MECHANICS OF SOLIDS

MECHANICS OF SOLIDS

(In S.I. Units)

For

B.E./B.Tech. 2nd YEAR

Strictly as per the latest syllabus prescribed by U.P. Technical University, Lucknow

By

Dr. R.K. BANSAL

B.Sc. Engg. (Mech.), M. Tech., Hons. (I.I.T., Delhi),
Ph.D., M.I.E. (India)

Formerly
Professor in Mechanical Engineering
Department of Mechanical Engineering
Delhi College of Engineering
Delhi
And
Ex. Dean (Academics)
Northern India Engineering College
Delhi

(An ISO 9001:2008 Company)

BENGALURU • CHENNAI • COCHIN • GUWAHATI • HYDERABAD
JALANDHAR • KOLKATA • LUCKNOW • MUMBAI • RANCHI • NEW DELHI
BOSTON (USA) • ACCRA (GHANA) • NAIROBI (KENYA)

MECHANICS OF SOLIDS

Compiled by : **Smt. Nirmal Bansal**

Typeset at Excellent Graphics, Delhi
First Edition: 2011; Reprint: 2013, 2014, 2015; Second Edition : 2016, Third : 2017
ISBN 978-81-318-0891-7

Branches

✆	Bengaluru	080-26 75 69 30	
✆	Chennai	044-24 34 47 26,	24 35 95 07
✆	Cochin	0484-237 70 04,	405 13 03
✆	Guwahati	0361-254 36 69,	251 38 81
✆	Hyderabad	040-27 55 53 83,	27 55 53 93
✆	Jalandhar	0181-222 12 72	
✆	Kolkata	033-22 27 43 84	
✆	Lucknow	0522-220 99 16	
✆	Mumbai	022-24 91 54 15,	24 92 78 69
✆	Ranchi	0651-220 44 64	

PUBLISHED IN INDIA BY

(An ISO 9001:2008 Company)
113, GOLDEN HOUSE, DARYAGANJ,
NEW DELHI - 110002, INDIA
Telephone : 91-11-4353 2500, 4353 2501
Fax : 91-11-2325 2572, 4353 2528
www.laxmipublications.com info@laxmipublications.com

Contents

Chapter *Pages*

Preface to the Second Edition

I am glad to present the Second Edition of the book entitled, 'Mechanics of Solids' to the engineering students of U.P. Technical University, Lucknow. The course contents have been planned in such a way that the book covers the complete course of second year of B. Tech. Mech. Engg./Production Engineering/Industrial and Production Engineering/Mechanical and Industrial Engineering/Manufacturing Technology/Automobile Engineering and Aeronautical Engineering students of U.P. Technical University, Lucknow.

During my long experience of teaching, I have observed that the students face difficulty in understanding clearly the basic principles, fundamental concepts and theory without adequate solved problems along with the text. To meet this very basic requirement of the students, a large number of the questions taken from different Universities have been solved along with the text in S.I. units at proper places.

The book contains the following five units:

Unit I: Compound Stress and Strains; 3-D Stresses, Theory of Failure, Equilibrium Equations and Impact Load.

Unit II: Stress in Beams, Deflection of Beams and Torsion.

Unit III: Helical and Leaf Springs, Columns and Struts.

Unit IV: Thin Cylinders and Spheres, Thick Cylinders.

Unit V: Curved Beams and Unsymmetrical Bending.

The U.P. Tech. University Exam paper for 2009–2010 has been included with solution at the end of the book. Objective Type Questions from Competitive Exams with explanation are also given.

The book has been written in a simple and easy-to-follow language, so that even an average student can grasp the subject by self-study. At the end of each chapter Highlights, Theoretical Questions and many Unsolved Numerical Problems with answers have been given for the students to solve them.

Though every care has been taken in checking the manuscripts and proofreading, yet claiming perfection is very difficult. I shall be very grateful to the readers and users of this book for pointing out any mistake that might have crept in. Suggestions for improvement are most welcome and would be incorporated in the next edition with a view to make the book more useful.

—R.K. BANSAL

SYLLABUS

(NME-302 : MECHANICS OF SOLIDS)

UNIT I

Compound Stress and Strains: Introduction, state of plane stress, Principal stress and strain, Mohr's stress circle. 3

3-D Stress, Theory of Failure, Equilibrium Equations and Impact Load: Three-dimensional state of stress and strain, equilibrium equations. Generalized Hook's Law. Theories of Failure. **Castigliano's** Theorem. Impact load and stresses. 5

UNIT II

Stresses in Beams: Pure Bending, Normal stresses in beams, shear stresses in beams due to transverse and axial loads, composite, beams. 2

Deflection of Beams: Equation of elastic curve, cantilever and simply supported beams, Macaulay's method, area moment method, fixed and continuous beams. 4

Torsion: Torsion, combined bending and torsion of solid and hollow shafts. Torsion of thin walled tubes. 2

UNIT III

Helical and Leaf Springs: Deflection of springs by energy method, helical springs under axial load and under axial twist (respectively for circular and square cross-sections) axial load and twisting moment acting simultaneously both for open and closed coiled springs, laminated springs. 4

Columns and Struts: Buckling and stability, slenderness ratio, combined bending and direct stress, middle third and middle quarter rules. Struts with different and conditions. Euler's theory for pin ended columns, effect of end conditions on column buckling. Ranking Gardon Formulae, Examples of columns in mechanical equipments and machines. 4

UNIT IV

Thin Cylinders and Spheres: Introduction, difference between thin walled and thick walled pressure vessels, thin walled spheres and cylinders. Hoop and axial stresses and strain. Volumetric strain. 2

Thick Cylinders: Radial, axial and circumferential stresses in thick cylinders subjected to internal or external pressures, compound cylinders. Stresses in rotating shaft and cylinders. Stresses due to interference fits. 4

UNIT V

Curved Beams: Bending of beams with large initial curvature, position of neutral axis for rectangular, trapezoidal and circular cross-sections, stress in crane hooks, stress in circular rings subjected to tension or compression. 4

Unsymmetrical Bending: Properties of beam cross-section, slope of neutral axis, stress and deflection in unsymmetrical bending, determination of shear centre and flexural axis (for symmetry about both axis and about one axis) for I-section and channel-section. 4

UNIT I

COMPOUND STRESS AND STRAINS

Chapters

1

Compound Stress and Strains

1.1. INTRODUCTION

In 1st and 2nd semesters, the concept and definition of stress, strain, types of stresses (*i.e.*, tensile, compressive and simple shear) and types of strain (*i.e.*, tensile, compressive, shear and volumetric strains etc.) were explained. These stresses were acting in a plane, which was at right angles to the line of action of the force. In many engineering problems both direct (tensile or compressive stress) and shear stresses are acting at the same time. In such situation the resultant stress across any section will be neither normal nor tangential to the plane. In this chapter the stresses, acting on an inclined plane (or oblique section) will be analysed.

1.2. NORMAL STRESS AND STRAIN

Stress is defined as the force of resistance per unit area offered by a body against deformation. The external force acting on the body is called the *load or force*. Load is applied on the body while the stress is induced in the material of the body. A loaded member remains in equilibrium when resistance offered by the body against deformation and applied load are equal.

Stress is written as, $\sigma = \dfrac{P}{A}$

where σ = Stress,
P = External load or force, and
A = Cross-section area.

Stress is expressed as N/m^2 or N/mm^2.

$$1\ N/m^2 = \frac{1\ N}{(1000\ mm)^2} = \frac{1\ N \times 10^{-6}}{mm^2} = \frac{10^{-6}\ N}{mm^2}$$

Strain is defined as the ratio of change of length to original length.

When a body is subjected to some external force, there is some change of dimension of the body. The ratio of change of dimension of the body to the original dimension in known as strain. Strain is dimensionless.

Strain may be :

1. Tensile strain,
2. Compressive strain,
3. Volumetric strain, and
4. Shear strain.

If there is some increase in length of a body due to external force, then the ratio of increase of length to the original length of the body is known as *tensile strain.* But if there is some decrease in length of the body, then the ratio of decrease of the length of the body to the original length is known as *compressive strain.* The ratio of change of volume of the body to the original volume is known as *volumetric strain.* The strain produced by shear stress is known as shear strain.

The stress may be normal stress or a shear stress.

Normal stress is the stress which acts in a direction perpendicular to the area. It is represented by σ (sigma). The normal stress is further divided into tensile stress and compressive stress.

1.2.1. Tensile Stress. The stress induced in a body, when subjected to two equal and opposite pulls as shown in Fig. 1.1 (*a*) as a result of which there is an increase in length, is known as tensile stress. The ratio of increase in length to the original length is known as *tensile strain.* The tensile stress acts normal to the area and it pulls on the area.

Let P = Pull (or force) acting on the body,
A = Cross-sectional area of the body,
L = Original length of the body,
dL = Increase in length due to pull P acting on the body,
σ = Stress induced in the body, and
e = Strain (*i.e.*, tensile strain).

Fig. 1.1 (*a*) shows a bar subjected to a tensile force P at its ends.

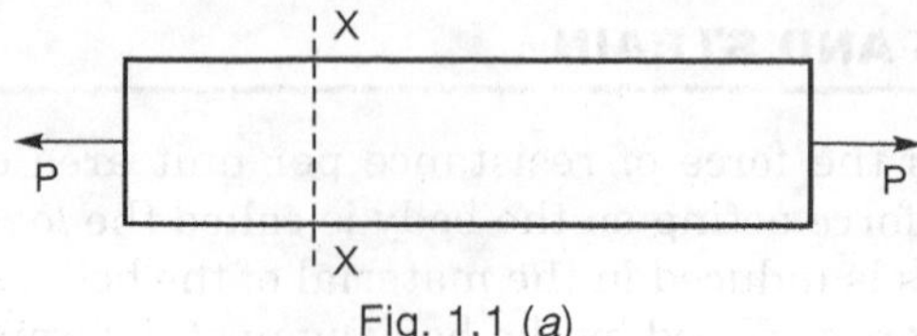

Fig. 1.1 (*a*)

$$\therefore \quad \text{Tensile stress,} \quad \sigma = \frac{\text{Tensile load } (P)}{A} \qquad (\because \quad P = R)$$

or
$$\sigma = \frac{P}{A}$$

And tensile strain is given by,

$$e = \frac{\text{Increase in length}}{\text{Original length}} = \frac{dL}{L}$$

1.2.2. Compressive Stress. The stress induced in a body, when subjected to two equal and opposite pushes as shown in Fig. 1.1 (*b*) as a result of which there is a decrease in length of the body, is known as compressive stress. And the ratio of decrease in length to the original length is known as *compressive strain.* The compressive stress acts normal to the area and it pushes on the area.

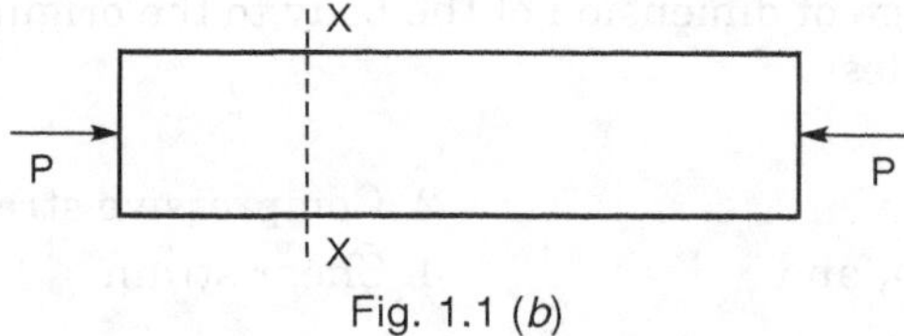

Fig. 1.1 (*b*)

Let an axial push P is acting on a body in cross-sectional area A. Due to external push P, let the original length L of the body decreases by dL.

Then compressive stress is given by,

$$\sigma = \frac{\text{Push } (P)}{\text{Area } (A)} = \frac{P}{A}.$$

And compressive strain is given by,

$$e = \frac{\text{Decrease in length}}{\text{Original length}} = \frac{dL}{L}.$$

1.3. SHEAR STRESS AND STRAIN

Shear stress is the stress which acts parallel (or tangential) to the area. The stress induced in a body, when subjected to two equal and opposite forces which are acting tangentially across the resisting section as shown in Fig 1.1 (*c*) as a result of which the body tends to shear off across the section, is known as shear stress. The corresponding strain is known as *shear strain*. The shear stress is the stress which acts tangential to the area. It is represented by τ.

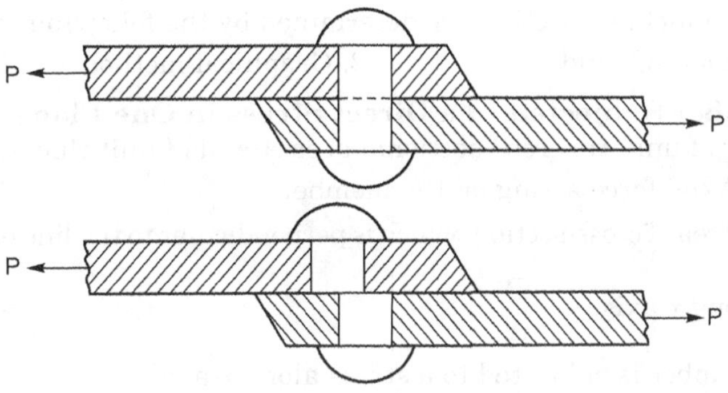

Fig. 1.1 (*c*)

Consider a rectangular block of height h, length L and width unity. Let the bottom face AB of the block be fixed to the surface as shown in Fig. 1.1 (*d*). Let a force P be applied tangentially along the top face CD of the block. Such a force acting tangentially along a surface is known as shear force. For the equilibrium of the block, the surface AB will offer a tangential reaction P equal and opposite to the applied tangential force P. And the shear force per unit area is known as shear stress which is represented by τ.

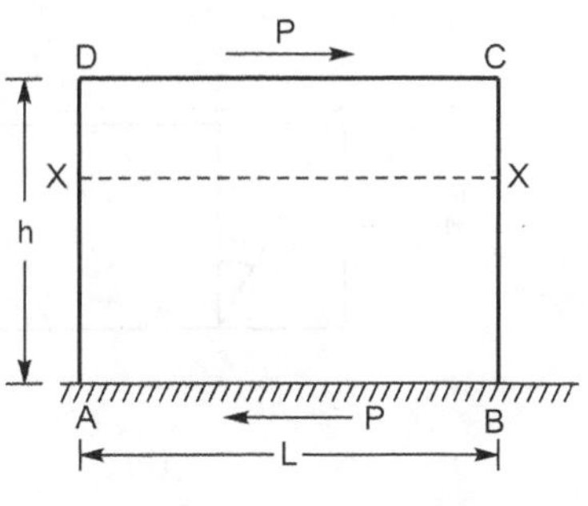

Fig. 1.1 (*d*)

$$\therefore \quad \text{Shear stress, } \tau = \frac{\text{Shear force}}{\text{Shear area}}$$

$$= \frac{P}{L \times 1} \qquad (\because \; A = L \times 1)$$

Note that shear stress is tangential to the area over which it acts.

As the bottom face of the block is fixed, the face $ABCD$ will be distorted to ABC_1D_1 through an angle ϕ as a result of force P as shown in Fig. 1.1 (*e*).

And shear strain (ϕ) is given by,

$$\phi = \frac{\text{Transversal displacement}}{\text{Distance } AD}$$

or $$\phi = \frac{DD_1}{AD} = \frac{dl}{h}$$

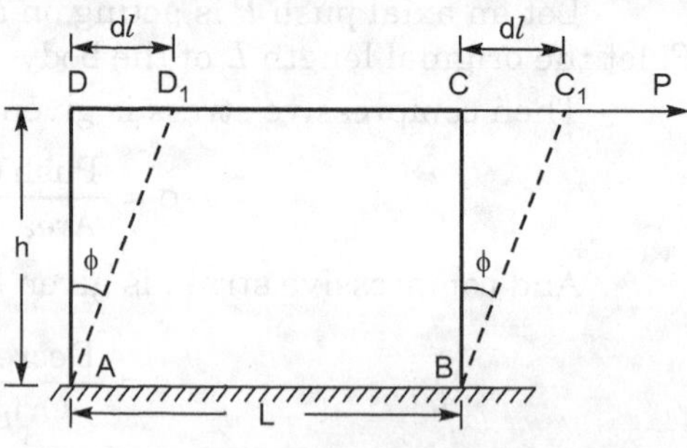

Fig. 1.1 (*e*)

1.3.1. Principal Planes and Principal Stresses. The planes, which have no shear stress, are known as principal planes. Hence principal planes are the planes of zero shear stress. These planes carry only normal stresses.

The normal stresses, acting on a principal plane, are known as principal stresses.

1.4. STRESSES ON INCLINED SECTIONS

The stresses on inclined section are determined by the following methods :

1. Analytical method, and 2. Graphical method.

1.4.1. A Member Subjected to a Direct Stress in One Plane. Fig. 1.2 (*a*) shows a rectangular member of uniform cross-sectional area *A* and of unit thickness.

Let P = Axial force acting on the member.

A = Area of cross-section, which is perpendicular to the line of action of the force P.

The stress along x-axis, $\sigma = \dfrac{P}{A}$

Hence, the member is subjected to a stress along x-axis.

Consider a cross-section EF which is perpendicular to the line of action of the force P.

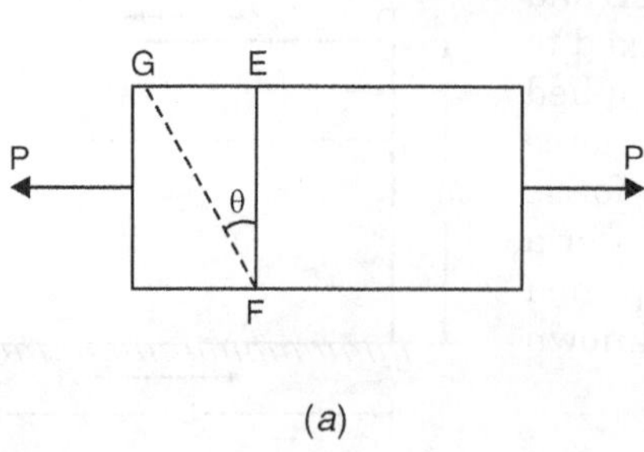

(*a*)

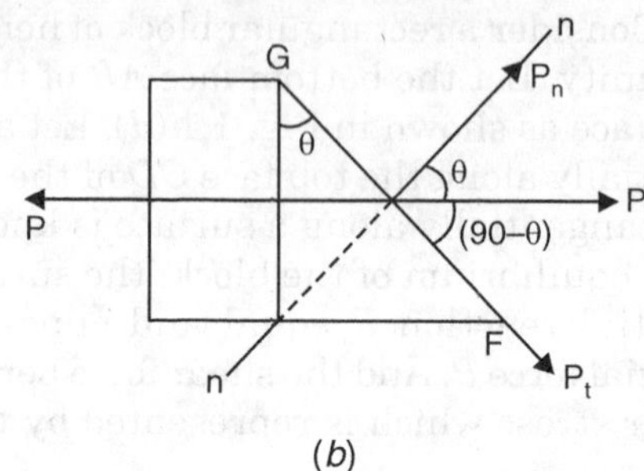

(*b*)

Fig. 1.2

Then area of section EF, $A = EF \times 1$

The stress on the section EF is given by

$$\sigma = \frac{\text{Force}}{\text{Area of } EF} = \frac{P}{A} \qquad \text{...(1.1)}$$

The stress on the section EF is entirely normal stress. There is no shear stress (or tangential stress) on the section EF.

Now consider a section FG at an angle θ with the normal cross-section EF as shown in Fig. 1.2 (a).

Area of section FG $= FG \times 1$ (member is having unit thickness)

$$= \frac{EF}{\cos\theta} \times 1 \qquad \left(\because \text{ In } \Delta\, EFG, \frac{EF}{FG} = \cos\theta \;\therefore\; FG = \frac{EF}{\cos\theta}\right)$$

$$= \frac{A}{\cos\theta} \qquad (\because\ EF \times 1 = A)$$

Let us find the stresses on the section FG.

The normal stress and tangential stress ($i.e.$, shear stress) on the section FG are obtained as given below [Refer to Fig. 1.2 (b)].

First, isolate the portion of the member to the left of the oblique section as a free-body. This portion will be in equilibrium, if a force equal, opposite and collinear with the original force P is acting on the oblique section as shown in Fig. 1.2 (b). This force will have components normal to and along the oblique section. Due to these components there will be normal stress and shear stress.

Let P_n = The component of the force P, normal to section FG

$= P\cos\theta$

P_t = The component of force P, along the surface of the section FG (or tangential to the surface FG)

$= P\sin\theta$

σ_n = Normal stress across the section FG

σ_t = Tangential stress ($i.e.$, shear stress) across the section FG.

$\therefore$ Normal stress and tangential stress across the section FG are obtained as,

Normal stress, $$\sigma_n = \frac{\text{Force normal to section } FG}{\text{Area of section } FG}$$

$$= \frac{P_n}{\left(\frac{A}{\cos\theta}\right)} = \frac{P\cos\theta}{\left(\frac{A}{\cos\theta}\right)} \qquad (\because\ P_n = P\cos\theta)$$

$$= \frac{P}{A}\cos\theta \,.\, \cos\theta = \frac{P}{A}\cos^2\theta$$

$$= \sigma\cos^2\theta \qquad \left(\because\ \frac{P}{A} = \sigma\right) \qquad ...(1.2)$$

Tangential stress ($i.e.$, shear stress),

$$\sigma_t = \frac{\text{Tangential force across section } FG}{\text{Area of section } FG}$$

$$= \frac{P_t}{\left(\frac{A}{\cos\theta}\right)} = \frac{P\sin\theta}{\left(\frac{A}{\cos\theta}\right)} \qquad (\because\ P_t = P\sin\theta)$$

$$= \frac{P}{A} \sin \theta \,.\, \cos \theta = \sigma \sin \theta .\, \cos \theta$$

$$= \frac{\sigma}{2} \times 2 \sin \theta \cos \theta \qquad \text{[Multiplying and dividing by 2]}$$

$$= \frac{\sigma}{2} \sin 2\theta \qquad (\because \quad 2 \sin \theta \cos \theta = \sin 2\theta) \qquad ...(1.3)$$

From equation (1.2), it is seen that the normal stress (σ_n) on the section *FG* will be maximum, when $\cos^2 \theta$ or $\cos \theta$ is maximum. And $\cos \theta$ will be maximum when $\theta = 0°$ as $\cos 0° = 1$. But when $\theta = 0°$, the section *FG* will coincide with section *EF*. But the section *EF* is normal to the line of action of the loading. This means the plane normal to the axis of loading will carry the maximum normal stress.

$$\therefore \text{ Maximum normal stress} = \sigma \cos^2 \theta = \sigma \cos^2 0° = \sigma \qquad ...(1.4)$$

From equation (1.3), it is observed that the tangential stress (*i.e.*, shear stress) across the section *FG* will be maximum when $\sin 2\theta$ is maximum. And $\sin 2\theta$ will be maximum when $\sin 2\theta = 1$ or $2\theta = 90°$ or $270°$

or $\qquad \theta = 45°$ or $135°$.

This means the shear stress will be maximum on two planes inclined at 45° and 135° to the normal section *EF*.

$$\therefore \quad \text{Max. value of shear stress} = \frac{\sigma}{2} \sin 2\theta = \frac{\sigma}{2} \sin 90° = \frac{\sigma}{2}. \qquad ...(1.5)$$

From equations (1.4) and (1.5) it is seen that maximum normal stress is equal to σ whereas the maximum shear stress is equal to $\sigma/2$ or equal to half the value of greatest normal stress.

Second Method

A member subjected to a direct stress in one plane. Fig. 1.2 (*c*) shows a rectangular member of uniform cross-sectional area *A* and of unit thickness. The bar is subjected to a principal tensile stress σ_1 on the faces *AD* and *BC*.

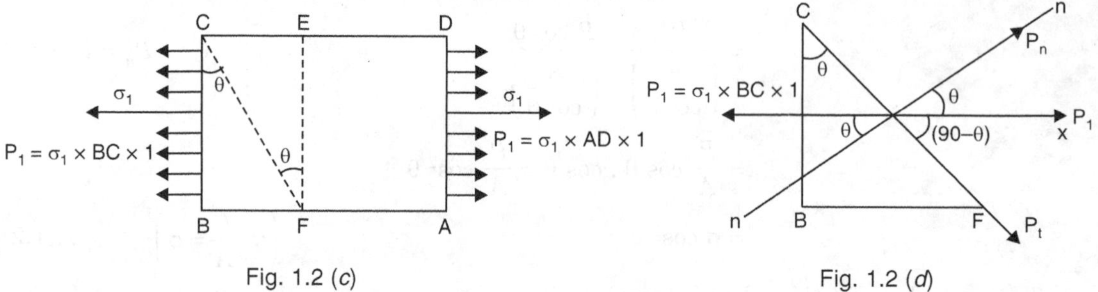

Fig. 1.2 (*c*) Fig. 1.2 (*d*)

Area of cross-section $= BC \times$ Thickness of bar

$= BC \times 1$ or $AD \times 1$

Let the stresses on the oblique plane *FC* are to be calculated. The plane *FC* is inclined at an angle θ with the normal cross-section *EF* (or *BC*). This can be done by converting the stress σ_1 acting on face *BC* into equivalent force. Then this force will be resolved along the inclined planes *FC* and perpendicular to *FC*. (Please note that it is force and not the stress which is to be resolved).

Tensile stress on face $BC = \sigma_1$

Now, the tensile force on BC,

$$P_1 = \text{Stress } (\sigma_1) \times \text{Area of cross-section}$$

$$= \sigma_1 \times BC \times 1 \qquad (\because \text{ Area} = BC \times 1)$$

The above tensile force P_1 is also acting on the inclined section FC, in the axial direction as shown in Fig. 1.2 (*d*). This force P_1 is resolved into two component, *i.e.*, one normal to the plane FC and other along the plane FC.

Let P_n = Component of the force P_1, normal to the section FC

$$= P_1 \cos\theta$$

$$= \sigma_1 \times BC \times 1 \times \cos\theta \qquad (\because P_1 = \sigma_1 \times BC \times 1)$$

P_t = Component of the force P_1, along the section FC

$$= P_1 \sin\theta$$

$$= \sigma_1 \times BC \times 1 \times \sin\theta$$

σ_n = Normal stress on the section FC

σ_t = Shear stress (or tangential stress) across the section FC.

Then normal stress, $\sigma_n = \dfrac{\text{Force normal to section } FC}{\text{Area of section } FC}$

$$= \frac{P_n}{FC \times 1} \qquad (\because \text{ bar is of unit thickness})$$

$$= \frac{\sigma_1 \times BC \times \cos\theta}{FC} \qquad (\because P_n = \sigma_1 \times BC \times \cos\theta)$$

$$= \sigma_1 \times \cos\theta \times \cos\theta \qquad \left(\because \text{ In triangle } FBC, \frac{BC}{FC} = \cos\theta\right)$$

$$= \sigma_1 \times \cos^2\theta \qquad \ldots(1.5A)$$

Similarly, tangential (or shear) stress,

$$\sigma_t = \frac{\text{Force along section } FC}{\text{Area of section } FC} = \frac{P_t}{FC \times 1}$$

$$= \frac{\sigma_1 \times BC \times 1 \times \sin\theta}{FC} \qquad (\because P_t = \sigma_1 \times BC \times 1)$$

$$= \sigma_1 \times \cos\theta \times \sin\theta \qquad \left(\because \text{ In triangle } FBC, \frac{BC}{FC} = \cos\theta\right)$$

$$= \sigma_1 \times \cos\theta \times \sin\theta$$

$$= \frac{\sigma_1}{2} \times 2 \times \cos\theta \times \sin\theta \qquad \text{(Multiplying and dividing by two)}$$

$$= \frac{\sigma_1}{2} \times \sin 2\theta \qquad \ldots(1.5B) \quad (\because 2\sin\theta\cos\theta = \sin 2\theta)$$

From equation (1.5*A*), it is seen that the normal stress (σ_n) on the section FC will be maximum, when $\cos^2\theta$ or $\cos\theta$ is maximum. And $\cos\theta$ will be maximum when $\theta = 0°$ as $\cos 0° = 1$. But when $\theta = 0°$, the section FC will coincide with section EF. But the section

EF is normal to the line of action of the loading. This means the plane normal to the axis of loading will carry the maximum normal stress.

$\therefore$ Maximum normal stress $= \sigma_1 \cos^2 \theta = \sigma_1 \cos^2 0° = \sigma_1$ ...(1.5*C*)

From equation (1.5*B*), it is observed that the tangential stress (*i.e.*, shear stress) across the section *FC* will be maximum when sin 2θ is maximum. And sin 2θ will be maximum when

$$\sin 2\theta = 1 \text{ or } 2\theta = 90° \text{ or } 270° \text{ or } \theta = 45° \text{ or } 135°.$$

This means the shear stress will be maximum on two planes inclined at 45° and 135° to the normal section *EF* or *BC*.

$\therefore$ Max. value of shear stress $= \dfrac{\sigma_1}{2} \sin 2\theta = \dfrac{\sigma_1}{2} \sin 90° = \dfrac{\sigma_1}{2}$ ...(1.5*D*)

From equations (1.5*C*) and (1.5*D*) it is seen that maximum normal stress is equal to σ_1 whereas the maximum shear stress is equal to $\dfrac{\sigma_1}{2}$ or equal to half the value of greatest normal stress.

Note. It is the force which is resolved in two components. The stress is not resolved.

Problem 1.1. *A rectangular bar of cross-sectional area 10000 mm² is subjected to an axial load of 20 kN. Determine the normal and shear stresses on a section which is inclined at an angle of 30° with normal cross-section of the bar.*

Sol. Given :

Cross-sectional area of the rectangular bar,

$$A = 10000 \text{ mm}^2$$

Axial load, $P = 20 \text{ kN} = 20{,}000 \text{ N}$

Angle of oblique plane with the normal cross-section of the bar,

$$\theta = 30°$$

Now direct stress, $\sigma = \dfrac{P}{A} = \dfrac{20000}{10000} = 2 \text{ N/mm}^2$

Let σ_n = Normal stress on the oblique plane

σ_t = Shear stress on the oblique plane.

Using equation (1.2) for normal stress, we get

$$\sigma_n = \sigma \cos^2 \theta$$

$= 2 \times \cos^2 30°$ $(\because \sigma = 2 \text{ N/mm}^2)$

$= 2 \times 0.866^2$ $(\because \cos 30° = 0.866)$

$=$ **1.5 N/mm². Ans.**

Using equation (1.3) for shear stress, we get

$$\sigma_t = \frac{\sigma}{2} \sin 2\theta = \frac{2}{2} \times \sin (2 \times 30°)$$

$= 1 \times \sin 60° =$ **0.866 N/mm². Ans.**

Problem 1.2. *Find the diameter of a circular bar which is subjected to an axial pull of 160 kN, if the maximum allowable shear stress on any section is 65 N/mm².*

Sol. Given :

Axial pull, $P = 160 \text{ kN} = 160000 \text{ N}$

Maximum shear stress $= 65 \text{ N/mm}^2$

Let D = Diameter of the bar

$\therefore$ Area of the bar $= \frac{\pi}{4} D^2$

$\therefore$ Direct stress, $\sigma = \frac{P}{A} = \frac{160000}{\frac{\pi}{4} D^2} = \frac{640000}{\pi D^2}$ N/mm^2

Maximum shear stress is given by equation (3.5).

$\therefore$ Maximum shear stress $= \frac{\sigma}{2} = \frac{640000}{2 \times \pi D^2}$.

But maximum shear stress is given as = 65 N/mm^2.

Hence equating the two values of maximum shear, we get

$$\therefore \quad 65 = \frac{640000}{2 \times \pi D^2}$$

$$\therefore \quad D^2 = \frac{640000}{2 \times \pi \times 65} = 1567$$

$$\therefore \quad D = \mathbf{39.58\ mm.} \quad \textbf{Ans.}$$

1.4.2. A Member Subjected to like Direct Stresses in Two Mutually Perpendicular Directions. Fig. 1.3 (*a*) shows a rectangular bar *ABCD* of uniform cross-sectional area *A* and of unit thickness. The bar is subjected to two direct tensile stresses (or two-principal tensile stresses) as shown in Fig. 1.3 (*a*).

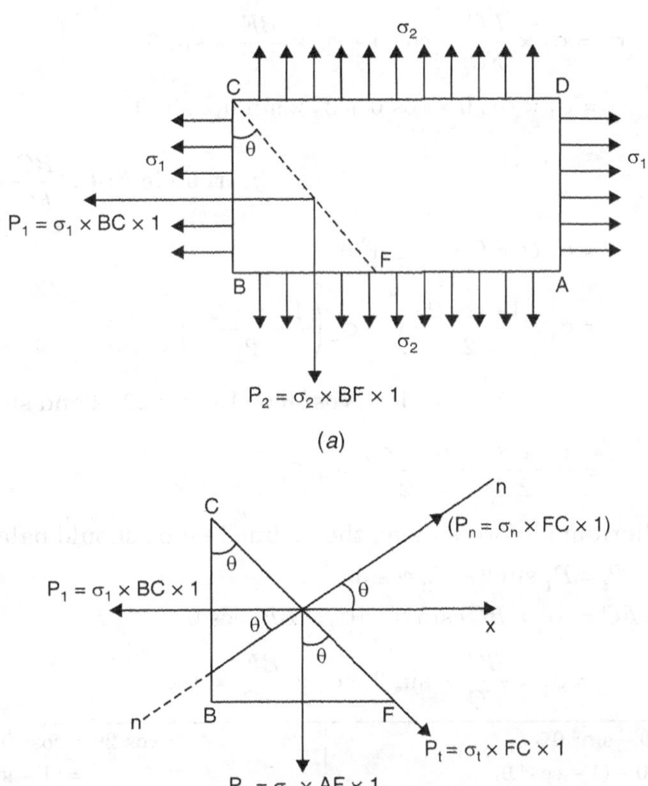

Fig. 1.3

Let FC be the oblique section on which stresses are to be calculated. This can be done by converting the stresses σ_1 (acting on face BC) and σ_2 (acting on face AB) into equivalent forces. Then these forces will be resolved along the inclined plane FC and perpendicular to FC. First, isolate the triangular portion CBF of the member as shown in Fig. 1.3 (*b*). The forces on the faces BC and BF will be $P_1 = \sigma_1 \times BC \times 1$ and $P_2 = \sigma_2 \times BF \times 1$. To balance these forces we need forces normal to and along the inclined section. These forces will be P_n and P_t.

Let θ = Angle made by oblique section FC with normal cross-section BC

σ_1 = Major tensile stress on face AD and BC *i.e.*, along x-axis

σ_2 = Minor tensile stress on face AB and CD *i.e.*, along y-axis

P_1 = Tensile force on face BC *i.e.*, along x-axis

P_2 = Tensile force on face FB *i.e.*, along y-axis.

The tensile force on face BC,

$$P_1 = \sigma_1 \times \text{Area of face } BC = \sigma_1 \times BC \times 1 \qquad (\because \quad \text{Area} = BC \times 1)$$

The tensile force on face FB,

$$P_2 = \text{Stress on } FB \times \text{Area of } FB = \sigma_2 \times FB \times 1.$$

For equilibrium, all forces normal to inclined plane should balance. Hence we get

$$P_n = P_1 \cos\theta + P_2 \sin\theta$$

or
$$\sigma_n \times FC = \sigma_1 \times BC \times \cos\theta + \sigma_2 \times BF \times \sin\theta \quad (\because \quad P_1 = \sigma_1 \times BC,\ P_2 = \sigma_2 \times BF)$$

or
$$\sigma_n = \sigma_1 \times \frac{BC}{FC} \times \cos\theta + \sigma_2 \times \frac{BF}{FC} \times \sin\theta$$

$$= \sigma_1 \times \cos\theta \times \cos\theta + \sigma_2 \times \sin\theta \times \sin\theta$$

$$\left(\because \quad \text{In triangle } FBC,\ \frac{BC}{FC} = \cos\theta,\ \frac{BF}{FC} = \sin\theta\right)$$

$$= \sigma_1 \cos^2\theta + \sigma_2 \sin^2\theta$$

$$= \sigma_1 \left(\frac{1 + \cos 2\theta}{2}\right)^{*} + \sigma_2 \left(\frac{1 - \cos 2\theta}{2}\right)^{**}$$

$$[\because \quad \cos^2\theta = (1 + \cos 2\theta)/2 \text{ and } \sin^2\theta = (1 - \cos 2\theta)/2]$$

$$= \frac{\sigma_1 + \sigma_2}{2} + \frac{\sigma_1 - \sigma_2}{2} \cos 2\theta \qquad \text{...(1.6)}$$

Also, for equilibrium all forces along the inclined plane should balance.

$$\therefore \quad P_t = P_1 \sin\theta - P_2 \cos\theta$$

or
$$\sigma_t \times FC = (\sigma_1 \times BC) \sin\theta - (\sigma_2 \times BF) \cos\theta$$

or
$$\sigma_t = \sigma_1 \times \frac{BC}{FC} \times \sin\theta - \sigma_2 \times \frac{BF}{FC} \times \cos\theta$$

---

* $\cos 2\theta = \cos^2\theta - \sin^2\theta$

$= \cos^2\theta - (1 - \cos^2\theta)$

$= 2\cos^2\theta - 1$

$\therefore \cos^2\theta = \dfrac{(1 + \cos 2\theta)}{2}$

** $\cos 2\theta = \cos^2\theta - \sin^2\theta$

$= (1 - \sin^2\theta) - \sin^2\theta$

$= 1 - 2\sin^2\theta$

$\therefore \quad \sin^2\theta = \dfrac{(1 - \cos 2\theta)}{2}$

$$= \sigma_1 \times \cos\theta \times \sin\theta - \sigma_2 \times \sin\theta \times \cos\theta$$

$$\left(\because \text{ In triangle } FBC, \frac{BC}{FC} = \cos\theta, \frac{BF}{FC} = \sin\theta\right)$$

$$= (\sigma_1 - \sigma_2)\cos\theta\sin\theta$$

$$= \frac{(\sigma_1 - \sigma_2)}{2} \times 2\cos\theta\sin\theta \qquad \text{(Multiplying and dividing by 2)}$$

$$= \frac{(\sigma_1 - \sigma_2)}{2}\sin 2\theta \qquad ...(1.7)$$

The resultant stress on the section FC will be given as

$$\sigma_R = \sqrt{\sigma_n^2 + \sigma_t^2} \qquad ...(1.8)$$

Obliquity [Refer to Fig. 1.4]. The angle made by the resultant stress with the normal of the oblique plane, is known as obliquity. It is denoted by ϕ. Mathematically,

$$\tan\phi = \frac{\sigma_t}{\sigma_n} \qquad ...[1.8\,(A)]$$

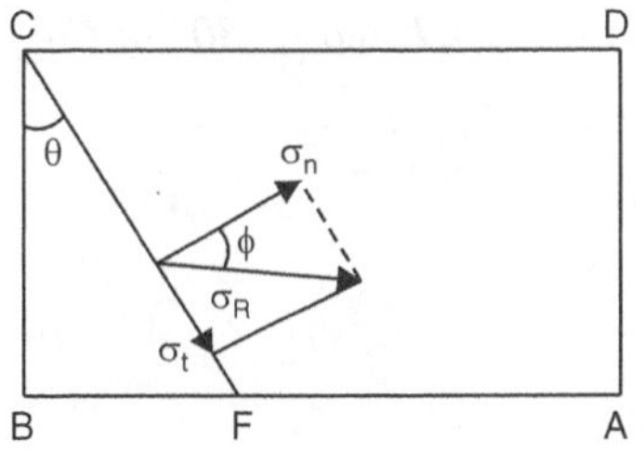

C D θ σ_n ϕ σ_R σ_t B F A

Fig. 1.4

Maximum shear stress. The shear stress is given by equation (1.7). The shear stress (σ_t) will be maximum when

$$\sin 2\theta = 1 \quad \text{or} \quad 2\theta = 90° \quad \text{or} \quad 270°$$

$$(\because \quad \sin 90° = 1 \text{ and also } \sin 270° = 1)$$

or $\qquad \theta = 45°$ or $135°$

And maximum shear stress, $(\sigma_t)_{max} = \dfrac{\sigma_1 - \sigma_2}{2}$ ...(1.9)

Principal planes. Principal planes are the planes on which shear stress is zero. To locate the position of principal planes, the shear stress given by equation (1.7) should be equated to zero.

$\therefore$ For principal planes,

$$\frac{\sigma_1 - \sigma_2}{2}\sin 2\theta = 0$$

or $\qquad \sin 2\theta = 0 \qquad [\because \ (\sigma_1 - \sigma_2)$ cannot be equal to zero$]$

or $\qquad 2\theta = 0$ or $180°$

$\therefore \qquad \theta = 0$ or $90°$

when $\theta = 0$, $\qquad \sigma_n = \dfrac{\sigma_1 + \sigma_2}{2} + \dfrac{\sigma_1 - \sigma_2}{2}\cos 2\theta$

$$= \frac{\sigma_1 + \sigma_2}{2} + \frac{\sigma_1 - \sigma_2}{2}\cos 0°$$

$$= \frac{\sigma_1 + \sigma_2}{2} + \frac{\sigma_1 - \sigma_2}{2} \times 1 \qquad (\because \quad \cos 0° = 1)$$

$$= \sigma_1$$

when $\theta = 90°$,
$$\sigma_n = \frac{\sigma_1 + \sigma_2}{2} + \frac{\sigma_1 - \sigma_2}{2} \cos 2 \times 90°$$

$$= \frac{\sigma_1 + \sigma_2}{2} + \frac{\sigma_1 - \sigma_2}{2} \cos 180°$$

$$= \frac{\sigma_1 + \sigma_2}{2} + \frac{\sigma_1 - \sigma_2}{2} \times (-1) \qquad (\because \quad \cos 180° = -1)$$

$$= \sigma_2.$$

Note. The relations, given by equations (1.6) to (1.9), also hold good when one or both the stresses are compressive.

Problem 1.3. *The tensile stresses at a point across two mutually perpendicular planes are 120 N/mm² and 60 N/mm². Determine the normal, tangential and resultant stresses on a plane inclined at 30° to the plane of stress 120 N/mm².*

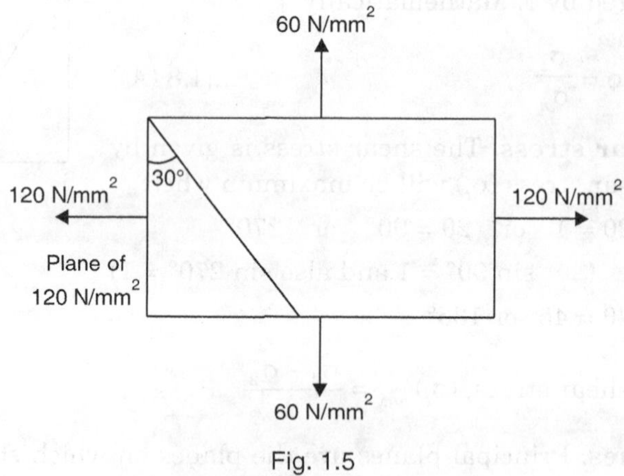

Fig. 1.5

Sol. Given :

Major principal stress, $\sigma_1 = 120$ N/mm²

Minor principal, $\sigma_2 = 60$ N/mm²

Angle of oblique plane with the plane of 120 N/mm²,

$$\theta = 30°.$$

Normal stress

The normal stress (σ_n) is given by equation (1.6),

$$\therefore \qquad \sigma_n = \frac{\sigma_1 + \sigma_2}{2} + \frac{\sigma_1 - \sigma_2}{2} \cos 2\theta$$

$$= \frac{120 + 60}{2} + \frac{120 - 60}{2} \cos 2 \times 30°$$

$$= 90 + 30 \cos 60° = 90 + 30 \times \tfrac{1}{2} = \mathbf{105\ N/mm^2}.$$ **Ans.**

Tangential stress

The tangential (or shear stress) σ_t is given by equation (1.7).

$$\therefore \quad \sigma_t = \frac{\sigma_1 - \sigma_2}{2} \sin 2\theta$$

$$= \frac{120 - 60}{2} \sin (2 \times 30°)$$

$$= 30 \times \sin 60° = 30 \times 0.866 = \mathbf{25.98\ N/mm^2}. \quad \textbf{Ans.}$$

Resultant stress

The resultant stress (σ_R) is given by equation (1.8)

$$\therefore \quad \sigma_R = \sqrt{\sigma_n^2 + \sigma_t^2} = \sqrt{105^2 + 25.98^2}$$

$$= \sqrt{11025 + 674.96} = \mathbf{108.16\ N/mm^2}. \quad \textbf{Ans.}$$

Problem 1.4. *The stresses at a point in a bar are 200 N/mm² (tensile) and 100 N/mm² (compressive). Determine the resultant stress in magnitude and direction on a plane inclined at 60° to the axis of the major stress. Also determine the maximum intensity of shear stress in the material at the point.*

Sol. Given :

Major principal stress, $\sigma_1 = 200\ N/mm^2$

Minor principal stress, $\sigma_2 = -100\ N/mm^2$

(Minus sign is due to compressive stress)

Angle of the plane, which it makes with the major principal stress = 60°

$\therefore$ Angle $\theta = 90° - 60° = 30°$.

Resultant stress in magnitude and direction

First calculate the normal and tangential stresses.

Using equation (1.6) for normal stress,

$$\sigma_n = \frac{\sigma_1 + \sigma_2}{2} + \frac{\sigma_1 - \sigma_2}{2} \cos 2\theta$$

$$= \frac{200 + (-100)}{2} + \frac{200 - (-100)}{2} \cos (2 \times 30°) \quad (\because \theta = 30°)$$

$$= \frac{200 - 100}{2} + \frac{200 + 100}{2} \cos 60°$$

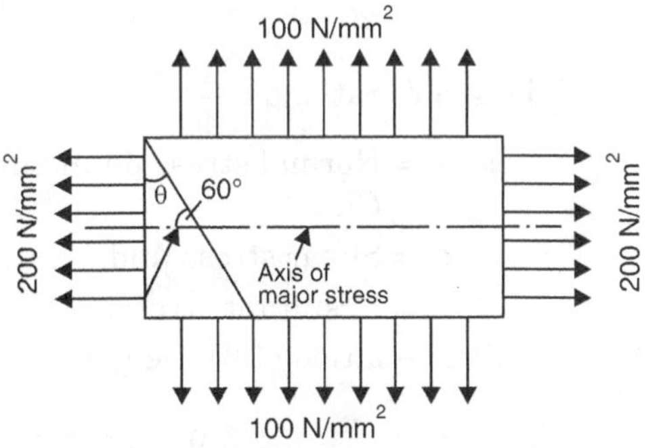

$$= 50 + 150 \times \tfrac{1}{2} \quad \left(\because \cos 60° = \tfrac{1}{2}\right)$$

$$= 50 + 75 = 125\ N/mm^2.$$

Fig. 1.6

Using equation (1.7) for tangential stress,

$$\sigma_t = \frac{\sigma_1 - \sigma_2}{2} \sin 2\theta = \frac{200 - (-100)}{2} \sin (2 \times 30°)$$

$$= \frac{200 + 100}{2} \sin 60° = 150 \times 0.866 = 129.9\ N/mm^2.$$

Using equation (1.8) for resultant stress,

$$\sigma_R = \sqrt{\sigma_n^2 + \sigma_t^2} = \sqrt{125^2 + 129.9^2}$$

$$= \sqrt{15625 + 16874} = \mathbf{180.27\ N/mm^2}.\quad \textbf{Ans.}$$

The inclination of the resultant stress with the normal of the inclined plane is given by equation [1.8 (*A*)] as

$$\tan\phi = \frac{\sigma_t}{\sigma_n} = \frac{129.9}{125} = 1.04$$

$$\therefore \quad \phi = \tan^{-1} 1.04 = \mathbf{46^\circ\ 6'}.\quad \textbf{Ans.}$$

Maximum shear stress

Maximum shear stress is given by equation (1.9)

$$\therefore \quad (\sigma_t)_{max} = \frac{\sigma_1 - \sigma_2}{2} = \frac{200 - (-100)}{2} = \frac{200 + 100}{2} = \mathbf{150\ N/mm^2}.\quad \textbf{Ans.}$$

Problem 1.5. *At a point in a strained material the principal tensile stresses across two perpendicular planes, are 80 N/mm² and 40 N/mm². Determine normal stress, shear stress and the resultant stress on a plane inclined at 20° with the major principal plane. Determine also the obliquity. What will be the intensity of stress, which acting alone will produce the same maximum strain if Poisson's ratio =* $\frac{1}{4}$*?*

Sol. Given :

Major principal stress, $\sigma_1 = 80$ N/mm²

Minor principal stress, $\sigma_2 = 40$ N/mm²

The plane *CE* is inclined at angle 20° with major principal plane (*i.e.*, plane *BC*).

$\therefore \quad \theta = 20°$

Poisson's ratio, $\mu = \frac{1}{4}$

Let σ_n = Normal stress on inclined plane *CE*

σ_t = Shear stress and

σ_R = Resultant stress.

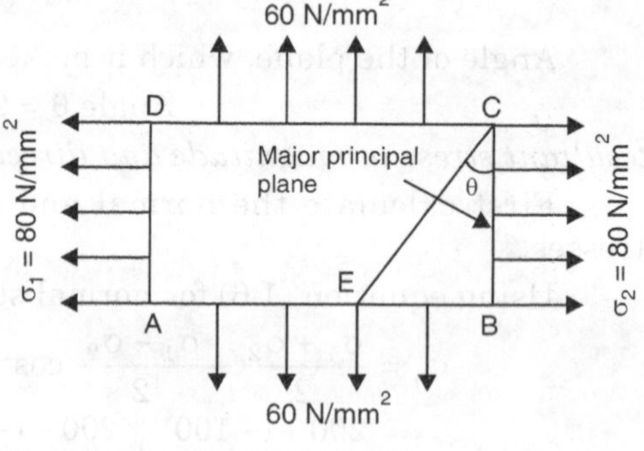

Fig. 1.7

Using equation (1.6), we get

$$\sigma_n = \frac{\sigma_1 + \sigma_2}{2} + \frac{\sigma_1 - \sigma_2}{2}\cos 2\theta = \frac{80 + 40}{2} + \frac{80 - 40}{2}\cos(2 \times 20^\circ)$$

$$= 60 + 20 \times \cos 40^\circ = \mathbf{75.32\ N/mm^2}.\quad \textbf{Ans.}$$

The shear stress is given by equation (1.7)

$$\therefore \quad \sigma_t = \frac{\sigma_1 - \sigma_2}{2}\sin 2\theta = \frac{80 - 40}{2}\sin(2 \times 20^\circ) = 20 \sin 40^\circ$$

$$= \mathbf{12.865\ N/mm^2}.\quad \textbf{Ans.}$$

The resultant stress is given by equation (1.8)

$$\therefore \quad \sigma_R = \sqrt{\sigma_n^2 + \sigma_t^2}$$

$$= \sqrt{75.32^2 + 12.856^2} = \mathbf{76.4\ N/mm^2}.\quad \textbf{Ans.}$$

Obliquity (ϕ) is given by equation [1.8 (*A*)]

$$\tan \phi = \frac{\sigma_t}{\sigma_n} = \frac{12.856}{75.32}$$

$$\therefore \qquad \phi = \tan^{-1} \frac{12.856}{75.32} = \mathbf{9° \; 41'}. \quad \textbf{Ans.}$$

Let σ = stress which acting alone will produce the same maximum strain. The maximum strain will be in the direction of major principal stress.

$$\therefore \quad \text{Maximum strain} = \frac{\sigma_1}{E} - \frac{\mu\sigma_2}{E} = \frac{1}{E}(\sigma_1 - \mu\sigma_2)$$

$$= \frac{1}{E}\left(80 - \frac{40}{4}\right) = \frac{70}{E}$$

The strain due to stress, $\sigma = \dfrac{\sigma}{E}$

Equating the two strains, we get $\dfrac{70}{E} = \dfrac{\sigma}{E}$

$$\therefore \qquad \sigma = \mathbf{70 \; N/mm^2}. \quad \textbf{Ans.}$$

Problem 1.6. *At a point in a strained material, the principal stresses are 100 N/mm² tensile and 40 N/mm² compressive. Determine the resultant stress in magnitude and direction on a plane inclined at 60° to the axis of the major principal stress. What is the maximum intensity of shear stress in the material at the point ?*

Sol. Given :

The major principal stress, σ_1 = 100 N/mm²

The minor principal stress, σ_2 = – 40 N/mm²

(Minus sign due to compressive stress)

Inclination of the plane with the axis of major principal stress = 60°

$\therefore$ Inclination of the plane with the axis of minor principal stress,

$$\theta = 90 - 60 = 30°.$$

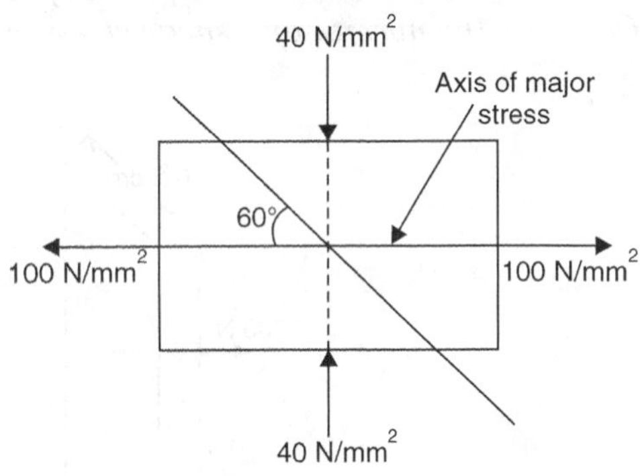

Fig. 1.8

Resultant stress in magnitude

The resultant stress (σ_R) is given by equation (1.8) as

$$\sigma_R = \sqrt{\sigma_n^2 + \sigma_t^2}$$

where σ_n = Normal stress and is given by equation (1.6) as

$$= \frac{\sigma_1 + \sigma_2}{2} + \frac{\sigma_1 - \sigma_2}{2} \cos 2\theta$$

$$= \frac{100 + (-40)}{2} + \frac{100 - (-40)}{2} \cos (2 \times 30°)$$

$$= \frac{100 - 40}{2} + \frac{100 + 40}{2} \cos 60°$$

$$= 30 + 70 \times 0.5 \qquad (\because \cos 60° = 0.5)$$

$$= 65 \text{ N/mm}^2$$

and σ_t = Shear stress and is given by equation (1.7) as

$$= \frac{\sigma_1 - \sigma_2}{2} \sin 2\theta = \frac{100 - (-40)}{2} \sin (2 \times 30°)$$

$$= \frac{100 + 40}{2} \sin 60° = 70 \times .866 = 60.62 \text{ N/mm}^2$$

$$\sigma_R = \sqrt{65^2 + 60.62^2} = \mathbf{88.9\ N/mm^2.\ \ Ans.}$$

Direction of resultant stress

Let the resultant stress is inclined at an angle ϕ to the normal of the oblique plane. Then using equation [1.8 (*A*)].

$$\tan \phi = \frac{\sigma_t}{\sigma_n} = \frac{60.62}{65}$$

$$\therefore \quad \phi = \tan^{-1} \frac{60.62}{65} = \mathbf{43°.\ \ Ans.}$$

Maximum shear stress

Using equation (1.9), $(\sigma_t)_{max} = \dfrac{\sigma_1 - \sigma_2}{2}$

$$= \frac{100 - (-40)}{2} = \frac{100 + 40}{2} = \mathbf{70\ N/mm^2.\ \ Ans.}$$

Problem 1.7. *A small block is 4 cm long, 3 cm high and 0.5 cm thick. It is subjected to uniformly distributed tensile forces of resultants 1200 N and 500 N as shown in Fig. 1.9 below. Compute the normal and shear stresses developed along the diagonal AB.*

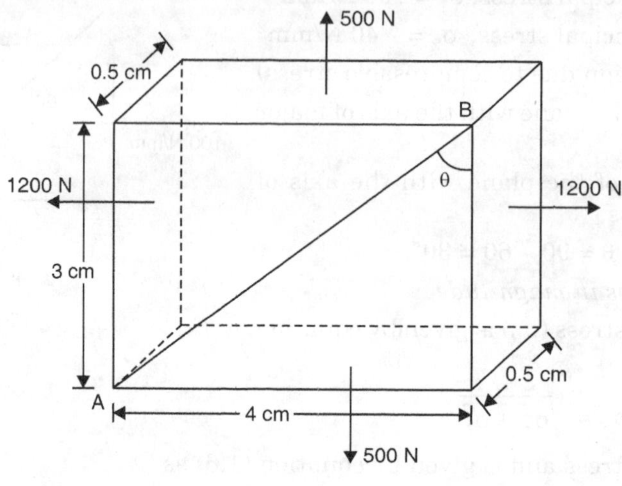

Fig. 1.9

Sol. Given :

Length = 4 cm, height = 3 cm and width = 0.5 cm

Force along x-axis = 1200 N

Force along y-axis = 500 N

Area of cross-section normal to x-axis = $3 \times 0.5 = 1.5$ cm^2

Area of cross-section normal to y-axis = $4 \times 0.5 = 2$ cm^2

$$\therefore \text{ Stress along } x\text{-axis} = \frac{\text{Force along } x\text{-axis}}{\text{Area normal to } x\text{-axis}}$$

$$= \frac{1200}{1.5} = 800 \text{ N/cm}^2$$

$$\therefore \qquad \sigma_1 = 800 \text{ N/cm}^2$$

$$\text{Stress along } y\text{-axis,} \quad \sigma_2 = \frac{\text{Force along } y\text{-axis}}{\text{Area normal to } y\text{-axis}}$$

$$= \frac{500}{2} = 250 \text{ N/cm}^2$$

Also, $\tan \theta = \frac{4}{3} = 1.33$

$\therefore$ $\theta = \tan^{-1} 1.33 = 53.06°$

Let σ_n = Normal stress on diagonal *AB*

σ_t = Shear stress on diagonal *AB*

$$\text{Using equation (1.6), } \sigma_n = \frac{\sigma_1 + \sigma_2}{2} + \frac{\sigma_1 - \sigma_2}{2} \cos 2\theta$$

$$= \frac{800 + 250}{2} + \frac{800 - 250}{2} \cos (2 \times 53.06)$$

$$= 525 + 275 \times \cos 106.12° = 525 + 275 \times (-0.2776)$$

$$= 525 - 76.35 = \mathbf{448.65 \text{ N/cm}^2}. \quad \textbf{Ans.}$$

$$\text{Now using equation (1.7), } \sigma_t = \frac{\sigma_1 - \sigma_2}{2} \sin 2\theta$$

$$= \frac{800 - 250}{2} \sin (2 \times 53.06°)$$

$$= 275 \sin 106.12° = 275 \times 0.96 = \mathbf{264.18 \text{ N/cm}^2}. \quad \textbf{Ans.}$$

1.4.3. A Member Subjected to a Simple Shear Stress. Fig. 1.10 shows a rectangular bar *ABCD* of uniform cross-sectional area *A* and of unit thickness. The bar is subjected to a simple shear stress (τ) across the faces *BC* and *AD*. Let *FC* be the oblique section on which normal and tangential stresses are to be calculated.

Let θ = Angle made by oblique section *FC* with normal cross-section *BC*,

τ = Shear stress across faces *BC* and *AD*.

It has already been proved that a shear stress is always accompanied by an equal shear stress at right angles to it. Hence the faces *AB* and *CD* will also be subjected to a shear stress τ as shown in Fig. 1.10. Now these stresses will be converted into equivalent forces. Then these forces will be resolved along the inclined surface and normal to inclined surface. Consider the forces acting on the wedge *FBC* of Fig. 1.11.

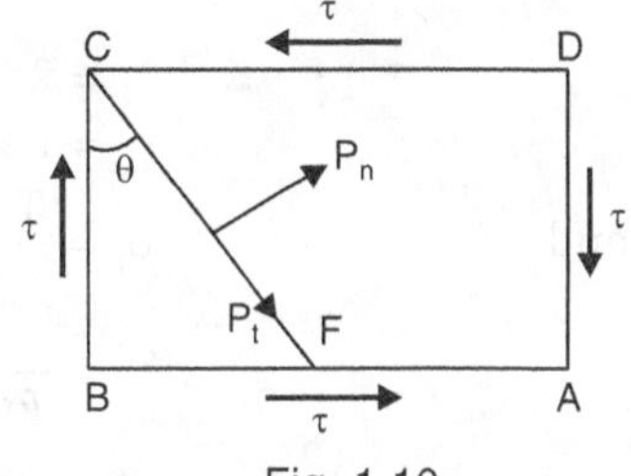

Fig. 1.10

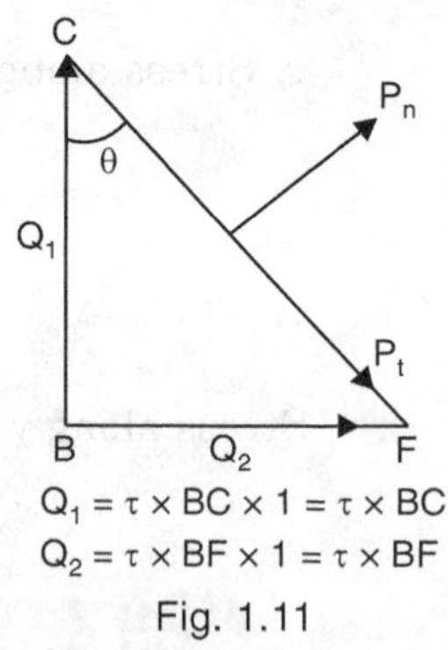

Fig. 1.11

Let Q_1 = Shear force on face BC

$= \text{Shear stress} \times \text{Area of face } BC$

$= \tau \times BC \times 1 \quad (\because \text{ Area of face } BC = BC \times 1)$

$= \tau \times BC$

Q_2 = Shear force on face FB

$= \tau \times \text{Area of } FB$

$= \tau \times FB \times 1 = \tau \,.\, FB$

P_n = Total normal force on section FC

P_t = Total tangential force on section FC.

The force Q_1 is acting along face CB as shown in Fig. 1.11. This force is resolved into two components *i.e.*, $Q_1 \cos \theta$ and $Q_1 \sin \theta$ along the plane CF and normal to the plane CF respectively.

The force Q_2 is acting along the face FB. This force is also resolved into two component *i.e.*, $Q_2 \sin \theta$ and $Q_2 \cos \theta$ along the plane FC and normal to the plane FC respectively.

∴ Total normal force on section FC,

$$P_n = Q_1 \sin \theta + Q_2 \cos \theta$$

$$= \tau \times BC \times \sin \theta + \tau \times FB \times \cos \theta. \quad (\because \; Q_1 = \tau \times BC \text{ and } Q_2 = \tau \times FB)$$

And total tangential force on section FC.

$$P_t = Q_2 \sin \theta - Q_1 \cos \theta. \quad \text{(–ve sign is taken due to opposite direction)}$$

$$= \tau \times FB \times \sin \theta - \tau \times BC \times \cos \theta \quad (\because \; Q_2 = \tau \,.\, FB \text{ and } Q_1 = \tau \,.\, BC)$$

Let σ_n = Normal stress on section FC

σ_t = Tangential stress on section FC

Then
$$\sigma_n = \frac{\text{Total normal force on section } FC}{\text{Area of section } FC}$$

$$= \frac{P_n}{FC \times 1}$$

$$= \frac{\tau \,.\, BC \,.\, \sin \theta + \tau \,.\, FB \,.\, \cos \theta}{FC \times 1} \quad (\because \; \text{Area} = FC \times 1)$$

$$= \tau \,.\, \frac{BC}{FC} \,.\, \sin \theta + \tau \,.\, \frac{FB}{FC} \,.\, \cos \theta$$

$$= \tau \,.\, \cos \theta \,.\, \sin \theta + \tau \,.\, \sin \theta \,.\, \cos \theta$$

$$\left(\because \; \text{In triangle } FBC, \frac{BC}{FC} = \cos \theta, \frac{FB}{FC} = \sin \theta\right)$$

$$= 2\tau \cos \theta \,.\, \sin \theta$$

$$= \tau \sin 2\theta \quad (\because \; 2 \sin \theta \cos \theta = \sin 2\theta) \quad \text{...(1.10)}$$

and
$$\sigma_t = \frac{\text{Total tangential force on section } FC}{\text{Area of section } FC}$$

$$= \frac{P_t}{FC \times 1}$$

$$= \frac{\tau \times FB \times \sin \theta - \tau \times BC \times \cos \theta}{FC \times 1}$$

$$= \tau \times \frac{FB}{FC} \times \sin \theta - \tau \times \frac{BC}{FC} \times \cos \theta$$

$$= \tau \times \sin\theta \times \sin\theta - \tau \times \cos\theta \times \cos\theta$$
$$= \tau \sin^2\theta - \tau\cos^2\theta = -\tau\,[\cos^2\theta - \sin^2\theta]$$
$$= -\tau\cos 2\theta \qquad (\because \cos^2\theta - \sin^2\theta = \cos 2\theta) \quad ...(1.11)$$

–ve sign shows that σ_t will be acting downwards on the plane *CF*.

1.4.4. A Member Subjected to Direct Stresses in Two Mutually Perpendicular Directions Accompanied by a Simple Shear Stress. Fig. 1.12 (*a*) shows a rectangular bar *ABCD* of uniform cross-sectional area *A* and of unit thickness. This bar is subjected to :

(*i*) tensile stress σ_1 on the face *BC* and *AD*

(*ii*) tensile stress σ_2 on the face *AB* and *CD*

(*iii*) a simple shear stress τ on face *BC* and *AD*.

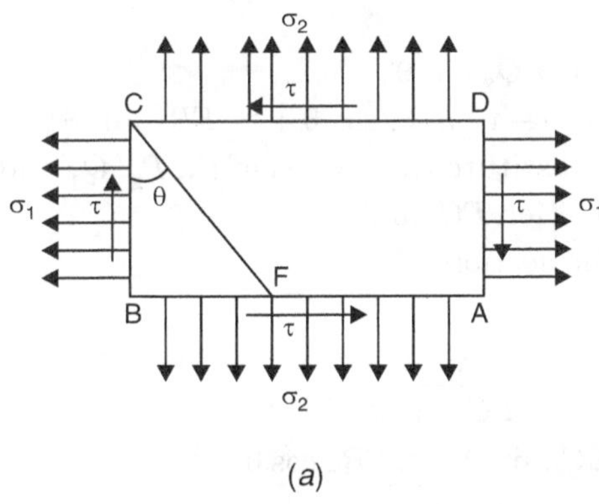

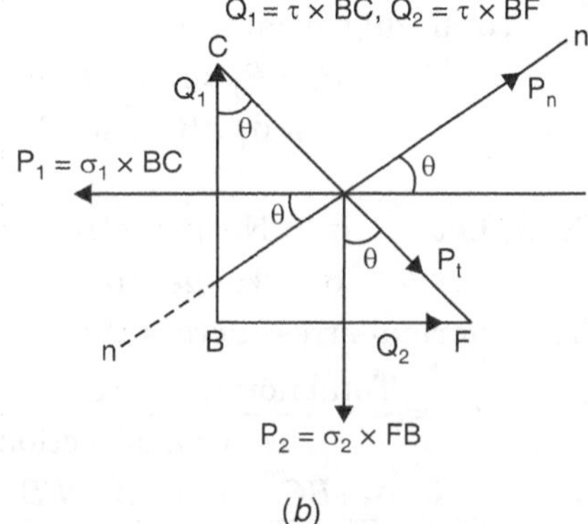

Fig. 1.12

But we know, a simple shear stress is always accompanied by an equal shear stress at right angles to it. Hence the faces *AB* and *CD* will also be subjected to a shear stress τ as shown in Fig. 1.12 (*a*).

We want to calculate normal and tangential stresses on oblique section *FC*, which is inclined at an angle θ with the normal cross-section *BC*. The given stresses are converted into equivalent forces.

Consider the equilibrium of wedge *FBC* shown in Fig. 1.12 (*b*).

The forces acting on the wedge *FBC* are :

P_1 = Tensile force on face *BC* due to tensile stress σ_1

$= \sigma_1 \times$ Area of *BC*

$= \sigma_1 \times BC \times 1 \qquad (\because \text{Area} = BC \times 1)$

$= \sigma_1 \times BC$

P_2 = Tensile force on face *FB* due to tensile stress σ_2

$= \sigma_2 \times$ Area of $FB = \sigma_2 \times FB \times 1$

$= \sigma_2 \times FB$

Q_1 = Shear force on face *BC* due to shear stress τ

$= \tau \times$ Area of *BC*

$= \tau \times BC \times 1 = \tau \times BC$

Q_2 = Shear force on face FB due to shear stress τ
$= \tau \times$ Area of FB
$= \tau \times FB \times 1 = \tau \times FB.$

Resolving the above four forces (*i.e.*, P_1, P_2, Q_1 and Q_2) normal to the oblique section FC, we get

Total normal force,

$$P_n = P_1 \cos\theta + P_2 \sin\theta + Q_1 \sin\theta + Q_2 \cos\theta$$

Substituting the values of P_1, P_2, Q_1 and Q_2, we get

$$P_n = \sigma_1 . BC . \cos\theta + \sigma_2 . FB . \sin\theta + \tau . BC . \sin\theta + \tau . FB . \cos\theta$$

Similarly, the total tangential force (P_t) is obtained by resolving P_1, P_2, Q_1 and Q_2 along the oblique section FC.

$\therefore$ Total tangential force,

$$P_t = P_1 \sin\theta - P_2 \cos\theta - Q_1 \cos\theta + Q_2 \sin\theta$$
$$= \sigma_1 . BC . \sin\theta - \sigma_2 . FB . \cos\theta - \tau . BC . \cos\theta + \tau . FB . \sin\theta$$

(substitute the values of P_1, P_2, Q_1 and Q_2)

Now, Let σ_n = Normal stress across the section FC, and
σ_t = Tangential stress across the section FC.

Then normal stress across the section FC,

$$\sigma_n = \frac{\text{Total normal force across section } FC}{\text{Area of section } FC} = \frac{P_n}{FC \times 1}$$
$$= \frac{\sigma_1 . BC . \cos\theta + \sigma_2 . FB . \sin\theta + \tau . BC . \sin\theta + \tau . FB . \cos\theta}{FC \times 1}$$
$$= \sigma_1 . \frac{BC}{FC} . \cos\theta + \sigma_2 . \frac{FB}{FC} . \sin\theta + \tau . \frac{BC}{FC} . \sin\theta + \tau . \frac{FB}{FC} . \cos\theta$$
$$= \sigma_1 . \cos\theta . \cos\theta + \sigma_2 \sin\theta . \sin\theta + \tau . \cos\theta . \sin\theta + \tau \sin\theta . \cos\theta$$
$$\left(\because \text{ In triangle } FBC, \frac{BC}{FC} = \cos\theta \text{ and } \frac{FB}{FC} = \sin\theta\right)$$
$$= \sigma_1 \cos^2\theta + \sigma_2 \sin^2\theta + 2\tau \cos\theta \sin\theta$$
$$= \sigma_1 \left(\frac{1+\cos 2\theta}{2}\right) + \sigma_2 \left(\frac{1-\cos 2\theta}{2}\right) + \tau \sin 2\theta$$
$$\left(\because \cos^2\theta = \frac{1+\cos 2\theta}{2}, \sin^2\theta = \frac{1-\cos 2\theta}{2} \text{ and } 2\cos\theta\sin\theta = \sin 2\theta\right)$$
$$= \frac{\sigma_1 + \sigma_2}{2} + \frac{\sigma_1 - \sigma_2}{2} \cos 2\theta + \tau \sin 2\theta \quad \text{...(1.12)}$$

and tangential stress (*i.e.*, shear stress) across the section FC,

$$\sigma_t = \frac{\text{Total tangential force across section } FC}{\text{Area of section } FC} = \frac{P_t}{FC \times 1}$$
$$= \frac{\sigma_1 . BC . \sin\theta - \sigma_2 . FB . \cos\theta - \tau . BC . \cos\theta + \tau . FB . \sin\theta}{FC \times 1}$$
$$= \sigma_1 . \frac{BC}{FC} . \sin\theta - \sigma_2 . \frac{FB}{FC} . \cos\theta - \tau . \frac{BC}{FC} . \cos\theta + \tau . \frac{FB}{FC} . \sin\theta$$

$$= \sigma_1 \cdot \cos\theta \cdot \sin\theta - \sigma_2 \cdot \sin\theta \cdot \cos\theta - \tau \cdot \cos\theta \cdot \cos\theta + \tau \cdot \sin\theta \cdot \sin\theta$$

$$\left(\because \text{ In triangle } FBC, \frac{BC}{FC} = \cos\theta \text{ and } \frac{FB}{FC} = \sin\theta\right)$$

$$= (\sigma_1 - \sigma_2) \cdot \cos\theta \sin\theta - \tau\cos^2\theta + \tau\sin^2\theta$$

$$= \left(\frac{\sigma_1 - \sigma_2}{2}\right) \cdot 2\cos\theta\sin\theta - \tau(\cos^2\theta - \sin^2\theta)$$

$$= \frac{\sigma_1 - \sigma_2}{2} \cdot \sin 2\theta - \tau\cos 2\theta \qquad (\because \cos^2\theta - \sin^2\theta = \cos 2\theta) \quad ...(1.13)$$

1.5. PRINCIPAL STRESS

The planes on which shear stress (*i.e.*, tangential stress) is zero, are known as principal planes. And the stresses acting on principal planes are known as **principal stresses**.

The position of principal planes are obtained by equating the tangential stress [given by equation (1.13)] to zero.

$\therefore$ For principal planes, $\sigma_t = 0$

or
$$\frac{\sigma_1 - \sigma_2}{2}\sin 2\theta - \tau\cos 2\theta = 0$$

or
$$\frac{\sigma_1 - \sigma_2}{2}\sin 2\theta = \tau\cos 2\theta$$

or
$$\frac{\sin 2\theta}{\cos 2\theta} = \frac{\tau}{\frac{(\sigma_1 - \sigma_2)}{2}} = \frac{2\tau}{(\sigma_1 - \sigma_2)}$$

or
$$\tan 2\theta = \frac{2\tau}{(\sigma_1 - \sigma_2)} \quad ...(1.14)$$

But the tangent of any angle in a right angled triangle

$$= \frac{\text{Height of right angled triangle}}{\text{Base of right angled triangle}}$$

$$\therefore \quad \frac{\text{Height of right angled triangle}}{\text{Base of right angled triangle}} = \frac{2\tau}{(\sigma_1 - \sigma_2)}$$

$\therefore$ Height of right angled triangle $= 2\tau$

Base of right angled triangle $= (\sigma_1 - \sigma_2)$.

Now diagonal of the right angled triangle

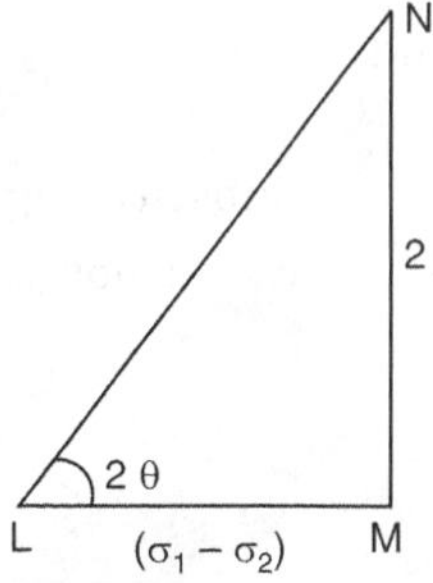

Fig. 1.13

$$= \pm\sqrt{(\sigma_1 - \sigma_2)^2 + (2\tau)^2} = \pm\sqrt{(\sigma_1 - \sigma_2)^2 + 4\tau^2}$$

$$= \sqrt{(\sigma_1 - \sigma_2)^2 + 4\tau^2} \quad \text{and} \quad -\sqrt{(\sigma_1 - \sigma_2)^2 + 4\tau^2}$$

1st Case. Diagonal $= \sqrt{(\sigma_1 - \sigma_2)^2 + 4\tau^2}$

Then
$$\sin 2\theta = \frac{\text{Height}}{\text{Diagonal}} = \frac{2\tau}{\sqrt{(\sigma_1 - \sigma_2)^2 + 4\tau^2}}$$

and
$$\cos 2\theta = \frac{\text{Base}}{\text{Diagonal}} = \frac{(\sigma_1 - \sigma_2)}{\sqrt{(\sigma_1 - \sigma_2)^2 + 4\tau^2}}.$$

The value of major principal stress is obtained by substituting the values of sin 2θ and cos 2θ in equation (1.12).

∴ Major principal stress

$$= \frac{\sigma_1 + \sigma_2}{2} + \frac{\sigma_1 - \sigma_2}{2} \cos 2\theta + \tau \sin 2\theta$$

$$= \frac{\sigma_1 + \sigma_2}{2} + \frac{\sigma_1 - \sigma_2}{2} \times \frac{(\sigma_1 - \sigma_2)}{\sqrt{(\sigma_1 - \sigma_2)^2 + 4\tau^2}} + \tau \times \frac{2\tau}{\sqrt{(\sigma_1 - \sigma_2)^2 + 4\tau^2}}$$

$$= \frac{\sigma_1 + \sigma_2}{2} + \frac{1}{2} \frac{(\sigma_1 - \sigma_2)^2}{\sqrt{(\sigma_1 - \sigma_2)^2 + 4\tau^2}} + \frac{2\tau^2}{\sqrt{(\sigma_1 - \sigma_2)^2 + 4\tau^2}}$$

$$= \frac{\sigma_1 + \sigma_2}{2} + \frac{(\sigma_1 - \sigma_2)^2 + 4\tau^2}{2\sqrt{(\sigma_1 - \sigma_2)^2 + 4\tau^2}}$$

$$= \frac{\sigma_1 + \sigma_2}{2} + \frac{1}{2}\sqrt{(\sigma_1 - \sigma_2)^2 + 4\tau^2}$$

$$= \frac{\sigma_1 + \sigma_2}{2} + \sqrt{\left(\frac{\sigma_1 - \sigma_2}{2}\right)^2 + \tau^2} \quad ...(1.15)$$

2nd Case. Diagonal $= -\sqrt{(\sigma_1 - \sigma_2)^2 + 4\tau^2}$

Then
$$\sin 2\theta = \frac{2\tau}{-\sqrt{(\sigma_1 - \sigma_2)^2 + 4\tau^2}}$$

and
$$\cos 2\theta = \frac{(\sigma_1 - \sigma_2)}{-\sqrt{(\sigma_1 - \sigma_2)^2 + 4\tau^2}}$$

Substituting these values in equation (1.12), we get minor principal stress.

∴ Minor principal stress

$$= \frac{\sigma_1 + \sigma_2}{2} + \frac{\sigma_1 - \sigma_2}{2} \cos 2\theta + \tau \sin 2\theta$$

$$= \frac{\sigma_1 + \sigma_2}{2} + \frac{\sigma_1 - \sigma_2}{2} \times \frac{\sigma_1 - \sigma_2}{-\sqrt{(\sigma_1 - \sigma_2)^2 + 4\tau^2}} + \tau \times \frac{2\tau}{-\sqrt{(\sigma_1 - \sigma_2)^2 + 4\tau^2}}$$

$$= \frac{\sigma_1 + \sigma_2}{2} - \frac{(\sigma_1 - \sigma_2)^2}{2\sqrt{(\sigma_1 - \sigma_2)^2 + 4\tau^2}} - \frac{2\tau^2}{\sqrt{(\sigma_1 - \sigma_2)^2 + 4\tau^2}}$$

$$= \frac{\sigma_1 + \sigma_2}{2} - \frac{(\sigma_1 - \sigma_2)^2 + 4\tau^2}{2\sqrt{(\sigma_1 - \sigma_2)^2 + 4\tau^2}}$$

$$= \frac{\sigma_1 + \sigma_2}{2} - \frac{1}{2}\sqrt{(\sigma_1 - \sigma_2)^2 + 4\tau^2}$$

$$= \frac{\sigma_1 + \sigma_2}{2} - \sqrt{\left(\frac{(\sigma_1 - \sigma_2)}{2}\right)^2 + \tau^2} \quad \text{...(1.16)}$$

Equation (1.15) gives the maximum principal stress whereas equation (1.16) gives minimum principal stress. These two principal planes are at right angles.

The position of principal planes is obtained by finding two values of θ from equation (1.14).

1.6. MAXIMUM SHEAR STRESS

The shear stress is given by equation (1.13). The shear stress will be maximum or minimum when

$$\frac{d}{d\theta}(\sigma_t) = 0$$

or
$$\frac{d}{d\theta}\left[\frac{\sigma_1 - \sigma_2}{2}\sin 2\theta - \tau\cos 2\theta\right] = 0$$

or
$$\frac{\sigma_1 - \sigma_2}{2}(\cos 2\theta) \times 2 - \tau(-\sin 2\theta) \times 2 = 0$$

$$(\sigma_1 - \sigma_2) \,.\, \cos 2\theta + 2\tau \sin 2\theta = 0$$

or
$$2\tau \sin 2\theta = -(\sigma_1 - \sigma_2)\cos 2\theta = (\sigma_2 - \sigma_1)\cos 2\theta$$

or
$$\frac{\sin 2\theta}{\cos 2\theta} = \frac{\sigma_2 - \sigma_1}{2\tau}$$

or
$$\tan 2\theta = \frac{\sigma_2 - \sigma_1}{2\tau} \quad \text{...(1.17)}$$

Equation (1.17) gives condition for maximum or minimum shear stress.

If $\tan 2\theta = \dfrac{\sigma_2 - \sigma_1}{2\tau}$

Then
$$\sin 2\theta = \pm\frac{\sigma_2 - \sigma_1}{\sqrt{(\sigma_2 - \sigma_1)^2 + 4\tau^2}}$$

and
$$\cos 2\theta = \pm\frac{2\tau}{\sqrt{(\sigma_2 - \sigma_1)^2 + 4\tau^2}}$$

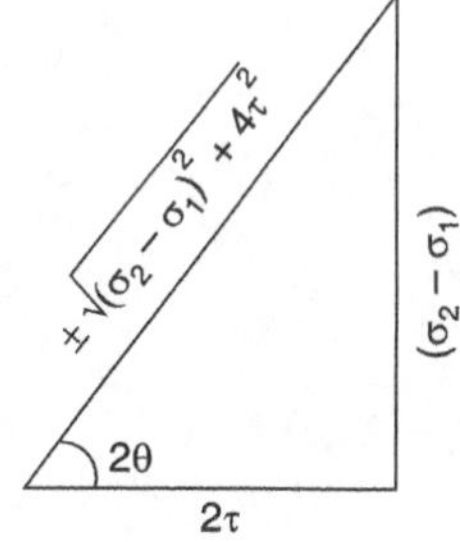

Fig. 1.14

Substituting the values of $\sin 2\theta$ and $\cos 2\theta$ in equation (1.13), the maximum and minimum shear stresses are obtained.

∴ Maximum shear stress is given by

$$(\sigma_t)_{max} = \frac{\sigma_1 - \sigma_2}{2}\sin 2\theta - \tau\cos 2\theta$$

$$= \pm\frac{\sigma_1 - \sigma_2}{2} \times \frac{(\sigma_2 - \sigma_1)}{\sqrt{(\sigma_2 - \sigma_1)^2 + 4\tau^2}} \pm \tau \times \frac{2\tau^2}{\sqrt{(\sigma_2 - \sigma_1)^2 + 4\tau^2}}$$

$$= \pm\frac{(\sigma_1 - \sigma_2)^2}{2\sqrt{(\sigma_2 - \sigma_1)^2 + 4\tau^2}} \pm \frac{2\tau^2}{\sqrt{(\sigma_2 - \sigma_1)^2 + 4\tau^2}}$$

$$= \pm \frac{(\sigma_2 - \sigma_1)^2 + 4\tau^2}{2\sqrt{(\sigma_2 - \sigma_1)^2 + 4\tau^2}} = \pm \frac{1}{2}\sqrt{(\sigma_2 - \sigma_1)^2 + 4\tau^2}$$

$$\therefore \quad (\sigma_t)_{max} = \frac{1}{2}\sqrt{(\sigma_2 - \sigma_1)^2 + 4\tau^2}$$

$$= \frac{1}{2}\sqrt{(\sigma_1 - \sigma_2)^2 + 4\tau^2} \quad \text{...(1.18)}$$

The planes on which maximum shear stress is acting, are obtained after finding the two values of θ from equation (1.17). These two values of θ will differ by 90°.

Note. The above relations hold good when one or both the stresses are compressive.

Problem 1.8. *At a point within a body subjected to two mutually perpendicular directions, the stresses are 80 N/mm² tensile and 40 N/mm² tensile. Each of the above stresses is accompanied by a shear stress of 60 N/mm². Determine the normal stress, shear stress and resultant stress on an oblique plane inclined at an angle of 45° with the axis of minor tensile stress.*

Sol. Given :

Major tensile stress, $\sigma_1 = 80 \text{ N/mm}^2$

Minor tensile stress, $\sigma_2 = 40 \text{ N/mm}^2$

Shear stress, $\tau = 60 \text{ N/mm}^2$

Angle of oblique plane, with the axis of minor tensile stress,

$\theta = 45°$.

(i) Normal stress (σ_n)

Using equation (1.12),

$$\sigma_n = \frac{\sigma_1 + \sigma_2}{2} + \frac{\sigma_1 - \sigma_2}{2}\cos 2\theta + \tau \sin 2\theta$$

$$= \frac{80 + 40}{2} + \frac{80 - 40}{2}\cos(2 \times 45°) + 60 \sin(2 \times 45°)$$

$$= 60 + 20 \cos 90° + 60 \sin 90°$$

$$= 60 + 20 \times 0 + 60 \times 1 \quad (\because \cos 90° = 0)$$

$$= 60 + 0 + 60 = \mathbf{120 \text{ N/mm}^2}.$$ **Ans.**

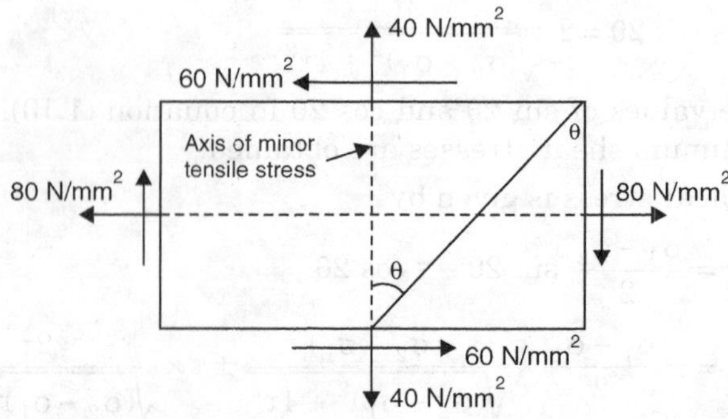

Fig. 1.15

(ii) Shear (or tangential) stress (σ_t)

Using equation (1.13),

$$\sigma_t = \frac{\sigma_1 - \sigma_2}{2} \sin 2\theta - \tau \cos 2\theta$$

$$= \frac{80 - 40}{2} \sin (2 \times 45°) - 60 \times \cos (2 \times 45°)$$

$$= 20 \times \sin 90° - 60 \cos 90°$$

$$= 20 \times 1 - 60 \times 0$$

$$= \mathbf{20\ N/mm^2.} \quad \textbf{Ans.}$$

(iii) Resultant stress (σ_R)

Using equation, $\sigma_R = \sqrt{\sigma_n{}^2 + \sigma_t{}^2}$

$$= \sqrt{120^2 + 20^2} = \sqrt{14400 + 400}$$

$$= \sqrt{14800} = \mathbf{121.655\ N/mm^2.} \quad \textbf{Ans.}$$

Problem 1.9. *A rectangular block of material is subjected to a tensile stress of 110 N/mm² on one plane and a tensile stress of 47 N/mm² on the plane at right angles to the former. Each of the above stresses is accompanied by a shear stress of 63 N/mm² and that associated with the former tensile stress tends to rotate the block anti-clockwise. Find :*

(i) the direction and magnitude of each of the principal stress and

(ii) magnitude of the greatest shear stress.

Sol. Given :

Major tensile stress, $\sigma_1 = 110$ N/mm²

Minor tensile stress, $\sigma_2 = 47$ N/mm²

Shear stress, $\tau = 63$ N/mm²

(i) Major principal stress is given by equation (1.15).

$$\therefore \text{ Major principal stress} = \frac{\sigma_1 + \sigma_2}{2} + \sqrt{\left(\frac{\sigma_1 - \sigma_2}{2}\right)^2 + \tau^2}$$

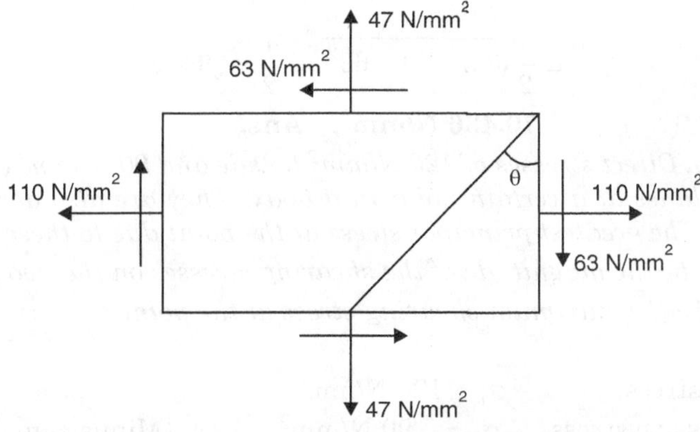

Fig. 1.16

$$= \frac{110+47}{2} + \sqrt{\left(\frac{110-47}{2}\right)^2 + 63^2} = \frac{157}{2} + \sqrt{\left(\frac{63}{2}\right)^2 + (63)^2}$$

$$= 78.5 + \sqrt{31.5^2 + 63^2} = 78.5 + \sqrt{992.25 + 3969}$$

$$= 78.5 + 70.436 = \mathbf{148.936\ N/mm^2}.\quad \textbf{Ans.}$$

Minor principal stress is given by equation (1.16).

$$\therefore \text{ Minor principal stress} = \frac{\sigma_1 + \sigma_2}{2} - \sqrt{\left(\frac{\sigma_1 - \sigma_2}{2}\right)^2 + \tau^2}$$

$$= \frac{110+47}{2} - \sqrt{\left(\frac{110-47}{2}\right)^2 + 63^2} = 78.5 - 70.436$$

$$= \mathbf{8.064\ N/mm^2}.\quad \textbf{Ans.}$$

The directions of principal stresses are given by equation (1.14).

$\therefore$ Using equation (1.14),

$$\tan 2\theta = \frac{2\tau}{\sigma_1 - \sigma_2} = \frac{2 \times 63}{110 - 47} = \frac{2 \times 63}{63} = 2.0$$

$$\therefore \quad 2\theta = \tan^{-1} 2.0 = 63°\ 26' \text{ or } 243°\ 26'$$

$$\therefore \quad \theta = \mathbf{31°\ 43' \text{ or } 121°\ 43'}.\quad \textbf{Ans.}$$

(ii) Magnitude of the greatest shear stress

Greatest shear stress is given by equation (1.18).

Using equation (1.18),

$$(\sigma_t)_{max} = \frac{1}{2}\sqrt{(\sigma_1 - \sigma_2)^2 + 4\tau^2}$$

$$= \frac{1}{2}\sqrt{(100-47)^2 + 4 \times 63^2}$$

$$= \frac{1}{2}\sqrt{63^2 + 4 \times 63^2} = \frac{1}{2} \times 63 \times \sqrt{5}$$

$$= \mathbf{70.436\ N/mm^2}.\quad \textbf{Ans.}$$

Problem 1.10. *Direct stresses of 120 N/mm² tensile and 90 N/mm² compression exist on two perpendicular planes at a certain point in a body. They are also accompanied by shear stress on the planes. The greatest principal stress at the point due to these is 150 N/mm².*

(a) What must be the magnitude of the shearing stresses on the two planes ?

(b) What will be the maximum shearing stress at the point ?

Sol. Given :

Major tensile stress, $\sigma_1 = 120$ N/mm²

Minor compressive stress, $\sigma_2 = -90$ N/mm² (Minus sign due to compression)

Greatest principal stress $= 150$ N/mm²

(*a*) Let τ = Shear stress on the two planes.

Using equation (1.15) for greatest principal stress, we get

$$\text{Greatest principal stress} = \frac{\sigma_1 + \sigma_2}{2} + \sqrt{\left(\frac{\sigma_1 - \sigma_2}{2}\right)^2 + \tau^2}$$

or $$150 = \frac{120 + (-90)}{2} + \sqrt{\left(\frac{120 - (-90)}{2}\right)^2 + \tau^2}$$

$$= \frac{120 - 90}{2} + \sqrt{\left(\frac{120 + 90}{2}\right)^2 + \tau^2} = 15 + \sqrt{105^2 + \tau^2}$$

or $$150 - 15 = \sqrt{105^2 + \tau^2}$$

or $$135 = \sqrt{105^2 + \tau^2}$$

Squaring both sides, we get

$$135^2 = 105^2 + \tau^2$$

or $$\tau^2 = 135^2 - 105^2 = 18225 - 11025 = 7200$$

$\therefore$ $$\tau = \sqrt{7200} = \mathbf{84.853\ N/mm^2.}\quad \textbf{Ans.}$$

(b) Maximum shear stress at the point

Using equation (1.18) for maximum shear stress,

$$(\sigma_t)_{max} = \frac{1}{2}\sqrt{(\sigma_1 - \sigma_2)^2 + 4\tau^2}$$

$$= \frac{1}{2}\sqrt{[120 - (-90)]^2 + 4 \times 7200} \qquad (\because\ \tau^2 = 7200)$$

$$= \frac{1}{2}\sqrt{210^2 + 28800} = \frac{1}{2}\sqrt{44100 + 28800} = \frac{1}{2} \times 270$$

$$= \mathbf{135\ N/mm^2.}\quad \textbf{Ans.}$$

Problem 1.11. *A point in a strained material is subjected to the stresses as shown in Fig. 1.17.*

Locate the principal planes, and evaluate the principal stresses.

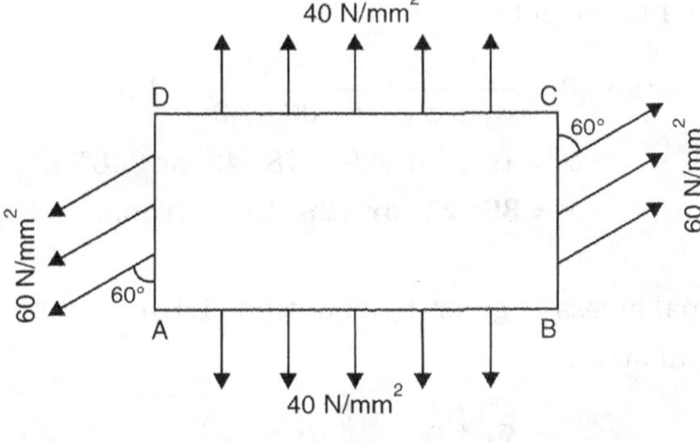

Fig. 1.17

Sol. Given :

The stress on the face *BC* or *AD* is not normal. It is inclined at an angle of 60° with face *BC* or *AD*. This stress can be resolved into two components *i.e.*, normal to the face *BC* (or *AD*) and along the face *BC* (or *AD*).

∴ Stress normal to the face *BC* or *AD*

$$= 60 \times \sin 60° = 60 \times 0.866 = 51.96 \text{ N/mm}^2$$

Stress along the face *BC* or *AD*

$$= 60 \times \cos 60° = 60 \times 0.5 = 30 \text{ N/mm}^2$$

The stress along the face *BC* or *AD* is known as shear stress. Hence $\tau = 30$ N/mm². Due to complementary shear stress the face *AB* and *CD* will also be subjected to shear stress of 30 N/mm². Now the stresses acting on the material are shown in Fig. 1.17 (*a*).

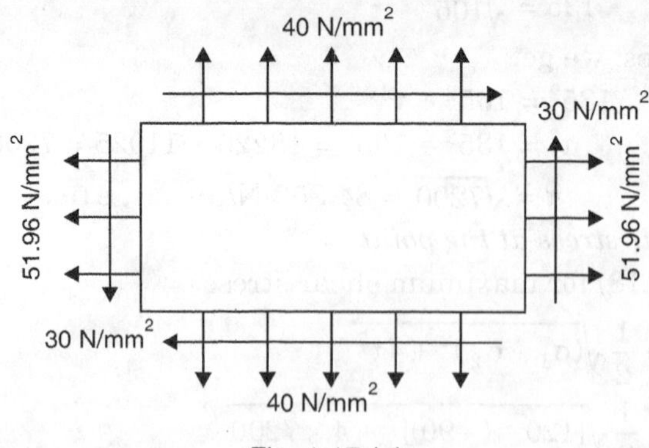

Fig. 1.17 (*a*)

Major tensile stress, $\sigma_1 = 51.96$ N/mm²

Minor tensile stress, $\sigma_2 = 40$ N/mm²

Shear stress, $\tau = 30$ N/mm²

Location of principal planes

Let θ = Angle, which one of the principal planes make with the stress of 40 N/mm².

The location of the principal planes is given by the equation (1.14).

Using equation (1.14), we get

$$\tan 2\theta = \frac{2\tau}{\sigma_1 - \sigma_2} = \frac{2 \times 30}{51.96 - 40} = 4.999$$

$$\therefore \quad 2\theta = \tan^{-1} 4.999 = 78° \, 42' \text{ or } 258° \, 42'$$

or θ = **39° 21′ or 129° 21′. Ans.**

Principal stress

The major principal stress is given by equation (1.15).

∴ Major principal stress

$$= \frac{\sigma_1 + \sigma_2}{2} + \sqrt{\left(\frac{\sigma_1 - \sigma_2}{2}\right)^2 + \tau^2}$$

$$= \frac{51.96 + 40}{2} + \sqrt{\left(\frac{51.96 - 40}{2}\right)^2 + 30^2}$$

$$= 45.98 + 30.6$$

$$= \mathbf{76.58\ N/mm^2.}\quad \textbf{Ans.}$$

The minor principal stress is given by equation (1.16).

∴ Minor principal stress

$$= \frac{\sigma_1 + \sigma_2}{2} - \sqrt{\left(\frac{\sigma_1 - \sigma_2}{2}\right)^2 + \tau^2}$$

$$= \frac{51.96 + 40}{2} - \sqrt{\left(\frac{51.96 - 40}{2}\right)^2 + 30^2}$$

$$= 45.98 - 30.6$$

$$= \mathbf{15.38\ N/mm^2.}\quad \textbf{Ans.}$$

Problem 1.12. *The normal stress in two mutually perpendicular directions are 600 N/mm² and 300 N/mm² both tensile. The complimentary shear stresses in these directions are of intensity 450 N/mm². Find the normal and tangential stresses on the two planes which are equally inclined to the planes carrying the normal stresses mentioned above.*

Sol. Given :

Major tensile stress, $\sigma_1 = 600\ N/mm^2$

Minor tensile stress, $\sigma_2 = 300\ N/mm^2$

Shear stress, $\tau = 450\ N/mm^2$

The normal and tangential stresses are to be calculated on the two planes which are equally inclined to the planes of major tensile stress and of minor tensile stress. This means $\theta = 45°$ and $135°$.

∴ Angle $\theta = 45°$ and $135°$.

(*i*) Normal stress (σ_n) is given by equation (1.12).

$$\therefore \qquad \sigma_n = \frac{\sigma_1 - \sigma_2}{2} + \frac{\sigma_1 - \sigma_2}{2} \cos 2\theta + \tau \sin 2\theta$$

(*a*) When $\theta = 45°$, the normal stresses (σ_n) becomes as

$$\sigma_n = \frac{600 + 300}{2} + \frac{600 - 300}{2} \cos (2 \times 45°) + 450 \sin (2 \times 45°)$$

$$= 450 + 150 \cos 90° + 450 \sin 90°$$

$$= 450 + 150 \times 0 + 450 \times 1 \qquad (\because \cos 90° = 0 \text{ and } \sin 90° = 1)$$

$$= \mathbf{900\ N/mm^2.}\quad \textbf{Ans.}$$

(*b*) When $\theta = 135°$, the normal stress (σ_n) becomes as

$$\sigma_n = \frac{600 + 300}{2} + \frac{600 - 300}{2} \cos (2 \times 135) + 450 \sin (2 \times 135°)$$

$$= 450 + 150 \cos (270°) + 450 \sin 270°$$

$$= 450 + 150 \times 0 + 450 \times (-1) \quad (\because \cos 270° = 0 \text{ and } \sin 270° = -1)$$

$$= \mathbf{450 - 450 = 0.}\quad \textbf{Ans.}$$

(*ii*) Tangential stress (σ_t) is given by equation (1.13)

$$\therefore \qquad \sigma_t = \frac{\sigma_1 - \sigma_2}{2} \sin 2\theta - \tau \cos 2\theta$$

(*a*) When θ = 45°, the tangential stress (σ_t) becomes as

$$\sigma_t = \frac{600 - 300}{2} \sin 90° - 450 \cos 90°$$

$$= 150 \times 1 - 450 \times 0 = \mathbf{150\ N/mm^2}. \quad \textbf{Ans.}$$

(*b*) When θ = 135°, the tangential stress (σ_t) becomes as

$$\sigma_t = \frac{600 - 300}{2} \sin 270° - 450 \cos 270°$$

$$= 150 \times (-1) - 450 \times 0 = \mathbf{-150\ N/mm^2}. \quad \textbf{Ans.}$$

Problem 1.13. *The intensity of resultant stress on a plane AB (Fig. 1.18) at a point in a material under stress is 800 N/cm² and it is inclined at 30° to the normal to that plane. The normal component of stress on another plane BC at right angles to plane AB is 600 N/cm².*

Determine the following :

(*i*) *the resultant stress on the plane BC,*

(*ii*) *the principal stresses and their directions,*

(*iii*) *the maximum shear stresses and their planes.*

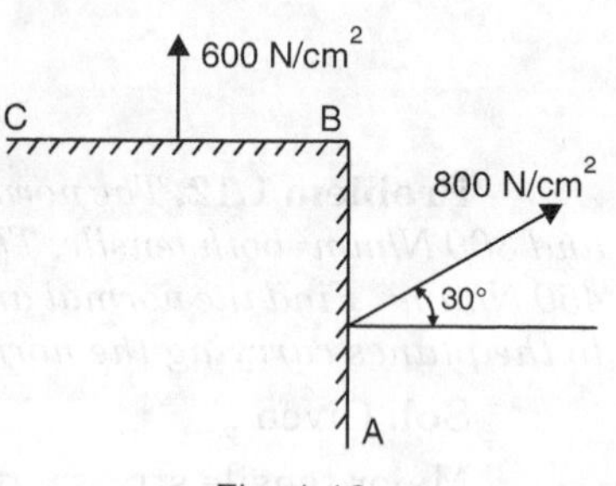

Fig. 1.18

Sol. Given :

Resultant stress on plane *AB* = 800 N/cm²

Angle of inclination of the above stress = 30°

Normal stress on plane *BC* = 600 N/cm²

The resultant stress 800 N/cm² on plane *AB* is resolved into normal stress and tangential stress.

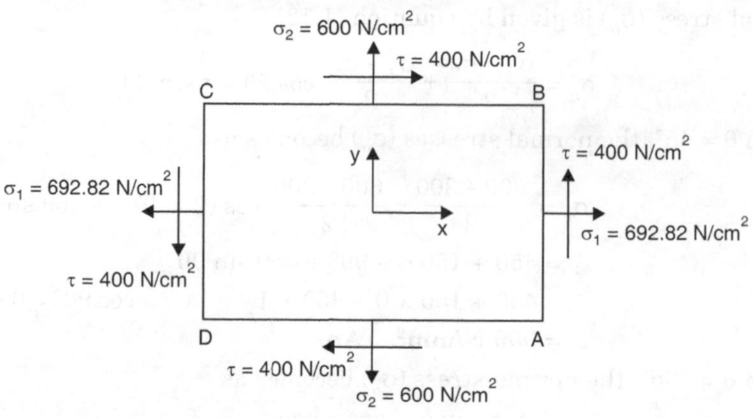

Fig. 1.18 (*a*)

The normal stress on plane *AB*

$$= 800 \times \cos 30° = 692.82\ \text{N/cm}^2.$$

The tangential stress on plane *AB*

$$= 800 \times \sin 30° = 400\ \text{N/cm}^2.$$

The shear stress on plane *AB* is, *i.e.*, τ_{AB} = 400 N/cm², then to maintain the equilibrium on the wedge *ABC*, another shear stress of the same magnitude, *i.e.*, τ_{BC} = 400 N/cm² must act on the plane *BC*. The free body diagram of the element *ABCD* is shown in Fig. 1.16 (*a*), showing normal and shear stresses acting on different faces.

(*i*) *Resultant stress on plane BC*

On plane *BC*, from Fig. 1.16 (*a*),

$$\sigma_2 = 600 \text{ N/cm}^2$$

Shear stress, $\tau = 400 \text{ N/cm}^2$

∴ Resultant stress on plane *BC*

$$= \sqrt{\sigma_2^{\,2} + \tau^2}$$

$$= \sqrt{600^2 + 400^2} = \mathbf{721\ N/cm^2.}\quad \textbf{Ans.}$$

The resultant will be inclined at an angle θ with the horizontal given by,

$$\tan\theta = \frac{\sigma_2}{\tau} = \frac{600}{400} = 1.5$$

∴ $\theta = \tan^{-1} 1.5 = \mathbf{56.3°}.$ **Ans.**

(*ii*) *Principal stresses and their directions*

The major principal stress is given by equation (1.15).

∴ Major principal stress

$$= \frac{\sigma_1 + \sigma_2}{2} + \sqrt{\left(\frac{\sigma_1 - \sigma_2}{2}\right)^2 + \tau^2}$$

$$= \frac{692.82 + 600}{2} + \sqrt{\left(\frac{692.82 - 600}{2}\right)^2 + 400^2}$$

$$= 646.41 - 402.68$$

$$= \mathbf{1049.09\ N/cm^2\ (Tensile).}\quad \textbf{Ans.}$$

The minor principal stress is given by equation (1.16)

∴ Minor principal stress

$$= \frac{\sigma_1 + \sigma_2}{2} - \sqrt{\left(\frac{\sigma_1 - \sigma_2}{2}\right)^2 + \tau^2}$$

$$= \frac{692.82 + 600}{2} - \sqrt{\left(\frac{692.82 - 600}{2}\right)^2 + 400^2}$$

$$= 646.41 - 402.68$$

$$= \mathbf{243.73\ N/cm^2\ (Tensile).}\quad \textbf{Ans.}$$

The directions of principal stresses are given by equation (1.14), as

$$\tan 2\theta = \frac{2\tau}{(\sigma_1 - \sigma_2)} = \frac{2 \times 400}{(692.82 - 600)} = \frac{800}{92.82} = 8.618$$

∴ $2\theta = \tan^{-1} 8.618 = 83.38° \text{ or } 263.38°$

∴ $\theta = \mathbf{41.69° \text{ or } 131.99°}.$ **Ans.**

(*iii*) *The maximum shear stress and their planes.*

The maximum shear stress is given by equation (1.18).

$$\therefore \quad (\sigma_t)_{max} = \frac{1}{2}\sqrt{(\sigma_1 - \sigma_2)^2 + 4\tau^2} = \sqrt{\left(\frac{\sigma_1 - \sigma_2}{2}\right)^2 + \tau^2}$$

$$= \sqrt{\left(\frac{692.82 - 600}{2}\right)^2 + 400^2}$$

$$= \mathbf{402.68\ N/cm^2.} \quad \textbf{Ans.}$$

Problem 1.14. *At a point in a strained material, on plane BC there are normal and shear stresses of 560 N/mm² and 140 N/mm² respectively. On plane AC, perpendicular to plane BC, there are normal and shear stresses of 280 N/mm² and 140 N/mm² respectively as shown in Fig. 1.19. Determine the following :*

(i) principal stresses and location of the planes on which they act,

(ii) maximum shear stress and the plane on which it acts.

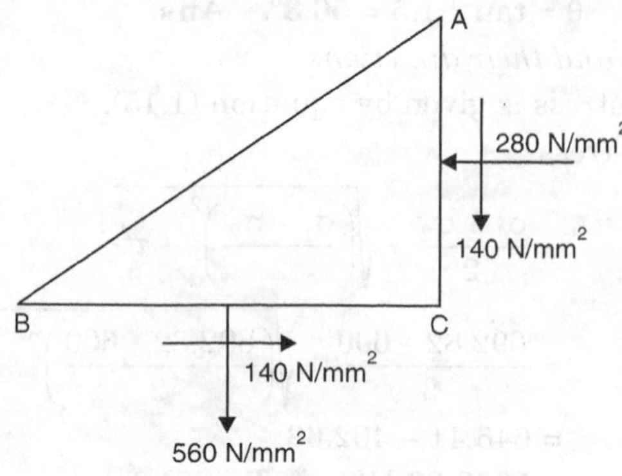

Fig. 1.19

Sol. Given :

On plane AC, $\sigma_1 = -280$ N/mm² (– ve sign due to compressive stress)

$\tau = 140$ N/mm²

On plane BC, $\sigma_2 = 560$ N/mm²

$\tau = 140$ N/mm²

(i) Principal stresses and location of the planes on which they act.

Principal stress are given by equations (1.15) and (1.16)

$$\therefore \quad \text{Principal stresses} = \frac{\sigma_1 + \sigma_2}{2} \pm \sqrt{\left(\frac{\sigma_1 - \sigma_2}{2}\right)^2 + \tau^2}$$

$$= \frac{-280 + 560}{2} \pm \sqrt{\left(\frac{-280 - 560}{2}\right)^2 + 140^2}$$

$$= 140 \pm 442.7$$

$$= 582.7 \text{ and } (140 - 442.7) \text{ N/mm}^2$$

$$= 582.7 \text{ and } -302.7 \text{ N/mm}^2$$

∴ Major principal stress $= 582.7 \text{ N/mm}^2$ **(Tensile). Ans.**

∴ Minor principal stress $= -302.7 \text{ N/mm}^2$. **Ans.**

The planes on which principal stresses act, are given by equation (1.14) as

$$\tan 2\theta = \frac{2\tau}{\sigma_1 - \sigma_2} = \frac{2 \times 140}{-280 - 560} = \frac{280}{-840} = -0.33$$

∴ $$2\theta = \tan^{-1} - 0.33 = -18.26°$$

– ve sign shows that 2θ is lying in 2nd and 4th quadrant

∴ $$2\theta = (180 - 18.26°) \text{ or } (360 - 18.26°)$$
$$= 161.34° \text{ or } 341.34°$$

∴ $\theta =$ **80.67° and 170.67°. Ans.**

(ii) Maximum shear stress and the plane on which it acts.

Maximum shear stress is given by equation (1.18).

∴ $$(\sigma_t)_{max} = \sqrt{\left(\frac{\sigma_1 - \sigma_2}{2}\right)^2 + \tau^2}$$

$$= \sqrt{\left(\frac{-280 - 560}{2}\right)^2 + 140^2}$$

$$= \sqrt{420^2 + 140^2} = \mathbf{442.7 \text{ N/mm}^2}. \textbf{ Ans.}$$

The plane on which maximum shear stress acts is given by equation (1.17) as

$$\tan 2\theta = \frac{\sigma_2 - \sigma_1}{2\tau}$$

$$= \frac{560 - (-280)}{2 \times 140} = \frac{840}{280} = 3.0$$

∴ $$2\theta = \tan^{-1} 3.0 = 71.56° \text{ or } 251.56°$$

∴ $\theta =$ **35.78° or 125.78°. Ans.**

1.7. MOHR'S STRESS CIRCLE

Mohr's stress circle is a graphical method of finding normal, tangential and resultant stresses on an oblique plane. Mohr's circle will be drawn for the following cases :

(i) A body subjected to two mutually perpendicular principal tensile stresses of unequal intensities.

(ii) A body subjected to two mutually perpendicular principal stresses which are unequal and unlike (*i.e.*, one is tensile and other is compressive).

(iii) A body subjected to two mutually perpendicular principal tensile stresses accompanied by a simple shear stress.

1.7.1. Mohr's Circle when a Body is Subjected to two Mutually Perpendicular Principal Tensile Stresses of Unequal Intensities. Consider a rectangular body subjected to two mutually perpendicular principal tensile stresses of unequal intensities. It is required to find the resultant stress on an oblique plane.

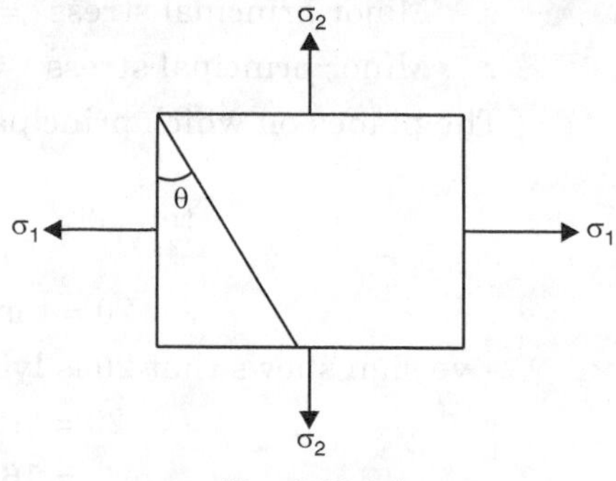

Fig. 1.20 (*a*)

Let σ_1 = Major tensile stress

σ_2 = Minor tensile stress, and

θ = Angle made by the oblique plane with the plane of stress σ_1.

Mohr's circle is drawn as : [See Fig. 1.20 (*b*)].

Take any point A and draw a horizontal line through A. Take $AB = \sigma_1$ and $AC = \sigma_2$ towards right from A to some suitable scale. With BC as diameter describe a circle. Let O is the centre of the circle. Now through O, draw a line OE marking an angle 2θ with OB in the same direction as the oblique plane makes with the plane of stress σ_1. In Fig. 1.20 (*a*), the oblique plane makes an angle θ with the plane of σ_1 in anticlockwise direction. The line OE is therefore drawn at an angle 2θ in anti-clockwise direction in Fig. 1.20 (*b*).

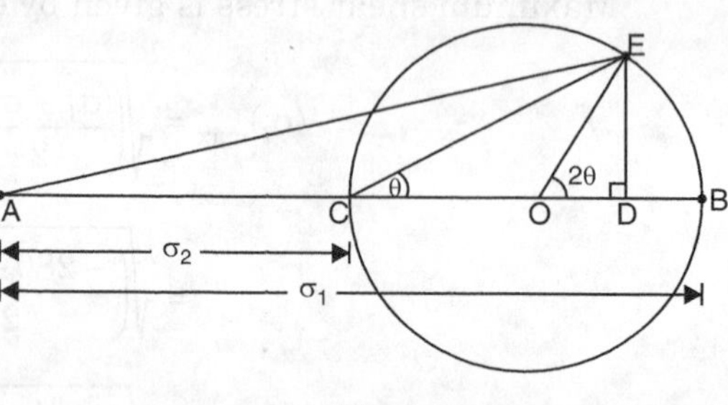

Fig. 1.20 (*b*)

From E, draw ED perpendicular on AB. Join AE. Then the normal and tangential stresses on the oblique plane are given by AD and ED respectively. The resultant stress on the oblique plane is given by AE.

From Fig. 1.20 (*b*), we have

Length AD = Normal stress on oblique plane

Length ED = Tangential stress on oblique plane

Length AE = Resultant stress on oblique plane.

Radius of Mohr's circle $= \dfrac{\sigma_1 - \sigma_2}{2}$

Angle ϕ = obliquity.

Proof. [See Fig. 1.20 (*b*)]

$$CO = OB = OE = \text{Radius of Mohr's circle} = \frac{\sigma_1 - \sigma_2}{2}$$

$$\therefore \quad AO = AC + CO$$

$$= \sigma_2 + \frac{\sigma_1 - \sigma_2}{2} = \frac{2\sigma_2 + \sigma_1 - \sigma_2}{2} = \frac{\sigma_1 + \sigma_2}{2}$$

$$OD = OE \cos 2\theta$$

$$= \frac{\sigma_1 - \sigma_2}{2} \cos 2\theta \qquad \left(\because \; OE = \frac{\sigma_1 - \sigma_2}{2}\right)$$

$\therefore$
$$AD = AO + OD$$
$$= \frac{\sigma_1 + \sigma_2}{2} + \frac{\sigma_1 - \sigma_2}{2} \cos 2\theta$$
$$= \sigma_n \text{ or Normal stress}$$

and
$$ED = OE \sin 2\theta$$
$$= \frac{\sigma_1 - \sigma_2}{2} \sin 2\theta$$
$$= \sigma_t \text{ or Tangential stress.}$$

Important points. [See Fig. 1.20 (*b*)]

(*i*) Normal stress is along the line *ACB*. Hence maximum normal stress will be when point *E* is at *B*. And minimum normal stress will be when point *E* is at *C*. Hence maximum normal stress = *AB* = σ_1 and minimum normal stress = *AB* = σ_2.

(*ii*) Tangential stress (or shear stress) is along a line which is perpendicular to line *CB*. Hence maximum shear stress will be when perpendicular to line *CB* is drawn from point *O*. Then maximum shear stress will be equal to the radius of the Mohr's circle.

$\therefore$
$$(\sigma_t)_{max} = \frac{\sigma_1 - \sigma_2}{2}.$$

(*iii*) When the point *E* is at *B* or at *C*, the shear stress will be zero.

(*iv*) The angle ϕ (which is known as angle of obliquity) will be maximum, when the line *AE* is tangent to the Mohr's circle.

Problem 1.15. *Solve problem 1.3 by using Mohr's circle method.*

Sol. The data given in problem 1.3, is

$$\sigma_1 = 120 \text{ N/mm}^2 \text{ (tensile)}$$
$$\sigma_2 = 60 \text{ N/mm}^2 \text{ (tensile)}$$
$$\theta = 30°.$$

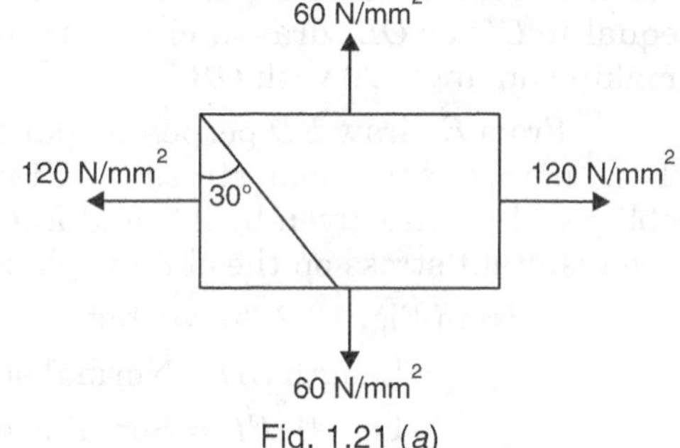

Fig. 1.21 (*a*)

Scale. Let $1 \text{ cm} = 10 \text{ N/mm}^2$

Then
$$\sigma_1 = \frac{120}{10} = 12 \text{ cm}$$

and
$$\sigma_2 = \frac{60}{10} = 6 \text{ cm}$$

Mohr's circle is drawn as : [See Fig. 1.21 (*b*)].

Take any point *A* and draw a horizontal line through *A*. Take *AB* = σ_1 = 12 cm and *AC* = σ_2 = 6 cm. With *BC* as diameter (*i.e.*, *BC* = 12 – 6 = 6 cm) describe a circle. Let *O* is the centre of the circle. Through *O*, draw a line *OE* making an angle 2θ (*i.e.*, 2 × 30 = 60°) with *OB*. From *E*, draw *ED* perpendicular to *CB*. Join *AE*. Measure lengths *AD*, *ED* and *AE*.

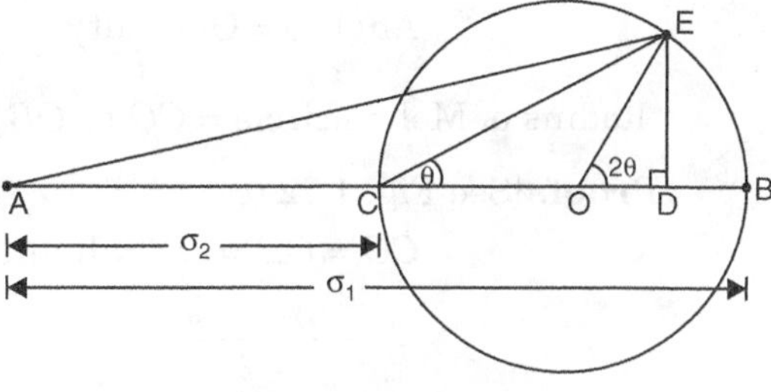

Fig. 1.21 (*b*)

By measurements :

Length *AD* = 10.50 cm

Length *ED* = 2.60 cm

Length *AE* = 10.82 cm

Then normal stress = Length $AD \times$ Scale
= 10.50×10 = **105 N/mm²**. **Ans.**

Tangential or shear stress = Length $ED \times$ Scale
= 2.60×10 = **26 N/mm²**. **Ans.**

Resultant stress = Length $AE \times$ Scale
= 10.82×10 = **108.2 N/mm²**. **Ans.**

1.7.2. Mohr's Circle when a Body is Subjected to two Mutually Perpendicular Principal Stresses which are Unequal and Unlike (*i.e.*, one is Tensile and other is Compressive). Consider a rectangular body subjected to two mutually perpendicular principal stresses which are unequal and one of them is tensile and the other is compressive. It is required to find the resultant stress on an oblique plane.

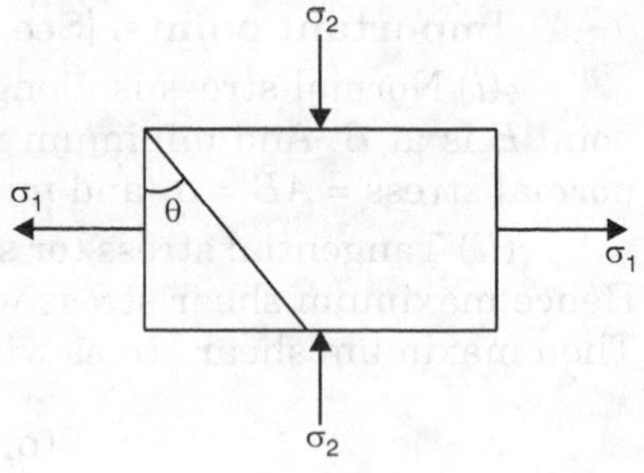

Fig. 1.22 (*a*)

Let σ_1 = Major principal tensile stress,

σ_2 = Minor principal compressive stress, and

θ = Angle made by the oblique plane with the plane of stress σ_1.

Mohr's circle is drawn as : [See Fig. 1.22 (*b*)]

Take any point A and draw a horizontal line through A on both sides of A as shown in Fig. 1.22 (*b*). Take $AB = \sigma_1(+)$ towards right of A and $AC = \sigma_2(-)$ towards left of A to some suitable scale. Bisect BC at O. With O as centre and radius equal to CO or OB, draw a circle. Through O draw a line OE making an angle 2θ with OB.

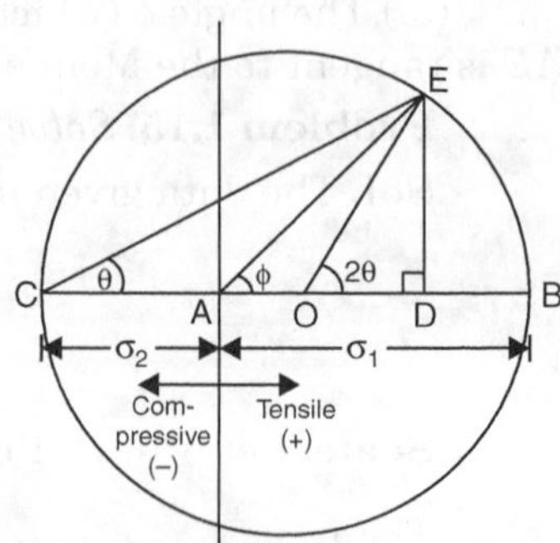

Fig. 1.22 (*b*)

From E, draw ED perpendicular to AB. Join AE and CE. Then normal and shear stress (*i.e.*, tangential stress) on the oblique plane are given by AD and ED. Length AE represents the resultant stress on the oblique plane.

∴ From Fig. 1.22 (*b*), we have

Length AD = Normal stress on oblique plane,

Length ED = Shear stress on oblique plane,

Length AE = Resultant stress on oblique plane, and

Angle ϕ = Obliquity.

$$\text{Radius of Mohr's circle} = CO \text{ or } OB = \frac{\sigma_1 + \sigma_2}{2}.$$

Proof. [See Fig. 1.22 (*b*)].

$$CO = OB = OE = \text{Radius of Mohr's circle}$$

$$= \frac{\sigma_1 + \sigma_2}{2}$$

$$AO = OC - AC$$

$$= \frac{\sigma_1 + \sigma_2}{2} - \sigma_2 = \frac{\sigma_1 + \sigma_2 - 2\sigma_2}{2} = \frac{\sigma_1 - \sigma_2}{2}$$

$\therefore \quad AD = AO + OD$

$= AO + OE \cos 2\theta \qquad (\because \; OD = OE \cos 2\theta)$

$= \frac{\sigma_1 - \sigma_2}{2} + \frac{\sigma_1 + \sigma_2}{2} \cos 2\theta \qquad \left(\because \; OE = \text{Radius} = \frac{\sigma_1 + \sigma_2}{2}\right)$

$= \sigma_n$ or Normal stress

and $\quad ED = OE \sin 2\theta$

$= \frac{\sigma_1 + \sigma_2}{2} \sin 2\theta \qquad \left(\because \; OE = \frac{\sigma_1 + \sigma_2}{2}\right)$

$= \sigma_t$ or Tangential (or shear) stress.

Problem 1.16. *Solve problem 1.4 by using Mohr's circle method.*

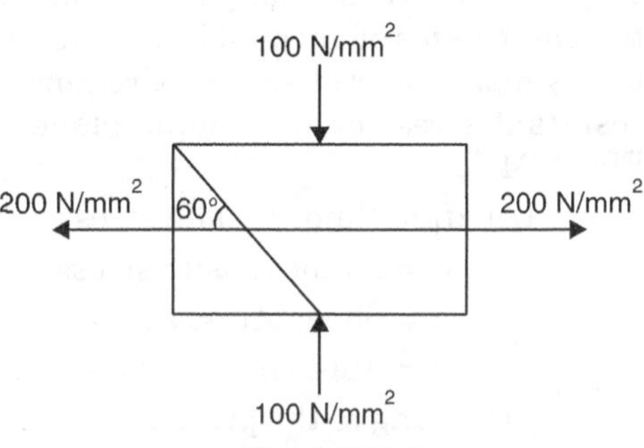

Fig. 1.23 (*a*)

Sol. Given : The data given in problem 1.4, is

$\sigma_1 = 200$ N/mm²

$\sigma_2 = -100$ N/mm² (compressive)

$\theta = 30°$.

It is required to determine the resultant stress and the maximum shear stress by Mohr's circle method. First choose a suitable scale.

Let 1 cm represents 20 N/mm².

Then $\quad \sigma_1 = \frac{200}{20} = 10$ cm

and $\quad \sigma_2 = \frac{-100}{20} = -5$ cm

Mohr's circle is drawn as given in Fig. 1.23 (*b*).

Take any point *A* and draw a horizontal line through *A* on both sides of *A*. Take $AB = \sigma_1 = 10$ cm towards right of *A* and $AC = \sigma_2 = -5$ cm towards left of *A*. Bisect *BC* at *O*. With *O* as centre and radius equal to *CO* or *OB*, draw a circle. Through *O* draw a line *OE* making an angle 2θ (*i.e.*, 2 × 30° = 60°) with *OB*. From *E*, draw *ED* perpendicular to *AB*. Join *AE* and *CE*. Then *AE* represents the resultant stress and angle φ represents the obliquity.

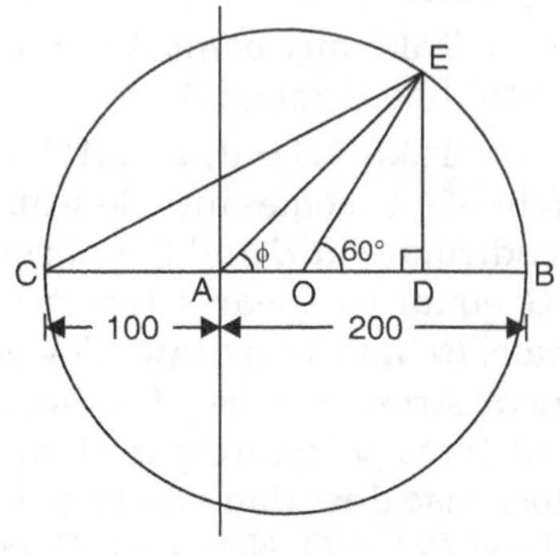

Fig. 1.23 (*b*)

By measurement from Fig. 1.23 (*b*), we have

Length *AE* = 9.0 cm

Length *AD* = 6.25 cm and length *ED* = 6.5 cm

Angle φ = 46°

$\therefore$ Resultant stress $= \text{Length } AE \times \text{Scale}$

$= 9.0 \times 20 =$ **180 N/mm². Ans.**

Angle made by the resultant stress with the normal of the inclined plane = φ = **46°. Ans.**

Normal stress $= \text{Length } AD \times 20$

$= 6.25 \times 20 = 125$ N/mm²

Shear stress $= \text{Length } ED \times 20$

$= 6.5 \times 20 = 130$ N/mm².

Maximum shear stress. Shear stress is along a line which is perpendicular to the line *AB*. Hence maximum shear stress will be when perpendicular to line *AB* is drawn from point *O*. Then maximum shear stress will be equal to the radius of Mohr's circle.

$\therefore$ Maximum shear stress = Radius of Mohr's circle

$$= \frac{\sigma_1 + \sigma_2}{2} = \frac{200 + 100}{2} = \mathbf{150\ N/mm^2.}\quad \textbf{Ans.}$$

1.7.3. Mohr's Circle when a Body is Subjected to two Mutually Perpendicular Principal Tensile Stresses Accompanied by a Simple Shear Stress. Consider a rectangular body subjected to two mutually perpendicular principal tensile stresses of unequal intensities accompanied by a simple shear stress. It is required to find the resultant stress on an oblique plane as shown in Fig. 1.24.

Fig. 1.24

Let σ_1 = Major tensile stress

σ_2 = Minor tensile stress

τ = Shear stress across face *BC* and *AD*

θ = Angle made by the oblique plane with the plane of major tensile stress.

According to the principal of shear stress, the faces *AB* and *CD* will also be subjected to a shear stress of τ.

We will consider clockwise shear stress as positive and anticlockwise shear stress as negative.

Mohr's circle is drawn as given in Fig. 1.25.

Take any point *A* and draw a horizontal line through *A*.

Take $AB = \sigma_1$ and $AC = \sigma_2$ towards right of *A* to some suitable scale. Draw perpendiculars at *B* and *C* and cut off *BF* and *CG* equal to shear stress τ to the same scale. *BF* has been drawn downward as the shear stress on face *AD* is acting upwards and trees to rotate the element in anti-clockwise direction and hence is negative. Bisect *BC* at *O*. Now with *O* as centre and radius equal to *OG* or *OF* draw a circle. Through *O*, draw a line *OE* making an angle of 2θ with *OF* as shown in Fig. 1.25. From *E*, draw *ED* perpendicular to *CB*. Join *AE*. Then length *AE* represents the resultant stress on the given oblique plane. And lengths *AD* and *ED* represents the normal stress and tangential stress respectively.

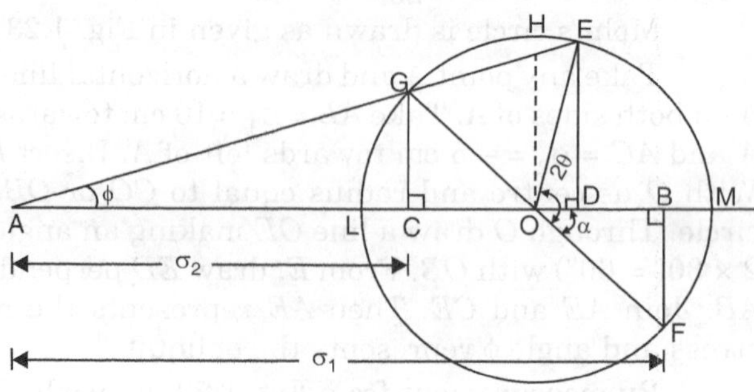

Fig. 1.25

Hence from Fig. 1.25, we have

Length *AE* = Resultant stress on the oblique plane

Length *AD* = Normal stress on the oblique plane

Length *ED* = Shear stress on the oblique plane.

Proof. (See Fig. 1.25).

$$CO = \frac{1}{2} CB = \frac{1}{2} [\sigma_1 - \sigma_2] \qquad (\because \; CB = \sigma_1 - \sigma_2)$$

$$AO = AC + CO = \sigma_2 + \frac{1}{2} [\sigma_1 - \sigma_2]$$

$$= \frac{2\sigma_2 + \sigma_1 - \sigma_2}{2} = \frac{\sigma_1 + \sigma_2}{2}$$

$$AD = AO + OD$$

$$= \frac{\sigma_1 + \sigma_2}{2} + OE \cos (2\theta - \alpha) \qquad [\because \; OD = OE \cos (2\theta - \alpha)]$$

$$= \frac{\sigma_1 + \sigma_2}{2} + OE [\cos 2\theta \cos \alpha + \sin 2\theta \sin \alpha]$$

$$= \frac{\sigma_1 + \sigma_2}{2} + OE \cos 2\theta \cos \alpha + OE \sin 2\theta \sin \alpha$$

$$= \frac{\sigma_1 + \sigma_2}{2} + OE \cos \alpha \,.\, \cos 2\theta + OE \sin \alpha \,.\, \sin 2\theta$$

$$= \frac{\sigma_1 + \sigma_2}{2} + OF \cos \alpha \,.\, \cos 2\theta + OF \sin \alpha \,.\, \sin 2\theta$$

$$(\because \; OE = OF = \text{Radius})$$

$$= \frac{\sigma_1 + \sigma_2}{2} + OB \cos 2\theta + BF \sin 2\theta$$

$$(\because \; OF \cos \alpha = OB, OF \sin \alpha = BF)$$

$$= \frac{\sigma_1 + \sigma_2}{2} + CO \cos 2\theta + \tau \sin 2\theta \qquad (\because \; OB = CO, BF = \tau)$$

$$= \frac{\sigma_1 + \sigma_2}{2} + \frac{\sigma_1 - \sigma_2}{2} \cos 2\theta + \tau \sin 2\theta \qquad \left(\because \; CO = \frac{\sigma_1 - \sigma_2}{2}\right)$$

$= \sigma_n$ or Normal stress

Now $ED = OE \sin (2\theta - \alpha) = OE (\sin 2\theta \cos \alpha - \cos 2\theta \sin \alpha)$

$$= OE \sin 2\theta \cos \alpha - OE \cos 2\theta \sin \alpha$$

$$= OE \cos \alpha \,.\, \sin 2\theta - OE \sin \alpha \,.\, \cos 2\theta$$

$$= OE \cos \alpha \,.\, \sin 2\theta - OE \sin \alpha \,.\, \cos 2\theta \qquad (\because \; OE = OF = \text{Radius})$$

$$= OB \,.\, \sin 2\theta - BF \cos 2\theta \qquad (\because \; OF \cos \alpha = OB, OF \sin \alpha = BF)$$

$$= CO \,.\, \sin 2\theta - \tau \cos 2\theta \qquad (\because \; OB = CO, BF = \tau)$$

$$= \frac{\sigma_1 - \sigma_2}{2} \sin 2\theta - \tau \cos 2\theta \qquad \left(\because \; CO = \frac{\sigma_1 - \sigma_2}{2}\right)$$

$= \sigma_t$ or Tangential stress.

Maximum and minimum value of normal stress. In Fig. 1.25, the normal stress is given by AD. Hence the maximum value of AD will be when D coincides with M and minimum value of AD will be when D coincides with L.

$\therefore$ Maximum value of normal stress,

$$(\sigma_n)_{max} = AM = AO + OM$$

$$= \frac{\sigma_1 + \sigma_2}{2} + OF \qquad \left(\because \; AO = \frac{\sigma_1 + \sigma_2}{2}, OM = OF = \text{Radius}\right)$$

$$= \frac{\sigma_1 + \sigma_2}{2} + \sqrt{OB^2 + BF^2} \qquad (\because \text{ In triangle } OBF, OF = \sqrt{OB^2 + BF^2})$$

$$= \frac{\sigma_1 + \sigma_2}{2} + \sqrt{\left(\frac{\sigma_1 - \sigma_2}{2}\right)^2 + \tau^2} \qquad \left(\because OB = \frac{\sigma_1 - \sigma_2}{2}, BF = \tau\right)$$

Minimum value of normal stress,

$$(\sigma_n)_{min} = AL = AO - LO$$

$$= \frac{\sigma_1 + \sigma_2}{2} - OF \qquad (\because LO = OF = \text{Radius})$$

$$= \frac{\sigma_1 + \sigma_2}{2} - \sqrt{\left(\frac{\sigma_1 - \sigma_2}{2}\right)^2 + \tau^2}$$

(*i*) For maximum normal stress, the point D coincides with M. But when the point D coincides with M, the point E also coincides with M. Hence for maximum value of normal stress,

Angle $2\theta = \alpha$ $\qquad (\because$ Line OE coincides with line $OM)$

$$\therefore \quad \theta = \frac{\alpha}{2} \qquad \text{...}(i)$$

Also $\tan 2\theta = \tan \alpha = \dfrac{BF}{OB} = \dfrac{\tau}{\dfrac{\sigma_1 - \sigma_2}{2}}$ $\qquad \left(\because BF = \tau, OB = \dfrac{\sigma_1 - \sigma_2}{2}\right)$

$$= \frac{2\tau}{\sigma_1 - \sigma_2}.$$

(*ii*) For maximum and minimum normal stresses, the shear stress is zero and hence the planes, on which maximum and minimum normal stresses act, are known as *principal planes* and the stresses are known as *principal stresses.*

(*iii*) For minimum normal stress, the point D coincides with point L. But when the point D coincides with L, the point E also coincides with L. Then

Angle $2\theta = \pi + \alpha$ $\qquad (\because$ Line OE coincides with line $OL)$

$$\therefore \quad \theta = \frac{\pi}{2} + \frac{\alpha}{2} \qquad \text{...}(ii)$$

From equations (*i*) and (*ii*), it is clear that the plane of minimum normal stress is inclined at an angle 90° to the plane of maximum normal stress.

Maximum value of shear stress. Shear stress is given by ED. Hence maximum value of ED will be when E coincides with G, and D coincides with O.

$\therefore$ Maximum shear stress,

$$(\sigma_t)_{max} = OH = OF \qquad (\because OH = OF = \text{radius})$$

$$= \sqrt{OB^2 + BF^2} \qquad (\because \text{ In triangle } OBF, OF = \sqrt{OB^2 + BF^2})$$

$$= \sqrt{\left(\frac{\sigma_1 - \sigma_2}{2}\right)^2 + \tau^2} \qquad \left(\because OB = \frac{\sigma_1 - \sigma_2}{2}, BF = \tau\right)$$

Problem 1.17. *A point in a strained material is subjected to stresses shown in Fig. 1.26. Using Mohr's circle method, determine the normal and tangential stresses across the oblique plane. Check the answer analytically.*

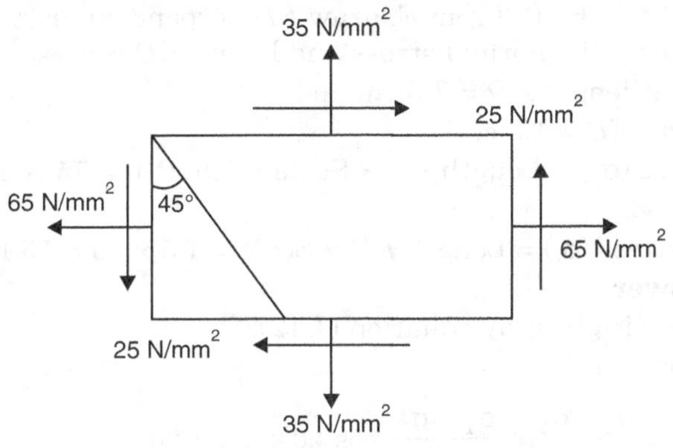

Fig. 1.26 (*a*)

Sol. Given :

Major principal stress, $\sigma_1 = 65\ \text{N/mm}^2$

Minor principal stress, $\sigma_2 = 35\ \text{N/mm}^2$

Shear stress, $\tau = 25\ \text{N/mm}^2$

Angle of oblique plane, $\theta = 45°$.

Mohr's circle method

Let 1 cm = 10 N/mm²

Then $\sigma_1 = \dfrac{65}{10} = 6.5$ cm,

$$\sigma_2 = \frac{35}{10} = 3.5 \text{ cm and } \tau = \frac{25}{10} = 2.5 \text{ cm}$$

Mohr's circle is drawn as given in Fig. 1.26 (*b*).

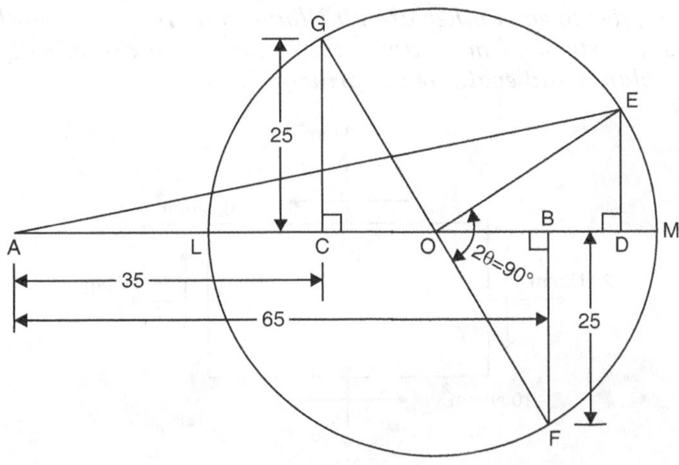

Fig. 1.26 (*b*)

Take any point A and draw a horizontal line through A. Take $AB = \sigma_1 = 6.5$ cm and $AC = \sigma_2 = 3.5$ cm towards right of A. Draw perpendicular at B and C and cut off BF and CG equal to shear stress $\tau = 2.5$ cm. Bisect BC at O. Now with O as centre and radius equal to OF (or OG) draw a circle. Through O, draw a line OE making an angle of 2θ (*i.e.*, $2 \times 45° = 90°$) with OF as shown in Fig. 1.26 (*b*). From E, draw ED perpendicular to AB produced. Join AE. Then length AD represents the normal stress and length ED represents the shear stress.

By measurements, length $AD = 7.5$ cm and

length $ED = 1.5$ cm.

$\therefore$ Normal stress (σ_n) = Length AD × Scale = 7.5 × 10 = **75 N/mm^2. Ans.**

($\because$ 1 cm = 10 N/mm^2)

And tangential stress (σ_t) = Length ED × Scale = 1.5 × 10 = **15 N/mm^2. Ans.**

Analytical Answers

Normal stress (σ_n) is given by equation (1.12).

$\therefore$ Using equation (1.12),

$$\sigma_n = \frac{\sigma_1 + \sigma_2}{2} + \frac{\sigma_1 - \sigma_2}{2} \cos 2\theta + \tau \sin 2\theta$$

$$= \frac{65 + 35}{2} + \frac{65 - 35}{2} \cos (2 \times 45°) + 25 \sin (2 \times 45°)$$

$$= 50 + 15 \cos 90° + 25 \sin 90°$$

$$= 50 + 15 \times 0 + 25 \times 1 \qquad (\because \cos 90° = 0, \sin 90° = 1)$$

$= 50 + 0 + 25 =$ **75 N/mm^2. Ans.**

Tangential stress is given by equation (1.13)

$\therefore$ Using equation (1.13),

$$\sigma_t = \frac{\sigma_1 - \sigma_2}{2} \sin 2\theta - \tau \cos 2\theta$$

$$= \frac{65 - 35}{2} \sin (2 \times 45) - 25 \cos (2 \times 45)$$

$$= 15 \sin 90° - 25 \cos 90° = 15 \times 1 - 25 \times 0 = 15 - 0$$

= **15 N/mm^2. Ans.**

Problem 1.18. *At a certain point in a strained material, the intensities of stresses on two planes at right angles to each other are 20 N/mm^2 and 10 N/mm^2 both tensile. They are accompanied by a shear stress of magnitude 10 N/mm^2. Find graphically or otherwise, the location of principal planes and evaluate the principal stresses.*

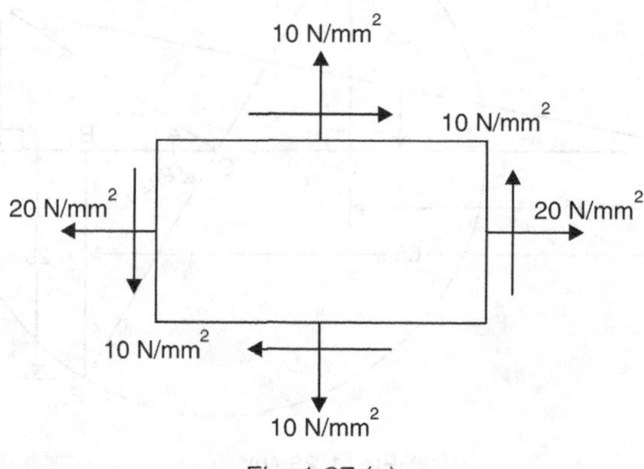

Fig. 1.27 (*a*)

Sol. Given :

Major tensile stress, $\sigma_1 = 20$ N/mm^2

Minor tensile stress, $\sigma_2 = 10$ N/mm^2

Shear stress, $\tau = 10$ N/mm^2

This problem may be solved analytically or graphically. Here we shall solve it graphically (*i.e.*, by Mohr's circle method).

Scale. Take 1 cm = 2 N/mm^2

Then $\sigma_1 = \dfrac{20}{2} = 10$ cm, $\sigma_2 = \dfrac{10}{2} = 5$ cm

and $\tau = \dfrac{10}{2} = 5$ cm.

Mohr's circle is drawn as given in Fig. 1.27 (*b*).

Take any point *A* and draw a horizontal line through *A*. Take $AB = \sigma_1 = 10$ cm and $AC = \sigma_2 = 5$ cm towards right side of *A*. Draw perpendiculars at *B* and *C* and cut off $BF = CG = \tau = 5$ cm. Bisect *BC* at *O*. Now with *O* as centre and radius equal to *OG* (or *OF*), draw a circle cutting the horizontal line through *A*, at *L* and *M* as shown in Fig. 1.27(*b*). Then *AM* and *AL* represent the major principal and minor principal stresses.

Fig. 1.27 (*b*)

By measurements, we have

Length *AM* = 13.1 cm and Length *AL* = 1.91 cm

$\angle FOB$ (or 2θ) = 63.7°.

$\therefore$ Major principal stress = Length *AM* × Scale

= 13.1 × 2 N/mm^2 ($\because$ 1 cm = 2 N/mm^2)

= **26.2 N/mm^2. Ans.**

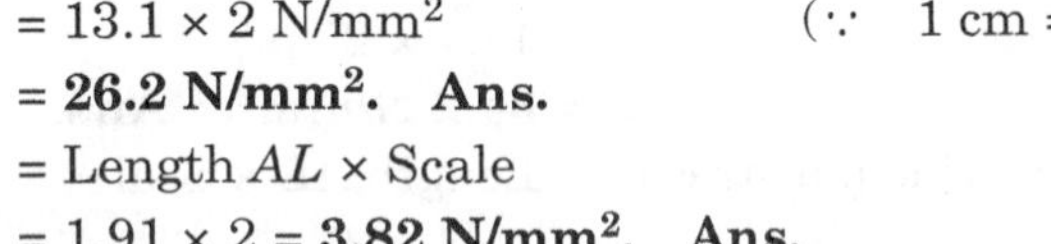

Minor principal stress = Length *AL* × Scale

= 1.91 × 2 = **3.82 N/mm^2. Ans.**

Location of principal planes (No shear stress)

$$2\theta = 63.7°$$

$$\therefore \quad \theta = \frac{63.7°}{2} = \mathbf{31.85°}. \quad \textbf{Ans.}$$

The second principal plane is given by

$\theta + 90°$ or 31.85° + 90° or **121.85°. Ans.**

Problem 1.19. *An elemental cube is subjected to tensile stresses of 30 N/mm^2 and 10 N/mm^2 acting on two mutually perpendicular planes and a shear stress of 10 N/mm^2 on these planes. Draw the Mohr's circle of stresses and hence or otherwise determine the magnitudes and directions of principal stresses and also the greatest shear stress.*

Sol. Given :

Major tensile stress, $\sigma_1 = 30$ N/mm^2

Minor tensile stress, $\sigma_2 = 10$ N/mm^2

Shear stress, $\tau = 10$ N/mm^2

Scale. Take $1 \text{ cm} = 2 \text{ N/mm}^2$

Then $\sigma_1 = \frac{30}{2} = 15 \text{ cm}$

$\sigma_2 = \frac{10}{2} = 5 \text{ cm}$ and $\tau = \frac{10}{2} = 5 \text{ cm}$

Mohr's circle of stresses is drawn as given in Fig. 1.28.

Take any point A and draw a horizontal line through A.

Take $AB = \sigma_1 = 15$ cm and $AC = \sigma_2 = 5$ cm towards right side of A. Draw perpendiculars at B and C and cut off $BF = CG = \tau = 5$ cm. Bisect BC at O. Now with O as centre and radius equal to OG (or OF), draw a circle cutting the horizontal line through A at L and M as shown in Fig. 1.28. Then AM and AL represents the major and minor principal stresses respectively. And OH represents the maximum shear stress.

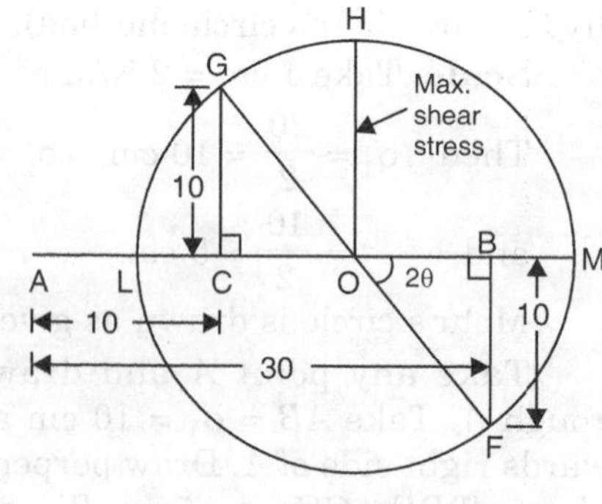

Fig. 1.28

By measurements, we have

Length $AM = 17.1$ cm

Length $AL = 2.93$ cm

Length OH = Radius of Mohr's circle $= 7.05$ cm

$\angle FOB$ (or 2θ) $= 45°$.

$\therefore$ Major principal stress

$= \text{Length } AM \times \text{Scale}$

$= 17.1 \times 2$ $\qquad (\because \ 1 \text{ cm} = 2 \text{ N/mm}^2)$

$= \mathbf{34.2 \text{ N/mm}^2}$. **Ans.**

Minor principal stress $= \text{Length } AL \times \text{Scale}$

$= 2.93 \times 2$ $\qquad (\because \ 1 \text{ cm} = 2 \text{ N/mm}^2)$

$= \mathbf{5.86 \text{ N/mm}^2}$. **Ans.**

$\angle FOB$ or $2\theta = 45°$

$\therefore \quad \theta = \frac{45}{2} = \mathbf{22.5°}$. **Ans.**

The second principal plane is given by $\theta + 90°$.

$\therefore$ Second principal plane $= 22.5 + 90 = \mathbf{112.5°}$. **Ans.**

The greatest shear stress $= \text{Length } OH \times \text{Scale}$

$= 7.05 \times 20 = \mathbf{14.1 \text{ N/mm}^2}$. **Ans.**

1.8. PRINCIPAL STRAIN ON AN INCLINED PLANE

To determine the strain on an inclined plane due to stresses σ_x, σ_y and τ, we will first consider strains due to each stress separately. After that we will combine the result.

1.8.1. Strain on an Inclined Plane due to Stress (σ_x). A rectangular bar $ABCD$ is subjected to a stress σ_x. Due to this stress there is an increase in length 'dx' as shown in Fig. 1.29. It is required to find the strain on the oblique plane AC (*i.e.*, strain in diagonal AC). Final position of the bar is shown by $AB'C'D$.

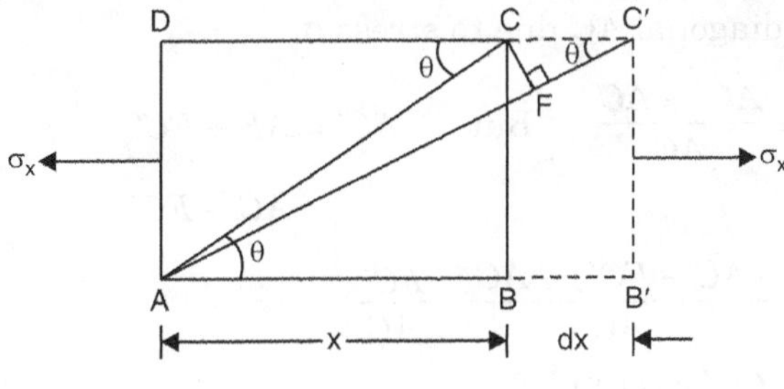

Fig. 1.29

Let dx = Increase in length x due to stress σ_x

e_x = Strain in x-direction

$= \dfrac{dx}{x}$

$\therefore \quad dx = x \times e_x$

But $\quad dx = BB' = CC' = x \times e_x$

Let e is the strain in diagonal AC.

$$\therefore \quad e = \frac{AC' - AC}{AC} \quad \text{But } AC' = AF + FC' = AC + FC'$$

$(\because \quad AF = AC$ and $FC' = CC' \cos \theta)$

$$= \frac{(AC + CC' \cos \theta) - AC}{AC}$$

$$= \frac{CC' \cos \theta}{AC}$$

$$= \frac{(x \times e_x) \times \cos \theta}{AC} \quad (\because \quad CC' = dx = x \times e_x) \quad \text{Also } x = AC \cos \theta$$

$$= \frac{AC \cos \theta \times e_x \times \cos \theta}{AC} = e_x \cos^2 \theta \qquad ...(1.19)$$

1.8.2. Strain on an Inclined Plane due to Stress (σ_y). Refer to Fig. 1.30.

Let dy = Increase in length 'y' due to stress σ_y

e_y = Strain in y-direction

e = Strain in diagonal AC

Now, $\quad CC' = DD' = dy$

But $\quad e_y$ = strain in y-direction

$= \dfrac{dy}{y}$

$\therefore \quad dy = y \times e_y$

or $\quad CC' = DD' = dy = y \times e_y$

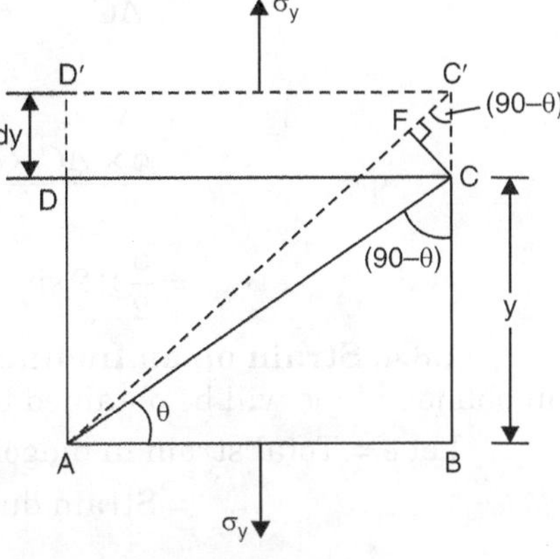

Fig. 1.30

Now e = strain in diagonal AC due to stress σ_y

$$= \frac{AC' - AC}{AC} \quad \text{But} \quad AC' = AF + FC'$$

$$= AC + FC' \qquad (\because \quad AF = AC)$$

$$= \frac{(AC + FC') - AC}{AC} = \frac{FC'}{AC}$$

But $FC' = CC' \cos (90 - \theta)$

$= CC' \sin \theta$

$= (y \times e_y) \sin \theta \qquad (\because \quad CC' = dy = y \times e_y)$

$= (AC \times \sin \theta \times e_y \times \sin \theta) \qquad (\because \quad y = AC \sin \theta)$

$$\therefore \quad e = \frac{FC'}{AC} = \frac{AC \times \sin^2 \theta \times e_y}{AC}$$

$$= e_y \times \sin^2 \theta \qquad \text{...(1.20)}$$

1.8.3. Strain on an Inclined Plane due to Shear Stress τ. Refer to Fig. 1.31.

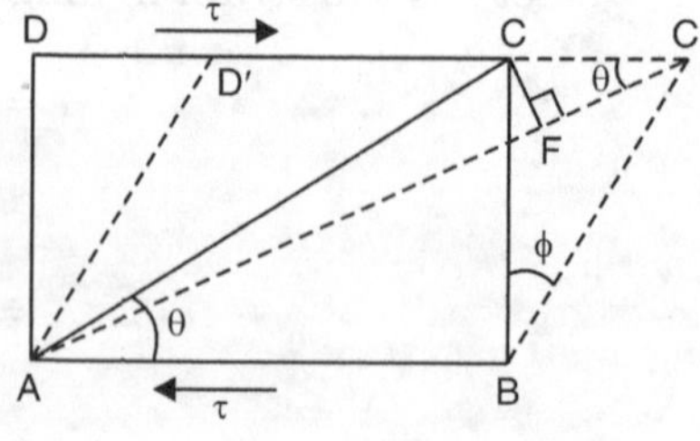

Fig. 1.31

Let ϕ = shear strain due to shear stress τ

$$= \frac{CC'}{CB}$$

$\therefore \quad CC' = \phi \times CB$

$= \phi \times AC \sin \theta \quad (\because \quad CB = AC \sin \theta)$

We know that the strain in diagonal AC is given by

$$e = \frac{AC' - AC}{AC} \quad \text{But} \quad AC' = AF + FC'$$

$$= AC + FC'$$

$$= \frac{(AC + FC') - AC}{AC}$$

$$= \frac{FC'}{AC} \quad \text{But} \quad FC' = CC' \times \cos \theta$$

$$= (\phi \times AC \sin \theta) \times \cos \theta$$

$$(\because \quad CC' = \phi \times AC \sin \theta)$$

$$= \frac{\phi \times AC \times \sin \theta \times \cos \theta}{AC} = \phi \times \sin \theta \times \cos \theta$$

$$= \frac{\phi}{2} \times 2 \sin \theta \times \cos \theta = \frac{\phi}{2} \times \sin 2\theta \qquad \text{...(1.21)}$$

1.8.4. Strain on an Inclined Plane due to Stresses σ_x, σ_y and τ. The total strain on an oblique plane will be obtained by adding the RHS of equations (1.19), (1.20) and (1.21).

Let e = Total strain in diagonal AC due to stresses σ_x, σ_y, and τ

= Strain due to σ_x + strain due to σ_y + strain due to τ

$$= e_x \cos^2 \theta + e_y \sin^2 \theta + \frac{\phi}{2} \sin 2\theta$$

$$= e_x \frac{(1+\cos 2\theta)}{2} + e_y \frac{(1-\cos 2\theta)}{2} + \frac{\phi}{2} \sin 2\theta$$

$$\left(\because \; \cos^2 \theta = \frac{1+\cos 2\theta)}{2}, \sin^2 \theta = \frac{1-\cos 2\theta}{2}\right)$$

$$= \frac{e_x + e_y}{2} + \frac{e_x - e_y}{2} \times \cos 2\theta + \frac{\phi}{2} \times \sin 2\theta \quad \text{...(1.22)}$$

1.8.5. Maximum or Minimum Value of Diagonal Strain. The diagonal strain 'e' given by equation (1.22) will by maximum or minimum if $\frac{de}{d\theta} = 0$

or $$\frac{d}{d\theta}\left[\frac{e_x + e_y}{2} + \left(\frac{e_x - e_y}{2}\right)\cos 2\theta + \frac{\phi}{2}\sin 2\theta\right] = 0$$

or $$0 + \left(\frac{e_x - e_y}{2}\right)(-\sin 2\theta) \times 2 + \frac{\phi}{2}(\cos 2\theta) \times 2 = 0$$

or $$-(e_x - e_y)\sin 2\theta + \phi \times \cos 2\theta = 0$$

or $$\frac{\sin 2\theta}{\cos 2\theta} = \frac{\phi}{(e_x - e_y)}$$

or $$\tan 2\theta = \frac{\phi}{(e_x - e_y)}$$

$\therefore$ $$2\theta = \tan^{-1} \frac{\phi}{(e_x - e_y)}$$

or $$\theta = \frac{1}{2} \tan^{-1} \frac{\phi}{(e_x - e_y)}$$

Fig. 1.31 (*a*)

From Fig. 1.31 (*a*), we have

$$\cos 2\theta = \pm \frac{(e_x - e_y)}{\sqrt{(e_x - e_y)^2 + \phi^2}}$$

and $$\sin 2\theta = \pm \frac{\phi}{\sqrt{(e_x - e_y)^2 + \phi^2}}$$

Hence the diagonal strain 'e' will be maximum or minimum when the value of $\tan 2\theta = \frac{\phi}{(e_x - e_y)}$. By substituting the +ve values of cos 2θ and sin 2θ in equation (1.22), we get maximum value of strain.

$\therefore$ $$e_{\text{max.}} = \frac{e_x + e_y}{2} + \frac{e_x - e_y}{2} \times \frac{(e_x - e_y)}{\sqrt{(e_x - e_y)^2 + \phi^2}} + \frac{\phi}{2} \times \frac{\phi}{\sqrt{(e_x - e_y)^2 + \phi^2}}$$

$$= \frac{e_x + e_y}{2} + \frac{(e_x - e_y)^2 + \phi^2}{2\sqrt{(e_x - e_y)^2 + \phi^2}}$$

$$= \frac{e_x + e_y}{2} + \frac{1}{2}\sqrt{(e_x - e_y)^2 + \phi^2} \quad ...(1.23)$$

Similarly by substituting the (–ve) values of cos 2θ and sin 2θ in equation (1.22), we get the minimum value of strain.

$$\therefore \quad e_{min} = \frac{e_x + e_y}{2} + \frac{e_x - e_y}{2} \times \frac{(-)(e_x - e_y)}{\sqrt{(e_x - e_y)^2 + \phi^2}} + \frac{\phi}{2} \times \frac{(-\phi)}{\sqrt{(e_x - e_y)^2 + \phi^2}}$$

$$= \frac{e_x + e_y}{2} - \frac{1}{2}\sqrt{(e_x - e_y)^2 + \phi^2} \quad ...(1.24)$$

The equations (1.23) and (1.24) gives the values of **principal strains.**

HIGHLIGHTS

1. The planes, which have no shear stress, are known as principal planes.
2. The stresses, acting on principal planes, are known as principal stresses.
3. Analytical and graphical methods are used for finding the stresses on an oblique section.
4. When a member is subjected to a direct stress (σ) in one plane, then the stresses on an oblique plane (which is inclined at an angle θ with the normal cross-section) are given by :

   Normal stress, $\sigma_n = \sigma \cos^2 \theta$

   Tangential stress, $\sigma_t = \frac{\sigma}{2} \sin 2\theta$

   Max. normal stress $= \sigma$

   Max. shear stress $= \frac{\sigma}{2}$.
5. When a member is subjected to two like direct stresses in two mutually perpendicular directions, then the stresses on an oblique plane inclined at an angle θ with the axis of the minor stress (or with the plane of major stress) are given by :

   Normal stress, $\sigma_n = \frac{\sigma_1 + \sigma_2}{2} + \frac{\sigma_1 - \sigma_2}{2} \cos 2\theta$.

   Tangential stress, $\sigma_t = \frac{\sigma_1 - \sigma_2}{2} \sin 2\theta$

   Resultant stress, $\sigma_R = \sqrt{\sigma_n^{\ 2} + \sigma_t^{\ 2}}$.
6. The angle made by the resultant stress with the normal of the oblique plane, is known as obliquity. It is denoted by φ. Mathematically, $\tan \phi = \frac{\sigma_t}{\sigma_n}$.
7. When a member is subjected to a simple shear stress (τ), then the stresses on an oblique plane are given as :

   Normal stress, $\sigma_n = \tau \sin 2\theta$

   Tangential stress, $\sigma_t = -\tau \cos 2\theta$.
8. When a member is subjected to two direct stresses (σ_1, σ_2) in two mutually perpendicular directions accompanied by a simple shear stress (τ), then the stresses, on an oblique plane inclined at an angle θ with the axis of minor stress, are given by :

   Normal stress, $\sigma_n = \frac{\sigma_1 + \sigma_2}{2} + \frac{\sigma_1 - \sigma_2}{2} \cos 2\theta + \tau \sin 2\theta$.

Tangential stress, $\sigma_t = \frac{\sigma_1 - \sigma_2}{2} \sin 2\theta - \tau \cos 2\theta$

(*a*) Position of principal planes is given by $\tan 2\theta = \frac{2\tau}{\sigma_1 - \sigma_2}$

(*b*) Major principal stress $= \frac{\sigma_1 + \sigma_2}{2} + \sqrt{\left(\frac{\sigma_1 - \sigma_2}{2}\right)^2 + \tau^2}$

(*c*) Minor principal stress $= \frac{\sigma_1 + \sigma_2}{2} - \sqrt{\left(\frac{\sigma_1 - \sigma_2}{2}\right)^2 + \tau^2}$

(*d*) Maximum shear stress $= \frac{1}{2}\sqrt{(\sigma_1 - \sigma_2)^2 + 4\tau^2}$

(*e*) Condition for maximum shear stress, $\tan 2\theta = \frac{\sigma_2 - \sigma_1}{2\tau}$.

9. Mohr's circle of stresses is a graphical method of finding normal, tangential and resultant stresses on an oblique plane.

10. Maximum shear stress by Mohr's circle method, is equal to the radius of the Mohr's circle.

11. The planes of maximum and minimum normal stresses are at an angle of 90° to each other.

EXERCISE 1

(A) Theoretical Questions

1. Define the terms : Principal planes and principal stresses.

2. A rectangular bar is subjected to a direct stress (σ) in one plane only. Prove that the normal and shear stresses on an oblique plane are given by

$$\sigma_n = \sigma \cos^2 \theta \quad \text{and} \quad \sigma_t = \frac{\sigma}{2} \sin 2\theta$$

where θ = Angle made by oblique plane with the normal cross-section of the bar,
σ_n = Normal stress, and
σ_t = Tangential or shear stress.

3. A rectangular bar is subjected to two direct stresses (σ_1 and σ_2) in two mutually perpendicular directions. Prove that the normal stress (σ_n) and shear stress (σ_t) on an oblique plane which is inclined at an angle θ with the axis of minor stress are given by

$$\sigma_n = \frac{\sigma_1 + \sigma_2}{2} + \frac{\sigma_1 - \sigma_2}{2} \cos 2\theta$$

and $$\sigma_t = \frac{\sigma_1 - \sigma_2}{2} \sin 2\theta.$$

4. Define the term 'obliquity' and how it is determined.

5. Derive an expression for the stresses on an oblique plane of a rectangular body, when the body is subjected to a simple shear stress.

6. A rectangular body is subjected to direct stresses in two mutually perpendicular directions accompanied by a shear stress. Prove that the normal stress and shear stress on an oblique plane inclined at an angle θ with the plane of major direct stress, are given by

$$\sigma_n = \frac{\sigma_1 + \sigma_2}{2} + \frac{\sigma_1 - \sigma_2}{2} \cos 2\theta + \tau \sin 2\theta$$

and $$\sigma_t = \frac{\sigma_1 - \sigma_2}{2} \sin 2\theta - \tau \cos 2\theta.$$

7. Derive an expression for the major and minor principal stresses on an oblique plane, when the body is subjected to direct stresses in two mutually perpendicular directions accompanied by a shear stress.

8. Write a note on Mohr's circle of stresses.

9. A body is subjected to direct stresses in two mutually perpendicular directions accompanied by a simple shear stress. Draw the Mohr's circle of stresses and explain how will you obtain the principal stresses and principal planes.

10. A body is subjected to direct stresses in two mutually perpendicular directions. How will you determine graphically the resultant stress on an oblique plane when :

(*i*) the stresses are unequal and unlike, and

(*ii*) the stresses are unequal and like.

(B) Numerical Problems

1. A rectangular bar of cross-sectional area 12000 mm^2 is subjected to an axial load of 360 N/mm^2. Determine the normal and shear stresses on a section which is inclined at an angle of 30° with the normal cross-section of the bar. [**Ans.** 9.25 N/mm^2, 1.3 N/mm^2]

2. Find the diameter of a circular bar which is subjected to an axial pull of 150 kN, if the maximum allowable shear stress on any section is 60 N/mm^2. [**Ans.** 3.989 cm]

3. A rectangular bar of cross-sectional area 10000 mm^2 is subjected to a tensile load P as shown in Fig. 1.32. The permissible normal and shear stresses on the oblique plane BC are given as 8 N/mm^2 and 4 N/mm^2 respectively. Determine the safe value of P. [**Ans.** 92.378 kN]

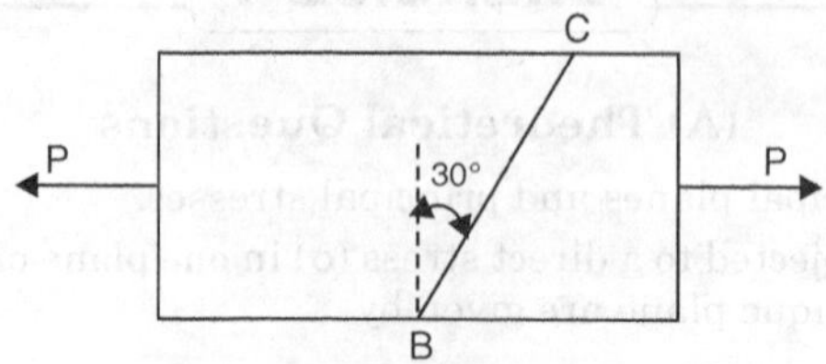

Fig. 1.32

4. The principal tensile stress at a point across two mutually perpendicular planes are 100 N/mm^2 and 50 N/mm^2. Determine the normal, tangential and resultant stresses on a plane inclined at 30° to the axis of the minor principal stress. [**Ans.** 0.875 N/mm^2, 21.65 N/mm^2, 90.138]

5. The principal stresses at a point in a bar are 160 N/mm^2 (tensile) and 80 N/mm^2 (compressive). Determine the resultant stress in magnitude and direction on a plane inclined at 60° to the axis of the major principal stress. Also determine the maximum intensity of shear stress in the material at the point. [**Ans.** 144.22 N/mm^2, ϕ = 46° 5.7, 120 N/mm^2]

6. At a point in a strained material, the principal stresses are 140 N/mm^2 (tensile) and 60 N/mm^2 (compressive). Determine the resultant stress in magnitude and direction on a plane inclined at 45° to the axis of the major principal stress. What is the maximum intensity of shear stress in the material at the point ? [**Ans.** 107.7 N/mm^2 , ϕ = 61° 11.9, 100 N/mm^2]

7. At a point within a body subjected to two mutually perpendicular directions, the stresses are 100 N/mm^2 (tensile) and 75 N/mm^2 (tensile). Each of the above stresses, is accompanied by a shear stress of 75 N/mm^2. Determine the normal, shear and resultant stresses on an oblique plane inclined at an angle of 45° with the axis of minor tensile stress.

[**Ans.** 150, 25, 152.07 N/mm^2]

8. For the problem 7, determine : (*i*) the direction and magnitude of each of the principal stress and (*ii*) magnitude of the greatest shear stress.

[**Ans.** 154.057, – 4.057 N/mm^2, θ = 35°, 468′ and 125° 46.8′ N/mm^2]

9. Direct stresses of 160 N/mm^2 tensile and 120 N/mm^2 compressive exist on two perpendicular planes at a certain point in a body. They are also accompanied by shear stresses on the planes. The greatest principal stress at the point due to these is 200 N/mm^2.

(*i*) What must be the magnitude of the shearing stresses on the two planes ?

(*ii*) What will be the maximum shearing stress at the point ?

[**Ans.** (*i*) 113.137 N/mm^2, (*ii*) 180 N/mm^2]

10. At a certain point in a strained material, the stresses on the two planes at right angles to each other are 40 N/mm^2 and 20 N/mm^2 both tensile. They are accompanied by a shear stress of magnitude 20 N/mm^2. Find graphically or otherwise, the location of principal planes and evaluate the principal stresses. [**Ans.** $\theta = 31° 43'$, $121° 43'$ and 52.36, 7.64 N/mm^2]

11. Solve problem 4, by graphical method.

12. Solve problem 5, by graphical method.

13. Solve problem 4, using Mohr's circle of stresses.

14. Solve problem 5, using Mohr's circle of stresses.

15. A point in a strained material is subjected to stresses shown in Fig. 1.33.

Using Mohr's circle method, determine the normal and tangential stresses across the oblique plane. Check the answer analytically. [**Ans.** 105 N/mm^2, 15 N/mm^2]

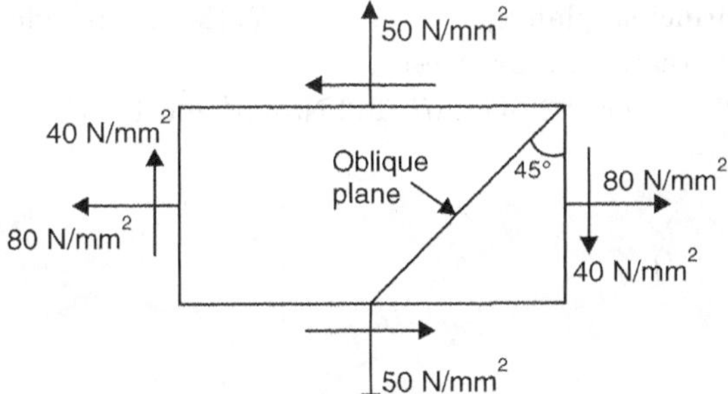

Fig. 1.33

16. An elemental cube is subjected to tensile stresses of 60 N/mm^2 and 20 N/mm^2 acting on two mutually perpendicular planes and a shear stress of 20 N/mm^2 on these planes. Draw the Mohr's circle of stresses and hence or otherwise determine the magnitudes and directions of principal stresses and also the greatest shear stress.

[**Ans.** 68.214, 11.72 N/mm^2, $\theta = 25.5°$ and 112.5°, 28.28 N/mm^2]

17. A strained material is subjected to two dimensional stresses. Prove that the sum of the normal components of stresses on any two mutually perpendicular planes is constant.

[**Hint.** Normal stresses on a plane inclined at θ with major principal plane is given by

$$\sigma_n = \frac{\sigma_1 + \sigma_2}{2} + \frac{\sigma_1 - \sigma_2}{2} \cos 2\theta \quad ...(i)$$

Normal stress on a plane inclined at $(\theta + 90°)$ is given by

$$\sigma_n^* = \frac{\sigma_1 + \sigma_2}{2} + \frac{\sigma_1 - \sigma_2}{2} \cos [2(\theta + 90°)]$$

$$= \frac{\sigma_1 + \sigma_2}{2} + \frac{\sigma_1 - \sigma_2}{2} \cos (180 + 2\theta)$$

$$= \frac{\sigma_1 + \sigma_2}{2} - \frac{\sigma_1 - \sigma_2}{2} \cos 2\theta \quad ...(ii)$$

Adding (*i*) and (*ii*), $\sigma_n + \sigma_n^* = \sigma_1 + \sigma_2$ = constant].

18. At a point in a two dimensional system, the normal stress on two mutually perpendicular planes are σ_1 and σ_2 (both alike) and shear stress is τ. Show that one of the principal stresses is zero if $\tau = \sqrt{\sigma_1 \times \sigma_2}$.

[**Hint.** Principal stresses $= \dfrac{\sigma_1 + \sigma_2}{2} \pm \sqrt{\left(\dfrac{\sigma_1 - \sigma_2}{2}\right)^2 + \tau^2}$

$$= \frac{\sigma_1 + \sigma_2}{2} \pm \sqrt{\left(\frac{\sigma_1 - \sigma_2}{2}\right)^2 + \sigma_1\sigma_2}$$

$$= \frac{\sigma_1 + \sigma_2}{2} \pm \sqrt{\left(\frac{\sigma_1 - \sigma_2}{2}\right)^2} = \frac{\sigma_1 + \sigma_2}{2} \pm \frac{\sigma_1 + \sigma_2}{2}$$

$= \sigma_1 + \sigma_2$ and zero].

19. A rectangular block of material is subjected to a tensile stress of 100 N/mm² on one plane and a tensile stress of 50 N/mm² on a plane at right angles, together with shear stresses of 60 N/mm² on the faces. Find :

(*i*) the direction of principal planes, (*ii*) the magnitude of principal stresses and (*iii*) magnitude of the greatest shear stress.

[**Ans.** (*i*) 33° 41′ or 123° 41′ , (*ii*) 140 N/mm² and 10 N/mm² tensiles, (*iii*) 65 N/mm²]

2

3-D Stresses, Theory of Failure, Equilibrium Equations and Impact Load

2.1. THREE DIMENSIONAL STATE OF STRESS

Fig. 2.1 shows a body on which a number of forces (P_1, P_2, P_3) are acting. Consider a section plane perpendicular to x-axis. On this section plane consider a small area ΔA, on which a small force ΔP is acting as shown in Fig. 2.1(a). This force ΔP is having three components ΔP_x, ΔP_y and ΔP_z. The component ΔP_x is perpendicular to the area ΔA and acting along x-axis whereas the components ΔP_y and ΔP_z are along the surface area ΔA in the direction of y and z. We know that force divided by area represents stress. The component of force normal to area will give normal stress whereas the components of the force along the surface area will give shear stresses. The normal stress will be represented by σ (sigma) and shear stress by τ (Tau).

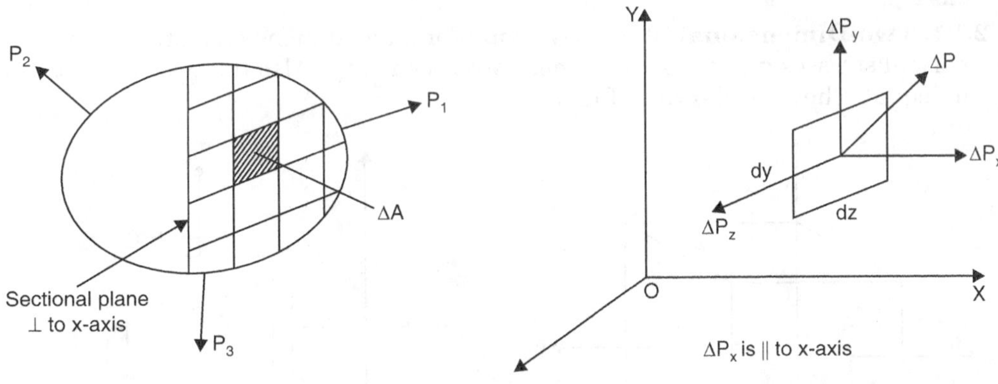

Fig. 2.1 Fig. 2.1(a)

$$\therefore \qquad \frac{\Delta P_x}{\Delta A} = \sigma_{xx} \;;\; \frac{\Delta P_y}{\Delta A} = \tau_{xy} \text{ and } \frac{\Delta P_z}{\Delta A} = \tau_{xz}.$$

The first subscript of stresses represents that the plane perpendicular to x-axis is considered and the second subscript represents the direction of the stress components.

If a section plane normal to y-axis is considered, then we will get

$$\frac{\Delta P_y}{\Delta A} = \sigma_{yy},\; \frac{\Delta P_x}{\Delta A} = \tau_{yx} \text{ and } \frac{\Delta P_z}{\Delta A} = \tau_{yz}$$

Similarly for a plane normal to z-axis, the three component of stresses will be σ_{zz}, τ_{zx} and τ_{zy}.

Now consider a cube of dimensions dx, dy and dz as shown in Fig. 2.2. The stresses acting on this cube are also shown.

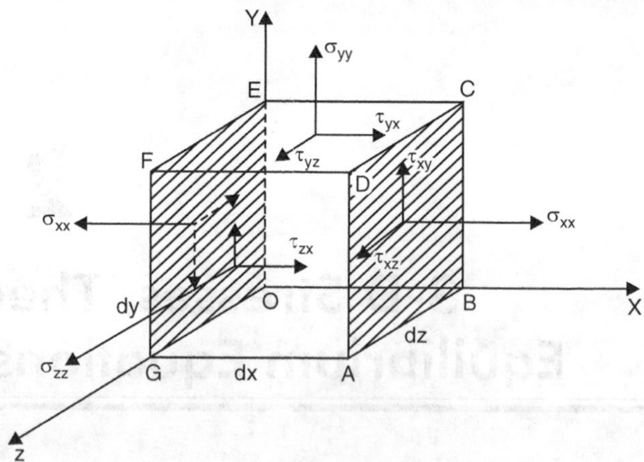

Fig. 2.2

On each face, there are three stresses, one of them is normal stress and the other two are shear stresses. The face *ABCD* in Fig. 2.2 is perpendicular to x-axis and stresses on this face are σ_{xx}, τ_{xy} and τ_{xz}. The first subscripts represents the plane perpendicular to the axis and second subscripts represents the direction of stress. On the face of the cube away from the origin, (such as face *ABCD*), the stresses are positive if they act in the positive direction of the axes. On the face of the cube coinciding with the origin (such as face OEFG), the stresses are positive if they act in the negative direction of the axes. The magnitude of stresses will be same due to action-reaction equilibrium concept. Similarly the stresses on the face *EFDC* (which is ⊥ to y-axis) are σ_{yy}, τ_{yx} and τ_{yz}. The face *ADFG* is ⊥ to z-axis and stresses are σ_{zz}, τ_{zx} and τ_{zy}. Hence from Fig. 2.2, it is clear that there are three normal stresses σ_{xx}, σ_{yy}, σ_{zz} and six shear stresses. τ_{xy}, τ_{xz}, τ_{yx}, τ_{yz}, τ_{zx} and τ_{zy}. The shear stresses τ_{xy} and τ_{yx}, τ_{xz} and τ_{zx}, τ_{yz} and τ_{zy} are equal. Hence for a three dimensional stress system, we require only six stresses (three normal stresses, σ_{xx}, σ_{yy}, σ_{zz} and three shear stresses τ_{xy}, τ_{xz}, and τ_{yz}).

2.1.1. Two-Dimensional Stress System. For a two-dimensional stress system, there are two normal stresses σ_{xx}, σ_{yy} and two shear stresses τ_{xy}, τ_{yx}. Also the two shear stresses τ_{xy} and τ_{yx} are equal. They are shown in Fig. 2.3.

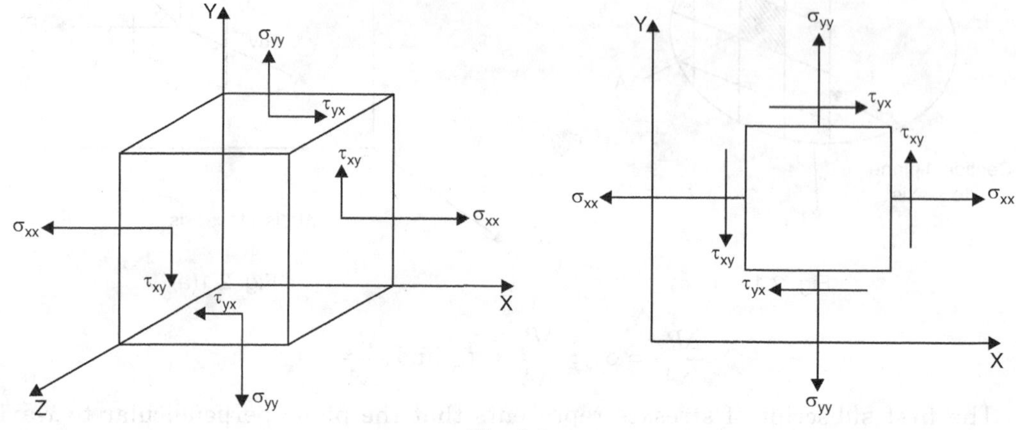

Fig. 2.3

The two-dimensional stress shown in Fig. 2.3 is known as **plane stress**. Hence for plane stress, there are two normal stresses σ_{xx} and σ_{yy} ($\sigma_{zz} = 0$) and two shear stresses (τ_{xy} and τ_{yx}). Also the shear stresses $\tau_{xy} = \tau_{yx} = \tau$. For simplicity the normal stresses σ_{xx} and σ_{yy} can be written as σ_x and σ_y.

Hence for plane stress, we have

$$\begin{pmatrix} \sigma_x & \tau & 0 \\ \tau & \sigma_y & 0 \\ 0 & 0 & 0 \end{pmatrix}$$

2.1.2. Triaxial Stress System. A system which is having three normal stresses only and no shear stresses is known as Triaxial stress system. Here σ_x, σ_y and σ_z are existing and τ_{xy}, τ_{xz}, τ_{yz} etc are zero. Hence for triaxial stress system. We have

$$\begin{pmatrix} \sigma_x & 0 & 0 \\ 0 & \sigma_y & 0 \\ 0 & 0 & \sigma_z \end{pmatrix}$$

Bi-axial stress ($\sigma_z = 0$) and no shear stress

$$\begin{pmatrix} \sigma_x & 0 & 0 \\ 0 & \sigma_y & 0 \\ 0 & 0 & 0 \end{pmatrix}$$

2.1.3. Uni-axial Stress System. Only σ_x will exists in uni-axial stress system

$$\begin{pmatrix} \sigma_x & 0 & 0 \\ 0 & 0 & 0 \\ 0 & 0 & 0 \end{pmatrix}$$

2.2. EQUILIBRIUM EQUATIONS

Consider a two-dimensional small element *ABCD* of dimensions $\Delta x \times \Delta y \times 1$ as shown in Fig. 2.4. The element has been assumed of unit thickness in the direction perpendicular to the plane of the paper. We have considered the possibility of an increment in stresses from one face to another face of the element. If the normal stress on face *AD* is σ_x, then $\frac{\partial \sigma_x}{\partial x}$ represents the rate of change of σ_x in the direction of x. For the length Δx, the increment in stress will be $\frac{\partial \sigma_x}{\partial x} \cdot \Delta x$. Hence the normal stress on face *BC* will be equal to $\sigma_x + \frac{\partial \sigma_x}{\partial x} \Delta x$.

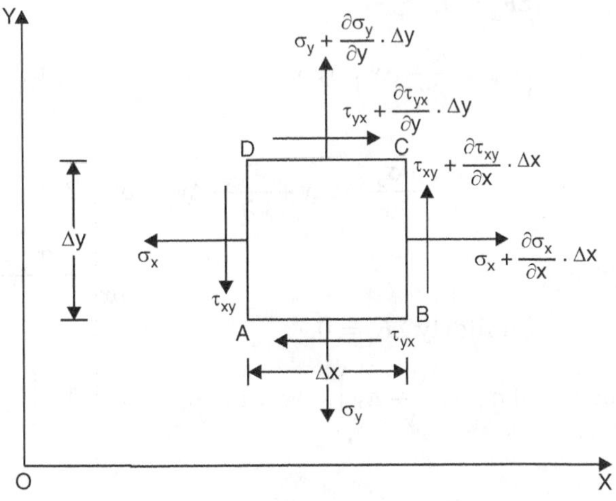

Fig. 2.4

The forces due to these stresses will be $\sigma_x \times dA$ on face *AD* and $\left(\sigma_x + \frac{\partial \sigma_x}{\partial x} \cdot \Delta x\right) dA$ on face *BC*.

Now the stresses and forces on the different faces of the element are :

Face AB,	Normal stress = σ_y Shear stress = τ_{yx}	Normal force = $\sigma_y \times AB \times 1 = \sigma_y \times \Delta x$ Shear force = $\tau_{yx} \times AB \times 1 = \tau_{yx} \times \Delta x$
Face CD,	Normal stress = $\sigma_y + \dfrac{\partial \sigma_y}{\partial y} \cdot \Delta y$ Shear stress = $\tau_{yx} + \dfrac{\partial \tau_{yx}}{\partial y} \Delta y$	Normal force = $\left(\sigma_y + \dfrac{\partial \sigma_y}{\partial y} \Delta y\right) \Delta x \times 1$ Shear force = $\left(\tau_{yx} + \dfrac{\partial \tau_{yx}}{\partial y} \Delta y\right) \Delta x \times 1$
Face AD,	Normal stress = σ_x Shear stress = τ_{xy}	Normal force = $\sigma_x \times \Delta y \times 1$ Shear force = $\tau_{xy} \times \Delta y \times 1$
Face BC,	Normal stress = $\sigma_x + \dfrac{\partial \sigma_x}{\partial x} \Delta x$ Shear stress = $\tau_{xy} + \dfrac{\partial \tau_{xy}}{\partial x} \Delta x$	Normal force = $\left(\sigma_x + \dfrac{\partial \sigma_x}{\partial x} \Delta x\right) \Delta y \times 1$ Shear force = $\left(\tau_{xy} + \dfrac{\partial \tau_{xy}}{\partial x} \Delta x\right) \Delta y$

Let B_x and B_y are the body force per unit volume due to weight or magnetic effect on the element in the direction of x and y.

Total body force in x-direction $= B_x \times$ volume of element

$= B_x \, (\Delta x \times \Delta y \times 1)$

Total body force in y-direction $= B_y \, (\Delta x \times \Delta y \times 1)$

Now the equations of equilibrium becomes as,

$\Sigma F_x = 0$,

$$\left(\sigma_x + \frac{\partial \sigma_x}{\partial x} \Delta x\right) \Delta y \times 1 - \sigma_x \times \Delta y \times 1 + \left(\tau_{yx} + \frac{\partial \tau_{yx}}{\partial y} \Delta y\right) \times \Delta x \times 1 - \tau_{yx} \times \Delta x \times 1 + B_x \, (\Delta x \times \Delta y \times 1) = 0$$

or
$$\frac{\partial \sigma_x}{\partial x} \Delta x \Delta y + \frac{\partial \tau_{yx}}{\partial y} \Delta y \times \Delta x + B_x \, (\Delta x \times \Delta y) = 0$$

or
$$\frac{\partial \sigma_x}{\partial x} + \frac{\partial \tau_{yx}}{\partial y} + B_x = 0 \qquad \text{...(2.1)}$$

Similarly $\Sigma F_y = 0$,

or
$$\left(\sigma_y + \frac{\partial \sigma_y}{\partial y} \Delta y\right) \times \Delta x \times 1 - \sigma_y \times \Delta x \times 1 + \left(\tau_{xy} + \frac{\partial \tau_{xy}}{\partial x} \Delta x\right) \Delta y \times 1 - \tau_{xy} \times \Delta y \times 1 + B_y \, (\Delta x \times \Delta y \times 1) = 0$$

or
$$\frac{\partial \sigma_y}{\partial y} \Delta y \times \Delta x + \frac{\partial \tau_{xy}}{\partial x} \Delta x \times \Delta y + B_y \times \Delta x \times \Delta y = 0$$

or
$$\frac{\partial \sigma_y}{\partial y} + \frac{\partial \tau_{xy}}{\partial x} + B_y = 0 \qquad \text{...(2.1A)}$$

Equations (2.1) and (2.1A) are the equilibrium equations for 2-D state of stress.

2.3. GENERALIZED HOOKE'S LAW

The six basic equations between a general state of stress and strain are referred as the generalized Hooke's Law. According to basic Hooke's law, the stress is directly proportional to the corresponding strain within elastic limit. This is true only for one dimensional stress system. But for two-dimensional or three dimensional system it is not true. Let us consider the following cases.

2.3.1. For One-Dimensional Stress System. The relationship between stress and strain for a unidirectional stress (*i.e.*, for normal stress in one direction only) is given by **Hooke's law,** which states that when a material is loaded within its elastic limit, the normal stress developed is proportional to the strain produced. This means that the ratio of the normal stress to the corresponding strain is a constant within the elastic limit. This constant is represented by E and is known as modulus of elasticity or Young's modulus of elasticity.

$$\therefore \qquad \frac{\text{Normal stress}}{\text{Corresponding strain}} = \text{Constant} \qquad \text{or} \qquad \frac{\sigma}{e} = E$$

where σ = Normal stress, e = Strain and E = Young's modulus

$$\text{or} \qquad e = \frac{\sigma}{E} \qquad \text{...(2.2)}$$

The above equation gives the stress and strain relation for the normal stress in one direction.

2.3.2. For Two-Dimensional Stress System. Before knowing the relationship between stress and strain for two-dimensional stress system, we shall have to define longitudinal strain, lateral strain, and Poisson's ratio.

1. Longitudinal strain. When a body is subjected to an axial tensile load, there is an increase in the length of the body. But at the same time there is a decrease in other dimensions of the body at right angles to the line of action of the applied load. Thus the body is having axial deformation and also deformation at right angles to the line of action of the applied load (*i.e.*, lateral deformation).

The ratio of axial deformation to the original length of the body is known as longitudinal (or linear) strain. The longitudinal strain is also defined as the deformation of the body per unit length in the direction of the applied load.

Let L = Length of the body

P = Tensile force acting on the body

δL = Increase in the length of the body in the direction of P.

Then, longitudinal strain $= \dfrac{\delta L}{L}$.

2. Lateral strain. The strain at right angles to the direction of applied load is known as lateral strain. Let a rectangular bar of length L, breadth b and depth d is subjected to an axial tensile load P as shown in Fig. 2.5. The length of the bar will increase while the breadth and depth will decrease.

Let δL = Increase in length,

δb = Decrease in breadth, and

δd = Decrease in depth.

Then longitudinal strain $= \dfrac{\delta L}{L}$ ...(2.3)

and lateral strain $= \dfrac{\delta b}{b}$ or $\dfrac{\delta d}{d}$ ...(2.4)

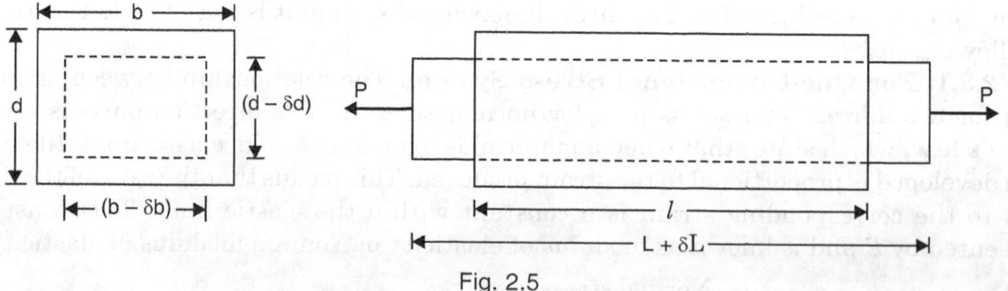

Fig. 2.5

Note. (*i*) If longitudinal strain is tensile, the lateral strains will be compressive.

(*ii*) If longitudinal strain is compressive then lateral strains will be tensile.

(*iii*) Hence every longitudinal strain in the direction of load is accompanied by lateral strains of the opposite kind in all directions perpendicular to the load.

3. Poisson's ratio. The ratio of lateral strain to the longitudinal strain is a constant for a given material, when the material is stressed within the elastic limit. This ratio is called **Poisson's ratio** and it is generally denoted by μ. Hence mathematically,

$$\text{Poisson's ratio, } \mu = \frac{\text{Lateral strain}}{\text{Longitudinal strain}} \quad ...(2.5)$$

or Lateral strain = μ × Longitudinal strain

As lateral strain is opposite in sign to longitudinal strain, hence algebraically, lateral strain is written as

Lateral strain = – μ × Longitudinal strain ...(2.5*A*)

4. Relationship between stress and strain. Consider a two-dimensional figure *ABCD*, subjected to two mutually perpendicular stresses σ_1 and σ_2.

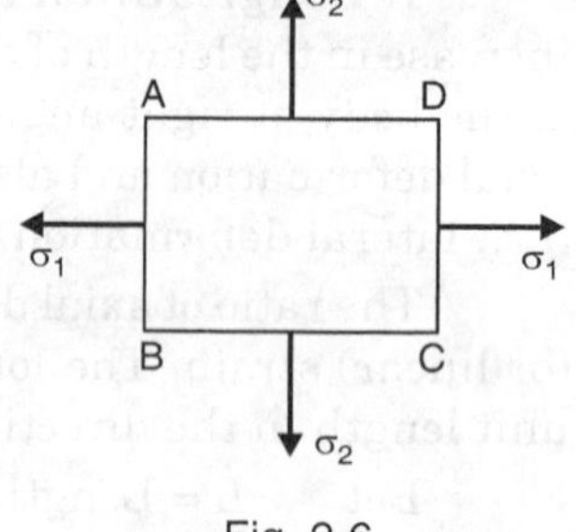

Fig. 2.6

Refer to Fig. 2.6.

Let σ_1 = Normal stress in *x*-direction

σ_2 = Normal stress in *y*-direction

Consider the strain produced by σ_1.

The stress σ_1 will produce strain in the direction of *x* and also in the direction of *y*. The strain in the direction of *x* will be longitudinal strain and $= \dfrac{\sigma_1}{E}$ whereas the strain in the direction of *y* will be lateral strain and $= -\mu \times \dfrac{\sigma_1}{E}$.

(∵ Lateral strain = – μ × longitudinal strain)

Now consider the strain produced by σ_2.

The stress σ_2 will produce strain in the direction of *y* and also in the direction of *x*. The strain in the direction of *y* will be longitudinal strain and $= \dfrac{\sigma_2}{E}$ whereas the strain in the direction of *x* will be lateral strain and $= -\mu \times \dfrac{\sigma_2}{E}$.

Let e_1 = Total strain in x-direction
e_2 = Total strain in y-direction

Now total strain in the direction of x due to stresses σ_1 and $\sigma_2 = \frac{\sigma_1}{E} - \mu \frac{\sigma_2}{E}$

Similarly total strain in the direction of y due to stresses σ_1 and $\sigma_2 = \frac{\sigma_2}{E} - \mu \frac{\sigma_1}{E}$

$$\therefore \quad e_1 = \frac{\sigma_1}{E} - \mu \frac{\sigma_2}{E} \quad ...(2.6)$$

$$e_2 = \frac{\sigma_2}{E} - \mu \frac{\sigma_1}{E} \quad ...(2.7)$$

The above two equations gives the stress and strain relationship for the two-dimensional stress system. In the above equations, tensile stress is taken to be positive whereas the compressive stress negative. The above equations shows that strains are not proportional to corresponding strains.

2.3.3. For Three-Dimensional Stress System. Fig. 2.7 shows a three-dimensional body subjected to three orthogonal normal stresses σ_1, σ_2, σ_3 acting in the directions of x, y and z respectively.

Consider the strains produced by each stress separately.

The stress σ_1 will produce strain in the direction of x and also in the directions of y and z. The strain in the direction of x will be $\frac{\sigma_1}{E}$ whereas the strains in the direction of y and z will be $-\mu \frac{\sigma_1}{E}$.

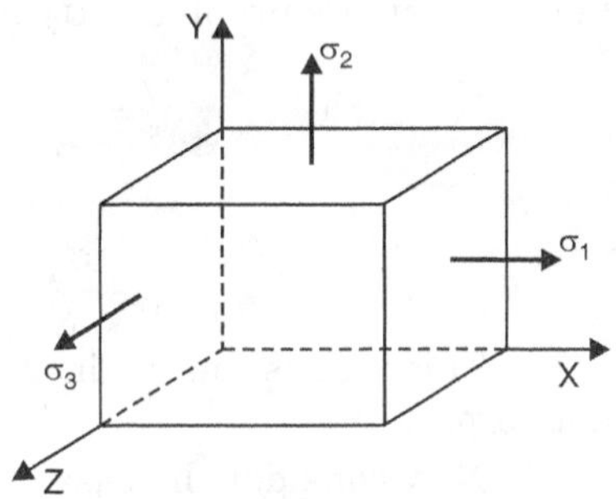

Fig. 2.7

Similarly the stress σ_2 will produce strain $\frac{\sigma_2}{E}$ in the direction of y and strain of $-\mu \frac{\sigma_2}{E}$ in the direction of x and y each.

Also the stress σ_3 will produce strain $\frac{\sigma_3}{E}$ in the direction of z and strain of $-\mu \times \frac{\sigma_3}{E}$ in the direction of x and y.

Total strain in the direction of x due to stresses σ_1, σ_2 and $\sigma_3 = \frac{\sigma_1}{E} - \mu \frac{\sigma_2}{E} - \mu \frac{\sigma_3}{E}$.

Similarly total strains in the direction of y due to stresses σ_1, σ_2 and σ_3

$$= \frac{\sigma_2}{E} - \mu \frac{\sigma_3}{E} - \mu \frac{\sigma_1}{E}$$

and total strains in the direction of z due to stresses σ_1, σ_2 and σ_3

$$= \frac{\sigma_3}{E} - \mu \frac{\sigma_1}{E} - \mu \frac{\sigma_2}{E}$$

Let e_1, e_2 and e_3 are total strains in the direction of x, y and z respectively. Then

$$e_1 = \frac{\sigma_1}{E} - \mu\frac{\sigma_2}{E} - \mu\frac{\sigma_3}{E} \quad ...(2.8A)$$

$$e_2 = \frac{\sigma_2}{E} - \mu\frac{\sigma_3}{E} - \mu\frac{\sigma_1}{E} \quad ...(2.8B)$$

and

$$e_3 = \frac{\sigma_3}{E} - \mu\frac{\sigma_1}{E} - \mu\frac{\sigma_2}{E} \quad ...(2.8C)$$

The above three equations give the stress and strain relationship for the three orthogonal normal stress system. If the body is subjected to shear stresses (τ_{xy}, τ_{yz}, τ_{zx}) then shear strains will also be produced. The relation between the shear stresses and shear strains are given as

$$\phi_{xy} = \frac{\tau_{xy}}{C} \text{ where } C = \text{Shear modulus} \quad ...(2.8D)$$

$$\phi_{yz} = \tau_{yz}/C \text{ and} \quad ...(2.8E)$$

$$\phi_{zx} = \tau_{zx}/C \quad ...(2.8F)$$

These six equations which give the relations between strains and stresses are referred as **generalised Hooke's Law.**

2.3.4. Plane Stress and Plane Strain Problems. From generalised Hooke's law equations, the difference between **plain stress** and **plain strain problems** can be explained. For plain stress problems, σ_x and σ_y may exist and $\sigma_z = 0$. Then the strains will be as

$$e_x = \frac{\sigma_x}{E} - \frac{\mu\sigma_y}{E},$$

[$\because$ σ_z or $\sigma_3 = 0$ in equations (1.8*A*), (1.8*B*) and (1.8*C*)]

$$e_y = \frac{\sigma_y}{E} - \frac{\mu\sigma_x}{E} \quad \text{and} \quad e_z = -\frac{\mu\sigma_y}{E} - \frac{\mu\sigma_x}{E}$$

This means the strain e_z is not zero. Hence for plain stress problems, σ_z is zero but e_z is not zero.

Now consider the case of **plain strain problems**. For plain strain problems strains e_x and e_y may exists. But strain e_z is zero. But e_z is given as

$$e_z = \frac{\sigma_z}{E} - \frac{\mu\sigma_y}{E} - \frac{\mu\sigma_x}{E}$$

If e_z is zero than σ_z should not be zero as first term on the R.H.S. of the above equation is positive and other two terms are negative. The R.H.S. will be zero only when σ_z is not zero. Hence for plane stress problems :

σ_x and σ_y may exist and $\sigma_z = 0$

But e_z is not zero

And for plane strain problems :

e_x and e_y may exist and $e_z = 0$

But σ_z is not zero.

2.4. THEORIES OF FAILURE

When some external load* is applied on a body, the stresses and strains are produced in the body. The stresses are directly proportional to the strains within the elastic limit. This

*The loading on the body is assumed gradual or static.

means when the load is removed, the body will return to its original shape. There is no permanent deformation in the body.

However, if the stress produced in the body due to the application of the load, is beyond the elastic limit, the permanent deformations occur in the body. This means if the load is removed, the body will not retain its original shape. There are some permanent deformations in the body. Whenever permanent deformations occur in the body, the body is said to have "*failed*". This should be clear that failure does not mean rupture of the body.

Let us consider the failure (or permanent deformation) of a bar in a simple tensile test. The tensile stress is directly proportional to the tensile strain upto elastic limit. This means that there is a definite value of tensile stress upto elastic limit. Beyond the elastic limit if the tensile stress increases, the failure of the bar will take place. At this stage, why the failure takes place ? It may be due to the increase of tensile stress or due to other quantities such as shear stress and strain energy also attain definite values, and anyone of these may be deciding factor of the failure of the bar. Certain theories have advanced to explain the cause of failure. According to the important theories, the failure takes place when a certain limiting value is reached by one of following :

1. The maximum principal stress,
2. The maximum principal strain,
3. The maximum shear stress,
4. The maximum strain energy,
5. The maximum shear strain energy.

In all the above cases,

$\sigma_1, \sigma_2, \sigma_3$ = principal stresses in any complex system

σ^* = tensile or compressive stress at the elastic limit.

2.4.1. Maximum Principal Stress Theory. According to this theory, the failure of a material will occur when the maximum principal tensile stress (σ_1) in the complex system reaches the value of the maximum stress at the elastic limit in simple tension or the maximum principal compressive stress reaches the value of the maximum stress at the elastic limit in simple compression.

Let in a complex three dimensional stress system,

σ_1, σ_2 and σ_3 = principal stresses at a point in three perpendicular directions. The stresses σ_1 and σ_2 are tensile and σ_3 is compressive. Also σ_1 is more than σ_2.

σ_t^* = tensile stress at elastic limit in simple tension.

Then according to this theory, the failure will take place if

$$\sigma_1 \geq \sigma_t^* \text{ in simple tension} \quad \text{...(2.9)}$$

This is the simplest and oldest theory of failure and is known as **Rankine's theory.** If the maximum principal stress (σ_1) is the design criterion, then maximum principal stress must not exceed the permissible stress (σ_t) for the given material.

Hence $$\sigma_1 = \sigma_t \quad \text{...(2.9}A\text{)}$$

where σ_t = permissible stress and is given by

$$\sigma_t = \sigma_t^*/\text{safety factor} \quad \text{...(2.9}B\text{)}$$

2.4.2. Maximum Principal Strain Theory. This theory is due to **Saint Venant.** According to this theory, the failure will occur in a material when the maximum principal strain reaches the strain due to yield stress in simple tension or when the minimum principal strain (*i.e.,* maximum compressive strain) reaches the strain due to yield stress in simple compression. Yield stress is the maximum stress at elastic limit. Consider a three dimensional stress system.

Principal strain in the direction of principal stress σ_1 is,

$$e_1 = \frac{\sigma_1}{E} - \frac{\mu\,\sigma_2}{E} - \frac{\mu\,\sigma_3}{E}$$

$$= \frac{1}{E}\,[\sigma_1 - \mu(\sigma_2 + \sigma_3)]$$

Principal strain in the direction of principal stress σ_3 is

$$e_3 = \frac{1}{E}\,[\sigma_3 - \mu(\sigma_1 + \sigma_2)]$$

Strain due to yield stress in simple tension $= \frac{1}{E} \times$ yield stress in tension

$$= \frac{1}{E} \times \sigma_t^*$$

According to this theory, the failure of the material will take place when

$$e_1 \geq \frac{\sigma_t^*}{E}$$

Substituting the values of e_1 and e_3, we get the conditions of failure as :

(*i*) $\frac{1}{E}\,[\sigma_1 - \mu(\sigma_2 + \sigma_3)] \geq \frac{1}{E} \times \sigma_t^*$

or $\quad \sigma_1 - \mu(\sigma_2 + \sigma_3) \geq \sigma_t^*$ ...(2.10)

Problem 2.1. *The principal stresses at a point in an elastic material are 100 N/mm² (tensile), 80 N/mm² (tensile) and 50 N/mm² (compressive). If the stress at the elastic limit in simple tension is 200 N/mm², determine whether the failure of material will occur according to maximum principal stress theory. If not, then determine the factor of safety.*

Sol. Given :

The three principal stresses are :

σ_1 = 100 N/mm² (tensile)

σ_2 = 80 N/mm² (tensile)

σ_3 = 50 N/mm² (compressive) = – 50 N/mm²

Stress at elastic limit in simple tension,

σ_t^* = 200 N/mm².

(*i*) *To determine whether failure of material will occur or not*

From the three given stresses, the maximum principal tensile stress is σ_1 = 100 N/mm². And the stress at elastic limit in simple tension is σ_t^* = 200 N/mm². As σ_1 is less than σ_t^*, the failure will not occur according to maximum principal stress theory. **Ans.**

(*ii*) *Factor of safety*

Using equation [2.9(*A*)], we get

$$\sigma_1 = \sigma_t \quad \therefore \quad \sigma_t = 100 \text{ N/mm}^2$$

From equation [2.9(*B*)],

$$\sigma_t = \frac{\sigma_t^*}{\text{Factor of safety}} \quad \therefore \quad \text{Factor of safety} = \frac{\sigma_t^*}{\sigma_t}$$

$$= \frac{200}{100} = \mathbf{2.0.} \quad \textbf{Ans.}$$

Problem 2.2. *The principal stresses at a point in an elastic material are 200 N/mm² (tensile), 100 N/mm² (tensile) and 50 N/mm² (compressive). If the stress at the elastic limit in simple tension is 200 N/mm², determine whether the failure of the material will occur according to maximum principal strain theory. Take Poisson's ratio = 0.3.*

Sol. Given :

The principal stresses are :

$\sigma_1 = 200$ N/mm² (tensile)

$\sigma_2 = 100$ N/mm² (tensile)

$\sigma_3 = 50$ N/mm² (compressive) = – 50 N/mm²

Stress at elastic limit in simple tension,

$\sigma_t^* = 200$ N/mm²

Poisson's ratio, $\mu = 0.3$

To determine whether failure of material will occur or not according to maximum principal strain theory.

Out of three principal stresses, the maximum principal tensile stress is σ_1. Hence the maximum principal strain will be in the direction of σ_1. Let this strain is e_1. Hence maximum principal strain in the direction of σ_1 will be,

$$e_1 = \frac{1}{E}\,[\sigma_1 - \mu(\sigma_2 + \sigma_3)]$$

$$= \frac{1}{E}\,[200 - 0.3\{100 + (-50)\}] \qquad [\because \ \sigma_3 = -50]$$

$$= \frac{1}{E}\,[200 - 30 + 15] = \frac{185}{E} \qquad \ldots(i)$$

Strain due to stress at elastic limit in simple tension,

$$e_t^* = \frac{\sigma_t^*}{E} = \frac{200}{E} \qquad \ldots(ii)$$

According to maximum principal strain theory, the failure of a material occurs if the maximum principal strain (e_1) reaches the strain due to stress at elastic limit in simple tension. Here $e_1 < e_t^*$, hence failure will not occur. **Ans.**

2.4.3. Maximum Shear Stress Theory. This theory is due to **Guest and Tresca** and therefore known as Guest's theory. According to this theory, the failure of a material will occur when the maximum shear stress in a material reaches the value of maximum shear stress in simple tension at the elastic limit. The maximum shear stress in the material is equal to half the difference between maximum and minimum principal stress.

If σ_1, σ_2 and σ_3 are principal stresses at a point in a material for which σ_t^* is the principal stress in simple tension at elastic limit, then

Max. shear stress in the material = Half of difference of maximum and minimum principal stresses

$$= \frac{1}{2}\,[\sigma_1 - \sigma_3]$$

In case of simple tension, at the elastic limit the principal stresses are σ_t^*, 0, 0

[In simple tension, the stress is existing in one direction only]

∴ Max. shear stress in simple tension at elastic limit

= Half of difference of maximum and minimum principal stresses

$$= \frac{1}{2}[\sigma_t^* - 0] = \frac{1}{2}\sigma_t^*$$

∴ For the failure of material,

$$\frac{1}{2}[\sigma_1 - \sigma_3] \geq \frac{1}{2}\sigma_t^* \quad \text{or} \quad (\sigma_1 - \sigma_3) \geq \sigma_t^* \qquad \text{...(2.11)}$$

Problem 2.3. *For the data given in problem 2.2, determine whether failure of material will occur or not according to maximum shear stress theory.*

Sol. Given : Data from problem 2.2 :

$\sigma_1 = 200$ N/mm^2 (tensile)

$\sigma_2 = 100$ N/mm^2 (tensile)

$\sigma_3 = 50$ N/mm^2 (compressive) = – 50 N/mm^2

$\sigma_t^* = 200$ N/mm^2.

Now max. shear stress developed in the material

= Half of difference of maximum and minimum principal stresses

$$= \frac{1}{2}[\sigma_1 - \sigma_3] = \frac{1}{2}[200 - (-50)]$$

$$= \frac{250}{2} = 125 \text{ N/mm}^2 \quad [\text{Here } \sigma_1 \text{ is maximum and } \sigma_3 \text{ is minimum}]$$

Max. shear stress at elastic limit in simple tension

$$= \frac{1}{2} \times \sigma_t^*$$

$$= \frac{1}{2} \times 200 = 100 \text{ N/mm}^2.$$

As maximum shear stress developed in the material is 125 N/mm^2 whereas maximum shear stress at the elastic limit in simple tension is 100 N/mm^2, hence failure will occur. **Ans.**

2.4.4. Maximum Strain Energy Theory. This theory is due to **Haigh** and is known as **Haigh's Theory**. According to this theory, the failure of a material occurs when the total strain energy per unit volume in the material reaches the strain energy per unit volume of the material at the elastic limit in simple tension.

We know that the strain energy in a body is equal to work done by the load (P) in straining the material and it is equal to $\frac{1}{2} \times P \times \delta L$

(∵ Load is gradually increased from 0 to P)

∴ U = Strain energy

$$= \frac{1}{2} \times P \times \delta L$$

$$= \frac{1}{2} \times (\sigma \times A) \times (e) \times L \qquad \left[\begin{array}{l} \because \ \sigma = \dfrac{P}{A} \quad \therefore \ P = \sigma \times A \\ \text{and } \ e = \dfrac{\delta L}{L} \quad \therefore \ \delta L = e \times L \end{array}\right]$$

$$= \frac{1}{2} \times \sigma \times e \times A \times L$$

$$= \frac{1}{2} \times \text{stress} \times \text{strain} \times \text{volume} \qquad (\because \; A \times L = \text{volume})$$

$\therefore$ Strain energy per unit volume

$$= \frac{1}{2} \times \text{stress} \times \text{strain}$$

$$= \frac{1}{2} \times \sigma \times e \qquad \text{...(2.12)}$$

For a three dimensional stress system, the principal stresses acting at a point are σ_1, σ_2 and σ_3. The corresponding strains are e_1, e_2 and e_3, where e_1 = principal strain in the direction of σ_1

Now, $$e_1 = \frac{\sigma_1}{E} - \frac{\mu}{E}(\sigma_2 + \sigma_3)$$

Similarly, $$e_2 = \frac{\sigma_2}{E} - \frac{\mu}{E}(\sigma_3 + \sigma_1)$$

and $$e_3 = \frac{\sigma_3}{E} - \frac{\mu}{E}(\sigma_1 + \sigma_2)$$

$\therefore$ Total strain energy per unit volume in three dimensional system,

$$U = \frac{1}{2} \times \sigma_1 \times e_1 + \frac{1}{2} \times \sigma_2 \times e_2 + \frac{1}{2}\sigma_3 \times e_3$$

$$= \frac{1}{2}\sigma_1 \times \left[\frac{\sigma_1}{E} - \frac{\mu}{E}(\sigma_2 + \sigma_3)\right] + \frac{1}{2}\sigma_2 \times \left[\frac{\sigma_2}{E} - \frac{\mu}{E}(\sigma_3 + \sigma_1)\right] + \frac{1}{2}\sigma_3 \times \left[\frac{\sigma_3}{E} - \frac{\mu}{E}(\sigma_1 + \sigma_2)\right]$$

$$= \frac{1}{2E}[\sigma_1^2 + \sigma_2^2 + \sigma_3^2 - 2\mu(\sigma_1\sigma_2 + \sigma_2\sigma_3 + \sigma_3\sigma_1)] \qquad \text{...(2.13)}$$

The strain energy per unit volume corresponding to stress at elastic limit in simple tension

$$= \frac{1}{2} \times \sigma_t^* \times e_t^* \qquad [\text{where } e_t^* = \text{strain due to } \sigma_t^*$$

$$= \frac{1}{2} \times \sigma_t^* \times \frac{\sigma_t^*}{E} \qquad \left.\therefore\; E = \frac{\sigma_t^*}{e_t^*} \text{ or } e_t^* = \frac{\sigma_t^*}{E}\right]$$

$$= \frac{1}{2E} \times (\sigma_t^*)^2 \qquad \text{...(2.14)}$$

For the failure of the material

$$\frac{1}{2E}[\sigma_1^2 + \sigma_2^2 + \sigma_3^2 - 2\mu(\sigma_1\sigma_2 + \sigma_2\sigma_3 + \sigma_3\sigma_1)] \geq \frac{1}{2E} \times (\sigma_t^*)^2$$

or $$[\sigma_1^2 + \sigma_2^2 + \sigma_3^2 - 2\mu(\sigma_1\sigma_2 + \sigma_2\sigma_3 + \sigma_3\sigma_1)] \geq (\sigma_t^*)^2 \qquad \text{...(2.15)}$$

For a two-dimensional stress system, $\sigma_3 = 0$. Hence above equation becomes as

$$\sigma_1^2 + \sigma_2^2 - 2\mu(\sigma_1\sigma_2) \geq (\sigma_t^*)^2 \qquad \text{...(2.16)}$$

Problem 2.4. *For the data given in problem 2.2, determine whether failure of material will occur or not according to maximum strain energy theory.*

Sol. Data from problem 2.2 are :

$\sigma_1 = 200$ N/mm^2 (tensile)

$\sigma_2 = 100$ N/mm^2 (tensile)

$\sigma_3 = 50$ N/mm^2 (compressive) $= -50$ N/mm^2

$\mu = 0.3$

Elastic limit in simple tension, $\sigma_t^* = 200$ N/mm^2

The total strain energy absorbed per unit volume in the material is given by equation (2.13).

$\therefore$ Total strain energy per unit volume in the material

$$= \frac{1}{2E}[\sigma_1^2 + \sigma_2^2 + \sigma_3^2 - 2\mu(\sigma_1\sigma_2 + \sigma_2\sigma_3 + \sigma_3\sigma_1)]$$

$$= \frac{1}{2E}[200^2 + 100^2 + (-50)^2 - 2 \times 0.3 \{200 \times 100 + 100 \times (-50) + (-50) \times 200\}]$$

$$= \frac{1}{2E}[40000 + 10000 + 2500 - 0.6(20000 - 5000 - 10000)]$$

$$= \frac{1}{2E}[52500 - 0.6 \times 5000] = \frac{1}{2E}[49500] \quad ...(i)$$

Strain energy per unit volume corresponding to stress at elastic limit in simple tension is given by equation (2.14).

$\therefore$ Strain energy per unit volume at elastic limit in simple tension

$$= \frac{1}{2E} \times \sigma_t^{*2}$$

$$= \frac{1}{2E} \times 200^2 \quad (\because \sigma_t^* = 200)$$

$$= \frac{40000}{2E} \quad ...(ii)$$

Now apply the theory of maximum strain energy.

If we compare equations (*i*) and (*ii*), we find the total strain energy per unit volume in the material is more than the strain energy per unit volume at elastic limit in simple tension. Hence failure will occur. **Ans.**

2.4.5. Maximum Shear Strain Energy Theory. This theory is due to Mises and Henky and is known as Mises-Henky theory. This theory is also called the *energy of distortion theory*. According to this theory, the failure of a material occurs when the total shear strain energy per unit volume in the stressed material reaches a value equal to the shear strain energy per unit volume at the elastic limit in the simple tension test.

The total shear strain energy* (or energy of distortion) per unit volume due to principal stresses σ_1, σ_2 and σ_3 in a stressed material is given as

$$= \frac{1}{12C}[(\sigma_1 - \sigma_2)^2 + (\sigma_2 - \sigma_3)^2 + (\sigma_3 - \sigma_1)^2] \quad ...(2.17)$$

The simple tension test is a uniaxial stress system which means the principal stresses are σ_1, 0, 0.

At the elastic limit the tensile stress in simple test is σ_t^*.

Hence at the elastic limit in simple tension test, the principal stresses are σ_t^*, 0, 0.

The shear strain energy per unit volume at the elastic limit in simple tension will be

$$= \frac{1}{12C}[(\sigma_t^* - 0)^2 + (0 - 0)^2 + (0 - \sigma_t^*)^2]$$

[Here $\sigma_1 = \sigma_t^*$, $\sigma_2 = 0$, $\sigma_3 = 0$]

$$= \frac{1}{12C}[2 \times \sigma_t^{*2}] \quad ...(2.18)$$

For the failure of the material,

$$\frac{1}{12C}[(\sigma_1 - \sigma_2)^2 + (\sigma_2 - \sigma_3)^2 + (\sigma_3 - \sigma_1)^2] \geq \frac{1}{12C}[2 \times (\sigma_t^*)^2]$$

or $$(\sigma_1 - \sigma_2)^2 + (\sigma_2 - \sigma_3)^2 + (\sigma_3 - \sigma_1)^2 \geq 2 \times (\sigma_t^*)^2 \quad ...(2.19)$$

For a two-dimensional stress system, $\sigma_3 = 0$.

$$(\sigma_1 - \sigma_2)^2 + (\sigma_2)^2 + (-\sigma_1)^2 = 2 \times (\sigma_t)^2$$

or $$\sigma_1^2 + \sigma_2^2 - 2\sigma_1\sigma_2 + \sigma_2^2 + \sigma_1^2 = 2 \times \sigma_t^2$$

or $$2(\sigma_1^2 + \sigma_2^2 - \sigma_1\sigma_2) = 2 \times \sigma_t^2$$

or $$\sigma_1^2 + \sigma_2^2 - \sigma_1\sigma_2 = \sigma_t^2.$$

2.5. STRAIN ENERGY

Whenever a body is strained, the energy is absorbed in the body. The energy which is absorbed in the body due to straining effect is known as **strain energy**. The straining effect may be due to gradually applied load, or suddenly applied or load with impact. Hence the strain energy will be stored in the body when load is applied gradually or suddenly or with an impact. The strain energy stored in the body is equal to work done by the applied load in stretching the body.

2.6. IMPACT LOAD AND STRESSES

Whenever a body is strained, the energy is absorbed in the body. The energy, which is absorbed in the body due to straining effect is known as *strain energy*. The straining effect may be due to gradually applied load or suddenly applied load or load with impact. Hence the strain energy will be stored in the body when the load is applied gradually or suddenly or with an impact. The strain energy stored in the body is equal to the work done by the applied load in stretching the body.

2.6.1. Expression for Strain Energy Stored in a Body when the Load is Applied Gradually. In Art. 2.6, we have mentioned that the strain energy stored in a body is equal to the work done by the applied load in stretching the body.

Fig. 2.8 shows load extension diagram of a body under tensile test upto elastic limit. The tensile load P increases gradually from zero to the value of P and the extension of the body increases from zero to the value of x.

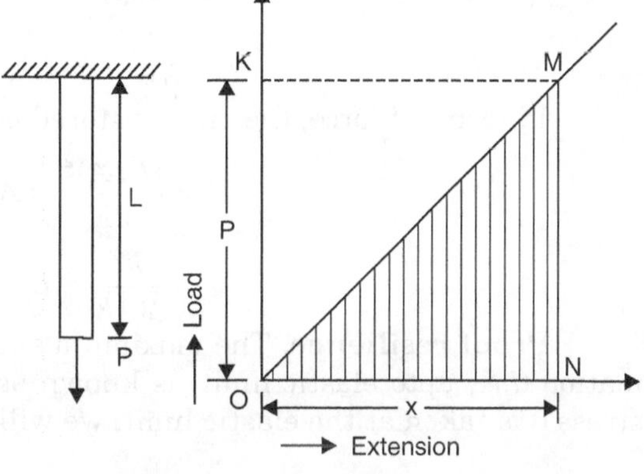

Fig. 2.8

The load P performs work in stretching the body. This work will be stored in the body as *strain energy* which is recoverable after the load P is removed.

Let P = Gradually applied load,
x = Extension of the body,
A = Cross-sectional area,
L = Length of the body,
V = Volume of the body,
E = Young's modulus,
U = Strain energy stored in the body, and
σ = Stress induced in the body.

Now work done by the load = Area of load extension curve (Shaded area in Fig. 2.8)

$$= \text{Area of triangle } ONM$$

$$= \frac{1}{2} \times P \times x. \qquad \text{...}(i)$$

But load, P = Stress × Area = $\sigma \times A$

and extension, x = Strain × Length $\left(\because \text{ Strain} = \frac{\text{Extension}}{\text{Length}} \quad \therefore \text{Extension} = \text{Strain} \times L\right)$

$$= \frac{\text{Stress}}{E} \times L \qquad \left(\because \text{ Strain} = \frac{\text{Stress}}{E}\right)$$

$$= \frac{\sigma}{E} \times L. \qquad \text{...(2.20)}$$

Substituting the values of P and x is equation (i), we get

$$\text{Work done by the load} = \frac{1}{2} \times \sigma \times A \times \frac{\sigma}{E} \times L = \frac{1}{2}\frac{\sigma^2}{E} \times A \times L$$

$$= \frac{\sigma^2}{2E} \times V \qquad (\because \text{ Volume } V = A \times L)$$

But the work done by the load in stretching the body is equal to the strain energy stored in the body.

$\therefore$ Energy stored in the body,

$$U = \frac{\sigma^2}{2E} \times V. \qquad \text{...(2.21)}$$

In terms of force, the energy stored becomes as

$$U = \frac{(F/A)^2}{2E} \times A \times L \qquad (\because \ \sigma = F/A \text{ and } V = A \times L)$$

$$= \frac{F^2}{2AE} \times L \qquad \text{...(2.21A)}$$

Proof resilience. The maximum energy stored in the body without permanent deformation (*i.e.*, upto elastic limit) is known as proof resilience. Hence if in equation (2.2), the stress σ is taken at the elastic limit, we will get proof resilience.

$\therefore$ Proof resilience $= \dfrac{\sigma^{*2}}{2E} \times$ Volume ...(2.22)

where σ^* = Stress at the elastic limit.

Modulus of resilience = Strain energy per unit volume

$$= \frac{\text{Total strain energy}}{\text{Volume}} = \frac{\frac{\sigma^2}{2E} \times V}{V} = \frac{\sigma^2}{2E} \qquad \text{...(2.23)}$$

2.6.2. Expression for Strain Energy Stored in a Body When the Load is Applied Suddenly. When the load is applied suddenly to a body, the load is constant throughout the process of the deformation of the body.

Consider a bar subjected to a sudden load.

Let P = Load applied suddenly,
L = Length of the bar,
A = Area of the cross-section,
V = Volume of the bar = $A \times L$,
E = Young's modulus,
x = Extension of the bar,
σ = Stress induced by the suddenly applied load, and
U = Strain energy stored.

As the load is applied suddenly, the load P is constant when the extension of the bar takes place.

$\therefore$ Work done by the load = Load × Extension = $P \times x$.

The maximum strain energy stored (*i.e.*, energy stored upto elastic limit) in a body is given by

$$U = \frac{\sigma^2}{2E} \times \text{Volume of the body}$$

$$= \frac{\sigma^2}{2E} \times A \times L. \qquad (\because \text{Volume} = A \times L)$$

Equating the strain energy stored in the body to the work done, we get

$$\frac{\sigma^2}{2E} \times A \times L = P \times x = P \times \frac{\sigma}{E} \times L. \qquad \left[\because \text{From equation (2.20), } x = \frac{\sigma}{E} \times L\right]$$

Cancelling $\dfrac{\sigma \times L}{E}$ to both sides, we get

$$\frac{\sigma \times A}{2} = P \quad \text{or} \quad \sigma = 2 \times \frac{P}{A}. \qquad \text{...(2.24)}$$

From the above equation it is clear that the maximum stress induced due to suddenly applied load is twice the stress induced when the same load is applied gradually.

After obtaining the value of stress (σ), the values of extension (x) and the strain energy stored in the body may be calculated easily.

Problem 2.5. *A tensile load of 60 kN is gradually applied to a circular bar of 4 cm diameter and 5 m long. If the value of E = 2.0 × 10^5 N/mm^2, determine :*

(i) stretch in the rod,

(ii) stress in the rod,

(iii) strain energy absorbed by the rod.

Sol. Given :

Gradually applied load,

$$P = 60 \text{ kN} = 60 \times 1000 \text{ N}$$

Dia. of rod, $d = 4 \text{ cm} = 40 \text{ mm}$

∴ Area, $A = \frac{\pi}{4} \times 40^2 = 400\ \pi \text{ mm}^2$

Length of rod, $L = 5 \text{ m} = 500 \text{ cm} = 5000 \text{ mm}$

∴ Volume of rod, $V = A \times L = 400\ \pi \times 5000 = 2 \times 10^6\ \pi \text{ mm}^3$

Young's modulus, $E = 2 \times 10^5 \text{ N/mm}^2$.

Let x = stretch or extension in the rod,

σ = stress in the rod, and

U = strain energy absorbed by the rod.

Now, stress, $\sigma = \frac{\text{Load}}{\text{Area}} = \frac{P}{A} = \frac{60000}{400\pi}$ = **47.746 N/mm². Ans.**

The stretch or extension is given by equation (4.1),

$$x = \frac{\sigma}{E} \times L = \frac{47.746}{2 \times 10^5} \times 5000 = \mathbf{1.19 \text{ mm}.} \quad \textbf{Ans.}$$

The strain energy absorbed by the rod is given by equation (2.21),

$$U = \frac{\sigma^2}{2E} \times V = \frac{47.476^2}{2 \times 2 \times 10^5} \times 2 \times 10^6\ \pi = 35810 \text{ Nmm} = \mathbf{35.81 \text{ Nm}.} \quad \textbf{Ans.}$$

Problem 2.6. *If in problem 2.5, the tensile load of 60 kN is applied suddenly determine* :

(i) Maximum instantaneous stress induced,

(ii) Instantaneous elongation in the rod, and

(iii) Strain energy absorbed in the rod.

Sol. Given :

The data given in problem 2.5 is $d = 40$ mm, Area = $400\ \pi \text{ mm}^2$, $L = 5000$ mm, Volume $= 2 \times 10^6\ \pi \text{ mm}^3$, $E = 2 \times 10^5 \text{ N/mm}^2$ and suddenly applied load, $P = 60000$ N.

(i) Maximum instantaneous stress induced

Using equation (2.24),

$$\sigma = 2 \times \frac{P}{A} = 2 \times \frac{60000}{400\pi} = \mathbf{95.493 \text{ N/mm}^2.} \quad \textbf{Ans.}$$

(ii) Instantaneous elongation in the rod

Let x = Instantaneous elongation

Then $$x = \frac{\sigma}{E} \times L = \frac{95.493}{2 \times 10^5} \times 5000 \qquad \text{[See equation (2.20)]}$$

$$= \mathbf{2.38 \text{ mm}.} \quad \textbf{Ans.}$$

(iii) Strain energy is given by,

$$U = \frac{\sigma^2}{2E} \times V = \frac{95.493^2}{2 \times 2 \times 10^5} \times 2 \times 10^6\ \pi = 143238 \text{ Nmm}$$

$$= \mathbf{143.238 \text{ Nm}.} \quad \textbf{Ans.}$$

Problem 2.7. *Calculate instantaneous stress produced in a bar 10 cm² in area and 3 m long by the sudden application of a tensile load of unknown magnitude, if the extension of the bar due to suddenly applied load is 1.5 mm. Also determine the suddenly applied load. Take E = 2 × 10⁵ N/mm².*

Sol. Given :

Area of bar, $A = 10 \text{ cm}^2 = 1000 \text{ mm}^2$

Length of bar, $L = 3 \text{ m} = 3000 \text{ mm}$

Extension due to suddenly applied load,

$x = 1.5 \text{ mm}$

Young's modulus, $E = 2 \times 10^5 \text{ N/mm}^2$.

Let σ = Instantaneous stress due to sudden load, and

P = Suddenly applied load.

The extension x is given by equation (2.20),

$$x = \frac{\sigma}{E} \times L \quad \text{or} \quad 1.5 = \frac{\sigma}{2 \times 10^5} \times 3000$$

$$\therefore \quad \sigma = \frac{1.5 \times 2 \times 10^5}{3000} = \mathbf{100 \ N/mm^2}. \quad \textbf{Ans.}$$

Suddenly applied load

The instantaneous stress produced by a sudden load is given by equation (2.24) as

$$\sigma = 2 \times \frac{P}{A} \quad \text{or} \quad 100 = 2 \times \frac{P}{1000}$$

$$\therefore \quad P = \frac{1000 \times 100}{2} = 50000 \text{ N} = \mathbf{50 \ kN}. \quad \textbf{Ans.}$$

Problem 2.8. *A steel rod is 2 m long and 50 mm in diameter. An axial pull of 100 kN is suddenly applied to the rod. Calculate the instantaneous stress induced and also the instantaneous elongation produced in the rod. Take E = 200 GN/m².*

Sol. Given :

Length, $L = 2 \text{ m} = 2 \times 1000 = 2000 \text{ mm}$

Diameter, $d = 50 \text{ mm}$

$\therefore$ Area, $A = \frac{\pi}{4} \times 50^2 = 625 \pi \text{ mm}^2$

Suddenly applied load,

$P = 100 \text{ kN} = 100 \times 1000 \text{ N}$

Value of $E = 200 \text{ GN/m}^2 = 200 \times 10^9 \text{ N/m}^2$ $\quad (\because G = \text{Giga} = 10^9)$

$$= \frac{200 \times 10^9}{10^6} \text{ N/mm}^2 \quad (\because 1 \text{ m} = 1000 \text{ mm} \quad \therefore \text{m}^2 = 10^6 \text{ mm}^2)$$

$$= 200 \times 10^3 \text{ N/mm}^2$$

Using equation (2.24) for suddenly applied load,

$$\sigma = 2 \times \frac{P}{A} = 2 \times \frac{100 \times 1000}{625 \pi} \text{ N/mm}^2 = \mathbf{101.86 \ N/mm^2}. \quad \textbf{Ans.}$$

Let dL = Elongation

Then $$dL = \frac{P}{E} \times L = \frac{101.86}{200 \times 10^3} \times 2000 = \mathbf{1.0186 \ mm}. \quad \textbf{Ans.}$$

Problem 2.9. *A uniform metal bar has a cross-sectional area of 700 mm² and a length of 1.5 m. If the stress at the elastic limit is 160 N/mm², what will be its proof resilience ? Determine also the maximum value of an applied load, which may be suddenly applied without exceeding the elastic limit. Calculate the value of the gradually applied load which will produce the same extension as that produced by the suddenly applied load above.*

Take $E = 2 \times 10^5$ *N/mm²*.

Sol. Given :

Area, $A = 700 \text{ mm}^2$

Length, $L = 1.5 \text{ m} = 1500 \text{ mm}$

$\therefore$ Volume of bar, $V = A \times L = 700 \times 1500 = 1050000 \text{ mm}^2$

Stress at elastic limit, $\sigma^* = 160 \text{ N/mm}^2$

Young's modulus, $E = 2 \times 10^5 \text{ N/mm}^2$

(*i*) Proof resilience is given by equation (2.22), as

$$\text{Proof resilience} = \frac{\sigma^{*2}}{2E} \times \text{Volume} = \frac{160^2}{2 \times 2 \times 10^5} \times 1050000$$

$$= 67200 \text{ Nmm} = \mathbf{67.2 \text{ Nm}}. \quad \textbf{Ans.}$$

(*ii*) Let P = Maximum value of suddenly applied load, and

P_1 = Gradually applied load.

Using equation (2.24) for suddenly applied load,

$$\sigma^* = 2 \times \frac{P}{A} \qquad \text{(Change } p \text{ to } p^*)$$

$$\therefore \quad P = \frac{\sigma^* \times A}{2} = \frac{160 \times 700}{2} = 56000 \text{ N} = \mathbf{56 \text{ kN}}. \quad \textbf{Ans.}$$

For gradually applied load,

$$\sigma^* = \frac{P_1}{A}$$

or $$P_1 = \sigma^* \times A = 160 \times 700 = 112000 \text{ N} = \mathbf{112 \text{ kN}}. \quad \textbf{Ans.}$$

Problem 2.10. *A tension bar 5 m long is made up of two parts, 3 metre of its length has a cross-sectional area of 10 cm² while the remaining 2 metre has a cross-sectional area of 20 cm². An axial load of 80 kN is gradually applied. Find the total strain energy produced in the bar and compare this value with that obtained in a uniform bar of the same length and having the same volume when under the same load. Take* $E = 2 \times 10^5$ *N/mm²*.

Sol. Given :

Total length of bar, $L = 5 \text{ m} = 5000 \text{ mm}$

Length of 1st part, $L_1 = 3 \text{ m} = 3000 \text{ mm}$

Area of 1st part, $A_1 = 10 \text{ cm}^2 = 10 \times 100 \text{ mm}^2 = 1000 \text{ mm}^2$

$\therefore$ Volume of 1st part, $V_1 = A_1 \times L_1 = 1000 \times 3000 = 3 \times 10^6 \text{ mm}^3$

Length of 2nd part, $L_2 = 2 \text{ m} = 2000 \text{ mm}$

Area of 2nd part, $A_2 = 20 \text{ cm}^2 = 20 \times 100 \text{ mm}^2 = 2000 \text{ mm}^2$

$\therefore$ Volume of 2nd part, $V_2 = 2000 \times 2000 = 4 \times 10^6 \text{ mm}^3$

Axial gradual load, $P = 80 \text{ kN} = 80 \times 1000 = 80000 \text{ N}$

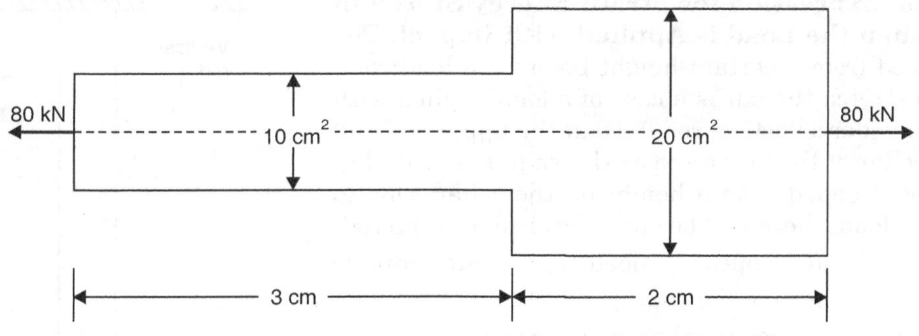

Fig. 2.9

Young's modulus, $E = 2 \times 10^5$ N/mm²

Stress in 1st part, $\sigma_1 = \dfrac{\text{Load}}{A_1} = \dfrac{80000}{1000} = 80$ N/mm²

Stress in 2nd part, $\sigma_2 = \dfrac{P}{A_2} = \dfrac{80000}{2000} = 40$ N/mm²

Strain energy in 1st part,

$$U_1 = \frac{\sigma_1^2}{2E} \times V_1 = \frac{80^2}{2 \times 2 \times 10^5} \times 3 \times 10^6 = 48000 \text{ Nmm} = 48 \text{ Nm}$$

Strain energy in 2nd part,

$$U_2 = \frac{\sigma_2^2}{2E} \times V_2 = \frac{40^2}{2 \times 2 \times 10^5} \times 4000000 = 16000 \text{ Nmm} = 16 \text{ Nm}$$

∴ Total strain energy produced in the bar,

$$U = U_1 + U_2 = 48 + 16 = \mathbf{64\ Nm.}\quad \textbf{Ans.}$$

Strain energy stored in a uniform bar

Volume of uniform bar, $V = V_1 + V_2 = 3000000 + 4000000 = 7000000$ mm²

Length of uniform bar, $L = 5$ m $= 5000$ mm

Let A = Area of uniform bar

Then $V = A \times L$ or $7000000 = A \times 5000$

∴ $A = \dfrac{7000000}{5000} = 1400$ mm²

Stress in uniform bar, $\sigma = \dfrac{P}{A} = \dfrac{80000}{5000} = 57.143$ N/mm²

∴ Strain energy stored in the uniform bar,

$$U = \frac{\sigma^2}{2E} \times V = \frac{57.143^2}{2 \times 2 \times 10^5} \times 7000000$$

$$= 57143 \text{ Nmm} = 57.143 \text{ Nm}$$

∴ $$\frac{\text{Strain energy in the given bar}}{\text{Strain energy in the uniform bar}} = \frac{64}{57.143} = \mathbf{1.12.}\quad \textbf{Ans.}$$

2.6.3. Expression for Strain Energy Stored in a Body when the Load is Applied with Impact. The load dropped from a certain height before the load commences to stretch the bar is a case of a load applied with impact. Consider a vertical rod fixed at the upper end and having a collar at the lower end as shown in Fig. 2.10. Let the load be dropped from a height on the collar. Due to this impact load, there will be some extension in the rod.

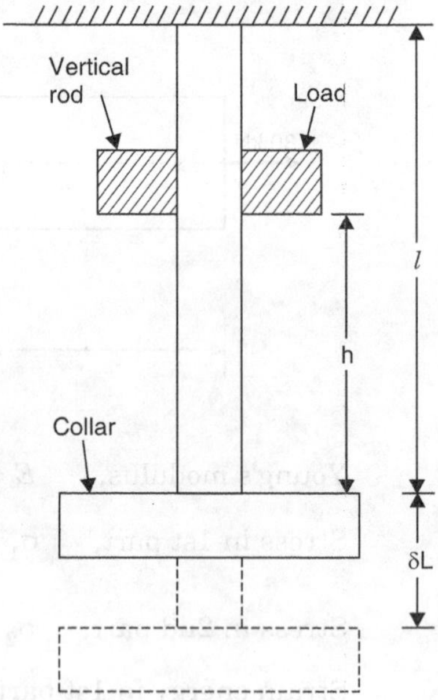

Let P = Load dropped (*i.e.,* load applied with impact)

L = Length of the rod,

A = Cross-sectional area of the rod,

V = Volume of rod = $A \times L$,

h = Height through which load is dropped,

δL = Extension of the rod due to load P,

E = Modulus of elasticity of the material of rod,

σ = Stress induced in the rod due to impact load.

Fig. 2.10

The strain in the bar is given by,

$$\text{Strain} = \frac{\text{Stress}}{E}$$

i.e.,

$$\frac{\delta L}{L} = \frac{\sigma}{E}$$

$$\therefore \quad \delta L = \frac{\sigma}{E} \times L \qquad \text{...(2.25)}$$

Work done by the load = Load × Distance moved

$$= P(h + \delta L) \qquad \text{...}(i)$$

The strain energy stored by the rod,

$$U = \frac{\sigma^2}{2E} \times V = \frac{\sigma^2}{2E} \times AL \qquad \text{...}(ii)$$

Equating the work done by the load to the strain energy stored, we get

$$P(h + \delta L) = \frac{\sigma^2}{2E} \,.\, AL$$

or

$$P\left(h + \frac{\sigma}{E} \,.\, L\right) = \frac{\sigma^2}{2E} \,.\, AL \qquad \left(\because \ \delta L = \frac{\sigma}{E} \,.\, L\right)$$

or

$$Ph + P \,.\, \frac{\sigma}{E} \,.\, L = \frac{\sigma^2}{2E} \,.\, AL$$

or

$$\frac{\sigma^2}{2E} \,.\, AL - P \,.\, \frac{\sigma}{E} \,.\, L - Ph = 0$$

Multiplying by $\frac{2E}{AL}$ to both sides, we get

$$\sigma^2 - P \,.\, \frac{\sigma}{E} \,.\, L \times \frac{2E}{A \,.\, L} - Ph \,.\, \frac{2E}{AL} = 0$$

or

$$\sigma^2 - \frac{2P}{A} \,.\, \sigma - \frac{2PEh}{A \,.\, L} = 0.$$

The above equation is a quadratic equation in 'σ',

$$\therefore \quad \sigma = \frac{\dfrac{2P}{A} \pm \sqrt{\left(\dfrac{2P}{A}\right)^2 + 4 \, . \, \dfrac{2PEh}{A \, . \, L}}}{2 \times 1} \qquad \left(\because \text{ roots} = \frac{b^2 \pm \sqrt{4ac}}{2a}\right)$$

$$= \frac{P}{A} \pm \sqrt{\frac{4P^2}{4A^2} + \frac{8 \, . \, PEh}{4 \, . \, A \, . \, L}} = \frac{P}{A} \pm \sqrt{\left(\frac{P}{A}\right)^2 + \frac{2PEh}{A \, . \, L}}$$

$$= \frac{P}{A} + \sqrt{\left(\frac{P}{A}\right)^2 + \frac{2PEh}{A \, . \, L}} \qquad \text{(Neglecting –ve root)}$$

$$= \frac{P}{A} + \frac{P}{A}\sqrt{1 + \frac{2PEh}{A \, . \, L} \times \frac{A^2}{P^2}} = \frac{P}{A} + \frac{P}{A}\sqrt{1 + \frac{2AEh}{P \, . \, L}}$$

$$= \frac{P}{A}\left(1 + \sqrt{1 + \frac{2AEh}{P \, . \, L}}\right) \qquad \text{...(2.26)}$$

After knowing the value of 'σ', the strain energy can be obtained.

Important Conclusions

(*i*) If δL is very small in comparison with h.

The work done by load $= P. \, h$

Equating the work done by the load to the strain energy stored in the rod, we get

$$P \, . \, h = \frac{\sigma^2}{2E} \, . \, AL$$

$$\therefore \quad \sigma^2 = \frac{2E \, . \, P \, . \, h}{A \, . \, L} \quad \text{and} \quad \sigma = \sqrt{\frac{2EPh}{A \, . \, L}} \qquad \text{...(2.27)}$$

(*ii*) In equation (2.26), if $h = 0$, we get

$$\sigma = \frac{P}{A}\,(1 + \sqrt{1 + 0}\,) = \frac{P}{A}\,(1 + 1) = \frac{2P}{A}$$

which is the case of suddenly applied load.

Once the stress p is known, the corresponding instantaneous extension (δL) and the strain energy (U) can be obtained.

Problem 2.11. *A weight of 10 kN falls by 30 mm on a collar rigidly attached to a vertical bar 4 m long and 1000 mm² in section. Find the instantaneous expansion of the bar. Take E = 210 GPa. Derive the formula you use.* (Bhavnagar University, Feb. 1992)

Sol. Given :

Falling weight, $P = 10$ kN $= 10{,}000$ N

Falling height, $h = 30$ mm

Length of bar, $L = 4$ m $= 4000$ mm

Area of bar, $A = 1000$ mm^2

Value of $E = 210$ GPa $= 210 \times 10^9$ N/m^2

($\because$ G = Giga = 10^9 and Pa = Pascal = 1 N/m^2)

$$= \frac{210 \times 10^9 \text{ N}}{10^6 \text{ mm}^2} \qquad (\because \; 1 \text{ m} = 1000 \text{ mm and m}^2 = 10^6 \text{ mm}^2)$$

$= 210 \times 10^3$ N/mm^2 $= 2.1 \times 10^5$ N/mm^2

Let dL = Instantaneous elongation due to falling weight

σ = Instantaneous stress produced due to falling weight

Using equation (2.26), we get

$$\sigma = \frac{P}{A}\left(1+\sqrt{1+\frac{2EAh}{P \times L}}\right)$$

$$= \frac{10000}{1000}\left(1+\sqrt{1+\frac{2 \times 2.1 \times 10^5 \times 1000 \times 30}{10000 \times 4000}}\right)$$

$$= 10\left(1+\sqrt{1+315}\right) = 10\left(1+\sqrt{316}\right)$$

$$= 10 \times 18.77 = 187.7 \text{ N/mm}^2$$

Now,
$$E = \frac{\text{Stress}}{\text{Strain}} = \frac{\sigma}{\left(\frac{\delta L}{L}\right)} \quad \text{or} \quad \frac{\delta L}{L} = \frac{\sigma}{E}$$

$\therefore$
$$\delta L = \frac{\sigma}{E} \times L = \frac{187.7 \times 4000}{2.1 \times 10^5} = \mathbf{3.575 \text{ mm.}} \quad \textbf{Ans.}$$

Problem 2.12. *A load of 100 N falls through a height of 2 cm on to a collar rigidly attached to the lower end of a vertical bar 1.5 m long and of 1.5 cm² cross-sectional area. The upper end of the vertical bar is fixed.*

Determine :

(*i*) *maximum instantaneous stress induced in the vertical bar,*

(*ii*) *maximum instantaneous elongation, and*

(*iii*) *strain energy stored in the vertical rod.*

Take $E = 2 \times 10^5$ *N/mm²*.

Sol. Given :

Impact load, $P = 100$ N

Height through which load falls,

$h = 2$ cm $= 20$ mm

Length of bar, $L = 1.5$ m $= 1500$ mm

Area of bar, $A = 1.5 \text{ cm}^2 = 1.5 \times 100 \text{ mm}^2 = 150 \text{ mm}^2$

$\therefore$ Volume, $V = A \times L = 150 \times 1500 = 225000 \text{ mm}^3$

Modulus of elasticity, $E = 2 \times 10^5 \text{ N/mm}^2$

Let σ = Maximum instantaneous stress induced in the vertical bar,

δL = Maximum elongation, and

U = Strain energy stored.

(*i*) Using equation (2.26),

$$\sigma = \frac{P}{A}\left(1+\sqrt{1+\frac{2AEh}{P.L}}\right) = \frac{100}{150}\left(1+\sqrt{1+\frac{2 \times 150 \times 2 \times 10^5 \times 20}{100 \times 1500}}\right)$$

$$= \frac{100}{150}\left(1+\sqrt{1+8000}\right) = \mathbf{60.23 \text{ N/mm}^2.} \quad \textbf{Ans.}$$

(*ii*) Using equation (2.25),

$$\delta L = \frac{\sigma}{E} \times L = \frac{60.23 \times 1500}{2 \times 10^5} = \textbf{0.452 mm. Ans.}$$

(*iii*) Strain energy is given by,

$$U = \frac{\sigma^2}{2E} \times V = \frac{60.23^2}{2 \times 2 \times 10^5} \times 225000 = 2045 \text{ Nmm}$$

$$= \textbf{2.045 Nm. Ans.}$$

Problem 2.13. *The maximum instantaneous extension, produced by an unknown falling weight through a height of 4 cm in a vertical bar of length 3 m and of cross-sectional area 5 cm², is 2.1 mm.*

Determine :

(*i*) *the instantaneous stress induced in the vertical bar, and*

(*ii*) *the value of unknown weight. Take* $E = 2 \times 10^5$ *N/mm².*

Sol. Given :

Instantaneous extension, $\delta L = 2.1$ mm

Length of bar, $L = 3 \text{ m} = 3000$ mm

Area of bar, $A = 5 \text{ cm}^2 = 500 \text{ mm}^2$

∴ Volume of bar, $V = 500 \times 3000 = 1500000 \text{ mm}^3$

Height through which weight falls, $h = 4 \text{ cm} = 40$ mm

Modulus of elasticity, $E = 2 \times 10^5 \text{ N/mm}^2$

Let σ = Instantaneous stress produced, and

P = Unknown weight.

We know $E = \frac{\text{Stress}}{\text{Strain}}$ or Stress = E × Strain

∴ Instantaneous stress $= E \times \text{Instantaneous strain} = E \times \frac{\delta L}{L}$

$$= 2 \times 10^5 \times \frac{2.1}{3000} \text{ N/mm}^2 = \textbf{140 N/mm}^2\textbf{. Ans.}$$

Equating the work done by the falling weight to the strain energy stored, we get

$$P(h + \delta L) = \frac{\sigma^2}{2E} \times V$$

or
$$P(40 + 2.1) = \frac{140^2}{2 \times 2 \times 10^5} \times 1500000 = 73500$$

or
$$P = \frac{73500}{42.1} = \textbf{1745.8 N. Ans.}$$

Note. The value of P can also be obtained by using equation (2.26).

Problem 2.14. *A bar 12 mm diameter gets stretched by 3 mm under a steady load of 8000 N. What stress would be produced in the same bar by a weight of 800 N, which falls vertically through a distance of 8 cm on to a rigid collar attached at its end ? The bar is initially unstressed. Take* $E = 2.0 \times 10^5$ *N/mm².* (AMIE, Winter 1986)

Sol. Given :

Dia. of bar, $d = 12$ mm

∵ Area of bar, $A = \frac{\pi}{4}(12)^2 = 113.1 \text{ mm}^2$

Increase in length, $\delta L = 3$ mm

Steady load, $W = 8000$ N

Falling weight, $P = 800$ N

Vertical distance, $h = 8$ cm $= 80$ mm

Young's modulus, $E = 2.0 \times 10^5 \text{ N/mm}^2$

Let L = Length of the bar, and

σ = Stress produced by the falling weight.

With steady load

We know
$$E = \frac{\text{Stress}}{\text{Strain}} = \frac{\left(\frac{\text{Steady load}}{\text{Area}}\right)}{\frac{\delta L}{L}}$$

or
$$2.0 \times 10^5 = \frac{\left(\frac{8000}{113.1}\right)}{\left(\frac{3}{L}\right)} = \frac{8000}{113.1} \times \frac{L}{3}$$

∴
$$L = \frac{2.0 \times 10^5 \times 113.1 \times 3}{8000} = 8482.5 \text{ mm}$$

Now using equation (2.26), we get
$$\sigma = \frac{P}{A}\left(1 + \sqrt{1 + \frac{2AEh}{PL}}\right)$$

$$= \frac{800}{113.1}\left(1 + \sqrt{1 + \frac{2 \times 113.1 \times 2.0 \times 10^5 \times 80}{8.0 \times 8482.5}}\right) \text{N/mm}^2$$

$$= 7.0734(1 + \sqrt{1 + 533.33}) = 7.0734 \times 24.1155$$

$= \mathbf{170.578}$ **N/mm²**. **Ans.**

Problem 2.15. *A rod 12.5 mm in diameter is stretched 3.2 mm under a steady load of 10 kN. What stress would be produced in the bar by a weight of 700 N, falling through 75 mm before commencing to stretch, the rod being initially unstressed ? The value of E may be taken as 2.1 × 10⁵ N/mm².* (AMIE, Winter 1988)

Sol. Given :

Dia. of rod, $d = 12.5$ mm

∴ Area of rod, $A = \frac{\pi}{4} \times 12.5^2 = 122.72 \text{ mm}^2$

Increase in length, $\delta L = 3.2$ mm

Steady load, $W = 10$ kN $= 10{,}000$ N

Falling load, $P = 700$ N

Falling height, $h = 75$ mm

Young's modulus, $E = 2.1 \times 10^5$ N/mm^2

Let L = Length of the rod,

σ = Stress produced by the falling weight.

We know
$$E = \frac{\text{Stress}}{\text{Strain}} = \frac{\left(\frac{\text{Steady load}}{\text{Area}}\right)}{\left(\frac{\delta L}{L}\right)}$$

or
$$2.1 \times 10^5 = \frac{\left(\frac{10{,}000}{122.72}\right)}{\left(\frac{3.2}{L}\right)} = \left(\frac{10{,}000}{122.72}\right) \times \left(\frac{L}{3.2}\right)$$

$$\therefore \quad L = \frac{2.1 \times 10^5 \times 122.72 \times 3.2}{10{,}000} = 8246.7 \text{ mm}$$

Now using equation (2.26), we get
$$\sigma = \frac{P}{A}\left(1 + \sqrt{1 + \frac{2AEh}{P \times L}}\right)$$
$$= \frac{700}{122.72}\left(1 + \sqrt{1 + \frac{2 \times 122.72 \times 2.1 \times 10^5 \times 75}{700 \times 8246.7}}\right)$$
$$= \mathbf{153.74 \text{ N/mm}^2}. \quad \textbf{Ans.}$$

Problem 2.16. *A vertical round steel rod 1.82 metre long is securely held at its upper end. A weight can slide freely on the rod and its fall is arrested by a stop provided at the lower end of the rod. When the weight falls from a height of 30 mm above the stop the maximum stress reached in the rod is estimated to be 157 N/mm². Determine the stress in the rod if the load had been applied gradually and also the minimum stress if the load had fallen from a height of 47.5 mm.*

Take $E = 2.1 \times 10^5$ N/mm^2.

Sol. Given :

Length of rod, $L = 1.82$ m $= 1.82 \times 1000 = 1820$ mm

Height through which load falls, $h = 30$ mm

Maximum stress induced in the rod, $\sigma = 157$ N/mm^2

Modulus of elasticity, $E = 2.1 \times 10^5$ N/mm^2

Let σ_1 = Stress induced in the rod if the load is applied gradually and

σ_2 = Maximum stress if the load had fallen from a height of 47.5 mm.

Strain energy stored in the rod when load falls through a height of 30 mm,

$$U = \frac{\sigma^2}{2E} \times \text{Volume} = \frac{157}{2 \times 2.1 \times 10^5} \times V$$

$$= 0.05868 \times V \text{ Nm}$$

The extension of the rod is given by equation (2.25),

$$\delta L = \frac{\sigma}{E} \times L$$

$$= \frac{157}{2.1 \times 10^5} \times 1820 = 1.36 \text{ mm}$$

∴ Total distance through which load falls

$$= h + \delta L = 30 + 1.36 = 31.36 \text{ mm}$$

∴ Work done by the falling load = Load × Total distance

$$= P \times 31.36$$

Equating the work done by the falling load to the strain energy stored, we get

$$P \times 31.36 = 0.05868 \times V$$

or $$\frac{P}{V} = \frac{0.05868}{31.36} = 0.001871$$

or $$\frac{P}{A.L} = 0.001871 \qquad (\because \; V = A.L)$$

or $$\frac{P}{A} = 0.001871 \times L = 0.001871 \times 1820 = 3.4$$

1st Case. If the load had been applied gradually, the stress induced is given by,

$$\sigma_1 = \frac{\text{Load}}{\text{Area}} = \frac{P}{A}$$

$$= \mathbf{3.4 \ N/mm^2}. \quad \textbf{Ans.}$$

2nd Case. If the load had fallen from a height of 47.5 mm.

Let σ_2 = Maximum stress induced.

Using equation (2.26), we get

$$\sigma_2 = \frac{P}{A}\left[1 + \sqrt{1 + \frac{2AEh}{P \times L}}\right] \qquad [\text{Here } \sigma = \sigma_2]$$

$$= 3.4\left[1 + \sqrt{1 + \frac{2 \times 2.1 \times 10^5 \times 47.5}{3.4 \times 1820}}\right] \qquad \left(\because \; \frac{P}{A} = 3.4, h = 47.5\right)$$

$$= 3.4\left(1 + \sqrt{1 + 3219.24}\right)$$

$$= \mathbf{196.64 \ N/mm^2}. \quad \textbf{Ans.}$$

HIGHLIGHTS

1. For a three-dimensional stress system subjected to principal stresses σ_1, σ_2 and σ_3 the equations for criterion of failure according to :
   (*a*) Maximum principal stress theory (or Rankine theory) are :
   Criterion of failure : $\sigma_1 \geq \sigma_t^*$ where σ_t^* = Elastic limit in simple tension
   (*b*) Maximum principal strain theory (or Saint Venant Theory)
   Criterion of failure : $\frac{1}{E}[\sigma_1 - \mu(\sigma_2 + \sigma_3)] \geq \frac{1}{E}\sigma_t^*$ or $\sigma_1 - \mu(\sigma_2 + \sigma_3) \geq \sigma_t^*$
   (*c*) Maximum shear stress theory (or Guest and Tresca Theory)
   At failure : $(\sigma_1 - \sigma_3) \geq \sigma_t^*$
   (*d*) Maximum strain energy theory (or Haigh's Theory)
   At failure : $\frac{1}{2E}[\sigma_1^2 + \sigma_2^2 + \sigma_3^2 - 2\mu(\sigma_1\sigma_2 + \sigma_2\sigma_3 + \sigma_3\sigma_1)] \geq \frac{1}{2E}(\sigma_t^*)^2$
   For two-dimensional stress-system :
   At failure : $\sigma_1^2 + \sigma_2^2 - 2\mu(\sigma_1 \times \sigma_2) \geq (\sigma_t^*)^2$
   (*e*) Maximum shear strain energy theory (or Mises and Henky Theory or Energy of distortion Theory)
   Criterion of failure : $(\sigma_1 - \sigma_2)^2 + (\sigma_2 - \sigma_3)^2 + (\sigma_3 - \sigma_1)^2 \geq 2 \times (\sigma_t^*)^2$
2. The energy stored in a body due to straining effect is known as strain energy.
3. Resilience is the total strain energy stored in a body. Resilience is also defined as the capacity of a strained body for doing work on the removal of the straining force.
4. The maximum strain energy stored in a body is known as proof resilience. The proof resilience is given by,

   Proof resilience = $\frac{\sigma^2}{2E} \times$ Volume

   where σ = Stress at the elastic limit.
5. The proof resilience of a body per unit volume is known as modulus of resilience.
6. The maximum stress induced in a body is given by

   $$\sigma = \frac{P}{A} \text{ .... if the load } P \text{ is applied gradually}$$
   $$= 2\frac{P}{A} \text{ ..... if the load } P \text{ is applied suddenly}$$
   $$= \frac{P}{A}\left(1 + \sqrt{1 + \frac{2AEh}{P.L}}\right) \text{ ..... if the load } P \text{ is applied with impact}$$

   where A = Cross-sectional area of the body,
   h = Height through which load falls,
   E = Modulus of rigidity,
   L = Length of the body.
7. The maximum stress induced in a body due to suddenly applied load is twice the stress induced when the same load is applied gradually.
8. If the extension produced in a rod due to impact load is very small in comparison with the height

through which the load falls, then the maximum stress induced in body is given by

$$\sigma = \sqrt{\frac{2E \cdot P \cdot h}{A \cdot L}}$$

where P = Impact load,
h = Height through which load falls.

9. To find the expression for the stress induced in a body either by suddenly applied load or by an impact load, the strain energy stored in a body is equated to the work done by the load.

EXERCISE 2

(A) Theoretical Questions

1. What do you understand by the term "Theories of failure" ? Name the important theories of failure.
2. Define and explain the following theories of failure :
   (*i*) Maximum principal stress theory.
   (*ii*) Maximum principal strain theory.
   (*iii*) Maximum shear stress theory.
   (*iv*) Maximum strain energy theory.
   (*v*) Maximum shear strain energy theory.
3. Explain with reasons which theory of failure is best suited for :
   (*i*) Ductile materials
   (*ii*) Brittle material.
4. Define the following terms :
   (*i*) Resilience (*ii*) Strain energy
   (*iii*) Impact loading, and (*iv*) Spring.
5. Define resilience, proof resilience and modulus of resilience.
6. Find an expression for the strain energy stored in a body when
   (*i*) the load is applied gradually (*ii*) the load is applied suddenly and
   (*iii*) the load is applied with an impact.
7. Prove that the maximum stress induced in a body due to suddenly applied load is twice the stress induced when the same load is applied gradually.
8. Derive an expression for the stress induced in a body due to suddenly applied load and hence find the value of extension produced in the body.
9. Prove that the maximum strain energy stored in a body is given by,

$$U = \frac{\sigma^2}{2E} \times \text{Volume}$$

where σ = Stress at the elastic limit.

10. Explain the terms : Gradually applied load, suddenly applied load, and load applied with an impact.
11. Prove that the stress induced in a body when the load is applied with impact is given by,

$$\sigma = \frac{P}{A}\left(1 + \sqrt{1 + \frac{2AEh}{P.L}}\right)$$

where P = Load applied with impact, A = Cross-sectional area of the body,
h = Height through which load falls, L = Length of the body, and
E = Modulus of elasticity.

12. If the extension produced in a rod due to impact load is very small in comparison with the height through which the load falls, prove that stress induced in the body will be given by

$$\sigma = \sqrt{\frac{2EPh}{A \cdot L}}.$$

(B) Numerical Problems

1. The principal stresses at a point in an elastic material are 22, N/mm^2 (tensile), 110 N/mm^2 (tensile) and 55 N/mm^2 (compressive).

If the elastic limit in simple tension is 220 N/mm^2 and μ = 0.3, then determine whether the failure of material will occur or not according to

(*i*) Maximum principal stress theory,
(*ii*) Maximum principal strain theory,
(*iii*) Maximum shear stress theory,
(*iv*) Maximum strain energy theory,
(*v*) Maximum shear strain energy theory. **[Ans.** (*i*) Yes ; (*ii*) No ; (*iii*) Yes ; (*iv*) Yes ; (*v*) Yes]

2. Determine the diameter of a bolt which is subjected to an axial pull of 12 kN together with a transverse shear force of 6 kN, when the elastic limit in tension is 300 N/mm^2, factor of safety = 3 and Poisson's ratio = 0.3 using :

(*i*) Maximum principal stress theory,
(*ii*) Maximum principal strain theory,
(*iii*) Maximum shear stress theory,
(*iv*) Maximum strain energy theory,
(*v*) Maximum shear strain energy theory

[Ans. (*i*) 13.59 mm ; (*ii*) 13.925 mm ; (*iii*) 14.7 mm ; (*iv*) 14.02 mm ; (*v*) 14.22 mm]

3. A bolt is under an axial thrust of 7.2 kN together with a transverse shear force of 3.6 kN. Calculate the diameter of the bolt according to :

(*i*) Maximum principal stress theory,
(*ii*) Maximum shear stress theory,
(*iii*) Maximum strain energy theory.

Take elastic limit in simple tension = 202 N/mm^2, factor of safety = 3 and Poisson's ratio = 0.3.

[Ans. (*i*) 12.8 mm ; (*ii*) 13.8 mm ; (*iii*) 13.33 mm]

4. A steel shaft is subjected to an end thrust producing a stress of 90 MPa and the maximum shearing stress on the surface arising from torsion is 60 MPa. The yield point of the material in simple tension was found to be 300 MPa. Calculate the factor of safety of the shaft according to the following theories :

(*i*) Maximum shear stress theory, *(AMIE, Summer 1977)*
(*ii*) Maximum distortion-energy theory. **[Ans.** (*i*) 2.5 ; (*ii*) 2.77]

5. At a section of a mild steel shaft of diameter 180 mm, the maximum torque is 67.5 kNm and maximum bending moment is 40.5 kNm. The elastic limit in simple tension is 220 N/mm^2 . Determine whether the failure of the material will occur or not according to maximum shear stress theory. If not, then find the factor of safety. **[Ans.** (*i*) Not ; (*ii*) 1.6]

6. A tensile load of 50 kN is gradually applied to a circular bar of 5 cm diameter and 4 m long. If the value of $E = 2.0 \times 10^5$ N/mm^2, determine : (*i*) stretch in the rod, (*ii*) stress in the rod, and (*iii*) strain energy absorbed by the rod. **[Ans.** (*i*) 0.0509 cm ; (*ii*) 25.465 N/mm^2 ; (*iii*) 12.73 Nm]

7. If in question 1, the tensile load of 50 kN is applied suddenly, determine : (*i*) maximum instantaneous stress induced, (*ii*) instantaneous elongation in the rod, and (*iii*) strain energy absorbed in the rod. **[Ans.** (*i*) 50.93 N/mm^2 ; (*ii*) 0.1018 cm ; (*iii*) 50.93 Nm]

8. Calculate instantaneous stress produced in a bar 10 cm^2 in area and 4 m long by the sudden application of a tensile load of unknown magnitude, if the extension of the bar due to suddenly applied load is 1.35 mm. Also determine the suddenly applied load. Take $E = 2 \times 10^5$ N/mm^2.

[**Ans.** 67.5 N/mm^2, 33.75 kN]

9. A uniform metal bar has a cross-sectional area of 6 cm^2 and a length of 1.4 m. If the stress at the elastic limit is 150 N/mm^2, find the proof resilience of the bar. Determine also the maximum value of an applied load, which may be suddenly applied without exceeding the elastic limit. Calculate the value of the gradually applied load which will produced the same extension as that produced by the suddenly applied load above. Take $E = 2 \times 10^5$ N/mm^2.

[**Ans.** 47.25 Nm, 45 kN, 90 kN]

10. A tension bar 6 m long is made up of two parts, 4 metre of its length has a cross-sectional area 12.5 cm^2 while the remaining 2 m length has a cross-sectional area of 25 cm^2. An axial load of 5 tonnes is gradually applied. Find the total strain energy produced in the bar and compare this value with that obtained in a uniform bar of the same length and having the same volume when under the same load. Take $E = 2 \times 10^6$ kgf/cm^2. [**Ans.** 242 kgf/cm, 1.054]

11. A load of 200 N falls through a height of 2.5 cm on to a collar rigidly attached to the lower end of a vertical bar 2 m long and of 3 cm^2 cross-sectional area. The upper end of the vertical bar is fixed. Determine : (*i*) maximum instantaneous stress induced in the vertical bar, (*ii*) maximum instantaneous elongation, and (*iii*) strain energy stored in the vertical rod. Take $E = 2 \times 10^6$ kgf/cm^2. [**Ans.** (*i*) 58.4 N/mm^2 ; (*ii*) 0.0584 cm ; (*iii*) 511.5 Nm]

12. The maximum instantaneous, produced by an unknown falling weight through a height of 4 cm in a vertical bar of length 5 m and of cross-sectional area 5 cm^2, is 1.80 mm. Determine : (*i*) the instantaneous stress induced in the vertical bar and (*ii*) the values of unknown weight.

Take $E = 2 \times 10^6$ kgf/cm^2. [**Ans.** (*i*) 72 N/mm^2 ; (*ii*) 775.1 N]

13. An unknown weight falls through a height of 20 mm on a collar rigidly attached to the lower end of a vertical bar 5 m long and 800 mm^2 in section. If the maximum extension of the rod is to be 2.5 mm, what is the corresponding stress and magnitude of the unknown weight ? Take $E = 2.0 \times 10^6$ kgf/cm^2. [**Ans.** 1000 kgf/cm^2, 444.44 kgf]

14. A bar 1.5 cm diameter gets stretched by 2.5 mm under a steady load of 100 kgf. What stress would be produced in the same bar by a weight of 120 kgf, which falls vertically through a distance 5 cm on to a rigid collar attached at its end ? The bar is initially unstressed. Take $E = 2.0 \times 10^6$ kgf/cm^2. [**Ans.** 1309.44 kgf/cm^2]

15. A vertical round steel rod 2 m long is securely held at its upper end. A weight can slide freely on the rod and its fall is arrested by a stop provided at the lower end of the rod. When the weight falls from a height of 2.5 cm above the stop, the maximum stress reached in the rod is estimated to be 1450 kgf/cm^2. Determine the stress in the rod if the load had been applied gradually and also the maximum stress if the load had fallen from a height of 4.5. Take $E = 2.0 \times 10^6$ kgf/cm^2.

[**Ans.** 39.743 kgf/cm^2, 193.42 kgf/cm^2]

UNIT II

STRESSES IN BEAMS, DEFLECTION IN BEAMS AND TORSION

Chapters

3

Stresses in Beams

(A) NORMAL OR BENDING STRESSES IN BEAMS

3.1. INTRODUCTION

When some external load acts on a beam, the shear force and bending moments are set up at all sections of the beam. Due to the shear force and bending moment, the beam undergoes certain deformation. The material of the beam will offer resistance or stresses against these deformations. These stresses with certain assumptions can be calculated. The stresses introduced by bending moment are known as *bending stresses* and the stresses due to shear force are known as shear stresses. In this chapter, the theory of pure bending, expression for bending stresses, bending stress in symmetrical and unsymmetrical sections, strength of a beam and composite beams will be discussed.

3.2. PURE BENDING

If a length of a beam is subjected to a constant bending moment and no shear force (*i.e.*, zero shear force), then the stresses will be set up in that length of the beam due to B.M.

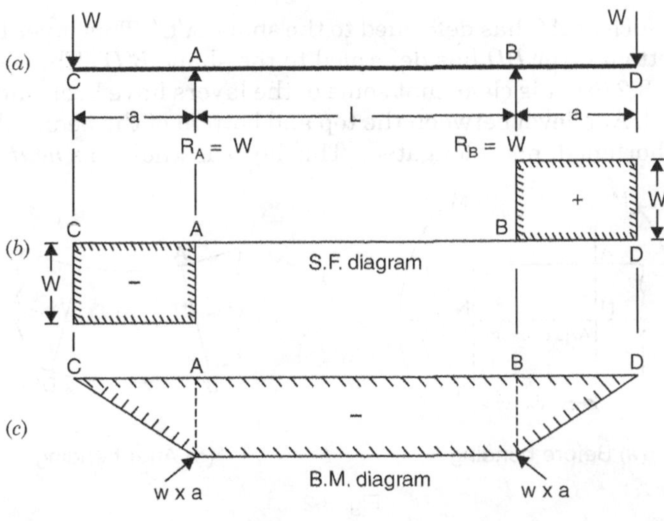

Fig. 3.1

only and that length of the beam is said to be in *pure bending* (or simple bending). The stresses set up in that length of beam are known as bending stresses.

A beam simply supported at A and B and overhanging by same length at each support is shown in Fig. 3.1. A point load W is applied at each end of the overhanging portion. The S.F. and B.M. for the beam are drawn as shown in Fig. 3.1 (*b*) and Fig. 3.1 (*c*) respectively. From these diagrams, it is clear that there is no shear force between A and B but the B.M. between A and B is constant.

This means that between A and B, the beam is subjected to a constant bending moment only. This condition of the beam between A and B is known as pure bending or simple bending.

3.2.1. Theory of Simple Bending with Assumptions Made. Before discussing the theory of simple bending, let us see the assumptions made in the theory of simple bending. The following are the important assumptions :

1. The material of the beam is homogeneous* and isotropic**.

2. The value of Young's modulus of elasticity is the same in tension and compression.

3. The transverse sections which were plane before bending, remain plane after bending also.

4. The beam is initially straight and all longitudinal filaments bend into circular arcs with a common centre of curvature.

5. The radius of curvature is large compared with the dimensions of the cross-section.

6. Each layer of the beam is free to expand or contract, independently of the layer, above or below it.

3.2.2. Theory of Simple Bending. Fig. 3.2 (*a*) shows a part of a beam subjected to simple bending. Consider a small length δx of this part of beam. Consider two sections AB and CD which are normal to the axis of the beam $N-N$. Due to the action of the bending moment, the part of length δx will be deformed as shown in Fig. 3.2 (*b*). From this figure, it is clear that all the layers of the beam, which were originally of the same length, do not remain of the same length any more.

The top layer such as AC has deformed to the shape $A'C'$. This layer has been shortened in its length. The bottom layer BD has deformed to the shape $B'D'$. This layer has been elongated. From the Fig. 3.2 (*b*), it is clear that some of the layers have been shortened while some of them are elongated. At a level between the top and bottom of the beam, there will be a layer which is neither shortened nor elongated. This layer is known as *neutral layer* or *neutral*

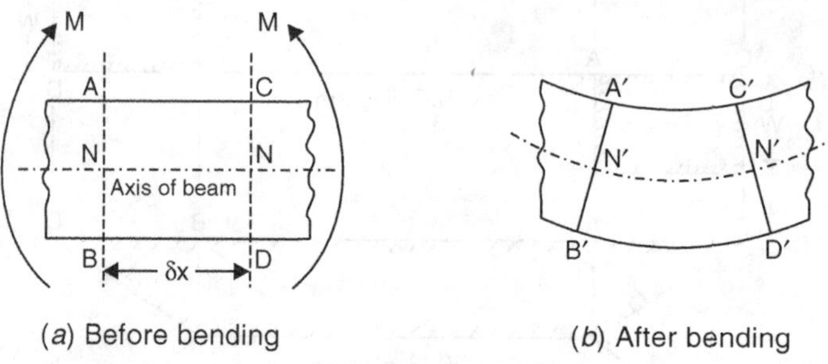

(*a*) Before bending (*b*) After bending

Fig. 3.2

*Homogeneous means the material is of the same kind throughout.

** Isotropic means that the elastic properties in all directions are equal.

surface. This layer in Fig. 3.2 (*b*) is shown by $N' - N'$ and in Fig. 3.2 (*a*) by $N - N$. The line of intersection of the neutral layer on a cross-section of a beam is known as *neutral axis* (written as N.A.).

The layers above $N - N$ (or $N' - N'$) have been shortened and those below, have been elongated. Due to the decrease in lengths of the layers above $N - N$, these layers will be subjected to compressive stresses. Due to the increase in the lengths of layers below $N - N$, these layers will be subjected to tensile stresses.

We also see that the top layer has been shortened maximum. As we proceed towards the layer $N - N$, the decrease in length of the layers decreases. At the layer $N - N$, there is no change in length. This means the compressive stress will be maximum at the top layer. Similarly the increase in length will be maximum at the bottom layer. As we proceed from bottom layer towards the layer $N - N$, the increase in length of layers decreases. Hence the amount by which a layer increases or decreases in length, depends upon the position of the layer with respect to $N - N$. This theory of bending is known as theory of simple bending.

3.2.3. Expression for Bending Stress. Fig. 3.3 (*a*) shows a small length δx of a beam subjected to a simple bending. Due to the action of bending, the part of length δx will be deformed as shown in Fig. 3.3 (*b*). Let $A'B'$ and $C'D'$ meet at O.

Let R = Radius of neutral layer $N'N'$

θ = Angle subtended at O by $A'B'$ and $C'D'$ produced.

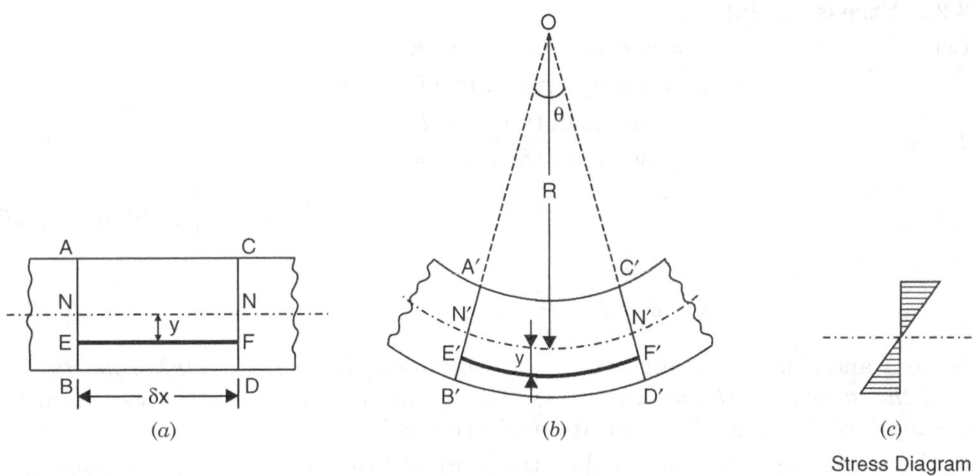

Fig. 3.3

3.2.4. Strain Variation Along the Depth of Beam. Consider a layer EF at a distance y below the neutral layer NN. After bending this layer will be elongated to $E'F'$.

Original length of layer $EF = \delta x$.

Also length of neutral layer $NN = \delta x$.

After bending, the length of neutral layer $N'N'$ will remain unchanged. But length of layer $E'F'$ will increase. Hence

$$N'N' = NN = \delta x.$$

Now from Fig. 3.3 (*b*),

$$N'N' = R \times \theta$$

and $$E'F' = (R + y) \times \theta \qquad (\because \text{ Radius of } E'F' = R + y)$$

But $$N'N' = NN = \delta x.$$

Hence $$\delta x = R \times \theta$$

∴ Increase in the length of the layer *EF*

$$= E'F' - EF = (R + y)\,\theta - R \times \theta \qquad (\because EF = \delta x = R \times \theta)$$

$$= y \times \theta$$

∴ Strain in the layer *EF*

$$= \frac{\text{Increase in length}}{\text{Original length}}$$

$$= \frac{y \times \theta}{EF} = \frac{y \times \theta}{R \times \theta} \qquad (\because EF = \delta x = R \times \theta)$$

$$= \frac{y}{R}$$

As R is constant, *hence the strain in a layer is proportional to its distance from the neutral axis.* The above equation shows the variation of strain along the depth of the beam. The variation of strain is linear.

3.2.5. Stress Variation

Let σ = Stress in the layer *EF*

E = Young's modulus of the beam

Then $$E = \frac{\text{Stress in the layer } EF}{\text{Strain in the layer } EF}$$

$$= \frac{\sigma}{\left(\frac{y}{R}\right)} \qquad \left(\because \text{ Strain in } EF = \frac{y}{R}\right)$$

∴ $$\sigma = E \times \frac{y}{R} = \frac{E}{R} \times y \qquad \text{...(3.1)}$$

Since E and R are constant, *therefore stress in any layer is directly proportional to the distance of the layer from the neutral layer.* The equation (3.1) shows the variation of stress along the depth of the beam. The variation of stress is linear.

In the above case, all layers below the neutral layer are subjected to tensile stresses whereas the layers above neutral layer are subjected to compressive stresses. The Fig. 3.3 (*c*) shows the stress distribution. These stresses are also known as **Normal stresses.**

The equation (3.1) can also be written as

$$\frac{\sigma}{y} = \frac{E}{R} \qquad \text{...(3.2)}$$

3.2.6. Neutral Axis and Moment of Resistance. The neutral axis of any transverse section of a beam is defined as the line of intersection of the neutral layer with the transverse section. It is written as N.A.

In Art. 3.2.3, we have seen that if a section of a beam is subjected to pure sagging moment, then the stresses will be compressive at any point above the neutral axis and tensile below the neutral axis. There is no stress at the neutral axis. The stress at a distance y from the neutral axis is given by equation (3.1) as

$$\sigma = \frac{E}{R} \times y.$$

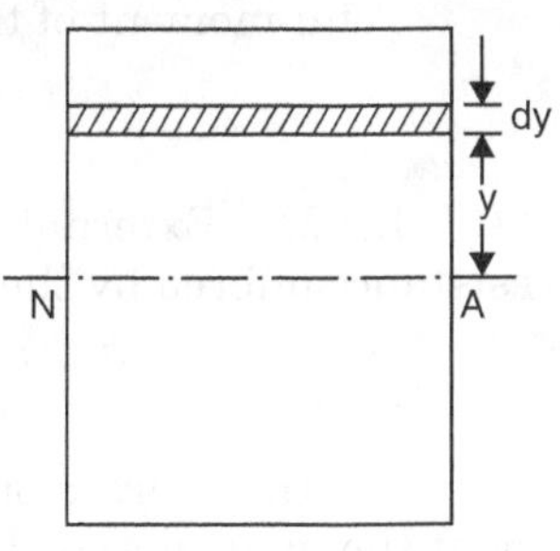

Fig. 3.4

Fig. 3.4 shows the cross-section of a beam. Let N.A. be the neutral axis of the section. Consider a small layer at a distance y from the neutral axis. Let dA = Area of the layer.

Now the force on the layer

$$= \text{Stress on layer} \times \text{Area of layer}$$
$$= \sigma \times dA$$
$$= \frac{E}{R} \times y \times dA \qquad \text{...}(i) \qquad \left(\because \ \sigma = \frac{E}{R} \times y\right)$$

Total force on the beam section is obtained by integrating the above equation.

∴ Total force on the beam section

$$= \int \frac{E}{R} \times y \times dA$$
$$= \frac{E}{R} \int y \times dA \qquad (\because \ E \text{ and } R \text{ is constant})$$

But for pure bending, there is no force on the section of the beam (or force is zero).

$$\therefore \qquad \frac{E}{R} \int y \times dA = 0$$

or
$$\int y \times dA = 0 \qquad \left(\text{as } \frac{E}{R} \text{ cannot be zero}\right)$$

Now $y \times dA$ represents the moment of area dA about neutral axis. Hence $\int y \times dA$ represents the moment of entire area of the section about neutral axis. But we know that moment of any area about an axis passing through its centroid, is also equal to zero. Hence neutral axis coincides with the centroidal axis. Thus the centroidal axis of a section gives the position of neutral axis.

3.2.7. Moment of Resistance. Due to pure bending, the layers above the N.A. are subjected to compressive stresses whereas the layers below the N.A. are subjected to tensile stresses. Due to these stresses, the forces will be acting on the layers. These forces will have moment about the N.A. The total moment of these forces about the N.A. for a section is known as moment of resistance of that section.

The force on the layer at a distance y from neutral axis in Fig. 3.4 is given by equation (i), as

$$\text{Force on layer} = \frac{E}{R} \times y \times dA$$

Moment of this force about N.A.

$$= \text{Force on layer} \times y$$
$$= \frac{E}{R} \times y \times dA \times y$$
$$= \frac{E}{R} \times y^2 \times dA$$

Total moment of the forces on the section of the beam (or moment of resistance)

$$= \int \frac{E}{R} \times y^2 \times dA = \frac{E}{R} \int y^2 \times dA$$

Let M = External moment applied on the beam section. For equilibrium the moment of resistance offered by the section should be equal to the external bending moment.

$$\therefore \quad M = \frac{E}{R} \int y^2 \times dA .$$

But the expression $\int y^2 \times dA$ represents the moment of inertia of the area of the section about the neutral axis. Let this moment of inertia be I.

$$\therefore \quad M = \frac{E}{R} \times I \quad \text{or} \quad \frac{M}{I} = \frac{E}{R} \qquad \text{...(3.3)}$$

But from equation (3.2), we have

$$\frac{\sigma}{y} = \frac{E}{R}$$

$$\therefore \quad \frac{M}{I} = \frac{\sigma}{y} = \frac{E}{R} \qquad \text{...(3.4)}$$

The equation (3.4) is known as bending equation.

In equation (3.4), the different quantities are expressed in consistent units as given below :

M is expressed in N mm ; I in mm^4

σ is expressed in N/mm^2 ; y in mm

and E is expressed in N/mm^2 ; R in mm.

3.2.8. Condition of Simple Bending. The equation (3.4) is applicable to a member which is subjected to a constant bending moment and the member is absolutely free from shear force. But in actual practice, a member is subjected to such loading that the B.M. varies from section to section and also the shear force is not zero. But shear force is zero at a section where bending moment is maximum. *Hence the condition of simple bending* may be assumed to be satisfied at such a section. Hence the stresses produced due to maximum bending moment, are obtained from equation (3.4) as the shear forces at these sections are generally zero. Hence the theory and equations discussed in the above articles are quite sufficient and give results which enables the engineers to design beams and structures and calculate their stresses and strains with a reasonable degree of approximation where B.M. is maximum.

3.3. BENDING STRESSES IN SYMMETRICAL SECTIONS

The neutral axis (N.A.) of a symmetrical section (such as circular, rectangular or square) lies at a distance of $d/2$ from the outermost layer of the section where d is the diameter (for a circular section) or depth (for a rectangular or a square section). There is no stress at the neutral axis. But the stress at a point is directly proportional to its distance from the neutral axis. The maximum stress takes place at the outermost layer. For a simply supported beam, there is a compressive stress above the neutral axis and a tensile stress below it. If we plot these stresses, we will get a figure as shown in Fig. 3.5.

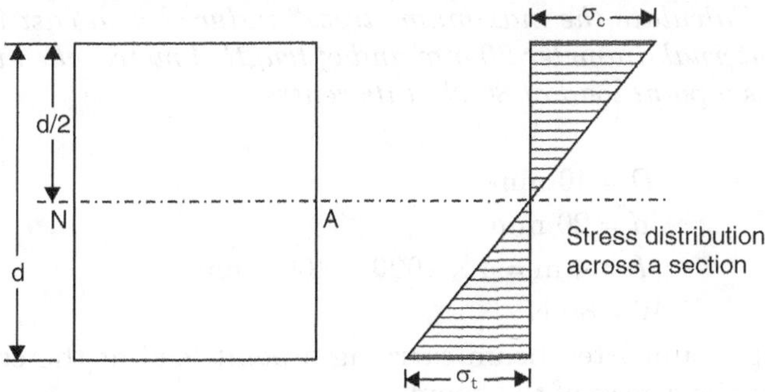

Fig. 3.5

Problem 3.1. *A steel plate of width 120 mm and of thickness 20 mm is bent into a circular arc of radius 10 m. Determine the maximum stress induced and the bending moment which will produce the maximum stress. Take $E = 2 \times 10^5$ N/mm².*

Sol. Given :

Width of plate, $b = 120$ mm

Thickness of plate, $t = 20$ mm

$\therefore$ Moment of inertia, $I = \dfrac{bt^3}{12} = \dfrac{120 \times 20^3}{12} = 8 \times 10^4 \text{ mm}^4$

Radius of curvature, $R = 10 \text{ m} = 10 \times 10^3$ mm

Young's modulus, $E = 2 \times 10^5 \text{ N/mm}^2$

Let σ_{max} = Maximum stress induced, and

M = Bending moment.

Using equation (3.2), $\dfrac{\sigma}{y} = \dfrac{E}{R}$

$$\therefore \quad \sigma = \frac{E}{R} \times y \qquad ...(i)$$

Equation (*i*) gives the stress at a distance *y* from N.A.

Stress will be maximum, when *y* is maximum. But *y* will be maximum at the top layer or bottom layer.

$$\therefore \quad y_{max} = \frac{t}{2} = \frac{20}{2} = 10 \text{ mm.}$$

Now equation (*i*) can be written as

$$\sigma_{max} = \frac{E}{R} \times y_{max}$$

$$= \frac{2 \times 10^5}{10 \times 10^3} \times 10 = \mathbf{200 \ N/mm^2}. \quad \textbf{Ans.}$$

From equation (3.4), we have

$$\frac{M}{I} = \frac{E}{R}$$

$$\therefore \quad M = \frac{E}{R} \times I = \frac{2 \times 10^5}{10 \times 10^3} \times 8 \times 10^4$$

$$= 16 \times 10^5 \text{ N mm} = \mathbf{1.6 \ kNm}. \quad \textbf{Ans.}$$

Problem 3.2. *Calculate the maximum stress* induced in a cast iron pipe of external diameter 40 mm, of internal diameter 20 mm and of length 4 metre when the pipe is supported at its ends and carries a point load of 80 N at its centre.*

Sol. Given :

External dia., $D = 40$ mm

Internal dia., $d = 20$ mm

Length, $L = 4 \text{ m} = 4 \times 1000 = 4000$ mm

Point load, $W = 80$ N

In case of simply supported beam carrying a point load at the centre, the maximum bending moment is at the centre of the beam.

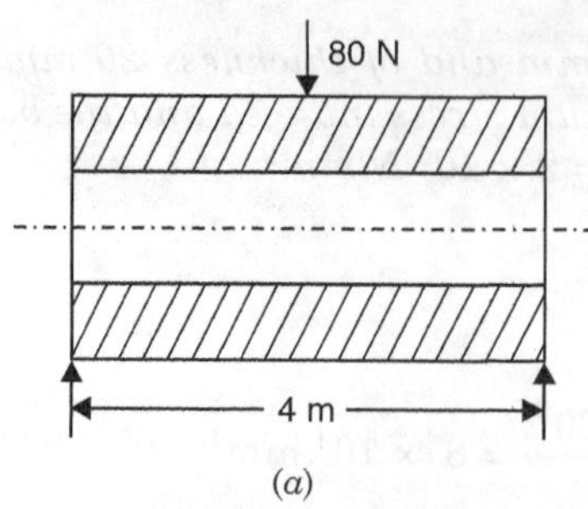

(a)

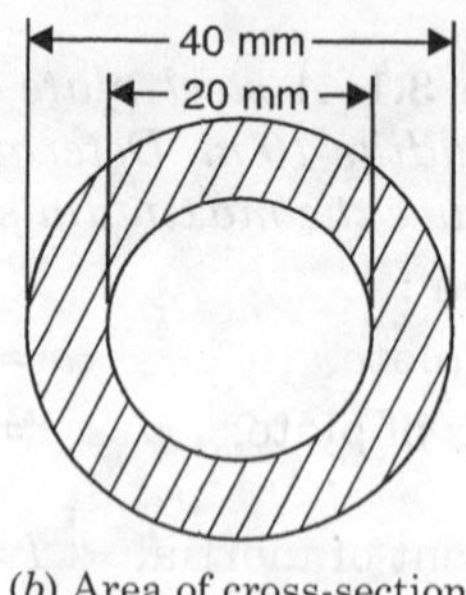

(b) Area of cross-section

Fig. 3.6

And maximum B.M. $= \dfrac{W \times L}{4}$

$\therefore$ Maximum B.M. $= \dfrac{80 \times 4000}{4} = 8 \times 10^4$ Nmm

$\therefore$ $M = 8 \times 10^4$ Nmm

Fig. 3.6 (*b*) shows the cross-section of the pipe.

Moment of inertia of hollow pipe,

$$I = \frac{\pi}{64}[D^4 - d^4]$$

$$= \frac{\pi}{64}[40^4 - 20^4] = \frac{\pi}{64}[2560000 - 160000]$$

$$= 117809.7 \text{ mm}^4$$

Now using equation (3.4),

$$\frac{M}{I} = \frac{\sigma}{y} \qquad ...(i)$$

when y is maximum, stress will be maximum. But y is maximum at the top layer from the N.A.

$$\therefore \quad y_{max} = \frac{D}{2} = \frac{40}{2} = 20 \text{ mm}$$

---

*The bending stress will be maximum at the section where B.M. is maximum. This is because $\dfrac{M}{I} = \dfrac{\sigma}{y}$ or $\sigma = \dfrac{M}{I} \times y$.

The above equation (i) can be written as

$$\frac{M}{I} = \frac{\sigma_{max}}{y_{max}}$$

$$\therefore \qquad \sigma_{max} = \frac{M}{I} \times y_{max}$$

$$= \frac{8 \times 10^4 \times 20}{117809.7} = \mathbf{13.58\ N/mm^2}. \quad \textbf{Ans.}$$

3.3.1. Section Modulus. Section modulus is defined as the ratio of moment of inertia of a section about the neutral axis to the distance of the outermost layer from the neutral axis. It is denoted by the symbol Z. Hence mathematically section modulus is given by,

$$Z = \frac{1}{y_{max}} \qquad ...(3.5)$$

where I = M.O.I. about neutral axis

and y_{max} = Distance of the outermost layer from the neutral axis.

From equation (3.4), we have

$$\frac{M}{I} = \frac{\sigma}{y}$$

The stress σ will be maximum, when y is maximum. Hence above equation can be written as

$$\frac{M}{I} = \frac{\sigma_{max}}{y_{max}}$$

$$\therefore \qquad M = \sigma_{max} \cdot \frac{I}{y_{max}}$$

But

$$\frac{I}{y_{max}} = Z$$

$$\therefore \qquad M = \sigma_{max} \cdot Z \qquad ...(3.6)$$

In the above equation, M is the maximum bending moment (or moment of resistance offered by the section). Hence moment of resistance offered the section is maximum when section modulus Z is maximum. Hence section modulus represent the strength of the section.

3.3.2. Section Modulus For Various Shapes or Beam Sections

1. *Rectangular Section*

Moment of inertia of a rectangular section about an axis through its C.G. (or through N.A.) is given by,

$$I = \frac{bd^3}{12}$$

Distance of outermost layer from N.A. is given by,

$$y_{max} = \frac{d}{2}$$

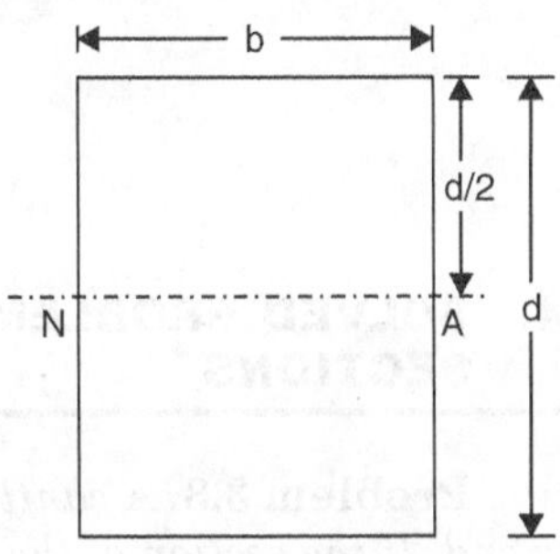

Fig. 3.7

∴ Section modulus is given by,

$$Z = \frac{I}{y_{max}} = \frac{bd^3}{12 \times \left(\frac{d}{2}\right)} = \frac{bd^3}{12} \times \frac{2}{d} = \frac{bd^2}{6} \qquad \text{...(3.7)}$$

2. *Hollow Rectangular Section*

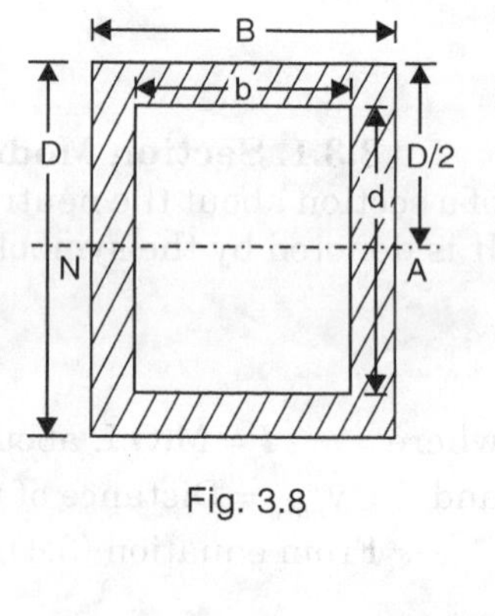

Fig. 3.8

Here
$$I = \frac{BD^3}{12} - \frac{bd^2}{12}$$

$$= \frac{1}{12}\,[BD^3 - bd^3]$$

$$y_{max} = \frac{D}{2}$$

∴
$$Z = \frac{I}{y_{max}}$$

$$= \frac{\frac{1}{12}[BD^3 - bd^3]}{\left(\frac{D}{2}\right)}$$

$$= \frac{1}{6D}\,[BD^3 - bd^3] \qquad \text{...(3.8)}$$

3. *Circular Section*
For a circular section,

$$I = \frac{\pi}{64}\,d^4 \quad \text{and} \quad y_{max} = \frac{d}{2}$$

∴
$$Z = \frac{I}{y_{max}} = \frac{\frac{\pi}{64}d^4}{\left(\frac{d}{2}\right)} = \frac{\pi}{32}\,d^3 \qquad \text{...(3.9)}$$

4. *Hollow Circular Section*

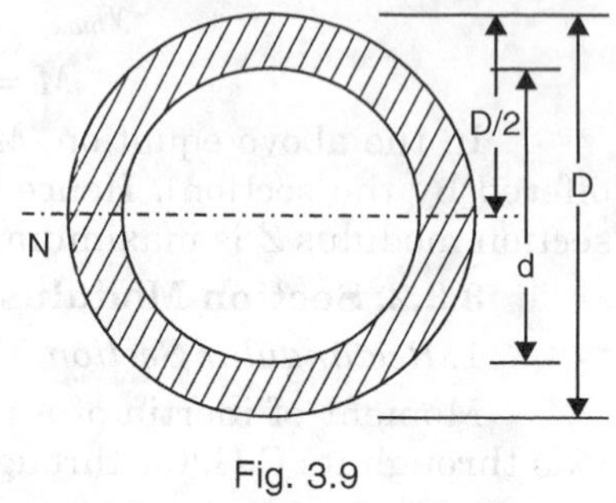

Fig. 3.9

Here
$$I = \frac{\pi}{64}\,[D^4 - d^4]$$

and
$$y_{max} = \frac{D}{2}$$

∴
$$Z = \frac{I}{y_{max}} = \frac{\frac{\pi}{64}[D^4 - d^4]}{\left(\frac{D}{2}\right)}$$

$$= \frac{\pi}{32D}\,[D^4 - d^4] \qquad \text{...(3.10)}$$

3.4. SOLVED PROBLEMS BASED ON BENDING STRESSES IN SYMMETRICAL SECTIONS

Problem 3.3. *A cantilever of length 2 metre fails when a load of 2 kN is applied at the free end. If the section of the beam is 40 mm × 60 mm, find the stress at the failure.*

Sol. Given :

Length, $L = 2 \text{ m} = 2 \times 10^3$ mm

Load, $W = 2$ kN $= 2000$ N

Section of beam is 40 mm × 60 mm.

∴ Width of beam, $b = 40$ mm

Depth of beam, $d = 60$ mm

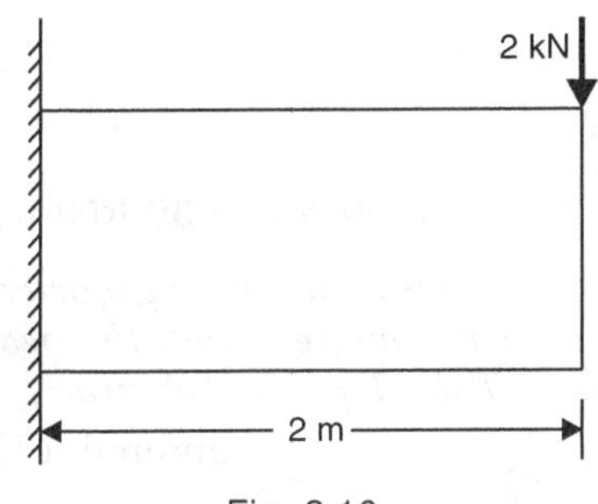

Fig. 3.10 Fig. 3.10 (*a*)

Fig. 3.10 (*a*) shows the section of the beam.

Section modulus of a rectangular section is given by equation (3.7).

$$\therefore \quad Z = \frac{bd^2}{6} = \frac{40 \times 60^2}{6} = 24000 \text{ mm}^3$$

Maximum bending moment for a cantilever shown in Fig. 3.10 is at the fixed end.

$$\therefore \quad M = W \times L = 2000 \times 2 \times 10^3 = 4 \times 10^6 \text{ Nmm}$$

Let σ_{max} = Stress at the failure

Using equation (3.6), we get

$$M = \sigma_{max} \, . \, Z$$

$$\therefore \quad \sigma_{max} = \frac{M}{Z} = \frac{4 \times 10^6}{24000} = \mathbf{166.67 \ N/mm^2}. \quad \textbf{Ans.}$$

Problem 3.4. *A rectangular beam 200 mm deep and 300 mm wide is simply supported over a span of 8 m. What uniformly distributed load per metre the beam may carry, if the bending stress is not to exceed 120 N/mm².*

Sol. Given :

Depth of beam, $d = 200$ mm

Width of beam, $b = 300$ mm

Length of beam, $L = 8$ m

Max. bending stress,

$$\sigma_{max} = 120 \text{ N/mm}^2$$

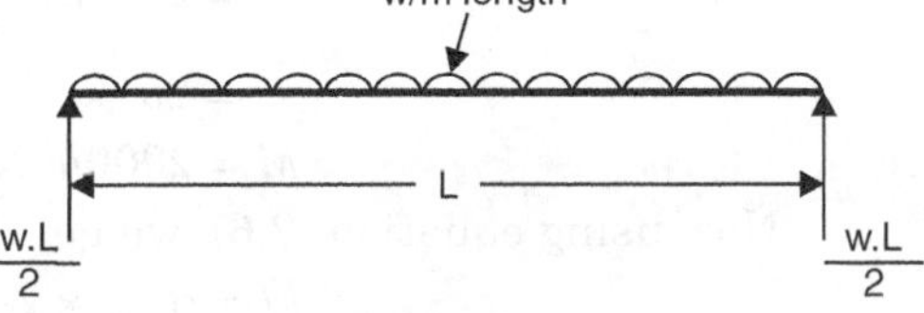

Fig. 3.11

Let w = Uniformly distributed load per m length over the beam

(Fig. 3.11 (*a*) shows the section of the beam).

Section modulus for a rectangular section is given by equation (3.7).

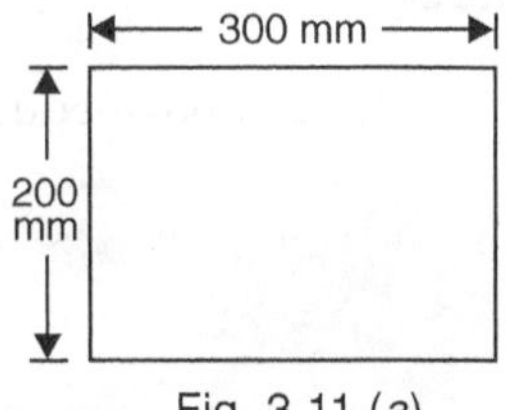

Fig. 3.11 (*a*)

$$\therefore \quad Z = \frac{bd^2}{6} = \frac{300 \times 200^2}{6} = 2000000 \text{ mm}^3$$

Max. B.M. for a simply supported beam carrying uniformly distributed load as shown in Fig. 3.11 is at the centre of the beam. It is given by

$$M = \frac{w \times L^2}{8} = \frac{w \times 8^2}{8} \qquad (\because \; L = 8 \text{ m})$$

$$= 8w \text{ Nm} = 8w \times 1000 \text{ Nmm}$$

$$= 8000w \text{ Nmm} \qquad (\because \; 1 \text{ m} = 1000 \text{ mm})$$

Now using equation (3.6), we get

$$M = \sigma_{max.} \,.\, Z$$

or $$8000w = 120 \times 2000000$$

$\therefore$ $$w = \frac{120 \times 2000000}{8000} = 30 \times 1000 \text{ N/m} = \mathbf{30 \text{ kN/m}}. \quad \textbf{Ans.}$$

Problem 3.5. *A rectangular beam 300 mm deep is simply supported over a span of 4 metres. Determine the uniformly distributed load per metre which the beam may carry, if the bending stress should not exceed 120 N/mm². Take I = 8 × 10⁶ mm⁴.*

(Annamalai University, 1991)

Sol. Given :

Depth, $d = 300$ mm

Span, $L = 4$ m

Max. bending stress, $\sigma_{max} = 120 \text{ N/mm}^2$

Moment of inertia, $I = 8 \times 10^6 \text{ mm}^4$

Let, w = U.D.L. per metre length over the beam in N/m.

The bending stress will be maximum, where bending moment is maximum. For a simply supported beam carrying U.D.L., the bending moment is maximum at the centre of the beam [*i.e.*, at point *C* of Fig. 3.11 (*b*)]

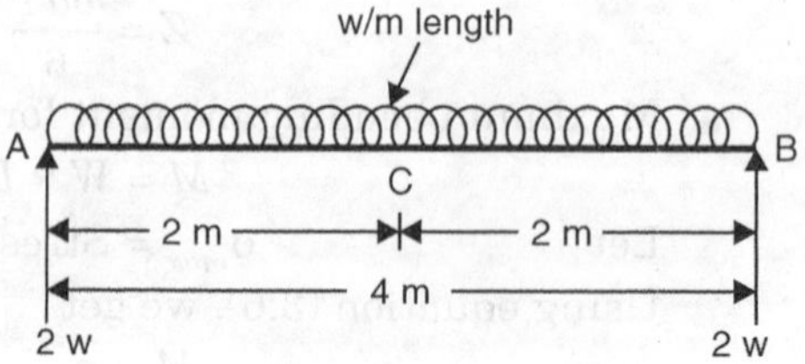

Fig. 3.11 (*b*)

$\therefore$ $$\text{Max. B.M.} = 2w \times 2 - 2w \times 1$$

$$= 4w - 2w$$

$$= 2w \text{ Nm} \qquad \left(\text{Also } M = \frac{w \times L^2}{8} = \frac{w \times 4^2}{8} = \frac{16w}{8} = 2w\right)$$

$$= 2w \times 1000 \text{ Nmm}$$

or $$M = 2000w \text{ Nmm}$$

Now using equation (3.6), we get

$$M = \sigma_{max} \times Z \qquad \ldots(i)$$

where $$Z = \frac{I}{y_{max}} = \frac{8 \times 10^6}{150} \qquad \left(\because \; y_{max} = \frac{d}{2} = \frac{300}{2} = 150 \text{ mm}\right)$$

Hence above equation (*i*) becomes as

$$2000w = 120 \times \frac{8 \times 10^6}{150}$$

or $$w = \frac{120 \times 8 \times 10^6}{2000 \times 150} = \mathbf{3200 \text{ N/m}}. \quad \textbf{Ans.}$$

Problem 3.6. *A square beam 20 mm × 20 mm in section and 2 m long is supported at the ends. The beam fails when a point load of 400 N is applied at the centre of the beam. What uniformly distributed load per metre length will break a cantilever of the same material 40 mm wide, 60 mm deep and 3 m long ?*

Sol. Given :

Depth of beam, $d = 20$ mm

Width of beam, $b = 20$ mm

Length of beam, $L = 2$ m

Point load, $W = 400$ N

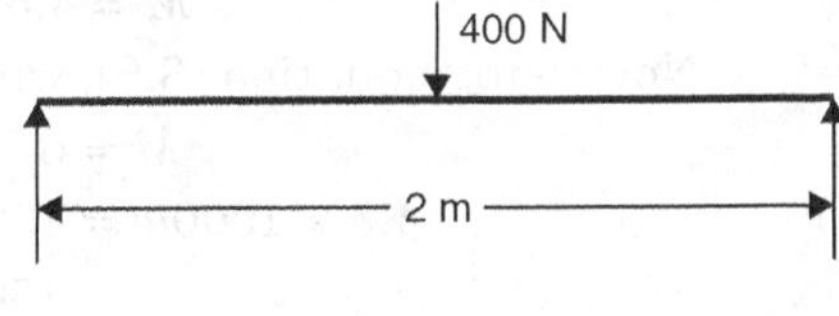

Fig. 3.12

In this problem, the maximum stress for the simply supported beam is to be calculated first. As the material of the cantilever is same as that of simply supported beam, hence maximum stress for the cantilever will also be same as that of simply supported beam.

Fig. 3.12 (*a*) shows the section of beam.

The section modulus for the rectangular section of simply supported beam is given by equation (3.7).

$$\therefore \quad Z = \frac{bd^2}{6} = \frac{20 \times 20^2}{6} = \frac{4000}{3} \text{ mm}^3$$

Fig. 3.12 (*a*)

Max. B.M. for a simply supported beam carrying a point load at the centre (as shown in Fig. 3.12) is given by,

$$M = \frac{w \times L}{4} = \frac{400 \times 2}{4} = 200 \text{ Nm}$$

$$= 200 \times 1000 = 200000 \text{ Nmm}$$

Let σ_{max} = Max. stress induced

Now using equation (3.6), we get

$$M = \sigma_{max} \cdot Z$$

or
$$200000 = \sigma_{max} \times \frac{4000}{3}$$

$$\therefore \quad \sigma_{max} = \frac{200000 \times 3}{4000} = 150 \text{ N/mm}^2$$

Now let us consider the cantilever as shown in Fig. 3.13.

Let w = Uniformly distributed load per m run.

Maximum stress will be same as in case of simply supported beam.

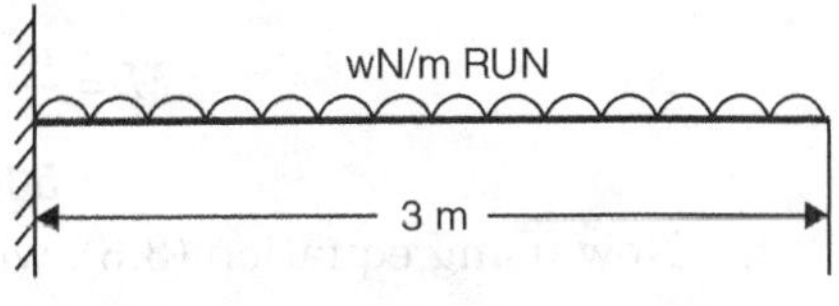

Fig. 3.13

$$\therefore \quad \sigma_{max} = 150 \text{ N/mm}^2$$

Width of cantilever, $b = 40$ mm

Depth of cantilever, $d = 60$ mm

Length of cantilever, $L = 3$ m

Fig. 3.13 (*a*) shows the section of cantilever beam.

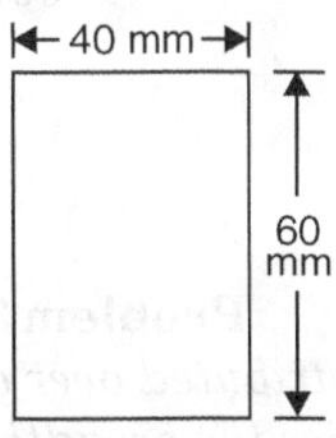

Fig. 3.13 (*a*)

Section modulus of rectangular section of cantilever $= \dfrac{bd^2}{6}$

$$\therefore \quad Z = \frac{40 \times 60^2}{6} = 24000 \text{ mm}^3$$

Maximum B.M. for a cantilever

$$= \frac{wL^2}{2} = \frac{w \times 3^2}{2} = 4.5w \text{ Nm} = 4.5 \times 1000w \text{ Nmm}$$

$\therefore$ $M = 4.5 \times 1000w$ Nmm

Now using equation (3.6), we get

$$M = \sigma_{max} . Z$$

or $4.5 \times 1000w = 150 \times 24000$

$\therefore$ $w = \frac{150 \times 24000}{4.5 \times 1000}$ = **800 N/m. Ans.**

Problem 3.7. *A beam is simply supported and carries a uniformly distributed load of 40 kN/m run over the whole span. The section of the beam is rectangular having depth as 500 mm. If the maximum stress in the material of the beam is 120 N/mm² and moment of inertia of the section is 7 × 10⁸ mm⁴, find the span of the beam.*

Sol. Given :

U.D.L., $w = 40$ kN/m $= 40 \times 1000$ N/m

Depth, $d = 500$ mm

Max. stress, $\sigma_{max} = 120$ N/mm^2

M.O.I. of section, $I = 7 \times 10^8$ mm^4

Let L = Span of simply supported beam.

Section modulus of the section is given by equation (3.5), as

$$Z = \frac{I}{y_{max}}$$

where $y_{max} = \frac{d}{2} = \frac{500}{2} = 250$ mm

$\therefore$ $Z = \frac{7 \times 10^8}{250} = 28 \times 10^5$ mm^3

The maximum B.M. for a simply supported beam, carrying a U.D.L. over the whole span is at the centre of the beam and is equal to $\frac{w . L^2}{8}$.

$$\therefore \quad M = \frac{w . L^2}{8} = \frac{40000 \times L^2}{8}$$

$$= 5000L^2 \text{ Nm} = 5000L^2 \times 1000 \text{ Nmm}$$

Now using equation (3.6), we get

$$M = \sigma_{max} . Z$$

or $5000 \times 1000 \times L^2 = 120 \times 28 \times 10^5$

or $L^2 = \frac{120 \times 28 \times 10^5}{5000 \times 1000} = 2.4 \times 28$

$\therefore$ $L = \sqrt{2.4 \times 28} = 8.197$ m say **8.20 m. Ans.**

Problem 3.8. *A timber beam of rectangular section is to support a load of 20 kN uniformly distributed over a span of 3.6 m when beam is simply supported. If the depth of section is to be twice the breadth, and the stress in the timber is not to exceed 7 N/mm², find the dimensions of the cross-section.*

How would you modify the cross-section of the beam, if it carries a concentrated load of 20 kN placed at the centre with the same ratio of breadth to depth ?

Sol. Given :

Total load, $W = 20 \text{ kN} = 20 \times 1000 \text{ N}$

Span, $L = 3.6 \text{ m}$

Max. stress, $\sigma_{max} = 7 \text{ N/mm}^2$

Let b = Breadth of beam in mm

Then depth, $d = 2b$ mm

Section modulus of rectangular beam $= \dfrac{bd^2}{6}$

$$\therefore \quad L = \frac{b \times (2b)^2}{6} = \frac{2b^3}{6} \text{ mm}^3$$

Maximum B.M., when the simply supported beam carries a U.D.L. over the entire span, is at the centre of the beam and is equal to $\dfrac{wL^2}{8}$ or $\dfrac{WL}{8}$.

$$\therefore \quad M = \frac{WL}{8} = \frac{20000 \times 3.6}{8} = 9000 \text{ Nm}$$

$$= 9000 \times 1000 \text{ Nmm}$$

Now using equation (3.6), we get

$$M = \sigma_{max} \cdot Z$$

or
$$9000 \times 1000 = 7 \times \frac{2b^3}{3}$$

or
$$b^3 = \frac{3 \times 9000 \times 1000}{7 \times 2} = 1.92857 \times 10^6$$

$$\therefore \quad b = (1.92857 \times 10^6)^{1/3}$$

$= 124.47$ mm say **124.5 mm. Ans.**

and $d = 2b = 2 \times 124.5 =$ **249 mm. Ans.**

Dimension of the section when the beam carries a point load at the centre.

B.M. is maximum at the centre and it is equal to $\dfrac{W \times L}{4}$ when the beam carries a point load at the centre.

$$\therefore \quad M = \frac{W \times L}{4} = \frac{20000 \times 3.6}{4} = 18000 \text{ Nm}$$

$$= 18000 \times 1000 \text{ Nmm}$$

$$\sigma_{max} = 7 \text{ N/mm}^2$$

and
$$Z = \frac{2b^3}{3} \qquad (\because \text{ In this case also } d = 2b)$$

Using equation (3.6), we get

$$M = \sigma_{max} . Z$$

or
$$18000 \times 1000 = 7 \times \frac{2b^2}{3}$$

$\therefore \quad b^3 = \frac{3 \times 18000 \times 1000}{7 \times 2} = 3.85714 \times 10^6$

$\therefore \quad b = (3.85714 \times 10^6)^{1/3} = $ **156.82 mm. Ans.**

and $\quad d = 2 \times 156.82 = $ **313.64 mm. Ans.**

Problem 3.9. *A timber beam of rectangular section of length 8 m is simply supported. The beam carries a U.D.L. of 12 kN/m run over the entire length and a point load of 10 kN at 3 metre from the left support. If the depth is two times the width and the stress in the timber is not to exceed 8 N/mm², find the suitable dimensions of the section.*

Sol. Given :

Length, $\quad L = 8$ m

U.D.L., $\quad w = 12$ kN/m $= 12000$ N/m

Point load, $\quad W = 10$ kN $= 10000$ N

Depth of beam $\quad = 2 \times$ Width of beam

$\therefore \quad d = 2b$

Stress, $\quad \sigma_{max} = 8$ N/mm²

First calculate the section where B.M. is maximum. Where B.M. is maximum, the shear force will be zero. Now the equations of pure bending can be used. For doing this, calculate the reactions R_A and R_B as shown in Fig. 3.14.

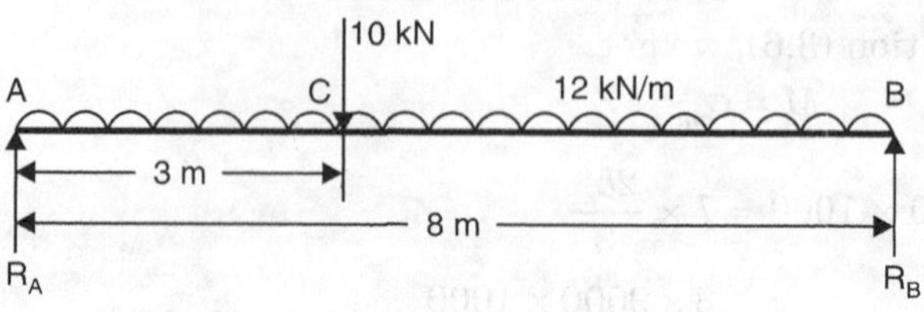

Fig. 3.14

Taking moments about A, we get

$$R_B \times 8 = 12000 \times 8 \times 4 + 10000 \times 3$$

$\therefore \quad R_B = \frac{12000 \times 32 + 30000}{8} = 51750$ N

$\therefore \quad R_A = \text{Total load} - R_B$

$= (12000 \times 8 + 10000) - 51750 = 54250$ N

Now $\quad$ S.F. at $A = + R_A = + 54250$ N

S.F. just L.H.S. at $C \quad = 54250 - 12000 \times 3 = + 18250$ N

S.F. just R.H.S. of $C \quad = 18250 - 10000 = 8250$ N

S.F. at $B \quad = - R_B = - 51750$ N

The S.F. is changing sign between section CB and hence at some section in C and B the S.F. will be zero.

Let S.F. is zero at x metre from B.

Equating the S.F. at this section to zero, we have

$$12000 \times x - R_B = 0$$

or $\quad 12000 \times x - 51750 = 0$

$\therefore \quad x = \frac{51750}{12000} = 4.3125$ m

$\therefore$ Maximum B.M. will occur at 4.3125 m from B.

$\therefore$ Maximum B.M. $= M = R_B \times 4.3125 - 12000 \times 4.3125 \times \dfrac{4.3125}{2}$

$= 51750 \times 4.3125 - 111585.9375$

$= 111585.9375$ Nm $= 111585.9375 \times 1000$ Nmm

Section modulus for rectangular beam is given by,

$$Z = \frac{bd^2}{6} = \frac{b \times (2b)^2}{6} = \frac{2b^3}{3}$$

Now using equation (3.6), we get

$$M = \sigma_{max} \cdot Z$$

or $$111585.9375 \times 1000 = 8 \times \frac{2b^3}{3}$$

$\therefore$ $$b^3 = \frac{3 \times 111585.9375 \times 1000}{16} = 20.9223 \times 10^6$$

$\therefore$ $b = (20.9223 \times 10^6)^{1/3} =$ **275.5 mm. Ans.**

and $d = 2 \times 275.5 =$ **551 mm. Ans.**

Problem 3.10. *A rolled steel joist of I section has the dimensions : as shown in Fig. 3.15. This beam of I section carries a u.d.l. of 40 kN/m run on a span of 10 m, calculate the maximum stress produced due to bending.*

Sol. Given :

u.d.l., $w = 40$ kN/m $= 40000$ N/m

Span, $L = 10$ m

Moment of inertia about the neutral axis

$$= \frac{200 \times 400^3}{12} - \frac{(200 - 10) \times 360^3}{12}$$

$= 1066666666 - 738720000$

$= 327946666$ mm^4

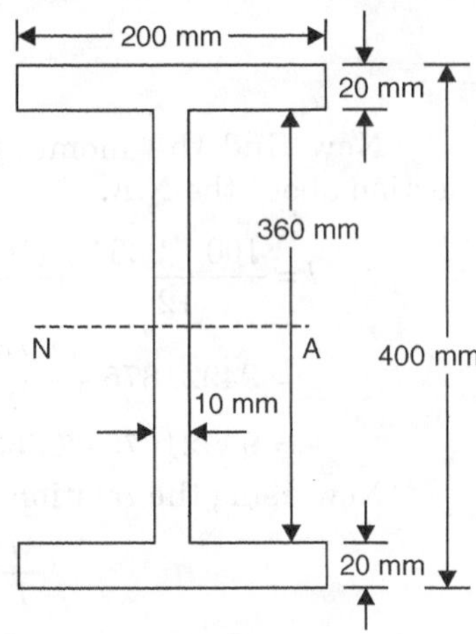

Fig. 3.15

Maximum B.M. is given by,

$$M = \frac{w \times L^2}{8} = \frac{40000 \times 10^2}{8}$$

$= 500000$ Nm

$= 500000 \times 1000$ Nmm

$= 5 \times 10^8$ Nmm

Now using the relation,

$$\frac{M}{I} = \frac{\sigma}{y}$$

$\therefore$ $$\sigma = \frac{M}{I} \times y$$

or $$\sigma_{max} = \frac{M}{I} \times y_{max} = \frac{5 \times 10^8}{327946666} \times 200 \qquad (\because \; y_{max} = 200 \text{ mm})$$

$=$ **304.92 N/mm². Ans.**

Problem 3.11. *An I-section shown in Fig. 3.16, is simply supported over a span of 12 m. If the maximum permissible bending stress is 80 N/mm², what concentrated load can be carried at a distance of 4 m from one support ?*

Sol. Given :

Bending stress, $\sigma_{max} = 80 \text{ N/mm}^2$

Let W = Concentrated load carried at a distance of 4 m from support B in Newton

To find the maximum bending moment (which will be at point C where concentrated load is acting), first calculate the reactions R_A and R_B.

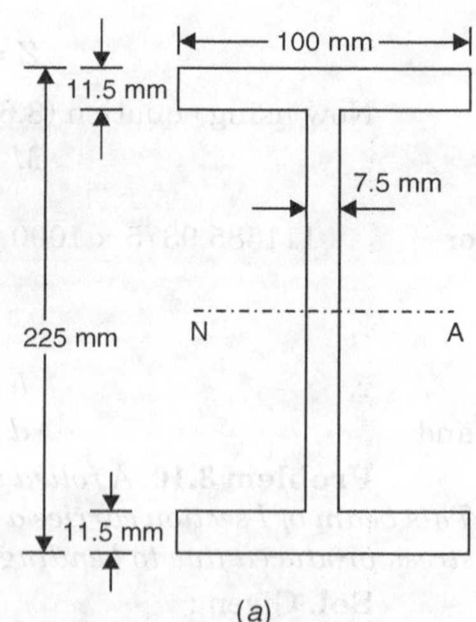

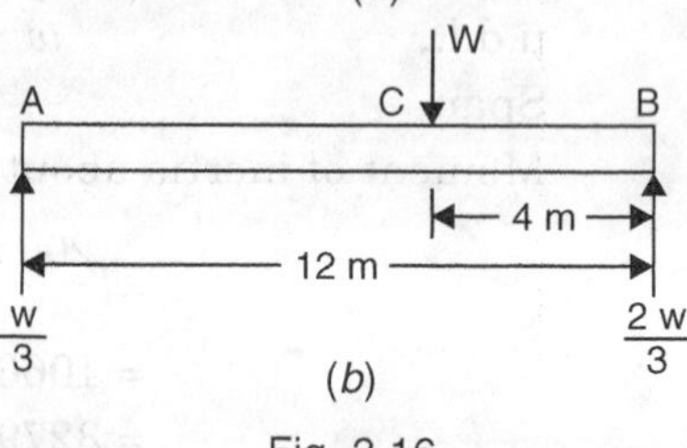

Fig. 3.16

Taking moments about point A, we get

$$R_B \times 12 = W \times 8$$

$$\therefore \quad R_B = \frac{8W}{12} = \frac{2}{3} W$$

and $$R_A = W - R_B = W - \frac{2}{3} W = \frac{W}{3}$$

B.M. at point C $$= R_A \times 8 = \frac{W}{3} \times 8 = \frac{8}{3} W \text{ Nm}$$

But B.M. at C is maximum

$\therefore$ Maximum B.M.,

$$M_{max} = \frac{8}{3} W \text{ Nm} = \frac{8}{3} W \times 1000 \text{ Nmm}$$

$$= \frac{8000}{3} W \text{ Nmm}$$

Now find the moment of inertia of the given I-section about the N.A.

$$\therefore \quad I = \frac{100 \times 225^3}{12} - \frac{(100 - 7.5) \times (225 - 2 \times 11.5)^3}{12}$$

$$= 94921875 - \frac{92.5 \times (202)^3}{12}$$

$$= 94921875 - 63535227.55 = 31386647.45 \text{ mm}^4$$

Now using the relation,

$$\frac{M}{I} = \frac{\sigma}{y}$$

or $$\frac{M}{I} = \frac{\sigma_{max}}{y_{max}}$$

where $y_{max} = \frac{225}{2} = 112.5$ mm.

Now substituting the known values, we get

$$\frac{\left(\frac{8000}{3} W\right)}{31386647.45} = \frac{80}{112.5}$$

or $$W = \frac{80}{112.5} \times 31386647.45 \times \frac{3}{8000} = \textbf{8369.77 N.} \quad \textbf{Ans.}$$

Problem 3.12. *Two circular beams where one is solid of diameter D and other is a hollow of outer dia. D_o and inner dia. D_i, are of the same length, same material and of same weight. Find the ratio of section modulus of these circular beams.*

Sol. Given :

Dia. of solid beam $= D$

Dias. of hollow beam $= D_o$ and D_i

Let L = Length of each beam (same length)

W = Weight of each beam (same weight)

ρ = Density of the material of each shaft (same material)

Now weight of solid beam $= \rho \times g \times$ Area of section $\times L$

$$= \rho \times g \times \frac{\pi}{4} D^2 \times L$$

Weight of hollow beam $= \rho \times g \times$ Area of section $\times L$

$$= \rho \times g \times \frac{\pi}{4} [D_0^2 - D_i^2] \times L$$

But the weights are same

$$\therefore \quad \rho \times g \times \frac{\pi}{4} D^2 \times L = \rho \times g \times \frac{\pi}{4} [D_0^2 - D_i^2] \times L$$

or
$$D^2 = D_0^2 - D_i^2 \quad \ldots(i)$$

Now section modulus of solid section,

$$Z = \frac{\pi}{32} D^3 \quad \text{[See equation (3.9)]}$$

And section modulus of hollow section,

$$Z_1 = \frac{\pi}{32 D_0} [D_0^4 - D_i^4] \quad \text{[See equation (3.10)]}$$

$$= \frac{\pi}{32 D_0} [D_0^2 + D_i^2] [D_0^2 - D_i^2]$$

$$\therefore \quad \frac{\text{Section modulus of solid section}}{\text{Section modulus of hollow section}}$$

$$= \frac{\frac{\pi}{32} D^3}{\frac{\pi}{32 D_0} [D_0^2 + D_i^2] [D_0^2 - D_i^2]}$$

$$= \frac{D^3 \times D_0}{[D_0^2 + D_i^2][D_0^2 - D_i^2]} = \frac{D \times D_0 \times D^2}{[D_0^2 + D_i^2][D_0^2 - D_i^2]}$$

$$= \frac{D \times D_0 \times [D_0^2 - D_i^2]}{[D_0^2 + D_i^2](D_0^2 - D_i^2)} \quad [\because \; D^2 = D_0^2 - D_i^2 \text{ from equation } (i)]$$

$$= \frac{D \times D_0}{(D_0^2 + D_i^2)} \quad \ldots(ii)$$

Also from equation (*i*),

$$D^2 = D_0^2 - D_i^2 \quad \text{or} \quad D_i^2 = D_0^2 - D^2$$

Substituting the value of D_i^2 in equation (*ii*), we get

$$\frac{\text{Section modulus of solid shaft}}{\text{Section modulus of hollow shaft}} = \frac{D \times D_0}{D_0^2 + D_0^2 - D^2} = \frac{D \times D_0}{(2D_0^2 - D^2)}$$

or

$$\frac{\text{Section modulus of hollow shaft}}{\text{Section modulus of solid shaft}} = \frac{2D_0^2 - D^2}{D \times D_0} = \frac{2D_0^2}{D \times D_0} - \frac{D^2}{D \times D_0}$$

$$= 2\frac{\mathbf{D_0}}{\mathbf{D}} - \frac{\mathbf{D}}{\mathbf{D_0}}. \quad \textbf{Ans.}$$

Problem 3.13. *A water main of 500 mm internal diameter and 20 mm thick is running full. The water main is of cast iron and is supported at two points 10 m apart. Find the maximum stress in the metal. The cast iron and water weigh 72000 N/m³ and 10000 N/m³ respectively.*

(Annamalai University, 1990)

Sol. Given :

Internal dia., $D_i = 500 \text{ mm} = 0.5 \text{ m}$

Thickness of pipe, $t = 20 \text{ mm}$

$\therefore$ Outer dia., $D_0 = D_i + 2 \times t = 500 + 2 \times 20 = 540 \text{ mm} = 0.54 \text{ m}$

Weight density of cast iron $= 72000 \text{ N/m}^3$

Weight density of water $= 10000 \text{ N/m}^3$

Internal area of pipe $= \frac{\pi}{4} D_i^2 = \frac{\pi}{4} \times 0.5^2 = 0.1960 \text{ m}^2$

This is also equal to the area of water section.

$\therefore$ Area of water section $= 0.196 \text{ m}^2$

Outer area of pipe $= \frac{\pi}{4} D_0^2 = \frac{\pi}{4} \times 0.54^2 \text{ m}^2$

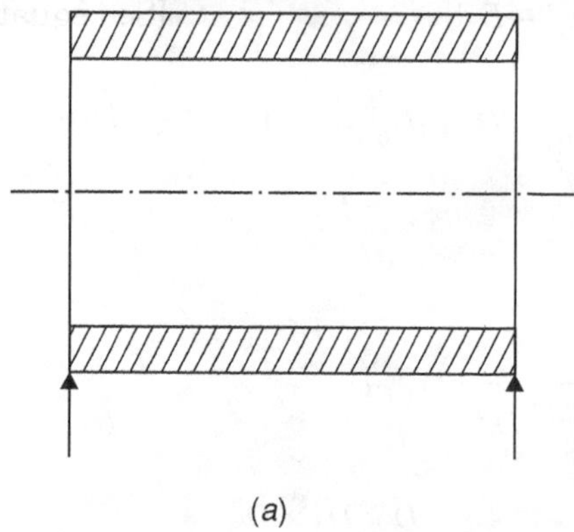

(*a*)

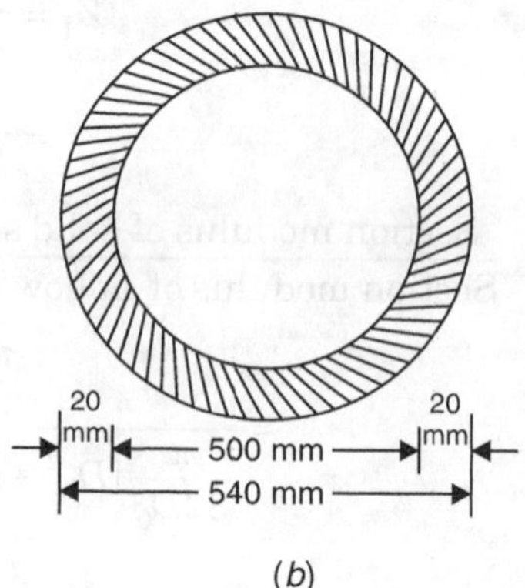

(*b*)

Fig. 3.17

$\therefore$ Area of pipe section $= \frac{\pi}{4} D_0^2 - \frac{\pi}{4} D_i^2$

$$= \frac{\pi}{4} [D_0^2 - D_i^2] = \frac{\pi}{4} [0.54^2 - 0.5^2] = 0.0327 \text{ m}^2$$

Moment of inertia of the pipe section about neutral axis,

$$I = \frac{\pi}{64} [D_0^4 - D_i^4] = \frac{\pi}{64} [540^4 - 500^4] = 1.105 \times 10^9 \text{ mm}^4$$

Let us now find the weight of pipe and weight of water for one metre length.

Weight of the pipe for one metre run

= Weight density of cast iron × Volume of pipe

= 72000 × [Area of pipe section × Length]

= 72000 × 0.0327 × 1 (∵ Length = 1 m)

= 2354 N

Weight of the water for one metre run

= Weight density of water × Volume of water

= 10000 × (Area of water section × Length)

= 10000 × 0.196 × 1 = 1960 N

∴ Total weight on the pipe for one metre run

= 2354 + 1960 = 4314 N

Hence the above weight is the U.D.L. (uniformly distributed load) on the pipe. The maximum bending moment due to U.D.L. is $w \times L^2/8$, where w = Rate of U.D.L. = 4314 N per metre run.

∴ Maximum bending moment due to U.D.L.,

$$M = \frac{w \times L^2}{8} = \frac{4314 \times 10^2}{8} \quad (\because L = 10 \text{ m})$$

$$= 53925 \text{ Nm} = 53925 \times 10^3 \text{ N mm}$$

Now using $\dfrac{M}{I} = \dfrac{\sigma}{y}$.

$$\therefore \quad \sigma = \frac{M}{I} \times y$$

The stress will be maximum, when y is maximum. But maximum value of

$$y = \frac{D_0}{2} = \frac{540}{2} = 270 \text{ mm.}$$

$$\therefore \quad y_{max} = 270 \text{ mm}$$

$$\therefore \text{ Maximum stress, } \sigma_{max} = \frac{M}{I} \times y_{max} = \frac{53925 \times 10^3}{1.105 \times 10^9} \times 270$$

$$= \mathbf{13.18 \ N/mm^2}. \quad \textbf{Ans.}$$

3.5. BENDING STRESS IN UNSYMMETRICAL SECTIONS

In case of symmetrical sections, the neutral axis passes through the geometrical centre of the section. But in case of unsymmetrical sections such as L, T sections, the neutral axis does not pass through the geometrical centre of the section. Hence the value of y for the topmost layer or bottom layer of the section from neutral axis will not be same. For finding the bending stress in the beam, the bigger value of y is used. As the neutral axis passes through the centre of gravity of the section, hence in unsymmetrical sections, first the centre of gravity is calculated.

3.6. SOLVED PROBLEMS FOR BENDING STRESSES IN UNSYMMETRICAL SECTIONS

Problem 3.14. *A cast iron bracket subject to bending has the cross-section of I-form with unequal flanges. The dimensions of the section are shown in Fig. 3.18. Find the position of the neutral axis and moment of inertia of the section about the neutral axis. If the maximum* *bending moment on the section is 40 MN mm, determine the maximum bending stress. What is the nature of the stress ?*

Sol. Given :

Max. B.M., M = 40 MN mm = 40 × 10^6 Nmm

Let us first calculate the C.G. of the section. Let $\bar{y}$ is the distance of the C.G. from the bottom face. The section is symmetrical about y-axis and hence $\bar{y}$ is only to be calculated.

Then,

$$\bar{y} = \frac{A_1 y_1 + A_2 y_2 + A_3 y_3}{(A_1 + A_2 + A_3)}$$

where A_1 = Area of bottom flange = 130 × 50 = 6500 mm^2

y_1 = Distance of C.G. of A_1 from bottom face

$$= \frac{50}{2} = 25 \text{ mm}$$

A_2 = Area of web = 200 × 50 = 10000 mm^2

y_2 = Distance of C.G. of A_2 from bottom face

$$= 50 + \frac{200}{2} = 150 \text{ mm}$$

A_3 = Area of top flange = 200 × 50 = 10000 mm^2

y_3 = Distance of C.G. of A_3 from bottom face

$$= 50 + 200 + \frac{50}{2} = 275 \text{ mm.}$$

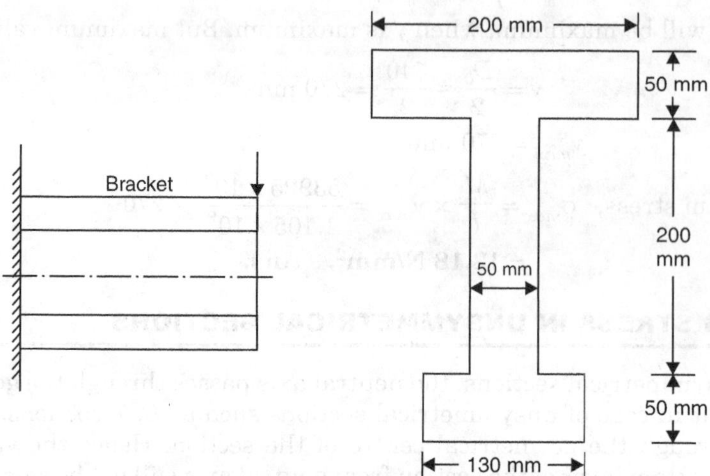

Fig. 3.18

$$\therefore \quad \bar{y} = \frac{6500 \times 25 + 10000 \times 150 + 10000 \times 275}{6500 + 10000 + 10000}$$

$$= \frac{162500 + 1500000 + 2750000}{26500}$$

$$= \frac{4412500}{26500} = 166.51 \text{ mm}$$

Hence neutral axis is at a distance of 166.51 mm from the bottom face. **Ans.**

Moment of inertia of the section about the N.A.

$$I = I_1 + I_2 + I_3$$

where I_1 = M.O.I. of bottom flange about N.A.

= M.O.I. of bottom flange about an axis passing through its C.G. + A_1 × (Distance of its C.G. from N.A.)^2

$$= \frac{130 \times 50^3}{12} + 6500 \times (166.51 - 25)^2$$

$$= 1354166.67 + 130163020 = 131517186.6 \text{ mm}^4$$

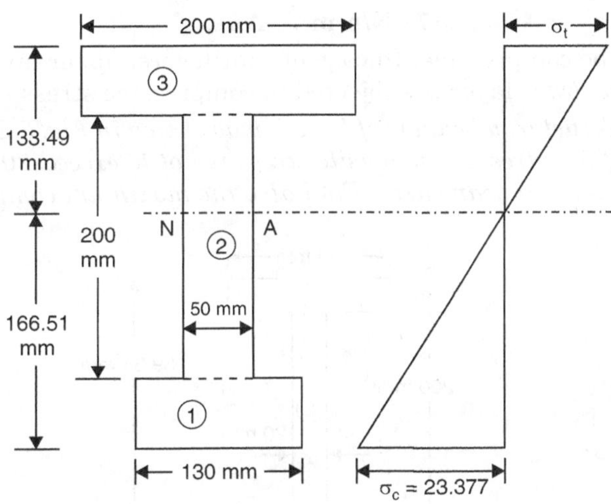

Fig. 3.19

Similarly I_2 = M.O.I. of web about N.A.

$$= \frac{50 \times 200^3}{12} + A_2 \,.\, (166.51 - y_2)^2$$

$$= \frac{50 \times 200^3}{12} + 10000\,(166.51 - 150)^2$$

$$= 33333333.33 + 272580.1$$

$$= 33605913.43 \text{ mm}^4$$

and I_3 = M.O.I. of top flange about N.A.

$$= \frac{200 \times 50^3}{12} + A_3 \,.\, (y_3 - 166.51)^2$$

$$= \frac{200 \times 50^3}{12} + 10000 \times (275 - 166.51)^2$$

$$= 2083333.33 + 117700801 = 119784134.3 \text{ mm}^4$$

∴ $I = I_1 + I_2 + I_3 = 131517186.6 + 33605913.43 + 119784134.3$

$= \mathbf{284907234.9 \text{ mm}^4}$. **Ans.**

Now distance of C.G. from the upper top fibre

$$= 300 - \bar{y} = 300 - 166.51 = 133.49 \text{ mm}$$

and the distance of C.G. from the bottom fibre

$$= \bar{y} = 166.51 \text{ mm}$$

Hence we shall take the value of y = 166.51 mm for maximum bending stress.

Now using the equation

$$\frac{M}{I} = \frac{\sigma}{y}$$

$$\therefore \qquad \sigma = \frac{M}{I} \times y = \frac{40 \times 10^6}{284907234.9} \times 166.51 = 23.377 \text{ N/mm}^2$$

∴ Maximum bending stress

$$= \mathbf{23.377\ N/mm^2.} \quad \textbf{Ans.}$$

This stress will be compressive. In case of cantilevers, upper layer is subjected to tensile stress, whereas the lower layer is subjected to compressive stress.

Problem 3.15. *A cast iron beam is of I-section as shown in Fig. 3.20. The beam is simply supported on a span of 5 metres. If the tensile stress is not to exceed 20 N/mm², find the safe uniformly load which the beam can carry. Find also the maximum compressive stress.*

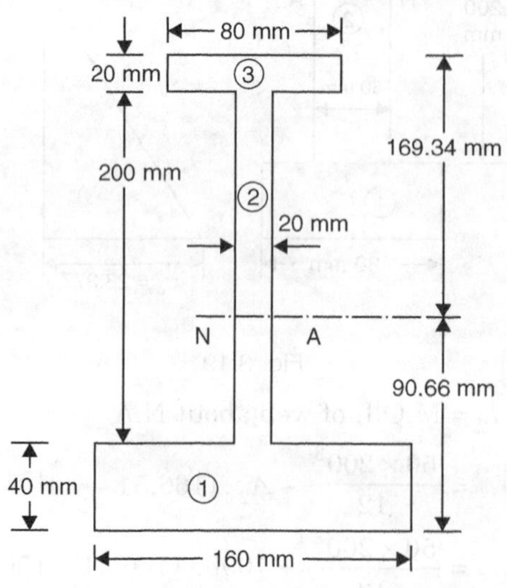

Fig. 3.20

Sol. Given :

Length of beam, L = 5 m

Maximum tensile stress, σ_t = 20 N/mm²

First calculate the C.G. of the section. Let $\bar{y}$ is the distance of the C.G. from the bottom face. As the section is symmetrical about y-axis, hence $\bar{y}$ is only to be calculated.

$$\text{Now} \qquad \bar{y} = \frac{A_1 y_1 + A_2 y_2 + A_3 y_3}{(A_1 + A_2 + A_3)}$$

$$= \frac{(160 \times 40) \cdot \dfrac{40}{2} + (200 \times 20)\left(40 + \dfrac{200}{2}\right) + (80 \times 20) \cdot \left(40 + 200 + \dfrac{20}{2}\right)}{160 \times 40 + 200 \times 20 + 80 \times 20}$$

$$= \frac{128000 + 560000 + 400000}{6400 + 4000 + 1600} = \frac{1088000}{12000} = 90.66 \text{ mm}$$

N.A. lies at a distance of 90.66 mm from the bottom face or 260 – 90.66 = 169.34 mm from the top face.

Now moment of inertia of the section about N-axis is given by,

$$I = I_1 + I_2 + I_3$$

where I_1 = M.O.I. of bottom flange about N.A.

= M.O.I. of bottom flange about its C.G. + A_1 × (Distance of its C.G. from N.A.)^2

$$= \frac{160 \times 40^3}{12} + 160 \times 40 \times (90.66 - 20)^2$$

= 853333.33 + 31954147.84 = 32807481.17 mm^4

I_2 = M.O.I. of web about N.A.

= M.O.I. of web about its C.G. + A_2 × (Distance of its C.G. from N.A.)^2

$$= \frac{20 \times 200^3}{12} + 200 \times 20 \times (140 - 90.66)^2$$

= 13333333.33 + 9737742.4 = 23071075.73 mm^4

I_3 = M.O.I. of top flange about N.A.

= M.O.I. of top flange about its C.G. + A_3 × (Distance of its C.G. from N.A.)^2

$$= \frac{80 \times 20^3}{12} + 80 \times 20 \times (250 - 90.66)^2$$

= 53333.33 + 40622776.96 = 40676110.29 mm^4

∴ I = 32807481.17 + 23071075.75 + 40676110.29 = 96554667.21 mm^4.

For a simply supported beam, the tensile stress will be at the extreme bottom fibre and compressive stress will be at the extreme top fibre.

Here maximum tensile stress = 20 N/mm^2

Hence for the maximum tensile stress,

$$y = 90.66 \text{ mm}$$

[*i.e.*, y is the distance of the extreme bottom fibre (where the tensile stress is maximum) from the N.A.]

Using the relation, $\frac{M}{I} = \frac{\sigma}{y}$

$$\therefore \quad M = \frac{\sigma}{y} \times I$$

$$= \frac{20}{90.66} \times 96554667.21 \qquad (\because \sigma = \sigma_t = 20 \text{ N/mm}^2)$$

= 21300389.85 Nmm ...(*i*)

Let w = Uniformly distributed load in N/m on the simply supported beam.

The maximum B.M. is at the centre and equal to $\frac{wL^2}{8}$

$\therefore \qquad M = \dfrac{w \times 5^2}{8}$ Nm $= \dfrac{w \times 25 \times 1000}{8}$ Nmm $= 3125\ w$ Nmm ...(*ii*)

Equating the two values of M, given by equation (*i*) and (*ii*), we get

$$3125w = 21300389.85$$

$\therefore \qquad w = \dfrac{21300389.85}{3125}$ = **6816.125 N/m. Ans.**

Maximum Compressive Stress

Distance of extreme top fibre from N.A.,

$$y_c = 169.34 \text{ mm}$$

$$M = 21300389.85$$

$$I = 96554667.21$$

Let $\qquad \sigma_c$ = Max. compressive stress

Using the relation, $\dfrac{M}{I} = \dfrac{\sigma}{y}$

$\therefore \qquad \sigma = \dfrac{M}{I} \times y$

or $\qquad \sigma_c = \dfrac{M}{I} \times y_c = \dfrac{21300389.85}{96554667.21} \times 169.34$ = **37.357 N/mm². Ans.**

Problem 3.16. *A cast iron beam is of T-section as shown in Fig. 3.21. The beam is simply supported on a span of 8 m. The beam carries a uniformly distributed load of 1.5 kN/m length on the entire span. Determine the maximum tensile and maximum compressive stresses.*

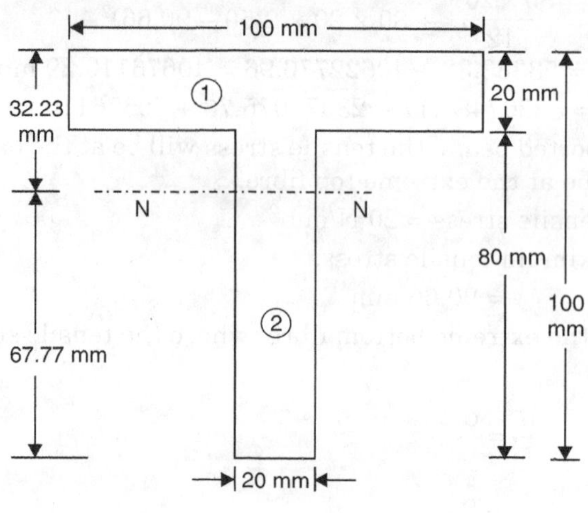

Fig. 3.21

Sol. Given :

Length, $\qquad L = 8$ m

U.D.L., $\qquad w = 1.5$ kN/m = 1500 N/m

To find the position of the N.A., the C.G. of the section is to be calculated first. The C.G. will be lying on the y-y axis.

Let $\qquad \bar{y}$ = Distance of the C.G. of the section from the bottom

$$\therefore \qquad \bar{y} = \frac{A_1y_1 + A_2y_2}{A_1 + A_2} = \frac{(100 \times 20) \times \left(80 + \frac{20}{2}\right) + 80 \times 20 \times \frac{80}{2}}{(100 \times 2) + (80 \times 20)}$$

$$= \frac{180000 + 64000}{2000 + 1600} = \frac{244000}{3600} = 67.77 \text{ mm}$$

$\therefore$ N.A. lies at a distance of 67.77 mm from the bottom face or 100 – 67.77 = 32.23 mm from the top face.

Now moment of inertia of the section about N.A. is given by,

$$I = I_1 + I_2$$

where I_1 = M.O.I. of top flange about N.A.

= M.O.I. of top flange about its C.G. + A_1 × (Distance of its C.G. from N.A.)^2

$$= \frac{100 \times 20^3}{12} + (100 \times 20) \times (32.23 - 10)^2$$

$$= 66666.7 + 988345.8 = 1055012.5 \text{ mm}^4$$

I_2 = M.O.I. of web about N.A.

= M.O.I. of web about its C.G. + A_2 × (Distance of its C.G. from N.A.)^2

$$= \frac{20 \times 80^3}{12} + (80 \times 20) \times (67.77 - 40)^2$$

$$= 853333.3 + 1233876.6 = 2087209.9 \text{ mm}^4$$

$$I = I_1 + I_2 = 1055012.5 + 2087209.9 = 3142222.4 \text{ mm}^4.$$

For a simply supported beam, the maximum tensile stress will be at the extreme bottom fibre and maximum compressive stress will be at the extreme top fibre.

Maximum B.M. is given by,

$$M = \frac{w \times L^2}{8} = \frac{1500 \times 8^2}{8} = 12000 \text{ Nm}$$

$$= 12000 \times 1000 = 12000000 \text{ Nmm}$$

Now using the relation

$$\frac{M}{I} = \frac{\sigma}{y} \quad \text{or} \quad \sigma = \frac{M}{I} \times y$$

(*i*) For maximum tensile stress,

y = Distance of extreme bottom fibre from N.A. = 67.77 mm

$$\therefore \qquad \sigma = \frac{12000000}{3142222.4} \times 67.77 = \mathbf{258.81 \text{ N/mm}^2}. \quad \textbf{Ans.}$$

(*ii*) For maximum compressive stress,

y = Distance of extreme top fibre from N.A. = 32.23 mm

$$\therefore \qquad \sigma = \frac{M}{I} \times y = \frac{12000000}{3142222.4} \times 32.23 = \mathbf{123.08 \text{ N/mm}^2}. \quad \textbf{Ans.}$$

Problem 3.17. *A simply supported beam of length 3 m carries a point load of 12 kN at a distance of 2 m from left support. The cross-section of the beam is shown in Fig. 3.22 (b). Determine the maximum tensile and compressive stress at X-X.*

Sol. Given :

Point load, W = 12 kN = 12000 N

First find the B.M. at *X-X*. And to do this first calculate reactions R_A and R_B.

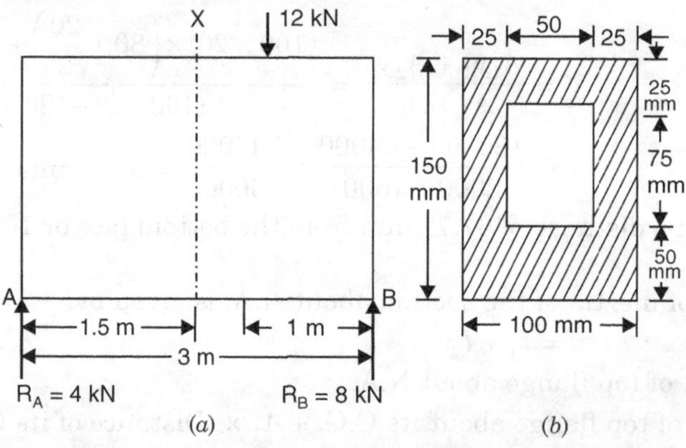

Fig. 3.22

Taking moments about A, we get

$$R_B \times 3 = 12 \times 2$$

$$\therefore \quad R_B = \frac{12 \times 2}{3} = 8 \text{ kN} \quad \text{and} \quad R_A = W - R_B = 12 - 8 = 4 \text{ kN}$$

B.M. at X-X

$$= R_A \times 1.5 = 4 \times 1.5 = 6 \text{ kNm}$$

$$= 6 \times 1000 \text{ Nm} = 6000 \times 1000 \text{ Nmm}$$

$$= 6000{,}000 \text{ Nmm}$$

$$\therefore \quad M = 6000{,}000 \text{ Nmm}$$

Now find the position of N.A. of the section of the beam. This can be obtained if we know the position of C.G. of the section.

Let $\bar{y}$ = Distance of the C.G. of the section from the bottom edge

$$= \frac{A_1 y_1 - A_2 y_2}{A_1 - A_2} \quad \text{(Negative sign is due to cut out part)}$$

$$= \frac{(150 \times 100) \times 75 - (75 \times 50) \times \left(50 + \frac{75}{2}\right)}{150 \times 100 - 75 \times 50}$$

$$= \frac{1125000 - 328125}{15000 - 3750} = \frac{796875}{11250} = 70.83 \text{ mm}$$

Hence N.A. will lie at a distance of 70.83 mm from the bottom edge or 150 – 70.83 = 79.17 mm from the top edge as shown in Fig. 3.23.

Now the moment of inertia of the section about N.A. is given by,

$$I = I_1 - I_2$$

where I_1 = M.O.I. of outer rectangle about N.A.

= M.O.I. of rectangle 100 × 150 about its C.G. + A_1 × (Distance of its C.G. from N.A.)^2

$$= \frac{100 \times 150^3}{12} + 100 \times 150 \times (75 - 70.83)^2$$

$$= 28125000 + 260833.5 = 28385833.5 \text{ mm}^4$$

I_2 = M.O.I. of cut out rectangular part about N.A.

= M.O.I. of cut out part about its C.G. + A_2 × (Distance of its C.G. from N.A.)^2

$$= \frac{50 \times 75^3}{12} + 50 \times 75 \times \left(50 + \frac{75}{2} - 70.83\right)^2$$

$= 1757812.5 + 1042083.375$

$= 2799895.875 \text{ mm}^4$

$\therefore \quad I = I_1 - I_2 = 28385833.5 - 2799895.875$

$= 25585937.63 \text{ mm}^4$

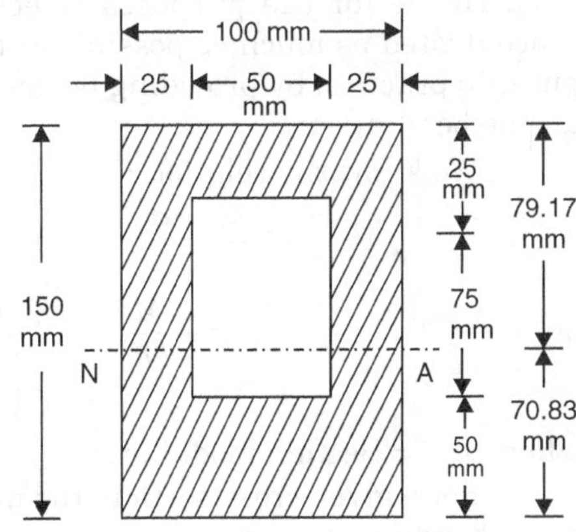

Fig. 3.23

The bottom edge of the section will be subjected to tensile stress whereas the top edge will be subjected to compressive stress. The top edge is at 79.17 mm from N.A. whereas bottom edge is 70.83 mm from N.A.

Now using the relation,

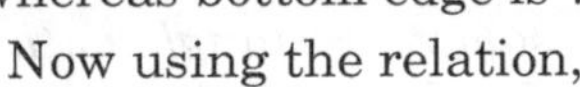

$$\frac{M}{I} = \frac{\sigma}{y}$$

$$\therefore \quad \sigma = \frac{M}{I} \times y$$

(*i*) For maximum tensile stress, $y = 70.83$ mm

$\therefore$ Maximum tensile stress,

$$\sigma = \frac{6000000}{25585937.63} \times 70.83 = \mathbf{16.60\ N/mm^2}. \quad \textbf{Ans.}$$

(*ii*) For maximum compressive stress,

$$y = 79.17 \text{ mm}.$$

$\therefore$ Maximum compressive stress,

$$\sigma = \frac{M}{I} \times y = \frac{6000000}{25585937.63} \times 79.17 = \mathbf{18.56\ N/mm^2}. \quad \textbf{Ans.}$$

3.7. STRENGTH OF A SECTION

The strength of a section means the greatest moment of resistance offered by the section and moment of resistance is given by,

$$M = \sigma \times Z \qquad \left(\because \frac{M}{I} = \frac{\sigma}{y} \text{ or } M = \frac{\sigma}{y} \times I = \sigma \times Z \text{ where } Z = \frac{I}{y}\right)$$

where M = Moment of resistance

σ = Bending stress, and

Z = Section modulus.

For a given value of allowable stress, the moment of resistance depends upon the section modulus. The section modulus, therefore, represents the strength of the section. Greater the value of *section modulus*, stronger will be the section.

The bending stress at any point in any beam section is proportional to its distance from the neutral axis. Hence the maximum tensile and compressive stresses in a beam section are proportional to the distances of the most distant tensile and compressive fibres from the neutral axis. Hence for the purposes of economy and weight reduction the material should be concentrated as much as possible at the greatest distance from the neutral axis. This idea is put into practice, by providing beams of I-section, where the flanges alone with-stand almost all the bending stress.

We know the relation :

$$\frac{M}{I} = \frac{\sigma}{y}$$

or
$$\sigma = \frac{M}{I} \times y = \frac{M}{\left(\frac{I}{y}\right)} = \frac{M}{Z}$$

where Z = Section modulus.

For a given cross-section, the maximum stress to which the section is subjected due to a given bending moment depends upon the section modulus of the section. If the section modulus is small, then the stress will be more.

Problem 3.18. *Three beams have the same length, same allowable bending stress and the same bending moment. The cross-section of the beams are a square, rectangle with depth twice the width and a circle. Find the ratios of weights of the circular and the rectangular beams with respect to square beams.*

Sol. Given :

Fig. 3.24 shows a square, a rectangular and a circular section.

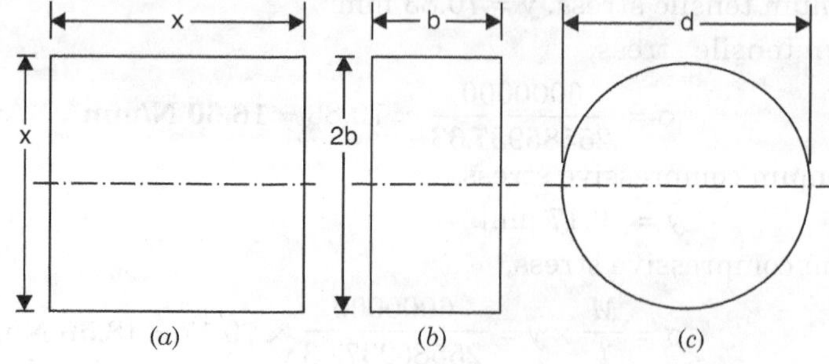

Fig. 3.24

Let x = Side of a square beam

b = Width of rectangular beam

$\therefore$ $2b$ = Depth of the rectangular beam

d = Diameter of a circular section.

The moment of resistance of a beam is given by,

$$M = \sigma \times Z$$

where Z = Section modulus.

As all the three beams have the same allowable bending stress (σ), and same bending moment (M), therefore the section modulus (Z) of the three beams must be equal.

Section modulus of a square beam

$$= \frac{I}{y} = \frac{\frac{bd^3}{12}}{\left(\frac{d}{2}\right)} = \frac{x.x^3}{12} \times \frac{2}{x} \qquad (\because \; b = d = x)$$

$$= \frac{x^2}{6}$$

Section modulus of a rectangular beam

$$= \frac{\frac{bd^3}{12}}{\frac{d}{2}} = \frac{\frac{b \times (2b)^3}{12}}{\left(\frac{2b}{2}\right)} \qquad (\because \; d = 2b)$$

$$= \frac{b \times 8b^3}{12} \times \frac{2}{2b} = \frac{2}{3} b^3$$

Section modulus of a circular beam

$$= \frac{\frac{\pi d^4}{64}}{\frac{d}{2}} = \frac{\pi d^4}{64} \times \frac{2}{d} = \frac{\pi d^3}{32}$$

Equating the section modulus of a square beam with that of a rectangular beam, we get

$$\frac{x^3}{6} = \frac{2}{3} b^3$$

$$\therefore \qquad b^3 = \frac{3x^3}{6 \times 2} = \frac{x^3}{4} = 0.25x^3$$

or $\qquad b = (0.25)^{1/3} x = 0.63x \qquad ...(i)$

Equating the section modulus of a square beam with that of a circular beam, we get

$$\frac{x^3}{6} = \frac{\pi d^3}{32}$$

$$\therefore \qquad d^3 = \frac{32x^3}{6\pi} \quad \text{or} \quad d = \left(\frac{32}{6\pi}\right)^{1/3} . x = 1.1927x$$

The weights of the beams are proportional to their cross-sectional areas. Hence

$$\frac{\text{Weight of rectangular beam}}{\text{Weight of square beam}} = \frac{\text{Area of rectangular beam}}{\text{Area of square beam}}$$

$$= \frac{b \times 2b}{x \times x} = \frac{0.63x \times 2 \times 0.63x}{x \times x}$$

$$= \mathbf{0.7938.} \quad \textbf{Ans.}$$

and

$$\frac{\text{Weight of circular beam}}{\text{Weight of square beam}} = \frac{\text{Area of circular beam}}{\text{Area of square beam}}$$

$$= \frac{\frac{\pi d^2}{4}}{x^2} = \frac{\pi d^2}{4x^2} = \frac{\pi \times (1.1927x)^2}{4x^2} \qquad (\because \; d = 1.1927x)$$

$$= \mathbf{1.1172.} \quad \textbf{Ans.}$$

Problem 3.19. *A beam is of square section of the side 'a'. If the permissible bending stress is 'σ', find the moment of resistance when the beam section is placed such that (i) two sides are horizontal, (ii) one diagonal is vertical. Find also the ratio of the moments of the resistance of the section in the two positions.* (Bangalore University, July 1988)

Sol. Given :

Bending stress = σ

1st Case

Fig. 3.25 (*a*) shows the square beam section when two sides are horizontal.

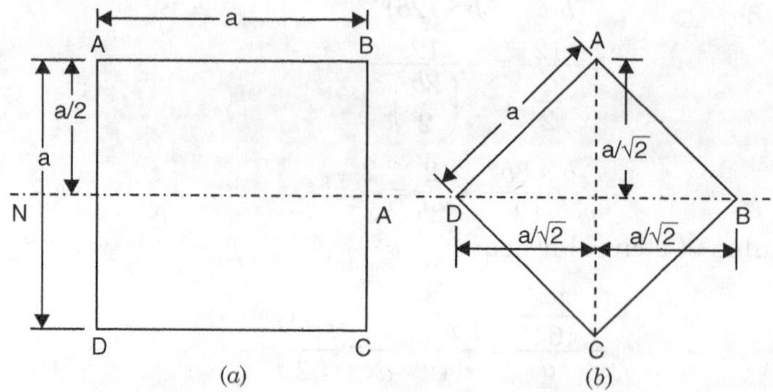

Fig. 3.25

Let M_1 = Moment of resistance of the square beam when two sides are horizontal.

Moment of resistance is given by,

$$M = \sigma \times Z$$

$$\therefore \quad M_1 = \sigma \times Z_1 \quad \quad ...(i)$$

where Z_1 = Section modulus

$$= \frac{I}{y_{max}} = \frac{\frac{a \times a^3}{12}}{a/2} = \frac{a^4}{12} \times \frac{2}{a} = \frac{a^3}{6}$$

$$\therefore \quad M_1 = \sigma \times \frac{\mathbf{a^3}}{\mathbf{6}}. \quad \textbf{Ans.} \quad \quad ...(ii)$$

2nd Case

Fig. 3.25 (*b*) shows the square beam section when one diagonal is vertical.

Let M_2 = Moment of resistance of the beam in this position

$\therefore$ $M_2 = \sigma \times Z_2$

where Z_2 = Section modulus for the section shown in Fig. 3.25 (*b*).

$$= \frac{I_2}{y_{max}} = \frac{2 \times \frac{bh^3}{12}}{\left(\frac{a}{\sqrt{2}}\right)}$$

$$\left(\because \text{ M.O.I. of a triangle about its base} = \frac{bh^3}{12}. \text{ There are two triangles}\right)$$

$$= \frac{\frac{2}{12} \times \sqrt{2}a \left(\frac{a}{\sqrt{2}}\right)^3}{\left(\frac{a}{\sqrt{2}}\right)} \qquad \left(\because \text{ Here base} = b = \sqrt{2}a \text{ and } h = \frac{a}{\sqrt{2}}\right)$$

$$= \frac{a^3}{6 \times \sqrt{2}}$$

$$\therefore \qquad M_2 = \sigma \times \frac{\mathbf{a^3}}{\mathbf{6 \times \sqrt{2}}}. \quad \textbf{Ans.}$$

Ratio of moment of resistance of the section in two positions

$$= \frac{M_1}{M_2} = \frac{\left(\frac{\sigma \times a^3}{6}\right)}{\left(\sigma \times \frac{a^3}{6 \times \sqrt{2}}\right)} = \sqrt{2} = \mathbf{1.414.} \quad \textbf{Ans.}$$

Problem 3.20. *Prove that the moment of a resistance of a beam of square section, with its diagonal in the plane of bending is increased by flatting top and bottom corners as shown in Fig. 3.26 and that the moment of resistance is a maximum when* $y = \frac{8a}{9}$. *Find the percentage increase in moment of resistance also.*

Sol. Given :

Fig. 3.26 (*a*) shows a square section with diagonal *AC* vertical. Let the portions *AEF* and *CGH* be cut off.

Let I_1 = M.O.I. of the square *ABCD* about diagonal B.D.

Z_1 = Section modulus of square *ABCD*

M_1 = Moment of resistance of the square *ABCD*

I_2 = M.O.I. of the new section with cut off portion (*i.e.*, M.O.I. of *DEFBHG* about diagonal *BD*)

Z_2 = Section modulus of new section

M_2 = Moment of resistance of the new section.

In Fig. 3.26 (*a*), diagonal $AC = 2a$

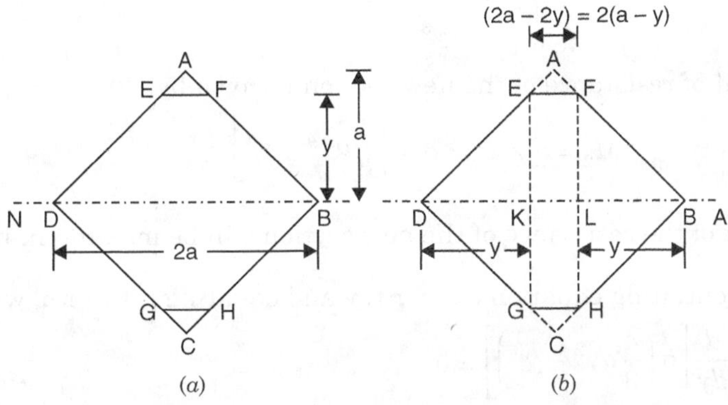

Fig. 3.26

$\therefore$ Diagonal $DB = AC = 2a$.

Now moment of inertia of the square $ABCD$ about N.A. (*i.e.*, diagonal BD) is given by

$\therefore$ I_1 = M.O.I. of two triangles ABD and BCD about their base BD

$$= 2 \times \frac{bh^3}{12} = 2 \times \frac{2a \times a^3}{12} \qquad \text{(Here } b = 2a \text{ and } h = a\text{)}$$

$$= \frac{a^4}{3}$$

Section modulus, $$Z_1 = \frac{I_1}{y_{max}} = \frac{\left(\frac{a^3}{3}\right)}{a} \qquad \text{(Here } y_{max} = a\text{)}$$

$$= \frac{a^4}{3} \times \frac{1}{a} = \frac{1}{3} a^3$$

Moment of resistance is given by,

$$M = \sigma \times Z$$

$\therefore$ $$M_1 = \sigma \times Z_1 = \sigma \times \frac{1}{3} a^3 = \sigma \times 0.3333a^3 \qquad ...(i)$$

Now the M.O.I. of the new section with cut off portion (*i.e.*, M.O.I. of $DEFBHG$) about the diagonal BD is given by [Refer to Fig. 3.26 (*b*)].

I_2 = M.O.I. of four triangles (*i.e.*, triangles DEK, FLB, DGH and HLB) plus M.O.I. of rectangle $EFHG$ about N.A. (*i.e.*, diagonal BD)

$$= \frac{4 \times bh^3}{12} + \frac{EF \times EG^3}{12} = \frac{4 \times y \times y^3}{12} + \frac{2(a-y) \times (2y)^3}{12}$$

$(\because$ Here $b = y$, $h = y$, $EF = 2(a - y)$ and $EG = 2y)$

$$= \frac{y^4}{3} + \frac{4}{3}(a - y) \times y^3 = \frac{y^4}{3} + \frac{4ay^3}{3} - \frac{4y^4}{3} = \frac{4}{3} ay^3 - y^4$$

and section modulus of new section is given by,

$$Z_2 = \frac{I_2}{y_{max}} = \frac{\frac{4}{3}ay^3 - y^4}{y} \qquad (\because \text{ Here } y_{max} = y)$$

$$= \frac{4}{3} ay^2 - y^3$$

Now moment of resistance of the new section is given by,

$$M_2 = \sigma \times Z_2 = \sigma \times \left[\frac{4}{3}ay^2 - y^3\right] \qquad ...(ii)$$

The moment of the resistance of the new section will be maximum, if $\frac{dM_2}{dy} = 0$.

Hence differentiating equation (*ii*) w.r.t. y and equating it to zero, we get

$$\frac{d}{dy}\left[\sigma\left(\frac{4}{3}ay^2 - y^3\right)\right] = 0$$

or $$\sigma\left(\frac{4}{3}a \times 2y - 3y^2\right) = 0 \qquad (\because \sigma \text{ and } a \text{ are constants})$$

or $$\frac{4}{3} a \times 2y - 3y^2 = 0 \qquad (\because \ \sigma \text{ cannot be zero})$$

$\therefore$ $$3y^2 = \frac{8}{3} a \times y$$

or $$y = \frac{8}{3} \frac{a \times y}{3 \times y} = \frac{8a}{9} \qquad ...(iii)$$

Substituting this value of y in equation (ii), we get

$$(M_2)_{max} = \sigma \times \left[\frac{4}{3} \times a \times \left(\frac{8a}{9}\right)^2 - \left(\frac{8a}{9}\right)^3\right] = \sigma \times \left[\frac{4 \times 64}{3 \times 81} a^3 - \frac{512}{729} a^3\right]$$

$$= \sigma \times [1.0535a^3 - 0.7023a^3] = \sigma \times 0.3512\, a^3 \qquad ...(iv)$$

But from equation (i), $M_1 = \sigma \times 0.3333\, a^3$

$\therefore$ M_2 is more than M_1. And from equation (iii), it is clear that M_2 is maximum when

$$y = \frac{\mathbf{8a}}{\mathbf{9}}. \quad \textbf{Ans.}$$

Now increase in moment of resistance

$$= (M_2)_{max} - M_1 = \sigma \times 0.3512\, a^3 - \sigma \times 0.3333\, a^2$$
$$= \sigma \times 0.0179\, a^3$$

Percentage increase in moment of resistance

$$= \frac{\text{Increase in moment of resistance}}{\text{Original moment of resistance}} \times 100$$

$$= \frac{\sigma \times 0.0719 \times a^3}{\sigma \times 0.3333 \times a^3} \times 100 = \mathbf{5.37\%}. \quad \textbf{Ans.}$$

Problem 3.21. *Prove that the ratio of depth to width of the strongest beam that can be cut from a circular log of diameter d is 1.414. Hence calculate the depth and width of the strongest beam that can be cut of a cylindrical log of wood whose diameter is 300 mm.*

Sol. Given :

Dia. of log = d

Let $ABCD$ be the strongest rectangular section which can be cut out of the cylindrical log.

Let b = Width of strongest section.

d = Depth of strongest section.

Now section modulus of the rectangular section

$$Z = \frac{I}{y} = \frac{\left(\frac{bh^3}{12}\right)}{\left(\frac{h}{2}\right)} = \frac{bh^2}{6} \qquad ...(i)$$

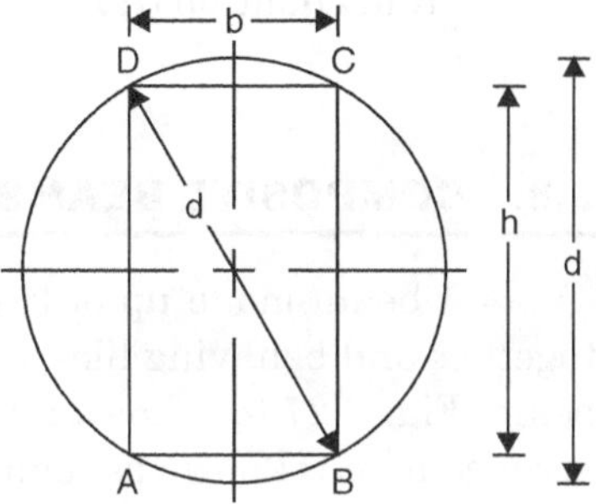

Fig. 3.27

In the above equation, b and h are variable.

From ΔBCD, $\quad b^2 + h^2 = d^2$

or $$h^2 = d^2 - b^2$$

Substituting the value of h^2 in equation (*i*), we get

$$Z = \frac{b}{6}[d^2 - b^2] = \frac{1}{6}[bd^2 - b^3] \quad ...(ii)$$

In the above equation, d is constant and hence only variable is b.

Now for the beam to be strongest, the section modulus should be maximum (or Z should be maximum).

For maximum value of Z,

$$\frac{dZ}{db} = 0$$

or $$\frac{d}{db}\left[\frac{bd^2 - b^3}{6}\right] = 0 \quad \text{or} \quad \frac{d^2 - 3b^2}{6} = 0$$

or $$d^2 - 3b^2 = 0 \quad \text{or} \quad d^2 = 3b^2 \quad ...(iii)$$

But from triangle BCD,

$$d^2 = b^2 + h^2$$

Substituting the value of d^2 in equation (*iii*), we get

$$b^2 + h^2 = 3b^2 \quad \text{or} \quad h^2 = 2b^2$$

or $$h = \sqrt{2} \times b \quad ...(iv)$$

or $$\frac{h}{b} = \sqrt{2} = \mathbf{1.414.} \quad \textbf{Ans.}$$

Numerical Part

Given, $d = 300$ mm

But for equation (*iii*), $d^2 = 3b^2$ or $3b^2 = d^2 = 300^2 = 90000$

or $$b^2 = \frac{90000}{3} = 30000$$

$\therefore$ $$b = (30000)^{1/2} = \mathbf{173.2\ mm.} \quad \textbf{Ans.}$$

From equation (*iv*),

$$h = \sqrt{2} \times b = 1.414 \times 173.2 = \mathbf{249.95\ mm.} \quad \textbf{Ans.}$$

3.8. COMPOSITE BEAMS (FLITCHED BEAMS)

A beam made up of two or more different materials assumed to be rigidly connected together and behaving like a single piece is known as a composite beam or a wooden flitched beam. Fig. 3.27 (*a*) shows a wooden beam (or timber beam) reinforced by steel plates. This arrangement is known as composite beam or a flitched beam. The strain at the common surfaces will be same for both materials. Also the total moment of resistance will be equal to the sum of the moments of individual sections.

When such a beam is subjected to bending, the bending stresses and hence strains due to bending stresses at a point are proportional to the distance of the point from the common neutral axis. Consider the composite beam as shown in Fig. 3.27 (*a*) and let at a distance y from the N.A., the stresses in steel and wood are f_1 and f_2 respectively.

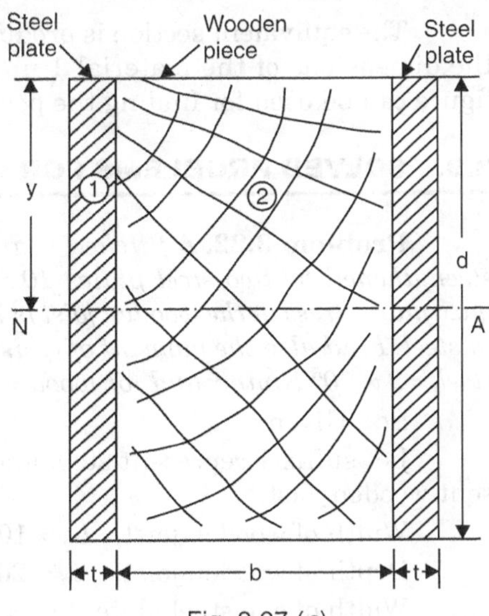

Fig. 3.27 (*a*)

Let E_1 = Young's modulus of steel plate

I_1 = Moment of inertia of steel about N.A.

M_1 = Moment of resistance of steel

E_2 = Young's modulus of wood

I_2 = M.O.I. of wood about N.A.

M_2 = Moment of resistance of wood.

Strain in steel at a distance y from N.A.

$$= \frac{\text{Stress}}{E} = \frac{\sigma_1}{E_1}$$

($\because$ Stress in steel = σ_1)

Strain in wood at a distance y from N.A.

$$= \frac{\sigma_2}{E_2}$$

But strain at the common surface is same

$$\therefore \quad \frac{\sigma_1}{E_1} = \frac{\sigma_2}{E_2} \quad \text{...(3.11)}$$

or

$$\sigma_1 = \frac{E_1}{E_2} \times \sigma_2$$

$$= m \times \sigma_2 \quad \text{...}(i)$$

where $m = \frac{E_1}{E_2}$ and is known as modular ratio between steel and wood.

Using the relation $\frac{M}{I} = \frac{\sigma}{y}$, we get

$$M = \frac{\sigma}{y} \times I$$

Hence moment of resistance of steel and wood are given by,

$$M_1 = \frac{\sigma_1}{y} \times I_1 \quad \text{and} \quad M_2 = \frac{\sigma_2}{y} \times I_2$$

$\therefore$ Total moment of resistance of the composite section,

$$M = M_1 + M_2$$

$$= \frac{\sigma_1}{y} \times I_1 + \frac{\sigma_2}{y} \times I_2$$

$$= \frac{m\sigma_2 \times I_1}{y} + \frac{\sigma_2}{y} \times I_2 \quad (\because \sigma_1 = m\sigma_2 \text{ from equation } i)$$

$$= \frac{\sigma_2}{y}[mI_1 + I_2] \quad \text{...(3.12)}$$

In equation (3.12) $I_2 + mI_1$ can be treated as equivalent moment of inertia of the cross-section, as if all made of material 2 (*i.e.*, wood) which will give the same amount of resistance as the composite beam. Let this be denoted by I.

$$\therefore \quad I = mI_1 + I_2 \quad \text{...(3.13)}$$

Then

$$M = \frac{\sigma_2}{y} \times I \quad \text{...(3.14)}$$

The equivalent section is produced by using $I = I_2 + mI_1$. This can be done by multiplying the dimensions of the material 1 in the direction parallel to the N.A. by m. The equivalent figure can be used for finding the position of N.A. and equivalent moment of inertia.

3.9. SOLVED PROBLEMS FOR COMPOSITE BEAMS

Problem 3.22. *A flitched beam consists of a wooden joist 10 cm wide and 20 cm deep strengthened by two steel plates 10 mm thick and 20 cm deep as shown in Fig. 3.28. If the maximum stress in the wooden joist is 7 N/mm², find the corresponding maximum stress attained in steel. Find also the moment of resistance of the composite section. Take Young's modulus for steel = 2 × 10⁵ N/mm² and for wood = 1 × 10⁴ N/mm².*

Sol. Given :

Let suffix 1 represent steel and suffix 2 represent wooden joist.

Width of wooden joist, $b_2 = 10$ cm

Depth of wooden joist, $d_2 = 20$ cm

Width of one steel plate, $b_1 = 1$ cm

Depth of one steel plate, $d_1 = 20$ cm

Number of steel plates $= 2$

Max. stress in wood, $\sigma_2 = 7\ \text{N/mm}^2$

E for steel, $E_1 = 2 \times 10^5\ \text{N/mm}^2$

E for wood, $E_2 = 1 \times 10^4\ \text{N/mm}^2$

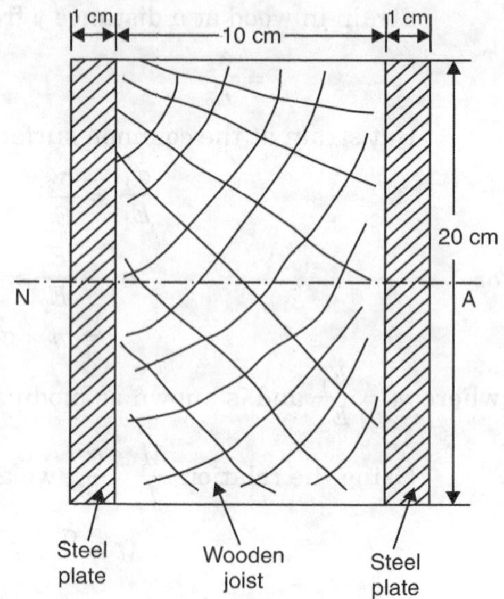

Fig. 3.28

Now M.O.I. of wooden joist about N.A.,

$$I_2 = \frac{b_2 d_2^3}{12} = \frac{10 \times 20^3}{12}$$

$$= 6666.66\ \text{cm}^4$$

$$= 6666.66 \times 10^4\ \text{mm}^4$$

M.O.I. of two steel plates about N.A.,

$$I_1 = \frac{2 \times b_1 d_1^3}{12} = \frac{2 \times 1 \times 20^3}{12}$$

$$= 1333.33\ \text{cm}^4 = 1333.33 \times 10^4\ \text{mm}^4.$$

Now modular ratio between steel and wood is given by,

$$m = \frac{E_1}{E_2} = \frac{2 \times 10^5}{1 \times 10^4} = 20$$

The equivalent moment of inertia (I) is given by equation (3.13).

$$\therefore \quad I = mI_1 + I_2$$

$$= 20 \times 1333.33 \times 10^4 + 6666.66 \times 10^4$$

$$= 10^4(26666.6 + 6666.66) = 10^4 \times 33333.2$$

Moment of resistance of the composite section is given by equation (3.14).

$$\therefore \quad M = \frac{\sigma_2}{y} \times I$$

$$= \frac{7 \times 10^4 \times 33333.2}{10 \times 10} \qquad (\because \ y = 10\ \text{cm} = 10 \times 10\ \text{mm})$$

$$= 233332.4 \times 10^2\ \text{N mm} = \mathbf{23333.24\ Nm.}$$ **Ans.**

Maximum Stress in Steel

Let σ_1 = Max. stress in steel.

Now using equation, we get

$$\frac{\sigma_1}{E_1} = \frac{\sigma_2}{E_2}$$

$$\therefore \quad \sigma_1 = \frac{E_1}{E_2} \times \sigma_2$$

$$= 20 \times 7 \qquad \left(\because \frac{E_1}{E_2} = m = 20 \text{ and } \sigma_2 = 7 \text{ N/mm}^2\right)$$

$$= \mathbf{140\ N/mm^2}. \quad \textbf{Ans.}$$

2nd Method

Total moment of resistance is equal to the sum of moment of resistance of individual member.

$$\therefore \quad M = M_1 + M_2 \qquad ...(i)$$

where $M_1 = \dfrac{\sigma_1}{y} \times I_1$ $\qquad \left(\because \dfrac{M}{I} = \dfrac{\sigma}{y}\right)$

$$= \frac{140}{100} \times 1333.33 \times 10^4 \qquad (\because \ y = 10 \times 10 = 100 \text{ mm})$$

$$= 18666620 \text{ Nmm} = 18666.620 \text{ Nm}$$

and $M_2 = \dfrac{\sigma_2}{y} \times I_2$

$$= \frac{7}{100} \times 6666.66 \times 10^4 \text{ Nmm}$$

$$= 46666.62 \text{ Nmm} = 4666.662 \text{ Nm}$$

$$\therefore \quad M = M_1 + M_2 = 18666.620 + 4666.662$$

$$= \mathbf{23333.282\ Nm}. \quad \textbf{Ans.}$$

3rd Method

The equivalent moment of inertia (I) is obtained by producing equivalent section.

(*a*) *The equivalent wooden section* is obtained by multiplying the dimension of steel plate in the direction parallel to the N.A. by the modular ratio between steel and wood $\left(i.e., \text{ by multiplying by } \dfrac{E_s}{E_w} = \dfrac{2 \times 10^5}{1 \times 10^4} = 20\right)$. But the width of one steel plate parallel to N.A. is 1 cm. Hence equivalent wooden width for this steel plate will be 20 × 1 = 20 cm. This is shown in Fig. 3.29.

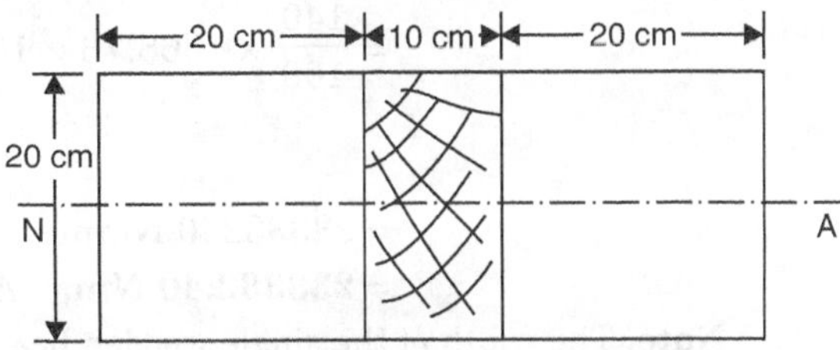

Fig. 3.29. Equivalent wooden section

$\therefore$ Equivalent M.O.I. is given by,

$$I = \frac{bd^3}{12}$$

$$= \frac{(20 + 10 + 20) \times 20^3}{12}$$

$= 33333.33 \text{ cm}^4 = 33333.33 \times 10^4 \text{ mm}^4$

$\therefore$ Total moment of resistance

= Moment of resistance of the equivalent wooden section

$$= \frac{\sigma}{y} \times I$$

$$= \frac{\text{Stress in wood}}{y} \times I$$

$$= \frac{7}{100} \times 33333.33 \times 10^4 = 23333333.33 \text{ Nmm}$$

= 23333.333 Nm. Ans.

(*b*) *The equivalent steel section* is obtained by multiplying the dimensions of wooden joist in the direction parallel to N.A. by the modular ratio between wood and steel $\left(\text{i.e., by multiplying by } \frac{E_w}{E_s} = \frac{1 \times 10^4}{2 \times 10^5} = \frac{1}{20}\right)$.

But the width of wooden joist parallel to N.A. is 10 cm. Hence equivalent steel width will be $10 \times \frac{1}{20} = 0.5$ cm. This is shown in Fig. 3.30.

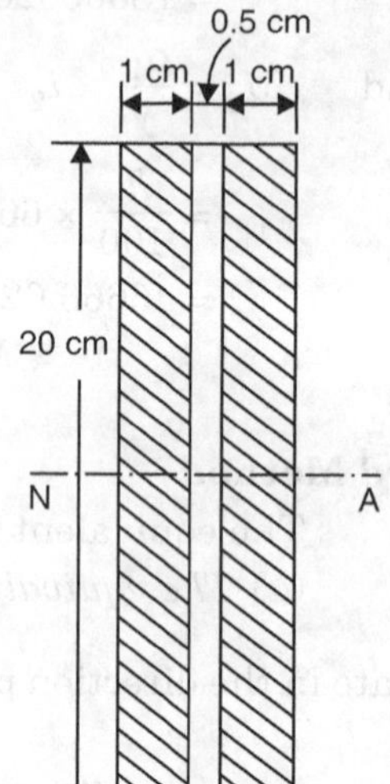

Fig. 3.30. Equivalent steel section

Hence equivalent M.O.I. is given by

$$I = \frac{bd^3}{12}$$

$$= \frac{(1 + 0.5 + 1) \times 20^3}{12}$$

$= 1666.66 \text{ cm}^4$

$= 1666.66 \times 10^4 \text{ mm}^4$

$$\therefore \quad M = \frac{\sigma}{y} \times I$$

$$= \frac{140}{100} \times 1666.76 \times 10^4$$

(Here σ is the stress in steel and = 140 N/mm²)

= 23333240 Nmm

= 23333.240 Nm. Ans.

Note. The width of the single wooden beam for the total moment of resistance of 23333.33 Nm should be 20 + 10 + 20 = 50 cm as shown in Fig. 3.29. But the width of flitched beam for the same moment of resistance is only 1 + 10 + 1 = 12 cm as shown in Fig. 3.28. Hence flitched beams require less space.

Problem 3.23. *A timber beam 100 mm wide and 200 mm deep is to be reinforced by bolting on two steel flitches each 150 mm by 12.5 mm in section. Calculate the moment of resistance in the following cases : (i) flitches attached symmetrically at the top and bottom ; (ii) flitches attached symmetrically at the sides. Allowable stress in timber is 6 N/mm². What is the maximum stress in the steel in each case ? Take* $E_s = 2 \times 10^5$ *N/mm² and* $E_t = 1 \times 10^4$ *N/mm².*

Sol. Given :

1st Case. *Flitches attached symmetrically at the top and bottom.*

(See Fig. 3.31).

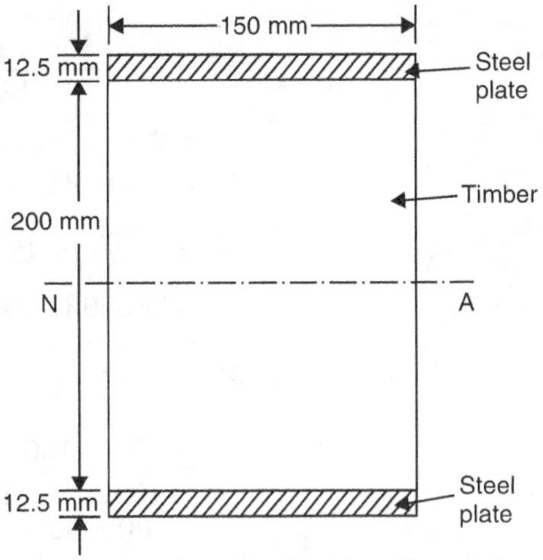

Fig. 3.31

Let suffix 1 represents steel and suffix 2 represents timber.

Width of steel, $b_1 = 150$ mm

Depth of steel, $d_1 = 12.5$ mm

Width of timber, $b_2 = 150$ mm

Depth of timber, $d_2 = 200$ mm

Number of steel plates = 2

Max. stress in timber, $\sigma_2 = 6$ N/mm²

E for steel, $E_1 = E_s = 2 \times 10^5$ N/mm²

E for timber, $E_2 = E_t = 1 \times 10^4$ N/mm²

Distance of extreme fibre of timber from N.A.,

$$y_2 = 100 \text{ mm}$$

Distance of extreme fibre of steel from N.A.,

$$y_1 = 100 + 12.5 = 112.5 \text{ mm.}$$

Let $\sigma_1^* =$ Max. stress in steel

$\sigma_1 =$ Stress in steel at a distance of 100 mm from N.A.

Now we know that strain at the common surface is same. The strain at a common distance of 100 mm from N.A. is steel and wood would be same. Hence using equation (3.11), we get

$$\frac{\sigma_1}{E_1} = \frac{\sigma_2}{E_2}$$

$$\therefore \quad \sigma_1 = \frac{E_1}{E_2} \times \sigma_2 = \frac{2 \times 10^5}{1 \times 10^4} \times 6 = 120 \text{ N/mm}^2.$$

But σ_1 is the stress in steel at a distance of 100 mm from N.A. Maximum stress in steel would be at a distance of 112.5 mm from N.A. As bending stresses are proportional to the distance from N.A.

Hence
$$\frac{\sigma_1}{100} = \frac{\sigma_1^*}{112.5}$$

$$\therefore \quad \sigma_1^* = \frac{112.5}{100} \times \sigma_1 = \frac{112.5}{100} \times 120 = \mathbf{135 \text{ N/mm}^2}. \quad \textbf{Ans.}$$

Now moment of resistance of steel is given by

$$M_1 = \frac{\sigma_1^*}{y_1} \times I_1 \text{ (where } \sigma_1^* \text{ is the maximum stress in steel)}$$

$$= \frac{135}{112.5} \times I_1$$

where I_1 = M.O.I. of two steel plates about N.A.

= 2 × [M.O.I. one steel plate about its C.G. + Area of one steel plate × (Distance between its C.G. and N.A.)²]

$$= 2 \times \left[\frac{b_1 d_1^3}{12} + b_1 d_1 \times \left(100 + \frac{d_1}{2}\right)^2\right]$$

$$= 2 \times \left[\frac{150 \times 12.5^3}{12} + 150 \times 12.5 \times \left(100 + \frac{12.5}{2}\right)^2\right]$$

$$= 2 \times [24414.06 + 21166992.18]$$

$$= 42382812.48 \text{ mm}^4$$

$$\therefore \quad M_1 = \frac{135}{112.5} \times 42382812.48$$

$$= 50859374.96 \text{ Nmm} = 50859.375 \text{ Nm}$$

Similarly, $M_2 = \dfrac{\sigma_2}{y_2} \times I_2$

$$= \frac{6}{100} \times \frac{150 \times 200^3}{12}$$

$$= 6000000 \text{ Nmm} = 6000 \text{ Nm}$$

∴ Total moment of resistance is given by,

$$M = M_1 + M_2$$

= 50859.375 + 6000 = **56859.375 Nm. Ans.**

2nd Case. *Flitches attached symmetrically at the sides* (See Fig. 3.32)

Here distance of the extreme fibre of steel from N.A.

$$= \frac{150}{2} = 75 \text{ mm.}$$

In the first case we have seen that stress in steel at a distance of 100 mm from N.A. is 120 N/mm².

Hence the stress in steel at a distance of 75 mm from N.A. is given by,

$$\sigma_1^{**} = \frac{120}{100} \times 75$$

(∵ Stress is proportional to the distance from N.A.)

= 90 N/mm²

∴ Maximum stress in steel

= σ_1^{**} = **90 N/mm². Ans.**

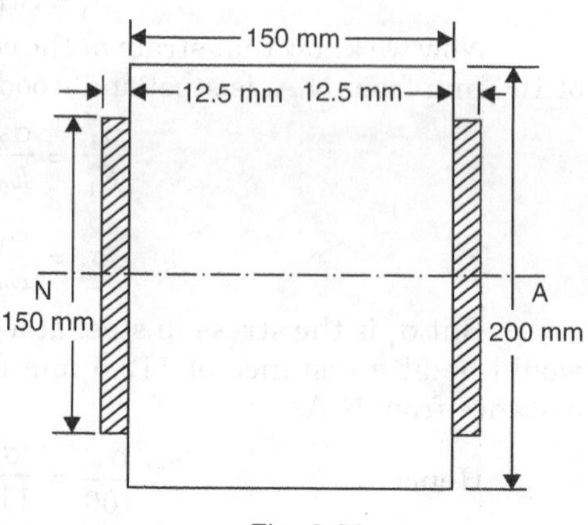

Fig. 3.32

Total moment of resistance is given by,

$$M = M_1 + M_2$$

where M_1 = Moment of resistance of two steel plates

$$= \frac{\sigma_1^{**}}{y_{max}} \times I_1$$

(Here σ_1^{**} = Maximum stress in steel = 90 N/mm²)

$$= \frac{90}{75} \times I_1 \qquad (y_{max} = 75 \text{ mm})$$

I_1 = M.O.I. of two steel plates about N.A.

$$= 2 \times \frac{12.5 \times 150^3}{12} = 7031250 \text{ mm}^4$$

$$\therefore \quad M_1 = \frac{90}{75} \times 7031250 \text{ Nmm} = 8437500 \text{ Nmm} = 8437.5 \text{ Nm}.$$

Similarly, M_2 = Moment of resistance of timber section

$$= \frac{\sigma_2}{y_2} \times I_2$$

$$= \frac{6}{100} \times \frac{150 \times 200^3}{12} \qquad \left(\because I_2 = \frac{150 \times 200^8}{12}\right)$$

$$= 6000000 \text{ Nmm} = 6000 \text{ Nm}$$

$\therefore$ Total moment of resistance,

$$M = M_1 + M_2$$

$$= 8437.5 + 6000 = \mathbf{14437.5 \text{ Nm}}. \quad \textbf{Ans.}$$

Problem 3.24. *Two rectangular plates, one of steel and the other of brass each 40 mm wide and 10 mm deep are placed together to form a beam 40 mm wide and 20 mm deep, on two supports 1 m apart, the brass plate being on the top of the steel plate. Determine the maximum load, which can be applied at the centre of the beam, if the plates are :*

(*i*) *separate and can bend independently,*

(*ii*) *firmly secured throughout their length.*

Maximum allowable stress in steel = 112.5 N/mm² and in brass = 75 N/mm². Take $E_s = 2 \times 10^5$ *N/mm² and* $E_b = 8 \times 10^4$ *N/mm².*

Sol. Given :

Width of plates, $b = 40$ mm

Depth of plates, $d = 10$ mm

Span, $L = 1$ m

Stress in steel, $\sigma_s = 112.5$ N/mm²

Stress in brass, $\sigma_b = 75$ N/mm²

Value of E for steel, $E_s = 2 \times 10^5$ N/mm²

Value of E for brass, $E_b = 8 \times 10^4$ N/mm².

Fig. 3.33

1st Case. *The plates are separate and can bend independently.*

Since the two materials bend independently, each will have its own neutral axis. It will be assumed that the radius of curvature R is the same for both the plates.

Using the relation $$\frac{\sigma}{y} = \frac{E}{R}$$

or $$R = \frac{E \times y}{\sigma}$$

or $$R = \frac{E_s \times y_s}{\sigma_s} = \frac{E_b \times y_b}{\sigma_b}$$

or $$\frac{\sigma_s}{\sigma_b} = \frac{E_s \times y_s}{E_b \times y_b}$$

But $y_s = y_b$ as the two plates are having their own N.A. The distance of the extreme fibre of brass from its own N.A. is 5 mm. Also the distance of extreme fibre of steel from its N.A. = 5 mm.

$$\therefore \quad \frac{\sigma_s}{\sigma_b} = \frac{E_s}{E_b} = \frac{2 \times 10^5}{8 \times 10^4} = 2.5$$

Now the allowable stress in steel is 112.5 N/mm²

i.e., $\sigma_s = 112.5 \text{ N/mm}^2$.

Then maximum stress in brass will be,

$$\sigma_b = \frac{\sigma_s}{2.5} = \frac{112.5}{2.5} = 45 \text{ N/mm}^2$$

This is less than the allowable stress of 75 N/mm².

Note. If maximum stress in brass is taken as 75 N/mm². Then the stress in steel will be $\sigma_s = 2.5 \times \sigma_b = 2.5 \times 187.5$ N/mm². This stress is more than the allowable stress in steel.

The total moment of resistance is given by,

$$M = M_s + M_b$$

where M_s = Moment of resistance of steel plate.

$$= \frac{\sigma_s}{y_s} \times I_s$$

$$= \frac{112.5}{5.0} \times \frac{40 \times 10^3}{12} \qquad \left(\because \; I_s = \text{M.O.I. of steel plate} = \frac{40 \times 10^3}{12} \right)$$

$$= 75000 \text{ Nmm} = 75 \text{ Nm}$$

and M_b = Moment of resistance of brass plate

$$= \frac{\sigma_b}{y_b} \times I_b$$

$$= \frac{45}{5.0} \times \frac{40 \times 10^3}{12} = 30000 \text{ Nmm} = 30 \text{ Nm}$$

$$\therefore \quad M = M_s + M_b = 75 + 30$$

$$= 105 \text{ Nm} \qquad \text{...(i)}$$

Let W = Maximum load applied at the centre in N to a simply supported beam.

Then maximum bending moment will be at the centre of the beam. And it is equal to,

$$M = \frac{W \times L}{4} = \frac{W \times 1.0}{4} \text{ Nm} \qquad \text{...(ii)}$$

Equation (*i*) and (*ii*), we get

$$\frac{W}{4} = 105 \quad \text{or} \quad W = 4 \times 105 = \textbf{420 N.} \quad \textbf{Ans.}$$

2nd Case. *The plates are firmly secured throughout their length.*

In this case, the two plates act as a single unit and thus will have a single N.A. Let us convert the composite section into an equivalent brass section as shown in Fig. 3.34 (*b*).

The equivalent brass section is obtained by multiplying the dimensions of steel plate in the direction parallel to the N.A. by the modular ratio between steel and brass (*i.e.*, by multiplying by $\left.\frac{E_s}{E_b} = 2.5\right)$. But the width of steel plate parallel to N.A. is 40 mm. Hence equivalent brass width for the steel plate will be 40 × 2.5 = 100 mm. This is shown in Fig. 3.34.

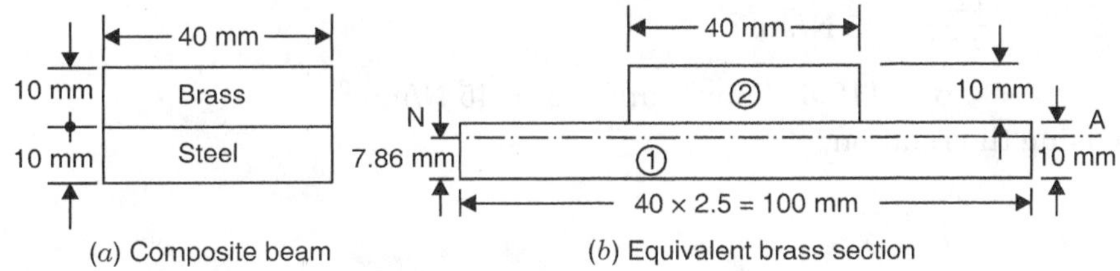

Fig. 3.34

Let $\bar{y}$ = Distance between C.G. of the equivalent brass section and bottom face.

$$= \frac{A_1 y_1 + A_2 y_2}{A_1 + A_2}$$

$$= \frac{100 \times 10 \times 5 + 40 \times 10 \times (10 + 5)}{100 \times 10 + 40 \times 10}$$

$$= \frac{5000 + 6000}{1000 + 400} = \frac{11000}{1400} = 7.86 \text{ mm.}$$

Hence N.A. of the equivalent brass section is at a distance of 7.86 mm from the bottom face.

Now the moment of inertia of the equivalent brass section about N.A. is given as

I = [M.O.I. of rectangle 100 × 10 about its C.G.
+ Area of rectangle 100 × 10 × (Distance between its C.G. and N.A.)^2]
+ [M.O.I. of rectangle 40 × 10 about its C.G. + Area of rectangle 40 × 10 × (Distance between its C.G. and N.A.)^2]

$$= \left[\frac{100 \times 10^3}{12} + 100 \times 10 \times (7.86 - 5)^2\right] + \frac{40 \times 10^3}{12} + 40 \times 10 \times [5 + (10 - 7.86)]^2$$

= 8333.33 + 8179.6 + 3333.33 + 20391.84

= 40238.1 mm^4.

Distance of upper extreme fibre from N.A.

= 20 – 7.86 = 12.14 mm

Distance of lower extreme fibre from N.A.

= 7.86 mm

Now allowable stress in brass is given 75 N/mm^2. As the upper plate is of brass.

Hence the upper extreme fibre will have a stress of 75 N/mm^2. Then the lowermost fibre will have the stress $= \frac{75}{12.14} \times 7.86 = 48.56$ N/mm^2. In Fig. 3.34 (b), the lowermost fibre is also of brass. Hence the actual stress in the lowermost fibre of steel will be

= 48.56 × 2.5 = 121.4 N/mm^2.

But the safe stress in steel is given as 112.5 N/mm². Hence the brass cannot be fully stressed.

If we take maximum stress in steel at the bottom to be 112.5 N/mm², then the corresponding stress in brass at the bottom fibre will be

$$\frac{112.5}{2.5} = 45 \text{ N/mm}^2.$$

$$\therefore \qquad \sigma_s = 112.5 \text{ N/mm}^2 \quad \text{and} \quad \sigma_b = 45 \text{ N/mm}^2.$$

Now using the relation,

$$\frac{M}{I} = \frac{\sigma}{y}$$

or

$$M = \frac{\sigma}{y} \times I$$

$$= \frac{45}{7.86} \times 40238.1 = 230370.8 \text{ Nmm}$$

$$= 230.3708 \text{ Nm} \qquad \text{...(iii)}$$

The maximum bending moment at the centre of a simply supported beam, carrying a point load W at the centre is given by,

$$M = \frac{W \times L}{4} = \frac{W \times 1.0}{4} \qquad \text{...(iv)}$$

Equating (*iii*) and (*iv*), we get

$$\frac{W}{4} = 230.3708$$

$$\therefore \qquad W = 4 \times 230.3708 = \mathbf{921.48 \text{ N.}} \quad \textbf{Ans.}$$

HIGHLIGHTS (3A)

1. The stresses produced due to constant bending moment (with zero shear force) are known as bending stresses.
2. The bending equation is given by,

$$\frac{M}{I} = \frac{\sigma}{y} = \frac{E}{R}$$

where M = Bending moment
σ = Bending stress
I = Moment of inertia about N.A.
y = Distance of the fibre from N.A.
R = Radius of curvature
E = Young's modulus of beam.

3. The bending stress in any layer is directly proportional to the distance of the layer from the neutral axis (N.A.).
4. The bending stress on the neutral axis is zero.
5. The neutral axis of a symmetrical section (such as circular, rectangular or square) lies at a distance of $\frac{d}{2}$ from the outermost layer of the section where d is the depth of the section.

6. If the top layer of the section is subjected to compressive stress then the bottom layer of the section will be subjected to tensile stress.

7. The ratio of moment of inertia of a section about the neutral axis to the distance of the outermost layer from the neutral axis is known as section modulus. It is denoted by Z.

$$\therefore \qquad Z = \frac{I}{y_{max}}$$

8. Section modulus for various sections are given as :

$$Z = \frac{bd^2}{6} \qquad \text{...For rectangular section}$$

$$= \frac{1}{6D}(BD^3 - bd^3) \qquad \text{...For a hollow rectangular section}$$

$$= \frac{\pi d^3}{32} \qquad \text{...For a circular section}$$

$$= \frac{\pi}{32D}[D^4 - d^4] \qquad \text{...For a hollow circular section.}$$

9. For finding bending stresses in unsymmetrical section, first their C.G. is to be obtained. This gives the position of N.A. The bigger value of y is to be used in bending equation.

10. The moment of resistance offered by the section is known as the strength of the section.

11. A beam made up of two or more different materials assumed to be rigidly connected together and behaving like a single unit, is known as a composite beam or flitched beam.

12. The strain at the common surface of a composite beam is same.

$$\therefore \qquad \frac{\sigma_1}{E_1} = \frac{\sigma_2}{E_2}.$$

13. The ratio of $\frac{E_1}{E_2}$ is known as modular ratio of first material to the second material.

14. Total moment of resistance of a composite beam is the sum of the moment of resistance of individual section.

EXERCISE (3A)

(A) Theoretical Questions

1. Define the terms : bending stress in a beam, neutral axis and section modulus.

2. What do you mean by 'simple bending' or 'pure bending' ? What are the assumptions made in the theory of simple bending ?

3. Derive an expression for bending stress at a layer in a beam.

4. What do you understand by neutral axis and moment of resistance ?

5. Prove that relation,

$$\frac{M}{I} = \frac{\sigma}{y} = \frac{E}{R}$$

where M = Bending moment, I = M.O.I.
σ = Bending stress, y = Distance from N.A.
E = Young's modulus, and R = Radius of curvature.

(Bangalore University, Jan. 1990)

6. What do you mean by section modulus ? Find an expression for section modulus for a rectangular, circular and hollow circular sections.
7. How would you find the bending stress in unsymmetrical section ?
8. What is the meaning of 'Strength of a section' ?
9. Define and explain the terms : modular ratio, flitched beams and equivalent section.
10. What is the procedure of finding bending stresses in case of flitched beams when it is of (*i*) a symmetrical section and (*ii*) an unsymmetrical section ?
11. Explain the terms : Neutral axis, section modulus, and moment of resistance. (*Bangalore University, July 1988*)
12. Show that for a beam subjected to pure bending, neutral axis coincides with the centroid of the cross-section. (*Bangalore University, March 1989*)
13. Prove that the bending stress in any fibre is proportional to the distance of that fibre from neutral layer in a beam. (*Bhavnagar University, 1992*)
14. Draw bending stress distribution for a T-section. (*Anna University, Nov. / Dec. 2004*)

(B) Numerical Problems

1. A steel plate of width 60 mm and of thickness 10 mm is bent into a circular arc of radius 10 m. Determine the maximum stress induced and the bending moment which will produce the maximum stress. Take $E = 2 \times 10^5$ N/mm^2. [**Ans.** 100 N/mm^2 ; 100 Nm]
2. A cast iron pipe of external diameter 60 mm, internal diameter of 40 mm, and of length 5 m is supported at its ends. Calculate the maximum bending stress induced in the pipe if it carries a point load of 100 N at its centre. [**Ans.** 7.34 N/mm^2]
3. A rectangular beam 300 mm deep is simply supported over a span of 4 m. What uniformly distributed load per metre, the beam may carry if the bending stress is not to exceed 120 N/mm^2 ? Take $I = 8 \times 10^6$ mm^4. [**Ans.** 3.2 kN/m]
4. A cast iron cantilever of length 1.5 metre fails when a point load W is applied at the free end. If the section of the beam is 40 mm × 60 mm and the stress at the failure is 120 N/mm^2, find the point load applied. [**Ans.** 1.92 kN]
5. A cast iron beam 20 mm × 20 mm in section and 100 cm long is simply supported at the ends. It carries a point load W at the centre. The maximum stress induced is 120 N/mm^2. What uniformly distributed load will break a cantilever of the same material 50 mm wide, 100 mm deep and 2 m long ? [**Ans.** 5 kN per m run]
6. A timber beam is 120 mm wide and 200 mm deep and is used on a span of 4 metres. The beam carries a uniformly distributed load of 2.8 kN/m run over the entire length. Find the maximum bending stress induced. [**Ans.** 7 N/mm^2]
7. A timber cantilever 200 mm wide and 300 mm deep is 3 m long. It is loaded with a U.D.L. of 3 kN/m over the entire length. A point load of 2.7 kN is placed at the free end of the cantilever. Find the maximum bending stress produced. [**Ans.** 7.2 N/mm^2]
8. A timber beam is freely supported on supports 6 m apart. It carries a uniformly distributed load of 12 kN/m run and a point load of 9 kN at 3.5 m from the right support. Design a suitable section of the beam making depth twice the width, if the stress in timber is not to exceed 8 N/mm^2. [**Ans.** 230 mm × 460 mm]
9. A beam of an I-section shown in Fig. 3.35 is simply supported over a span of 4 metres. Determine the load that the beam can carry per metre length, if the allowable stress in the beam is 30.82 N/mm^2. [**Ans.** 2.5 kN/m run]

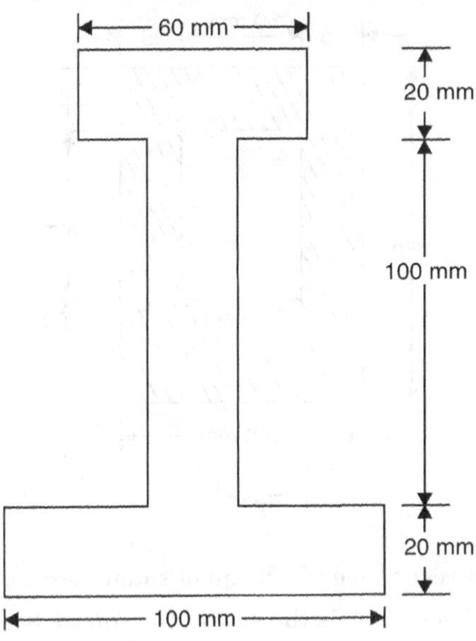

Fig. 3.35

10. A beam is of T-section as shown in Fig. 3.36. The beam is simply supported over a span of 4 m and carries a uniformly distributed load of 1.7 kN/m run over the entire span. Determine the maximum tensile and maximum compressive stress. [**Ans.** 8 N/mm^2 and 4.8 N/mm^2]

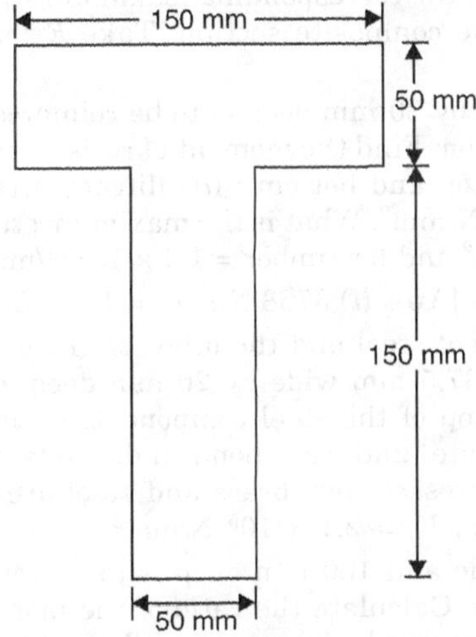

Fig. 3.36

11. A simply supported beam of length 4 m carries a point load of 16 kN at a distance of 3 m from left support. The cross-section of the beam is shown in Fig. 3.37. Determine the maximum tensile and compressive stress at a section which is at a distance of 2.25 m from the left support.

[**Ans.** 24.9 N/mm^2 ; 27.84 N/mm^2]

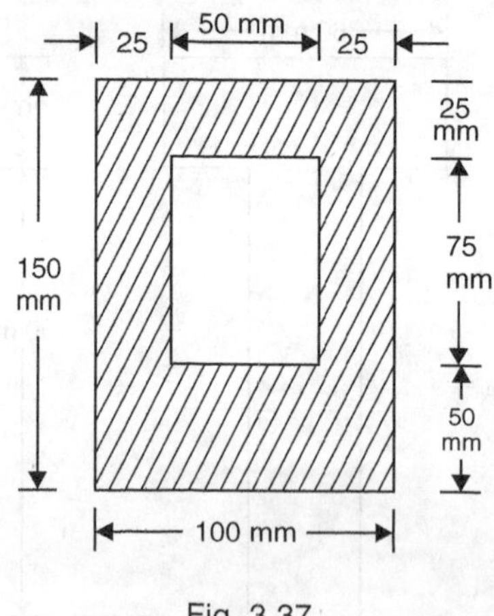

Fig. 3.37

12. Prove that the moment of resistance of a beam of square section is equal to $\sigma \times \frac{x^3}{6}$ where 'σ' is the permissible stress in bending, x is the side of the square beam and beam is placed such that its two sides are horizontal.

13. Find the moment of resistance of the above beam, if it is placed such that its one diagonal is vertical, the permissible bending stress is same (*i.e.*, equal to 'σ'). [**Ans.** $x^3 \times \sigma/6 \times \sqrt{2}$]

14. A flitched beam consists of a wooden joist 150 mm wide and 300 mm deep strengthened by a steel plate 12 mm thick and 300 mm deep on either side of the joist. If the maximum stress in the wooden joist is 7 N/mm², find the corresponding maximum stress attained in steel. Find also the moment of resistance of the composite section. Take E for steel = 2×10^5 N/mm² and for wood = 1×10^4 N/mm². [**Ans.** 140 N/mm², 66150 Nm]

15. A timber beam 60 mm wide by 80 mm deep is to be reinforced by bolting on two steel flitches, each 60 mm by 5 mm in section. Find the moment of resistance in the following cases : (*i*) flitches attached symmetrically at top and bottom ; (*ii*) flitches attached symmetrically at the sides. Allowable timber stress is 8 N/mm². What is the maximum stress in the steel in each case ? Take E for steel = 2.1×10^5 N/mm² and for timber = 1.4×10^4 N/mm².

[**Ans.** (*i*) 3768 Nm, σ_s = 135 N/mm² ; (*ii*) 1052 Nm, σ_s = 90 N/mm²]

16. Two rectangular plates, one of steel and the other of brass each 37.5 mm by 10 are placed to either to from a beam 37.5 mm wide by 20 mm deep, on two supports 75 cm apart, the brass component being on top of the steel component. Determine the maximum central load if the plates are (*i*) separate and can bend independently, (*ii*) firmly secured throughout their length. Permissible stresses for brass and steel are 70 N/mm² and 100 N/mm². Take $E_b = 0.875 \times 10^5$ N/mm² and $E_s = 2.1 \times 10^5$ N/mm². [**Ans.** (*i*) 472.2 N ; (*ii*) 1043.5 N]

17. A timber beam 150 mm wide and 100 mm deep is to be reinforced by two steel flitches each 150 mm × 10 mm in section. Calculate the ratio of the moments of the resistance in the two-mentioned cases : (*i*) Flitches attached symmetrically on the sides (*ii*) Flitches attached at top and bottom. [**Ans.** 0.31]

(B) SHEAR STRESSES IN BEAMS

3.10. SHEAR STRESSES IN BEAMS

In the earlier articles, we have seen that when a part of a beam is subjected to a constant bending moment and zero shear force, then there will be only bending stresses in the beam. The shear stress will be zero as shear stress is equal to shear force divided by the area. As shear force is zero, the shear stress will also be zero.

But in actual practice, a beam is subjected to a bending moment which varies from section to section. Also the shear force acting on the beam is not zero. It also varies from section to section. Due to these shear forces, the beam will be subjected to shear stresses. These shear stresses will be acting across transverse sections of the beam. These transverse shear stresses will produce a complimentary horizontal shear stresses, which will be acting on longitudinal layers of the beam. Hence beam will also be subjected to shear stresses. In this chapter, the distribution of the shear stress across the various sections (such as Rectangular section, Circular section, I-section, T-sections etc.) will be determined.

3.10.1. Shear Stress at a Section. Fig. 3.38 (*a*) shows a simply supported beam carrying a uniformly distributed load. For a uniformly distributed load, the shear force and bending moment will vary along the length of the beam. Consider two sections AB and CD of this beam at a distance dx apart.

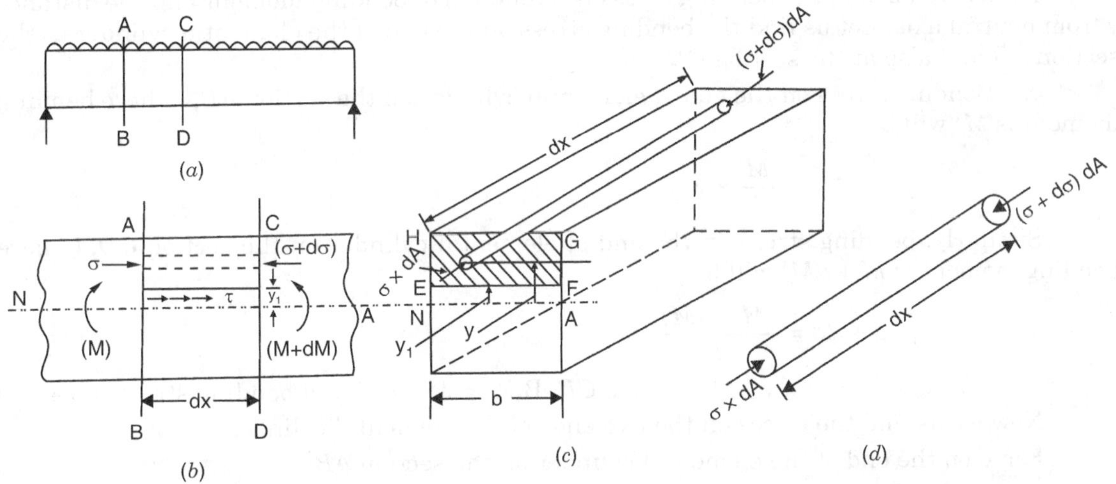

Area, A = Area of $EFGH$

Fig. 3.38

Let at the section AB,

$$F = \text{Shear force}$$

$$M = \text{Bending moment}$$

and at section CD, $F + dF$ = Shear force

$$M + dM = \text{Bending moment}$$

$$I = \text{Moment of inertia of the section about the neutral axis.}$$

Let it is required to find the shear stress on the section *AB* at a distance y_1 from the neutral axis. Fig. 3.38 (*c*) shows the cross-section of the beam. On the cross-section of the beam, let *EF* be a line at a distance y_1 from the neutral axis. Now consider the part of the beam above the level *EF* and between the sections *AB* and *CD*. This part of the beam may be taken to consists of an infinite number of elemental cylinders each of area *dA* and length *dx*. Consider one such elemental cylinder at a distance *y* from the neutral axis. The elemental cylinder is also shown in Fig. 3.38 (*d*).

$\therefore$ dA = Area of elemental cylinder

dx = Length of elemental cylinder

y = Distance of elemental cylinder from neutral axis

Let σ = Intensity of bending stress* on the end of the elemental cylinder on the section *AB*

$\sigma + d\sigma$ = Intensity of bending stress on the end of the elemental cylinder on the section *CD*.

The bending stress at distance *y* from the neutral axis is given by equation (3.6) as

$$\frac{M}{I} = \frac{\sigma}{y}$$

$$\therefore \quad \sigma = \frac{M}{I} \times y$$

For a given beam, the bending stress is a function of bending moment and the distance *y* from neutral axis. Let us find the bending stress on the end of the elemental cylinder at the section *AB* and also at the section *CD*.

$\therefore$ Bending stress on the end of elemental cylinder on the section *AB*, (where bending moment is *M*) will be

$$\sigma = \frac{M}{I} \times y$$

Similarly, bending stress on the end of elemental cylinder on the section *CD*, (where bending moment is *M* + *dM*) will be

$$\sigma + d\sigma = \frac{(M + dM)}{I} \times y$$

($\because$ On section *CD*, B.M. = $M + dM$ and bending stress = $\sigma + d\sigma$)

Now let us find the forces on the two ends of the elemental cylinder.

Force on the end of the elemental cylinder on the section *AB*

$$= \text{Stress} \times \text{Area of elemental cylinder}$$

$$= \sigma \times dA$$

$$= \frac{M}{I} \times y \times dA \qquad \left(\because \sigma = \frac{M}{I} \times y\right)$$

Similarly, force on the end of the elemental cylinder on the section *CD*

$$= (\sigma + d\sigma)\, dA$$

$$= \frac{(M + dM)}{I} \times y \times dA \qquad \left[\because \sigma + d\sigma = \frac{(M + dM)}{I} \times y\right]$$

---

*Bending stresses are acting normal to the cross-section.

At the two ends of the elemental cylinder, the forces are different. They are acting along the same line but are in opposite direction. Hence there will be unbalanced force on the elemental cylinder.

∴ Net unbalanced force on the elemental cylinder

$$= \frac{(M + dM)}{I} \times y \times dA - \frac{M}{I} \times y \times dA$$

$$= \frac{dM}{I} \times y \times dA \qquad ...(i)$$

The total unbalanced force above the level EF and between the two sections AB and CD may be found out by considering all the elemental cylinders between the sections AB and CD and above the level EF (*i.e.*, by integrating the above equation (*i*)).

∴ Total unbalanced force

$$= \int \frac{dM}{I} \times y \times dA = \frac{dM}{I} \int y \times dA$$

$$= \frac{dM}{I} \times A \times \bar{y} \qquad (\because \int y \times dA = A \times \bar{y})$$

where A = Area of the section above the level EF (or above y_1)
= Area of $EFGH$ as shown in Fig. 3.38 (*c*)

$\bar{y}$ = Distance of the C.G. of the area A from the neutral axis.

Due to the total unbalanced force acting on the part of the beam above the level EF and between the sections AB and CD as shown in Fig. 3.39 (*a*), the beam may fail due to shear. Hence in order the above part may not fail by shear, the horizontal section of the beam at the level EF must offer a shear resistance. This shear resistance at least must be equal to total unbalanced force to avoid failure due to shear.

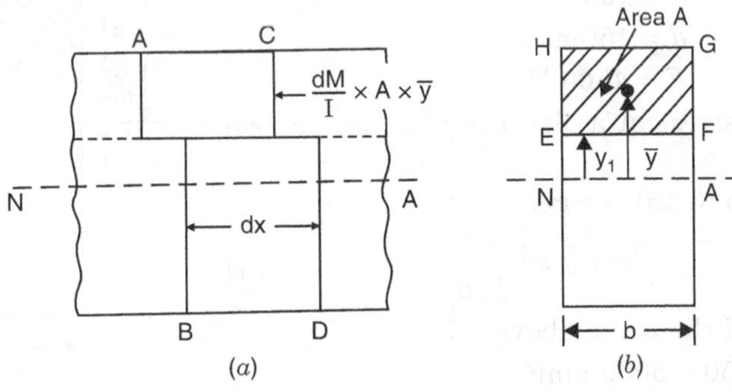

Fig. 3.39

∴ Shear resistance (or shear force) at the level EF

= Total unbalanced force

$$= \frac{dM}{I} \times A \times \bar{y} \qquad ...(ii)$$

Let τ = Intensity of horizontal shear at the level EF

b = Width of beam at the level EF

$\therefore$ Area on which τ is acting

$$= b \times dx$$

$\therefore$ Shear force due to τ

$$= \text{Shear stress} \times \text{Shear area}$$

$$= \tau \times b \times dx \qquad ...(iii)$$

Equating the two values of shear force given by equation (*ii*) and (*iii*), we get

$$\tau \times b \times dx = \frac{dM}{I} \times A \times \bar{y}$$

$$\therefore \qquad \tau = \frac{dM}{I} \times \frac{A \times \bar{y}}{b \times dx}$$

$$= \frac{dM}{dx} \cdot \frac{A\bar{y}}{I \times b}$$

$$= F \times \frac{A\bar{y}}{I \times b} \qquad \left(\because \frac{dM}{dx} = F = \text{Shear force}\right) \qquad ...(3.15)$$

The shear stress given by equation (3.15) is the horizontal shear stress at the distance y_1 from the neutral axis. But by the *principal of complementary shear*, the horizontal shear stress is accompanied by a vertical shear stress τ of the same quantity.

Sometimes $A \times \bar{y}$ is also expressed as the moment of area A about the neutral axis.

Note. In equation (3.15), b is the actual width at the level EF (Though here b is same at all levels, in many cases b may not be same at all levels) and I is the total moment of inertia of the section about N.A.

Problem 3.25. *A wooden beam 100 mm wide and 150 mm deep is simply supported over a span of 4 metres. If shear force at a section of the beam is 4500 N, find the shear stress at a distance of 25 mm above the N.A.*

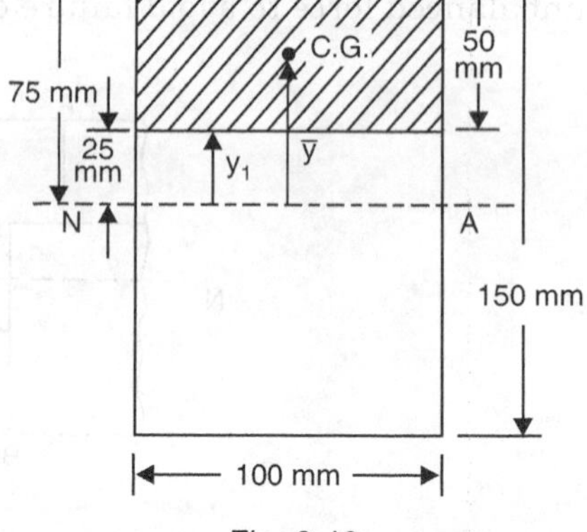

Fig. 3.40

Sol. Given :

Width, $b = 100$ mm

Depth, $d = 150$ mm

Shear force, $F = 4500$ N

Let τ = Shear stress at a distance of 25 mm above the neutral axis.

Using equation (3.15), we get

$$\tau = F.\frac{A\bar{y}}{I\,.\,b} \qquad ...(i)$$

where A = Area of the beam above y_1

$= 100 \times 50 = 5000 \text{ mm}^2$

(Shaded area of Fig. 3.40)

$\bar{y}$ = Distance of the C.G. of the area A from neutral axis

$$= 25 + \frac{50}{2} = 50 \text{ mm}$$

$$I = \text{M.O.I. of the total section}$$

$$= \frac{bd^3}{12}$$

$$= \frac{100 \times 150^3}{12} = 28125000 \text{ mm}^4$$

b = Actual width of section at a distance y_1 from N.A. = 100 m

Substituting these values in the above equation (*i*), we get

$$\tau = \frac{4500 \times 5000 \times 50}{28125000 \times 100} = \mathbf{0.4\ N/mm^2.} \quad \textbf{Ans.}$$

Problem 3.26. *A beam of cross-section of an isosceles triangle is subjected to a shear force of 30 kN at a section where base width = 150 mm and height = 450 mm. Determine :*

(*i*) *horizontal shear stress at the neutral axis,*

(*ii*) *the distance from the top of the beam where shear stress is maximum, and*

(*iii*) *value of maximum shear stress.*

Sol. Given :

Shear force at the section, F = 30 kN = 30,000 N

Base width, CD = 150 mm

Height, h = 450 mm.

(*i*) *Horizontal shear stress at the neutral axis*

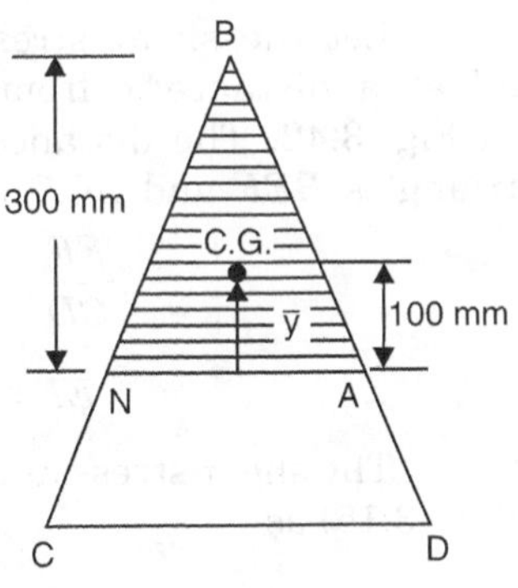

Fig. 3.41

The neutral axis of the triangle is at a distance of $\frac{h}{3}$ from base or $\frac{2h}{3}$ from the apex B. Hence distance of neutral axis from B will be $\frac{2 \times 450}{3}$ = 300 mm as shown in Fig. 3.41. The width of the section at neutral axis is obtained from similar triangles BCD and BNA as

$$\frac{NA}{CD} = \frac{300}{450}$$

or

$$NA = \frac{300}{450} \times CD = \frac{300}{450} \times 150 = 100 \text{ mm.}$$

The shear stress at any section is given by equation (3.15) as

$$\tau = F \times \frac{A \times \bar{y}}{I \times b} \qquad \text{...}(i)$$

where τ = Shear stress at the section

F = Shear force = 30,000 N

A = Area above the axis at which shear stress is to be obtained [*i.e.*, shaded area of Fig. 3.41]

$$= \frac{NA \times 300}{2} = \frac{100 \times 300}{2} = 15000 \text{ mm}^2$$

$\bar{y}$ = Distance of the C.G. of the area A from neutral axis

$$= \frac{1}{3} \times 300 = 100 \text{ mm}$$

I = M.O.I. of the total section about neutral axis

$$= \frac{\text{Base width} \times \text{Height}^3}{36} \qquad \left(\because \frac{B \times h^3}{36} \text{ where } B = \text{Base Width of Triangle} \right)$$

$$= \frac{150 \times 450^3}{36} \text{ mm}^4$$

b = Actual width of the section at which shear stress is to be obtained
= NA = 100 mm.

Substituting these values in equation (*i*), we get

$$\tau = 30{,}000 \times \frac{15000 \times 100}{\left(\frac{150 \times 450}{36}\right)^3 \times 100} \text{ N/mm}^2$$

$$= \mathbf{1.185 \text{ N/mm}^2.} \quad \textbf{Ans.}$$

(*ii*) *The distance from the top of the beam where shear stress is maximum*

Let the shear stress is maximum at the section *EF* at a distance *x* from the top of the beam as shown in Fig. 3.42. The distance *EF* is obtained from similar triangles *BEF* and *BCD* as

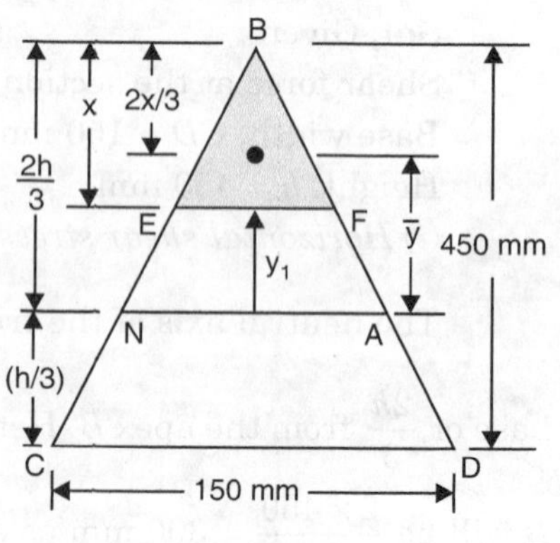

Fig. 3.42

$$\frac{EF}{CD} = \frac{x}{450}$$

$$\therefore \quad EF = \frac{x}{450} \times CD = \frac{x}{450} \times 150 = \frac{x}{3}.$$

The shear stress at the section *EF* is given by equation (3.15) as

$$\tau = F \times \frac{A \times \bar{y}}{I \times b} \qquad \ldots(ii)$$

where F = 30,000 N

A = Area of section above *EF* *i.e.*, Area of shaded triangle *BEF*

$$= \frac{EF \times x}{2} = \frac{x}{3} \times \frac{x}{2} \qquad \left(\because EF = \frac{x}{3} \right)$$

$$= \frac{x^2}{6}$$

$\bar{y}$ = Distance of C.G. of the Area A from neutral axis

$$= \frac{2h}{3} - \frac{2x}{3} = \frac{2 \times 450}{3} - \frac{2x}{3} = \left(300 - \frac{2x}{3}\right)$$

I = M.O.I. of ΔBCD about neutral axis

$$= \frac{150 \times 450^3}{36} \text{ mm}^4$$

b = Width of section $EF = \dfrac{x}{3}$.

Substituting these values in equation (*ii*), we get

$$\tau = \frac{30{,}000 \times \left(\frac{x^2}{6}\right) \times \left(300 - \frac{2x}{3}\right)}{\left(\frac{150 \times 450^3}{36}\right) \times \frac{x}{3}} = 0.0000395x\left(300 - \frac{2x}{3}\right)$$

$$= 0.0000395\left(300x - \frac{2x^2}{3}\right) \qquad ...(iii)$$

For maximum shear stress $\frac{d\tau}{dx} = 0$

or $$300 - \frac{2}{3} \times 2x = 0 \quad \text{or} \quad 300 = \frac{4x}{3}$$

or $$x = \frac{300 \times 3}{4} = \textbf{225 mm.} \quad \textbf{Ans.}$$

Hence, shear stress is maximum at a distance of 225 mm from the top of the beam.

(*iii*) *Value of Maximum Shear Stress*

The value of maximum shear stress will be obtained by substituting $x = 225$ mm in equation (*iii*).

$\therefore$ Maximum shear stress $= 0.0000395\left(300 \times 225 - \frac{2}{3} \times 225^2\right)$

$= \textbf{1.333 N/mm}^2$**. Ans.**

3.11. SHEAR STRESS DISTRIBUTION FOR DIFFERENT SECTIONS

The following are the important sections over which the shear stress distribution is to be obtained :

1. Rectangular Section,
2. Circular Section,
3. I-Section,
4. T-Sections, and
5. Miscellaneous Sections.

3.11.1. Rectangular Section. Fig. 3.43 shows a rectangular section of a beam of width b and depth d. Let F is the shear force acting at the section. Consider a level EF at a distance y from the neutral axis.

The shear stress at this level is given by equation (3.15) as

$$\tau = F \cdot \frac{A\bar{y}}{b \times I}$$

where A = Area of the section above y (*i.e.*, shaded area $ABFE$)

$$= \left(\frac{d}{2} - y\right) \times b$$

$\bar{y}$ = Distance of the C.G. of area A from neutral axis

$$= y + \frac{1}{2}\left(\frac{d}{2} - y\right) = y + \frac{d}{4} - \frac{y}{2} = \frac{y}{2} + \frac{d}{4} = \frac{1}{2}\left(y + \frac{d}{2}\right)$$

b = Actual width of the section at the level EF

I = M.O.I. of the whole section about N.A.

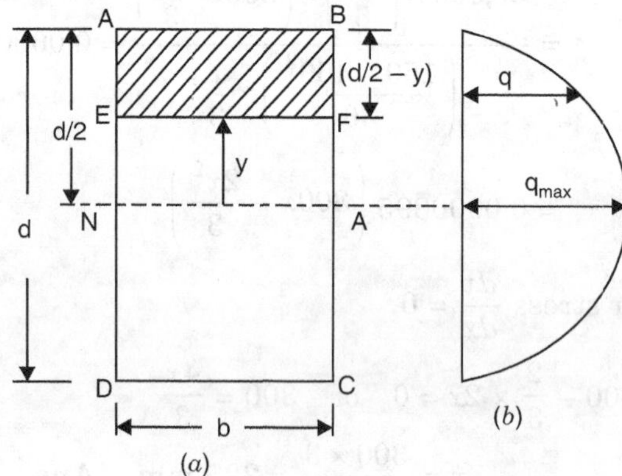

Fig. 3.43

Substituting these values in the above equation, we get

$$\tau = \frac{F.\left(\frac{d}{2} - y\right) \times b \times \frac{1}{2}\left(y + \frac{d}{2}\right)}{b \times I}$$

$$= \frac{F}{2I}\left(\frac{d^2}{4} - y^2\right) \qquad ...(3.16)$$

From equation (3.16), we see that τ increases as y decreases. Also the variation of τ with respect to y is a parabola. Fig. 3.43 (*b*) shows the variation of shear stress across the section.

At the top edge, $y = \frac{d}{2}$ and hence

$$\tau = \frac{F}{2I}\left[\frac{d^2}{4} - \left(\frac{d}{2}\right)^2\right] = \frac{F}{2I} \times 0 = 0$$

At the neutral axis, $y = 0$ and hence

$$\tau = \frac{F}{2I}\left(\frac{d^2}{4} - 0\right) = \frac{F}{2I} \times \frac{d^2}{4}$$

$$= \frac{Fd^2}{8I} = \frac{Fd}{8 \times \frac{bd^3}{12}} \qquad \left(\because I = \frac{bd^3}{12}\right)$$

$$= \frac{12}{8} \cdot \frac{F}{bd} = 1.5\,\frac{F}{bd} \qquad ...(i)$$

Now average shear stress, $\tau_{avg} = \frac{\text{Shear force}}{\text{Area of section}} = \frac{F}{b \times d}$.

Substituting the above value in equation (*i*), we get

$$\tau = 1.5 \times \tau_{avg} \qquad ...(3.17)$$

Equation (3.17) gives the shear stress at the neutral axis where $y = 0$. This stress is also the maximum shear stress.

$$\therefore \qquad \tau_{max} = 1.5\tau_{avg} \qquad ...(3.18)$$

From equation (3.15), $\tau = \dfrac{A\bar{y}}{Ib}$. In this equation the value of $A\bar{y}$ can also be calculated as given below :

$A\bar{y}$ = Moment of shaded area of Fig. 3.43 (*a*) about N.A.

Consider a strip of thickness dy at a distance y from N.A. Let dA is the area of this strip.

Then $\qquad dA$ = Area of strip = $b \times dy$

Moment of the area dA about N.A.

$$= dA \,.\, y \quad \text{or} \quad y \times dA$$

$$= y \times bdy \qquad (\because \; dA = b \times dy)$$

The moment of the shaded area about N.A. is obtained by integrating the above equation between the limits y to $\dfrac{d}{2}$.

$\therefore$ Moment of shaded area about N.A.

$$= \int_y^{d/2} y \times b \times dy$$

$$= b \int_y^{d/2} y \times dy \qquad \text{(as } b \text{ is constant)}$$

$$= b\left[\frac{y^2}{2}\right]_y^{d/2} = \frac{b}{2}\left[\left(\frac{d}{2}\right)^2 - y^2\right] = \frac{b}{2}\left[\frac{d^2}{4} - y^2\right]$$

But moment of shaded area about N.A. is also equal to $A\bar{y}$

$$\therefore \qquad A\bar{y} = \frac{b}{2}\left[\frac{d^2}{4} - y^2\right]$$

Substituting the value of $A\bar{y}$ in equation (3.15), we get

$$\tau = \frac{F \times \dfrac{b}{2}\left(\dfrac{d^2}{4} - y^2\right)}{I \times b} = \frac{F}{2I}\left(\frac{d^2}{4} - y^2\right)$$

This equation is same as equation (3.16).

Problem 3.27. *A rectangular beam 100 mm wide and 250 mm deep is subjected to a maximum shear force of 50 kN. Determine :*

(i) Average shear stress, *(ii) Maximum shear stress, and*

(iii) Shear stress at a distance of 25 mm above the neutral axis.

Sol. Given :

Width, $\qquad b = 100$ mm

Depth, $\qquad d = 250$ mm

Maximum shear force, $\quad F = 50 \text{ kN} = 50{,}000$ N.

(*i*) *Average shear stress is given by,*

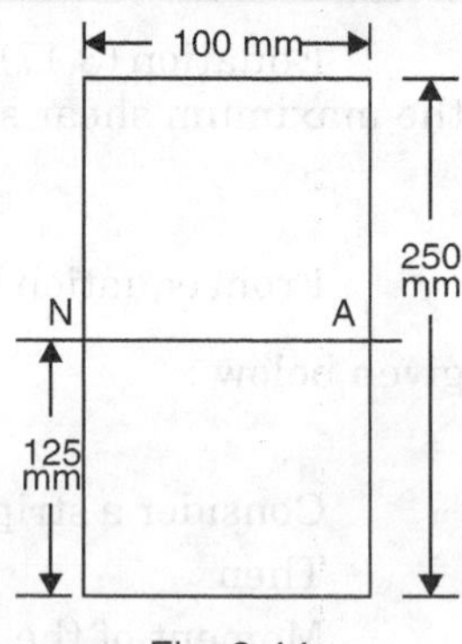

Fig. 3.44

$$\tau_{avg} = \frac{F}{\text{Area}} = \frac{50,000}{b \times d}$$

$$= \frac{50,000}{100 \times 250} = \mathbf{2\ N/mm^2.} \quad \textbf{Ans.}$$

(*ii*) *Maximum shear stress is given by equation (3.18)*

$$\therefore \quad \tau_{max} = 1.5 \times \tau_{avg}$$

$$= 1.5 \times 2 = \mathbf{3\ N/mm^2.} \quad \textbf{Ans.}$$

(*iii*) The shear stress at a distance *y from N.A. is given by equation (3.16).*

$$\therefore \quad \tau = \frac{F}{2I}\left(\frac{d^2}{4} - y^2\right)$$

$$= \frac{50000}{2I}\left(\frac{250^2}{4} - 25^2\right) \quad (\because \ y = 25 \text{ mm})$$

$$= \frac{50000}{2 \times \frac{bd^3}{12}}\left(\frac{62500}{4} - 625\right) = \frac{50000 \times 12}{2 \times 100 \times 250^3} \times 15000 \text{ N/mm}^2$$

$$= \mathbf{2.88\ N/mm^2.} \quad \textbf{Ans.}$$

Alternate Method [See Fig. 3.44 (*a*)]

The shear stress at a distance 25 mm from neutral axis is given by equation (3.15) as

$$\tau = F \times \frac{A \times \bar{y}}{I \times b}$$

where F = 50,000 N

A = Area of beam above 25 mm (*i.e.*, shaded area in Fig. 3.44 (*a*))

= 100 × 100 = 10000 mm^2

$\bar{y}$ = Distance of the C.G. of the area A from neutral axis

$$= 25 + \frac{100}{2} = 75 \text{ mm}$$

I = M.O.I. of total section about neutral axis

$$= \frac{bd^3}{12} = \frac{100 \times 250^3}{12}$$

b = Actual width of the section at a distance 25 mm from neutral axis = 100.

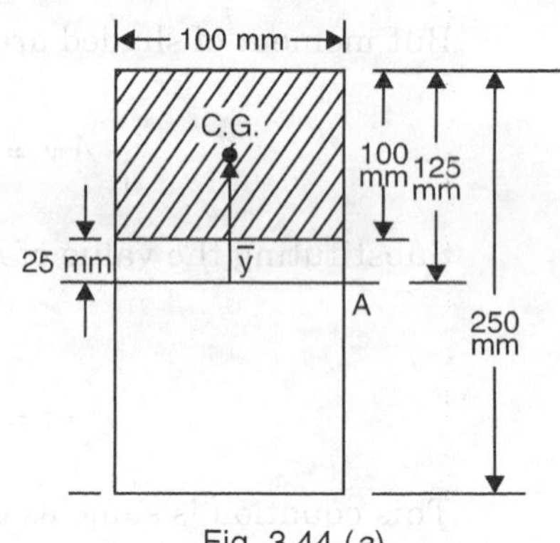

Fig. 3.44 (*a*)

Substituting these values in equation (3.15), we get

$$\tau = 50,000 \times \frac{10000 \times 75}{\left(\frac{100 \times 250^3}{12}\right) \times 100}$$

$$= \frac{50000 \times 10000 \times 75 \times 12}{100 \times 250^3 \times 100} = \mathbf{2.88\ N/mm^2.} \quad \textbf{Ans.}$$

Problem 3.28. *A timber beam of rectangular section is simply supported at the ends and carries a point load at the centre of the beam. The maximum bending stress is 12 N/mm² and maximum shearing stress is 1 N/mm², find the ratio of the span to the depth.*

Sol. Given :

Maximum bending stress, $\sigma_{max} = 12 \text{ N/mm}^2$

Maximum shear stress, $\tau_{max} = 1 \text{ N/mm}^2$.

Let b = Width of the beam,

d = Depth of the beam,

L = Span of the beam,

W = Point load at the centre.

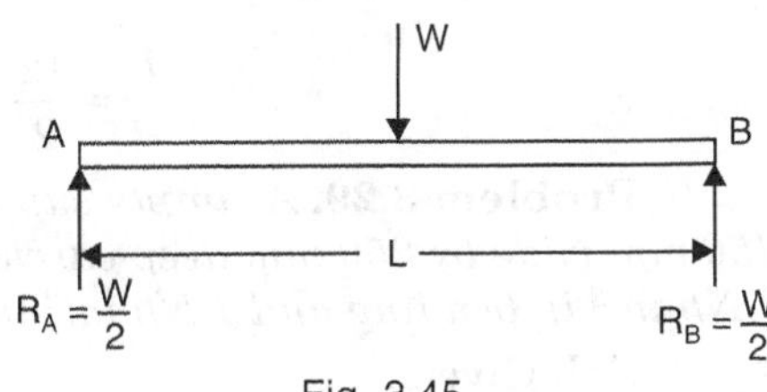

Fig. 3.45

Maximum shear force, $F = \frac{W}{2}$

and maximum B.M., $M = \frac{W \times L}{4}$

Now average shear stress, $\tau_{avg} = \frac{\text{Shear force}}{\text{Area}} = \frac{\left(\frac{W}{2}\right)}{b \times d} = \frac{W}{2bd}$.

Maximum shear stress is given by equation (3.18)

$$\therefore \quad \tau_{max} = 1.5 \times \tau_{avg}$$

or $$1 = 1.5 \times \frac{W}{2bd} \qquad \left(\because \tau_{max} = 1, \ \tau_{avg} = \frac{W}{2bd}\right)$$

or $$\frac{W}{bd} = \frac{2}{1.5} = \frac{4}{3} \qquad \ldots(i)$$

Now using bending equation

$$\frac{M}{I} = \frac{\sigma}{y} \quad \text{or} \quad \sigma = \frac{M}{I} \times y$$

$\therefore$ Maximum bending stress,

$$\sigma_{max} = \frac{M}{I} \times y_{max}$$

$$= \frac{\frac{W \times L}{4} \times \frac{d}{2}}{\frac{bd^3}{12}} \qquad \left(\because y_{max} = \frac{d}{2}\right)$$

$$= \frac{12}{8} \times \frac{W.L.d}{bd^3} = 1.5 \frac{W.L}{bd^2}$$

or $$12 = 1.5 \frac{W.L}{bd^2} \qquad (\because \sigma_{max} = 12)$$

$$= 1.5 \frac{W}{bd} \cdot \frac{L}{d}$$

$$= 1.5 \times \frac{4}{3} \times \frac{L}{d} \qquad \left[\because \frac{W}{bd} = \frac{4}{3} \text{ from equation } (i)\right]$$

$$\therefore \qquad = 2 \times \frac{L}{d}$$

$$\therefore \qquad \frac{L}{d} = \frac{12}{2} = \mathbf{6.}\ \textbf{Ans.}$$

Problem 3.29. *A simply supported wooden beam of span 1.3 m having a cross-section 150 mm wide by 250 mm deep carries a point load W at the centre. The permissible stress are 7 N/mm² in bending and 1 N/mm² in shearing. Calculate the safe load W.*

Sol. Given :

Span, $L = 1.30$ mm

Width, $b = 150$ mm

Depth, $d = 250$ mm

Bending stress, $\sigma = 7\ \text{N/mm}^2$

Shearing stress, $\tau = 1\ \text{N/mm}^2$

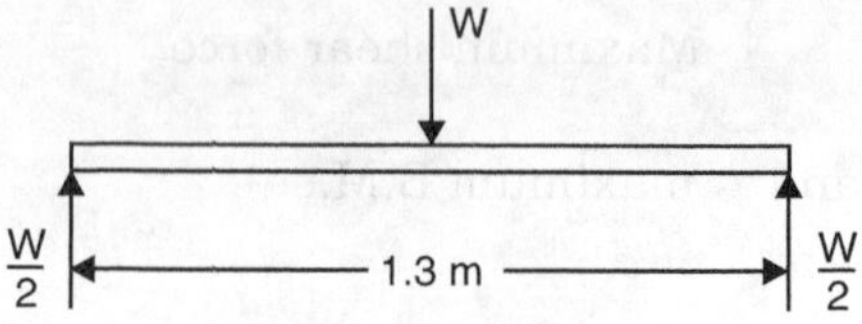

Fig. 3.46

Maximum B.M., $M = \frac{W \times L}{4} = \frac{W}{2} \times 1.3$ Nm

$$= \frac{W}{4} \times 1.3 \times 1000 \text{ Nmm} = 325\ W \text{ Nmm}$$

Maximum S.F. $= \frac{W}{2}$ N.

(*i*) *Value of W for bending stress consideration*

Using bending equation

$$\frac{M}{I} = \frac{\sigma}{y} \qquad \ldots(i)$$

where $M = 325\ W$ Nmm

$$I = \frac{bd^3}{12} = \frac{150 \times 250^3}{12} = 195312500 \text{ mm}^4$$

$\sigma = 7\ \text{N/mm}^2$

and $y = \frac{d}{2} = \frac{205}{2} = 125.$

Substituting these values in the above equation (*i*), we get

$$\frac{325W}{195312500} = \frac{7}{125}$$

$$\therefore \qquad W = \frac{7 \times 195312500}{325 \times 125} = 33653.8 \text{ N.}$$

(*ii*) *Value of W for shear stress consideration*

Average shear stress,

$$\tau_{avg} = \frac{\text{Shear force}}{\text{Area}} = \frac{\left(\frac{W}{2}\right)}{b \times d} = \frac{W}{2 \times 150 \times 250}$$

Maximum shear stress is given by equation (3.18)

$$\therefore \qquad \tau_{max} = \frac{3}{2} \times \tau_{avg}$$

But $\qquad \tau_{max} = 1 \text{ N/mm}^2$

$$\therefore \qquad 1 = \frac{3}{2} \times \frac{W}{2 \times 150 \times 250}$$

or $$W = \frac{2 \times 2 \times 150 \times 250}{3} = 50000 \text{ N.}$$

Hence, the safe load is minimum of the two values (*i.e.*, 33653.8 and 50000 N) of *W*. Hence safe load is 33653.8 N. **Ans.**

3.11.2. Circular Section. Fig. 3.47 shows a circular section of a beam. Let *R* is the radius of the circular section of *F* is the shear force acting on the section. Consider a level *EF* at a distance *y* from the neutral axis.

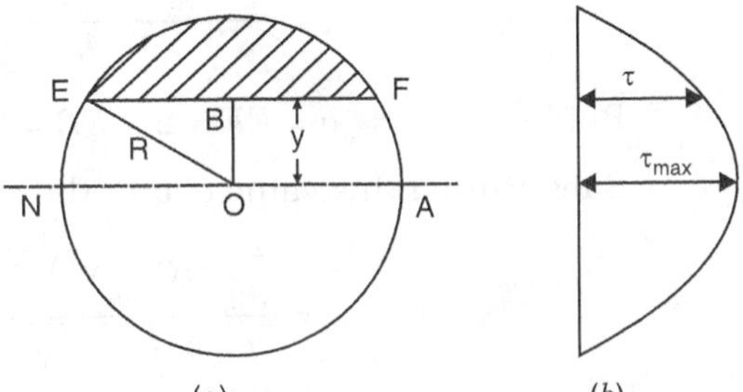

The shear stress at this level is given by equation (3.15) as

$$\tau = \frac{F \times A \times \bar{y}}{I \times b} \qquad \text{...}(i)$$

where $A\bar{y}$ = Moment of the shaded area about the neutral axis (N.A.)

I = Moment of inertia of the whole circular section

b = Width of the beam at the level *EF*.

(a) (b)

Fig. 3.47

Consider a strip of thickness *dy* at a distance *y* from N.A. Let *dA* is the area of strip.

Then $\qquad dA = b \times dy = EF \times dy \qquad (\because \; b = EF)$

$= 2 \times EB \times dy \qquad (\because \; EF = 2 \times EB)$

$= 2 \times \sqrt{R^2 - y^2} \times dy$

($\because$ In rt. angled triangle *OEB*, side $EB = \sqrt{R^2 - y^2}$)

Moment of this area *dA* about N.A.

$= y \times dA$

$= y \times 2\sqrt{R^2 - y^2} \times dy \qquad (\because dA = 2\sqrt{R^2 - y^2}\, dy)$

$= 2y\sqrt{R^2 - y^2}\, dy.$

Moment of the whole shaded area about the N.A. is obtained by integrating the above equation between the limits *y* and *R*

$$\therefore \qquad A\bar{y} = \int_y^R 2y\sqrt{R^2 - y^2}\, dy$$

$$= -\int_y^R (-2y)\sqrt{R^2 - y^2}\, dy.$$

Now $(-2y)$ is the differential of $(R^2 - y^2)$. Hence, the integration of the above equation becomes as

$$A\bar{y} = -\left[\frac{(R^2 - y^2)^{3/2}}{3/2}\right]_y^R$$

$$= -\frac{2}{3}\left[(R^2 - R^2)^{3/2} - (R^2 - y^2)^{3/2}\right]$$

$$= -\frac{2}{3}\left[0 - (R^2 - y^2)^{3/2}\right] = \frac{2}{3}(R^2 - y^2)^{3/2}$$

Substituting the value of $A\bar{y}$ in equation (*i*), we get

$$\tau = \frac{F \times \frac{2}{3}(R^2 - y^2)^{3/2}}{I \times b}$$

But $\qquad b = EF = 2 \times EB = 2 \times \sqrt{R^2 - y^2}$

Substituting this value of b in the above equation, we get

$$\tau = \frac{\frac{2}{3}F(R^2 - y^2)^{3/2}}{I \times 2\sqrt{R^2 - y^2}} = \frac{F}{EI}(R^2 - y^2) \qquad \text{...(3.19)}$$

Equation (3.19) shows that shear stress distribution across a circular section is parabolic. Also it is clear from this equation that with the increase of y, the shear stress decreases. At $y = R$, the shear stress, $\tau = 0$. Hence, shear stress will be maximum when $y = 0$ *i.e.*, at the neutral axis.

$\therefore$ At $y = 0$ *i.e.*, at the neutral axis, the shear stress is maximum and is given by

$$\tau_{max} = \frac{F}{3I}R^2$$

But $\qquad I = \frac{\pi}{64}D^4 = \frac{\pi}{64} \times (2R)^4 \qquad (\because \; D = 2R)$

$$= \frac{\pi}{4}R^4$$

$$\therefore \quad \tau_{max} = \frac{F \times R^2}{3 \times \frac{\pi}{4}R^4} = \frac{4}{3} \times \frac{F}{\pi R^2} \qquad \text{...(3.20)}$$

But average shear stress,

$$\tau_{avg} = \frac{\text{Shear force}}{\text{Area of circular section}} = \frac{F}{\pi R^2}$$

Hence equation (3.20) becomes as,

$$\tau_{max} = \frac{4}{3} \times \tau_{avg} \qquad \text{...(3.21)}$$

Problem 3.30. *A circular beam of 100 mm diameter is subjected to a shear force of 5 kN. Calculate :*

(i) Average shear stress, *(ii) Maximum shear stress, and*

(iii) Shear stress at a distance of 40 mm from N.A.

Sol. Given :

Diameter, $D = 100$ mm

$\therefore$ Radius, $R = \frac{100}{2} = 50$ mm

Shear force, $F = 5$ kN $= 5000$ N.

(i) Average shear stress is given by,

$$\tau_{avg} = \frac{\text{Shear force}}{\text{Area of circular section}}$$

$$= \frac{5000}{\pi(50)^2} = \mathbf{0.6366\ N/mm^2}. \quad \textbf{Ans.}$$

(ii) Maximum shear stress for a circular section is given by equation (3.21).

$$\therefore \quad \tau_{max} = \frac{4}{3} \times \tau_{avg}$$

$$= \frac{4}{3} \times 0.6366 = \mathbf{0.8488\ N/mm^2}. \quad \textbf{Ans.}$$

(iii) The shear stress at a distance 40 mm from N.A. is given by equation (3.19).

$$\therefore \quad \tau = \frac{F}{3I}(R^2 - y^2)$$

$$= \frac{5000}{3 \times \frac{\pi}{64} \times 100^4}(50^2 - 40^2) \qquad \left(\because y = 40 \text{ and } I = \frac{\pi}{64} 100^4\right)$$

$$= \frac{5000 \times 64}{3 \times \pi \times 100000000}(2500 - 1600)$$

$$= \mathbf{0.3055\ N/mm^2}. \quad \textbf{Ans.}$$

3.11.3. I-Section. Fig. 3.48 shows the I-section of a beam.

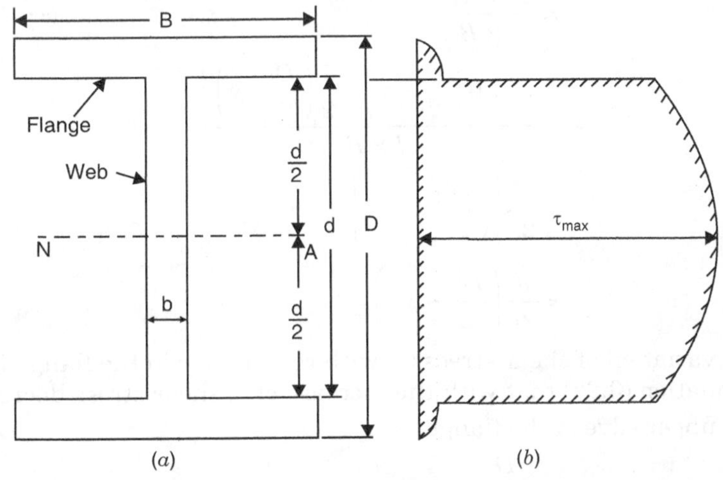

Fig. 3.48

Let B = Overall width of the section,

D = Overall depth of the section,

b = Thickness of the web, and

d = Depth of web.

The shear stress at a distance y from the N.A., is given by equation (3.1) as

$$\tau = F \times \frac{A\bar{y}}{I \times b}.$$

In this case the shear stress distribution in the web and shear stress distribution in the flange are to be calculated separately. Let us first calculate the shear stress distribution in the flange.

(*i*) *Shear stress distribution in the flange*

Consider a section at a distance y from N.A. in the flange as shown in Fig. 3.48 (*c*).

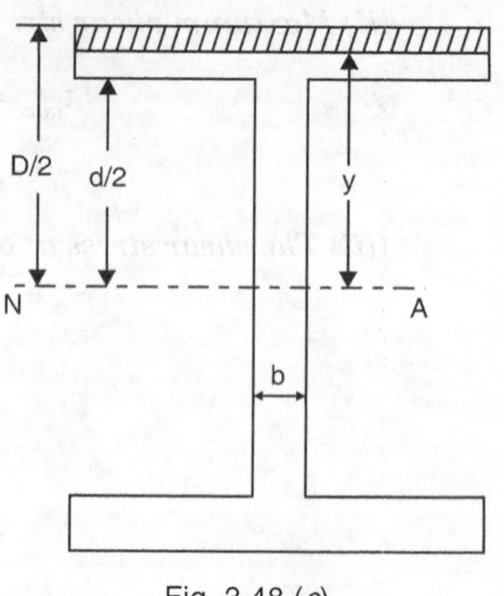

Fig. 3.48 (*c*)

Width of the section = B

Shaded area of flange, $A = B\left(\frac{D}{2} - y\right)$

Distance of the C.G. of the shaded area from neutral axis is given as

$$\bar{y} = y + \frac{1}{2}\left(\frac{D}{2} - y\right)$$

$$= y + \frac{D}{4} - \frac{y}{2}$$

$$= \frac{D}{4} + \frac{y}{2} = \frac{1}{2}\left(\frac{D}{2} + y\right)$$

Hence shear stress in the flange becomes,

$$\tau = \frac{F \times A\bar{y}}{I \times B} \qquad (\because \text{ Here width} = B)$$

$$= \frac{F \times B\left(\frac{D}{2} - y\right) \times \frac{1}{2}\left(\frac{D}{2} + y\right)}{I \times B}$$

$$= \frac{F}{2I}\left[\left(\frac{D}{2}\right)^2 - y^2\right]$$

$$= \frac{F}{2I}\left(\frac{D^2}{2} - y^2\right) \qquad \text{...(3.22)}$$

Hence, the variation of shear stress (τ) with respect to y in the flange is parabolic. It is also clear from equation (3.22) that with the increase of y, shear stress decreases.

(*a*) For the upper edge of the flange,

$$y = \frac{D}{2}$$

Hence shear stress, $\tau = \frac{F}{2I}\left[\frac{D^2}{4} - \left(\frac{D}{2}\right)^2\right] = 0.$

(*b*) For the lower edge of the flange,

$$y = \frac{d}{2}$$

Hence $$\tau = \frac{F}{2I}\left[\frac{D^2}{4} - \left(\frac{d}{2}\right)^2\right] = \frac{F}{2I}\left(\frac{D^2}{4} - \frac{d^2}{4}\right)$$

$$= \frac{F}{8I}(D^2 - d^2) \qquad \text{...(3.23)}$$

(*ii*) *Shear stress distribution in the web*

Consider a section at a distance y in the web from the N.A. as shown in Fig. 3.49.

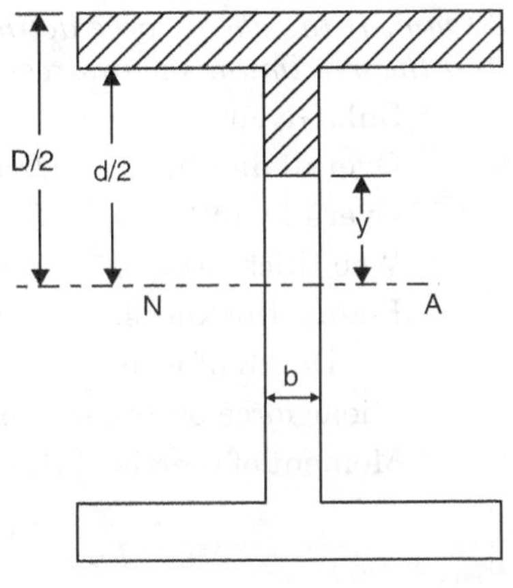

Fig. 3.49

Width of the section = b.

Here $A\bar{y}$ is made up of two parts *i.e.*, moment of the flange area about N.A. plus moment of the shaded area of the web about the N.A.

$\therefore A\bar{y}$ = Moment of the flange area about N.A. + moment of the shaded area of web about N.A.

$$= B\left(\frac{D}{2} - \frac{d}{2}\right) \times \frac{1}{2}\left(\frac{D}{2} + \frac{d}{2}\right) + b\left(\frac{d}{2} - y\right) \times \frac{1}{2}\left(\frac{d}{2} + y\right)$$

$$= \frac{B}{8}(D^2 - d^2) + \frac{b}{2}\left(\frac{d^2}{4} - y^2\right)$$

Hence the shear stress in the web becomes as

$$\tau = \frac{F \times A\bar{y}}{I \times b} = \frac{F}{I \times b} \times \left[\frac{B}{8}(D^2 - d^2) + \frac{b}{2}\left(\frac{d^2}{4} - y^2\right)\right] \qquad \text{...(3.24)}$$

From equation (3.24), it is clear that variation of τ with respect to y is parabolic. Also with the increase of y, τ decreases.

At the neutral axis, $y = 0$ and hence shear stress is maximum.

$$\therefore \qquad \tau_{max} = \frac{F}{I \times b}\left[\frac{B}{8}(D^2 - d^2) + \frac{b}{2} \times \frac{d^2}{4}\right]$$

$$= \frac{F}{I \times b}\left[\frac{B(D^2 - d^2)}{8} + \frac{bd^2}{8}\right] \qquad \text{...(3.25)}$$

At the junction of top of the web and bottom of flange,

$$y = \frac{d}{2}$$

Hence shear stress is given by,

$$\tau = \frac{F}{I \times b}\left[\frac{B}{8}(D^2 - d^2) + \frac{b}{2}\left(\frac{d^2}{4} - \left(\frac{d}{2}\right)^2\right)\right]$$

$$= \frac{F \times B \times (D^2 - d^2)}{8I \times b} \qquad \text{...(3.26)}$$

The shear stress distribution for the web and flange is shown in Fig. 3.48 (*b*). The shear stress at the junction of the flange and the web changes abruptly. The equation (3.23) gives the stress at the junction of the flange and the web when stress distribution is considered in the flange. But equation (3.26) gives the stress at the junction when stress distribution is considered in the web. From these two equations it is clear that the stress at the junction changes abruptly from $\frac{F}{8I}(D^2 - d^2)$ to $\frac{B}{b} \times \frac{F}{8I}(D^2 - d^2)$.

Problem 3.31. *An I-section beam 350 mm × 15 mm has a web thickness of 10 mm and a flange thickness of 20 mm. If the shear force acting on the section is 40 kN, find the maximum shear stress developed in the I-section.*

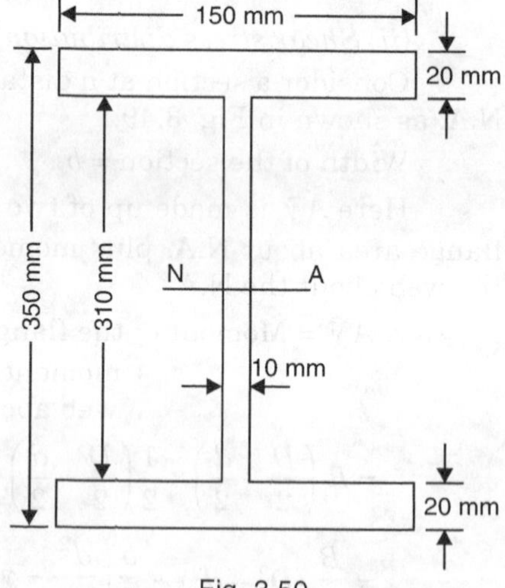

Fig. 3.50

Sol. Given :

Overall depth, $D = 350$ mm

Overall width, $B = 150$ mm

Web thickness, $b = 10$ mm

Flange thickness, $= 20$ mm

$\therefore$ Depth of web, $d = 350 - (2 \times 20) = 310$ mm

Shear force on the section, $F = 40$ kN $= 40{,}000$ N.

Moment of inertia of the section about neutral axis,

$$I = \frac{150 \times 350^3}{12} - \frac{140 \times 310^3}{12} \text{ mm}^4$$

$$= 535937500 - 347561666.6$$

$$= 188375833.4 \text{ mm}^4.$$

Maximum shear stress is given by equation (3.25)

$$\therefore \quad \tau_{max} = \frac{F}{I \times b}\left[\frac{B(D^2 - d^2)}{8} + \frac{bd^2}{8}\right]$$

$$= \frac{40000}{188375833.4 \times 10}\left[\frac{150(350^2 - 310^2)}{8} + \frac{10 \times 310^2}{8}\right]$$

$$= 0.000021234\left[\frac{150}{8}(122500 - 96100) + 120125\right]$$

$$= \mathbf{13.06 \ N/mm^2}. \quad \textbf{Ans.}$$

Alternate Method

The maximum shear stress developed in the I-section will be at the neutral axis. This shear stress is given by,

$$\tau_{max} = \frac{F \times A \times \bar{y}}{I \times b}$$

where $F = 40{,}000$ N

$A \times \bar{y}$ = Moment of the area above the neutral axis about the neutral axis

= Area of flange × Distance of C.G. of the area of flange from neutral axis + Area of web above neutral axis × Distance of the C.G. of this area from neutral axis

$$= (150 \times 20) \times \left(\frac{310}{2} + \frac{20}{2}\right) + \left(\frac{310}{2} \times 10\right) \times \left(\frac{310}{2} \times \frac{1}{2}\right)$$

$= 3000 \times 165 + 1550 \times 77.5$

$= 495000 + 120125 = 615125 \text{ mm}^3$

I = Moment of inertia of the whole section about neutral axis

$= 188375833.4 \text{ mm}^4$ (Already Calculated)

b = Width of the web at neutral axis

$= 10$ mm

$$\therefore \qquad \tau_{max} = \frac{40{,}000 \times 615125}{188375833.4 \times 10} = \mathbf{13.06\ N/mm^2}. \quad \textbf{Ans.}$$

Problem 3.32. *For the problem 3.31, sketch the shear stress distribution across the section. Also calculate the total shear force carried by the web.*

Sol. Given :

From problem 3.31, we have

$B = 150$ mm ; $D = 350$ mm

$d = 310$ mm ; $b = 10$ mm

$F = 40000$ N ; $I = 188.375 \times 10^6$ mm^4

$\tau_{max} = 13.06$ N/mm^2.

Shear stress distribution in the flange

The shear stress at the upper edge of the flange is zero.

Actually shear stress distribution in the flange is given by equation (3.22) as

$$\tau = \frac{F}{2I}\left(\frac{D^2}{4} - y^2\right) \qquad ...(i)$$

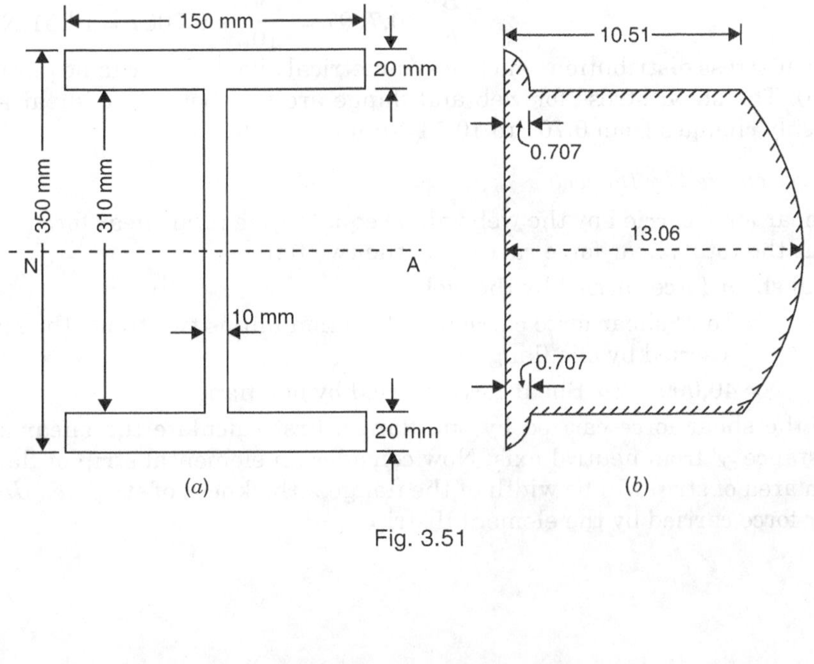

150 mm
20 mm
350 mm
310 mm
N A
10 mm
20 mm
(a)

10.51
0.707
13.06
0.707
(b)

Fig. 3.51

For the upper edge of the flange,

$$y = \frac{D}{2}$$

$$\therefore \qquad \tau = \frac{F}{2I}\left(\frac{D^2}{4} - \left(\frac{D}{2}\right)^2\right) = \frac{F}{2I}\left(\frac{D^2}{4} - \frac{D^2}{4}\right) = 0$$

For the lower edge of the upper flange (*i.e.,*) at the joint of web and flange,

$$y = \frac{d}{2}$$

$\therefore$ Substituting this value in equation (*i*), we get

$$\tau = \frac{F}{2I}\left[\frac{D^2}{4} - \left(\frac{d}{2}\right)^2\right] = \frac{F}{2I}\left(\frac{D^2}{4} - \frac{d^2}{4}\right)$$

$$= \frac{F}{8I}(D^2 - d^2) = \frac{40000}{8 \times 188.375 \times 10^6}(350^2 - 310^2)$$

$$= 0.7007 \text{ N/mm}^2.$$

Shear stress distribution in the web

The shear stress is maximum at N.A. and it is given by,

$$\tau_{max} = 13.06 \text{ N/mm}^2 \qquad \text{(calculate in problem 3.31)}$$

The shear stress at the junction of web and flange is given by equation (3.26) as

$$\tau = \frac{F \times B}{8I \times b}(D^2 - d^2)$$

$$= \frac{40000 \times 150}{8 \times 188.375 \times 10^6 \times 10}(350^2 - 310^2) = 10.51 \text{ N/mm}^2$$

(The shear stress at the junction can also be obtained as equal to

$$\frac{B}{b} \times 0.7007 = \frac{150}{10} \times 0.7007 = 10.51 \text{ N/mm}^2)$$

Now shear stress distribution which is symmetrical about N.A., can be plotted as shown in Fig. 3.51 (*b*). The shear stress for web and flange are parabolic. The shear stress at the junction suddenly changes from 0.707 to 10.51 N/mm^2.

Total Shear force carried by the web

Total shear force carried by the web will be equal to the total shear force carried by the I-section minus the total shear force carried by the two flanges.

$\therefore$ Total shear force carried by the web

= Total shear force carried by I-section minus two times the shear force carried by one flange

= 40,000 – 2 × Shear force carried by one flange ...(*ii*)

To find the shear force carried by one flange, first calculate the shear stress in the flange at a distance '*y*' from neutral axis. Now consider an elemental strip of flange of thickness '*dy*'. Then area of strip will be width of the flange × thickness of strip *i.e.*, $dA = 150 \times dy$. Now the shear force carried by the elemental strip

= Shear stress at a distance y in the flange × Area of the strip
$= \tau \times 150 \times dy$

Total shear force carried by the flange will be obtained by integrating the above equation from $\frac{310}{2}$ to $\frac{350}{2}$ (*i.e.*, from 155 to 175).

∴ Total shear force carried by one flange

$$= \int_{155}^{175} \tau \times 150 \times dy \qquad \text{...(iii)}$$

The value of 'τ' (*i.e.*, shear stress) in the flange at a distance y from neutral axis is given by

$$\tau = \frac{F \times A \times \bar{y}}{I \times b}$$

where $F = 40{,}000$

$A\bar{y}$ = Moment of area of the flange above y, about neutral axis
[*i.e.*, shaded area of Fig. 3.48 (*c*) on page 162]

$$= B\left(\frac{D}{2} - y\right) \times \frac{1}{2}\left(\frac{D}{2} + y\right)$$

$$= 150\left(\frac{350}{2} - y\right) \times \frac{1}{2}\left(\frac{350}{2} + y\right) \qquad (\because \text{ Here } B = 150, D = 350)$$

$$= 150\,(175 - y) \times \frac{1}{2}\,(175 + y)$$

$$= 75\,(175^2 - y^2) = 75\,(30625 - y^2)$$

I = Moment of inertia of the whole section about neutral axis
$= 188.375 \times 10^6$ mm^4 (Already calculated)

b = Width of flange
= 150 mm.

Substituting the above values, we get

$$\therefore \qquad \tau = \frac{40{,}000 \times 75\,(30625 - y^2)}{188.375 \times 10^6 \times 150} = 0.000106\,(30625 - y^2)$$

Substituting this value of τ in equation (*iii*), we get

Total shear force carried by one flange

$$= \int_{155}^{175} 0.000106(30625 - y^2) \times 150 \times dy$$

$$= 0.000106 \times 150 \int_{155}^{175} (30625 - y^2)\, dy$$

$$= 0.0159 \left[30625y - \frac{y^3}{3}\right]_{155}^{175}$$

$$= 0.0159 \left[30625\,(175 - 155) - \frac{1}{3}\,(175^3 - 155^3)\right]$$

$$= 0.0159\left[612500 - \frac{1}{3}(5359375 - 3723875)\right]$$

$$= 0.0159\,[612500 - 545166.66]$$

$$= 1070.61 \text{ N}$$

Substituting this value in equation (*ii*), we get

Total shear force carried by web

$$= 40{,}000 - 2 \times 1070.61$$

$$= 37858.78 \text{ N} = \mathbf{37.858\ kN.}\quad \textbf{Ans.}$$

3.11.4. T-Section. The shear stress distribution over a T-section is obtained in the same manner as over an I-section. But in this case the position of neutral axis (*i.e.*, position of C.G.) is to be obtained first, as the section is not symmetrical about *x-x* axis. The shear stress distribution diagram will also not be symmetrical.

Problem 3.33. *The shear force acting on a section of a beam is 50 kN. The section of the beam is of T-shaped of dimensions 100 mm × 100 mm × 20 mm as shown in Fig. 3.52. The moment of inertia about the horizontal neutral axis is 314.221 × 10⁴ mm⁴. Calculate the shear stress at the neutral axis and at the junction of the web and the flange.*

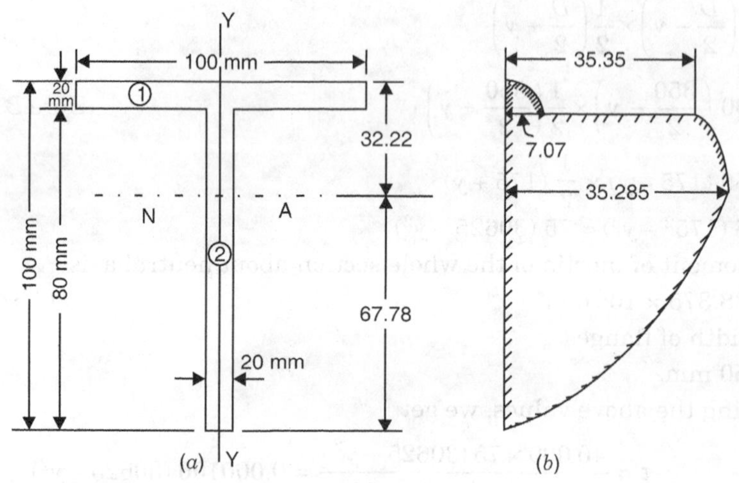

Fig. 3.52

Sol. Given :

Shear force, $F = 50 \text{ kN} = 50000 \text{ N}$

Moment of inertia about N.A.,

$$I = 314.221 \times 10^4 \text{ mm}^4.$$

First calculate the position of neutral axis. This can be obtained if we know the position of C.G. of given T-section. The given section is symmetrical about the axis *Y-Y* and hence the C.G. of the section will lie on *Y-Y* axis.

Let y^* = Distance of the C.G. of the section from the top of the flange.

Then $$y^* = \frac{A_1 y_1 + A_2 y_2}{(A_1 + A_2)}$$

$$= \frac{(100 \times 20) \times 10 + (20 \times 80) \times \left(20 + \frac{80}{2}\right)}{(100 \times 10) + (10 \times 90)}$$

$$= \frac{20000 + 96000}{2000 + 1600} = 32.22.$$

Hence, neutral axis will be at a distance of 32.22 mm from the top of the flange as shown in Fig. 3.52 (*a*).

Shear stress distribution in the flange

Now the shear stress at the top edge of the flange, and bottom of the web is zero.

Shear stress in the flange just at the junction of the flange and web is given by,

$$\tau = \frac{F \times A\bar{y}}{I \times b}$$

where $A = 100 \times 20 = 2000 \text{ mm}^2$

$\bar{y}$ = Distance of C.G. of the area of flange from N.A.

$$= 32.22 - \frac{20}{2} = 22.22 \text{ mm}$$

b = Width of flange = 100 mm

$$\therefore \qquad \tau = \frac{50000 \times 2000 \times 22.22}{314.221 \times 10^4 \times 100} = 7.07 \text{ N/mm}^2.$$

Shear stress distribution in the web

The shear stress in the web just at the junction of the web and flange will suddenly increase from 7.07 N/mm^2 to $7.07 \times \frac{100}{20} = 35.35$ N/mm^2. The shear stress will be maximum at N.A. Hence shear stress at the N.A. is given by

$$\tau = \frac{F \times A\bar{y}}{I \times b}$$

where $A\bar{y}$ = Moment of the above N.A. about N.A.

= Moment of area of flange about N.A. + Moment of area of web about N.A.

$$= 20 \times 100 \times (32.22 - 10) + 20 \times (32.22 - 10) \times \frac{22.22}{2}$$

$$= 44440 + 4937.28 = 49377.284 \text{ mm}^2$$

b = 20 mm

$$\therefore \qquad \tau = \frac{50000 \times 49377.284}{314.221 \times 10^4 \times 20} = 39.285 \text{ N/mm}^2$$

Now the shear stress distribution diagram can be drawn as shown in Fig. 3.52 (*b*).

3.11.5. Miscellaneous Sections. The shear stress distribution over miscellaneous sections is obtained in the same manner as over a T-section. Here also the position of neutral axis is obtained first.

Problem 3.34. *The shear force acting on a beam at an I-section with unequal flanges is 50 kN. The section is shown in Fig. 3.53. The moment of inertia of the section about N.A. is 2.849 × 10^4. Calculate the shear stress at the N.A. and also draw the shear stress distribution over the depth of the section.*

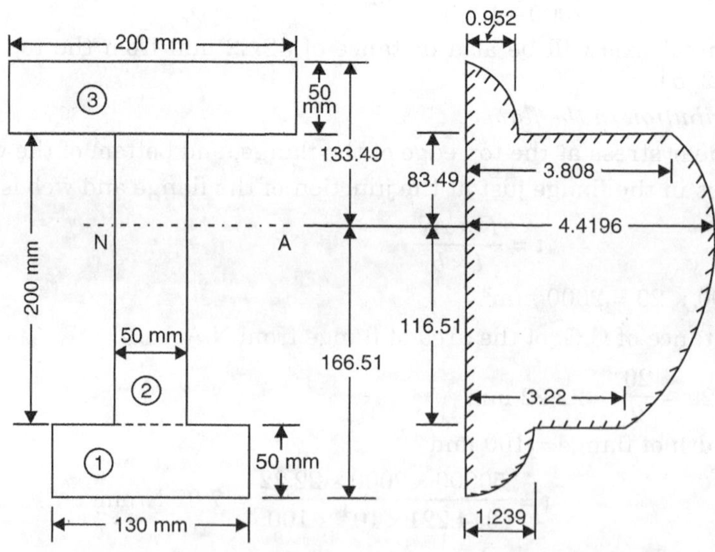

Fig. 3.53

Sol. Given :

Shear force, $F = 50 \text{ kN} = 50{,}000 \text{ N}$

Moment of inertia about N.A.,

$$I = 2.849 \times 10^8 \text{ mm}^4.$$

Let us first calculate the position of N.A. This is obtained if we know the position of the C.G. of the given I-section. Let y^* is the distance of the C.G. from the bottom face. Then

$$y^* = \frac{A_1 y_1 + A_2 y_2 + A_3 y_3}{(A_1 + A_2 + A_3)}$$

where A_1 = Area of bottom flange

$= 130 \times 50 = 6500 \text{ mm}^2$

A_2 = Area of web $= 200 \times 50 = 10000 \text{ mm}^2$

A_3 = Area of top flange $= 200 \times 50 = 10000 \text{ mm}^2$

y_1 = Distance of C.G. of A_1 from bottom face

$= \frac{50}{2} = 25$ mm

y_2 = Distance of C.G. of A_2 from bottom face

$= 50 + \frac{200}{2} = 150$ mm

y_3 = Distance of C.G. of A_3 from bottom face

$= 50 + 200 + \frac{50}{2} = 275$ mm

$$\therefore \qquad y^* = \frac{6500 \times 25 \times 10000 \times 150 + 10000 \times 275}{6500 + 10000 + 10000} = 166.51 \text{ mm}$$

Hence N.A. is at a distance of 166.51 mm from the bottom face (or 300 – 166.51 = 133.49 mm from upper top fibre).

Shear stress distribution

(*i*) Shear stress at the extreme edges of the flanges is zero.

(*ii*) The shear stress in the upper flange just at the junction of upper flange and web is given by,

$$\tau = \frac{F \times A\overline{y}}{I \times b}$$

where $A\overline{y}$ = Moment of the area of the upper flange about N.A.

= Area of upper flange × Distance of the C.G. of upper flange from N.A.

= (200 × 50) × (133.49 – 25) = 1084900

b = Width of upper flange = 200 mm

$$\therefore \qquad \tau = \frac{50000 \times 1084900}{2.849 \times 108 \times 200} = 0.9520 \text{ N/mm}^2.$$

(*iii*) The shear stress in the web just at the junction of the web and upper flange will suddenly increase from 0.952 to $0.952 \times \frac{200}{50} = 3.808$ N/mm^2.

(*iv*) The shear stress will be maximum at the N.A. This is given by

$$\tau_{max} = \frac{F \times A\overline{y}}{I \times b}$$

where $A\overline{y}$ = Moment of total area (about N.A.) about N.A.

= Moment of area of upper flange about N.A. + Moment of area of web about N.A.

$= 200 \times 50 \times (133.49 - 25) + (133.49 - 50) \times 50 \times \frac{(133.49 - 50)}{2}$

= 1084900 + 174264.5 = 1259164.5

and b = 50 mm

$$\therefore \qquad \tau_{max} = \frac{50000 \times 1259164.5}{2.849 \times 10^8 \times 50} = 4.4196 \text{ N/mm}^2.$$

(*v*) The shear stress in the lower flange just at the junction of the lower flange and the web is given by

$$\tau = \frac{F \times A\overline{y}}{I \times b}$$

where $A\overline{y}$ = Moment of the area of the lower flange about N.A.

= 130 × 50 × (166.51 – 25) = 918125

b = Width of lower flange = 130 mm

$$\therefore \qquad \tau = \frac{50000 \times 918125}{2.849 \times 10^8 \times 130} = 1.239 \text{ N/mm}^2.$$

(*vi*) The shear stress in the web just at the junction of the web and lower flange will suddenly increase from 1.239 to $\frac{1.239 \times 130}{50}$ = 3.22 N/mm^2.

The shear stress diagram is shown in Fig. 3.53.

Problem 3.35. *The shear force acting on a beam at a section is F. The section of the beam is triangular base b and of an altitude h. The beam is placed with its base horizontal. Find the maximum shear stress and the shear stress at the N.A.*

Sol. Given :

$$\text{Base} = b$$

$$\text{Altitude} = h$$

The N.A. of the triangle *ABC* will lie at the C.G. of the triangle. But the C.G. of the triangle will be at a distance of $\frac{2h}{3}$ from the top.

∴ Neutral axis will be at a distance of $\frac{2h}{3}$ from the top.

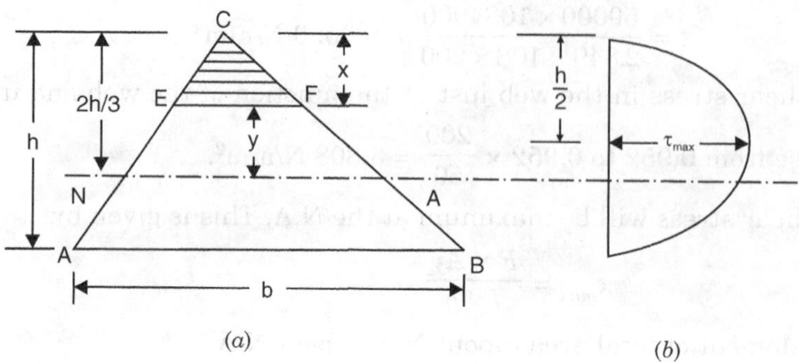

(*a*) (*b*)

Fig. 3.54

Consider a level *EF* at a distance *y* from the N.A. The shear stress at this level is given by,

$$\tau = \frac{F \times A\bar{y}}{I \times b} \quad ...(i)$$

where $A\bar{y}$ = Moment of the shaded area about the neutral axis

= Area of triangle *CEF* × Distance of C.G. of triangle *CEF* from N.A.

$$= \left(\frac{1}{2} \times EF \times x\right) \times \left(\frac{2h}{3} - \frac{2x}{3}\right)$$

$$= \left(\frac{1}{2} \times \frac{bx}{h} \times x\right) \times \frac{2}{3}(h - x)$$

(∵ As triangles *CEF* and *CAB* are similar. Hence $\frac{EF}{AB} = \frac{x}{h}$ or $EF = \frac{x}{h} \times AB = \frac{x \times b}{h}$)

$$= \frac{1}{2} \times \frac{bx^2}{h} \times \frac{2}{3}(h - x)$$

$$= \frac{1}{3} \times \frac{bx^2}{h} \times (h - x)$$

I = Moment of inertia of the whole triangular section CAB about N.A.

b = Actual width at the level EF

$$= EF = \frac{x \times b}{h}.$$

Substituting these values in equation (i), we get

$$\tau = \frac{F \times \frac{1}{3} \cdot \frac{bx^2}{h} \cdot (h - x)}{I \times \frac{x \times b}{h}} = \frac{1}{3} \cdot \frac{F \cdot x\,(h - x)}{I}$$

$$= \frac{F}{3I}(xh - x^2) \qquad \text{...}(ii)$$

From equation (ii), it is clear that variation of τ with respect to x is parabolic. At the top, $x = 0$ and hence τ is also zero. At the bottom $x = h$ and τ is also zero.

At the N.A., $x = \frac{2h}{3}$, and hence the shear at the N.A. becomes as,

$$\tau = \frac{F}{3I}\left[\frac{2h}{3} \times h - \left(\frac{2h}{3}\right)^2\right]$$

$$= \frac{F}{3I}\left[\frac{2h^2}{3} - \frac{4h^2}{9}\right] = \frac{F}{3I} \times \frac{(6h^2 - 4h^2)}{9} = \frac{F}{3I} \times \frac{2h^2}{9} = \frac{2}{27} \times \frac{Fh^2}{I}$$

But $$I = \frac{bh^3}{36}$$

$$\therefore \qquad \tau = \frac{2}{27} \times \frac{Fh^2}{\left(\frac{bh^3}{36}\right)} = \frac{2}{27} \times \frac{36 \times Fh^2}{bh^3}$$

$$= \frac{8}{3} \cdot \frac{F}{bh} \qquad \text{...(3.27)}$$

Maximum shear stress

The shear stress of any depth x from the top is given by equation (ii). The maximum shear stress will be obtained by differentiating equation (ii) with respect to x and equating to zero.

$$\therefore \qquad \frac{d}{dx}\left[\frac{F}{3I}(xh - x^2)\right] = 0$$

or $$\frac{F}{3I}(h - 2x) = 0$$

or $$h - 2x = 0 \qquad (\because \; F \text{ and } I \text{ are constants and cannot be zero})$$

or $$x = \frac{h}{2}$$

Now substituting this value of x in equation (ii), we get

$$\tau_{max} = \frac{F}{3I}\left[\frac{h}{2} \times h - \left(\frac{h}{2}\right)^2\right]$$

$$= \frac{F}{3I}\left[\frac{h^2}{2} - \frac{h^2}{4}\right] = \frac{F}{3I} \times \frac{h^2}{4} = \frac{Fh^2}{12I}$$

$$= \frac{Fh^3}{12 \times \frac{bh^3}{36}} \qquad \left(\because I = \frac{bh^3}{12}\right)$$

$$= \frac{36}{12} \times \frac{Fh^2}{bh^3}$$

$$= \frac{3F}{bh} \qquad \text{...(3.28)}$$

Now draw the shear stress diagram as shown in Fig. 3.54 (*b*).

Note. In the above case, the shear stress is not maximum at the N.A., but it is maximum at a depth of $h/2$ from the top. In all other cases, the shear stress was maximum at the N.A.

Problem 3.36. *A beam of triangular cross-section is subjected to a shear force of 50 kN. The base width of the section is 250 mm and height 200 mm. The beam is placed with its base horizontal. Find the maximum shear stress and the shear stress at the N.A.*

Sol. Given :

Shear force, $F = 50 \text{ kN} = 50000 \text{ N}$

Base width, $b = 250 \text{ mm}$

Height, $h = 200 \text{ mm}$

Maximum shear stress is given by equation (3.28).

$$\therefore \quad \tau_{max} = \frac{3F}{bh} = \frac{3 \times 50000}{250 \times 200} = \mathbf{3 \ N/mm^2}. \quad \textbf{Ans.}$$

Shear stress at N.A. is given by equation (3.27).

$$\therefore \quad \tau = \frac{8F}{3bh} = \frac{8 \times 50000}{3 \times 250 \times 200} = \mathbf{2.67 \ N/mm^2}. \quad \textbf{Ans.}$$

Problem 3.37. *A beam of square section is used as a beam with one diagonal horizontal. The beam is subjected to a shear force F, at a section. Find the maximum shear in the cross-section of the beam and draw the shear distribution diagram for the section.*

Sol. Given :

Fig. 3.55 shows a square beam *ABCD*, having diagonal *AC* horizontal.

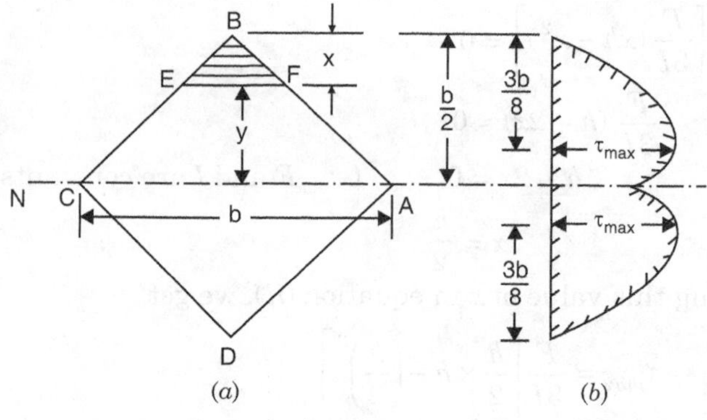

Fig. 3.55

Let b = Length of diagonal AC. This is also length of diagonal BD.

The N.A. of the beam shown in Fig. 3.55 (a), passes through diagonal AC.

Consider a level EF at a distance y from the N.A. The shear stress at this level is given by,

$$\tau = \frac{F \times A\bar{y}}{I \times b} \qquad \ldots(i)$$

where $A\bar{y}$ = Moment of the shaded area about N.A.

= Area of triangle BEF × Distance of C.G. of triangle BEF from N.A.

$$= \left(\frac{1}{2} \times EF \times x\right)\left(\frac{b}{2} - \frac{2}{3}x\right)$$

$$= \left(\frac{1}{2} \times 2x \times x\right)\left(\frac{b}{2} - \frac{2}{3}x\right) \qquad \left(\because \frac{EF}{CA} = \frac{x}{(b/2)}, \therefore EF = \frac{x}{(b/2)} \times b = 2x\right)$$

$$= x^2\left(\frac{b}{2} - \frac{2x}{3}\right)$$

I = Moment of inertia of the whole section about N.A.

$$= 2 \times \frac{b \times \left(\frac{b}{2}\right)^3}{12} = \frac{b^4}{48}$$

and b = Actual width at the level $EF = 2x$

Substituting these values in equation (i), we get

$$\tau = \frac{F \times x^2\left(\frac{b}{2} - \frac{2x}{3}\right)}{\left(\frac{b^4}{48}\right) \times 2x} = \frac{F \times 24x}{b^4}\left[\frac{3b - 4x}{6}\right]$$

$$= \frac{4F}{b^4}\, x\,(3b - 4x) \qquad \ldots(ii)$$

At the top, $x = 0$ hence $\tau = 0$

At the N.A., $x = \frac{b}{2}$ hence $\tau = \frac{4F}{b^2} \cdot \frac{b}{2}\left(3b - 4 \times \frac{b}{2}\right)$

$$= \frac{4F}{b^4} \cdot \frac{b}{2} \cdot b = \frac{2F}{b^2}$$

Maximum shear stress

Maximum shear stress will be obtained by differentiating equation (ii) with respect to x and equating to zero.

$$\therefore \qquad \frac{d}{dx}\left[\frac{4F}{b^4}(3bx - 4x^2)\right] = 0$$

or

$$\frac{4F}{b^4}(3b - 8x) = 0$$

or $$3b - 8x = 0 \qquad \left(\because \frac{4F}{b^4} \text{ cannot be zero}\right)$$

or $$x = \frac{3b}{8}$$

Substituting this value of x in equation (*ii*), we get maximum shear stress.

$$\therefore \quad \tau_{max} = \frac{4F}{b^4} \times \frac{3b}{8}\left(3b - 4 \times \frac{3b}{8}\right) = \frac{4F}{b^4} \times \frac{3b}{8} \times \frac{3b}{2} = \frac{9}{4} \times \frac{F}{b^2}.$$

The shear stress distribution is shown in Fig. 3.55 (*b*).

Problem 3.38. *Fig. 3.56 shows a section, which is subjected to a shear force of 100 kN. Determine the shear stresses at A, B, C and D. Sketch the shear stress distribution also.*

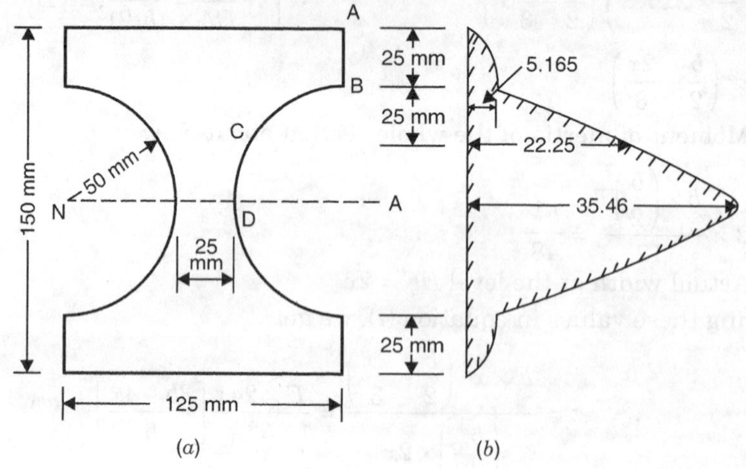

Fig. 3.56

Sol. Given :

Shear force, F = 100 kN = 100000 N.

The neutral axis will be at a distance of $\frac{125}{2}$ = 62.5 mm from the top, as the given section is symmetrical about *X-X* and *Y-Y* axis.

Moment of inertia of the given section about N.A. is given by,

I = M.O.I. of rectangle 125 × 150 about N.A.

– M.O.I. two semi-circle (or one circular hole) about N.A.

$$= \frac{125 \times 150^3}{12} - \frac{\pi}{64} \times 100^4 \text{ mm}^4 = 3.025 \times 10^7 \text{ mm}^4$$

The shear stress is given by,

$$\tau = \frac{F \times A\bar{y}}{I \times b}$$

At *A*, $\quad A\bar{y} = 0$ and hence $\tau = 0$

At *B*, $\quad A\bar{y}$ = Moment of area (125 × 25) about N.A.

$$= (125 \times 25) \times \left(50 + \frac{25}{2}\right) \qquad \left(\because A = 125 \times 25 \text{ and } \bar{y} = 50 + \frac{25}{2}\right)$$

$$= 125 \times 25 \times 62.5 = 1.953 \times 10^5 \text{ mm}^3$$

$$b = 125 \text{ mm}$$

$$\therefore \qquad \tau = \frac{100000 \times 1.9531 \times 10^5}{3.025 \times 10^7 \times 125} = \mathbf{5.165\ N/mm^2}. \quad \textbf{Ans.}$$

At C, $\quad A\bar{y}$ = Moment of area above an horizontal line passing through $C.C.$, about N.A.

= Moment of area of rectangle 125 × 50 about N.A.

– Moment of area of circular portion between C and B about N.A.

$$= (125 \times 50) \times \left(25 + \frac{50}{2}\right) - \int_{y=25}^{y=50} 2x \,.\, dy.y$$

$$= 3.125 \times 10^5 - \int_{25}^{50} 2 \times \sqrt{2500 - y^2} \times y \times dy \qquad \left(\because x = \sqrt{R^2 - y^2}\right)$$

$$= 3.125 \times 10^5 - \int_{25}^{50} -\sqrt{2500 - y^2}\,(-2y) \times dy$$

$$= 3.125 \times 10^5 + \left[\frac{(2500 - y^2)^{3/2}}{3/2}\right]_{25}^{60}$$

$$= 3.125 \times 10^5 + \frac{2}{3}\left[(2500 - 50^2)^{3/2} - (2500 - 25^2)^{3/2}\right]$$

$$= 3.125 \times 10^5 + \frac{2}{3}\left[0 - (2500 - 625)^{3/2}\right]$$

$$= 3.125 \times 10^5 + \frac{2}{3}(-81189)$$

$$= 3.125 \times 10^5 - 54126 = 258374 \text{ mm}^3$$

b = Width of beam at C (*i.e.*, length C-C)

= Width of complete section – 2 × Width of circular portion at C (*i.e.*, length EC)

$$= 125 - 2 \times \sqrt{R^2 - 25^2}$$

$$= 125 - 2 \times \sqrt{50^2 - 25^2}$$

$$= 125 - 86.6$$

$$= 38.4 \text{ mm}$$

$$\therefore \quad \tau = \frac{100000 \times 258374}{3.025 \times 10^7 \times 38.4}$$

$$= \mathbf{22.25\ N/mm^2}. \quad \textbf{Ans.}$$

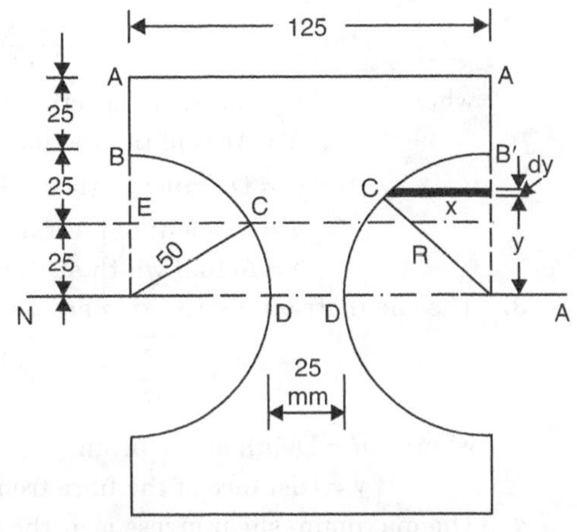

Fig. 3.57

At D, $\quad A\bar{y}$ = Moment of area above N.A., about N.A.

= Moment of area of rectangle 125 × 75 about N.A.

– Moment of area of circular portion between D and B about N.A.

$$= 125 \times 75 \times \frac{75}{2} - \int_{y=50}^{y=50} 2.x.dy.y$$

$$= 351500 - \int_0^{50} 2 \times \sqrt{2500 - y^2} \times y \times dy \qquad \left(\because x = \sqrt{2500 - y^2}\right)$$

$$= 351500 - \int_0^{50} -\sqrt{2500 - y^2}\,(-2y)dy$$

$$= 351500 + \left[\frac{(2500 - y^2)^{3/2}}{3/2}\right]_0^{50}$$

$$= 351500 + \frac{2}{3}\left[(2500 - 50^2)^{3/2} - (2500 - 0)^{3/2}\right]$$

$$= 351500 + \frac{2}{3}[0 - 125000]$$

$$= 351500 - 83333.33 \text{ mm}^3$$

$$= 268166.67 \text{ mm}^3$$

b = Width of beam at D (*i.e.*, length D-D)

= 25 mm

$$\therefore \qquad \tau = \frac{100000 \times 268166.67}{3.025 \times 10^7 \times 25} = \mathbf{35.46\ N/mm^2}. \quad \textbf{Ans.}$$

The variation of shear stress is shown in Fig. 3.56 (*b*).

HIGHLIGHTS (3B)

1. The stresses produced in a beam, which is subjected to shear force is known as shear stresses.
2. The shear stress at a fibre in a section of a beam is given by,

$$\tau = \frac{F \times A\bar{y}}{I \times b}$$

where F = Shear force acting at the given section.

A = Area of the section above the fibre.

$\bar{y}$ = Distance of the C.G. of the area A from the N.A.

I = Moment of inertia of whole section about N.A.

b = Actual width at the fibre.

3. The shear stress distribution across a rectangular section is parabolic and is given by,

$$\tau = \frac{F}{2I}\left(\frac{d^2}{4} - y^2\right)$$

where d = Depth of the beam

y = Distance of the fibre from N.A.

4. The maximum shear stress is at the N.A. for a rectangular section and is given by,

$$\tau_{max} = 1.5\ \tau_{avg.}$$

5. The shear stress distribution across a circular section is parabolic and is given by,

$$\tau = \frac{F}{3I}(R^2 - y^2).$$

6. The shear stress is maximum at the N.A. for a circular section and is given by,

$$\tau_{max} = \frac{4}{3} \times \tau_{avg}.$$

7. The shear stress distribution in I-section is parabolic. But at the junction of web and flange, the shear stress changes abruptly. The shear stress at the junction of the flange and the web changes from $\frac{F}{8I}(D^2 - d^2)$ to $\frac{B}{b} \times \frac{F}{8I}(D^2 - d^2)$ abruptly,

where D = Overall depth of the section,
d = Depth of web,
b = Thickness of web,
B = Overall width of the section.

8. The shear stress distribution for unsymmetrical sections is obtained after calculating the position of N.A.

9. In case of triangular section, the shear stress is not maximum at the N.A. The shear stress is maximum at a height of $h/2$.

10. The shear stress distribution diagram for a composite section, should be drawn by calculating the shear stress at important points.

EXERCISE (3B)

(A) Theoretical Questions

1. What do you mean by shear stresses in beams ?

2. Prove that the shear stress at any point (or in a fibre) in the cross-section of a beam which is subjected to a shear force F, is given by

$$\tau = F \times \frac{A\bar{y}}{b \times I}$$

where A = Area of the section above the fibre,
$\bar{y}$ = Distance of the C.G. of the area A from N.A.,
b = Actual width at the fibre, and
I = Moment of inertia of the section about N.A.

3. Show that for a rectangular section of the maximum shear stress is 1.5 times the average stress.

4. Prove that the shear stress distribution in a rectangular section of a beam which is subjected to a shear force F is given by

$$\tau = \frac{F}{2I}\left(\frac{d^2}{4} - y^2\right).$$

5. Prove that the maximum shear stress in a circular section of a beam is 4/3 times the average shear stress.

6. Derive an expression for the shear stress at any point in a circular section of a beam, which is subjected to a shear force F.

7. How will you draw the shear stress distribution diagram for composite section ?

8. How will you prove that the shear stress changes abruptly at the junction of the flange and the web of an I-section ?

9. The shear stress is not maximum at the N.A. in case of a triangular section. Prove this statement.
10. Prove that the maximum shear stress in a triangular section of a beam is given by

$$\tau_{max} = \frac{3F}{bh}$$

where b = Base width, and

h = Height.

11. Show that the ratio of maximum shear stress to mean shear stress in a rectangular cross-section is equal to 1.50 when it is subjected to a transverse shear force F. Plot the variation of shear stress across the section. *(Bangalore University, March 1989)*
12. Sketch the distribution of shear stress across the depth of the beams of the following cross-sections :
(*i*) T-section, and
(*ii*) Square section with diagonal vertical. *(Bangalore University, March 1989)*

(B) Numerical Problems

1. A rectangular beam 100 mm wide and 150 mm deep is subjected to a shear force of 30 kN. Determine : (*i*) average shear stress and (*ii*) maximum shear stress.
[**Ans.** (*i*) 2 N/mm^2 ; (*ii*) 3 N/mm^2]
2. A rectangular beam 100 mm wide is subjected to a maximum shear force of 100 kN. Find the depth of the beam if the maximum shear stress is 6 N/mm^2. [**Ans.** 250 mm]
3. A timber beam of rectangular section is simply supported at the ends and carries a point load at the centre of the beam. The length of the beam is 6 m and depth of beam is 1 m. Determine the maximum bending stress and the maximum shear stress. [**Ans.** 12 N/mm^2 ; 1 N/mm^2]
4. A timber beam 100 mm wide and 150 mm deep supports a uniformly distributed load of intensity w kN/m length over a span of 2 m.
If the safe stresses are 28 N/mm^2 in bending and 2 N/mm^2 in shear, calculate the safe intensity of the load which can be supported by the beam. [**Ans.** 20 kN/m]
5. A circular beam of 105 mm diameter is subjected to a shear force of 5 kN. Calculate : (*i*) average shear stress, and (*ii*) maximum shear stress. Also sketch the variation of the shear stress along the depth of the beam. [**Ans.** (*i*) 0.577 N/mm^2 ; (*ii*) 0.769 N/mm^2]
6. The maximum shear stress in a beam of circular section of diameter 150 mm, is 5.28 N/mm^2. Find the shear force to which the beam is subjected. [**Ans.** 70 kN]
7. A beam of I-section is having overall depth as 500 mm and overall width as 190 mm. The thickness of flanges is 25 mm whereas the thickness of the web is 15 mm. The moment of inertia about N.A. is given as 6.45×10^8 mm^4. If the section carries a shear force of 40 kN, calculate the maximum shear stress. Also sketch the shear stress distribution across the section.
[**Ans.** 62.33 N/mm^2]
8. An I-section has flanges of width b and the overall depth is $2b$. The flanges and web are of uniform thickness t. Find the ratio of the maximum shear stress to the average shear stress.
[**Ans.** 2.25]
9. An I-section has the following dimensions :
flanges : 150 mm × 20 mm
web : 30 mm × 10 mm
The maximum shear stress developed in the beam is 16.8 N/mm^2. Find the shear force to which the beam is subjected. [**Ans.** 50 kN]
10. A 12 cm by 5 cm I-section is subjected to a shearing force of 10 kN. Calculate the shear stress at the neutral axis and at the top of the web. What percentage of shearing force is carried by the web ? Given $I = 220 \times 10^4$ mm^4, area = 9.4×10^2 mm^2, web thickness = 3.5 mm and flange thickness = 5.5 mm. [**Ans.** 27.2 N/mm^2 ; 20.1 N/mm^2 ; 9.5 kN. *i.e.*, 95% of the total]

11. The shear force acting on a section of a beam is 100 kN. The section of the beam is of T-shaped of dimensions 200 mm × 250 mm × 50 mm. The flange thickness and web thickness are 50 mm. Moment of inertia about the horizontal neutral axis is 1.134×10^8 mm^4. Find the shear stress at the neutral axis and at the junction of the web and the flange.

[**Ans.** 11.64 N/mm^2 ; 2.76 N/mm^2 and 11.04 N/mm^2]

12. A beam is of T-section, flange 12 cm by 1 cm, web 10 cm by 1 cm. What percentage of the shearing force at any section is carried by the web ? [**Ans.** 93.5%]

13. For the section shown in Fig. 3.58, determine the average shearing stresses at *A*, *B*, *C* and *D* for a shearing force of 20 kN. Draw also the shear stress distribution across the section.

[**Ans.** 0 ; 6.47 N/mm^2 ; 27.7 N/mm^2 ; 44.4. N/mm^2]

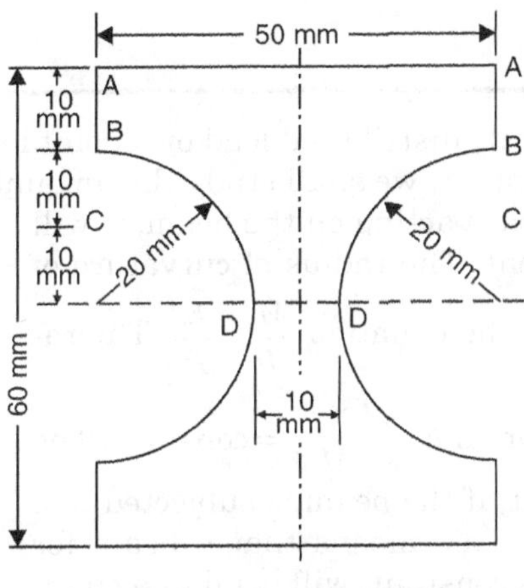

Fig. 3.58

14. A rectangular beam is simply supported at the ends and carries a point load at the centre. Prove that the ratio of span to depth

$$= \frac{\text{Maximum bending stress}}{2 \times \text{Maximum shear stress}}.$$

[**Hint.** Let W = Point load at centre,

b = width, and d = Depth.

Max. Shear force $= \dfrac{W}{2}$, Max. bending moment $= \dfrac{WL}{4}$.

$$\therefore \quad \text{Max. bending stress} = \frac{M}{Z} = \frac{\left(\dfrac{WL}{4}\right)}{\left(\dfrac{bd^2}{6}\right)} = \frac{WL}{4} \times \frac{6}{bd^2} = \frac{3}{2}\frac{WL}{bd^2}$$

$$\text{Max. shear stress} = \frac{3}{2} \times \text{Average shear stress}$$

$$= \frac{3}{2} \times \frac{\text{Max. shear force}}{\text{Area of section}} = \frac{3}{2} \times \frac{W}{2} \times \frac{1}{b \times d} = \frac{3}{4} \times \frac{W}{b \times d}$$

$$\therefore \quad \frac{\text{Max. bending stress}}{2 \times \text{Max. shear stress}} = \frac{\left(\dfrac{3}{2}\dfrac{WL}{bd^2}\right)}{\left(2 \times \dfrac{3}{4}\dfrac{W}{b \times d}\right)} = \frac{L}{d}.$$]

4

Deflection of Beams

4.1. INTRODUCTION

If a beam carries uniformly distributed load or a point load, the beam is deflected from its original position. In this chapter, we shall study the amount by which a beam is deflected from its position. Due to the loads acting on the beam, it will be subjected to bending moment. The radius of curvature of the deflected beam is given by the equation $\frac{M}{I} = \frac{E}{R}$. The radius of curvature will be constant if $R = \frac{I \times E}{M}$ = constant. The term $(I \times E)/M$ will be constant, if the beam is subjected to a constant bending moment M. This means that a beam for which, the value of $(E \times I)/M$ is constant, will bend in a circular arc.

Fig. 4.1 (*a*) shows the beam position before any load is applied on the beam whereas Fig. 4.1 (*b*) shows the beam position after loading.

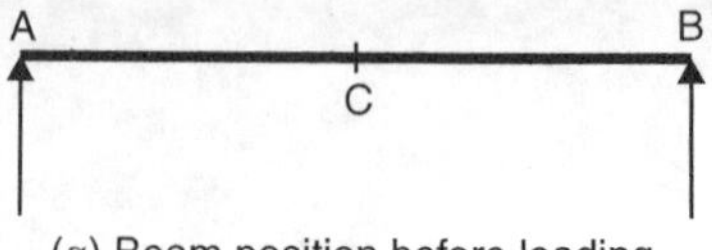

(*a*) Beam position before loading

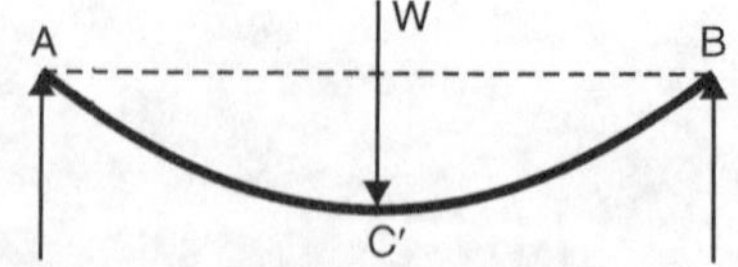

(*b*) Beam position after loading

Fig. 4.1

4.2. DEFLECTION AND SLOPE OF A BEAM SUBJECTED TO UNIFORM BENDING MOMENT

A beam AB of length L is subjected to a uniform bending moment M as shown in Fig. 4.1 (*c*). As the beam is subjected to a constant bending moment, hence it will bend into a circular arc. The initial position of the beam is shown by ACB, whereas the deflected position is shown by $AC'B$.

Let R = Radius of curvature of the deflected beam,

y = Deflection of the beam at the centre (*i.e.*, distance CC'),

I = Moment of inertia of the beam section,

E = Young's modulus for the beam material, and

θ = Slope of the beam at the end A (*i.e.*, the angle made by the tangent at A with the beam AB). For a practical beam the deflection y is a small quantity.

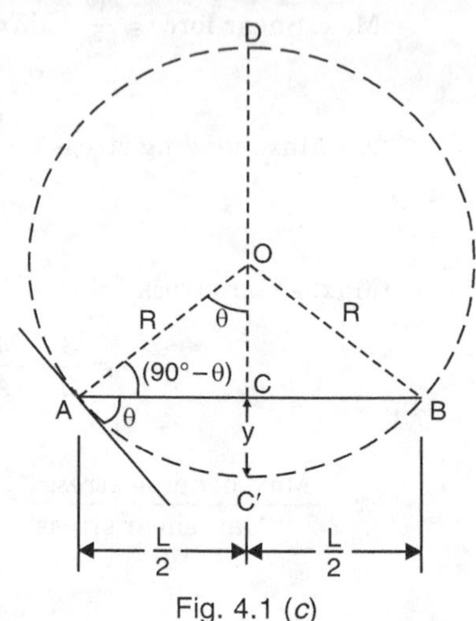

Fig. 4.1 (*c*)

Hence, $\tan\theta = \theta$ where θ is in radians. Hence θ becomes the slope as slope is

$$\frac{dy}{dx} = \tan\theta = \theta.$$

Now, $$AC = BC = \frac{L}{2}$$

Also, from the geometry of a circle, we know that

$$AC \times CB = DC \times CC'$$

$$\frac{L}{2} \times \frac{L}{2} = (2R - y) \times y \qquad (\because \; DC = DC' - CC' = 2R - y)$$

or $$\frac{L^2}{4} = 2Ry - y^2$$

For a practical beam, the deflection y is a small quantity. Hence, the square of a small quantity will be negligible. Hence neglecting y^2 in the above equation, we get

$$\frac{L^2}{4} = 2Ry$$

$\therefore$ $$y = \frac{L^2}{8R} \qquad \text{...}(i)$$

But from bending equation, we have

$$\frac{M}{I} = \frac{E}{R}$$

or $$R = \frac{E \times I}{M} \qquad \text{...}(ii)$$

Substituting the value of R in equation (i), we get

$$y = \frac{L^2}{8 \times \frac{EI}{M}}$$

or $$y = \frac{ML^2}{8EI} \qquad \text{...(4.1)}$$

The equation (4.1) gives the central deflection of a beam which bends in a circular arc.

Value of Slope (θ)

From triangle AOB, we know that

$$\sin\theta = \frac{AC}{AO} = \frac{\left(\frac{L}{2}\right)}{R} = \frac{L}{2R}$$

Since, the angle θ is very small, hence $\sin\theta = \theta$ (in radians)

$\therefore$ $$\theta = \frac{L}{2R}$$

$$= \frac{L}{2 \times \frac{EI}{M}} \qquad \left(\because \; R = \frac{EI}{M} \text{ from equation } (ii)\right)$$

$$= \frac{M \times L}{2EI} \qquad \text{...(4.2)}$$

Equation (4.2) gives the slope of the deflected beam at A or at B.

4.3. EQUATION OF ELASTIC CURVE

Let the curve AB represents the deflection of a beam as shown in Fig. 4.2 (a). Consider a small portion PQ of this beam. Let the tangents at P and Q make angle ψ and $\psi + d\psi$ with x-axis as shown in Fig. 4.2 (b). Normal at P and Q will meet at C such that

$$PC = QC = R$$

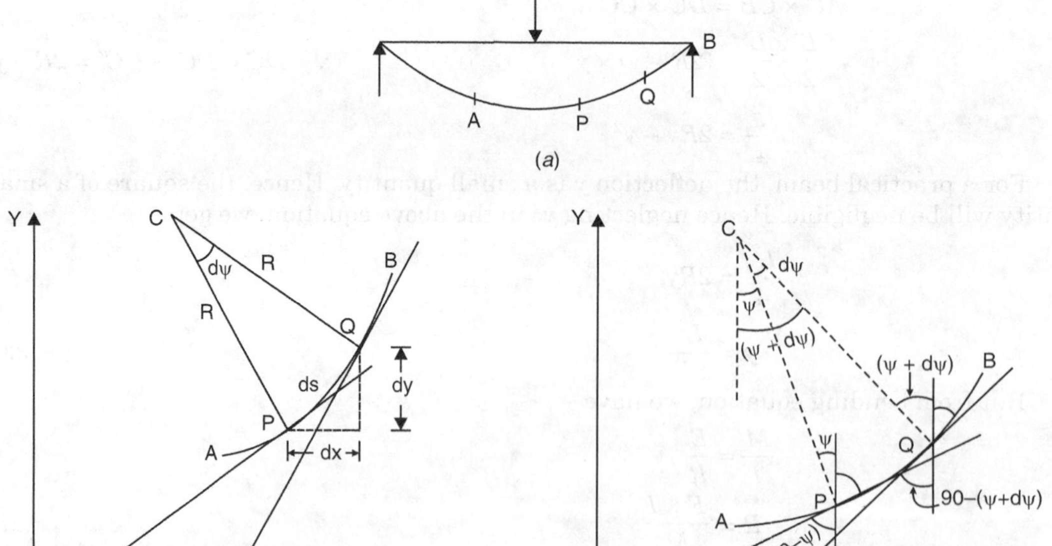

Fig. 4.2

The point C is known as centre of curvature of the curve PQ.

Let the length of PQ is equal to ds.

From Fig. 4.2 (c), we see that

$$\text{Angle } PCQ = d\psi$$

$\therefore$ $$PQ = ds = R.d\psi$$

or $$R = \frac{ds}{d\psi} \qquad \text{...(i)}$$

Let x and y be the co-ordinates of P. The length ds is having components dx and dy in x- and y-directions, then

$$\tan\psi = \frac{dy}{dx} \qquad \text{...(ii)}$$

$$\sin\psi = \frac{dy}{ds}$$

and $$\cos\psi = \frac{dx}{ds}$$

Now, equation (i) can be written as

$$R = \frac{ds}{d\psi} = \frac{ds}{dx} \times \frac{dx}{d\psi}$$

$$= \left(\frac{1}{\cos\psi}\right) \times \frac{dx}{d\psi}$$

or $$R = \frac{\sec\psi}{\left(\frac{d\psi}{dx}\right)} \quad ...(iii)$$

Let us find the value of $\frac{d\psi}{dx}$. As $\tan\psi = \frac{dy}{dx}$, differentiation of this equation w.r.t. x, will give

$$\sec^2\psi \cdot \frac{d\psi}{dx} = \frac{d^2y}{dx^2}$$

or $$\frac{d\psi}{dx} = \frac{\left(\frac{d^2y}{dx^2}\right)}{\sec^2\psi}$$

Substituting this value of $\frac{d\psi}{dx}$ in equation (iii), we get

$$R = \frac{\sec\psi}{\left(\frac{\frac{d^2y}{dx^2}}{\sec^2\psi}\right)} = \frac{\sec\psi \cdot \sec^2\psi}{\frac{d^2y}{dx^2}} = \frac{\sec^3\psi}{\left(\frac{d^2y}{dx^2}\right)}$$

Taking the reciprocal to both sides, we get

$$\frac{1}{R} = \frac{\frac{d^2y}{dx^2}}{\sec^3\psi} = \frac{\frac{d^2y}{dx^2}}{(\sec^2\psi)^{3/2}}$$

$$= \frac{\frac{d^2y}{dx^2}}{(1+\tan^2\psi)^{3/2}}$$

For a practical beam, the slope $\tan\psi$ at any point is a small quantity. Hence $\tan^2\psi$ can be neglected in comparison to 1.

$\therefore$ $$\frac{1}{R} = \frac{d^2y}{dx^2} \quad ...(iv)$$

From the bending equation, we have

$$\frac{M}{I} = \frac{E}{R}$$

or $$\frac{1}{R} = \frac{M}{EI} \quad ...(v)$$

Equating equations (*iv*) and (*v*), we get

$$\frac{M}{EI} = \frac{d^2y}{dx^2}$$

or
$$\frac{d^2y}{dx^2} = \frac{M}{EI}$$

The above equation is a second order linear differential equation. It is the governing differential equation for elastic curve.

Now,
$$M = EI\frac{d^2y}{dx^2} \quad ...(4.3)$$

Differentiating the above equation w.r.t. x, we get

$$\frac{dM}{dx} = EI\frac{d^3y}{dx^3}$$

But,
$$\frac{dM}{dx} = F \text{ shear force}$$

$$F = EI\frac{d^3y}{dx^3} \quad ...(4.4)$$

Differentiating equation (4.4) w.r.t. x, we get

$$\frac{dF}{dx} = EI\frac{d^4y}{dx^4}$$

But
$$\frac{dF}{dx} = w \text{ the rate of loading}$$

$\therefore$
$$w = EI\frac{d^4y}{dx^4} \quad ...(4.5)$$

Hence, the relation between curvature, slope, deflection etc., at a section is given by :

Deflection $= y$

Slope $= \dfrac{dy}{dx}$

Bending moment $= EI\dfrac{d^2y}{dx^2}$

Shearing force $= EI\dfrac{d^3y}{dx^3}$

The rate of loading $= EI\dfrac{d^4y}{dx^4}$.

Units. In the above equations, E is taken in N/mm^2

I is taken in mm^4, y is taken in mm,

M is taken in Nm and x is taken in m.

4.3.1. Methods of Determining Slope and Deflection at a Section in a Loaded Beam. The followings are the important methods for finding the slope and deflection at a section in a loaded beam :

(*i*) Double integration method

(*ii*) Moment area method, and

(*iii*) Macaulay's method

Incase of double integration method, the equation used is

$$M = EI\frac{d^2y}{dx^2} \quad \text{or} \quad \frac{d^2y}{dx^2} = \frac{M}{EI}$$

First integration of the above equation gives the value of $\frac{dy}{dx}$ or slope. The second integration gives the value of y or deflection.

The first two methods are used for a single load whereas the third method is used for several loads.

(A) DEFLECTION OF SIMPLY SUPPORTED BEAM

4.4. DEFLECTION OF A SIMPLY SUPPORTED BEAM

The following cases will be considered :

(*i*) Simply supported beam with a point load at the centre.

(*ii*) Simply supported beam with a uniformly distributed load.

4.4.1. Beam with a Point Load at the Centre. A simply supported beam *AB* of length *L* and carrying a point load *W* at the centre is shown in Fig. 4.3.

As the load is symmetrically applied the reactions R_A and R_B will be equal. Also the maximum deflection will be at the centre.

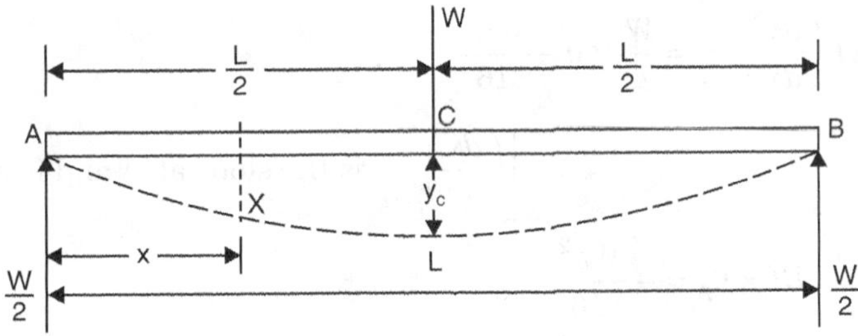

Fig. 4.3

Now, $$R_A = R_B = \frac{W}{2}$$

Consider a section *X* at a distance *x* from *A*. The bending moment at this section is given by,

$$M_x = R_A \times x$$

$$= \frac{W}{2} \times x$$ (Plus sign is as B.M. for left portion at *X* is clockwise)

But B.M. at any section is also given by equation (4.3) as

$$M = EI\frac{d^2y}{dx^2}$$

Equating the two values of B.M., we get

$$EI\frac{d^2y}{dx^2} = \frac{W}{2} \times x \qquad ..(i)$$

On integration, we get

$$EI\frac{dy}{dx} = \frac{W}{2} \times \frac{x^2}{2} + C_1 \qquad ...(ii)$$

where C_1 is the constant of integration. And its value is obtained from boundary conditions. The boundary condition is that at $x = \frac{L}{2}$, slope $\left(\frac{dy}{dx}\right) = 0$ (As the maximum deflection is at the centre, hence slope at the centre will be zero). Substituting this boundary condition in equation (*ii*), we get

$$O = \frac{W}{4} \times \left(\frac{L}{2}\right)^2 + C_1$$

or
$$C_1 = -\frac{WL^2}{16}$$

Substituting the value of C_1 in equation (*ii*), we get

$$EI\frac{dy}{dx} = \frac{Wx^2}{4} - \frac{WL^2}{16} \qquad ...(iii)$$

The above equation is known the *slope equation.* We can find the slope at any point on the beam by substituting the values of x. Slope is maximum at A. At A, $x = 0$ and hence slope at A will be obtained by substituting $x = 0$ in equation (*iii*).

$$\therefore \quad EI\left(\frac{dy}{dx}\right)_{\text{at } A} = \frac{W}{4} \times 0 - \frac{WL^2}{16}$$

$$\left[\left(\frac{dy}{dx}\right)_{\text{at } A} \text{ is the slope at } A \text{ and is represented by } \theta_A\right]$$

or
$$EI \times \theta_A = -\frac{WL^2}{16}$$

$$\therefore \quad \theta_A = -\frac{WL^2}{16EI}$$

The slope at point B will be equal to θ_A, since the load is symmetrically applied.

$$\therefore \quad \theta_B = \theta_A = -\frac{WL^2}{16EI} \qquad ...(4.6)$$

Equation (4.6) gives the slope in radians.

Deflection at any point

Deflection at any point is obtained by integrating the slope equation (*iii*). Hence integrating equation (*iii*), we get

$$EI \times y = \frac{W}{4} \cdot \frac{x^3}{3} - \frac{WL^2}{16}x + C_2 \qquad ...(iv)$$

where C_2 is another constant of integration. At A, $x = 0$ and the deflection (y) is zero.

Hence, substituting these values in equation (*iv*), we get

$$EI \times 0 = 0 - 0 + C_2$$

or $$C_2 = 0$$

Substituting the value of C_2 in equation (*iv*), we get

$$EI \times y = \frac{Wx^3}{12} - \frac{WL^2 \cdot x}{16} \quad \ldots(v)$$

The above equation is known as *the deflection equation.* We can find the deflection at any point on the beam by substituting the values of x. The deflection is maximum at centre point C, where $x = \frac{L}{2}$. Let y_c represents the deflection at C. Then substituting $x = \frac{L}{2}$ and $y = y_c$ in equation (*v*), we get

$$EI \times y_c = \frac{W}{12}\left(\frac{L}{2}\right)^3 - \frac{WL^2}{16} \times \left(\frac{L}{2}\right)$$

$$= \frac{WL^3}{96} - \frac{WL^3}{32} = \frac{WL^3 - 3WL^3}{96}$$

$$= -\frac{2WL^3}{96} = -\frac{WL^3}{48}$$

$$\therefore \quad y_c = -\frac{WL^3}{48EI}$$

(Negative sign shows that deflection is downwards)

$$\therefore \quad \text{Downward deflection, } y_c = \frac{WL^3}{48EI} \quad \ldots(4.7)$$

Problem 4.1. *A beam 6 m long, simply supported at its ends, is carrying a point load of 50 kN at its centre. The moment of inertia of the beam (i.e., I) is given as equal to 78 × 10⁶ mm⁴. If E for the material of the beam = 2.1 × 10⁵ N/mm², calculate : (i) deflection at the centre of the beam and (ii) slope at the supports.*

Sol. Given :

Length, $L = 6 \text{ m} = 6 \times 1000 = 6000 \text{ mm}$

Point load, $W = 50 \text{ kN} = 50{,}000 \text{ N}$

M.O.I., $I = 78 \times 10^6 \text{ mm}^4$

Value of $E = 2.1 \times 10^5 \text{ N/mm}^2$

Let y_c = Deflection at the centre and

θ_A = Slope at the support.

(*i*) Using equation (4.7) for the deflection at the centre, we get

$$y_c = \frac{WL^3}{48EI}$$

$$= \frac{50000 \times 6000^3}{48 \times 2.1 \times 10^5 \times 78 \times 10^6} = \mathbf{13.736 \text{ mm.}} \quad \textbf{Ans.}$$

(*ii*) Using equation (4.6) for the slope at the supports, we get

$$\theta_B = \theta_A = -\frac{WL^2}{16EI}$$

$$= \frac{WL^2}{16EI} \qquad \text{(Numerically)}$$

$$= \frac{50000 \times 6000^2}{16 \times 2.1 \times 10^5 \times 78 \times 10^6} \text{ radians}$$

$$= 0.06868 \text{ radians}$$

$$= 0.06868 \times \frac{180}{\pi} \text{ degree} \qquad \left(\because \ 1 \text{ radian} = \frac{180}{\pi} \text{ degree}\right)$$

$$= \mathbf{3.935°.} \quad \textbf{Ans.}$$

Problem 4.2. *A beam 4 metre long, simply supported at its ends, carries a point load W at its centre. If the slope at the ends of the beam is not to exceed 1°, find the deflection at the centre of the beam.*

Sol. Given :

Length, $L = 4 \text{ m} = 4000 \text{ mm}$

Point load at centre = W

Slope at the ends, $\theta_A = \theta_B = 1° = \frac{1 \times \pi}{180} = 0.01745$ radians

Let y_c = Deflection at the centre

Using equation (4.6), for the slope at the supports, we get

$$\theta_A = \frac{WL^2}{16EI} \qquad \text{(Numerically)}$$

or

$$0.01745 = \frac{WL^2}{16EI} \qquad ...(i)$$

Now, using equation (4.7), we get

$$y_c = \frac{WL^3}{48EI}$$

$$= \frac{WL^2}{16EI} \times \frac{L}{3} \qquad \left(\because \ \frac{WL^3}{48EI} = \frac{WL^2}{16EI} \times \frac{L}{3}\right)$$

$$= 0.01745 \times \frac{4000}{3} \qquad \left[\because \ \frac{WL^2}{16EI} = 0.01745 \text{ from equation } (i)\right]$$

$$= \mathbf{23.26 \ mm.} \quad \textbf{Ans.}$$

Problem 4.3. *A beam 3 m long, simply supported at its ends, is carrying a point load W at the centre. If the slope at the ends of the beam should not exceed 1°, find the deflection at the centre of the beam.* (Annamalai University, 1991)

Sol. Given :

Length, $L = 3 \text{ m} = 3 \times 1000 = 3000$ mm

Point load at centre = W

Slope at the ends, $\theta_A = \theta_B = 1°$

$$= \frac{1 \times \pi}{180} = 0.01745 \text{ radians}$$

Let y_c = Deflection at the centre

Using equation (4.6), we get

$$\theta_A = \frac{WL^2}{16EI} \quad \text{or} \quad 0.01745 = \frac{WL^2}{16EI} \qquad ...(i)$$

Now, using equation (4.7), we get

$$y_c = \frac{WL^3}{48EI} = \frac{WL^2}{16EI} \times \frac{L}{3}$$

$$= 0.01745 \times \frac{L}{3} \qquad \left(\because \frac{WL^2}{16EI} = 0.01745\right)$$

$$= 0.01745 \times \frac{3000}{3} \qquad (\because L = 3000 \text{ mm})$$

$$= \mathbf{17.45\ mm.} \quad \textbf{Ans.}$$

4.4.2. Deflection of a Simply Supported Beam with a Uniformly Distributed Load. A simply supported beam *AB* of length *L* and carrying a uniformly distributed load of *w* per unit length over the entire length is shown in Fig. 4.4. The reactions at *A* and *B* will be equal. Also the maximum deflection will be at the centre. Each vertical reaction = $\frac{w \times L}{2}$.

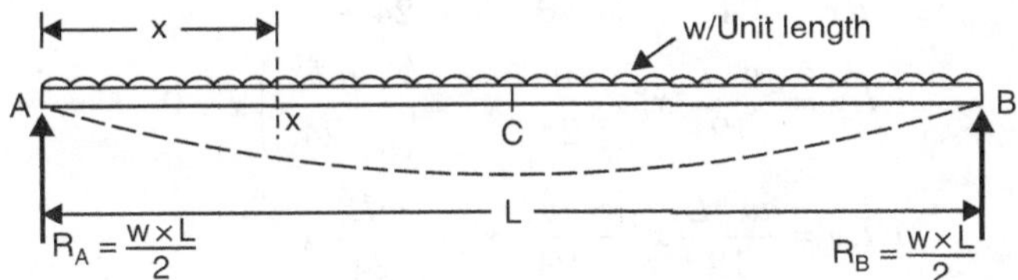

Fig. 4.4

$$\therefore \quad R_A = R_B = \frac{w \times L}{2}$$

Consider a section *X* at a distance *x* from *A*. The bending moment at this section is given by,

$$M_x = R_A \times x - w \times x \times \frac{x}{2} = \frac{w\,.\,L}{2}\,.\,x - \frac{w\,.\,x^2}{2}$$

But B.M. at any section is also given by equation (4.3), as

$$M = EI\,\frac{d^2y}{dx^2}$$

Equating the two values of B.M., we get

$$EI\,\frac{d^2y}{dx^2} = \frac{w\,.\,L}{2}x - \frac{w\,.\,x^2}{2}$$

Integrating the above equation, we get

$$EI\,\frac{dy}{dx} = \frac{w\,.\,L}{2}\,.\,\frac{x^2}{2} - \frac{w}{2}\,.\,\frac{x^3}{3} + C_1 \qquad ...(i)$$

where C_1 is a constant of integration.

Integrating the above equation again, we get

$$EI.y = \frac{w.L}{4}.\frac{x^3}{3} - \frac{w}{6}.\frac{x^4}{4} + C_1 x + C_2 \quad ...(ii)$$

where C_2 is another constant of integration. Thus, two constants of integration (*i.e.*, C_1 and C_2) are obtained from boundary conditions. The boundary conditions are :

(*i*) at $x = 0, y = 0$ and (*ii*) at $x = L, y = 0$

Substituting first boundary condition *i.e.*, $x = 0, y = 0$ in equation (*ii*), we get

$$0 = 0 - 0 + 0 + C_2 \quad \text{or} \quad C_2 = 0$$

Substituting the second boundary condition *i.e.*, at $x = L, y = 0$ in equation (*ii*), we get

$$0 = \frac{w.L}{4}.\frac{L^3}{3} - \frac{w}{6}.\frac{L^4}{4} + C_1.L \quad (C_2 \text{ is already zero})$$

$$= \frac{w.L^4}{12} - \frac{w.L^4}{24} + C_1.L$$

or

$$C_1 = -\frac{wL^3}{12} + \frac{wL^3}{24} = -\frac{wL^3}{24}$$

Substituting the value of C_1 in equations (*i*) and (*ii*), we get

$$EI\frac{dy}{dx} = \frac{w.L}{4}.x^2 - \frac{w}{6}x^3 - \frac{wL^3}{24} \quad ...(iii)$$

and

$$EI.y = \frac{w.L}{12}x^3 - \frac{w}{24}.x^4 + \left(-\frac{wL^3}{24}\right)x + 0 \quad (\because \ C_2 = 0)$$

or

$$EI.y = \frac{w.L}{12}x^3 - \frac{w}{24}.x^4 - \frac{wL^3}{24}x \quad ...(iv)$$

The equation (*iii*) is known as *slope equation.* We can find the slope $\left(i.e., \text{ the value of } \frac{dy}{dx}\right)$ at any point on the beam by substituting the different values of x in this equation.

The equation (*iv*) is known as *deflection equation.* We can find the deflection (*i.e.*, the value of y) at any point on the beam by substituting the different values of x in this equation.

Slope at the supports

Let θ_A = Slope at support A. This is equal to $\left(\frac{dy}{dx}\right)_{at\ A}$

and θ_B = Slop at support $B = \left(\frac{dy}{dx}\right)_{at\ B}$

At A, $x = 0$ and $\frac{dy}{dx} = \theta_A$.

Substituting these values in equation (*iii*), we get

$$EI.\theta_A = \frac{wL}{4} \times 0 - \frac{w}{6} \times 0 - \frac{wL^3}{24}$$

$$= \frac{wL^3}{24} = -\frac{WL^2}{24} \quad (\because \ w.L = W = \text{Total load})$$

$$\therefore \qquad \theta_A = -\frac{WL^2}{24EI} \qquad \text{...(4.8)}$$

(Negative sign means that tangent at A makes an angle with AB in the anti-clockwise direction)

By symmetry,
$$\theta_B = -\frac{WL^2}{24EI} \qquad \text{...(4.9)}$$

Maximum Deflection

The maximum deflection is at the centre of the beam *i.e.*, at point C, where $x = \frac{L}{2}$. Let y_C = deflection at C which is also maximum deflection. Substituting $y = y_C$ and $x = \frac{L}{2}$ in the equation (*iv*), we get

$$EI.y_C = \frac{w.L}{12}\cdot\left(\frac{L}{2}\right)^3 - \frac{w}{24}\cdot\left(\frac{L}{2}\right)^4 - \frac{wL^3}{24}\cdot\left(\frac{L}{2}\right)$$

$$= \frac{w.L^4}{96} - \frac{wL^4}{384} - \frac{wL^4}{48} = -\frac{5w.L^4}{384}$$

$$\therefore \qquad y_C = -\frac{5}{384}\cdot\frac{wL^4}{EI} = -\frac{5}{384}\cdot\frac{W.L^3}{EI} \qquad (\because \; w.L = W = \text{Total load})$$

Negative sign indicates that deflection is downwards.

$\therefore$ Downward deflection,

$$y_C = \frac{5}{384}\cdot\frac{WL^3}{EI} \qquad \text{...(4.10)}$$

4.5. SOLVED PROBLEMS BASED ON DEFLECTION OF SIMPLY SUPPORTED BEAM

Problem 4.4. *A beam of uniform rectangular section 200 mm wide and 300 mm deep is simply supported at its ends. It carries a uniformly distributed load of 9 kN/m run over the entire span of 5 m. If the value of E for the beam material is 1 × 10⁴ N/mm², find :*

(i) the slope at the supports and *(ii) maximum deflection.*

Sol. Given :

Width, $b = 200$ mm

Depth, $d = 300$ mm

M.O.I., $I = \frac{bd^3}{12} = \frac{200 \times 300^3}{12} = 4.5 \times 10^8 \text{ mm}^4$

U.d.l., $w = 9$ kN/m $= 9000$ N/m

Span, $L = 5$ m $= 5000$ mm

$\therefore$ Total load, $W = w\,.\,L^* = 9000 \times 5 = 45{,}000$ N

Value of $E = 1 \times 10^4$ N/mm²

Let θ_A = Slope at the support

and y_C = Maximum deflection.

*Here L should be taken in metre. Hence for calculating total load, L must be in metre and in other calculations L is taken in mm.

(*i*) Using equation (4.8), we get

$$\theta_A = -\frac{W \cdot L^2}{24EI}$$

$$= -\frac{45000 \times 5000^2}{24 \times 1 \times 10^4 \times 4.5 \times 10^8} \text{ radians}$$

$$= \mathbf{0.0104 \text{ radians.}} \quad \textbf{Ans.}$$

(*ii*) Using equation (4.10), we get

$$y_C = \frac{5}{384} \cdot \frac{W \cdot L^3}{EI}$$

$$= \frac{5}{384} \times \frac{45000 \times 5000^3}{1 \times 10^4 \times 4.5 \times 10^8}$$

$$= \mathbf{16.27 \text{ mm.}} \quad \textbf{Ans.}$$

Problem 4.5. *A beam of length 5 m and of uniform rectangular section is simply supported at its ends. It carries a uniformly distributed load of 9 kN/m run over the entire length. Calculate the width and depth of the beam if permissible bending stress is 7 N/mm² and central deflection is not to exceed 1 cm.*

Take E for beam material = 1 × 10⁴ N/mm².

Sol. Given :

Length, $L = 5 \text{ m} = 5000 \text{ mm}$

U.d.l., $w = 9 \text{ kN/m}$

∴ Total load, $W = w.L = 9 \times 5 = 45 \text{ kN} = 45,000 \text{ N}$

Bending stress, $f = 7 \text{ N/mm}^2$

Central deflection, $y_C = 1 \text{ cm} = 10 \text{ mm}$

Value of $E = 1 \times 10^4 \text{ N/mm}^2$

Let b = Width of beam is mm

and d = Depth of beam in mm

∴ M.O.I., $I = \frac{bd^3}{12}$

Using equation (4.10), we get

$$y_C = \frac{5}{384} \cdot \frac{W \cdot L^3}{EI}$$

or

$$10 = \frac{5}{384} \times \frac{45000 \times 5000^3}{1 \times 10^4 \times \left(\frac{bd^3}{12}\right)}$$

or

$$bd^3 = \frac{5}{384} \times \frac{45000 \times 5000^3 \times 12}{1 \times 10^4 \times 10}$$

$$= 878.906 \times 10^7 \text{ mm}^4 \quad \ldots(i)$$

The maximum bending moment for a simply supported beam carrying a uniformly distributed load is given by,

$$M = \frac{w \cdot L^2}{8} = \frac{W \cdot L}{8} \quad (\because \; W = w.L = \text{Total load})$$

$$= \frac{45000 \times 5}{8} \text{ Nm} = \frac{45000 \times 5}{8} \times 1000 \text{ Nmm}$$

$$= 28125000 \text{ Nmm}$$

Now using the bending equation as

$$\frac{M}{I} = \frac{f}{y}$$

or $$\frac{28125000}{\left(\frac{bd^3}{12}\right)} = \frac{7}{\left(\frac{d}{2}\right)} \qquad \left(\because \text{ Here } y = \frac{d}{2}\right)$$

or $$\frac{28125000 \times 12}{bd^3} = \frac{14}{d}$$

or $$bd^2 = \frac{28125000 \times 12}{14} = 24107142.85 \text{ mm}^3 \qquad ...(ii)$$

Dividing equation (*i*) by equation (*ii*), we get

$$d = \frac{838.906 \times 10^7}{24107142.85} = \mathbf{364.58 \text{ mm.}} \quad \textbf{Ans.}$$

Substituting this value of 'd' in equation (*ii*), we get

$$b \times (364.58)^2 = 24107142.85$$

$$\therefore \qquad b = \frac{24107142.85}{364.58^2} = \mathbf{181.36 \text{ mm.}} \quad \textbf{Ans.}$$

Problem 4.6. *A beam of length 5 m and of uniform rectangular section is supported at its ends and carries uniformly distributed load over the entire length. Calculate the depth of the section if the maximum permissible bending stress is 8 N/mm² and central deflection is not to exceed 10 mm.*

Take the value of E = 1.2 × 10⁴ N/mm².

Sol. Given :

Length, $L = 5 \text{ m} = 5000 \text{ mm}$

Bending stress, $f = 8 \text{ N/mm}^2$

Central deflection, $y_C = 10 \text{ mm}$

Value of $E = 1.2 \times 10^4 \text{ N/mm}^2$

Let W = Total load

and d = Depth of beam

The maximum bending moment for a simply supported beam carrying a uniformly distributed load is given by,

$$M = \frac{w.L^2}{8} = \frac{W.L}{8} \qquad (\because \quad W = w.L) \qquad ...(i)$$

Now using the bending equation,

$$\frac{M}{I} = \frac{f}{y}$$

or $$M = \frac{f \times I}{y} = \frac{8 \times I}{(d/2)} \qquad \left(\because \ y = \frac{d}{2}\right)$$

$$\therefore \qquad M = \frac{16I}{d} \qquad \text{...(ii)}$$

Equating the two values of B.M., we get

$$\frac{W.L}{8} = \frac{16I}{d}$$

or

$$W = \frac{16 \times 8I}{L \times d} = \frac{128I}{L \times d} \qquad \text{...(iii)}$$

Now, using equation (4.10), we get

$$y_C = \frac{5}{384} \times \frac{WL^3}{EI}$$

or

$$10 = \frac{5}{384} \times \frac{128I}{L \times d} \times \frac{L^3}{EI} \qquad \left(\because \ y_C = 10 \text{ mm and } W = \frac{128I}{L \times d}\right)$$

$$= \frac{5}{384} \times \frac{128 \times L^2}{d \times E}$$

or

$$d = \frac{5}{384} \times \frac{128 \times L^2}{10 \times E} = \frac{5}{384} \times \frac{128 \times 5000^2}{10 \times 1.2 \times 10^4}$$

$= 347.2$ mm $=$ **34.72 cm. Ans.**

4.6. MACAULAY'S METHOD

The procedure of finding slope and deflection for a simply supported beam with an eccentric point load as mentioned in Art. 4.5, is a very laborious. There is a convenient method for determining the deflections of the beam subjected to point loads.

This method was devised by Mr. M.H. Macaulay and is known as Macaulay's method. This method mainly consists in the special manner in which the bending moment at any section is expressed and in the manner in which the integrations are carried out.

4.6.1. Deflection of a Simply Supported Beam with an Eccentric Point Load. A simply supported beam AB of length L and carrying a point load W at a distance 'a' from left support and at a distance 'b' from right support is shown in Fig. 4.5. The reactions at A and B are given by,

$$R_A = \frac{W.b}{L} \quad \text{and} \quad R_B = \frac{W.a}{L}$$

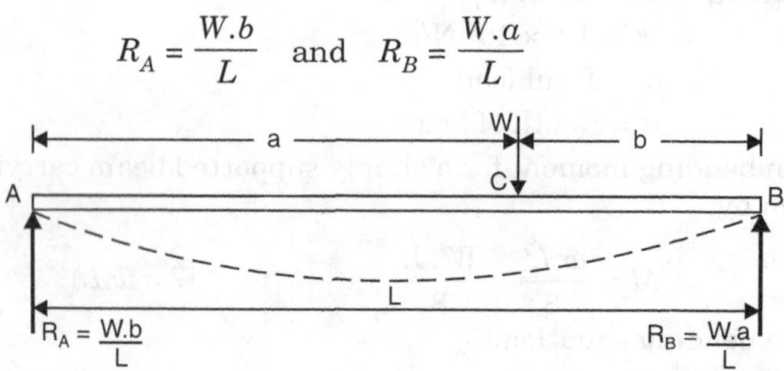

Fig. 4.5

The bending moment at any section between A and C at a distance x from A is given by,

$$M_x = R_A \times x = \frac{W.b}{L} \times x$$

The above equation of B.M. holds good for the values of x between 0 and 'a'. The B.M. at any section between C and B at a distance x from A is given by,

$$M_x = R_A.x - W \times (x - a)$$

$$= \frac{W.b}{L}.x - W(x - a)$$

The above equation of B.M. holds good for all values of x between $x = a$ and $x = b$.

The B.M. for all sections of the beam can be expressed in a single equation written as

$$M_x = \frac{W.b}{L}x \quad \vdots \quad - W(x - a) \qquad ...(i)$$

Stop at the dotted line for any point in section AC. But for any point in section CB, add the expression beyond the dotted line also.

The B.M. at any section is also given by equation (4.3) as

$$M = EI\frac{d^2y}{dx^2} \qquad ...(ii)$$

Hence, equating (*i*) and (*ii*), we get

$$EI\frac{d^2y}{dx^2} = \frac{W.b}{L}.x \quad \vdots \quad - W(x - a) \qquad ...(iii)$$

Integrating the above equation, we get

$$EI\frac{dy}{dx} = \frac{W.b}{L}\frac{x^2}{2} + C_1 \quad \vdots \quad - \frac{W(x-a)^2}{2} \qquad ...(iv)$$

where C_1 is a constant of integration. This constant of integration should be written after the first term. Also *the brackets are to be integrated as a whole*. Hence the integration of $(x - a)$ will be $\frac{(x-a)^2}{2}$ and not $\frac{x^2}{2} - ax$.

Integrating equation (*iv*) once again, we get

$$EI.y = \frac{W.b}{2L}.\frac{x^3}{3} + C_1x + C_2 \quad \vdots \quad - \frac{W}{2}\frac{(x-a)^3}{3} \qquad ...(v)$$

where C_2 is another constant of integration. This constant is written after C_1x. The integration of $(x - a)^2$ will be $\left(\frac{x-a}{3}\right)^3$. This type of integration is justified as the constant of integrations C_1 and C_2 are valid for all values of x.

The values of C_1 and C_2 are obtained from boundary conditions. The two boundary conditions are :

(*i*) At $x = 0$, $y = 0$ and (*ii*) At $x = L$, $y = 0$

(*i*) At A, $x = 0$ and $y = 0$. Substituting these values in equation (*v*) upto dotted line only, we get

$$0 = 0 + 0 + C_2$$

$$\therefore \quad C_2 = 0$$

(*ii*) At B, $x = L$ and $y = 0$. Substituting these values in equation (*v*), we get

$$0 = \frac{W.b}{2L}.\frac{L^3}{3} + C_1 \times L + 0 - \frac{W}{2}\frac{(L-a)^3}{3}$$

($\because$ $C_2 = 0$. Here complete Eq. (*v*) is to be taken)

$$= \frac{W.b.L^2}{6} + C_1 \times L - \frac{W}{2}\frac{b^3}{3} \qquad (\because \; L - a = b)$$

$$\therefore \quad C_1 \times L = \frac{W}{6} . b^3 - \frac{W.b.L^2}{6} = - \frac{W.b}{6}(L^2 - b^2)$$

$$\therefore \quad C_1 = - \frac{W.b}{6L}(L^2 - b^2) \qquad ...(vi)$$

Substituting the value of C_1 in equation (*iv*), we get

$$EI \frac{dy}{dx} = \frac{W.b}{L}\frac{x^2}{2} + \left[- \frac{W.b}{6L}\left(L^2 - b^2\right) \right] \; \vdots \; - \frac{W(x-a)^2}{2}$$

$$= \frac{W.b.x^2}{2L} - \frac{W.b}{6L}\left(L^2 - b^2\right) \; \vdots \; - \frac{W(x-a)^2}{2} \qquad ...(vii)$$

The equation (*vii*) gives the slope at any point in the beam. Slope is maximum at A or B. To find the slope at A, substitute $x = 0$ in the above equation upto dotted line as point A lies in AC.

$$\therefore \quad EI.\theta_A = \frac{W.b}{2L} \times 0 - \frac{Wb}{6L}(L^2 - b^2) \qquad \left(\because \; \frac{dy}{dx} \text{ at } A = \theta_A \right)$$

$$= - \frac{Wb}{6L}(L^2 - b^2)$$

$$\therefore \quad \theta_A = - \frac{Wb}{6EIL}(L^2 - b^2) \qquad \text{(as given before)}$$

Substituting the values of C_1 and C_2 in equation (*v*), we get

$$EI.y = \frac{W.b}{6L}.x^3 + \left[- \frac{Wb}{6L}(L^2 - b^2) \right] x + 0 \; \vdots \; - \frac{W}{6}(x-a)^3 \qquad ...(viii)$$

The equation (*viii*) gives the deflection at any point in the beam. To find the deflection y_c under the load, substitute $x = a$ in equation (*viii*) and consider the equation upto dotted line (as point C lies in AC). Hence, we get

$$EI.y_c = \frac{W.b}{6L}.a^3 - \frac{W.b}{6L}(L^2 - b^2)a = \frac{W.b}{6L}.a\,(a^2 - L^2 + b^2)$$

$$= - \frac{W.a.b}{6L}(L^2 - a^2 - b^2)$$

$$= - \frac{W.a.b}{6L}[(a+b)^2 - a^2 - b^2] \qquad (\because \; L = a + b)$$

$$= - \frac{W.a.b}{6L}[a^2 + b^2 + 2ab - a^2 - b^2]$$

$$= - \frac{W.a.b}{6L}[2ab] = - \frac{Wa^2.b^2}{3L}$$

$$\therefore \quad y_c = - \frac{Wa^2.b^2}{3EIL} \qquad \text{...(same as before)}$$

Note. While using Macaulay's Method, the section x is to be taken in the last portion of the beam.

4.7. SOLVED PROBLEMS BASED ON MACAULAY'S METHOD

Problem 4.7. *A beam of length 6 m is simply supported at its ends and carries a point load of 40 kN at a distance of 4 m from the left support. Find the deflection under the load and maximum deflection. Also calculate the point at which maximum deflection takes place. Given M.O.I. of beam = 7.33×10^7 mm⁴ and E = 2×10^5 N/mm².*

Sol. Given :

Length, $L = 6 \text{ m} = 6000 \text{ mm}$

Point load, $W = 40 \text{ kN} = 40{,}000 \text{ N}$

Distance of point load from left support, $a = 4 \text{ m} = 4000 \text{ mm}$

$\therefore$ $b = L - a = 6 - 4 = 2 \text{ m} = 2000 \text{ mm}$

Let y_c = Deflection under the load

y_{max} = Maximum deflection

Using equation $$y_c = -\frac{W \cdot a^2 \cdot b^2}{3EIL}$$

$$\therefore \quad y_c = -\frac{40000 \times 4000^2 \times 2000^2}{3 \times 2 \times 10^5 \times 7.33 \times 10^7 \times 6000}$$

$= -$ **9.7 mm. Ans.**

Problem 4.8. *A beam of length 6 m is simply supported at its ends and carries two point loads of 48 kN and 40 kN at a distance of 1 m and 3 m respectively from the left support. Find :*

(i) deflection under each load,

(ii) maximum deflection, and

(iii) the point at which maximum deflection occurs.

Given E = 2×10^5 N/mm² and I = 85×10^6 mm⁴.

Sol. Given :

$I = 85 \times 10^5 \text{ mm}^4 \; ; E = 2 \times 10^5 \text{ N/mm}^2$

First calculate the reactions R_A and R_B.

Taking moments about A, we get

$$R_B \times 6 = 48 \times 1 + 40 \times 3 = 168$$

$$\therefore \quad R_B = \frac{168}{6} = 28 \text{ kN}$$

$$\therefore \quad R_A = \text{Total load} - R_B = (48 + 40) - 28 = 60 \text{ kN}$$

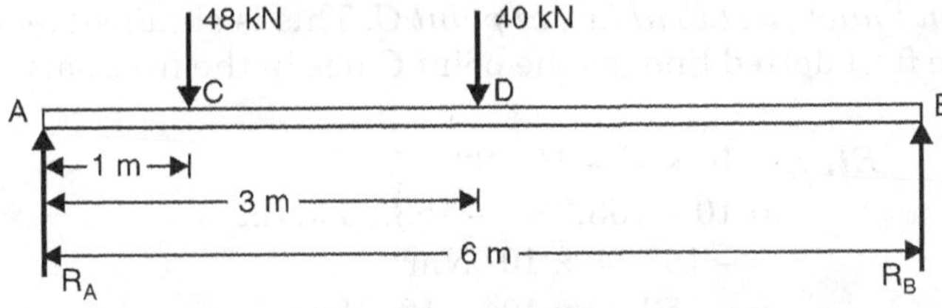

Fig. 4.6

Consider the section X in the last part of the beam (*i.e.,* in length DB) at a distance x from the left support A. The B.M. at this section is given by,

$$EI\frac{d^2y}{dx^2} = R_A.x \quad \vdots \quad -48(x-1) \quad \vdots \quad -40(x-3)$$

$$= 60x \quad \vdots \quad -48(x-1) \quad \vdots \quad -40(x-3)$$

Integrating the above equation, we get

$$EI\frac{dy}{dx} = \frac{60x^2}{2} + C_1 \quad \vdots \quad -48\frac{(x-1)^2}{2} \quad \vdots \quad -40\frac{(x-3)^2}{2}$$

$$= 30x^2 + C_1 \quad \vdots \quad -24(x-1)^2 \quad \vdots \quad -20(x-3)^2 \qquad ...(i)$$

Integrating the above equation again, we get

$$EI.y = \frac{30x^3}{3} + C_1x + C_2 \quad \vdots \quad \frac{-24(x-1)^3}{3} \quad \vdots \quad \frac{-20(x-3)^3}{3}$$

$$= 10x^3 + C_1x + C_2 \quad \vdots \quad -8(x-1)^3 \quad \vdots \quad -\frac{20}{3}(x-3)^3 \qquad ...(ii)$$

To find the values of C_1 and C_2, use two boundary conditions. The boundary conditions are :

(*i*) at $x = 0, y = 0$, and (*ii*) at $x = 6$ m, $y = 0$.

(*i*) Substituting the first boundary condition *i.e.,* at $x = 0$, $y = 0$ in equation (*ii*) and considering the equation upto first dotted line (as $x = 0$ lies in the first part of the beam), we get

$$0 = 0 + 0 + C_2 \quad \therefore \quad C_2 = 0$$

(*ii*) Substituting the second boundary condition *i.e.,* at $x = 6$ m, $y = 0$ in equation (*ii*) and considering the complete equation (as $x = 6$ lies in the last part of the beam), we get

$$0 = 10 \times 6^3 + C_1 \times 6 + 0 - 8(6-1)^3 - \frac{20}{3}(6-3)^3 \qquad (\because \quad C_2 = 0)$$

or

$$0 = 2160 + 6C_1 - 8 \times 5^3 - \frac{20}{3} \times 3^3$$

$$= 2160 + 6C_1 - 1000 - 180 = 980 + 6C_1$$

$$\therefore \qquad C_1 = \frac{-980}{6} = -163.33$$

Now, substituting the values of C_1 and C_2 in equation (*ii*), we get

$$EI.y = 10x^3 - 163.33x \quad \vdots \quad -8(x-1)^3 \quad \vdots \quad -\frac{20}{3}(x-3)^3 \qquad ...(iii)$$

(*i*) (*a*) *Deflection under first load i.e., at point C*. This is obtained by substituting $x = 1$ in equation (*iii*) upto the first dotted line (as the point C lies in the first part of the beam). Hence, we get

$$EI.\,y_c = 10 \times 1^3 - 163.33 \times 1$$

$$= 10 - 163.33 = -153.33 \text{ kNm}^3$$

$$= -153.33 \times 10^3 \text{ Nm}^3$$

$$= -153.33 \times 10^3 \times 10^9 \text{ Nmm}^3$$

$$= -153.33 \times 10^{12} \text{ Nmm}^3$$

$$\therefore \quad y_c = \frac{-153.33 \times 10^{12}}{EI} = \frac{-153.33 \times 10^{12}}{2 \times 10^5 \times 85 \times 10^6} \text{ mm}$$

$$= -\mathbf{9.019\ mm.} \quad \textbf{Ans.}$$

(Negative sign shows that deflection is downwards).

(*b*) *Deflection under second load i.e., at point D.* This is obtained by substituting $x = 3$ m in equation (*iii*) upto the second dotted line (as the point D lies in the second part of the beam). Hence, we get

$$EI.y_D = 10 \times 3^3 - 163.33 \times 3 - 8(3-1)^3$$

$$= 270 - 489.99 - 64 = -283.99 \text{ kNm}^3$$

$$= -283.99 \times 10^{12} \text{ Nmm}^3$$

$$\therefore \quad y_D = \frac{-283.99 \times 10^{12}}{2 \times 10^5 \times 85 \times 10^6} = -\mathbf{16.7\ mm.} \quad \textbf{Ans.}$$

(*ii*) and (*iii*) *Maximum Deflection.* The deflection is likely to be maximum at a section between C and D. For maximum deflection, $\frac{dy}{dx}$ should be zero. Hence, equate the equation (*i*) equal to zero upto the second dotted line.

$$\therefore \quad 30x^2 + C_1 - 24(x-1)^2 = 0$$

or $\quad 30x^2 - 163.33 - 24(x^2 + 1 - 2x) = 0 \qquad (\because \ C_1 = -163.33)$

or $\quad 6x^2 + 48x - 187.33 = 0$

The above equation is a quadratic equation. Hence its solution is

$$x = \frac{-48 \pm \sqrt{48^2 + 4 \times 6 \times 187.33}}{2 \times 6} = 2.87 \text{ m.}$$

(Neglecting –ve root)

Now substituting $x = 2.87$ m in equation (*iii*) upto the second dotted line, we get maximum deflection as

$$EI.y_{max} = 10 \times 2.87^3 - 163.33 \times 2.87 - 8(2.87-1)^3$$

$$= 236.39 - 468.75 - 52.31$$

$$= 284.67 \text{ kNm}^3 = -284.67 \times 10^{12} \text{ Nmm}^3$$

$$\therefore \quad y_{max} = \frac{-284.67 \times 10^{12}}{2 \times 10^5 \times 85 \times 10^6} = -\mathbf{16.745\ mm.} \quad \textbf{Ans.}$$

Problem 4.9. *A beam of length 8 m is simply supported at its ends. It carries a uniformly distributed load of 40 kN/m as shown in Fig. 4.7. Determine the deflection of the beam at its mid-point and also the position of maximum deflection and maximum deflection. Take* $E = 2 \times 10^5$ *N/mm*^{2} *and* $I = 4.3 \times 10^8$ *mm*^{4}.

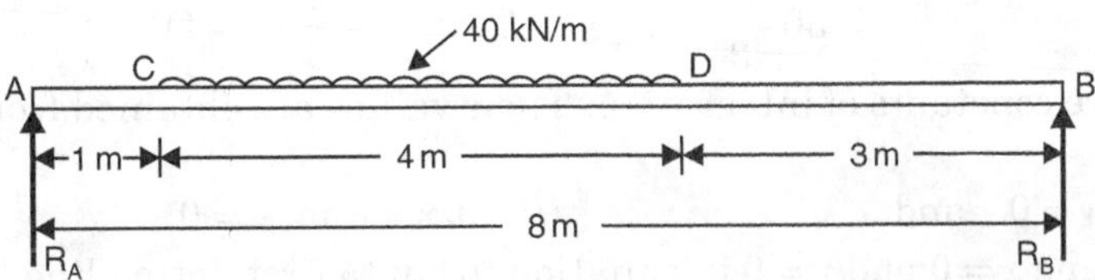

Fig. 4.7

Sol. Given :

Length, $L = 8$ m

U.d.l., $W = 40$ kN/m

Value of $E = 2 \times 10^5$ N/mm^2

Value of $I = 4.3 \times 10^8$ mm^4

First calculate the reactions R_A and R_B.

Taking moments about A, we get

$$R_B \times 8 = 40 \times 4 \times \left(1 + \frac{4}{2}\right) = 480 \text{ kN}$$

$$\therefore \quad R_B = \frac{480}{8} = 60 \text{ kN}$$

$$\therefore \quad R_A = \text{Total load} - R_B = 40 \times 4 - 60 = 100 \text{ kN}$$

In order to obtain the general expression for the bending moment at a distance x from the left end A, which will apply for all values of x, it is necessary to extend the uniformly distributed load upto the support B, compensating with an equal upward load of 40 kN/m over the span DB as shown in Fig. 4.8. Now, Macaulay's method can be applied.

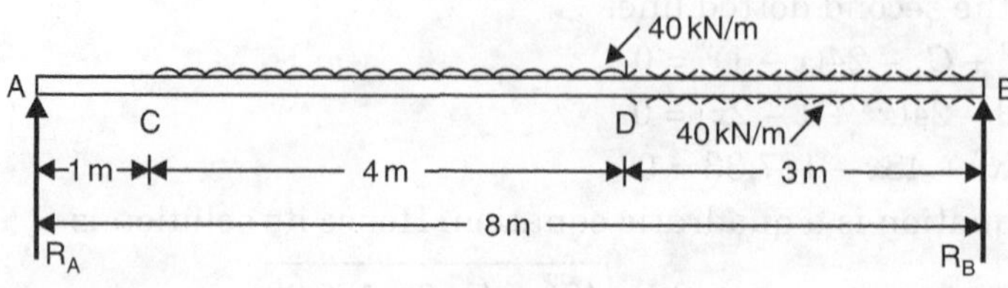

Fig. 4.8

The B.M. at any section at a distance x from end A is given by,

$$EI\frac{d^2y}{dx^2} = R_A.x \quad \vdots \quad -40(x-1) \times \frac{(x-1)}{2} \quad \vdots \quad +40 \times (x-5) \times \frac{(x-5)}{2}$$

or

$$EI\frac{d^2y}{dx^2} = 100x \quad \vdots \quad -20(x-1)^2 \quad \vdots \quad +20(x-5)^2$$

Integrating the above equation, we get

$$EI\frac{dy}{dx} = \frac{100x^2}{2} + C_1 \quad \vdots \quad -\frac{20(x-1)^3}{3} \quad \vdots \quad +20\frac{(x-5)^3}{3} \qquad ...(i)$$

Integrating again, we get

$$EI.y = 50\frac{x^3}{3} + C_1x + C_2 \quad \vdots \quad -\frac{20}{3}\frac{(n-1)^4}{4} \quad \vdots \quad +\frac{20}{3}\frac{(x-5)^4}{3}$$

$$= 50\frac{x^3}{3} + C_1x + C_2 \quad \vdots \quad -\frac{5}{3}(x-1)^4 \quad \vdots \quad +\frac{5}{3}(x-5)^4 \qquad ...(ii)$$

where C_1 and C_2 are constants of integration. Their values are obtained from boundary conditions which are :

(*i*) at $x = 0, y = 0$ and (*ii*) at $x = 8$ m, $y = 0$

(*i*) Substituting $x = 0$ and $y = 0$ in equation (*ii*) upto first dotted line (as $x = 0$ lies in the first part AC of the beam), we get

$$0 = 0 + C_1 \times 0 + C_2 \qquad \therefore C_2 = 0$$

(*ii*) Substituting $x = 8$ and $y = 0$ in complete equation (*ii*) (as point $x = 8$ lies in the last part *DB* of the beam), we get

$$0 = \frac{50}{3} \times 8^3 + C_1 \times 8 + 0 - \frac{5}{3}(8-1)^4 + \frac{5}{3}(8-5)^4 \qquad (\because \; C_2 = 0)$$

$$= 8533.33 + 8C_1 - 4001.66 + 135$$

or $$8C_1 = -4666.67$$

or $$C_1 = \frac{-4666.67}{8} = -583.33$$

Substituting the value of C_1 and C_2 in equation (*ii*), we get

$$EI.y = \frac{50}{3}x^3 - 583.33x \; \vdots \; -\frac{5}{3}(x-1)^4 \; \vdots \; +\frac{5}{3}(x-4)^4 \qquad \text{...(iii)}$$

(*a*) *Deflection at the centre*

By substituting $x = 4$ m in equation (*iii*) upto second dotted line, we get the deflection at the centre. [The point $x = 4$ lies in the second part (*i.e., CD*) of the beam].

$$\therefore \quad EI.y = \frac{50}{3} \times 4^3 - 583.33 \times 4 - \frac{5}{3}(4-1)^4$$

$$= 1066.66 - 2333.32 - 135 = -1401.66 \text{ kNm}^3$$

$$= -1401.66 \times 1000 \text{ Nm}^3$$

$$= -1401.66 \times 1000 \times 10^9 \text{ Nmm}^3$$

$$= -1401.66 \times 10^{12} \text{ Nmm}^3$$

$$\therefore \quad y = \frac{-1401.66 \times 10^{12}}{EI} = \frac{-1401.66 \times 10^{12}}{2 \times 10^5 \times 4.5 \times 10^8}$$

$$= \mathbf{-16.29 \text{ mm downward.}} \quad \textbf{Ans.}$$

(*b*) *Position of maximum deflection*

The maximum deflection is likely to lie between *C* and *D*. For maximum deflection the slope $\frac{dy}{dx}$ should be zero. Hence, equating the slope given by equation (*i*) upto second dotted line to zero, we get

$$0 = 100\frac{x^2}{2} + C_1 - \frac{20}{3}(x-1)^3$$

$$0 = 50x^2 - 583.33 - 6.667(x-1)^3 \qquad \text{...(iv)}$$

The above equation is solved by trial and error method.

Let $x = 1$, then R.H.S. of equation (*iv*)

$$= 50 - 583.33 - 6.667 \times 0 = -533.33$$

Let $x = 2$, then R.H.S. $= 50 \times 4 - 583.33 - 6.667 \times 1 = -390.00$

Let $x = 3$, then R.H.S. $= 50 \times 9 - 583.33 - 6.667 \times 8 = -136.69$

Let $x = 4$, then R.H.S. $= 50 \times 16 - 583.33 - 6.667 \times 27 = +36.58$

In equation (*iv*), when $x = 3$ then R.H.S. is negative but when $x = 4$ then R.H.S. is positive. Hence exact value of x lies between 3 and 4.

Let $x = 3.82$, then R.H.S. $= 50 \times 3.82 - 583.33 - 6.667(3.82-1)^3$

$$= 729.63 - 583.33 - 149.51 = -3.22$$

Let $x = 3.83$, then R.H.S. $= 50 \times 3.83^2 - 583.33 = 6.667(3.83-1)^3$

$$= 733.445 - 583.33 - 151.1 = -0.99$$

The R.H.S. is approximately zero in comparison to the three terms (*i.e.*, 733.445, 583.33 and 151.1).

$\therefore$ Value of x = **3.83. Ans.**

Hence, maximum deflection will be at a distance of 3.83 m from support A.

(*c*) *Maximum deflection*

Substituting x = 3.83 m in equation (*iii*) upto second dotted line, we get the maximum deflection [the point x = 3.83 lies in the second part *i.e.*, *CD* of the beam.]

$$\therefore \quad EI.y_{max} = \frac{50}{3} \times 3.83^3 - 583.33 \times 3.83 - \frac{5}{3}(3.83-1)^4$$

$$= 936.36 - 2234.15 - 106.9 = -1404.69 \text{ kNm}^3$$

$$= -1404.69 \times 10^{12} \text{ Nmm}^3$$

$$\therefore \quad y_{max} = \frac{-1404.69 \times 10^{12}}{2 \times 10^5 \times 4.3 \times 10^8} = \mathbf{-16.33 \text{ mm}.} \quad \textbf{Ans.}$$

Problem 4.10. *An overhanging beam ABC is loaded as shown in Fig. 4.9. Find the slopes over each support and at the right end. Find also the maximum upward deflection between the supports and the deflection at the right end.*

Take $E = 2 \times 10^5$ *N/mm*^{2} *and* $I = 5 \times 10^8$ *mm*^{4}.

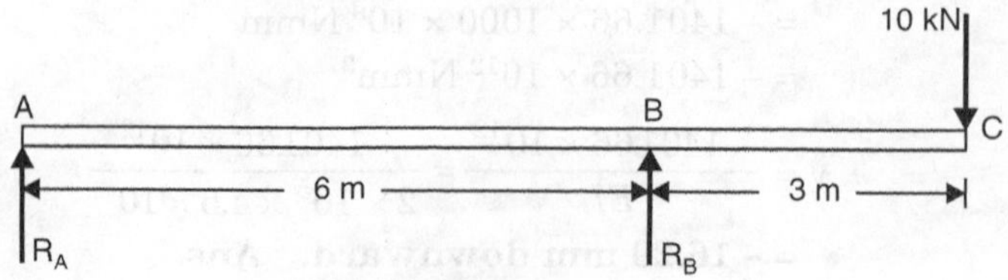

Fig. 4.9

Sol. Given :

Point load, W = 10 kN

Value of $E = 2 \times 10^5$ N/mm^2

Value of $I = 5 \times 10^8$ mm^4

First calculate the reaction R_A and R_B.

Taking moments about A, we get

$$R_B \times 6 = 10 \times 9$$

$$\therefore \quad R_B = \frac{10 \times 9}{6} = 15 \text{ kN}$$

$$\therefore \quad R_A = \text{Total load} - R_B = 10 - 15 = -5 \text{ kN}$$

Hence the reaction R_A will be in the downward direction. Hence Fig. 4.10 will be modified as shown in Fig. 4.10. Now, write down an expression for the B.M. in the last section of the beam.

The B.M. at any section at a distance x from the support A is given by,

$$EI\frac{d^2y}{dx^2} = -R_A \times x \;\vdots\; + R_B \times (x-6)$$

$$= -5x \;\vdots\; + 15(x-6) \qquad (\because \; R_A = 5)$$

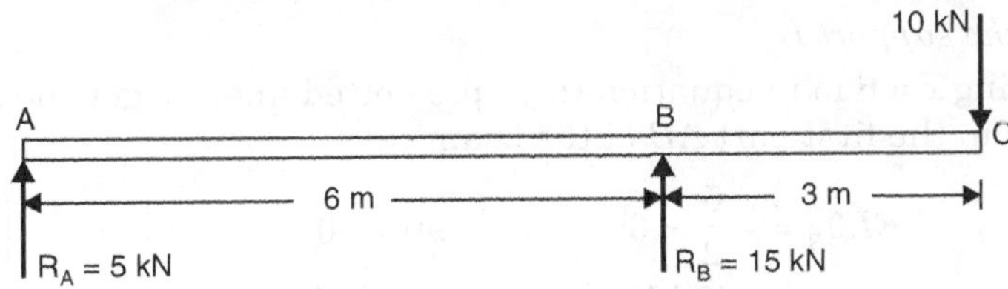

Fig. 4.10

Integrating the above equation, we get

$$EI\frac{dy}{dx} = \frac{-5x^2}{2} + C_1 \;\vdots\; + \frac{15(x-6)^2}{2} \quad ...(i)$$

Integrating again, we get

$$EI.y = -\frac{5}{2}\frac{x^3}{3} + C_1x + C_2 \;\vdots\; + \frac{15}{2}\frac{(x-6)^3}{3}$$

$$= -\frac{5}{6}x^3 + C_1x + C_2 \;\vdots\; + \frac{5}{2}(x-6)^3 \quad ...(ii)$$

where C_1 and C_2 are constant of integration. Their values are obtained from boundary conditions which are :

(*i*) at $x = 0$, $y = 0$ and (*ii*) at $x = 6$ m, $y = 0$.

(*i*) Substituting $x = 0$ and $y = 0$ in equation (*ii*) upto dotted line (as $x = 0$ lies in the first part *AB* of the beam), we get

$$0 = 0 + C_1 \times 0 + C_2 \qquad \therefore C_2 = 0$$

(*ii*) Substituting $x = 6$ m and $y = 0$ in equation (*ii*) upto dotted line (as $x = 6$ lies in the first part *AB* of the beam), we get

$$0 = \frac{-5}{6} \times 6^3 + C_1 \times 6 + 0 \qquad (\because C_2 = 0)$$

$$\therefore \quad = -5 \times 36 + 6C_1$$

$$\therefore \quad C_1 = \frac{+5 \times 36}{6} = 30$$

Substituting the values of C_1 and C_2 in equations (*i*) and (*ii*), we get

$$EI\frac{dy}{dx} = -\frac{5}{2}x^2 + 30 \;\vdots\; + \frac{15}{2}(x-6)^2 \quad ...(iii)$$

and

$$EI.y = -\frac{5}{6}x^3 + 30x \;\vdots\; + \frac{5}{2}(x-6)^3 \quad ...(iv)$$

(*a*) *Slope over the support A*

By substituting $x = 0$ in equation (*iii*) upto dotted line, we get the slope at support *A* (the point $x = 0$ lies in the first part *AB* of the beam).

$$\therefore \quad EI.\theta_A = -\frac{5}{2} \times 0 + 30 = 30 \text{ kNm}^2 = 30 \times 1000 \text{ Nm}^2 \left(\because \frac{dy}{dx} \text{ at } A = \theta_A\right)$$

$$= 30 \times 1000 \times 10^6 \text{ Nmm}^2 = 30 \times 10^9 \text{ Nmm}^2$$

$$\therefore \quad \theta_A = \frac{30 \times 10^9}{E \times I} = \frac{30 \times 10^9}{2 \times 10^5 \times 5 \times 10^8}$$

$$= \mathbf{0.0003 \text{ radians.}} \quad \textbf{Ans.}$$

(*b*) *Slope at the support B*

By substituting $x = 6$ m in equation (*iii*) upto dotted line, we get the slope at support B (the point $x = 6$ lies in the first part AB of the beam).

$$EI.\theta_B = -\frac{5}{2} \times 6^2 + 30 = -90 + 30 \qquad \left(\because \frac{dy}{dx} \text{ at } B = \theta_B\right)$$

$$= -60 \text{ kNm}^2 = -60 \times 10^9 \text{ Nmm}^2$$

$$\therefore \qquad \theta_B = \frac{-60 \times 10^9}{E \times I} = \frac{-60 \times 10^9}{2 \times 10^5 \times 5 \times 10^8}$$

$$= \mathbf{-0.0006 \text{ radians.}} \quad \mathbf{Ans.}$$

(*c*) *Slope at the right end i.e., at C*

By substituting $x = 9$ m in equation (*iii*), we get the slope at C. In this case, complete equation is to be taken as point $x = 9$ m lies in the last part of the beam.

$$\therefore \qquad EI.\theta_C = -\frac{5}{2} \times 9^2 + 30 + \frac{15}{2}(9-6)^2 \qquad \left(\because \frac{dy}{dx} \text{ at } C = \theta_C\right)$$

$$= -202.5 + 30 + 67.5 = -105 \text{ kNm}^2$$

$$= -105 \times 10^9 \text{ Nmm}^2$$

$$\therefore \qquad \theta_C = \frac{-105 \times 10^9}{E \times I} = \frac{-105 \times 10^9}{2 \times 10^5 \times 5 \times 10^8}$$

$$= \mathbf{-0.00105 \text{ radians.}} \quad \mathbf{Ans.}$$

(*d*) *Maximum upward deflection between the supports*

For maximum deflection between the supports, $\frac{dy}{dx}$ should be zero. Hence, equating the slope given by the equation (*iii*) to be zero upto dotted line, we get

$$0 = -\frac{5}{2}x^2 + 30 = -5x^2 + 60$$

or $$5x^2 = 60 \qquad \text{or} \quad x = \sqrt{\frac{60}{5}} = \sqrt{12} = 3.464 \text{ m}$$

Now, substituting $x = 3.464$ m in equation (*iv*) upto dotted line, we get the maximum deflection as

$$EI.y_{max} = -\frac{5}{6} \times 3.464^3 + 30 \times 3.464$$

$$= -34.638 + 103.92 = 69.282 \text{ kNm}^3$$

$$= 69.282 \times 1000 \times 10^9 \text{ Nmm}^3 = 69.282 \times 10^{12} \text{ mm}^3$$

$$\therefore \qquad y_{max} = \frac{69.282 \times 10^{12}}{2 \times 10^5 \times 5 \times 10^8}$$

$$= \mathbf{0.6928 \text{ mm (upward).}} \quad \mathbf{Ans.}$$

(*e*) *Deflection at the right end i.e., at point C*

By substituting $x = 9$ m in equation (*iv*), we get the deflection at point C. Here complete equation is to be taken as point $x = 9$ m lies in the last part of the beam.

$$\therefore \quad EI.y_C = -\frac{5}{6} \times 9^3 + 30 \times 9 + \frac{5}{2}(9-6)^3$$

$$= -607.5 + 270 + 67.5$$

$$= -270 \text{ kNm}^3 = -270 \times 10^{12} \text{ Nmm}^3$$

$$\therefore \quad y_C = \frac{-270 \times 10^{12}}{2 \times 10^5 \times 5 \times 10^8}$$

$$= \mathbf{-2.7 \text{ mm (downwards).}} \quad \textbf{Ans.}$$

Problem 4.11. *A beam ABC of length 9 m has one support of the left end and the other support at a distance of 6 m from the left end. The beam carries a point load of 1 kN at right end and also carries a uniformly distributed load of 4 kN/m over a length of 3 m as shown in Fig. 4.11. Determine the slope and deflection at point C.*

Take $E = 2 \times 10^5$ *N/mm*^{2} *and* $I = 5 \times 10^8$ *mm*^{4}*.*

Sol. Given :

Point load, $W = 12$ kN

U.d.l., $w = 4$ kN/m

Value of $E = 2 \times 10^5$ N/mm^2

Value of $I = 5 \times 10^8$ mm^4

First calculate the reactions R_A and R_B.

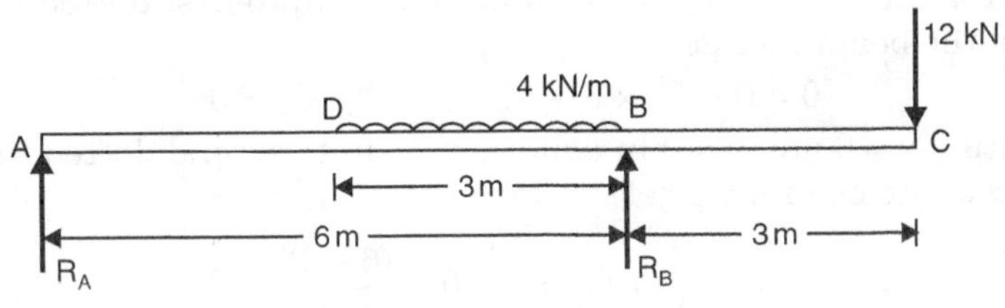

Fig. 4.11

Taking moments about A, we get

$$R_B \times 6 = 4 \times 3 \times \left(3 + \frac{3}{2}\right) + 12 \times 9$$

$$= 54 + 108 = 162$$

$$\therefore \quad R_B = \frac{162}{6} = 27 \text{ kN } (\uparrow)$$

and $\quad R_A = \text{Total load} - R_B = 24 - 27 = -3 \text{ kN } (\downarrow)$

Negative sign shows that R_A will be acting downwards. In order to obtain general expression for the bending moment at a distance x from the left end A, which will apply for all values of x, it is necessary to extend the uniformly distributed load upto point C, compensating with an equal upward load of 4 kN/m over the span BC as shown in Fig. 4.12. Now Macaulay's method can be applied.

The B.M. at any section at a distance x from the support A is given by,

$$EI\frac{d^2y}{dx^2} = -R_A \times x \;\;\vdots\;\; -4(x-3)\frac{(x-3)}{2} \;\;\vdots\;\; +R_B(x-6) \;\;\vdots\;\; +4(x-6)\frac{(x-6)}{2}$$

$$= -3x \;\;\vdots\;\; -2(x-3)^2 \;\;\vdots\;\; +27(x-6) \;\;\vdots\;\; +2(x-6)^2$$

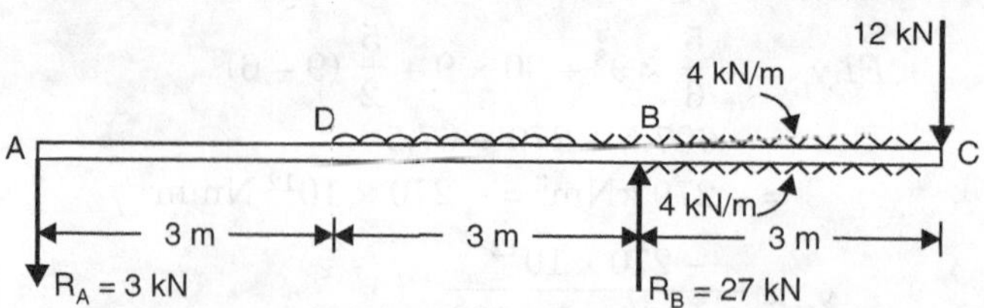

Fig. 4.12

Integrating the above equation, we get

$$EI\,\frac{dy}{dx} = -\frac{3x^2}{2} + C_1 \;\vdots\; -\frac{2(x-3)^3}{3} \;\vdots\; + \frac{27(x-6)^2}{2} + \frac{2(x-6)^3}{3} \quad \text{...}(i)$$

Integrating again, we get

$$EI.y = -\frac{3}{2}\frac{x^3}{3} + C_1x + C_2 \;\vdots\; -\frac{2}{3}\frac{(x-3)^4}{4} \;\vdots\; + \frac{27}{2}\frac{(x-6)^3}{3} \;\vdots\; + \frac{2}{3}\frac{(x-6)^4}{4}$$

or
$$EI.y = -\frac{x^3}{2} + C_1x + C_2 \;\vdots\; -\frac{(x-3)^4}{6} \;\vdots\; + \frac{9}{2}(x-6)^3 \;\vdots\; + \frac{1}{6}(x-6)^4 \quad \text{..}(ii)$$

where C_1 and C_2 are constant of integration. Their values are obtained from boundary conditions which are :

(*i*) at $x = 0$, $y = 0$ and (*ii*) at $x = 6$ m, $y = 0$.

(*i*) Substituting the $x = 0$ and $y = 0$ in equation (*ii*) upto first dotted line (as $x = 0$ lies in the first part *AD* of the beam), we get

$$0 = 0 + C_1 \times 0 + C_2 \qquad \therefore C_2 = 0$$

(*ii*) Substituting $x = 6$ and $y = 0$ in equation (*ii*) upto second dotted line (as $x = 6$ lies in the second part *DB* of the beam), we get

$$0 = -\frac{6^3}{2} + C_1 \times 6 + 0 - \frac{(6-3)^4}{6}$$
$$= -108 + 6C_1 - 13.5 = -121.5 + 6C_1$$

or
$$C_1 = \frac{121.5}{6} = 20.25$$

Substituting the values of C_1 and C_2 in equations (*i*) and (*ii*), we get

$$EI\,\frac{dy}{dx} = -\frac{3}{2}x^2 + 20.25 \;\vdots\; -\frac{2(x-3)^3}{3} \;\vdots\; + \frac{27}{2}(x-6)^2 \;\vdots\; + \frac{2}{3}(x-6)^3 \quad \text{...}(iii)$$

and
$$EI.y = -\frac{x^3}{2} + 20.2 \times x \;\vdots\; -\frac{1}{6}(x-3)^4 \;\vdots\; + \frac{9}{2}(x-6)^3 \;\vdots\; + \frac{1}{6}(x-6)^4 \quad \text{...}(iv)$$

(*a*) *Slope at the point C*

By substituting $x = 9$ m in equation (*iii*), we get the slope at *C*. Here, complete equation is to be taken as point $x = 9$ m lies in the last part of the beam.

$$\therefore \qquad EI.\theta_C = -\frac{3}{2} \times 9^2 + 20.25 - \frac{2}{3}(9-3)^3 + \frac{27}{2}(9-6)^2 + \frac{2}{3}(9-6)^3$$

$$\left(\because \frac{dy}{dx} \text{ at } C = \theta_C\right)$$

$$= -121.5 + 20.25 - 144 + 121.5 + 18 = -105.75 \text{ kNm}^2$$
$$= -105.75 \times 10^3 \times 10^6 \text{ Nmm}^2 = -105.75 \times 10^9 \text{ Nmm}^2$$

$$\therefore \quad \theta_C = -\frac{105.75 \times 10^9}{2 \times 10^5 \times 5 \times 10^8} = \mathbf{-0.0010575 \text{ radians.}} \quad \textbf{Ans.}$$

(b) Deflection at the point C

By substituting $x = 9$ m in complete equation (*iv*), we get the deflection at C.

$$\therefore \quad EI \times y_C = -\frac{9^3}{2} + 20.25 \times 9 - \frac{1}{6}(9-3)^4 + \frac{9}{2}(9-6)^3 + \frac{1}{6}(9-6)^4$$
$$= -364.5 + 182.25 - 216 + 121.5 + 13.5$$
$$= -263.25 \text{ kNm}^3 = -263.25 \times 10^{12} \text{ Nmm}^3$$

$$\therefore \quad y_C = -\frac{263.25 \times 10^{12}}{2 \times 10^5 \times 5 \times 10^8} = \mathbf{-2.6325 \text{ mm.}} \quad \textbf{Ans.}$$

Problem 4.12. *A horizontal beam AB is simply supported at A and B, 6 m apart. The beam is subjected to a clockwise couple of 300 kNm at a distance of 4 m from the left end as shown in Fig. 4.13. If* $E = 2 \times 10^5$ *N/mm*^{2} *and* $I = 2 \times 10^8$ *mm*^{4}*, determine :*

(*i*) *deflection at the point where couple is acting and*

(*ii*) *the maximum deflection.*

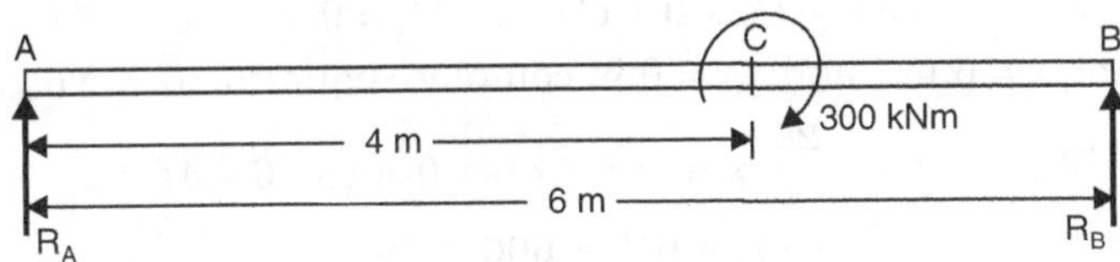

Fig. 4.13

Sol. Given :

Length, $L = 6$ m

Couple $= 300$ kNm

Value of $E = 2 \times 10^5$ N/mm^2

Value of $I = 2 \times 10^8$ mm^4

First calculate the reactions R_A and R_B.

Taking moments about A, we get

$$R_B \times 6 = 300$$

$$\therefore \quad R_B = \frac{300}{6} = 50 \text{ kN } (\uparrow)$$

and $\quad R_A = \text{Total load} - R_B = 0 - 50 \text{ kN} \quad (\because$ There is no load on beam)

$$= -50 \text{ kN}$$

Negative sign shows that R_A is acting downwards as shown in Fig. 4.14.

The B.M. at any section at a distance x from A, is given by

$$EI \frac{d^2y}{dx^2} = -50x \quad \vdots \quad + 300$$

$$= -50x \quad \vdots \quad + 300(x-4)^0$$

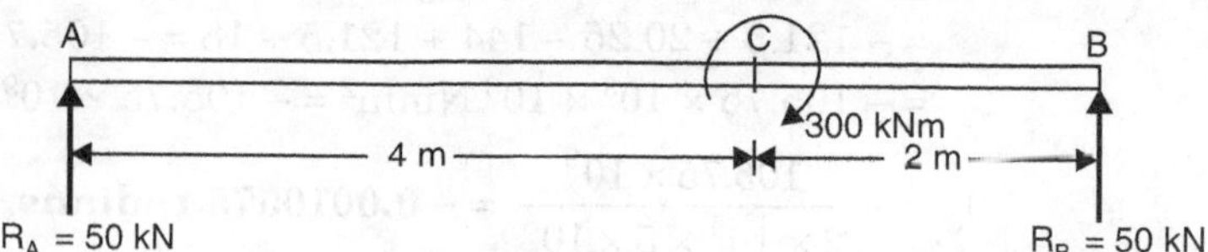

Fig. 4.14

Integrating the above equation, we get

$$EI\frac{dy}{dx} = -\frac{-50x^2}{2} + C_1 \quad \vdots \quad + 300(x-4) \qquad \text{...}(i)$$

Integrating again, we get

$$EIy = -\frac{50}{2} \times \frac{x^3}{3} + C_1x + C_2 \quad \vdots \quad + \frac{300\,(x-4)^2}{2}$$

$$= -\frac{25}{3}x^3 + C_1x + C_2 \quad \vdots \quad + 150(x-4)^2 \qquad \text{...}(ii)$$

where C_1 and C_2 are constants of integration. Their values are obtained from boundary conditions which are :

(*i*) at $x = 0, y = 0$ and (*ii*) at $x = 6$ m and $y = 0$.

(*i*) Substituting $x = 0$ and $y = 0$ in equation (*ii*) upto dotted line, we get

$$0 = 0 + C_1 \times 0 + C_2 \quad \therefore \quad C_2 = 0$$

(*ii*) Substituting $x = 6$ m and $y = 0$ in complete equation (*ii*), we get

$$0 = -\frac{25}{3} \times 6^3 + C_1 \times 6 + 0 + 150(6-4)^2$$

$$= -1800 + 6C_1 + 600$$

$$\therefore \quad C_1 = \frac{1800 - 600}{6} = 200$$

Substituting the values of C_1 and C_2 in equation (*ii*), we get

$$EI.y = -\frac{25}{3}x^3 + 200x \quad \vdots \quad + 150(x-4)^2 \qquad (\because C_2 = 0) \text{ ...}(iii)$$

(*i*) *Deflection at C (i.e., y_C)*

By substituting $x = 4$ in equation (*iii*) upto dotted line, we get the deflection at *C*.

$$\therefore \quad EI.y_C = -\frac{25}{3} \times 4^3 + 200 \times 4$$

$$= -533.33 + 800 = +266.67 \text{ kNm}^3$$

$$= 266.67 \times 10^{12} \text{ Nmm}^3$$

$$\therefore \quad y_C = \frac{266.67 \times 10^{12}}{2 \times 10^5 \times 2 \times 10^8} = \textbf{6.66 mm upwards.} \quad \textbf{Ans.}$$

(*ii*) *Maximum deflection*

First find the point where maximum deflection takes place. The maximum deflection is likely to occur in the larger segment *AC* of the beam. For maximum deflection $\frac{dy}{dx}$ should be zero. Hence equating the slope given by equation (*i*) upto dotted line to zero, we get

$$-\frac{50}{2}x^2 + 200 = 0 \qquad (\because \; C_1 = 200)$$

or $$-25x^2 + 200 = 0$$

or $$x = \sqrt{\frac{200}{25}} = 2 \times \sqrt{2} \text{ m}$$

Now, substituting $x = 2 \times \sqrt{2}$ in equation (*iii*) upto dotted line, we get the maximum deflection.

$$\therefore \quad EI \times y_{max} = -\frac{25}{3} \times (2 \times \sqrt{2})^3 + 200(2 \times \sqrt{2})$$

$$= -188.56 + 565.68$$

$$= 377.12 \text{ kNm}^3 = 377.12 \times 10^{12} \text{ Nmm}^3$$

$$\therefore \quad y_{max} = \frac{377.12 \times 10^{12}}{2 \times 10^5 \times 2 \times 10^8} = \textbf{9.428 mm upwards.} \quad \textbf{Ans.}$$

4.8. MOMENT AREA METHOD

Fig. 4.15 shows a beam *AB* carrying some type of loading, and hence subjected to bending moment as shown in Fig. 4.15 (*a*). Let the beam bent into AQ_1P_1B as shown in Fig. 4.15 (*b*) due to the load acting on the beam. Let *A* be a point of zero slope and zero deflection.

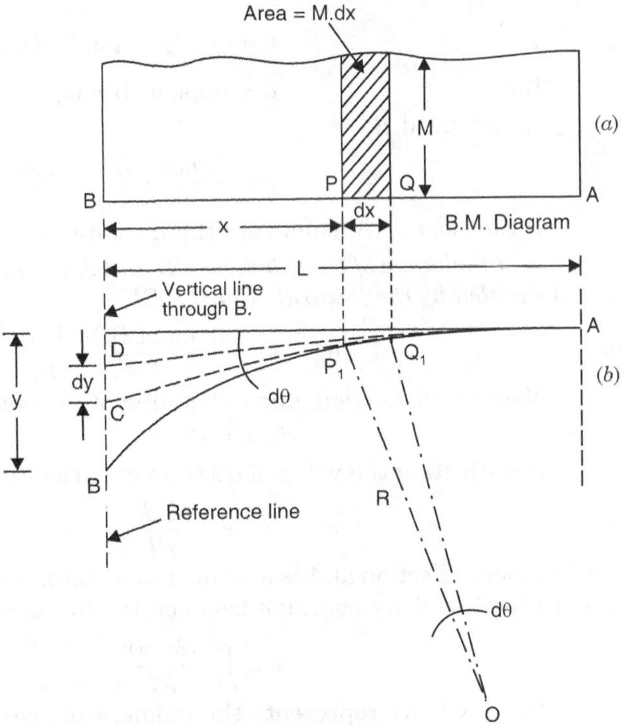

Consider an element *PQ* of small length *dx* at a distance *x* from *B*. The corresponding points on the deflected beam are P_1Q_1 as shown in Fig. 4.15 (*b*).

Let R = Radius of curvature of deflected part P_1Q_1

$d\theta$ = Angle subtended by the arc P_1Q_1 at the centre O

M = Bending moment between *P* and *Q*

P_1C = Tangent at point P_1

Q_1D = Tangent at point Q_1.

Fig. 4.15

The tangent at P_1 and Q_1 are cutting the vertical line through *B* at points *C* and *D*. The vertical line through point *B* which is not having zero slope is known as reference line. The angle between the normals at P_1 and Q_1 will be equal to the angle between the tangents at P_1 and Q_1. Hence the angle between the lines CP_1 and DQ_1 will be equal to $d\theta$.

For the deflected part P_1Q_1 of the beam, we have

$$P_1Q_1 = R.d\theta$$

But, $P_1Q_1 \approx dx \quad P_1Q_1 = ds \simeq dx$

$\therefore$ $dx = R.d\theta$

$$\therefore \quad d\theta = \frac{dx}{R} \qquad \text{...}(i)$$

But, for a loaded beam, we have

$$\frac{M}{I} = \frac{E}{R} \text{ or } R = \frac{EI}{M}$$

Substituting the values of R in equation (i), we get

$$d\theta = \frac{dx}{\left(\frac{EI}{M}\right)} = \frac{M\,dx}{EI} \qquad \text{...}(ii)$$

Since the slope at point A is assumed zero, hence slope at B is obtained by integrating the above equation between the limits 0 and L.

$$\therefore \quad \theta = \int_0^L \frac{M.dx}{EI} = \frac{1}{EI}\int_0^L M.dx$$

But $M.dx$ represents the area of B. M. diagram of length dx. Hence $\int_0^L M.dx$ represents the area of B. M. diagram between A and B.

$$\therefore \quad \theta = \frac{1}{EI} \text{ [Area of B. M. diagram between } A \text{ and } B]$$

But θ = slope at $B = \theta_B$

$\therefore$ Slope at B,

$$\theta_B = \frac{\text{Area of } B.M. \text{ diagram between } A \text{ and } B}{EI} \qquad \text{...(4.11)}$$

If the slope at A is not zero then, we have

"Total change of slope between B and A is equal to the area of B. M. diagram between B and A divided by the flexural rigidity EI"

or
$$\theta_B - \theta_A = \frac{\text{Area of B.M. between } A \text{ and } B}{EI} \qquad \text{...(4.12)}$$

Now, the deflection, due to bending of the portion P_1Q_1 is given by

$$dy = x.d\theta$$

Substituting the value of $d\theta$ from equation (ii), we get

$$dy = x \,.\, \frac{M.dx}{EI} \qquad \text{...}(iii)$$

Since deflection at A is assumed to be zero, hence the total deflection at B is obtained by integrating the above equation between the limits zero and L.

$$\therefore \quad y = \int_0^L \frac{xM.dx}{EI} = \frac{1}{EI}\int_0^L xM.dx$$

But $x \times M.dx$ represents the moment of area of the B.M. diagram of length dx about point B. *i.e.,* the point where slope is not zero.

Hence $\int_0^L xM.dx$ represents the moment of area of the B.M. diagram between B and A about B. This is equal to the total area of B.M. diagram between B and A multiplied by the distance of the C.G. of the area of B.M. diagram from B.

$$\therefore \qquad y = \frac{1}{EI} \times A \times \bar{x} = \frac{A\bar{x}}{EI} \qquad ...(4.13)$$

where, A = Area of B.M. diagram between A and B

$\bar{x}$ = Distance of C.G. of the area A from B *i.e.,* the point at which the slope is not zero.

4.8.1. Mohr's Theorems. The results given by equation (4.11) for slope and (4.13) for deflection are known as Mohr's theorems. They are state as :

I. The change of slope between any two points is equal to the net area of the B.M. diagram between these points divided by EI.

II. The total deflection between any two points is equal to the moment of the area of B.M. diagram between the two points about the point (at which slope is not zero. *i.e.,* B) divided by EI.

The Mohr's theorems is conveniently used for following cases :

1. Problems on cantilevers (zero slope at fixed end).
2. Simply supported beams carrying symmetrical loading (zero slope at the centre).
3. Beams fixed at both ends (zero slope at each end).

The B.M. diagram is a parabola for uniformly distributed loads. The following properties of area and centroids or parabola are given as :

Let $BC = d$

$AB = b$

In Fig. 4.16, ABC is a parabola and $ABCD$ is a surrounding rectangle.

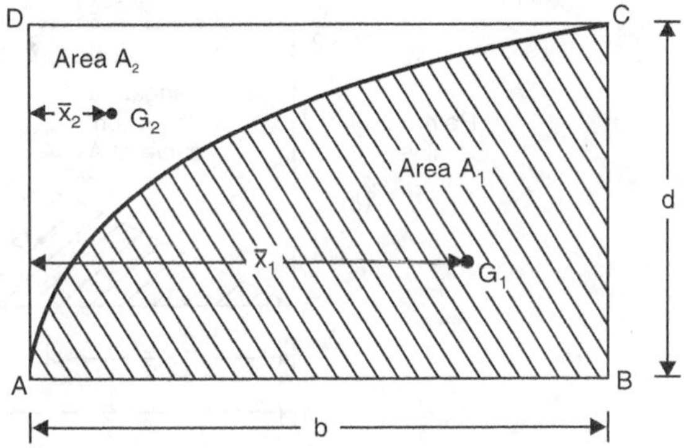

Let A_1 = Area of ABC

$\bar{x}_1$ = Distance of C.G. of A_1 from AD

A_2 = Area of ACD

$\bar{x}_2$ = Distance of C.G. of A_2 from AD

G_1 = C.G. of area A_1

G_2 = C.G. of area A_2.

Fig. 4.16

Then,

$$A_1 = \text{Area of parabola } ABC = \frac{2}{3}bd$$

$$A_2 = \text{Area } ACD = \text{Area } ABCD - \text{Area } ABC = b \times d - \frac{2}{3}bd = \frac{1}{3}bd$$

$$\bar{x}_1 = \frac{5}{8}b$$

$$\bar{x}_2 = \frac{1}{4}b.$$

4.8.2. Slope and Deflection of a Simply Supported Beam Carrying a Point Load at the Centre by Mohr's Theorem. Fig. 4.17 (*a*) shows a simply supported *AB* of length *L* and carrying a point load *W* at the centre of the beam *i.e.*, at point *C*. The B.M. diagram is shown in Fig. 4.17 (*b*). This is a case of symmetrical loading, hence slope is zero at the centre *i.e.*, at point *C*.

But the deflection is maximum at the centre.

For drawing B.M. diagram,

$$R_A = R_B = \frac{W}{2}$$

B.M. at *A* and *C* are :

$$M_A = 0,$$

$$M_C = R_A \times \frac{L}{2} = \frac{W}{2} \times \frac{L}{2} = \frac{WL}{4}$$

Take $CD = \frac{W \times L}{4}$. Join *A* to *D* and *D* to *B*. We get B.M. diagram

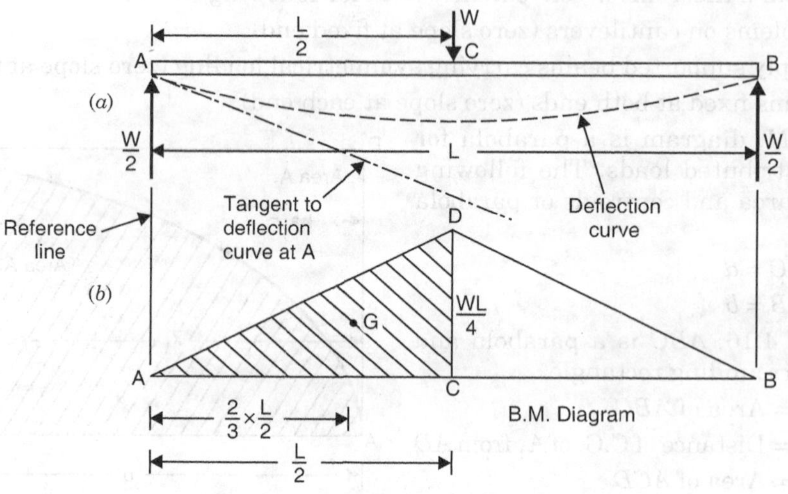

Fig. 4.17

(*i*) *Slope at A* (θ_A)

Now using Mohr's theorem for slope, we get

Slope at *A*, $\theta_A = \frac{\text{Area of B.M. diagram between } A \text{ and } C}{EI}$

But area of B.M. diagram between *A* and *C*

= Area of triangle *ACD*

$$= \frac{1}{2} \times \frac{L}{2} \times \frac{WL}{4} = \frac{WL^2}{16}$$

$\therefore$ Slope at *A* or $\theta_A = \frac{WL^2}{EI}$

(*ii*) *Deflection at C* (y_c)

Now using Mohr's theorem for deflection*, we get from equation (4.13) as

$$y_c = \frac{A\bar{x}}{EI}$$

where A = Area of B.M. Diagram between A and C

$$= \frac{WL^2}{16}$$

$\bar{x}$ = Distance of C.G. of area of B.M. diagram about the reference line. The reference line is a vertical line through the point where slope is not zero. Slope is not zero at point A. Hence distance of C.G. should be taken from point A.

$$= \frac{2}{3} \times \frac{L}{2} = \frac{L}{3}$$

$$\therefore \qquad y = \frac{\frac{WL^2}{16} \times \frac{L}{3}}{EI} = \frac{WL^3}{48EI}.$$

4.8.3. Slope and Deflection of a simply supported beam carrying A uniformly distributed load by Mohr's Theorem

Fig. 4.18 (*a*) shows a simply supported beam *AB* of length *L* and carrying a uniformly distributed load of *w*/unit length over the entire span. The B.M. diagram is shown in Fig. 4.18 (*b*). This is a case of symmetrical loading, hence slope is zero at the centre *i.e.*, at point *C*.

The B.M diagram is a parabola.

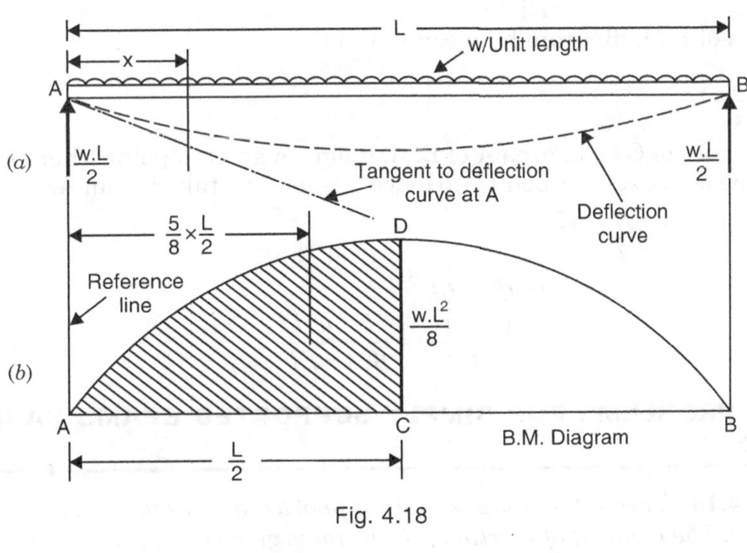

Fig. 4.18

Here $\qquad R_A = R_B = \dfrac{W \times L}{2}$

$$M_A = 0$$

*Deflection at *A* with respect to *C* is equal to deflection of *C* with respect to *A*. The deflection of *A* with respect to *C* (point *C* is having zero slope) is equal to moment of area of B.M. diagram between *A* and *C* about point *A* divided by *EI*.

$$M_x = R_A \times x - w \times x \times \frac{x}{2}$$

$$M_C = R_A \times \frac{L}{2} - w \times \frac{L}{2} \times \frac{L}{2 \times 2}$$

$$= \frac{wL}{2} \times \frac{L}{2} - \frac{wL^2}{8}$$

$$= \frac{wL^2}{4} - \frac{wL^2}{8} = \frac{wL^2}{8}$$

(*i*) Now using Mohr's theorem for slope, we get

Slope at A, $\theta_A = \dfrac{\text{Area of B.M. diagram between } A \text{ and } C}{EI}$

But area of B.M. diagram between A and C

$$= \text{Area of parabola } ACD$$

$$= \frac{2}{3} \times AC \times CD$$

$$= \frac{2}{3} \times \frac{L}{2} \times \frac{wL^2}{8} = \frac{w \, . \, L^3}{24}$$

$\therefore$ Slope at A, $\theta_A = \dfrac{w \, . \, L^3}{24EI}$

(*ii*) Now using Mohr's theorem for deflection, we get from equation (4.13) as

$$y_c = \frac{A\bar{x}}{EI}$$

where A = Area of B.M. diagram between A and C

$$= \frac{w \, . \, L^3}{24}$$

and $\bar{x}$ = Distance of C.G. of area A of B.M. diagram about a point where slope is not zero. Slope is not zero at point A. Hence $\bar{x}$ shows be taken from A.

$$= \frac{5}{8} \times AC = \frac{5}{8} \times \frac{L}{2} = \frac{5L}{16}$$

$$\therefore \quad y = \frac{\frac{w.L^3}{24} \times \frac{5L}{16}}{EI} = \frac{5}{384} \frac{w.L^4}{EI}.$$

4.9. SOLVED PROBLEMS FOR SIMPLY SUPPORTED BEAMS BASED ON AREA METHOD

Problem 4.13. *A beam 6 m long, simply supported at its ends, is carrying point load of 50 kN at its centre. The moment of inertia of the beam is given as equal to 78 × 10^6 mm^4. If E for the material of the beam = 2.1 × 10^5 N/mm^2, calculate using moment area theorem :*

(*i*) *Deflection at the centre of the beam and*

(*ii*) *Slope at the supports.*

Sol. Given :

$$L = 6 \text{ m}, \; W = 50 \text{ kN}$$

$$E = 2.1 \times 10^5 \text{ N/mm}^2$$

$$I = 78 \times 10^6 \text{ mm}^4$$

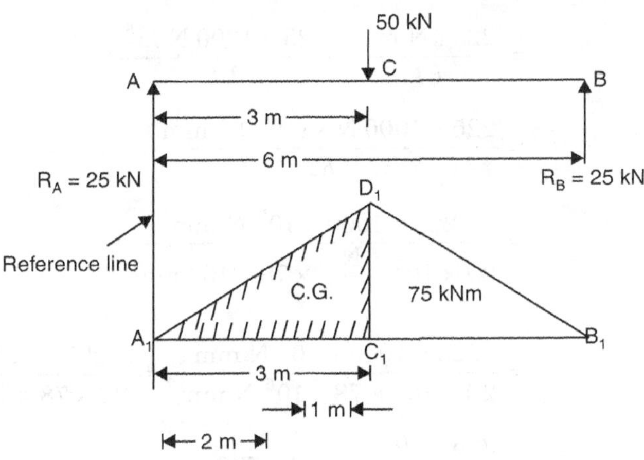

Fig. 4.19

(i) Deflection (y) at the centre

According to moment area theorem, the total deflection between any two points is equal to moment of the area of bending moment diagram between the two points about the point where slope is not zero divided by *EI*. The deflection at the support A is zero. Hence deflection at point *C* is equal to the moment of the area of B.M. diagram (between *A* and *C*) about point *A* divided by *EI*. The Fig. 4.19 shows the B.M. diagram for the given simply supported beam.

$$\left(\text{Reactions at } A \text{ and } B \text{ are equal } i.e.\ R_A = R_B = \frac{50}{2} = 25 \text{ kN.}\right)$$

The B.M. diagram is a triangle with max B.M. at the centre equal to

$$R_A \times 3 = 25 \times 3 = 75 \text{ kNm.}$$

Area of B.M. diagram between AC = Area of triangle $A_1C_1D_1$

$$= \frac{1}{2} \times A_1C_1 \times C_1D_1 = \frac{1}{2} \times 3 \times 75$$

$$= 112.5 \text{ kN m}^2$$

The slope is zero at the middle point *C*. Hence moment is taken about point *A*.

Moment of this area about point *A*

= Area of B.M. diagram × distance of C.G. of area from point A

$$= 112.5 \times \left(\frac{2}{3} \times 3\right) = 225 \text{ kN m}^3$$

$$\left(\because \text{ Distance of C.G. of triangle } A_1C_1D_1 \text{ from point } A_1 = \frac{2}{3} \text{ of } A_1C_1 = \frac{2}{3} \times 3 = 2 \text{ m}\right)$$

∴ Deflection (*y*) at point *c*

$$= \frac{\text{Moment of B.M. diagram about point } A}{EI}$$

$$= \frac{225 \text{ kN m}^3}{EI} = \frac{225 \times 1000 \text{ N m}^3}{EI} \qquad (\because \; 1 \text{ kN} = 1000 \text{ N})$$

$$= \frac{225 \times 1000 \text{ N} \times (1000)^3 \text{ mm}^3}{EI} \qquad (\because \; 1 \text{ m} = 1000 \text{ mm})$$

$$= \frac{225 \times 1000 \times 10^9 \text{ N mm}^3}{2.1 \times 10^5 \dfrac{\text{N}}{\text{mm}^2} \times 78 \times 10^6 \text{ mm}^4}$$

$$= \frac{225 \times 1000 \times 10^9 \text{ N mm}^3}{2.1 \times 10^5 \times 78 \times 10^6 \text{ N mm}^2} = \frac{225 \times 10^{12}}{2.1 \times 78 \times 10^{11}} \text{ mm}$$

$$= \frac{225 \times 10}{2.1 \times 78} \text{ mm} = \mathbf{13.736 \text{ mm}} \textbf{ Ans.}$$

(ii) Slope at the support (θ)

According to moment area theorem, the change of slope between any two points is equal to the net area of the B.M. diagram between these points divided by EI. The slope at the centre (*i.e.*, at point C) in this case is zero. Hence slope at the supports *i.e.*, at point A is equal to area of B.M. diagram between A and C.

But area of B.M. diagram between A and C

$= 112.5 \text{ kN m}^2$ (already calculated)

$\therefore$ Slope at the support A is given by,

$$\theta_A = \frac{\text{Area of B.M. diagram}}{EI}$$

$$= \frac{112.5 \text{ kN m}^2}{EI} = \frac{112.5 \times 1000 \text{ N} \times 10^6 \text{ mm}^2}{2.1 \times 10^5 \dfrac{\text{N}}{\text{mm}^2} \times 78 \times 10^6 \text{ mm}^4}$$

$$= \frac{112.5 \times 10^9 \text{ N mm}^2}{2.1 \times 10^5 \times 78 \times 10^6 \text{ N mm}^2} = \frac{112.5 \times 10^9}{2.1 \times 78 \times 10^{11}} \text{ radians}$$

$$= \frac{112.5}{2.1 \times 78 \times 10^2} \text{ radians}$$

$= 0.006868$ radians

$$= 0.006868 \times \frac{180}{\pi} \text{ degree} \qquad \left(\because \; 1 \text{ radian} = \frac{180}{\pi} \text{ degrees}\right)$$

$= \mathbf{0.3935°}$ **Ans.**

Problem 4.14. *A beam 4 m long, simply supported at its ends, carries a point load W at its centre. If the slope at the ends of the beam is not to exceed 1°, find the deflection at the centre of the beam using moment area theorem.*

Sol. Given

$L = 4$ m ; load at centre $= W$; slope at the ends $= 1° = \dfrac{\pi}{180}$ radians $= 0.01745$ radians.

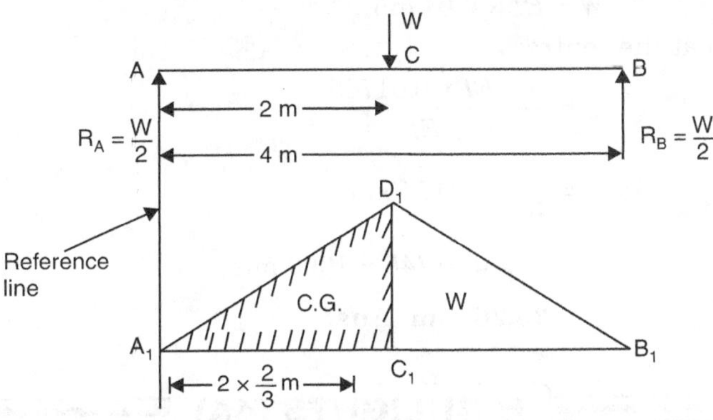

Fig. 4.20

The reactions R_A and R_B, at will be equal to $\frac{W}{2}$.

The B.M. diagram in this case will be triangle having max B.M. at the centre equal to

$$= R_A \times 2 \text{ Nm} = \frac{W}{2} \times 2 = W \text{ Nm}$$

Area of B.M. diagram between CA

$$= \text{Area of triangle } A_1C_1D_1$$

$$= \frac{1}{2} \times A_1C_1 \times C_1D_1$$

$$= \frac{1}{2} \times 2 \text{ m} \times W \text{ Nm}$$

$$= W \text{ Nm}^2$$

Moment of this area about A

$$= \text{Area of B.M.} \times \text{Distance of C.G. of area from point } A_1$$

$$= W \times \frac{2}{3} \times 2 = \frac{4W}{3} \text{ Nm}^3$$

According to moment area theorem,

$$\text{Slope at the support} = \frac{\text{Area of B.M. diagram}}{EI}$$

or $$0.01745 = \frac{W}{EI}$$

or $$W = EI \times 0.01745$$

Also according to moment area theorem,

Deflection at the centre

$$= \frac{\text{Moment of area of B.M. diagram}}{EI}$$

$$= \frac{\left(\frac{4W}{3}\right)}{EI} = \frac{4W}{3\,EI}$$

But $W = EI \times 0.01745$

$\therefore$ Deflection at the centre

$$= \frac{4 \times EI \times 0.01745}{3\ EI}$$

$$= \frac{4}{3} \times 0.01745 \text{ m}$$

$$= \frac{4}{3} \times 0.01745 \times 1000 \text{ mm}$$

$$= \mathbf{23.26 \text{ mm}} \quad \textbf{Ans.}$$

HIGHLIGHTS (4A)

1. The relation between curvature, slope, deflection etc. at a section is given by :

$$\text{Deflection} = y$$

$$\text{Slope} = \frac{dy}{dx}$$

$$\text{B.M.} = EI\,\frac{d^2y}{dx^2}$$

$$\text{S.F.} = EI\,\frac{d^3y}{dx^3}$$

$$w = EI\,\frac{d^4y}{dx^4}$$

As deflection is very small, hence slope is also given by $\frac{dy}{dx} = \tan\theta = \theta$.

2. Slope at the supports of a simply supported beam carrying a point load at the centre is given by :

$$\theta_A = \theta_B = -\frac{WL^2}{16EI}$$

where W = Point load at the centre, L = Length of beam
E = Young's modulus, I = M.O.I.

3. The deflection at the centre of a simply supported beam carrying a point load at the centre is given by $y_C = -\frac{WL^3}{48EI}$.

4. The slope and deflection of a simply supported beam, carrying a uniformly distributed load of w/unit length over the entire span, are given by,

$$\theta_A = \theta_B = \frac{WL^2}{24EI} \quad \text{and} \quad y_C = \frac{5}{384}\frac{WL^2}{EI}.$$

5. Macaulay's method is used in finding slopes and deflections at any point of a beam. In this method :
   (*i*) Brackets are to be integrated as a whole.
   (*ii*) Constants of integrations are written after the first term.
   (*iii*) The section, for which B.M. equation is to be written, should be taken in the last part of the beam.

6. For maximum deflection, the slope $\frac{dy}{dx}$ is zero.

7. The slope at point B if slope of A is zero by moment-area method is given by,

$$\theta_B = \frac{\text{Area of B.M. diagram between } A \text{ and } B}{EI}.$$

8. The deflection by moment area method is given by

$$y = \frac{A\bar{x}}{EI}$$

where A = Area of B.M. diagram between A and B

$\bar{x}$ = Distance of C.G. of area from B.

EXERCISE (4A)

(A) Theoretical Questions

1. Derive an expression for the slope and deflection of a beam subjected to uniform bending moment.
2. Prove that the relation that $M = EI \dfrac{d^2y}{dx^2}$

   where M = Bending moment, E = Young's modulus, I = M.O.I.
3. Find an expression for the slope at the supports of a simply supported beam, carrying a point load at the centre.
4. Prove that the deflection at the centre of a simply supported beam, carrying a point load at the centre, is given by $y_C = \dfrac{WL^3}{48EI}$

   where W = Point load, L = Length of beam.
5. Find an expression for the slope and deflection of a simply supported beam, carrying a point load W at a distance 'a' from left support and at a distance 'b' from right support where $a > b$.
6. Prove that the slope and deflection of a simply supported beam of length L and carrying a uniformly distributed load of w per unit length over the entire length are given by

   Slope at supports $= -\dfrac{WL^2}{24EI}$, and Deflection at centre $= \dfrac{5}{384}\dfrac{WL^3}{EI}$

   where W = Total load = $w \times L$.
7. What is a Macaulay's method ? Where is it used ? Find an expression for deflection at any section of a simply supported beam with an eccentric point load, using Macaulay's method.
8. What is moment-area method ? Where is it conveniently used ? Find the slope and deflection of a simply supported beam carrying a (*i*) point load at the centre and (*ii*) uniformly distributed load over the entire length using moment-area method.

(B) Numerical Problems

1. A wooden beam 4 m long, simply supported at its ends, is carrying a point load of 7.25 kN at its centre. The cross-section of the beam is 140 mm wide and 240 mm deep. If E for the beam = 6×10^3 N/mm^2, find the deflection at the centre. [**Ans.** 10 mm]
2. A beam 5 m long, simply supported at its ends, carries a point load W at its centre. If the slope at the ends of the beam is not to exceed 1°, find the deflection at the centre of the beam. [**Ans.** 29.08 mm]

3. Determine : (*i*) slope at the left support, (*ii*) deflection under the load and (*iii*) maximum deflection of a simply supported beam of length 10 m, which is carrying a point load of 10 kN at a distance 6 m from the left end.

Take $E = 2 \times 10^5$ N/mm² and $I = 1 \times 10^8$ mm⁴. [**Ans.** 0.00028 rad., 0.96 mm and 0.985 mm]

4. A beam of uniform rectangular section 100 mm wide and 240 mm deep is simply supported at its ends. It carries a uniformly distributed load of 9.125 kN/m run over the entire span of 4 m. Find the deflection at the centre if $E = 1.1 \times 10^4$ N/mm². [**Ans.** 6.01 mm]

5. A beam of length 4.8 m and of uniform rectangular section is simply supported at its ends. It carries a uniformly distributed load of 9.375 kN/m run over the entire length. Calculate the width and depth of the beam if permissible bending stress is 7 N/mm² and maximum deflection is not to exceed 0.95 cm.

Take E for beam material = 1.05×10^4 N/mm². [**Ans.** b = 240 mm and d = 336.8 mm]

6. Solve problem 3, using Macaulay's method.

7. A beam of length 10 m is simply supported at its ends and carries two point loads of 100 kN and 60 kN at a distance of 2 m and 5 m respectively from the left support. Calculate the deflections under each load. Find also the maximum deflection.

Take $I = 18 \times 10^8$ mm⁴ and $E = 2 \times 10^5$ N/mm².

[**Ans.** (*i*) – 4.35 mm ; (*ii*) – 6.76 mm ; (*iii*) y_{max} = – 6.78 mm]

8. A beam of length 20 m is simply supported at its ends and carries two point loads of 4 kN and 10 kN at a distance of 8 m and 12 m from left end respectively. Calculate : (*i*) deflection under each load (*ii*) maximum deflection.

Take $E = 2 \times 10^6$ N/mm² and $I = 1 \times 10^9$ mm⁴.

[**Ans.** (*i*) 10.3 mm and 10.6 downwards ; (*ii*) 11 mm]

9. A beam of length 6 m is simply supported at its ends. It carries a uniformly distributed load of 10 kN/m as shown in Fig. 4.21. Determine the deflection of the beam at its mid-point and also the position and the maximum deflection.

Take $EI = 4.5 \times 10^8$ N/mm². [**Ans.** – 2.578 mm, x = 2.9 m, y_{max} = – 2.582 mm]

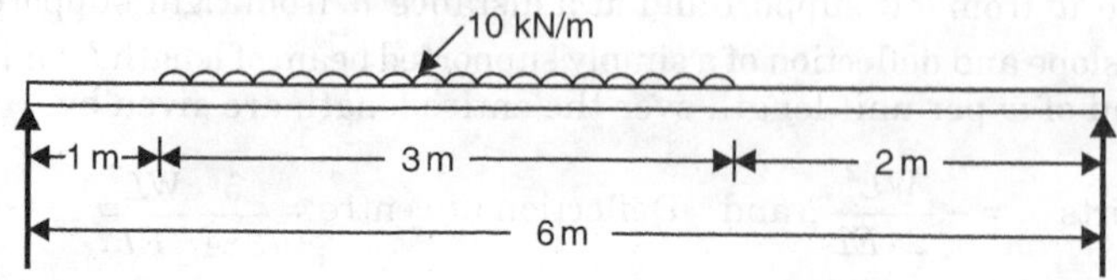

Fig. 4.21

10. A beam *ABC* of length 12 metre has one support at the left end and other support at a distance of 8 m from the left end. The beam carries a point load of 12 kN at the right end as shown in Fig. 4.22. Find the slopes over each support and at the right end. Find also the deflection at the right end.

Take $E = 2 \times 10^5$ N/mm² and $I = 5 \times 10^8$ mm⁴.

[**Ans.** θ_A = 6.00364, θ_B = – 0.00128, θ_C = – 0.00224, y_C = – 7.68 mm]

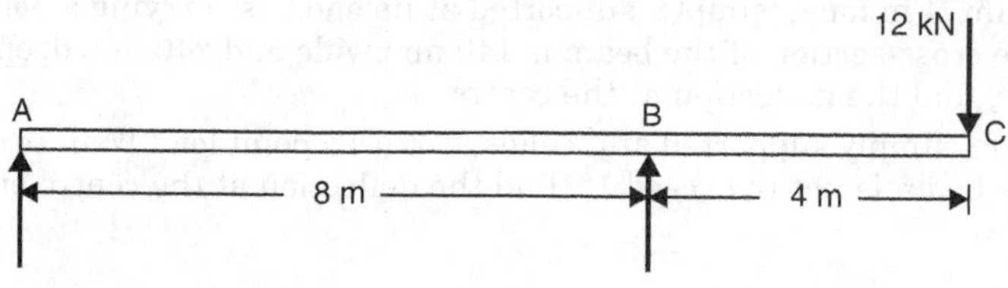

Fig. 4.22

11. An overhanging beam *ABC* is loaded as shown in Fig. 4.23. Determine the deflection of the beam at point *C*.

Take $E = 2 \times 10^5$ N/mm^2 and $I = 5 \times 10^8$ mm^4. **[Ans.** $y_c = -4.16$ mm]

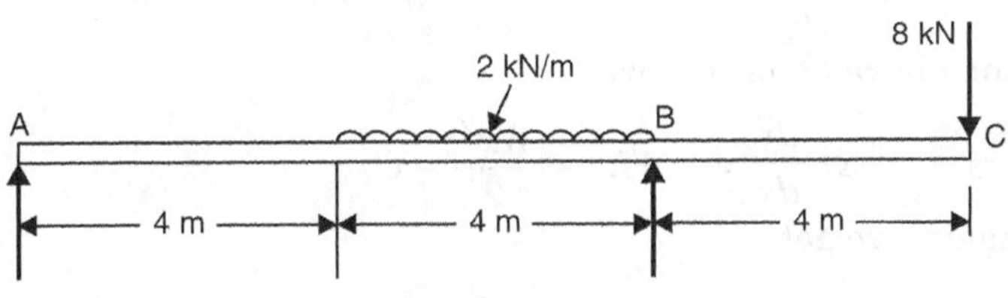

Fig. 4.23

12. A beam of span 8 m and of uniform flexural rigidity EI = 40 MN-m^2, is simply supported at its ends. It carries a uniformly distributed load of 15 kN/m run over the entire span. It is also subjected to a clockwise moment of 160 kNm at a distance of 3 m from the left support. Calculate the slope of the beam at the point of application of the moment. **[Ans.** 0.0061 radian]

(B) DEFLECTION OF CANTILEVER

4.10. DEFLECTION OF CANTILEVERS

Cantilever is a beam whose one end is fixed and other end is free. In this article we shall discuss the methods of finding slope and deflection for the cantilevers when they are subjected to various types of loading. The important methods are (*i*) Double integration method (*ii*) Macaulay's method and (*iii*) Moment-area-method. These methods have also been used for finding deflections and slope of the simply supported beams.

4.10.1. Deflection of a Cantilever with a Point Load at the Free end by Double Integration Method. A cantilever *AB* of length *L* fixed at the point *A* and free at the point *B* and carrying a point load at the free end *B* is shown in Fig. 4.24. *AB* shows the position of cantilever before any load is applied whereas *AB′* shows the position of the cantilever after loading.

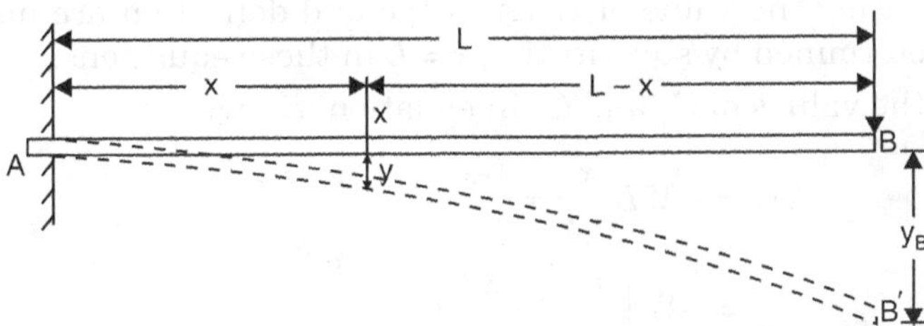

Fig. 4.24

Consider a section *X*, at a distance *x* from the fixed end *A*. The B.M. at this section is given by,

$$M_x = -W(L-x) \qquad \text{(Minus sign due to hogging)}$$

But B.M. at any section is also given by equation (4.3) as

$$M = EI \frac{d^2y}{dx^2}$$

Equating the two values of B.M., we get

$$EI\frac{d^2y}{dx^2} = -W(L-x) = -WL + W.x$$

Integrating the above equation, we get

$$EI\frac{dy}{dx} = -WLx + \frac{Wx^2}{2} + C_1 \quad ...(i)$$

Integration again, we get

$$EIy = -WL\frac{x^2}{2} + \frac{W}{2}\frac{x^3}{3} + C_1x + C_2 \quad ...(ii)$$

where C_1 and C_2 are constant of integrations. Their values are obtained from boundary conditions, which are : (*i*) at $x = 0, y = 0$ (*ii*) $x = 0, \frac{dy}{dx} = 0$

[At the fixed end, deflection and slopes are zero]

(*i*) By substituting $x = 0, y = 0$ in equation (*ii*), we get

$$0 = 0 + 0 + 0 + C_2 \quad \therefore \quad C_2 = 0$$

(*ii*) By substituting $x = 0, \frac{dy}{dx} = 0$ in equation (*i*), we get

$$0 = 0 + 0 + C_1 \quad \therefore \quad C_1 = 0$$

Substituting the value of C_1 in equation (*i*), we get

$$EI\frac{dy}{dx} = -WLx + \frac{Wx^2}{2}$$

$$= -W\left(Lx - \frac{x^2}{2}\right) \quad ...(iii)$$

The equation (*iii*) is known as *slope equation*. We can find the slope at any point on the cantilever by substituting the value of x. The slope and deflection are maximum at the free end. These can be determined by substituting $x = L$ in these equations.

Substituting the values of C_1 and C_2 in equation (*ii*), we get

$$EIy = -WL\frac{x^2}{2} + \frac{Wx^3}{6} \quad (\because \quad C_1 = 0, C_2 = 0)$$

$$= -W\left(\frac{Lx^2}{2} - \frac{x^3}{6}\right) \quad ...(iv)$$

The equation (*iv*) is known as *deflection equation*.

Let θ_B = slope at the free end B *i.e.*, $\left(\frac{dy}{dx}\right)$ at $B = \theta_B$ and

y_B = Deflection at the free end B

(*a*) Substituting θ_B for $\frac{dy}{dx}$ and $x = L$ in equation (*iii*), we get

$$EI.\theta_B = -W\left(L.L - \frac{L^2}{2}\right) = -W.\frac{L^2}{2}$$

$$\therefore \qquad \theta_B = -\frac{WL^2}{2EI} \qquad ...(4.14)$$

Negative sign shows that tangent at B makes an angle in the anti-clockwise direction with AB

$$\therefore \qquad \theta_B = \frac{WL^2}{2EI} \qquad ...(4.14A)$$

(b) Substituting y_B for y and $x = L$ in equation (iv), we get

$$EI.y_B = -W\left(L.\frac{L^2}{2}-\frac{L^3}{6}\right) = -W\left(\frac{L^3}{2}-\frac{L^3}{6}\right) = -W.\frac{L^3}{3}$$

$$\therefore \qquad y_B = -\frac{WL^3}{3EI} \qquad ...(4.15)$$

(Negative sign shows that deflection is downwards)

$$\therefore \quad \text{Downward deflection, } y_B = \frac{WL^3}{3EI} \qquad ...(4.15A)$$

4.10.2. Deflection of a Cantilever with a Point Load at a Distance a from the Fixed End. A cantilever AB of length L fixed at point A and free at point B and carrying a point load W at a distance 'a' from the fixed end A, is shown in Fig. 4.25.

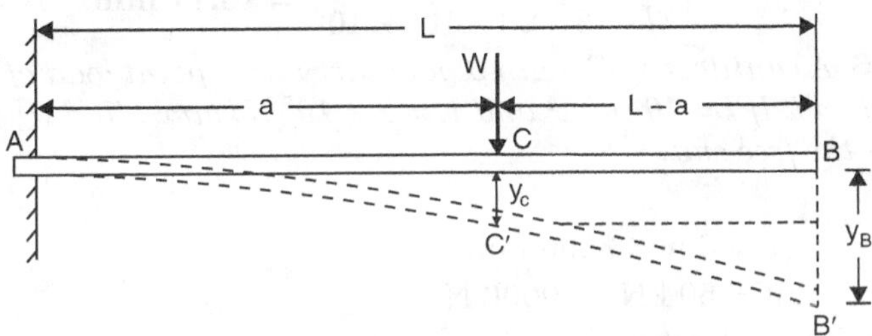

Fig. 4.25

Let θ_C = Slope at point C *i.e.*, $\left(\frac{dy}{dx}\right)$ at C

y_C = Deflection at point C

y_B = Deflection at point B

The portion AC of the cantilever may be taken as similar to a cantilever in Art. 4.10 (*i.e.*, load at the free end).

$$\therefore \qquad \theta_C = +\frac{Wa^2}{2EI} \qquad \text{[In equation (4.14 A) change } L \text{ to } a]$$

and

$$y_C = \frac{Wa^3}{3EI} \qquad \text{[In equation (4.15 A) change } L \text{ to } a]$$

The beam will bend only between A and C, but from C to B it will remain straight since B.M. between C and B is zero.

Since the portion CB of the cantilever is straight, therefore

Slope at C = slope at B

or

$$\theta_C = \theta_B = \frac{Wa^2}{2EI} \qquad ...(4.16)$$

Now from Fig. 4.25, we have

$$y_B = y_C + \theta_C(L - a)$$

$$= \frac{Wa^3}{3EI} + \frac{Wa^2}{2EI}(L - a) \qquad \left(\because \theta_C = \frac{Wa^2}{2EI}\right) \qquad \text{...(4.17)}$$

Problem 4.15. *A cantilever of length 3 m is carrying a point load of 25 kN at the free end. If the moment of inertia of the beam = 10^8 mm^4 and value of E = 2.1×10^5 N/mm^2, find (i) slope of the cantilever at the free end and (ii) deflection at the free end.*

Sol. Given :

Length, $L = 3 \text{ m} = 3000 \text{ mm}$

Point load, $W = 25 \text{ kN} = 25000 \text{ N}$

M.O.I., $I = 10^8 \text{ mm}^4$

Value of $E = 2.1 \times 10^5 \text{ N/mm}^2$

(*i*) Slope at the free end is given by equation (4.14 A).

$$\therefore \quad \theta_B = \frac{WL^2}{2EI} = \frac{25000 \times 3000^2}{2 \times 2.1 \times 10^5 \times 10^8} = \mathbf{0.005357 \text{ rad.}} \quad \textbf{Ans.}$$

(*ii*) Deflection at the free end is given by equation (4.15 A),

$$y_B = \frac{WL^3}{3EI} = \frac{25000 \times 3000^3}{3 \times 2.1 \times 10^5 \times 10^8} = \mathbf{10.71 \text{ mm.}} \quad \textbf{Ans.}$$

Problem 4.16. *A cantilever of length 3 m is carrying a point load of 50 kN at a distance of 2 m from the fixed end. If I = 10^8 mm^4 and E = 2×10^5 N/mm^2, find (i) slope at the free end and (ii) deflection at the free end.*

Sol. Given :

Length, $L = 3 \text{ m} = 3000 \text{ mm}$

Point load, $W = 50 \text{ kN} = 50000 \text{ N}$

Distance between the load and the fixed end,

$a = 2 \text{ m} = 2000 \text{ mm}$

M.O.I., $I = 10^8 \text{ mm}^4$

Value of $E = 2 \times 10^5 \text{ N/mm}^2$

(*i*) Slope at the free end is given by equation (4.16) as

$$\theta_B = \frac{Wa^2}{2EI} = \frac{50000 \times 2000^2}{2 \times 2 \times 10^5 \times 10^8} = \mathbf{0.005 \text{ rad.}} \quad \textbf{Ans.}$$

(*ii*) Deflection at the free end is given by equation (4.17) as

$$y_B = \frac{Wa^3}{3EI} + \frac{Wa^2}{2EI}(L - a)$$

$$= \frac{50000 \times 2000^3}{3 \times 2 \times 10^5 \times 10^8} + \frac{50000 \times 2000^2}{2 \times 2 \times 10^5 \times 10^8}(3000 - 2000)$$

$$= 6.67 + 5.0 = \mathbf{11.67 \text{ mm.}} \quad \textbf{Ans.}$$

4.10.3. Deflection of a Cantilever with a Uniformly Distributed Load. A cantilever *AB* of length *L* fixed at the point *A* and free at the point *B* and carrying a uniformly distributed load of *w* per unit length over the whole length, is shown in Fig. 4.26.

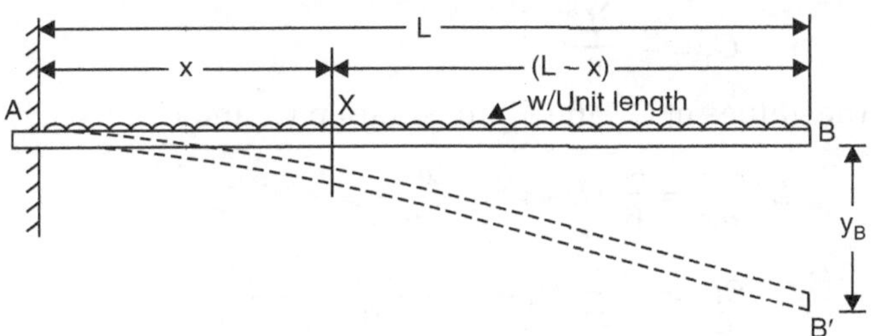

Fig. 4.26

Consider a section X, at a distance x from the fixed end A. The B.M. at this section is given by,

$$M_x = -w(L-x) \cdot \frac{(L-x)}{2} \qquad \text{(Minus sign due to hogging)}$$

But B.M. at any section is also given by equation (4.3) as

$$M = EI \frac{d^2y}{dx^2}$$

Equating the two values of B.M., we get

$$EI \frac{d^2y}{dx^2} = -\frac{w}{2}(L-x)^2$$

Integrating the above equation, we get

$$EI \frac{dy}{dx} = -\frac{w}{2}\frac{(L-x)^3}{3}(-1) + C_1$$

$$= \frac{w}{6}(L-x)^3 + C_1 \qquad \text{...(i)}$$

Integrating again, we get

$$EIy = \frac{w}{6} \cdot \frac{(L-x)^4}{4}(-1) + C_1x + C_2$$

$$= -\frac{w}{24}(L-x)^4 + C_1x + C_2 \qquad \text{...(ii)}$$

where C_1 and C_2 are constant of integrations. Their values are obtained from boundary conditions, which are : (*i*) at $x = 0, y = 0$ and (*ii*) at $x = 0, \frac{dy}{dx} = 0$ (as the deflection and slope at fixed end A are zero).

(*i*) By substituting $x = 0, y = 0$ in equation (*ii*), we get

$$0 = -\frac{w}{24}(L-0)^4 + C_1 \times 0 + C_2 = -\frac{wL^4}{24} + C_2$$

$$\therefore \qquad C_2 = \frac{wL^4}{24}$$

(*ii*) By substituting $x = 0$ and $\frac{dy}{dx} = 0$ in equation (*i*), we get

$$0 = \frac{w}{6}(L-0)^3 + C_1 = \frac{wL^3}{6} + C_1$$

$$\therefore \qquad C_1 = -\frac{wL^3}{6}$$

Substituting the values of C_1 and C_2 in equation (*i*) and (*ii*), we get

$$EI\frac{dy}{dx} = \frac{w}{6}(L-x)^3 - \frac{wL^3}{6} \qquad \text{...(iii)}$$

and

$$EIy = -\frac{w}{24}(L-x)^4 - \frac{wL^3}{6}x + \frac{wL^4}{24} \qquad \text{...(iv)}$$

The equation (*iii*) is known as *slope equation* and equation (*iv*) as *deflection equation.* From these equations the slope and deflection can be obtained at any sections. To find the slope and deflection at point B, the value of $x = L$ is substituted in these equations.

Let θ_B = Slope at the free end B *i.e.*, $\left(\frac{dy}{dx}\right)$ at B

y_B = Deflection at the free end B.

From equation (*iii*), we get slope at B as

$$EI.\theta_B = \frac{w}{6}(L-L)^3 - \frac{wL^3}{6} = -\frac{wL^3}{6}$$

$$\therefore \qquad \theta_B = -\frac{wL^3}{6EI} = -\frac{WL^2}{6EI} \qquad (\because W = \text{Total load} = w.L) \text{ ...(4.18)}$$

From equation (*iv*), we get the deflection at B as

$$EI.y_B = -\frac{w}{24}(L-L)^4 - \frac{wL^3}{6}\times L + \frac{wL^4}{24}$$

$$= -\frac{wL^4}{6} + \frac{wL^4}{24} = -\frac{3}{24}wL^4 = -\frac{wL^4}{8}$$

$$\therefore \qquad y_B = -\frac{wL^4}{8EI} = -\frac{WL^3}{8EI} \qquad (\because W = w.L)$$

$\therefore$ Downward deflection at B,

$$y_B = \frac{wL^4}{8EI} = \frac{WL^3}{8EI} \qquad \text{...(4.19)}$$

Problem 4.17. *A cantilever of length 2.5 m carries a uniformly distributed load of 16.4 kN per metre length over the entire length. If the moment of inertia of the beam = 7.95 × 10⁷ mm⁴ and value of E = 2 × 10⁵ N/mm², determine the deflection at the free end.*

Sol. Given :

Length, L = 2.5 mm = 2500 mm

U.d.l., w = 16.4 kN/m

$\therefore$ Total load, $W = w \times L = 16.4 \times 2.5 = 41$ kN = 41000 N

Value of $I = 7.95 \times 10^7$ mm^4

Value of $E = 2 \times 10^5$ N/mm^2

Let y_B = Deflection at the free end,

Using equation (4.19), we get

$$y_B = \frac{WL^3}{8EI} = \frac{41000 \times 2500^3}{8 \times 2 \times 10^5 \times 7.95 \times 10^7}$$

= **5.036 mm. Ans.**

Problem 4.18. *A cantilever of length 3 m carries a uniformly distributed load over the entire length. If the deflection at the free end is 40 mm, find the slope at the free end.*

Sol. Given :

Length, $L = 3 \text{ m} = 3000 \text{ mm}$

Deflection at free end, $y_B = 40$ mm

Let θ_B = slope at the free end

Using equation (4.19), we get

$$y_B = \frac{WL^3}{8EI}$$

or
$$40 = \frac{WL^2 \times L}{8EI} = \frac{WL^2 \times 3000}{8EI}$$

or
$$\frac{WL^2}{EI} = \frac{40 \times 8}{3000} \quad ...(i)$$

Slope at the free end is given by equation (4.18),

$$\therefore \quad \theta_B = -\frac{WL^2}{6EI} = -\frac{WL^2}{EI} \times \frac{1}{6} = -\frac{40 \times 8}{3000} \times \frac{1}{6}$$

$$\left[\because \text{ From equation } (i), \frac{WL^2}{EI} = \frac{40 \times 8}{3000}\right]$$

$= \mathbf{0.01777}$ **rad. Ans.**

Problem 4.19. *A cantilever 120 mm wide and 200 mm deep is 2.5 m long. What is the uniformly distributed load which the beam can carry in order to produce a deflection of 5 mm at the free end ? Take E = 200 GN/m².* (Annamalai University, 1991)

Sol. Given :

Width, $b = 120$ mm

Depth, $d = 200$ mm

Length, $L = 2.5 \text{ m} = 2.5 \times 1000 = 2500$ mm

Deflection at free end, $y_B = 5$ mm

Value of $E = 200 \text{ GN/m}^2 = 200 \times 10^9 \text{ N/m}^2$ $\quad (\because \; G = \text{Giga} = 10^9)$

$$= \frac{200 \times 10^9 \text{ N}}{(1000)^2 \text{ mm}^2} \quad [\because \; 1 \text{ m}^2 = (1000 \text{ mm})^2]$$

$$= 2 \times 10^5 \text{ N/mm}^2$$

Moment of inertia, $I = \dfrac{bd^3}{12} = \dfrac{120 \times 200^3}{12} = 8 \times 10^7 \text{ mm}^4$

Let w = uniformly distributed load per m length in N

W = Total load

$= w \times L$ (Here L is in metre)

$= w \times 2.5 = 2.5 \times w$ N

Using equation (4.19), we get

$$y_B = \frac{WL^3}{8EI}$$

or $$5 = \frac{2.5\,w \times 2500^3}{8 \times 2 \times 10^5 \times 8 \times 10^7}$$

or $$w = \frac{5 \times 8 \times 2 \times 10^5 \times 8 \times 10^7}{2.5 \times 2500^3} = 16384 \text{ N/m}$$

$$= \mathbf{16.384\ kN/m.} \quad \textbf{Ans.}$$

4.10.4. Deflection of a Cantilever with a Uniformly Distributed Load for a Distance 'a' from the Fixed end. A cantilever AB of length L fixed at the point A and free at the point B and carrying a uniformly distributed load of w/m length for a distance 'a' from the fixed end, is shown in Fig. 4.27.

The beam will bend only between A and C, but from C to B it will remain straight since B.M. between C and B is zero. The deflected shape of the cantilever is shown by $AC'B'$ in which portion $C'B'$ is straight.

Let θ_C = Slope at C, *i.e.*, $\left(\frac{dy}{dx}\right)$ at C

y_C = Deflection at point C, and

y_B = Deflection at point B.

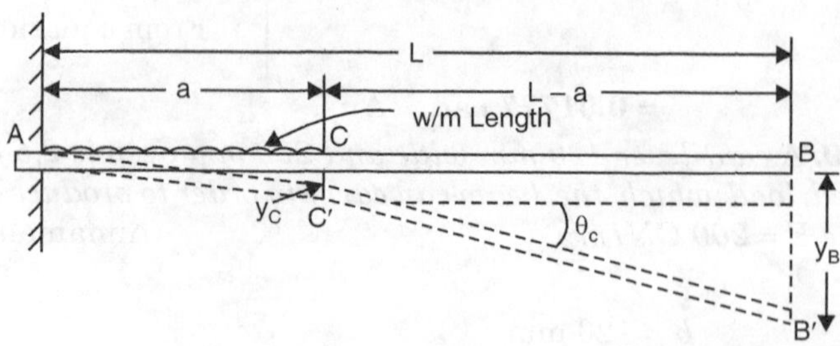

Fig. 4.27

The portion AC of the cantilever may be taken as similar to a cantilever in Art. 4.10.3.

$\therefore$ $$\theta_C = \frac{w\,.\,a^3}{6EI} \quad \text{[In equation (4.18) put } L = a\text{]}$$

and $$y_C = \frac{w\,.\,a^4}{8EI} \quad \text{[In equation (4.19) put } L = a\text{]}$$

Since the portion $C'B'$ of the cantilever is straight, therefore slope at C = slope at B

or $$\theta_C = \theta_B = \frac{wa^3}{6EI} \quad \text{...(4.20)}$$

Now from Fig. 4.27, we have

$$y_B = y_C + \theta_C\,(L - a)$$

$$= \frac{wa^4}{8EI} + \frac{w\,.\,a^3}{6EI}\,(L - a) \quad \text{...(4.21)}$$

4.10.5. Deflection of a Cantilever with a Uniformly Distributed Load for a Distance 'a' from the Free End. A cantilever AB of length L fixed at the point A and free at the point B and carrying a uniformly distributed load of w/m length for a distance 'a' from the free end is shown in Fig. 4.28 (a).

The slope and deflection at the point B is determined by considering :

(*i*) the whole cantilever AB loaded with a uniformly distributed load of w per unit length as shown in Fig. 4.28 (*b*).

(*ii*) a part of cantilever from A to C of length $(L - a)$ loaded with an upward uniformly distributed load of w per unit length as shown in Fig. 4.28 (*c*).

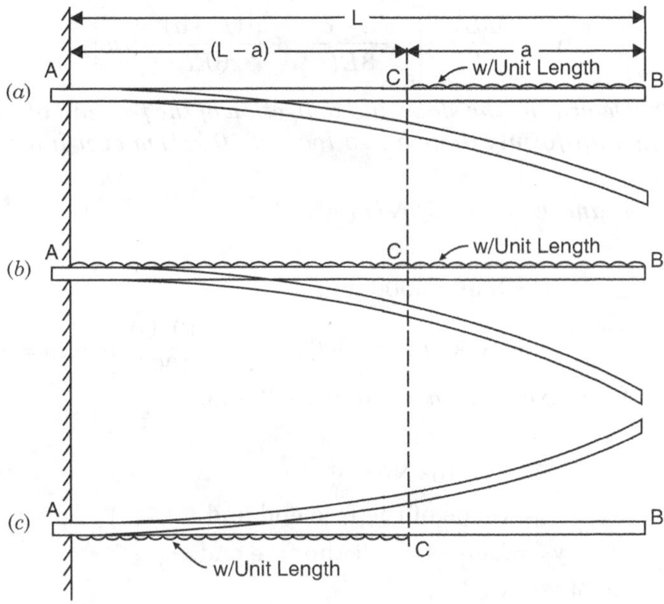

Fig. 4.28

Then slope at B = Slope due to downward uniform load over the whole length – slope due to upward uniform load from A to C

and deflection at B = Deflection due to downward uniform load over the whole length – deflection due to upward uniform load from A to C.

(*a*) Now slope at B due to downward uniformly distributed load over the whole length

$$= \frac{wL^3}{6EI}$$

(*b*) Slope at B or at C due to upward uniformly distributed load over the length $(L - a)$

$$= \frac{w(L-a)^3}{6EI}$$

Hence net slope at B is given by,

$$\theta_B = \frac{wL^3}{6EI} - \frac{w(L-a)^3}{6EI} \qquad \text{...(4.22)}$$

The downward deflection of point B due to downward distributed load over the whole length AB

$$= \frac{wL^4}{8EI}$$

The upward deflection of point B due to upward uniformly distributed load acting on the portion AC = upward deflection of C + slope at $C \times CB$

$$= \frac{w(L-a)^4}{8EI} + \frac{w.(L-a)^3}{6EI} \times a \qquad (\because \; CB = a)$$

$\therefore$ Net downward deflection of the free end B is given by

$$y_B = \frac{wL^4}{8EI} - \left[\frac{w(L-a)^4}{8EI} + \frac{w(L-a)^3}{6EI} \times a\right] \qquad \text{...(4.23)}$$

Problem 4.20. *Determine the slope and deflection of the free end of a cantilever of length 3 m which is carrying a uniformly distributed load of 10 kN/m over a length of 2 m from the fixed end.*

Take $I = 10^8$ mm^4 and $E = 2 \times 10^5$ N/mm^2.

Sol. Given :

Length, $L = 3$ m $= 3000$ mm

U.d.l., $w = 10$ kN/m $= 10000$ N/m $= \frac{10000}{1000}$ N/mm $= 10$ N/mm

Length of u.d.l. from fixed end, $a = 2$ m $= 2000$ mm.

Value of $I = 10^8$ mm^4

Value of $E = 2 \times 10^5$ N/mm^2

Let θ_B = Slope of the free end and

y_B = Deflection at the free end.

(*i*) Using equation (4.20), we have

$$\theta_B = \frac{wa^3}{6EI} = \frac{10 \times 2000^3}{6 \times 2 \times 10^5 \times 10^8} = \textbf{0.00066.} \quad \textbf{Ans.}$$

(*ii*) Using equation (4.21), we get

$$y_B = \frac{wa^4}{8EI} + \frac{w.a^3}{6EI}(L-a)$$

$$= \frac{10 \times 2000^4}{8 \times 2 \times 10^5 \times 10^8} + \frac{10 \times 2000^3}{6 \times 2 \times 10^5 \times 10^8} \times (3000 - 2000)$$

$$= 1 + 0.67 = \textbf{1.67 mm.} \quad \textbf{Ans.}$$

Problem 4.21. *A cantilever of length 3 m carries a uniformly distributed load of 10 kN/m over a length of 2 m from the free end. If $I = 10^8$ mm^4 and $E = 2 \times 10^5$ N/mm^2 ; find : (i) slope at the free end, and (ii) deflection at the free end.*

Sol. Given :

Length, $L = 3$ m $= 3000$ mm

U.d.l., $w = 10$ kN/m $= 10000$ N/m $= \frac{10000}{1000}$ N/mm $= 10$ N/mm

Length of u.d.l. from free end, $a = 2$ m $= 2000$ mm

Value of $I = 10^8$ mm^4

Value of $E = 2 \times 10^5$ N/mm^2

Let θ_B = Slope at the free end *i.e.*, $\left(\frac{dy}{dx}\right)$ at B and

y_B = Deflection at the free end.

(*i*) Using equation (4.22), we get

$$\theta_B = \frac{wL^3}{6EI} - \frac{w(L-a)^3}{6EI}$$

$$= \frac{10 \times 3000^3}{6 \times 2 \times 10^5 \times 10^8} - \frac{10(3000-2000)^3}{6 \times 2 \times 10^5 \times 10^8}$$

$$= 0.00225 - 0.000083 = \mathbf{0.002167\ rad.}\quad \textbf{Ans.}$$

(*ii*) Using equation (4.23), we get

$$y_B = \frac{wL^4}{8EI} - \left[\frac{w(L-a)^4}{8EI} + \frac{w(L-a)^3}{6EI} \times a\right]$$

$$= \frac{10 \times 3000^4}{8 \times 2 \times 10^5 \times 10^8} - \left[\frac{10(3000-2000)^4}{8 \times 2 \times 10^5 \times 10^8} + \frac{10(3000-2000)^3}{6 \times 2 \times 10^5 \times 10^8} \times 2000\right]$$

$$= 5.0625 - [0.0625 + 0.1667] = \mathbf{4.8333\ mm.}\quad \textbf{Ans.}$$

Problem 4.22. *A cantilever of length 3 m carries two point loads of 2 kN at the free end and 4 kN at a distance of 1 m from the free end. Find the deflection at the free end. Take E = 2 × 10⁵ N/mm² and I = 10⁸ mm⁴.*

Sol. Given :

Length, $L = 3$ m = 3000 mm

Load at free end, $W_1 = 2$ kN = 2000 N

Load at a distance one m from free end,

$W_2 = 4$ kN = 4000 N

Distance AC, $a = 2$ m = 2000 mm

Value of $E = 2 \times 10^5$ N/mm²

Value of $I = 10^8$ mm⁴

Let y_1 = Deflection at the free end due to load 2 kN alone

y_2 = Deflection at the free end due to load 4 kN alone.

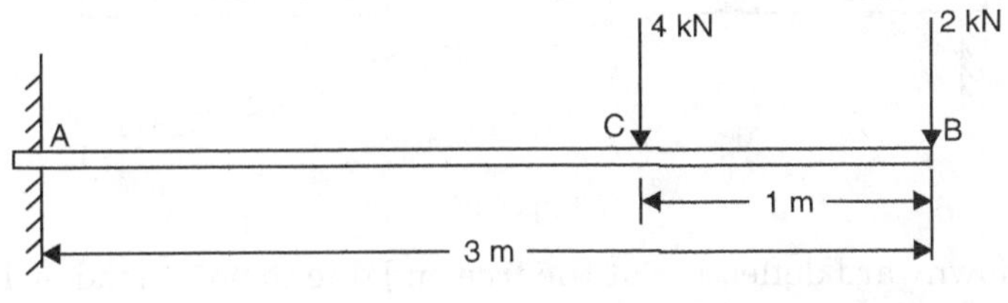

Fig. 4.29

Downward deflection due to load 2 kN alone at the free end is given by equation (4.15 *A*) as

$$y_1 = \frac{WL^3}{3EI} = \frac{2000 \times 3000^3}{3 \times 2 \times 10^5 \times 10^8} = 0.9 \text{ mm.}$$

Downward deflection at the free end due to load 4 kN (*i.e.*, 4000 N) alone at a distance 2 m from fixed end is given by (4.17) as

$$y_2 = \frac{Wa^3}{3EI} + \frac{Wa^2}{2EI}(L - a)$$

$$= \frac{4000 \times 2000^3}{3 \times 2 \times 10^5 \times 10^8} + \frac{4000 \times 2000^2}{2 \times 2 \times 10^5 \times 10^8}(3000 - 2000)$$

$$= 0.54 + 0.40 = 0.94 \text{ mm}$$

∴ Total deflection at the free end

$$= y_1 + y_2$$

$$= 0.9 + 0.94 = \mathbf{1.84\ mm.}\quad \textbf{Ans.}$$

Problem 4.23. *A cantilever of length 2 m carries a uniformly distributed load of 2.5 kN/m run for a length of 1.25 m from the fixed end and a point load of 1 kN at the free end. Find the deflection at the free end if the section is rectangular 12 cm wide and 24 cm deep and E = 1 × 10⁴ N/mm².* (Annamalai University, 1990)

Sol. Given :

Length, $L = 2 \text{ m} = 2000 \text{ mm}$

U.d.l., $w = 2.5 \text{ kN/m} = 2.5 \times 1000 \text{ N/m}$

$$= \frac{2.5 \times 1000}{1000} \text{ N/mm} = 2.5 \text{ N/mm}$$

Point load at free end, $W = 1 \text{ kN} = 1000 \text{ N}$

Distance AC, $a = 1.25 \text{ m} = 1250 \text{ mm}$

Width, $b = 12 \text{ mm}$

Depth, $d = 24 \text{ mm}$

Value of $I = \frac{bd^3}{12} = \frac{12 \times 24^3}{12}$

$= 13824 \text{ cm}^4 = 13824 \times 10^4 \text{ mm}^4 = 1.3824 \times 10^4 \text{ mm}^4$

Value of $E = 1 \times 10^8 \text{ N/mm}^2$

Let y_1 = Deflection at the free end due to point load 1 kN alone

y_2 = Deflection at the free end due to u.d.l. on length AC.

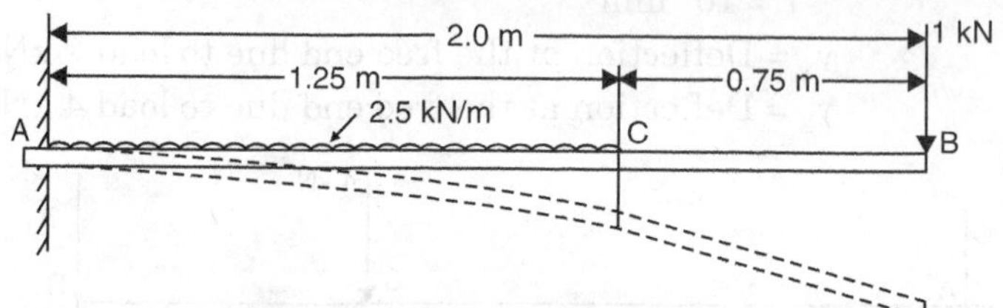

Fig. 4.30

(*i*) Now the downward deflection at the free end due to point load of 1 kN (or 1000 N) at the free end is given by equation (4.15 *A*) as

$$y_1 = \frac{WL^3}{3EI} = \frac{1000 \times 2000^3}{3 \times 10^4 \times 1.3824 \times 10^8} = 1.929 \text{ mm.}$$

(*ii*) The downward deflection at the free end due to uniformly distributed load of 2.5 N/mm on a length of 1.25 m (or 1250 mm) is given by equation (4.21) as

$$y_2 = \frac{wa^4}{8EI} + \frac{w \cdot a^3}{6EI}(L - a)$$

$$= \frac{2.5 \times 1250^4}{8 \times 10^4 \times 1.3824 \times 10^8} + \frac{2.5 \times 1250^3}{6 \times 10^4 \times 1.3824 \times 10^8}(2000 - 1250)$$

$$= 0.5519 + 0.4415 = 0.9934$$

∴ Total deflection at the free end due to point load and u.d.l.

$$= y_1 + y_2 = 1.929 + 0.9934 = \mathbf{2.9224\ mm.}\quad \textbf{Ans.}$$

Problem 4.24. *A cantilever of length 2 m carries a uniformly distributed load 2 kN/m over a length of 1 m from the free end, and a point load of 1 kN at the free end. Find the slope and deflection at the free end if E = 2.1 × 10⁵ N/mm² and I = 6.667 × 10⁷ mm⁴.*

Sol. Given : (See Fig. 4.31)

Length, $L = 2\text{ m} = 2000\text{ mm}$

U.d.l. $w = 2\text{ kN/m} = \frac{2 \times 1000}{1000}\text{ N/mm} = 2\text{ N/mm}$

Length BC, $a = 1\text{ m} = 1000\text{ mm}$

Point load, $W = 1\text{ kN} = 1000\text{ N}$

Value of $E = 2.1 \times 10^5\text{ N/mm}^2$

Value of $I = 6.667 \times 10^7\text{ mm}^4$.

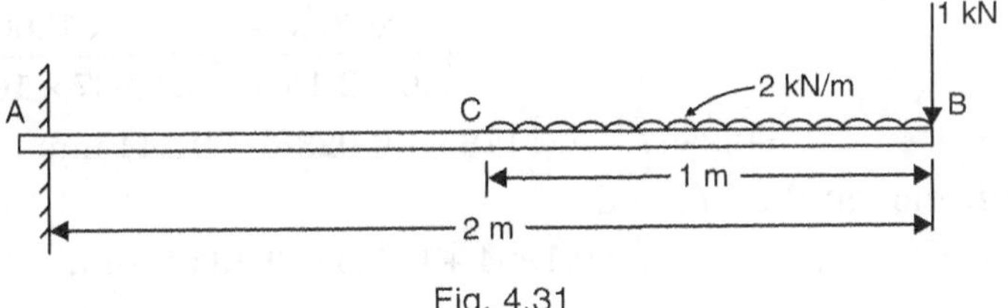

Fig. 4.31

(*i*) *Slope at the free end*

Let θ_1 = Slope at the free end due to point load of 1 kN *i.e.,* 1000 N

θ_2 = Slope at the free end due to u.d.l. on length BC.

The slope at the free end due to a point load of 1000 N at B is given by equation (4.14*A*) as

$$\theta_1 = \frac{WL^2}{2EI} \qquad (\because\ \theta_B = \theta_1 \text{ here})$$

$$= \frac{1000 \times 2000^2}{2 \times 2.1 \times 10^5 \times 6.667 \times 10^7} = 0.0001428\text{ rad.}$$

The slope at the free end due to u.d.l. of 2 kN/m over a length of 1 m from the free end is given by equation (4.22) as

$$\theta_2 = \frac{wL^3}{6EI} - \frac{w(L-a)^3}{6EI} \qquad (\because\ \theta_B = \theta_2 \text{ here})$$

$$= \frac{2 \times 2000^3}{6 \times 2.1 \times 10^5 \times 6.667 \times 10^7} - \frac{2 \times (2000 - 1000)^3}{6 \times 2.1 \times 10^5 \times 6.667 \times 10^7}$$

$$= 0.0001904 - 0.000238 = 0.0001666\text{ rad.}$$

∴ Total slope at the free end

$$= \theta_1 + \theta_2 = 0.0001428 + 0.0001666 = \mathbf{0.0003094\ rad.}\quad \textbf{Ans.}$$

(ii) Deflection at the free end

Let y_1 = Deflection at the free end due to point load of 1000 N

y_2 = Deflection at the free end due to u.d.l. on length *BC*.

The deflection at the free end due to point load of 1000 N is given by equation (4.15*A*) as

$$y_1 = \frac{WL^3}{3EI} \qquad (\because \text{ Here } y_1 = y_B)$$

$$= \frac{1000 \times 2000^3}{3 \times 2.1 \times 10^5 \times 6.667 \times 10^7} = 0.1904 \text{ mm.}$$

The deflection at the free end due to u.d.l. of 2 N/mm over a length of 1 m from the free end is given by equation (4.23) as

$$y_2 = \frac{wL^4}{8EI} - \left[\frac{w(L-a)^4}{8EI} + \frac{w(L-a)^3}{6EI} \times a\right]$$

$$= \frac{2 \times 2000^4}{8 \times 2.1 \times 10^5 \times 6.667 \times 10^7} - \left[\frac{2(2000-1000)^4}{8 \times 2.1 \times 10^5 \times 6.667 \times 10^7} + \frac{2(2000-1000)^3 \times 1000}{6 \times 2.1 \times 10^5 \times 6.667 \times 10^7}\right]$$

$$= 0.2857 - [0.01785 + 0.0238] = 0.244 \text{ mm}$$

$\therefore$ Total deflection at the free end

$$= y_1 + y_2 = 0.1904 + 0.244 = \mathbf{0.4344 \text{ mm.}} \quad \textbf{Ans.}$$

4.11. DEFLECTION AND SLOPE OF A CANTILEVER BY MOMENT AREA METHOD

The moment area method is discussed in Art. 12.8, where this method was applied to a simply supported beam. Let us apply this method to a cantilever. According to this method the change of slope between any two points is equal to the net area of the B.M. diagram between these two points divided by *EI*. If one of the points is having zero slope, then we can obtain the slope at the other point.

Similarly if the deflection at a point *A* is zero, then the deflection at the point *B* according to this method is given by

$$y = \frac{A\bar{x}}{EI}$$

where A = Area of B.M. diagram between *A* and *B*, and

$\bar{x}$ = Distance of C.G. of the area of B.M. diagram from the point where slope is not zero. Slope is not zero at point *B*, hence $\bar{x}$ should be taken from point *B*.

4.11.1. Cantilever Carrying a Point Load at the Free end. Fig. 4.32 (*a*) shows a cantilever of length *L* fixed at end *A* and free at the end *B*. It carries a point load *W* at *B*.

The B.M. will be zero at *B* and will be *W.L* at *A*. The variation of B.M. between *A* and *B* is linear as shown in Fig. 4.32 (*b*).

At the fixed end *A*, the slope and deflection are zero.

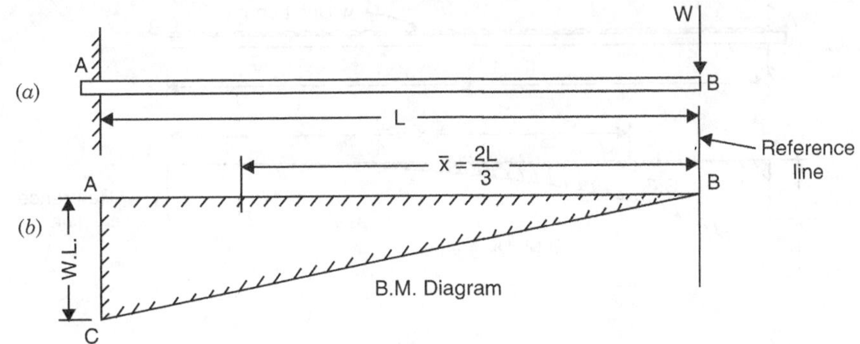

Fig. 4.32

Let θ_B = Slope at B *i.e.*, $\left(\frac{dy}{dx}\right)$ at B and

y_B = Deflection at B

Then according to moment area method,

$$\theta_B = \frac{\text{Area of B.M. Diagram between } A \text{ and } B}{EI}$$

$$= \frac{\frac{1}{2} \times AB \times AC}{EI} \quad \text{(Area of triangle } ABC\text{)}$$

$$= \frac{\frac{1}{2} \times L \times W \,.\, L}{EI} = \frac{WL^2}{2EI}$$

and $$y_B = \frac{A\bar{x}}{EI} \quad ...(i)$$

where A = Area of B.M. diagram between A and $B = \frac{W \,.\, L^2}{2}$

$\bar{x}$ = Distance of C.G. of area of B.M. diagram from $B = \frac{2L}{3}$

$$\therefore \quad y_B = \frac{\frac{W \,.\, L^2}{2} \times \frac{2L}{3}}{EI} = \frac{W \,.\, L^3}{3EI}.$$

4.11.2. Cantilever Carrying a Uniformly Distributed load. Fig. 4.33 (*a*) shows a cantilever of length L fixed at end A and free at the end B. It carries a uniformly distributed load of w/unit length over the entire length.

The B.M. will be zero at B and will be $\frac{w \,.\, L^2}{2}$ at A. The variation of B.M. between A and B is parabolic as shown in Fig. 4.33 (*b*). At the fixed end A, the slope and deflection are zero.

Area of B.M. diagram (ABC), $A = \frac{1}{3} \,.\, L \,.\, \frac{w \,.\, L^2}{2} = \frac{w \,.\, L^3}{6}$

and the distance of the C.G. of the B.M. diagram from B,

$$\bar{x} = \frac{3L}{4}$$

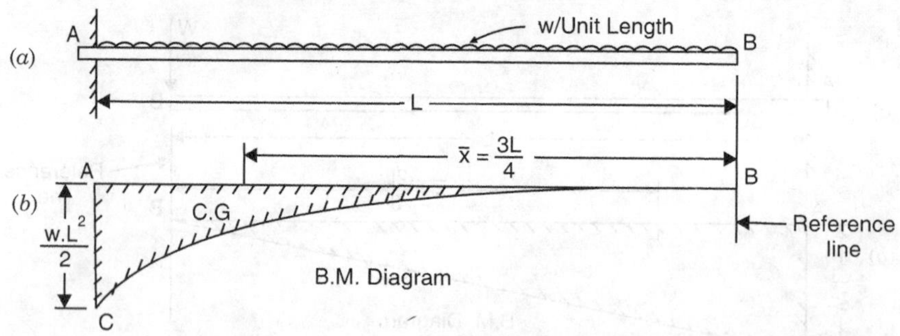

Fig. 4.33

Let $\qquad \theta_B$ = Slope at B, *i.e.*, $\left(\frac{dy}{dx}\right)$ and at B

y_B = Deflection at B.

Then according to moment area method,

$$\theta_B = \frac{\text{Area of B. M. diagram}}{EI} = \frac{wL^3}{6EI}$$

and
$$y_B = \frac{A\bar{x}}{EI} = \frac{w\,.\,L^3}{6EI} \times \frac{3L}{4} = \frac{w\,.\,L^4}{8EI}.$$

4.11.3. Cantilever Carrying a Uniformly Distributed Load upto a Length 'a' from the Fixed end. Fig. 4.34 (*a*) shows a cantilever of length *L* fixed at end *A* and free at the end *B*. It carries a uniformly distributed load of *w*/unit length over a length '*a*' from the fixed end.

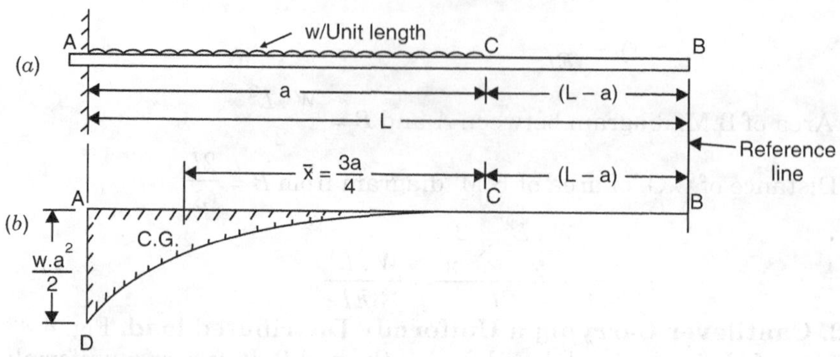

Fig. 4.34

The B.M. will be zero at *B* and *C*. But B.M. at *A* will be $\frac{w\,.\,a^2}{2}$. The variation of B.M. between *C* and *A* will be parabolic as shown in Fig. 4.34 (*b*). At the fixed end the slope and deflection are zero.

Area of B.M. diagram $\quad A = \frac{1}{3}\,.\,a\,.\,\frac{w\,.\,a^2}{2} = \frac{w\,.\,a^3}{6}$

and the distance of the C.G. of B.M. diagram from B, for cantilevers

$$\bar{x} = (L - a) + \frac{3a}{4}$$

Let θ_B = Slope at B *i.e.*, $\left(\frac{dy}{dx}\right)$ at B and

y_B = Deflection at B.

Then according to moment area method,

$$\theta_B = \frac{A}{EI} = \frac{w \cdot a^3}{6EI}$$

and

$$y_B = \frac{A\bar{x}}{EI} = \frac{w \cdot a^3}{6EI} \times \left[(L - a) + \frac{3a}{4}\right] = \frac{w \cdot a^3}{6EI}(L - a) + \frac{w \cdot a^4}{8EI}.$$

4.12. SOLVED PROBLEMS BASED ON MOMENT AREA METHOD

Problem 4.25. *A cantilever of length 2 m carries a point load of 20 kN at the free end and another load of 20 kN at its centre. If $E = 10^5$ N/mm^2 and $I = 10^8$ mm^4 for the cantilever then determine by moment area method, the slope and deflection of the cantilever at the free end.*

Sol. Given :

Length, $L = 2$ m

Load at free end, $W_1 = 20$ kN $= 20000$ N

Load at centre, $W_2 = 20$ kN $= 20000$ N

Value of $E = 10^5$ N/mm^2

Value of $I = 10^8$ mm^4

First draw the B.M. diagram,

B.M. at B $= 0$

B.M. at C $= -20 \times 1 = -20$ kNm $= -20 \times 10^3 \times 10^3$ Nmm

B.M. at A $= -20 \times 1 - 20 \times 2 = -60$ kNm $= -60 \times 10^3 \times 10^3$ Nmm

B.M. diagram is shown in Fig. 13.14 (*b*).

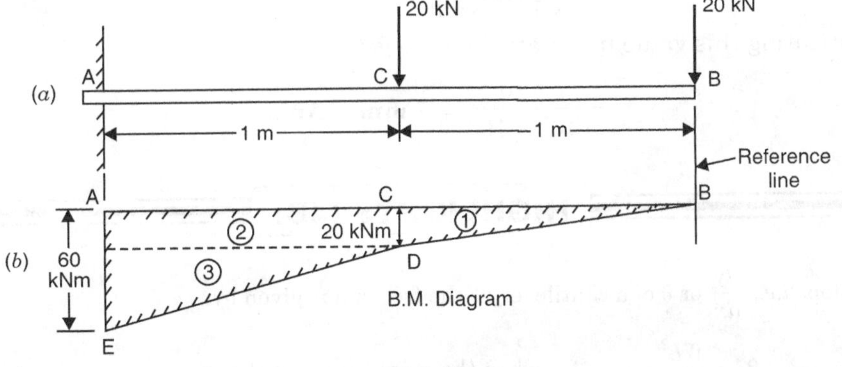

Fig. 4.35

To find the area of B.M. diagram, divide the Fig. 4.35 (*b*) into two triangles and one rectangle.

Now $\quad$ area $A_1 = \frac{1}{2} \times CD \times BC = \frac{1}{2} \times 20 \times 1$

$= 10 \text{ kNm}^2 = 10 \times 10^3 \times 10^6 \text{ Nmm}^2 \qquad (\because \; \text{m}^2 = 10^6 \text{ mm}^2)$

$= 10^{10} \text{ Nmm}^2$

Similarly area $\quad A_2 = CD \times AC = 20 \times 1 = 20 \text{ kNm}^2$

and $\quad$ area $A_3 = \frac{1}{2} \times \text{FD} \times \text{EF} = \frac{1}{2} \times 1 \times 40 = 20 \text{ kNm}^2$

$\therefore$ Total area of B.M. diagram,

$$A = A_1 + A_2 + A_3 = 10 + 20 + 20 = 50 \text{ kNm}^2$$

$$= 50 \times 10^3 \times 10^6 \text{ Nmm}^2 \qquad (\because \; \text{m}^2 = 10^6 \text{ mm}^2)$$

Slope and deflection at the fixed end is zero.

Let $\quad \theta_B$ = Slope at the free end B.

Then according to the moment area method,

$$\theta_B = \frac{\text{Area of B.M. diagram}}{EI}$$

$$= \frac{50 \times 10^3 \times 10^6}{10^5 \times 10^8} = \mathbf{0.005 \text{ radians.}} \quad \textbf{Ans.}$$

Let $\quad y_B$ = Deflection at the free end B.

Then according to moment area method,

$$y_B = \frac{A\bar{x}}{EI} \qquad \ldots(i)$$

Now let us find $\bar{x}$ or $A\bar{x}$.

Then total moment of the bending moment diagram about B is given by

$$A \,.\, \bar{x} = A_1\bar{x}_1 + A_2\bar{x}_2 + A_3\bar{x}_3$$

$$= 10 \times \left(\frac{2}{3} \times 1\right) + 20 \times \left(1 + \frac{1}{2}\right) + 20 \times \left(1 + \frac{2}{3} \times 1\right)$$

$$= \frac{20}{3} + 30 + \frac{100}{3} = 70 \text{ kNm}^3$$

$$= 70 \times 10^3 \times 10^9 \text{ Nmm}^3 \qquad (\because \; \text{m}^3 = 10^9 \text{ mm}^3)$$

$$= 7 \times 10^{13} \text{ Nmm}^3$$

Substituting this value in equation (*i*), we get

$$y_B = \frac{7 \times 10^{13}}{10^5 \times 10^8} = \mathbf{7 \text{ mm.}} \quad \textbf{Ans.}$$

HIGHLIGHTS (4B)

1. The slope *i.e.*, $\frac{dy}{dx}$ or θ of a cantilever at the free end is given by,

$\theta_B = \frac{WL^2}{2EI}$ $\quad$ when the point load is at the free end

$\theta_B = \theta_C = \frac{Wa^2}{2EI}$ $\quad$ when the point load is at a distance of '*a*' from the fixed end

$$\theta_B = \frac{w \cdot L^2}{6EI}$$ when it carries a uniformly distributed load over the whole length.

$$\theta_B = \theta_C = \frac{wa^3}{6EI}$$ when it carries a uniformly distributed load over a length '*a*' from the fixed end.

$$\theta_B = \frac{w \cdot L^3}{6EI} - \frac{w \cdot (L-a)^3}{6EI}$$ when it carries a uniformly distributed load over a distance '*a*' from the free end

$$\theta_B = \frac{w \cdot L^3}{24EI}$$ when it carries a gradually varying load from zero at the free end to *w*/m run at fixed end.

where W = Point load,

w = Uniformly distributed load,

L = Length of beam,

I = Moment of inertia, and

E = Young's modulus.

2. The deflection *i.e.*, y of a cantilever of length L, at the free end is given by,

$$y_B = \frac{WL^3}{3EI}$$ when the point load is at the free end

$$y_B = \frac{Wa^3}{3EI} + \frac{Wa^2}{2EI}(L-a)$$ when the point load is at a distance of 'a' from the fixed end

$$y_B = \frac{wL^4}{8EI}$$ when it carries a uniformly distributed load over the whole length

$$y_B = \frac{wa^4}{8EI} + \frac{w \cdot a^3}{6EI}(L-a)$$ when it carries a uniformly distributed load over a length '*a*' from the fixed end.

$$y_B = \frac{wL^4}{8EI} - \left[\frac{w(L-a)^4}{8EI} + \frac{w(L-a)^3}{6EI} \times a\right]$$ when it carries a uniformly distributed load over a length '*a*' from the free end

$$y_B = \frac{wL^4}{30EI}$$ when it carries a gradually varying load from zero at the free end to *w*/m run at the fixed end.

3. For a cantilever, at the fixed end slope and deflections are zero. Hence moment area method can be easily applied for finding slope and deflections of cantilevers. The slope (θ_B) and deflection (y_B) at the free end is given by,

$$\theta_B = \frac{\text{Area of B.M. diagram between free end and fixed end}}{EI}$$

and
$$y_B = \frac{A\bar{x}}{EI}$$

where A = Area of B.M. Diagram and

$\bar{x}$ = Distance of C.G. of B.M. diagram from free end.

4. Area of B.M. diagram sometimes is found easily by splitting the combined areas into triangles and rectangles.

EXERCISE (4B)

(A) Theoretical Questions

1. What is a cantilever ? What are the different methods of finding of slope and deflection of a cantilever ?
2. Derive an expression for the slope and deflection of a cantilever of length L, carrying a point load W at the free end by double integration method.
3. Solve question 2, by moment area method.
4. Prove that the slope and deflection of a cantilever carrying uniformly distributed load over the whole length are given by,

$$\theta_B = \frac{wL^3}{6EI} \quad \text{and} \quad y_B = \frac{wL^4}{8EI}$$

where w = Uniformly distributed load and
EI = Flexural rigidity.

5. Find the expression for the slope and deflection of a cantilever of length L which carries a uniformly distributed load over a length 'a' from the fixed end by
   (*i*) Double integration method and (*ii*) Moment area method.
6. Prove that the slope and deflection of a cantilever length L, which carries a gradually varying load from zero at the free end to w/m run at the fixed end are given by :

$$\theta_B = \frac{wL^3}{24EI} \quad \text{and} \quad y_B = \frac{wL^4}{30EI}$$

where EI = Flexural rigidity.

(B) Numerical Problems

1. A cantilever of length 2 m carries a point load of 30 kN at the free end. If moment of inertia $I = 10^8$ mm^4 and value of $E = 2 \times 10^5$ N/mm^2, then find :
   (*i*) slope of the cantilever at the free end and
   (*ii*) deflection at the free end. [**Ans.** (*i*) 0.003 rad. ; (*ii*) 4 mm]
2. A cantilever of length 3 m carries a point load of 60 kN at a distance of 2 m from the fixed end. If $E = 2 \times 10^5$ and $I = 10^8$, find :
   (*i*) slope at the free end and (*ii*) deflection at the free end.
   [**Ans.** (*i*) 0.006 rad. ; (*ii*) 14 mm]
3. A cantilever of length 30 m carries a uniformly distributed load of 24 kN/m length over the entire length. If moment of inertia of the beam = 10^8 mm^3 and value of $E = 2 \times 10^5$ N/mm^2, determine the slope and deflection at the free end. [**Ans.** 0.0054 rad. ; 12.15 mm]
4. A cantilever of length 3 m carries a uniformly distributed load over the entire length. If the slope at the free end is 0.01777 radians, find the deflection at the free end. [**Ans.** 39.99 mm]
5. Determine the slope and deflection at the free end of a cantilever of length 4 m which is carrying a uniformly distributed load of 12 kN/m over a length of 3 m from the fixed end. Take $EI = 2 \times 10^{13}$ N/mm^2. [**Ans.** 0.0027 rad., 8.775 mm]
6. A cantilever of length 3 m carries a uniformly distributed load of 15 kN/m over a length of 2 m from the free end. If $I = 10^8$ mm^4 and $E = 2 \times 10^5$ N/mm^2, find :
   (*i*) slope at the free end and (*ii*) deflection at the free end.
   [**Ans.** (*i*) 0.00326 rad. ; (*ii*) 7.25 mm]
7. A cantilever of length 2 m carries a load of 20 kN at the free end and 30 kN at a distance 1 m from the end. Find the slope and deflection at the free end. Take $E = 2.0 \times 10^5$ N/mm^2 and $I = 1.5 \times 10^8$ mm^4. [**Ans.** 0.00183 rad., 2.6 mm]

8. Determine the deflection at the free end of a cantilever which is 2 m long and carries a point load of 9 kN at the free end and a uniformly distributed load of 8 kN/m over a length of 1 m from the fixed end.
Take $I = 2.25 \times 10^7$ mm³ and $E = 2.2 \times 10^5$ N/mm². [**Ans.** 6.54 mm]

9. A cantilever of length 2 m carries a uniformly varying load of zero intensity at the free end, and 45 kN/m at the fixed end. If $E = 2 \times 10^5$ N/mm² and $I = 10^8$ mm⁴, find the slope and deflection of the free end. [**Ans.** 0.00075 rad. ; 1.2 mm]

10. A cantilever of length 2 m carries a point load of 30 kN at the free end and another load of 30 kN at its centre. If $EI = 10^{13}$ N/mm² for the cantilever then determine by moment area method, the slope and deflection at the free end of cantilever. [**Ans.** 0.0075 rad. ; 10.50 mm]

11. A cantilever of length 'L' carries a U.D.L. of w per unit for a length of $\frac{L}{2}$ from the fixed end. Determine the slope and deflection at the free end using area moment method.
(Bangalore University, 1990)

[**Hint.** See Article 4.11.3 on page 238. Here $a = \frac{L}{2}$. Hence slope at free end,

$$\theta_B = \frac{w \times a^3}{6EI} = \frac{w \times \left(\frac{L}{2}\right)^3}{6EI} = \frac{w \times L^3}{48EI}.$$

and deflection at free end,

$$y_B = \frac{w \times a^3}{6EI}(L - a) + \frac{w \times a^4}{8EI}$$

$$= \frac{w \times \left(\frac{L}{2}\right)^3}{6EI}\left(L - \frac{L}{2}\right) + \frac{w \times \left(\frac{L}{2}\right)^4}{8EI} = \frac{w \times L^3}{48EI} \times \frac{L}{2} + \frac{w \times L^4}{128EI}$$

$$= \frac{w \times L^4}{96EI} + \frac{wL^4}{128EI} = \frac{w \times L^4}{EI}\left(\frac{1}{96} + \frac{1}{128}\right) = \frac{wL^4}{EI}\left(\frac{4+3}{128 \times 3}\right)$$

$$= \frac{WL^4}{EI} \times \frac{7}{384} = \frac{7}{384}\frac{w \times L^4}{EI}$$ **Ans.**]

(C) FIXED AND CONTINUOUS BEAMS

4.13. FIXED BEAM

A beam whose both ends are fixed is known as a fixed beam. Fixed beam is also called a *built-in* or *encaster beam.* In case of a fixed beam both its ends are rigidly fixed and the slope and deflection at the fixed ends are zero. But the fixed ends are subjected to end moments. Hence end moments are not zero in case of a fixed beam.

In case of simply supported beam, the deflection is zero at the ends. But the slope is not zero at the ends as shown in Fig. 4.36 (*a*).

In case of fixed beam, the deflection and slope are zero at the fixed ends as shown in Fig. 4.36 (*b*). The slope will be zero at the ends if the deflection curve is horizontal at the ends. To bring the slope back to zero (*i.e.,* to make the deflection curve horizontal at the fixed ends), the end moments M_A and M_B will be acting in which M_A will be acting anti-clockwise and M_B will be acting clockwise as shown in Fig. 4.36 (*b*).

In case of fixed beams the B.M. diagram, slope and deflection for various types of loading such as point loads, uniformly distributed load and combination of point load and u.d.1., are discussed.

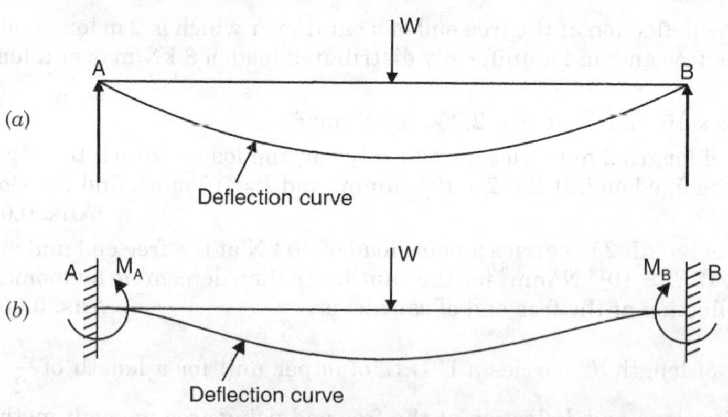

Fig. 4.36

4.13.1. Bending Moment Diagram for Fixed Beams. Fig. 4.37 shows a fixed beam *AB* of length *L* subjected to two loads *W* and 2*W* at a distance of $\frac{L}{4}$ from each ends.

Let R_A = Reaction at *A*
R_B = Reaction at *B*
M_A = Fixed end moment at *A*
M_B = Fixed end moment at *B*

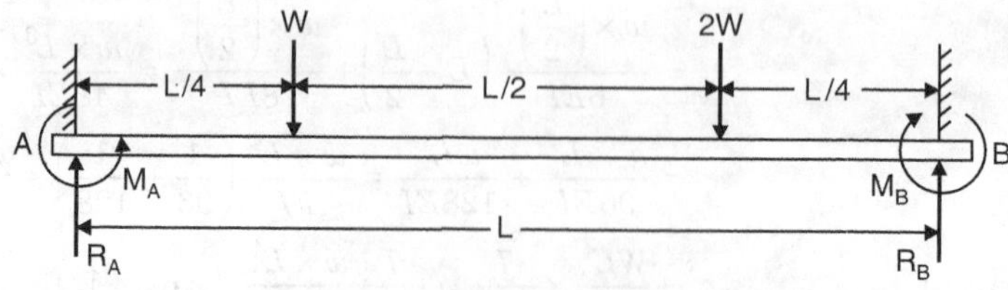

Fig. 4.37

The above four quantities *i.e.*, R_A, R_B, M_A and M_B are unknown.

The values of R_A, R_B, M_A and M_B are calculated by analysing the given beam in the following two stages :

(i) A simply supported beam subjected to given vertical loads as shown in Fig. 4.38.

Consider the beam *AB* as simply supported.

Let R_A^* = Reaction at *A* due to vertical loads
R_B^* = Reaction at *B* due to vertical loads.

Taking moments about *A*, we get

$$R_B^* \times L = W \times \frac{L}{4} + 2W \times \frac{3L}{4}$$

$$\therefore \quad R_B^* = \frac{W}{4} + \frac{6W}{4} = \frac{7W}{4}$$

and

$$R_A^* = \text{Total load} - R_B^*$$

$$= 3W - \frac{7W}{4} = \frac{5W}{4}$$

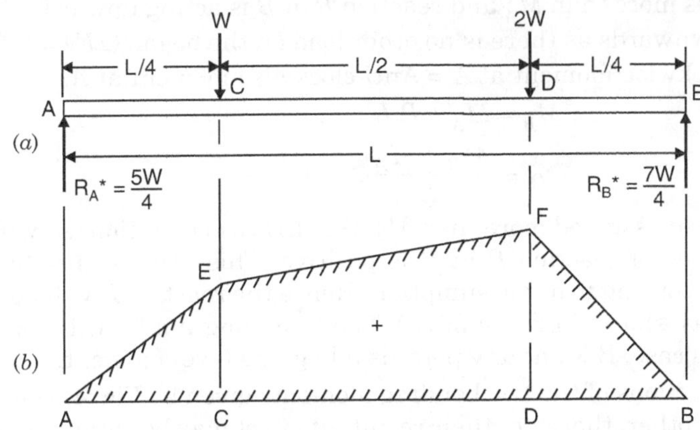

B.M. diagram considering beam as simply supported

Fig. 4.38

B.M. at $A = 0$, B.M. at $B = 0$.

B.M. at $C = \frac{5W}{4} \times \frac{L}{4} = \frac{5WL}{16}$

B.M. at $D = \frac{7W}{4} \times \frac{L}{4} = \frac{7WL}{16}$

Now B.M. diagram can be drawn as shown in Fig. 4.38 (*b*). In this case, B.M. at any point is a sagging (+ve) moment.

(ii) A simply supported beam subjected to end moments only (without given loading) as shown in Fig. 4.39.

Let M_A = Fixed end moment at A

M_B = Fixed end moment at B

R = Reaction* at each end due to these moments.

As the vertical loads acting on the beam are not symmetrical (they are W at distance $L/4$ from A and $2W$ at a distance $L/4$ from B), the fixed end moments will be different.

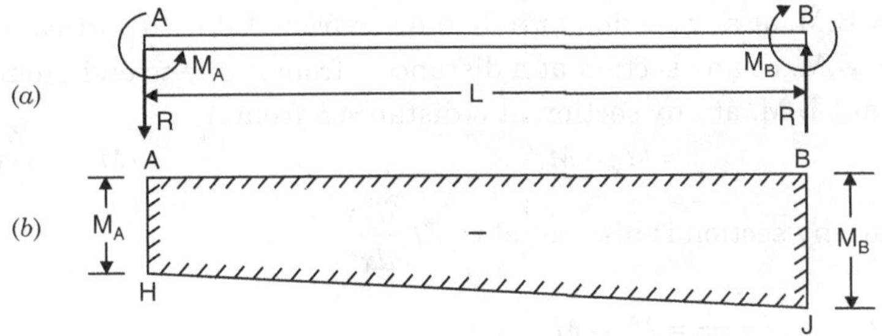

B.M. diagram due to end moments

Fig. 4.39

*The reaction at each end will be equal. There is no vertical load on the beam hence reaction at A + reaction at $B = O$. Or reaction at A = – reaction at B.

Suppose M_B is more than M_A and reaction R at B is acting upwards. Then reaction R at A will be acting downwards as there is no other load on the beam. ($\Sigma Fy = 0$). Taking moments about A, we get clockwise moment at A = Anti-clockwise moment at A.

$$M_B = M_A + R.L$$

$$\therefore \quad R = \frac{M_B - M_A}{L} \qquad \text{...(A)}$$

As M_B has been assumed more than M_A, the R.H.S. of equation (A) will be positive. This means the magnitude of reaction R at B is positive. This also means that the direction of reaction R at B is according to our assumption. Hence the reaction R will be upwards at B and downwards at A as shown in Fig. 4.39 (*a*). The B.M. diagram for this condition is shown in Fig. 4.39 (*b*). In this case, B.M. at any point is a hogging (–ve) moment.

Since the directions of the two bending moments given by Fig. 4.38 (*b*) and Fig. 4.39 (*b*) are opposite to each other, therefore their resultant effect may be obtained by drawing the two moments on the same side of the base AB, as shown in Fig. 4.40.

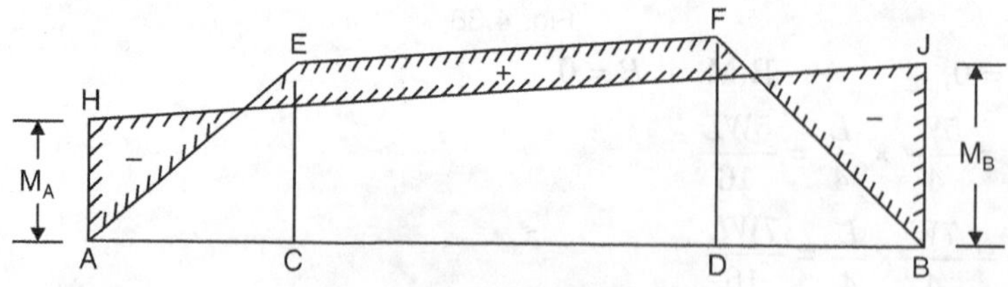

Fig. 4.40

Now the final reactions R_A and R_B are given by

$$R_A = R_A{}^* - R$$

and

$$R_B = R_B{}^* + R$$

In the above two equations, $R_A{}^*$ and $R_B{}^*$ are already calculated. They are : $R_A{}^* = 5W/4$ and $R_B{}^* = 7W/4$. But the value of R is in terms of M_B and M_A. It is given by $R = (M_B - M_A)/L$. Hence to find the value of R, we must calculate the value of M_B and M_A first.

To find the values of M_A and M_B

Let M_x = B.M. at any section at a distance x from A due to vertical loads

$M_x{}'$ = B.M. at any section at a distance x from A due to end moments.

The resultant B.M. at any section at a distance x from A

$$= M_x - M_x{}' \qquad (M_x \text{ is +ve but } M_x{}' \text{ is –ve})$$

But B.M. at any section is also equal to $EI \dfrac{d^2y}{dx^2}$

$$\therefore \quad EI\frac{d^2y}{dx^2} = M_x - M_x{}' \qquad \text{...(i)}$$

Integrating the above equation for the entire length, we get

$$EI\left[\frac{dy}{dx}\right]_0^L = \int_0^L M_x dx - \int_0^L M_x{}' dx$$

But $\frac{dy}{dx}$ represents the slope. And slope at the fixed ends *i.e.*, at A and B are zero. The above equation can be written as

$$EI\left[\left(\frac{dy}{dx}\text{ at } x=L\right)-\left(\frac{dy}{dx}\text{ at } x=0\right)\right]$$

$$=\int_0^L M_x.dx-\int_0^L M_x'dx$$

$$EI\,[0-0]=\int_0^L M_x.dx-\int_0^L M_x'dx$$

or

$$0=\int_0^L M_x.dx-\int_0^L M_x'dx \quad ...(ii)$$

Now $\int_0^L M_x.dx$ represents the area of B.M. diagram due to vertical loads and $\int_0^L M_x'.dx$ represents the area of B.M. diagram due to end moments.

Let a = Area of B.M. diagram due to vertical loads

a' = Area of B.M. diagram due to end moments.

Then

$$\int_0^L M_x.dx = a$$

and

$$\int_0^L M_x'.dx = a'$$

Substituting these values in equation (*ii*), we get

$$0 = a - a'$$

or

$$a = a' \quad ...(4.24)$$

The above equation shows that area of B.M. diagram due to vertical loads is equal to the area of B.M. diagram due to end moments.

Again consider the equation (*i*)

$$EI\frac{d^2y}{dx^2}=M_x-M_x'$$

Multiplying the above equation by x, we get

$$EI.x.\frac{d^2y}{dx^2}=x.M_x-x.M_x'$$

Integrating for the whole length of the beam *i.e.*, from 0 to L, we get

$$\int_0^L EI.x\frac{d^2y}{dx^2}.dx=\int_0^L x.M_x.dx-\int_0^L x.M_x'dx$$

$$EI\int_0^L x.\frac{d^2y}{dx^2}.dx=\int_0^L x.M_x.dx-\int_0^L x.M_x'.dx \quad ...(iii)$$

In the above equation, $M_x.dx$ represents the area of B.M. diagram due to vertical loads at a distance x from the end A. And the term $(x.M_x.dx)$ represents the moment of area of B.M.

diagram about the end A. Hence $\int_0^L x.M_x.dx$ represents the moment of the total area of B.M. diagram due to vertical loads about A, and it is equal to total area of B.M. diagram due to vertical loads multiplied by the distance of C.G. of area from A,

$$\therefore \qquad \int_0^L x.M_x dx = a\bar{x}$$

where $\bar{x}$ = Distance of the C.G. of B.M. diagram due to vertical loads.

$$\text{Similarly} \qquad \int_0^L x.M_x'.dx = a'\bar{x}'$$

where $\bar{x}'$ = Distance of the C.G. of B.M. diagram due to end moments.

Substituting the above values in equation (*iii*), we get

$$EI\int_0^L x.\frac{d^2y}{dx^2}.dx = a\bar{x} - a'\bar{x}'$$

or
$$EI\left[x\frac{dy}{dx} - y\right]_0^L = a\bar{x} - a'\bar{x}' \qquad \left[\because \frac{d}{dx}\left(x\frac{dy}{dx} - y\right) = \left(x\frac{d^2y}{dx^2} + \frac{dy}{dx}\right) - \frac{dy}{dx} = x\frac{d^2y}{dx^2}\right]$$

or
$$EI\left[\left(\frac{xdy}{dx} - y\right)_{\text{at } x = L} - \left(x\frac{dy}{dx} - y\right)_{\text{at } x = 0}\right] = a\bar{x} - a'\bar{x}'$$

or
$$EI\,[(L\theta_B - y_B) - (0 \times \theta_A - y_A)] = a\bar{x} - a'\bar{x}'.$$

Since slope and deflection at A and B are zero, hence θ_A, θ_B, y_A and y_B are zero.

$$\therefore \qquad 0 = a\bar{x} - a'\bar{x}'$$

or
$$a\bar{x} = a'\bar{x}' \qquad \text{...(4.25)}$$

But from equation (4.24), we have

$$a = a'$$

$$\therefore \qquad \bar{x} = \bar{x}' \qquad \text{...(4.26)}$$

Hence the distance of C.G. of B.M. diagram due to vertical loads from A is equal to the distance of C.G. of B.M. diagram due to end moments from A.

Now by using equations (4.24) and (4.26) the unknowns M_A and M_B can be calculated. This also means that M_A and M_B can be calculated by

(*i*) equating the area of B.M. diagram due to vertical loads to the area of B.M. diagram due to end moments.

(*ii*) equating the distance of C.G. of B.M. diagram due to vertical loads to the distance of C.G. of B.M. diagram due to end moments. The distance of C.G. must be taken from the same end in both cases.

4.13.2. Slope and Deflection for a Fixed Beam carrying a Point Load at the Centre

Fig. 4.41 (*a*) shows a fixed beam AB of length L, carrying a point load W at the centre C of the beam.

Let M_A = Fixed end moment at A

M_B = Fixed end moment at B

R_A = Reaction at A

R_B = Reaction at B.

The above four are unknown *i.e.*, R_A, R_B, M_A and M_B are unknown.

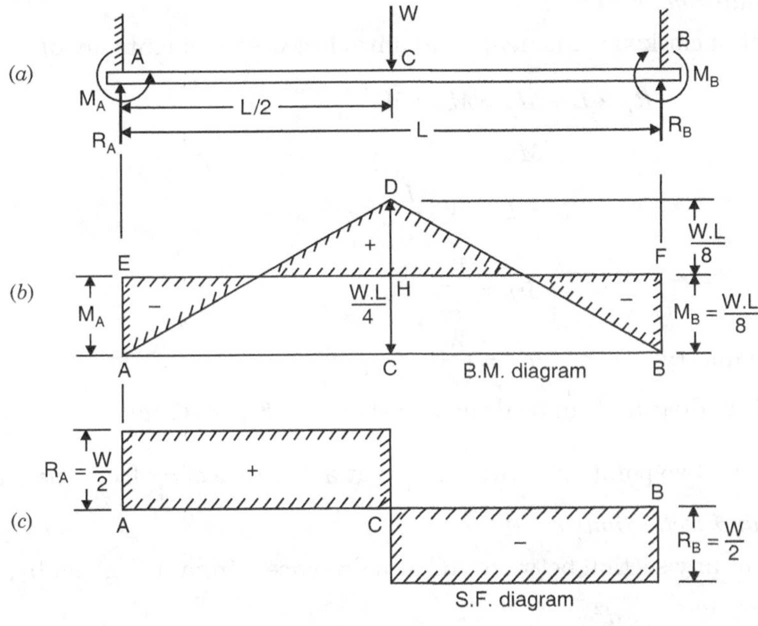

Fig. 4.41

(i) B.M. Diagram

Due to symmetry, the end moments M_A and M_B will be equal. Hence the B.M. diagram due to end moments will be a rectangle as shown in Fig. 4.41 (*b*) by *AEFB*. Here the magnitude of M_A and M_B are unknown. The bending moment diagram for a simply supported beam carrying a point load at the centre will be a triangle with the maximum B.M. at the centre equal to $\frac{W.L}{4}$. The B.M. diagram for this case is shown in Fig. 4.41 (*b*) by a triangle *ADB* in which

$$CD = \frac{W.L}{4}.$$

Now according to equation (4.24), area of B.M. diagram due to vertical loads should be equal to the area of B.M. diagram due to end moments.

∴ Equating the areas of the two bending moment diagrams, we get

Area of triangle *ADC* = Area of rectangle *AEFB*

or
$$\frac{1}{2} \times AB \times CD = AB \times AE$$

or
$$\frac{1}{2} \times L \times \frac{W.L}{4} = L \times M_A$$

∴
$$M_A = \frac{W.L}{8}$$

Also $$M_B = M_A = \frac{W.L}{8} \quad ...(4.27)$$

Now the B.M. diagram can be drawn as shown in Fig. 4.41 (*b*).

(*ii*) *S.F. Diagram*

Equating the clockwise moments and anti-clockwise moments about A, we get

$$R_B \times L + M_A = M_B + W \cdot \frac{L}{2}$$

But $$M_A = M_B$$

$\therefore$ $$R_B \times L = W \cdot \frac{L}{2}$$

or $$R_B = \frac{W}{2}$$

Due to symmetry, $$R_A = \frac{W}{2}$$

Now the S.F. diagram can be drawn as shown in Fig. 4.41 (*c*).

There will be two points of contraflexure at a distance of $\frac{L}{4}$ from the ends.

(*iii*) *Slope and Deflection*

The B.M. at any section between AC at a distance x from A is given by,

$$EI\frac{d^2y}{dx^2} = R_A \times x - M_A$$

$$= \frac{W.x}{2} - \frac{W.L}{8} \quad \left(\because M_A = \frac{W.L}{8},\ R_A = \frac{W}{2}\right)$$

Integrating the above equation, we get

$$EI\frac{dy}{dx} = \frac{W}{2} \cdot \frac{x^2}{2} - \frac{W.L}{8} \cdot x + C_1$$

where C_1 is a constant of integration.

At $x = 0$, $\frac{dy}{dx} = 0$. Hence $C_1 = 0$

Therefore the above equation becomes as

$$EI\frac{dy}{dx} = \frac{Wx^2}{4} - \frac{W.L}{8}x \quad ...(i)$$

The above equation (*i*) gives the *slope* of the beam at any point :

Integrating equation (*i*) again, we get

$$EIy = \frac{W}{4} \cdot \frac{x^3}{3} - \frac{WL}{8} \cdot \frac{x^2}{2} + C_2$$

where C_2 is another constant of integration. At $x = 0$, $y = 0$. Hence $C_2 = 0$.

Therefore the above equation becomes as

$$EIy = \frac{W.x^3}{12} - \frac{W.Lx^2}{16} \quad ...(ii)$$

The above equation (*ii*) gives the *deflection* of the beam at any point. The deflection is maximum at the centre of the beam, where $x = \frac{L}{2}$. Hence substituting $x = \frac{L}{2}$ in equation (*ii*), we get

$$EIy_{max} = \frac{W}{12}\left(\frac{L}{2}\right)^3 - \frac{W.L}{16}.\left(\frac{L}{2}\right)^2$$

$$= \frac{WL^3}{96} - \frac{WL^3}{64} = \frac{2WL^2 - 3WL^3}{192} = -\frac{WL^3}{192}$$

or

$$y_{max} = \frac{-WL^3}{192\,EI}$$

Minus sign means that the deflection is downwards.

$$\therefore \quad \text{Downward deflection, } y_{max} = \frac{WL^3}{192\,EI} \quad ...(4.28)$$

Note. The deflection at the centre of a simply supported beam carrying a point load W at the centre is $\frac{WL^3}{48EI}$. Hence the deflection of the simply supported beam is four times the deflection of the fixed beam.

Or in other words, the deflection of a fixed beam is one fourth times the deflection of the simply supported beam. Hence when fixed beams are used, the deflection will be less.

Problem 4.26. *A fixed beam AB, 6 m long, is carrying a point load of 50 kN at its centre. The moment of inertia of the beam is 78 × 10⁶ mm⁴ and value of E for beam material is 2.1 × 10⁵ N/mm². Determine :*

(i) Fixed end moments at A and B, and

(ii) Deflection under the load.

Sol. Given :

Length, $L = 6 \text{ m} = 6000 \text{ mm}$

Point load, $W = 50 \text{ kN} = 50{,}000 \text{ N}$

M.O.I., $I = 78 \times 10^8 \text{ mm}^4$

Value of $E = 2.1 \times 10^5 \text{ N/mm}^2$

Let M_A = Fixed end moment at A,

M_B = Fixed end moment at B,

y_{max} = Deflection under the central point load.

Using equation (4.27), we get

$$M_A = M_B = \frac{W.L}{8}$$

$$= \frac{50 \times 6}{8} = \textbf{37.5 kNm.} \quad \textbf{Ans.}$$

Using equation (4.28), we get

$$y_{max} = \frac{WL^3}{192\ EI}$$

$$= \frac{50000 \times 6000^3}{192 \times 2.1 \times 10^5 \times 78 \times 10^8} = \mathbf{3.434\ mm.}\quad \textbf{Ans.}$$

Alternate Method

Fig. 4.42 (*b*) shows the simply supported beam, which is having Max. B.M. at the centre equal to $R_A^* \times 3 = 25 \times 3 = 75$ kNm. Fig. 4.42 (*c*) shows the B.M. diagram for simply supported beam.

Fig. 4.42 (*d*) shows the fixed beam with end moments only. Due to symmetry end moments are equal. Hence $M_A = M_B$. Fig. 4.42 (*e*) shows the B.M. diagram due to end moments only. This diagram is a rectangle.

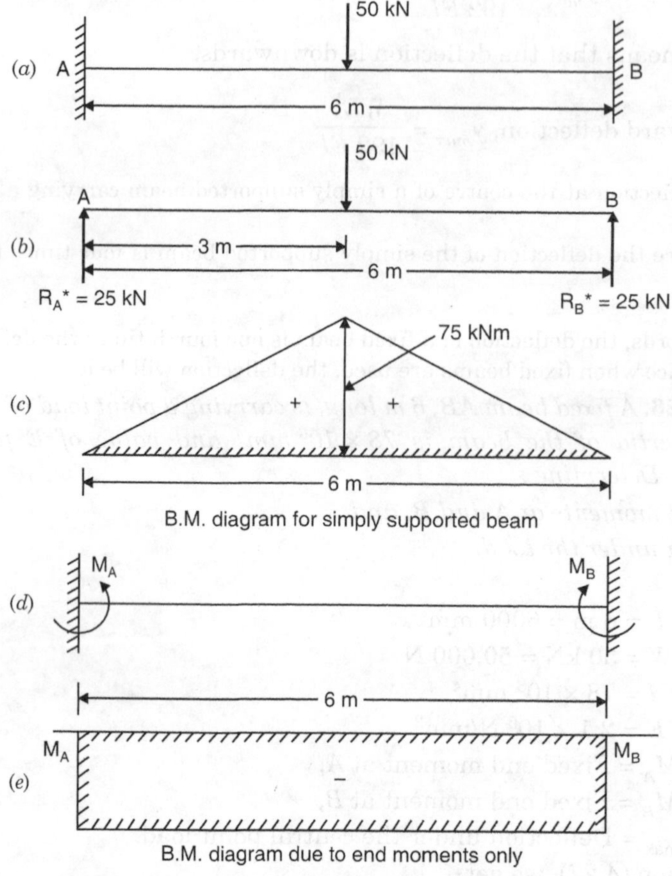

Fig. 4.42

Equating the areas of two B.M. diagrams, we get

Area of B.M. diagram for simply supported beam

= Area of B.M. diagram due to end moments.

i.e., $$\frac{75 \times 6}{2} = M_A \times 6$$

or $$M_A = \frac{75}{2} = 37.5 \text{ kNm}$$

But $M_A = M_B$

$\therefore$ $M_A = M_B$

= **37.5 kNm. Ans.**

4.13.3. Slope and Deflection for a Fixed Beam carrying an Eccentric Point Load. Fig. 4.43 (*a*) shows a fixed beam *AB* of length *L*, carrying a point load *W* at *C* at a distance of '*a*' from *A* and at a distance of '*b*' from *B*. The fixed end moments M_A and M_B and also reactions at *A* and *B i.e.*, R_A and R_B are shown in the same figure.

(*i*) *B.M. Diagram*

As the load is not acting symmetrically, therefore M_A and M_B will be different. In this case M_B will be more than M_A as the load is nearer to point *B*. The B.M. diagram due to end moments will be trapezium as shown in Fig. 4.43 (*b*) by *AEFB*. Here the length *AE* (*i.e.*, M_A) and *BF* (*i.e.*, M_B) are unknown.

The B.M. diagram for a simply supported beam carrying an eccentric point load will be triangle with maximum B.M. under the point load equal to $\frac{W.a.b}{L}$. The B.M. diagram for this case is shown in Fig. 4.43 (*b*) by a triangle *ADB* in which $CD = \frac{W.a.b.}{L}$.

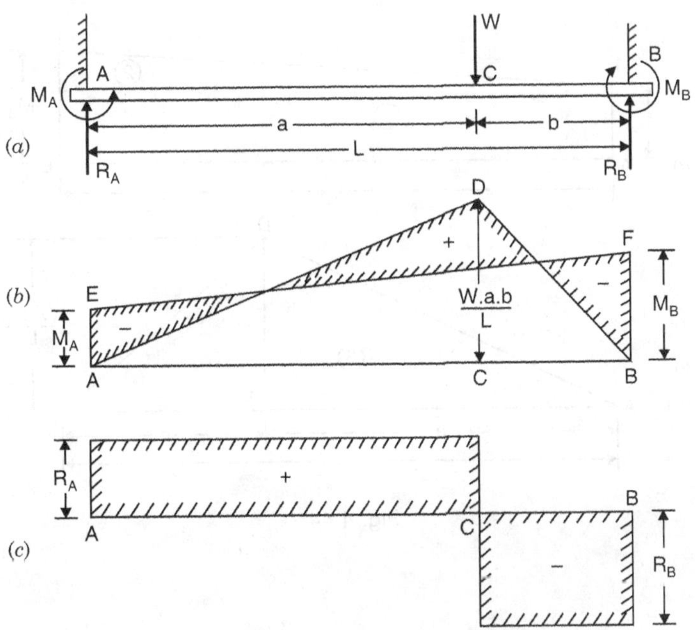

Fig. 4.43

Equating the areas of the two bending moment diagrams, we get

Area of trapezium $AEFB$ = Area of triangle ADB

$$\frac{1}{2}(AE + BF).AB = \frac{1}{2} \times AB \times CD$$

$$\frac{1}{2}(M_A + M_B).L = \frac{1}{2} \times L \frac{W.a.b}{L}$$

or $$M_A + M_B = \frac{W.a.b}{L} \quad \ldots(i)$$

Now using equation (4.3),

$$\bar{x} = \bar{x}'$$

or Distance of C.G. of B.M. diagram due to vertical loads from A = Distance of C.G. of B.M. diagram due to end moments from A.

Now $$\bar{x}' = \frac{A_1x_1 + A_2x_2}{(A_1 + A_2)} \quad \text{[See Fig. 4.44 } (a)]$$

$$= \frac{(M_A \, . \, L) \, . \, \frac{L}{2} + \frac{1}{2} \, . \, L(M_B - M_A) \times \frac{2L}{3}}{M_A \, . \, L + \frac{1}{2} \, . \, L(M_B - M_A)}$$

$$= \frac{M_A \, . \frac{L}{2} + (M_B - M_A) \, . \, \frac{L}{3}}{M_A + \frac{1}{2}(M_B - M_A)} = \frac{3M_A \, . \, L + 2M_B \, . \, L - 2M_A . L}{3(2M_A + M_B - M_A)}$$

$$= \frac{M_A \, . \, L + 2M_B \, . \, L}{3(M_A + M_B)} = \frac{(M_A + 2M_B) \, . \, L}{3(M_A + M_B)}$$

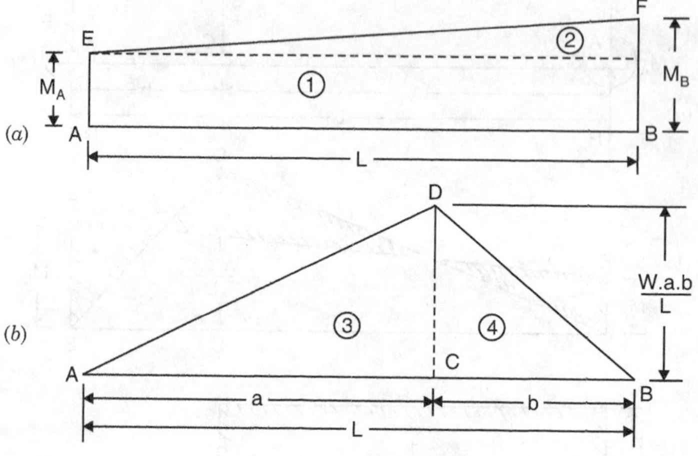

Fig. 4.44

and $$\bar{x} = \frac{A_3 x_3 + A_4 x_4}{(A_3 + A_4)}$$ [See Fig. 4.44 (*b*)]

$$= \frac{\left(\frac{1}{2} \times a \times CD\right) \times \frac{2a}{3} + \frac{1}{2} . b . CD \times \left(a + \frac{b}{3}\right)}{\frac{1}{2} . a . CD + \frac{1}{2} . b . CD}$$

$$= \frac{\frac{2a^2}{3} + b\left(a + \frac{b}{3}\right)}{a + b}$$ $\left(\text{Cancelling } \frac{CD}{2}\right)$

$$= \frac{2a^2 + 3ab + b^2}{3(a+b)} = \frac{2a^2 + 2ab + ab + b^2}{3(a+b)}$$

$$= \frac{2a(a+b) + b(a+b)}{3(a+b)} = \frac{(2a+b)(a+b)}{3(a+b)}$$

$$= \frac{2a+b}{3} = \frac{a + (a+b)}{3} = \frac{a+L}{3}$$ $(\because \quad a + b = L)$

But $$\bar{x}' = \bar{x}$$

$\therefore$ $$\frac{(M_A + 2M_B) . L}{3(M_A + M_B)} = \frac{a+L}{3}$$

or $$M_A + 2M_B = \frac{(a+L)(M_A + M_B)}{L}$$

$$= \frac{(a+L)}{L} . \frac{W.a.b}{L}$$ $\left[\because \; M_A + M_B = \frac{W.a.b}{L} \text{ from equation } (i)\right]$

$$= (a + L) . \frac{W.a.b}{L^2} \quad \text{...(ii)}$$

Subtracting equation (*i*) from equation (*ii*), we get

$$M_B = (a + L) . \frac{W.a.b}{L^2} - \frac{W.a.b}{L}$$

$$= \frac{W.a.b}{L}\left(\frac{a+L}{L} - 1\right)$$

$$= \frac{W.a.b}{L}\left(\frac{a+L-L}{L}\right) = \frac{W.a^2.b}{L^2} \quad \text{...(iii)}$$

Substituting the value of M_B in equation (*i*), we get

$$M_A + \frac{W.a^2.b}{L^2} = \frac{W.a.b}{L}$$

$\therefore$ $$M_A = \frac{W.a.b}{L} - \frac{Wa^2b}{L^2}$$

$$= \frac{W.a.b}{L^2}(L - a) = \frac{W.a.b.b}{L^2}$$ $(\because \quad L - a = b)$

$$= \frac{W.a.b^2}{L^2} \quad \text{...(iv)}$$

Now M_A and M_B are known and hence bending moment diagram can be drawn. From equations (*iii*) and (*iv*), it is clear that if $a > b$ than $M_B > M_A$.

(*ii*) *S. F. Diagram*

Equating the clockwise moments and anti-clockwise about A,

$$R_B \times L + M_A = M_B + W.a$$

$$\therefore \quad R_B = \frac{(M_B - M_A) + W.a}{L}$$

Similarly $$R_A = \frac{(M_A - M_B) + W.b}{L}$$

By substituting the values of M_A and M_B from equations (*iii*) and (*iv*), in the above equations, we shall get R_A and R_B. Now S.F. can be drawn as shown in Fig. 4.43 (*c*).

(*iii*) *Slope and Deflection*

The B.M. at any section between AC at a distance x is given by

$$EI\frac{d^2y}{dx^2} = R_A \times x - M_A$$

Substituting the value of R_A in the above equation, we get

$$EI\frac{d^2y}{dx^2} = \left[\frac{(M_A - M_B) + W.b}{L}\right].x - M_A$$

$$= \frac{(M_A - M_B)}{L}x + \frac{W.b}{L}.x - M_A$$

$$= \frac{W.b}{L}.x - \left[M_A + (M_B - M_A)\frac{x}{L}\right]$$

Substituting the values of M_A and M_B, we get

$$EI\frac{d^2y}{dx^2} = \frac{W.b}{L}.x - \left[\frac{W.a.b^2}{L^2} + \left(\frac{W.a^2.b}{L^2} - \frac{W.a.b^2}{L^2}\right)\frac{x}{L}\right]$$

$$= \frac{W.b.x}{L} - \frac{W.a.b^2}{L^2} - \left(\frac{W.a^2.b}{L^2} - \frac{W.a.b^2}{L^2}\right)\frac{x}{L}$$

$$= \frac{W.b.x}{L} - \frac{W.a.b^2}{L^2} - \frac{W.a.b}{L^2}(a-b).\frac{x}{L}$$

$$= \frac{W.b.x}{L} - \frac{W.a.b}{L^2}(a-b).\frac{x}{L} - \frac{W.a.b^2}{L^2}$$

$$= \frac{W.b}{L^3}[L^2x - a(a-b)x] - \frac{W.a.b^2}{L^2}$$

$$= \frac{W.b}{L^3}(L^2 - a^2 + ab)x - \frac{W.a.b^2}{L^2}$$

But $$L = a + b$$

$$\therefore \quad L^2 = (a+b)^2 = a^2 + b^2 + 2ab.$$

Substituting the value of L^2 in the above equation, we get

$$\therefore \quad EI\frac{d^2y}{dx^2} = \frac{W.b}{L^3}(a^2 + b^2 + 2ab - a^2 + ab)x - \frac{W.a.b^2}{L^2}$$

$$= \frac{W.b}{L^3}(b^2 + 3ab)x - \frac{W.a.b^2}{L^2}$$

$$= \frac{W.b^2}{L^3}(b + 3a)x - \frac{W.a.b^2}{L^2}$$

Integrating the above equation, we get

$$EI\frac{dy}{dx} = \frac{W.b^2}{L^3}(b + 3a).\frac{x^2}{2} - \frac{W.a.b^2}{L^2}x + C_1$$

where C_1 is a constant of integration.

At $x = 0$, $\frac{dy}{dx} = 0$. Hence $C_1 = 0$.

$$\therefore \quad EI\frac{dy}{dx} = \frac{W.b^2}{2L^3}(b + 3a).x^2 - \frac{W.a.b^2}{L^2}.x \qquad ...(v)$$

Integrating again, we get

$$EI.y = \frac{W.b^2}{2L^2}(b + 3a).\frac{x^3}{3} - \frac{W.a.b^2}{L^2}.\frac{x^2}{2} + C_2$$

where C_2 is another constant of integration. At $x = 0$, $y = 0$. Hence $C_2 = 0$.

$$\therefore \quad EI.y = \frac{W.b^2}{6L^3}(b + 3a)x^3 - \frac{W.a.b^2}{2L^2}.x^2 \qquad ...(vi)$$

The deflection under the load is obtained by substituting $x = a$ in the above equation. Let y_c is the deflection under the load, then

$$EI.y_c = \frac{W.b^2}{6L^3}(b + 3a).a^3 - \frac{W.a.b^2}{2L^2}.a^2$$

$$= \frac{W.b^2.a^3}{6L^3}.(b + 3a - 3L)$$

$$= -\frac{W.a^3.b^2}{6L^3}(3L - 3a - b)$$

$$= -\frac{W.a^3.b^2}{6L^3}[3(L - a) - b]$$

$$= -\frac{W.a^3.b^2}{6L^3}(3b - b) \qquad (\because \; L - a = b)$$

$$= -\frac{W.a^3.b^3}{3L^3}$$

$$\therefore \quad y_c = -\frac{W.a^3.b^3}{3EIL^3} \qquad ...(4.29)$$

Maximum deflection

Since $a > b$, hence maximum deflection will take place between A and C. For maximum deflection, $\frac{dy}{dx}$ should be zero. Hence substituting $\frac{dy}{dx} = 0$ in equation (v), we get

$$0 = \frac{W.b^2}{2L^3}(b + 3a).x^2 - \frac{W.a.b^2}{L^2}.x$$

$$\frac{W.a.b^2}{L^2}x = \frac{W.b^2}{2L^3}(b + 3a)x^2$$

$$\therefore \quad x = \frac{W.a.b^2}{L^2} \times \frac{2L^3}{W.b^2(b + 3a)} = \frac{2aL}{(b + 3a)} \quad \text{...(4.30)}$$

Substituting this value of x in equation (vi), we get maximum deflection. If y_{max} represents the maximum deflection, then

$$EI.y_{max} = \frac{W.b^2}{6L^3}(b + 3a)\left(\frac{2aL}{b + 3a}\right)^3 - \frac{W.a.b^2}{2L^2} \times \left(\frac{2aL}{b + 3a}\right)^2$$

$$= \frac{W.b^2}{6L^3}.\left(\frac{2aL}{b + 3a}\right)^2\left[(b + 3a).\frac{2aL}{(b + 3a)} - 3aL\right]$$

$$= -\frac{Wb^2}{6L^3}.\left(\frac{2aL}{b + 3a}\right)^2.aL$$

$$= -\frac{Wb^2}{6L^3}.\frac{4a^2L^2}{(b + 3a)^2}.aL = -\frac{2}{3}.\frac{Wa^3b^2}{(b + 3a)^2}$$

$$\therefore \quad y_{max} = -\frac{2}{3EI}.\frac{Wa^3b^2}{(b + 3a)^2}. \quad \text{...(4.31)}$$

Problem 4.27. *A fixed beam AB of length 3 m carries a point load of 45 kN at a distance of 2 m from A. If the flexural rigidity (i.e., EI) of the beam is 1 × 10⁴ kNm², determine :*

(i) Fixed end moments at A and B,

(ii) Deflection under the load,

(iii) Maximum deflection, and

(iv) Position of maximum deflection.

Sol. Given :

Length, $L = 3$ m

Point load, $W = 45$ kN

Flexural rigidity, $EI = 1 \times 10^4$ kNm^2

Distance of load from A, $a = 2$ m

$\therefore$ Distance of load from B, $b = 1$ m

Let M_A and M_B = Fixed end moments,

y_c = Deflection under the load

y_{max} = Maximum deflection and

x = Distance of maximum deflection from A.

(*i*) The fixed end moments at *A* and *B* are given by

$$M_A = \frac{W.a.b^2}{L^2} = \frac{45 \times 2 \times 1^2}{3^2} = \textbf{10 kNm.} \quad \textbf{Ans.}$$

and
$$M_B = \frac{W.a^2.b}{L^2} = \frac{45 \times 2^2 \times 1}{3^2} = \textbf{20 kNm.} \quad \textbf{Ans.}$$

(*ii*) Deflection under load is given by equation (4.29) as

$$y_c = -\frac{W.a^3.b^3}{3EIL^3} = -\frac{45 \times 2^3 \times 1^3}{3 \times 1 \times 10^4 \times 3^3} = -0.000444 \text{ m}$$

$$= \textbf{– 0.444 mm.} \quad \textbf{Ans.}$$

–ve sign means the deflection is downwards.

(*iii*) Maximum deflection is given by equation (4.31) as

$$y_{max} = -\frac{2}{3EI} \times \frac{Wa^3.b^2}{(b+3a)^2}$$

$$= -\frac{2}{3 \times 1 \times 10^4} \cdot \frac{45 \times 2^3 \times 1^2}{(1+3 \times 2)^2} = -\frac{16 \times 45}{3 \times 10^4 \times 49}$$

$$= -0.00049 \text{ m} = \textbf{– 0.49 m.} \quad \textbf{Ans.}$$

(*iv*) The distance of maximum deflection from point *A* is given by equation (4.30) as

$$x = \frac{2a.L}{(b+3a)}$$

$$= \frac{2 \times 2 \times 3}{1 + 3 \times 2} = \frac{12}{7} = \textbf{1.714 m.} \quad \textbf{Ans.}$$

Alternate Method

Fig. 4.45 (*b*) shows the simply supported beam with vertical load of 45 kN at a distance 2 m from *A*.

The reactions R_A* and R_B* due to vertical load will be :

$$3R_B^* = 45 \times 2 \quad \text{or} \quad R_B^* = 90/3 = 30 \text{ kN and } R_A^* = 45 - 30 = 15 \text{ kN.}$$

Fig. 4.45 (*c*) shows the B.M. diagram with max. B.M.at *C* and equal to $R_A^* \times 2 = 15 \times 2$ = 30 kNm.

Fig. 4.45 (*d*) shows the fixed beam with end moments and reactions. As the vertical load is not acting symmetrically, therefore M_A and M_B will be different. In this case M_B will be more than M_A, as load is nearer to point *B*. The B.M. diagram is shown in Fig. 4.45 (*e*).

(*i*) *Fixed end moments at A and B.* To find the value of M_A and M_B, equate the areas of two B.M. diagrams.

∴ Area of B.M. diagram due to vertical loads

= Area of B.M. diagram due to end moments

$$\therefore \quad A_1 + A_2 = A_3 + A_4 \text{ where } A_1 = \frac{30 \times 2}{2} = 30, A_2 = \frac{30 \times 1}{2} = 15$$

$$A_3 = 3M_A, A_4 = \frac{(M_B - M_A) \times 3}{2}$$

$$= 1.5\,(M_B - M_A)$$

or $$30 + 15 = 3M_A + 1.5M_B - 1.5\,M_A$$

or $$45 = 1.5M_A + 1.5M_B$$

or $$\frac{45}{1.5} = M_A + M_B \quad \text{or} \quad M_A + M_B = 30 \qquad ...(i)$$

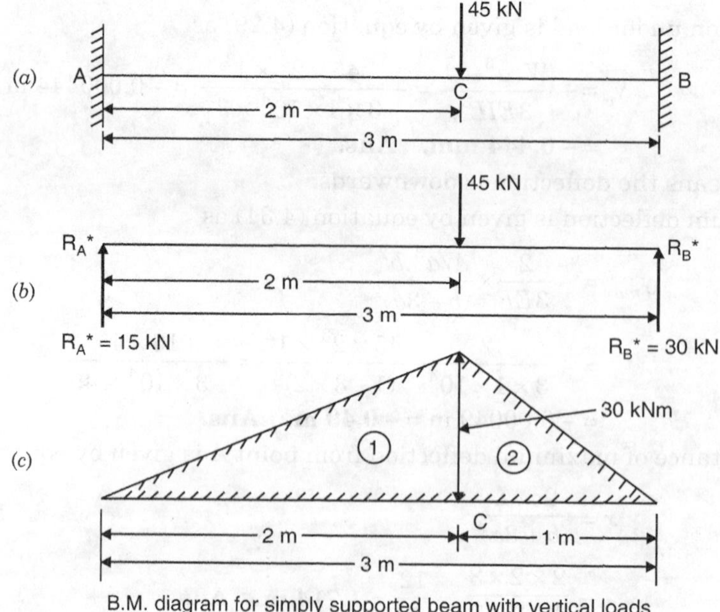

B.M. diagram for simply supported beam with vertical loads

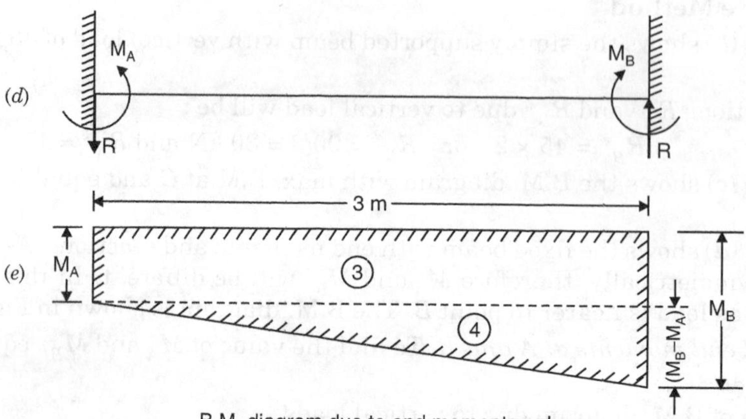

B.M. diagram due to end moments only

Fig. 4.45

Now equating the distance of C.G. of B.M. diagram due to vertical load to the distance of C.G. of B.M. diagram due to end moments from the some end (*i.e.*, from end *A*)

or $$\bar{x} = \bar{x}'$$

or $$\frac{A_1x_1 + A_2x_2}{A_1 + A_2} = \frac{A_3 \times x_3 + A_4x_4}{A_3 + A_4}$$

or $$\frac{30\times\frac{4}{3}+15\times\left(2+\frac{1}{3}\right)}{30+15}=\frac{3M_A\times\frac{3}{2}+1.5\,(M_B-M_A)\times 2}{3M_A+1.5M_B-1.5M_A}$$

or $$\frac{40+35}{45}=\frac{4.5M_A+3M_B-3M_A}{1.5M_A+1.5\,M_B}=\frac{1.5M_A+3M_B}{1.5M_A+1.5M_B}$$

or $$\frac{75}{45}=\frac{1.5(M_A+2M_B)}{1.5(M_A+M_B)}\quad\text{or}\quad\frac{5}{3}=\frac{M_A+2M_B}{M_A+M_B}$$

or $$5M_A+5M_B=3M_A+6M_B$$

or $$2M_A=M_B \qquad \text{...(ii)}$$

Solving equations (*i*) and (*ii*), we get

M_A = **10 kNm** and M_B = **20 kNm. Ans.**

Let us now find the reaction R due to end moments only. As the end moments are different, hence there will be reaction at A and B. Both the reactions will be equal and opposite in direction, as there is no vertical load, when we consider end moments only. As M_B is more, the reaction R will be upwards at B and downwards at A as shown in Fig. 4.45 (*d*).

Taking the moments about A for Fig. 4.45 (*d*), we get clockwise moment at A = Anti-clockwise moments at A

$$M_B=M_A+R\times 3$$

$$\therefore \qquad R=\frac{M_B-M_A}{3}=\frac{20-10}{3}=\frac{10}{3}\text{kN}$$

Now the total reaction at A and B will be,

$$R_A=R_A{}^{*}-R=15-\frac{10}{3}=\frac{35}{3}\text{kN}$$

and $$R_B=R_B{}^{*}+R=30+\frac{10}{3}=\frac{100}{3}\text{ kN}$$

Now, consider the fixed beam as shown in Fig. 4.46.

The B.M. at any section between AC at a distance x from A is given by $R_A\times x-M_A$

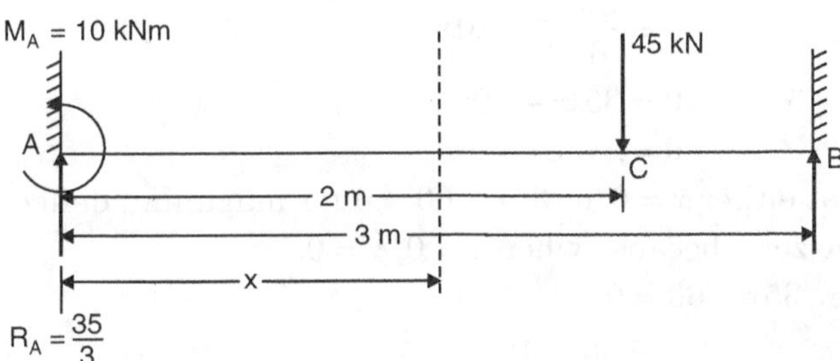

Fig. 4.46

or $$EI\frac{d^2y}{dx^2}=R_A\times x-M_A=\frac{35}{3}\times x-10$$

Integrating, we get

$$EI\frac{dy}{dx}=\frac{35}{3}\times\frac{x^2}{2}-10x+C_1$$

at $x = 0, \frac{dy}{dx} = 0 \quad \therefore \quad C_1 = 0$

$$\therefore \qquad EI\frac{dy}{dx} = \frac{35}{6}x^2 - 10x \qquad \text{...(iii)}$$

Integrating again, we get

$$EI \times y = \frac{35}{6} \times \frac{x^3}{3} - \frac{10x^2}{2} + C_2$$

at $x = 0, y = 0, \quad \therefore \quad C_2 = 0$

$$\therefore \qquad EI \times y = \frac{35}{18}x^3 - 5x^2 \qquad \text{...(iv)}$$

(ii) Deflection under the load

From equation (*iv*), we have

$$y = \frac{1}{EI}\left[\frac{35}{18}x^3 - 5x^2\right]$$

To find the deflection under the load, substitute $x = 2$ m in the above equation.

$$\therefore \qquad y = \frac{1}{EI}\left[\frac{35}{18} \times 2^3 - 5 \times 2^2\right]$$

$$= \frac{1}{1 \times 10^4}\left[\frac{35 \times 8}{18} - 20\right] \qquad (\because \quad EI = 1 \times 10^4)$$

$= -\,0.000444$ m $=$ **– 0.444 mm. Ans.**

(– ve sign means the deflection is downwards).

(iii) Maximum deflection

Deflection (y) will be maximum when $\frac{dy}{dx} = 0$.

Hence substituting the value of $\frac{dy}{dx} = 0$ in equation (*iii*), we get

$$0 = \frac{35}{6}x^2 - 10x$$

or $\qquad 0 = 35x^2 - 60x$

or $\qquad 0 = x\,(35x - 60)$

This means that either $x = 0$ or $35x - 60 = 0$ for maximum deflection.

But x cannot be zero, because when $x = 0$, $y = 0$.

$\therefore \qquad 35x - 60 = 0$

or $\qquad x = \frac{60}{35} = \frac{12}{7} = 1.714$ m

Substituting $x = 1.714$ m in equation (*iv*), we get maximum deflection.

$$\therefore \qquad EIy_{max} = \frac{35}{18}(1.714)^3 - 5(1.714)^2$$

or $$y_{max} = \frac{1}{EI}\left[\frac{35}{18}(1.714)^3 - 5(1.714)^2\right]$$

$$= \frac{1}{1 \times 10^4}[9.79 - 14.69]$$

$$= 0.00049 \text{ m} = \mathbf{0.49 \text{ mm}}. \quad \textbf{Ans.}$$

(*iv*) *Position of maximum deflection*

The maximum deflection will be at a distance of 1.714 m (*i.e.*, x = 1.714 m) from end A. **Ans.**

Problem 4.28. *A fixed beam AB of length 6 m carries point loads of 160 kN and 120 kN at a distance of 2 m and 4 m from the left end A. Find the fixed end moments and the reactions at the supports. Draw B.M. and S.F. diagrams.*

Sol. Given :

Length = 6 m

Load at C, W_C = 160 kN

Load at D, W_D = 120 kN

Distance AC = 2 m

Distance AD = 4 m

For the sake of convenience, let us first calculate the fixed end moments due to loads at C and D and then add up the moments.

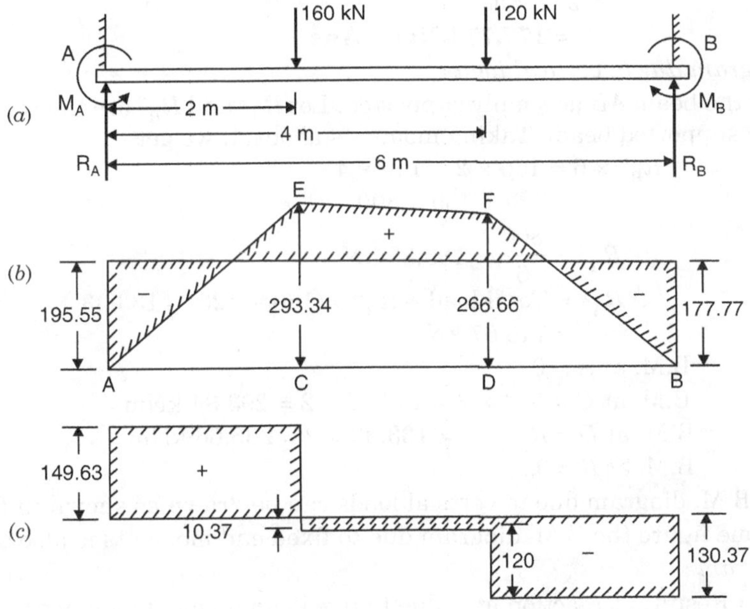

Fig. 4.47

(*i*) Fixed end moments due to load at *C*.

For the load at C, a = 2 m and b = 4 m

$$\therefore \qquad M_{A_1} = \frac{W_C . a . b^2}{L^2}$$

$$= \frac{160 \times 2 \times 4^2}{6^2} = 142.22 \text{ kNm}$$

$$M_{B_1} = \frac{W_C.a^2.b}{L^2} = \frac{160 \times 2^2 \times 4}{6^2} = 71.11 \text{ kNm}$$

(*ii*) Fixed end moments due to load at D.

Similarly for the load at D, $a = 4$ m and $b = 2$ m

$$\therefore \quad M_{A_2} = \frac{W_D.a.b^2}{L^2}$$

$$= \frac{120 \times 4 \times 2^2}{6^2} = 53.33 \text{ kNm}$$

and $$M_{B_2} = \frac{W_D.a^2.b}{L^2} = \frac{160 \times 4^2 \times 2}{6^2} = 106.66 \text{ kNm}$$

$\therefore$ Total fixing moment at A,

$$M_A = M_{A_1} + M_{A_2} = 142.22 + 53.33$$

$$= \mathbf{195.55 \text{ kNm}}. \quad \textbf{Ans.}$$

and total fixing moment at B,

$$M_B = M_{B_1} + M_{B_2} = 71.11 + 106.66$$

$$= \mathbf{177.77 \text{ kNm}}. \quad \textbf{Ans.}$$

B.M. diagram due to vertical loads

Consider the beam AB as simply supported. Let R_A* and R_B* are the reactions at A and B due to simply supported beam. Taking moments about A, we get

$$R_B^* \times 6 = 160 \times 2 + 120 \times 4$$

$$= 320 + 480 = 800$$

$$\therefore \quad R_B^* = \frac{800}{6} = 133.33 \text{ kN}$$

and $$R_A^* = \text{Total load} - R_B^* = (160 + 120) - 133.33$$

$$= 146.67 \text{ kN}$$

B.M. at $A = 0$

B.M. at $C = R_A^* \times 2 = 146.67 \times 2 = 293.34$ kNm

B.M. at $D = R_B^* \times 2 = 133.33 \times 2 = 266.66$ kNm

B.M. at $B = 0$.

Now the B.M. diagram due to vertical loads can be drawn as shown in Fig. 4.47 (*b*). In the same figure the B.M. diagram due to fixed end moments is also shown.

S.F. Diagram

Let R_A = Resultant reaction at A due to fixed end moments and vertical loads

R_B = Resultant reaction at B.

Equating the clockwise moments and anti-clockwise moments about A, we get

$$R_B \times 6 + M_A = 160 \times 2 + 120 \times 4 + M_B$$

or $$R_B \times 6 + 195.55 = 320 + 480 + 177.77$$

$$\therefore \quad R_B = \frac{800 + 177.77 - 195.55}{6} = 130.37 \text{ kN}$$

and $$R_A = \text{Total load} - R_B$$

$$= (160 + 120) - 130.37 = 149.63 \text{ kN}$$

$\therefore$ S.F. at $A = R_A = 149.63$ kN

S.F. at $C = 149.63 - 160 = -10.37$ kN

S.F. at $D = -10.37 - 120 = -130.37$ kN

S.F. at $B = -130.37$ kN

Now S.F. diagram can be drawn as shown in Fig. 4.47 (*c*).

Alternate Method

Fig. 4.47 (*b*) shows the simply supported beam with vertical loads.

Let $R_A{}^*$ and $R_B{}^*$ are the reactions at A and B due to vertical loads. Taking moments about A, we get

$$R_B{}^* \times 6 = 160 \times 2 + 120 \times 4 = 320 + 480 = 800$$

$$\therefore \quad R_B{}^* = \frac{800}{6} = \frac{400}{3} = 133.33 \text{ kN}$$

and
$$R_A{}^* = \text{Total load} - R_B{}^*$$
$$= (160 + 120) - 133.33 = 146.67 \text{ kN}$$

B.M. at $A = 0$

B.M. at $C = R_A{}^* \times 2 = 146.67 \times 2 = 293.34$ kNm

B.M. at $D = R_B{}^* \times 2 = 133.33 \times 2 = 266.66$ kNm

Now the B.M. diagram due to vertical loads can be drawn as shown in Fig. 4.48 (*c*).

Fig. 4.48 (*d*) shows the fixed beam with end moments only. As the load 160 kN is nearer to end A, hence M_A will be more than M_B. The B.M. diagram due to end moments is shown in Fig. 4.48 (*c*).

To find the values of M_A and M_B, equate the areas of two B.M. diagrams.

$\therefore$ Area of B.M. diagram due to vertical loads

= Area of B.M. diagram due to end moments

$$\therefore \quad A_1 + A_2 + A_3 + A_4 = A_5 + A_6 \quad ...(i)$$

where $A_1 = \dfrac{AC \times CE}{2} = \dfrac{2 \times 293.33}{2} = 293.33$

$A_2 = CD \times DF = 2 \times 266.67 = 533.34$

$A_3 = \dfrac{GF \times GE}{2} = \dfrac{2 \times 26.66}{2} = 66.66$

$A_4 = \dfrac{DB \times DF}{2} = \dfrac{2 \times 266.67}{2} = 266.67$

$A_5 = M_B \times 6 = 6M_B,\ A_6 = \dfrac{6 \times (M_A - M_B)}{2} = 3\,(M_A - M_B) = 3M_A - 3M_B$

Substituting these values in equation (*i*), we get

$$293.33 + 533.34 + 66.66 + 266.67 = 6M_B + 3M_A - 3M_B$$

or
$$1119.98 = 3M_B + 3M_A = 3\,(M_B + M_A)$$

$$\therefore \quad M_B + M_A = \frac{1119.98}{3} = 373.33 \quad ...(ii)$$

To get the other equation between M_A and M_B, equate the distance of C.G. of B.M. diagram due to vertical loads to the distance of C.G. of B.M. diagram due to end moments from end A.

or
$$\bar{x} = \bar{x}'$$

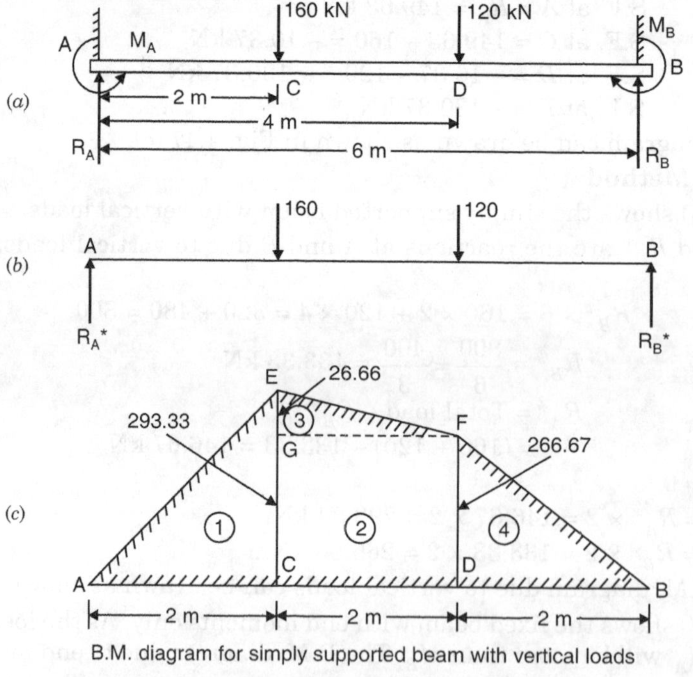

B.M. diagram for simply supported beam with vertical loads

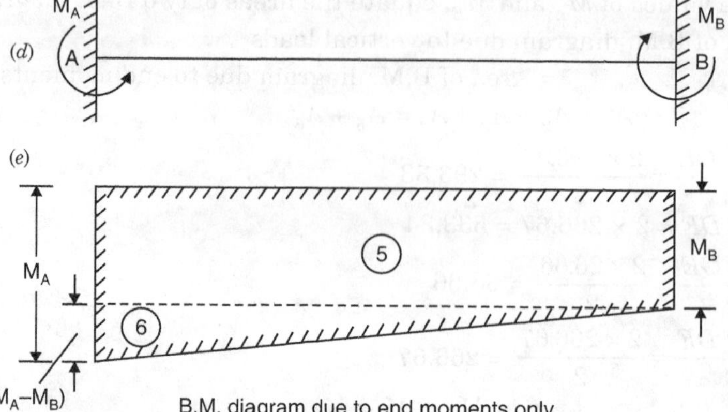

B.M. diagram due to end moments only

Fig. 4.48

or
$$\frac{A_1x_1 + A_2x_2 + A_3x_3 + A_4x_4}{A_1 + A_2 + A_3 + A_4} = \frac{A_5x_5 + A_6x_6}{A_5 + A_6}$$

or
$$\frac{293.33 \times \frac{4}{3} + 533.34 \times 3 + 26.66 \times \left(2 + \frac{2}{3}\right) + 266.66 \times \left(4 + \frac{2}{3}\right)}{293.33 + 533.34 + 26.66 + 266.66}$$

$$= \frac{6M_B \times 3 + 3(M_A - 3M_B) \times \frac{1}{3} \times 6}{6M_B + 3M_A - 3M_B}$$

or $$\frac{391.1 + 1600 + 70.91 + 1245.35}{1119.98} = \frac{3(6M_B + 2M_A - 2M_B)}{3(M_B + M_A)}$$

or $$2.95 = \frac{4M_B + 2M_A}{M_B + M_A}$$

or $$2.95M_B + 2.95M_A = 4M_B + 2M_A$$

or $$2.95M_A - 2M_A = 4M_B - 2.95M_B$$

or $$0.95M_A = 1.05M_B$$

or $$M_A = \frac{1.05}{0.95}M_B = 1.1M_B \quad ...(iii)$$

Substituting this value of M_A in equation (*ii*), we get

$$M_B + 1.1\,M_B = 373.33$$

or $$M_B = \frac{373.33}{2.1} = \mathbf{177.77\ kNm.}\quad \textbf{Ans.}$$

From equation (*iii*), $M_A = 1.1 \times 177.77 =$ **195.55 kNm. Ans.**

Combined B.M. Diagram

$M_A = 195.55$ kNm and $M_B = 177.77$ kNm. Now the combined B.M. diagram can be drawn as shown in Fig. 4.47 (*b*).

To draw the S.F. diagram, let us first find the values of resultant reactions due to vertical loads and fixed end moments R_A and R_B. Refer to Fig. 4.48 (*a*). Taking moments about *A*, we get clockwise moments at *A* = Anti-clockwise moments at *A*

$$\therefore \quad 160 \times 2 + 120 \times 4 + M_B = M_A + R_B \times 6$$

or $$320 + 480 + M_B = M_A + 6R_B$$

or $$800 + 177.77 = 195.55 + 6R_B$$

$$\therefore \quad R_B = \frac{800 + 177.77 - 195.55}{6} = 130.37 \text{ kN}$$

and $$R_A = \text{Total load} - R_B = (160 + 120) - 130.37 = 149.63 \text{ kN}$$

S.F. Diagram

S.F. at $A = R_A = 149.63$ kN

S.F. at $C = 149.63 - 160 = -10.37$ kN

S.F. at $D = -10.37 - 120 = -130.37$ kN

S.F. at $B = -130.37$ kN

Now S.F. diagram can be drawn as shown in Fig. 4.47(*c*).

Problem 4.29. *A fixed beam of length 6 m carries two point loads of 30 kN each at a distance of 2 m from both ends. Determine the fixed end moments and draw the B.M. diagram.*

Sol. Given :

Length, $L = 6$ m

Point load at *C*, $W_1 = 30$ kN

Point load at *D*, $W_2 = 30$ kN

Distance $AC = 2$ m

Distance $AD = 4$ m

The fixing moment at A due to loads at C and D is given by

M_A = Fixing moment due to load at C + Fixing moment due to load at D

$$= \frac{W_1 a_1 b_1^2}{L^2} + \frac{W_2 a_2 . b_2^2}{L^2}$$

$$= \frac{30 \times 2 \times 4^2}{6^2} + \frac{30 \times 4 \times 2^2}{6^2} = \frac{80}{3} + \frac{40}{3} = 40 \text{ kNm.}$$

Since the beam and loading is symmetrical, therefore fixing moments at A and B should be equal.

$\therefore$ $M_B = M_A$ = **40 kNm. Ans.**

To draw the B.M. diagram due to vertical loads, consider the beam AB as simply supported. The reactions at the simply supported beam will be equal to 30 kN each.

B.M. at A and B = 0

B.M. at C = 30 × 2 = 60 kNm

B.M. at D = 30 × 2 = 60 kNm.

Now the B.M. diagram due to vertical loads and due to end moments can be drawn as shown in Fig. 4.49 (*b*).

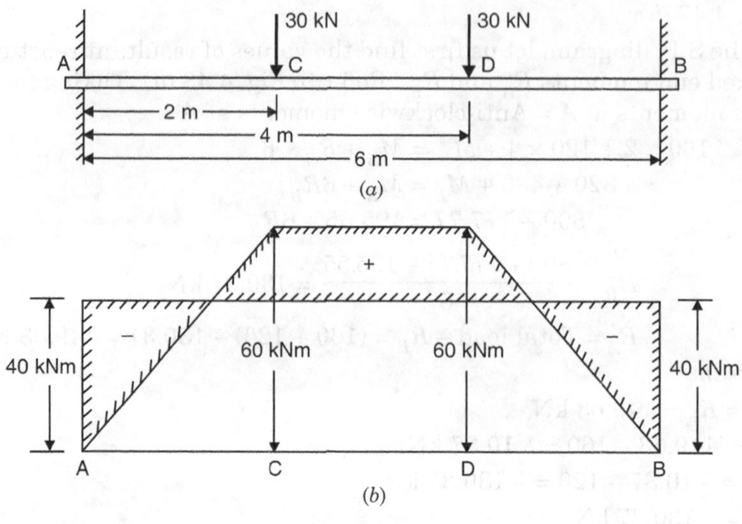

Fig. 4.49

4.13.4. Slope and Deflection for a Fixed Beam Carrying a Uniformly distributed Load over the Entire Length. Fig. 4.50 (*a*) shows a fixed beam of length L, carrying uniformly distributed load of w/unit length over the entire length.

Let M_A = Fixed end moment at A

M_B = Fixed end moment at B

R_A = Reaction at A

R_B = Reaction at B.

(i) B.M. Diagram

Since the loading on the beam is symmetrical, hence $M_A = M_B$. The B.M. diagram due to end moments will be a rectangle as shown in Fig. 4.50 (*b*) by *AEFB*. The magnitude of M_A or M_B is unknown.

The B.M. diagram for a simply supported beam carrying a uniformly distributed load will be parabola whose central ordinate will be $w.L^2/8$. The B.M. diagram for this case is shown in Fig. 4.50 (*b*) by parabola *ADB* in which $CD = \frac{w.L^2}{8}$.

Equating the areas of the two bending moment diagrams, we get

Area of rectangle *AEFB* = Area of parabola *ADB*

$$AB \times AE = \frac{2}{3} \times [AB \times CD]$$

$$L \times M_A = \frac{2}{3} \times L \times \frac{w.L^2}{8} \quad \text{or} \quad M_A = \frac{w.L^2}{12}$$

$$\therefore \quad M_B = M_A = \frac{w.L^2}{12} \quad \text{...(4.32)}$$

Now the B.M. diagram can be drawn as shown in Fig. 4.50 (*b*).

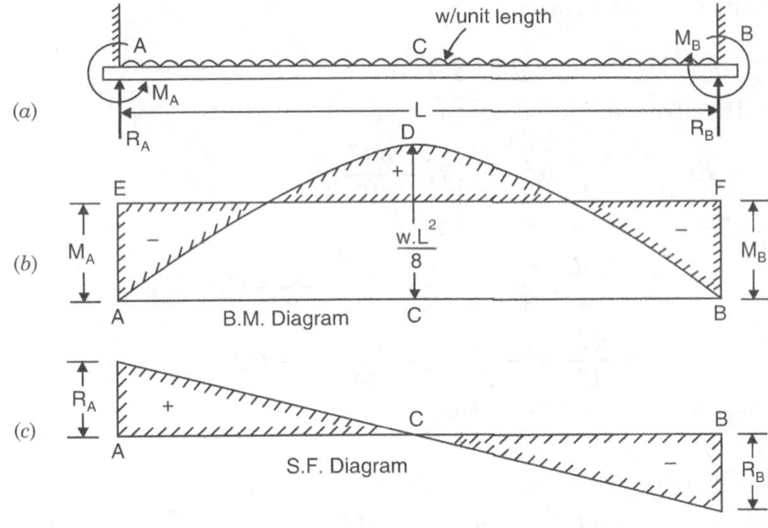

Fig. 4.50

(ii) S.F. Diagram

Equating the clockwise moments and anti-clockwise moments about *A*, we get

$$R_B \times L + M_A = w.L.\frac{L}{2} + M_B$$

But $\quad M_A = M_B$

$$\therefore \quad R_B \times L = w.L.\frac{L}{2} \quad \text{or} \quad R_B = \frac{w.L}{2}$$

Due to symmetry,

$$R_A = R_B = \frac{w.L}{2} \quad \text{...(4.33)}$$

Now the S.F. diagram can be drawn as shown in Fig. 4.50 (*c*).

(iii) Slope and deflection

The B.M. at any section at a distance x from A is given by,

$$EI\frac{d^2y}{dx^2} = R_A \times x - M_A - w.x\frac{x}{2}$$

$$= \frac{w.L}{2}.x - \frac{wL^2}{12} - \frac{wx^2}{2} \qquad \left(\because\ R_A = \frac{w.L}{2}, M_A = \frac{wL^2}{12}\right)$$

$$= \frac{wL.x}{2} - \frac{wx^2}{2} - \frac{wL^2}{12} \quad \text{...(i)}$$

Integrating the above equation, we get

$$EI\frac{dy}{dx} = \frac{w.L}{2}.\frac{x^2}{2} - \frac{w}{2}\frac{x^3}{3} - \frac{wL^2}{12}.x + C_1$$

$$= \frac{wL}{4}.x^2 - \frac{w}{6}x^3 - \frac{wL^2}{12}.x + C_1$$

where C_1 is a constant of integration.

At $x = 0$, $\frac{dy}{dx} = 0$. Hence $C_1 = 0$.

Therefore, the above equation becomes as

$$EI\frac{dy}{dx} = \frac{w.L}{4}.x^2 - \frac{w}{6}.x^3 - \frac{wL^2}{12}.x \quad \text{...(ii)}$$

Integrating the above equation, we get

$$EIy = \frac{wL}{4}.\frac{x^3}{3} - \frac{w}{6}.\frac{x^4}{4} - \frac{wL^2}{12}.\frac{x^2}{2} + C_2$$

$$= \frac{wL}{12}.x^3 - \frac{w}{24}.x^4 - \frac{wL^2}{24}.x^2 - C_2$$

where C_2 is another constant of integration.

At $x = 0$, $y = 0$. Hence $C_2 = 0$.

Therefore, the above equation becomes as

$$EIy = \frac{wL}{12}.x^3 - \frac{w}{24}.x^4 - \frac{wL^2}{24}.x^2 \quad \text{...(iii)}$$

The deflection at the centre is obtained by substituting $x = \frac{L}{2}$ in the above equation. Let the deflection at the centre is y_c

$$\therefore \quad EIy_c = \frac{wL}{12}.\left(\frac{L}{2}\right)^3 - \frac{w}{24}\left(\frac{L}{2}\right)^4 - \frac{wL^2}{24}\left(\frac{L}{2}\right)^2$$

$$= \frac{w.L^4}{96} - \frac{wL^4}{384} - \frac{wL^4}{96} = -\frac{wL^4}{384}$$

$$\therefore \qquad y_c = -\frac{wL^4}{384EI} \qquad \ldots(4.34)$$

Minus sign means that the deflection is downwards.

Note. The deflection at the centre of a simply supported beam carrying a uniformly distributed load over the entire length is 5/384, wL^4. This means that the central deflection for the fixed beam is one-fifth of the central deflection of the simply supported beam.

(*iv*) *Points of contraflexures*

For the points of contraflexures, B.M. given by equation (*i*) should be zero. Hence equating equation (*i*) to zero, we get

$$0 = \frac{wLx}{2} - \frac{wx^2}{2} - \frac{wL^2}{12}$$

or

$$0 = wLx - wx^2 - \frac{wL^2}{6} = Lx - x^2 - \frac{L^2}{6}$$

or

$$x^2 - Lx + \frac{L^2}{6} = 0$$

Solving the above quadratic equation, we get

$$x = \frac{+L \pm \sqrt{L^2 - \frac{4 \times L^2}{6}}}{2} = \frac{L \pm \sqrt{\frac{L^2}{3}}}{2} = \frac{L}{2} \pm \frac{L}{2\sqrt{3}}.$$

As $L/2$ represents the centre of the beam. Hence the two points of contraflexures occur at a distance of $L/2\sqrt{3}$ from the centre of the beam.

Problem 4.30. *A fixed beam of length 5 m carries a uniformly distributed load of 9 kN/m run over the entire span. If $I = 4.5 \times 10^{-4}$ m⁴ and $E = 1 \times 10^7$ kN/m², find the fixing moments at the ends and the deflection at the centre.*

Sol. Given :

Length, $L = 5$ m

U.d.l. $w = 9$ kN/m

Value of $I = 4.5 \times 10^{-4}$ m^4

Value of $E = 1 \times 10^7$ kN/m^2.

(*i*) The fixing moments at the ends is given by equation (4.32) as

$$M_A = M_B = \frac{w.L^2}{12} = \frac{9 \times 5^2}{12} = \mathbf{18.75\ kNm.} \quad \textbf{Ans.}$$

(*ii*) The deflection at the centre is given by equation (4.34) as

$$y_c = -\frac{wL^4}{384EI} = -\frac{9 \times 5^4}{384 \times 1 \times 10^7 \times 4.5 \times 10^{-4}}$$

$$\equiv 0.003254 \text{ m} = \mathbf{-\ 3.254\ mm.} \quad \textbf{Ans.}$$

Problem 4.31. *Find the fixing moments and support reactions of a fixed beam AB of length 6 m, carrying a uniformly distributed load of 4 kN/m over the left half of the span.*

Sol. Given :

Length, $L = 6$ m

U.d.l., $w = 4$ kN/m

(i) B.M. diagram due to end moments

Let M_A = Fixing moment at A

M_B = Fixing moment at B.

The value of M_A will be more than M_B as load due u.d.l. is nearer to point A.

The B.M. diagram due to end moments will be trapezium as shown in Fig. 4.51 (*b*) by *AEFB*.

The area of B.M. diagram due to end moments is given by,

$$a' = \frac{1}{2}(M_A + M_B) \times 6 = 3(M_A + M_B) \qquad ...(i)$$

(ii) B.M. diagram due to vertical loads

Now draw the B.M. diagram due to u.d.l. for a simply supported beam.

Let $R_A{}^*$ = Reaction at A for a simply supported beam

$R_B{}^*$ = Reaction at B for a simply supported beam.

Taking moments about A for a simply supported beam, we get

$$R_B{}^* \times 6 = 4 \times 3 \times 1.5 = 18$$

$$\therefore \qquad R_B{}^* = \frac{18}{6} = 3 \text{ kN}$$

and

$$R_A{}^* = \text{Total load} - R_B{}^*$$
$$= 4 \times 3 - 3 = 9 \text{ kN}$$

The B.M. at A and B are zero.

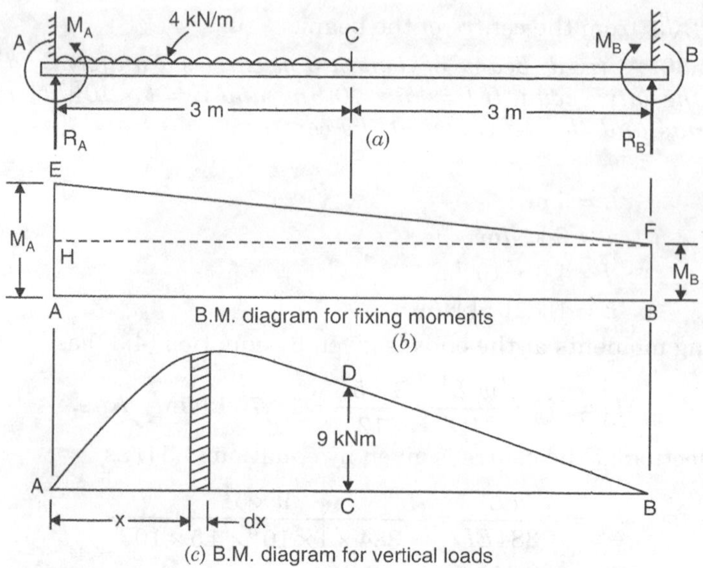

Fig. 4.51

B.M. at $C = R_B{}^* \times 3 = 3 \times 3 = 9$ kNm.

The B.M. diagram from A to C will be parabolic and from C to A the B.M. diagram will follow a straight line law as shown in Fig. 4.51 (*c*).

The area of the B.M. due to vertical loads is given by

$$a = \text{Area of parabola } ACD + \text{Area of triangle } BCD$$

$$= \text{Area of parabola } ACD + \tfrac{1}{2} \times 9 \times 3 \qquad ...(ii)$$

To find the area of the parabola ACD, consider a strip of length 'dx' at a distance x from A in portion AC.

The B.M. at a distance x from A is given by

$$M_x = R_A{}^* \times x - 4 \times x \,.\, \frac{x}{2} = 9x - 2x^2 \qquad (\because \; R_A{}^* = 9)$$

Area of B.M. diagram of length dx

$$= M_x.dx = (9x - 2x^2).dx$$

Total area of parabola from A to C is obtained by integrating the above equation between the limits of 0 and 3.

∴ Area of parabola ACD

$$= \int_0^3 (9x - 2x^2).dx$$

$$= \left[\frac{9x^2}{2} - \frac{2x^3}{3}\right]_0^3 = \frac{9\times 3^2}{2} - \frac{2\times 3^3}{3} = 40.5 - 18 = 22.5$$

Substituting this value in equation (ii), we get

$$a = 22.5 + \frac{1}{2} \times 9 \times 3 = 36.0 \qquad ...(iii)$$

Equating the two areas given by equations (i) and (iii), we get

$$3(M_A + M_B) = 36.0$$

or

$$M_A + M_B = 12.0 \qquad ...(iv)$$

Now moment of B.M. diagram due to vertical loads about A is given by

$$a\bar{x} = \int_0^3 x.M_x.dx + \text{Area of triangle } BCD$$

$$\times \text{ Distance of C.G. of BCD from } A$$

$$= \int_0^3 x(9x - 2x^2)\,.\,dx + \frac{1}{2}\times 9\times 3\times\left(3+\frac{1}{3}\times 3\right)$$

$$= \int_0^3 (9x^2 - 2x^3)\,.\,dx + 54$$

$$= \left[\frac{9x^3}{3} - \frac{2x^4}{4}\right]_0^3 + 54 = \left[3\times 3^3 - \frac{1}{2}\times 3^4\right] + 54$$

$$= (81 - 40.5) + 54 = 94.5 \qquad ...(v)$$

Moment of B.M. diagram due to end moments about A is given by [see Fig. 4.51 (b)].

$$a'\bar{x}' = \text{Area } ABFH \times \text{Distance of C.G. of } ABFH \text{ from } A$$

$$+ \text{ Area } HFE \times \text{Distance of its C.G. from } A$$

$$= (M_B \times L) \times \frac{L}{2} + \frac{1}{2} \times L \times (M_A - M_B) \times \frac{1}{3} \times L$$

$$= M_B \times 6 \times \frac{6}{2} + \frac{1}{2} \times 6 \times (M_A - M_B) \times \frac{6}{3}$$
$$= 18M_B + 6M_A - 6M_B$$
$$= 6M_A + 12M_B = 6(M_A + 2M_B) \qquad ...(vi)$$

But $\qquad a\bar{x} = a'\bar{x}'$

$\therefore \qquad 94.5 = 6(M_A + 2M_B)$

or $$M_A + 2M_B = \frac{94.5}{6} = 15.75 \qquad ...(vii)$$

Substracting equation (*iv*) from (*vii*), we get

$$M_B = 15.75 - 12.0 = \mathbf{3.75\ kNm.}\quad \textbf{Ans.}$$

Substituting this value in equation (*iv*), we get

$$M_A = 12 - 3.75 = \mathbf{8.25\ kNm.}\quad \textbf{Ans.}$$

Support reactions

Let R_A = Resultant reaction at A

R_B = Resultant reaction at B.

Equating the anti-clockwise moments and clockwise moments about A,

$$R_B \times 6 + M_A = 4 \times 3 \times 1.5 + M_B$$

or $$R_B \times 6 + 8.25 = 18 + 3.75 = 21.75$$

$\therefore$ $$R_B = \frac{21.75 - 8.25}{6} = \frac{13.50}{6} = \mathbf{2.25\ kN.}\quad \textbf{Ans.}$$

and $$R_A = \text{Total load} - R_B$$
$$= 4 \times 3 - 2.25 = \mathbf{9.75\ kN.}\quad \textbf{Ans.}$$

Second Method for Problem 4.31

Macaulay's method can be used and directly the fixing moments and end reactions can be calculated. This method is used where the areas of B.M. diagrams cannot be determined conveniently.

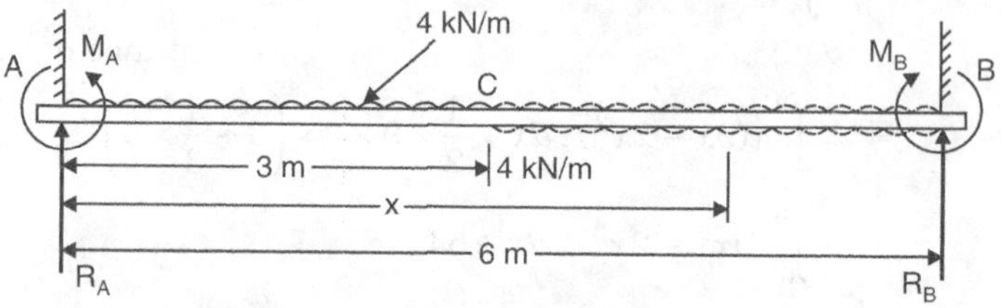

Fig. 4.52

For this method it is necessary that u.d.l. should be extended upto B and then compensated for upward u.d.l. for length BC as shown in Fig. 4.52.

The B.M. at any section at a distance x from A is given by

$$EI\frac{d^2y}{dx^2} = R_A.x - M_A - w \times x \times \frac{x}{2} \;\vdots\; + w \times (x-3) \times \frac{(x-3)}{2}$$
$$= R_A.x - M_A - \frac{4 \times x^2}{2} \;\vdots\; + \frac{4(x-3)^2}{2}$$

$$= R_A.x - M_A - 2x^2 \;\vdots\; + 2(x-3)^2 \qquad ...(A)$$

Integrating, we get

$$EI\frac{dy}{dx} = R_A.\frac{x^2}{2} - M_A.x - \frac{2x^3}{3} + C_1 \;\vdots\; + \frac{2(x-3)^3}{3} \qquad ...(i)$$

when $x = 0$, $\frac{dy}{dx} = 0$.

Substituting this value in the above equation upto dotted line, we get

$$C_1 = 0.$$

Therefore equation (i) becomes as

$$EI\frac{dy}{dx} = R_A.\frac{x^2}{2} - M_A.x - \frac{2x^3}{3} \;\vdots\; + \frac{2(x-3)^3}{3} \qquad ...(ii)$$

Integrating again, we get

$$EI\,y = \frac{R_A}{2}.\frac{x^3}{3} - \frac{M_A.x^2}{2} - \frac{2}{3}\frac{x^4}{4} + C_2 \;\vdots\; + \frac{2}{3}\frac{(x-3)^4}{4} \qquad ...(iii)$$

when $x = 0$, $y = 0$.

Substituting this value upto dotted line, we get

$$C_2 = 0$$

Therefore equation (iii) becomes as

$$EI\,y = \frac{R_A.x^3}{6} - \frac{M_A.x^2}{2} - \frac{1}{6}x^4 \;\vdots\; + \frac{1}{6}(x-3)^4 \qquad ...(iv)$$

when $x = 6$, $y = 0$.

Substituting this value in equation (iv) [Here complete equation is taken], we get

$$0 = \frac{R_A \times 6^3}{6} - \frac{M_A \times 6^2}{2} - \frac{1}{6}\times 6^4 + \frac{1}{6} \times (6-3)^4$$

$$= 36R_A - 18M_A - 216 + 13.5$$

or $$202.50 = 36R_A - 18M_A$$

$$101.25 = 18R_A - 9M_A \qquad ...(v)$$

At $x = 6$ m, $\frac{dy}{dx} = 0$.

Substituting these values in the complete equation (ii), we get

$$0 = R_A \times \frac{6^2}{2} - M_A \times 6 - \frac{2}{3}\times 6^3 + \frac{2}{3}(6-3)^3$$

$$= 18R_A - M_A \times 6 - 144 + 18$$

$$126 = 18R_A - 6M_A \qquad ...(vi)$$

Substracting equation (v) from equation (vi), we get

$$126 - 101.25 = 9M_A - 6M_A$$

or $$24.75 = 3M_A$$

$\therefore$ $$M_A = \frac{24.75}{3} = \textbf{8.25 kNm.} \quad \textbf{Ans.}$$

Substituting this value in equation (*vi*), we get

$$126 = 18R_A - 6 \times 8.25$$

$$\therefore \quad R_A = \frac{126 + 6 \times 8.25}{18} = \mathbf{9.75\ kN.} \quad \textbf{Ans.}$$

Now
$$R_B = \text{Total load} - R_A$$
$$= 4 \times 3 - 9.75 = \mathbf{2.25\ kN.} \quad \textbf{Ans.}$$

To find the value of M_B, we must equate the clockwise moments and anti-clockwise moments about B. Hence

Clockwise moments about B = Anti-clockwise moments about B.

$$M_B + R_A \times 6 = M_A + 4 \times 3 \times (4.5)$$

or
$$M_B + 9.75 \times 6 = 8.25 + 54 \qquad (\because\ R_A = 9.75 \text{ and } M_A = 8.25)$$

or
$$M_B + 58.50 = 62.25$$

$$\therefore \quad M_B = 62.25 - 58.50 = \mathbf{3.75\ kNm.} \quad \textbf{Ans.}$$

Problem 4.32. *A fixed beam of length 20 m, carries a uniformly distributed load of 8 kN/m on the left hand half together with a 120 kN load at 15 m from the left hand end. Find the end reactions and fixing moments and magnitude and the position of the maximum deflection. Take E = 2 × 10⁸ kN/m³ and I = 4 × 10⁸ mm⁴.*

Sol. Given :

Length, $L = 20$ m

U.d.l., $w = 8$ kN/m

Point load, $W = 120$ kN

Value of $E = 2 \times 10^8$ kN/m^2

Value of $I = 4 \times 10^8$ mm^4 $= 4 \times 10^{-4}$ m^4

Lengths, $AC = 10$ m, $AD = 15$ m

Fig. 4.53 shows the loading on the fixed beam.

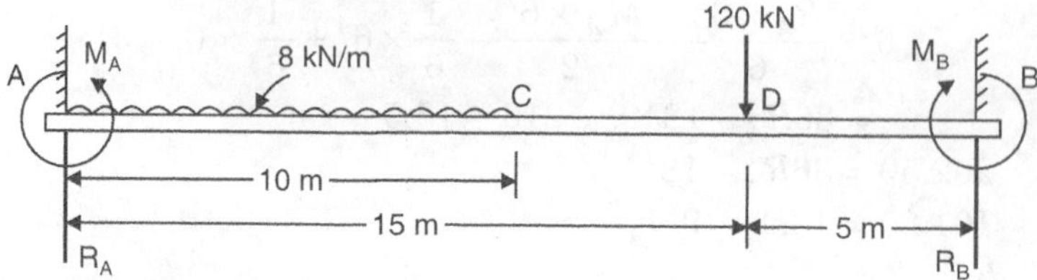

Fig. 4.53

Let R_A and R_B = End reactions at A and B

M_A and M_B = Fixing moments at A and B

Let us apply Macaulay's method for this case. Hence it is necessary that the u.d.l. should be extended upto B and then compensated for upward u.d.l. for length BC as shown in Fig. 4.54.

The B.M. at any section at a distance x from A is given by,

$$EI\frac{d^2y}{dx^2} = R_A.x - M_A - w \times x \times \left(\frac{x}{2}\right) \Big| - 120(x - 15) \Big| + w \times (x - 10) \times \left(\frac{x - 10}{2}\right)$$

$$= R_A \times x - M_A - 8 \times \frac{x^2}{2} \; \vdots \; - 120(x - 15) \; \vdots \; + \frac{8 \times (x - 10)^2}{2}$$

$$= R_A.x - M_A - 4x^2 \; \vdots \; - 120(x - 15) \; \vdots \; + 4(x - 10)^2$$

Integrating the above equation, we get

$$EI \frac{dy}{dx} = R_A.\frac{x^2}{2} - M_A.x - 4.\frac{x^3}{3} + C_1 \; \vdots \; - \frac{120(x - 15)^2}{2} \; \vdots \; + \frac{4(x - 10)^3}{3} \quad ...(i)$$

when $x = 0$, $\frac{dy}{dx} = 0$. Substituting this value in the above equation upto first dotted line, we get $C_1 = 0$. Therefore, equation (i) becomes as

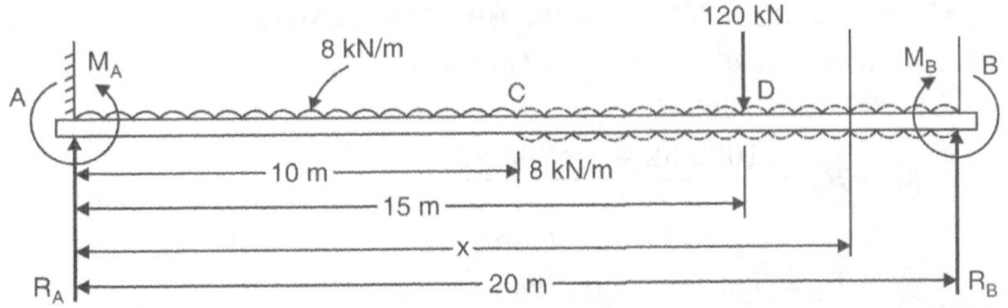

Fig. 4.54

$$EI \frac{dy}{dx} = \frac{R_A}{2}.x^2 - M_A.x - \frac{4}{3} x^3 \; \vdots \; - 60(x - 15)^2 \; \vdots \; + \frac{4}{3}(x - 10)^3 \quad ...(ii)$$

Integrating again, we get

$$EIy = \frac{R_A.x^3}{6} - \frac{M_A.x^2}{2} - \frac{4x^4}{3 \times 4} + C_2 \; \vdots \; - \frac{60(x - 15)^3}{3} \; \vdots \; + \frac{4}{3}\frac{(x - 10)^4}{4} \quad ...(iii)$$

when $x = 0$, $y = 0$. Substituting this value in the above equation upto first dotted line, we get $C_2 = 0$. Therefore equation (iii) becomes as

$$EIy = \frac{R_A.x^3}{6} - \frac{M_A.x^2}{2} - \frac{x^4}{3} \; \vdots \; + \frac{1}{3}(x - 10)^4 \quad ...(iv)$$

when $x = 20$, $y = 0$. Substituting these values in complete equation (iv), we get

$$0 = \frac{R_A \times 20^3}{6} - \frac{M_A \times 20^2}{2} - \frac{20^4}{3} - 20(20 - 15)^3 + \frac{1}{3}(20 - 10)^4$$

$$= \frac{20}{6} R_A - \frac{M_A}{2} - \frac{20^2}{3} - \frac{125}{20} + \frac{1}{3} \times \frac{10^4}{400} \qquad \text{(Dividing by } 20^2\text{)}$$

$$= \frac{20}{6} R_A - \frac{M_A}{2} - \frac{400}{3} - \frac{12.5}{2} + \frac{25}{3}$$

$$= \frac{20R_A - 3M_A - 800 - 37.5 + 50}{6}$$

or $$20R_A - 3M_A = 800 + 37.5 - 50 = 787.5 \quad ...(v)$$

At $x = 20$, $\frac{dy}{dx} = 0$. Substituting these values in complete equation (*ii*), we get

$$0 = \frac{R_A}{2} \times 20^2 - M_A \times 20 - \frac{4}{3} \times 20^3 - 60(20 - 15)^2 + \frac{4}{3}(20 - 10)^3$$

$$= 10R_A - M_A - \frac{4 \times 400}{3} - 3 \times 25 + \frac{4}{3} \times \frac{1000}{20} \qquad \text{(Dividing by 20)}$$

$$= 10R_A - M_A - \frac{1600}{3} - 75 + \frac{200}{3}$$

or $$10R_A - M_A = \frac{1600}{3} + 75 - \frac{200}{3} = \frac{1400}{3} + 75$$

or $$10R_A - M_A = 541.66$$

or $$20R_A - 2M_A = 1083.32 \qquad \text{(Multiplying by 2 both sides)} \quad ...(vi)$$

Substracting equation (*v*) from equation (*vi*), we get

$M_A = 1083.32 - 787.50 =$ **295.82 kNm. Ans.**

Substituting this values of M_A in equation (*vi*), we get

$$20R_A - 2 \times 295.82 = 1083.32$$

$\therefore$ $$R_A = \frac{1083.32 + 2 \times 295.82}{20}$$

$=$ **83.748 kN. Ans.**

Now $R_B =$ Total load on beam $- R_A$

$= (10 \times 8 + 120) - 83.748 =$ **116.252 kN. Ans.**

Equating the clockwise moment and anti-clockwise moment about *B*, we get

$$M_B + R_A \times 20 = M_A + 120 \times 5 + 8 \times 10 \times 15$$

or $$M_B + 83.748 \times 20 = 295.82 + 600 + 1200$$

or $M_B = 2095.82 - 83.748 \times 20 =$ **420.86 kNm. Ans.**

Maximum deflection and position of maximum deflection

Since the point load is more than the toal distributed load and acts at an equal distance from the nearest end, hence maximum deflection will be in the portion *AD*. For maximum deflection, $\frac{dy}{dx}$ should be zero. Substituting the value of $\frac{dy}{dx} = 0$ in equation (*ii*) [the term $- 60(x - 15)^2$ in equation (*ii*) should be ignored as this term is for the portion *DB*], we get

$$0 = \frac{R_A \times x^2}{2} - M_A . x - \frac{4}{3}x^3 + \frac{4}{3}(x - 10)^3$$

$$= \frac{83.748}{2} \times x^2 - 295.82x - \frac{4x^3}{3} + \frac{4(x-10)^3}{3}$$

$$= 41.874x^2 - 295.82x - \frac{4}{3}x^3 + \frac{4}{3}[x^3 - 1000 - 3x \times 10(x - 10)]$$

$$= 41.874x^2 - 295.82x - \frac{4000}{3} - \frac{4}{3} \times 3x \times 10 \times x + \frac{4}{3} \times 3x \times 10 \times 10$$

$$= 41.874x^2 - 295.82x - \frac{4000}{3} - 40x^2 + 400x$$

$$= 1.874x^2 + 104.18x - 1333.33$$

This is a quadratic equation. Hence its solution is

$$x = \frac{-104.18 \pm \sqrt{104.18^2 + 4 \times 1.874 \times 1333.33}}{2 \times 1.874}$$

$$= \frac{-104.81 + 144.387}{2 \times 1.874} \qquad \text{(Neglecting –ve root)}$$

$$= \mathbf{10.727\ m.} \quad \textbf{Ans.}$$

Hence maximum deflection occurs at a distance of 10.27 m from A. Maximum deflection is obtained by substituting $x = 10.727$ m in equation (*iv*) [neglecting the term $- 20(x - 15)^3$]

$$\therefore \quad EIy_{max} = \frac{R_A . x^3}{6} - \frac{M_A . x^2}{2} - \frac{x^4}{3} + \frac{1}{3}(x - 10)^4$$

$$= \frac{83.748 \times 10.727^3}{6} - \frac{295.82 \times 10.727^2}{2} - \frac{10.727^4}{3} + \frac{1}{3}(10.727 - 10)^4$$

$$= 17228.9 - 17019.8 - 4413.6 + 0.09 = - 4204.5$$

$$\therefore \quad y_{max} = \frac{-4204.5}{EI} = \frac{-4204.5}{2 \times 10^8 \times 4 \times 10^{-4}} = 0.05255 \text{ m} = \mathbf{52.56\ mm.} \quad \textbf{Ans.}$$

4.13.5. Advantages of Fixed Beams. The following are the advantages of a fixed beam over a simply supported beam :

(*i*) For the same loading, the maximum deflection of a fixed beam is less than that of a simply supported beam.

(*ii*) For the same loading, the fixed beam is subjected to a lesser maximum bending moment.

(*iii*) The slope at both ends of a fixed beam is zero.

(*iv*) The beam is more stable and stronger.

4.14. CONTINUOUS BEAM

Continuous beam is a beam which is supported on more than two supports. Fig. 4.55 shows such a beam, which is subjected to some external loading (here a uniformly distributed load). The deflection curve for the beam is shown by dotted line. The deflection curve is having convexity upwards over the intermediate supports, and concavity upwards over the mid of the span. Hence there will be hogging moments (*i.e.*, negative) over the intermediate supports and sagging moments (*i.e.*, positive) over the mid of the span. The end supports of a simply supported continuous beam will not be subjected to any bending moment. But the end support of fixed continuous beam will be subjected to fixing moments. If the moments over the intermediate supports are known, then the B.M. diagram can be drawn.

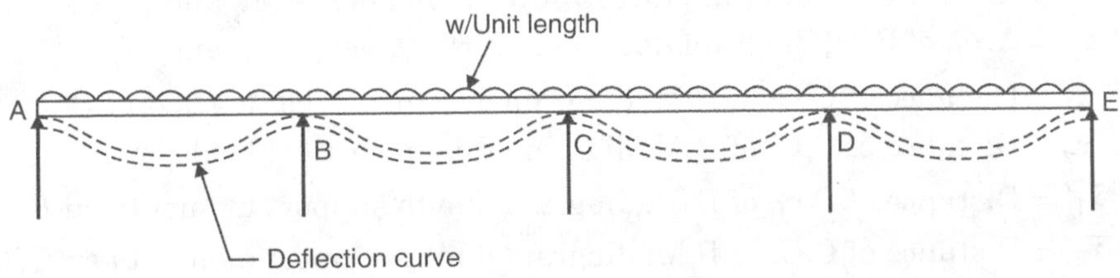

Fig. 4.55

The Fig. 4.55 shows a simply supported continuous beam. In this figure the end supports at A and E will not be subjected to any bending moment. Hence in this case $M_A = M_E = 0$.

Fig. 4.55 (a) shows a continuous beam with fixed ends at A and E. Here the end supports at A and E will be subjected to fixing moments. Hence M_A and M_E will not be zero.

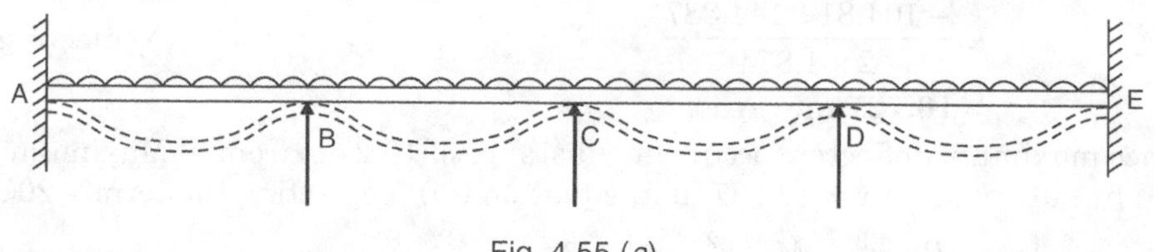

Fig. 4.55 (a)

In case of continuous beam, Clapeyron's equation of three moments and application of this equation to continuous beam of simply supportedends and fixed ends are explained.

4.14.1. Bending Moment Diagram for Continuous Beams. In Art. 4.14 it is mentioned that if the moments over the intermediate supports of a continuous beam are known, then the B.M. diagram can be drawn easily. The moments over the intermediate supports are determined by using *Clapeyron's theorem of three moments* which states that :

If BC and CD are any two consecutive span of a continuous beam subjected to an external loading, then the moments M_B, M_C and M_D at the supports B, C and D are given by,

$$M_B . L_1 + 2M_C(L_1 + L_2) + M_D . L_2 = \frac{6a_1\bar{x}_1}{L_1} + \frac{6a_2\bar{x}_2}{L_2} \qquad ...(4.35)$$

where L_1 = Length of span BC

L_2 = Length of span CD

a_1 = Area of B.M. diagram due to vertical loads on span BC

a_2 = Area of B.M. diagram due to vertical loads on span CD

$\bar{x}_1$ = Distance of C.G. of the B.M. diagram due to vertical loads on BC from B

$\bar{x}_2$ = Distance of C.G. of the B.M. diagram due to vertical loads on CD from D.

The equation (4.35) is known as the *equation of three moments* or Clapeyron's equation.

4.14.2. Derivation of Clapeyron's Equation of three Moments. Fig. 4.56 shows the length BCD (two consecutive spans) of a continuous beam which is shown in Fig. 4.55. Let M_B, M_C and M_D are the support moments at B, C and D respectively.

Let L_1 = Length of span BC

L_2 = Length of span CD

a_1 = Area of B.M. diagram due to vertical loads on span BC

a_2 = Area of B.M. diagram due to vertical loads on span CD

a_1' = Area of B.M. diagram due to support moments M_B and M_C

a_2' = Area of B.M. diagram due to support moments M_C and M_D

$\bar{x}_1$ = Distance of C.G. of B.M. diagram due to vertical loads on BC

$\bar{x}_2$ = Distance of C.G. of B.M. diagram due to vertical loads on CD

$\bar{x}_1'$ = Distance of C.G. of B.M. diagram due to support moments on BC

$\bar{x}_2'$ = Distance of C.G. of B.M. diagram due to support moments on CD.

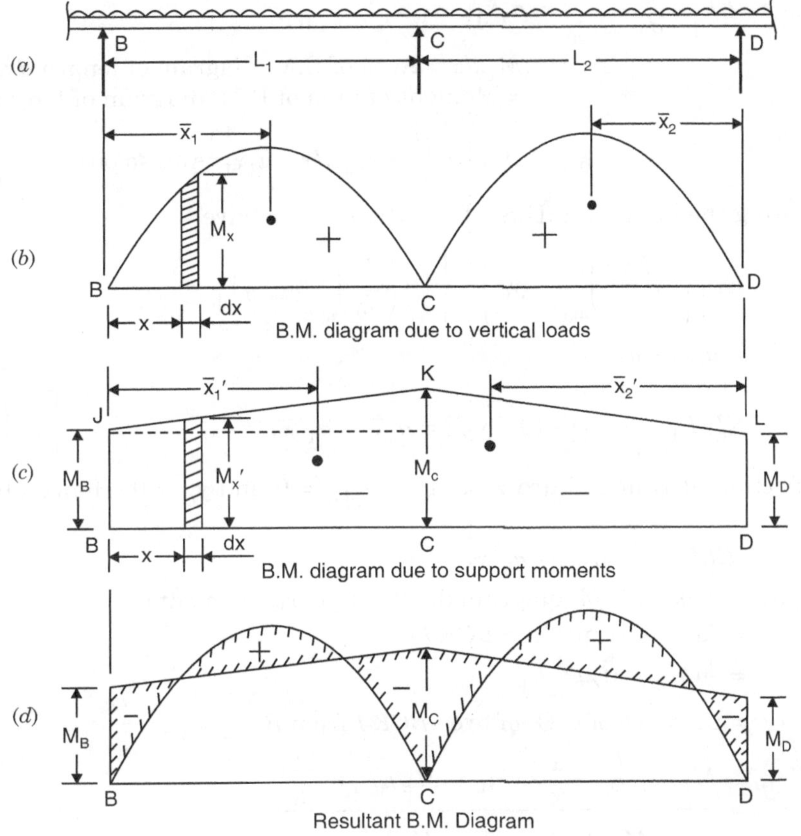

Fig. 4.56

Fig. 4.56 (*b*) and (*c*) shows the B.M. diagrams due to vertical loads and due to supports moments respectively.

(*i*) *Consider the span BC*

Let M_x = B.M. due to vertical loads at a distance x from B (sagging)

M_x' = B.M. due to support moments at a distance x from B (hogging)

$\therefore$ Net B.M. at a distance x from B is given by,

$$EI\,\frac{d^2y}{dx^2} = M_x - M_x'$$

Multiplying by x to both sides, we get

$$EI.x.\,\frac{d^2y}{dx^2} = x.M_x - x.M'x$$

Integrating from zero to L_1, we get

$$\int_0^{L_1} EI.x.\frac{d^2y}{dx^2}.dx = \int_0^{L_1} x.M_x.dx - \int_0^{L_1} x.M_x'.dx$$

or $$EI\left[x\frac{dy}{dx}-y\right]_0^{L_1}=a_1\bar{x}_1-a_1'\bar{x}_1' \quad ...(i)$$

($\because$ $M_x.dx$ = Area of B.M. diagram of length dx. And $x.M_x.dx$ = Moment of area of B.M. diagram of length dx about B.

Hence $\int_0^{L_1} x.M_x.dx = a_1\bar{x}_1$. And so on)

Substituting the limits in L.H.S. of equation (*i*), we have

$$EI\left[\left\{L_1\left(\frac{dy}{dx}\right)_{\text{at }C}-y_C\right\}-\left\{0\times\left(\frac{dy}{dx}\right)_{\text{at }B}-y_B\right\}\right]$$

$$= a_1\bar{x}_1-a_1'\bar{x}_1'$$

or $$EI[(L_1.\theta_C-y_C)-(0-y_B)]=a_1\bar{x}_1-a_1'\bar{x}_1'. \qquad \left[\because\ \left(\frac{dy}{dx}\right)_{\text{at }C}=\theta_C\right]$$

But deflection at B and C are zero. Hence $y_B = 0$ and $y_C = 0$. Hence above equation becomes as

$$[EI.L_1.\theta_C = a_1\bar{x}_1-a_1'\bar{x}_1' \quad ...(ii)$$

But a_1' = Area of B.M. diagram due to supports moments

= Area of trapezium $BCKJ$

$= \frac{1}{2}(M_B+M_C)\times L_1$

and $\bar{x}_1'$ = Distance of C.G. of area $BCKJ$ from B

$$=\frac{M_B.L_1.\frac{L_1}{2}+\frac{1}{2}\times(M_C-M_B).L_1\times\frac{2L_1}{3}}{M_B.L_1+\frac{1}{2}(M_C-M_B).L_1}$$

$$=\frac{M_B.\frac{L_1}{2}+(M_C-M_B)\times\frac{L_1}{3}}{M_B+(M_C-M_B).\frac{1}{2}}=\frac{\frac{3M_BL_1+2L_1(M_C-M_B)}{6}}{\frac{2M_B+M_C-M_B}{2}}$$

$$=\frac{\frac{L_1}{3}[3M_B+2M_C-2M_B]}{M_B+M_C}=\left(\frac{M_B+2M_C}{M_B+M_C}\right)\times\frac{L_1}{3}$$

Substituting the values of $\bar{a}_1$ and $\bar{x}_1'$ in equation (*ii*), we get

$$EI.L_1.\theta_C = a_1\bar{x}_1-\frac{1}{2}(M_B+M_C).L_1\times\left(\frac{M_B+2M_C}{M_B+M_C}\right)\times\frac{L_1}{3}$$

$$= a_1\bar{x}_1-\frac{L_1^2}{6}(M_B+2M_C)$$

or $$6EI.\theta_C=\frac{6a_1\bar{x}_1}{L_1}-L_1(M_B+2M_C) \quad ...(iii)$$

(ii) Consider the span CD

Similarly considering the span CD and taking D as origin and x positive to the left, it can be shown that

$$6EI.(-\theta_C) = \frac{6a_2\bar{x}_2}{L_2} - L_2(M_D + 2M_C)$$

[In the above case the slope at C (*i.e.*, θ_C) will have opposite sign than that given by equation (*iii*). The reason is that the direction of x from B for the span BC, and from D for span CD are in the opposite direction].

Hence the above equation becomes as

$$\therefore \quad -6EI\theta_c = \frac{6a_2\bar{x}_2}{L_2} - L_2(M_D + 2M_C) \quad \ldots(iv)$$

Adding equation (*iii*) and (*iv*), we get

$$0 = \frac{6a_1\bar{x}_1}{L_1} - L_1(M_B + 2M_C) + \frac{6a_2\bar{x}_2}{L_2} - L_2(M_D + 2M_C)$$

$$= \frac{6a_1\bar{x}_1}{L_1} + \frac{6a_2\bar{x}_2}{L_2} - L_1M_B - 2L_1M_C - L_2M_D - 2L_2M_C$$

or
$$L_1.M_B + L_2M_D + 2M_C(L_1 + L_2) = \frac{6a_1\bar{x}_1}{L_1} + \frac{6a_2\bar{x}_2}{L_2}$$

or
$$M_BL_1 + 2M_C(L_1 + L_2) + M_DL_2 = \frac{6a_1\bar{x}_1}{L_1} + \frac{6a_2\bar{x}_2}{L_2}$$

4.14.3. Application of Clapeyron's Equation of Three Moments to Continuous Beam with Simply Supported Ends. The fixing moments on the ends of a simply supported beam is zero. The continuous beam with simply supported ends may carry uniformly distributed load or point loads as given in the following problems:

Problem Based on Continuous Beams

Problem 4.33. *A continuous beam ABC covers two consecutive span AB and BC of lengths 4 m and 6 m, carrying uniformly distributed loads of 6 kN/m and 10 kN/m respectively. If the ends A and C are simply supported, find the support moments at A, B and C. Draw also B.M. and S.F. diagrams.*

Sol. Given :

Length AB, $L_1 = 4$ m

Length BC, $L_2 = 6$ m

U.d.l. on AB, $w_1 = 6$ kN/m

U.d.l. on BC, $w_2 = 10$ kN/m

Since the ends A and C are simply supported, the support moments at A and C will be zero.

$$\therefore \quad M_A = M_C = 0$$

To find the support moment at B (*i.e.*, M_B), Clapeyron's equation of three moments should be applied. Hence, we get

$$M_A.L_1 + 2M_B(L_1 + L_2) + M_C.L_2 = \frac{6a_1\bar{x}_1}{L_1} + \frac{6a_2\bar{x}_2}{L_2}$$

or
$$0 \times 4 + 2M_B(4 + 6) + 0 \times L_2 = \frac{6a_1\bar{x}_2}{4} + \frac{6a_2\bar{x}_2}{6}$$

or $$20M_B = \frac{3a_1\bar{x}_1}{2} + a_2\bar{x}_2 \qquad ...(i)$$

The B.M. diagram on a simply supported beam carrying u.d.l. is a parabola having an altitude of $\frac{wL^2}{8}$. And area of B.M. diagram = $\frac{2}{3}$ × Span × Altitude. The distance of C.G. of this area from one end = $\frac{\text{Span}}{2}$.

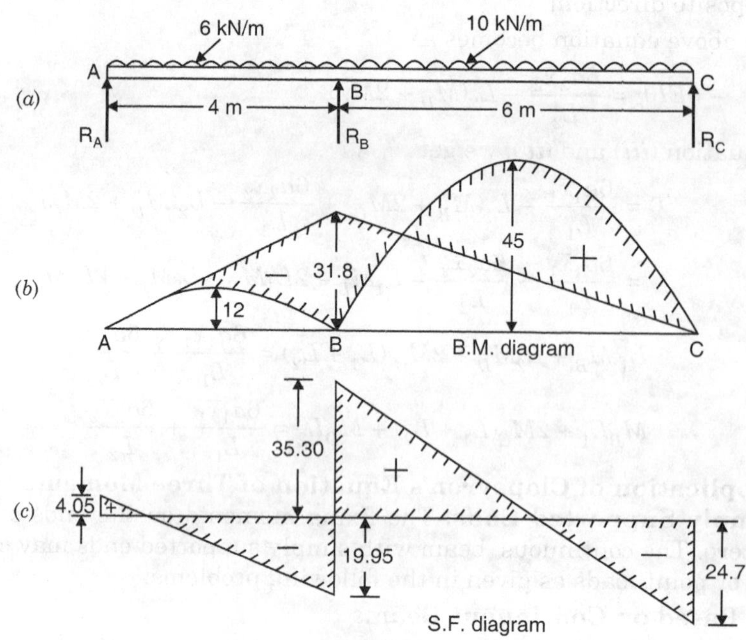

Fig. 4.57

Now a_1 = Area of B.M. diagram due to u.d.l. on AB

$$= \frac{2}{3} \times AB \times \text{Altitude} = \frac{2}{3} \times AB \times \frac{w_1 {L_1}^2}{8}$$

$$= \frac{2}{3} \times 4 \times \frac{6 \times 4^2}{8} = 32$$

$$\bar{x}_1 = \frac{L_1}{2} = \frac{4}{2} = 2 \text{ m}$$

a_2 = Area of B.M. diagram due to u.d.l. on BC

$$= \frac{2}{3} \times BC \times \frac{w_2 {L_2}^2}{8} = \frac{2}{3} \times 6 \times \frac{10 \times 6^2}{8} = 180$$

and $$\bar{x}_2 = \frac{L_2}{2} = \frac{6}{2} = 3 \text{ m.}$$

Substituting these values in equation (i), we get

$$20M_B = \frac{3 \times 32 \times 2}{2} + 180 \times 3$$

$$= 96 + 540 = 636$$

$\therefore \qquad M_B = \dfrac{636}{20} = 31.8$ kNm.

Now B.M. diagram due to supports moments is drawn as shown in Fig. 4.57 (*b*) in which

$$M_A = 0, M_C = 0 \text{ and } M_B = 31.8 \text{ kNm}.$$

The B.M. diagram due to vertical loads (here u.d.l.) on span *AB* and span *BC* are also shown by parabolas of altitudes

$$\frac{w_1 L_1^{\,2}}{8} = \frac{6 \times 4^2}{8} = 12 \text{ kNm and } \frac{w_2 L_2^{\,2}}{8} = \frac{10 \times 6^2}{8} = 45 \text{ kNm}$$

respectively in Fig. 4.57 (*b*).

S.F. Diagram

First calculate the reactions R_A, R_B and R_C at *A*, *B* and *C* respectively. For the span *AB*, taking moments about *B*, we get

$$R_A \times 4 - 6 \times 4 \times \frac{4}{2} = M_B \qquad \text{(The support } B \text{ has moment } M_B\text{)}$$

$$= -31.8$$

($\because$ $M_B = 31.8$. Negative sign is taken as the moment at *B* is hogging)

or $\qquad 4R_A - 48 = -31.8$

or $\qquad R_A = \dfrac{-31.8 + 48}{4} = 4.05$ kN.

Similarly for the span *BC*, taking moments about *B*, we get

$$R_C \times 6 - 6 \times 10 \times \frac{6}{2} = M_B = -31.8$$

or $\qquad 6R_C - 180 = -31.8$

or $\qquad R_C = \dfrac{180 - 31.8}{6} = 24.7$ kN.

Now, $\qquad R_B = \text{Total load on } ABC - (R_A + R_C)$

$$= (6 \times 4 + 10 \times 6) - (4.05 + 24.7) = 55.25 \text{ kN}.$$

Now complete the S.F. diagram as shown in Fig. 4.57 (*c*).

Problem 4.34. *A continuous beam ABCD of length 15 m rests on four supports covering 3 equal spans and carries a uniformly distributed load of 1.5 kN/m length. Calculate the moments and reactions at the supports. Draw the S.F. and B.M. diagrams also.*

Sol. Given :

Length *AB*, $\quad L_1 = 5$ m

Length *BC*, $\quad L_2 = 5$ m

Length *CD*, $\quad L_3 = 5$ m

U.d.l., $\quad w_1 = w_2 = w_3 = 1.5$ kN/m.

Since ends *A* and *D* are simply supported, the support moments at *A* and *D* will be zero.

$\therefore \qquad M_A = 0$ and $M_D = 0$

From symmetry $\qquad M_B = M_C$

To find the support moments at *B* and *D*, Clapeyron's equation of three moments is applied for *ABC* and for *BCD*.

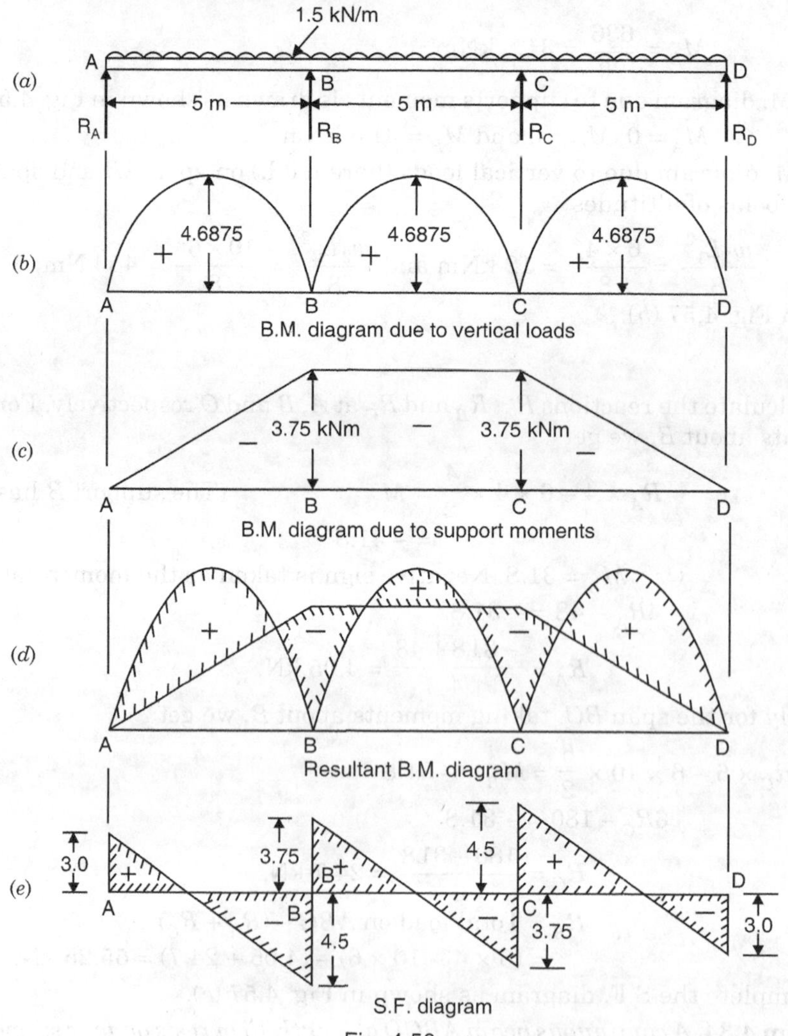

Fig. 4.58

For ABC, we get

$$M_A L_1 + 2M_B(L_1 + L_2) + M_C \,.\, L_2 = \frac{6a_1\bar{x}_1}{L_1} + \frac{6a_2\bar{x}_2}{L_2}$$

or
$$0 \times 5 + 2M_B(5 + 5) + M_C \times 5 = \frac{6a_1\bar{x}_1}{5} + \frac{6a_2\bar{x}_2}{5}$$

or
$$20M_B + 5M_C = \frac{6}{5}\,(a_1\bar{x}_1 + a_2\bar{x}_2) \qquad \ldots(i)$$

Now $\quad a_1$ = Area of B.M. Diagram due to u.d.l. on AB when AB is considered as simply supported beam

$$= \frac{2}{3} \times AB \times \text{Altitude of parabola}$$

$$= \frac{2}{3} \times 5 \times \frac{w_1 L_1}{8} = \frac{2}{3} \times 5 \times \frac{1.5 \times 5^2}{8} = 15.625$$

$$\bar{x}_2 = \frac{L_1}{2} = \frac{5}{2} = 2.5 \text{ m}$$

Due to symmetry $a_2 = a_1 = 15.625$ and $\bar{x}_2 = \bar{x}_1 = 2.5$

Substituting these values in equation (*i*), we get

$$20M_B + 5M_C = \frac{6}{5}(15.625 \times 2.5 + 15.625 \times 2.5)$$

$$= \frac{6}{5} \times 2 \times 15.625 \times 2.5 = 93.750$$

or $20M_B + 5M_B = 93.750$ $(\because M_B = M_C$ due to symmetry)

or $M_B = \dfrac{93.750}{25} = 3.75$ kNm

$\therefore$ $M_B = M_C =$ **3.75 kNm. Ans.**

Now the B.M. diagram due to supports moments is drawn as shown in Fig. 4.58 (*c*), in which

$$M_A = 0,\ M_D = 0,\ M_B = M_C = 3.75 \text{ kNm}.$$

The B.M. diagram due to vertical loads (here u.d.l.) on span *AB*, *BC* and *CD* (considering each span as simply supported) are shown by parabolas of altitudes $\dfrac{w_1 L_1^2}{8} = \dfrac{1.5 \times 5^2}{8} = 4.6875$ kNm each in Fig. 4.58 (*b*). Resultant B.M. diagram is shown in Fig. 4.58 (*d*).

Support Reactions

Let R_A, R_B, R_C and R_D are the support reactions at *A*, *B*, *C* and *D* respectively.

Due to symmetry, $R_A = R_D$

$R_B = R_C$

For the span *AB*, taking moments about *B*, we get

$$M_B = R_A \times 5 - 1.5 \times 5 \times \frac{5}{2}$$

or $-3.75 = R_A \times 5 - 18.75$ $(\because M_B = -3.75)$

or $5R_A = 18.75 - 3.75 = 15$

$\therefore$ $R_A = \dfrac{15}{5} =$ **3.0 kN. Ans.**

$\therefore$ Due to symmetry, $R_D = R_A =$ **3.0 kN. Ans.**

Now, $R_A + R_B + R_C + R_D =$ Total load on *ABCD*

or $R_A + R_B + R_B + R_A = 1.5 \times 15$ $(\because R_C = R_B, R_D = R_A)$

or $2(R_A + R_B) = 22.5$

or $R_A + R_B = \dfrac{22.5}{2} = 11.25$

or $R_B = 11.25 - R_A = 11.25 - 3.00 = 8.25$ $(\because R_A = 3.0)$

$\therefore$ $R_B = R_C =$ **8.25 kN. Ans.**

Now the S.F. diagram can be drawn as shown in Fig. 4.58 (*e*).

Problem 4.35. *A continuous beam ABCD, simply supported at A, B, C and D is loaded as shown in Fig. 4.59 (a). Find the moments over the beam and draw B.M. and S.F. diagrams.*

Sol. Given :

Length AB, $L_1 = 6$ m

Length BC, $L_2 = 5$ m

Length CD, $L_3 = 4$ m

Point load in BD, $W_1 = 9$ kN

Point load in BC, $W_2 = 8$ kN

U.d.l. on CD, $w = 3$ kN/m.

(i) B.M. diagram due to vertical loads taking each span as simply supported

Consider beam AB as simply supported

$$\text{B.M. at point load at } E = \frac{W_1 \times a \times b}{L_1} = \frac{9 \times 2 \times 4}{6} \quad (\because \text{ Here } a = 2 \text{ m}, b = 4 \text{ m})$$

$$= 12 \text{ kNm}$$

Similarly B.M. at F, considering beam BC as simply supported

$$= \frac{W_2 \,.\, a \,.\, b}{L_2} = \frac{8 \times 2 \times 3}{5} \quad (\because \text{ Here } a = 2, b = 3 \text{ and } L_2 = 5)$$

$$= 9.6 \text{ kNm}$$

The B.M. at the centre of a simply supported beam CD, carrying u.d.l.

$$= \frac{w \times {L_3}^2}{8} = \frac{3 \times 4^2}{8} = 6 \text{ kNm.}$$

Now the B.M. diagram due to vertical loads taking each span as simply supported can be drawn as shown in Fig. 4.59 (*b*).

(ii) B.M. diagram due to support moments

Let M_A, M_B, M_C and M_D are the supports moments at A, B, C and D respectively. But the end supports of a simply supported beam are not subjected to any bending moment. Hence the support moments at A and D will be zero.

$$\therefore \quad M_A = 0 \quad \text{and} \quad M_D = 0$$

To find the support moments at B and C, Clapeyron's equation of three moments in applied for ABC and for BCD.

(*a*) For spans AB and BC from equation of three moments, we have

$$M_A . L_1 + 2M_B (L_1 + L_2) + M_C \,.\, L_2 = \frac{6a_1\bar{x}_1}{L_1} + \frac{6a_2\bar{x}_2}{L_2}$$

or
$$0 + 2M_B (6 + 5) + M_C \times 5 = \frac{6a_1\bar{x}_1}{6} + \frac{6a_2\bar{x}_2}{5}$$

or
$$22M_B + 5M_C = a_1\bar{x}_1 + \frac{6}{5} a_2\bar{x}_2 \quad \ldots(i)$$

Now, $a_1\bar{x}_1$ = Moment of area of B.M. diagram due to vertical load on AB when AB is considered as simply supported beam about point A.

$$= \frac{1}{2} \times 2 \times 12 \times \frac{2 \times 2}{3} + \frac{1}{2} \times 4 \times 12 \times \left(2 + \frac{1}{3} \times 4\right)$$

$$= 16 + 80 = 96$$

$a_2\bar{x}_2$ = Moment of area of B.M. diagram due to vertical load on BC when BC is considered as simply supported beam about point C

$$= \frac{1}{2} \times 3 \times 9.6 \times \frac{2}{3} \times 3 + \frac{1}{2} \times 2 \times 9.6 \times \left(3 + \frac{1}{3} \times 2\right)$$

$$= 28.8 + 35.2 = 64.0$$

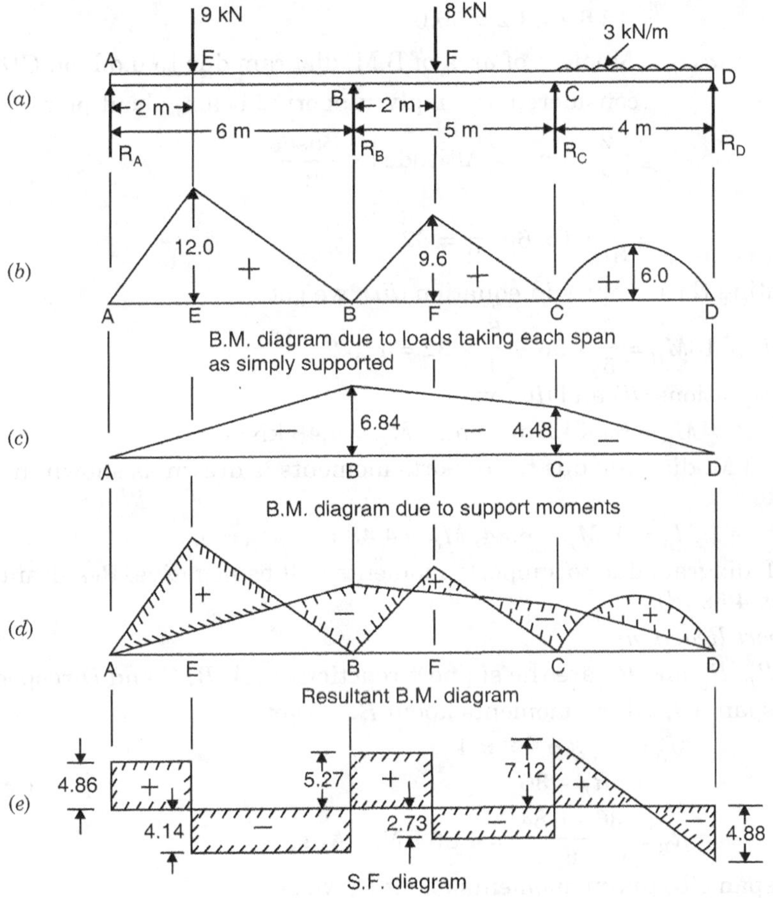

Fig. 4.59

Substituting these values in equation (*i*), we get

$$22M_B + 5M_C = 96 + \frac{6}{5} \times 64$$

$$= 172.8 \qquad \ldots(ii)$$

(*b*) For spans BC and CD from equation of three moments, we have

$$M_B \cdot L_2 + 2M_C(L_2 + L_3) + M_D \cdot L_3 = \frac{6a_2\bar{x}_2}{L_2} + \frac{6a_3\bar{x}_3}{L_3}$$

or

$$M_B \times 5 + 2M_C(5 + 4) + 0 = \frac{6a_2\bar{x}_2}{5} + \frac{6a_3\bar{x}_3}{4} \qquad (\because \; M_D = 0)$$

or $$5M_B + 18M_C = \frac{6}{5}a_2\bar{x}_2 + \frac{6}{5}a_3\bar{x}_3 \quad ...(iii)$$

where $a_2\bar{x}_2$ = Moment of area of B.M. diagram due to vertical load on BC when BC is considered as simply supported beam, about point B

$$= \frac{1}{2} \times 2 \times 9.6 \times \frac{2}{3} \times 2 + \frac{1}{2} \times 3 \times 9.6 \times \left(2 + \frac{1}{3} \times 3\right)$$

$$= 12.8 + 43.2 = 56.0$$

and $a_3\bar{x}_3$ = Moment of area of B.M. diagram due to u.d.l. on CD, when CD is considered as simply supported beam, about point D

$$= \left(\frac{2}{3} \times \text{Base} \times \text{Altitude}\right) \times \frac{\text{Base}}{2}$$

$$= \frac{2}{3} \times 4 \times 6 \times \frac{4}{2} = 32$$

Substituting these values in equation (*iii*), we get

$$5M_B + 18M_C = \frac{6}{5} \times 56 + \frac{6}{4} \times 32 = 115.2 \quad ...(iv)$$

Solving equations (*ii*) and (*iv*), we get

$$M_B = 6.84 \text{ kNm} \quad \text{and} \quad M_C = 4.48 \text{ kNm}.$$

Now the B.M. diagram due to supports moments is drawn as shown in Fig. 4.59 (*c*), in which

$$M_A = 0,\ M_B = 6.84,\ M_C = 4.48 \text{ and } M_D = 0.$$

The B.M. diagram due to supports moments will be negative. Resultant B.M. diagram is shown in Fig. 4.59 (*d*).

(*iii*) *Support Reactions*

Let R_A, R_B, R_C and R_D are the support reactions at A, B, C and D respectively,

For the span AB, taking moments about B, we get

$$M_B = R_A \times 6 - 9 \times 4$$

or $$-6.84 = 6R_B - 36 \quad (\because \ M_B = -6.84)$$

or $$R_B = \frac{36 - 6.84}{6} = \textbf{4.86 kN.} \quad \textbf{Ans.}$$

For the span CD, taking moments about C, we get

$$M_C = R_D \times 4 - 3 \times 4 \times \frac{4}{2}$$

or $$-4.48 = 4R_D - 24 \quad (\because \ M_C = -4.48)$$

$\therefore$ $$R_D = \frac{24 - 4.48}{4} = \textbf{4.88 kN.} \quad \textbf{Ans.}$$

Now taking moments about C for ABC, we get

$$M_C = R_A \times (6 + 5) - 9\,(5 + 4) + R_B \times 5 - 8 \times 3$$

or $$-4.48 = 4.86 \times 11 - 9 \times 9 + R_B \times 5 - 24 \quad (\because \ M_C = -4.48,\ R_A = 4.86)$$

$\therefore$ $$5R_B = 81 + 24 - 4.86 \times 11 - 4.48 = 47.06$$

$\therefore$ $$R_B = \frac{47.06}{5} = \textbf{9.41 kN.} \quad \textbf{Ans.}$$

Now, R_C = Total load on $ABCD - (R_A + R_B + R_D)$
$= (9 + 8 + 4 \times 3) - (4.86 + 9.41 + 4.88)$
= **9.85 kN. Ans.**

Now complete the S.F. diagram as shown in Fig. 4.59 (*e*).

HIGHLIGHTS (4C)

1. A beam whose both ends are fixed is known as fixed beam. And a beam which is supported on more than two supports is known as a continuous beam.
2. In case of a fixed beam :
   (*i*) $a = a'$ (*ii*) $a\bar{x} = a'\bar{x}'$ and $\bar{x} = \bar{x}'$

   Or

   (*i*) The area of B.M. diagram due to vertical loads is equal to the area of B.M. diagram due to end moments.
   (*ii*) Distance of C.G. of B.M. diagram due to vertical loads is equal to the distance of C.G. of B.M. diagram due to end moments from the same point.
3. The deflection at the centre of a fixed beam carrying a point load at the centre is given by
$$y_c = \frac{WL^3}{192EI}$$
   where W = Point load,
   L = Length of beam.
4. The deflection at the centre of a fixed beam carrying a point load at the centre is one-fourth of the deflection of a simply supported beam.
5. The deflection of a fixed beam with an eccentric load, under the point load is given by,
$$y_c = \frac{Wa^3b^3}{3EIL^3}.$$
6. (*a*) For a fixed beam carrying uniformly distributed load over the whole length :
$$\text{End moments} = \frac{W \times L^2}{12}$$
$$\text{Max. deflection} = \frac{w \,.\, L^4}{384EI}$$
   (*b*) The deflection at the centre of a fixed beam carrying uniformly distributed load over the whole span is one-fifth of the deflection of a simply supported beam.
7. The end moments of a fixed beam due to sinking of a support is given by
$$M_A = M_B \frac{6EI\delta}{L^2}$$
   where δ = Sinking of one support with respect to the other. At the higher end this moment is –ve whereas at the lower end it is positive.
8. Clapeyron's theorem of three moments for a continuous beam ABC is given by,
$$M_A.L_1 + 2M_B(L_1 + L_2) + M_C \times L_2 = \frac{6a_1\bar{x}_1}{L_1} + \frac{6a_2\bar{x}_2}{L_2}$$
   where a_1 = Area of B.M. diagram due to vertical loads on span AB
   a_2 = Area of B.M. diagram due to vertical loads on span BC
   $\bar{x}_1$ = Distance of C.G. of B.M. diagram due to vertical loads on AB from A
   $\bar{x}_2$ = Distance of C.G. of B.M. diagram due to vertical loads on BC from point C.

9. To apply the theorem of three moments to a fixed continuous beam, an imaginary support of zero span is introduced.

EXERCISE (4C)

(A) Theoretical Questions

1. What do you mean by a fixed beam and a continuous beam ?
2. Prove that for a fixed beam :
   (*i*) Area of B.M. diagram due to vertical loads is equal to the area of B.M. diagram due to end moments.
   (*ii*) Distance of C.G. of B.M. diagram due to vertical loads is equal to the distance of C.G. of B.M. diagram due to end moment from the same point.
3. Find an expression for the deflection for a fixed beam carrying a point load at the centre. Also obtain the value of maximum deflection.
4. Prove that the deflection at the centre of a fixed beam is one-fourth the deflection of a simply supported beam of the same length, when they carry a point load W at the centre.
5. Draw the S.F. and B.M. diagrams for a fixed beam, carrying an eccentric load.
6. Prove that the deflection at the centre of a fixed beam, carrying a uniformly distributed load is given by

$$y_c = \frac{wL^4}{384EI}$$

   Determine the position of points of contraflexures also.
7. Derive an expression for the fixing moments, when one of the supports of a fixed beam sinks down by δ from its original position.
8. What are advantages and disadvantages of a fixed beam over a simply supported beam ?
9. What is the Clapeyron's theorem of three moments ? Derive an expression for Clapeyron's theorem of three moments.
10. How will you apply Clapeyron's theorem of three moments to a
    (*i*) continuous beam with simply supported ends
    (*ii*) continuous beam with fixed end supports ?

(B) Numerical Problems

1. A fixed beam *AB*, 5 m long, carries a point load of 48 kN at its centre. The moment of inertia of the beam is 5×10^7 mm^4 and value of E for the beam material is 2×10^5 N/mm^2. Determine :
   (*i*) Fixed end moments at *A* and *B*, and
   (*ii*) Deflection under the load. **[Ans.** (*i*) $M_A = M_B = 30$ kNm, (*ii*) 3.125 mm]
2. A fixed beam of length 5 m carries a point load of 20 kN at a distance of 2 m from *A*. Determine the fixed end moments and deflection under the load, if the flexural rigidity of the beam is 1×10^4 kNm^2. **[Ans.** $M_A = 14.4$ kNm, $M_B = 9.6$ kNm, $y_c = 1.15$ mm]
3. A fixed beam of length 6 m carries point loads of 20 kN and 15 kN at distances 2 m and 4 m from the left end *A*. Find the fixed end moments and the reactions at the supports. Draw B.M. and S.F. diagrams. **[Ans.** $M_A = 24.44$ kNm, $M_B = 22.22$ kNm, $R_A = 18.70$ kN, $R_B = 16.30$ kN]
4. A fixed beam of length 3 m carries two point loads of 30 kN each at a distance of 1 m from both the ends. Determine the fixing moments and draw the B.M. diagram.
   [Ans. $M_A = M_B = 20$ kNm]

5. A fixed beam AB of length 6 m carries a uniformly distributed load of 3 kN/m over the left half of the span together with a point load of 4 kN at a distance of 4.5 m from the left end. Determine the fixing end moments and the support reactions.

[**Ans.** $M_A = 7.3$ kNm, $M_B = 6.2$ kNm, $R_A = 7.93$ kN, $R_B = 5.07$ kN]

6. A fixed beam AB of length 6 m is having moment of intertia $I = 5 \times 10^6$ mm^4. The support B sinks down by 6 mm. If $E = 2 \times 10^5$ N/mm^2 find the fixing moments. [**Ans.** $M_A = M_B = 1000$ Nm]

7. A continuous beam ABC of length 10 m rests on three supports A, B and C at the same level in which span $AB = 6$ m and span $BC = 4$ m. In span AB, there is a point load of 3 kN at a distance of 2 m from the end A, whereas in the span BC, there is a uniformly distributed load of 1 kN/m run over the whole length. Determine the support moments and support reactions. Draw S.F. and B.M. diagrams also. [**Ans.** (*i*) $M_A = M_C = 0$, $M_B = 2.4$ kNm,

(*ii*) $R_A = 1.6$ kN, $R_B = 4$ kN, $R_C = 1.4$ kN]

8. A continuous beam consists of three successive span of 8 m, 10 m and 6 m and carries loads of 6 kN/m, 4 kN/m and 8 kN/m respectively on the spans. Determine the bending moments and reactions at the supports. [**Ans.** (*i*) $M_A = M_D = 0$, $M_C = 32.2$ kNm, $M_B = 40.16$ kNm,

(*ii*) $R_A = 18.98$ kN, $R_B = 49.82$ kN, $R_C = 48.57$ kN, $R_D = 18.63$ kN]

9. A continuous beam ABC consists of two consecutive spans AB and BC of length 8 m and 6 m respectively. The beam carries a uniformly distributed load of 1 kN/m throughout its length. The end A is fixed and the end C is simply supported. Find the support moments and the reactions. Also draw the S.F. and B.M. diagrams. [**Ans.** (*i*) $M_A = 5.75$ kNm, $M_B = 4.5$ kNm, $M_C = 0$,

(*ii*) $R_A = 4.15$ kN, $R_B = 7.6$ kN, $R_C = 2.25$ kN]

10. Draw the S.F. and B.M. diagram of a continuous beam ABC of length 10 m which is fixed at A and is supported on B and C. The beam carries a uniformly distributed load of 2 kN/m length over the entire length. The spans AB and BC are equal to 5 m each.

[**Ans.** (*i*) $M_A = 3.57$ kNm, $M_B = 5.357$ kNm, $M_C = 0$,

(*ii*) $R_A = 5.357$ kN, $R_B = 8.571$ kN, $R_C = 6.071$ kN]

5

Torsion

5.1. TORSION

A shaft is said to be in torsion, when equal and opposite torques are applied at the two ends of the shaft. The torque is equal to the product of the force applied (tangentially to the ends of a shaft) and radius of the shaft. Due to the application of the torques at the two ends, the shaft is subjected to a twisting moment. This causes the shear stresses and shear strains in the material of the shaft.

The followings are the main points of torsion :

1. A shaft is in torsion, when equal and opposite torques are applied at the two ends of a shaft.

2. The relation of maximum shear stress induced in a shaft subjected to twisting moment is given by

$$\frac{\tau}{R} = \frac{C\theta}{L}$$

where τ = Maximum shear stress,

R = Radius of shaft,

C = Modulus of rigidity,

θ = Angle of twist in radian, and

L = Length of the shaft.

3. When a circular shaft is subjected to torsion, the shear stress at any point varies linearly from the axis to the surface *i.e.,*

$$\frac{\tau}{R} = \frac{q}{r}$$

where τ = Maximum shear stress on the surface of the shaft,

R = Radius of surface,

q = Shear stress at a point which is at a radius 'r'.

4. The shear stress is maximum on the surface of the shaft and is zero at the axis of the shaft.

5. The torque transmitted by a solid shaft is given by

$$T = \frac{\pi}{16} \tau D^3$$

where D = Dia. of solid shaft and

τ = Max. shear stress.

6. The torque transmitted by a hollow circular shaft is given by

$$T = \frac{\pi}{16} \tau \left(\frac{D_0^4 - D_i^4}{D_0} \right)$$

where D_0 = External diameter, and

D_i = Internal diameter.

7. Relation between torque, polar moment of inertia and shear stress is given as

$$\frac{T}{J} = \frac{\tau}{R} = \frac{C\theta}{L}$$

where I = Polar moment of inertia

$$= \frac{\pi}{32} D^4 \quad \text{... For a solid shaft}$$

$$= \frac{\pi}{32} [D_0^4 - D_1^4] \quad \text{... For a hollow shaft.}$$

8. The ratio of the polar moment of inertia to the radius of the shaft is known as polar modulus or torsional section modulus. It is denoted by Z_p.

$$\therefore \quad Z_p = \frac{J}{R}$$

$$= \frac{\pi}{16} D^3 \quad \text{... For a solid shaft}$$

$$= \frac{\pi}{16 D_0} [D_0^4 - D_i^4] \quad \text{... For a hollow shaft.}$$

9. Strength of a shaft means the maximum torque or maximum power the shaft can transmit.

10. The product of modulus of rigidity and polar moment of inertia of a shaft is known as torsional rigidity or stiffness of the shaft. Mathematically,

Torsional rigidity $= C \times J$

$$= \frac{T \times L}{\theta} \quad \left(\because \frac{T}{J} = \frac{C_\theta}{L} \quad \therefore \quad C \times J = \frac{T \times L}{\theta} \right)$$

$= T$ if $L = 1$ metre and θ = one radian.

11. The power transmitted by a shaft is given by

$$P = \frac{2\pi NT}{60}.$$

12. The strain energy stored in a shaft due to torsion is given by,

$$U = \frac{\tau^2}{4C} \times V \quad \text{... For a solid shaft}$$

$$= \frac{\tau}{4C \cdot D^2} (D^2 + d^2) \times V \quad \text{... For a hollow shaft}$$

where D = External diameter of shaft,

d = Internal diameter of shaft,

T = Modulus of rigidity, and

τ = Shear stress on the surface of the shaft.

13. Polar moment of inertia (J) is given by

$$J = \frac{\pi}{32} D^4 \quad \text{... For a solid shaft}$$

$$= \frac{\pi}{32} (D^4 - d^4) \quad \text{... For hollow shaft.}$$

5.2. DERIVATION OF SHEAR STRESS PRODUCED IN A CIRCULAR SHAFT SUBJECTED TO TORSION

When a circular shaft is subjected to torsion, shear stresses are set up in the material of the shaft. To determine the magnitude of shear stress at any point on the shaft, consider a shaft fixed at one end *AA* and free at the end *BB* as shown in Fig. 5.1. Let *CD* is any line on the outer surface of the shaft. Now let the shaft is subjected to a torque *T* at the end *BB* as shown in Fig. 5.2. As a result of this torque *T*, the shaft at the end *BB* will rotate clockwise and every cross-section of the shaft will be subjected to shear stresses. The point *D* will shift to *D′* and hence line *CD* will be deflected to *CD′* as shown in Fig. 5.2 (*a*). The line *OD* will be shifted to *OD′* as shown in Fig. 5.2 (*b*).

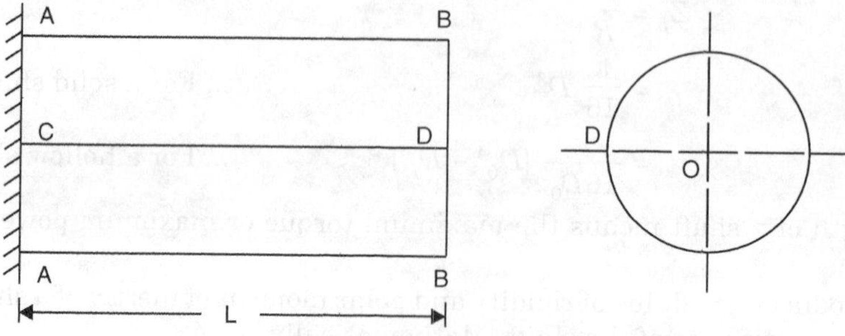

Fig. 5.1. Shaft fixed at one end *AA* before torque *T* is applied.

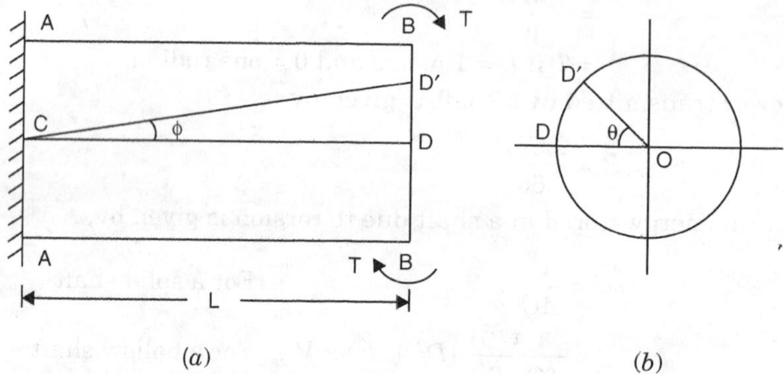

Fig. 5.2. Shaft fixed at *AA* and subjected to torque *T* at *BB*.

Let R = Radius of shaft

L = Length of shaft

T = Torque applied at the end BB

τ = Shear stress induced at the surface of the shaft due to torque T

C = Modulus of rigidity of the material of the shaft

$\phi = \angle DCD'$ also equal to shear strain

$\theta = \angle DOD'$ and is also called angle of twist.

Now distortion at the outer surface due to torque T

$$= DD'$$

$\therefore$ Shear strain at outer surface

$$= \text{Distortion per unit length}$$

$$= \frac{\text{Distortion at the outer surface}}{\text{Length of shaft}} = \frac{DD'}{L}$$

$$= \frac{DD'}{CD} = \tan \phi$$

$$= \phi \qquad (\text{if } \phi \text{ is very small then } \tan \phi \simeq \phi)$$

$\therefore$ Shear strain at outer surface,

$$\phi = \frac{DD'}{L} \qquad \text{...}(i)$$

Now from Fig. 5.2 (*b*).

Arc $DD' = OD \times \theta = R\theta$ $\qquad (\because \; OD = R = \text{Radius of shaft})$

Substituting the value of DD' in equation (*i*), we get

Shear strain at outer surface

$$\phi = \frac{R \times \theta}{L} \qquad \text{...}(ii)$$

Now the modulus of rigidity (C) of the material of the shaft is given as

$$C = \frac{\text{Shear stress induced}}{\text{Shear strain produced}} = \frac{\text{Shear stress at the outer surface}}{\text{Shear strain at outer surface}}$$

$$= \frac{\tau}{\left(\frac{R\theta}{L}\right)} \qquad \left(\because \text{ From equation } (ii), \text{ shear strain} = \frac{R\theta}{L}\right)$$

$$= \frac{\tau \times L}{R\theta}$$

$$\therefore \qquad \frac{C\theta}{L} = \frac{\tau}{R} \qquad \text{...(5.1)}$$

$$\therefore \qquad \tau = \frac{R \times C \times \theta}{L}$$

Now for a given shaft subjected to a given torque (T), the values of C, θ and L are constant. Hence shear stress produced is proportional to the radius R.

$$\therefore \qquad \tau \propto R \quad \text{or} \quad \frac{\tau}{R} = \text{constant} \qquad \text{...}(iii)$$

If q is the shear stress induced at a radius 'r' from the centre of the shaft then

$$\frac{\tau}{R} = \frac{q}{r} \qquad \text{...(5.2)}$$

But $\qquad \frac{\tau}{R} = \frac{C\theta}{L}$ from equation (5.1)

$$\therefore \qquad \frac{\tau}{R} = \frac{C\theta}{L} = \frac{q}{r}. \qquad \text{...(5.3)}$$

From equation (*iii*), it is clear that shear stress at any point in the shaft is proportional to the distance of the point from the axis of the shaft. Hence the shear stress is maximum at the outer surface and shear stress is zero at the axis of the shaft.

5.2.1. Assumptions Made in the Derivation of Shear Stress Produced in a Circular Shaft Subjected to Torsion. The derivation of shear stress produced in a circular shaft subjected to torsion, is based on the following assumptions :

1. The material of the shaft is uniform throughout.
2. The twist along the shaft is uniform.
3. The shaft is of uniform circular section throughout.
4. Cross-sections of the shaft, which are plane before twist remain plain after twist.
5. All radii which are straight before twist remain straight after twist.

5.2.2. Maximum Torque Transmitted by a Circular Solid Shaft. The maximum torque transmitted by a circular solid shaft, is obtained from the maximum shear stress induced at the outer surface of the solid shaft. Consider a shaft subjected to a torque T as shown in Fig. 5.3.

Let τ = Maximum shear stress induced at the outer surface

R = Radius of the shaft

q = Shear stress at a radius 'r' from the centre.

Consider an elementary circular ring of thickness 'dr' at a distance 'r' from the centre as shown in Fig. 5.3. Then the area of the ring,

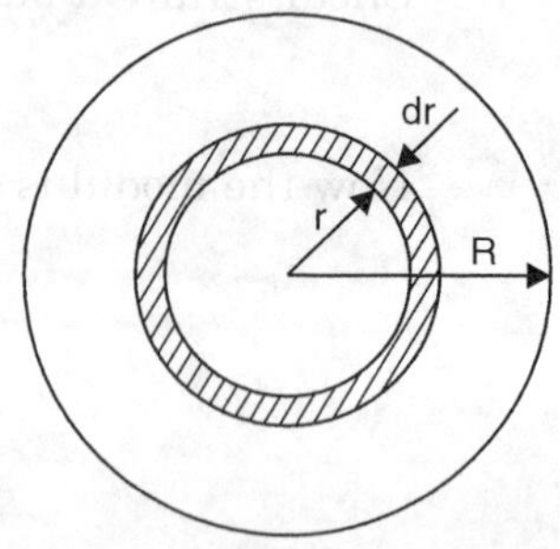

Fig. 5.3

$$dA = 2\pi r dr$$

From equation (5.2), we have

$$\frac{\tau}{R} = \frac{q}{r}$$

$\therefore$ Shear stress at the radius r,

$$q = \frac{\tau}{R} r = \tau \frac{r}{R}$$

$\therefore$ Turning force on the elementary circular ring

= Shear stress acting on the ring × Area of ring

$= q \times dA$

$$= \tau \times \frac{r}{R} \times 2\pi r dr \qquad \left(\because \; q = \tau \times \frac{r}{R}\right)$$

$$= \frac{\tau}{R} \times 2\pi r^2 dr$$

Now turning moment due to the turning force on the elementary ring,

dT = Turning force on the ring × Distance of the ring from the axis

$$= \frac{\tau}{R} \times 2\pi r^2 dr \times r$$

$$= \frac{\tau}{R} \times 2\pi r^3 dr \qquad \text{...[5.3 (A)]}$$

∴ The total turning moment (or total torque) is obtained by integrating the above equation between the limits O and R

$$\therefore \qquad T = \int_0^R dT = \int_0^R \frac{\tau}{R} \times 2\pi r^3 dr$$

$$= \frac{\tau}{R} \times 2\pi \int_0^R r^3 dr = \frac{\tau}{R} \times 2\pi \left[\frac{r^4}{4}\right]_0^R$$

$$= \frac{\tau}{R} \times 2\pi \times \frac{R^4}{4} = \tau \times \frac{\pi}{2} \times R^3$$

$$= \tau \times \frac{\pi}{2} \times \left(\frac{D}{2}\right)^3 \qquad \left(\because R = \frac{D}{2}\right)$$

$$= \tau \times \frac{\pi}{2} \times \frac{D^3}{8} = \tau \times \frac{\pi D^3}{16} = \frac{\pi}{16} \tau D^3 \qquad \text{...(5.4)}$$

Problem 5.1. *A solid shaft of 150 mm diameter is used to transmit torque. Find the maximum torque transmitted by the shaft if the maximum shear stress induced to the shaft is 45 N/mm².*

Sol. Given :

Diameter of the shaft, $D = 150$ mm

Maximum shear stress, $\tau = 45$ N/mm²

Let T = Maximum torque transmitted by the shaft.

Using equation (5.4), $T = \frac{\pi}{16} \tau D^3 = \frac{\pi}{16} \times 45 \times 150^3$

$= 29820586$ Nmm $=$ **29820.586 Nm. Ans.**

Problem 5.2. *The shearing stress is a solid shaft is not to exceed 40 N/mm² when the torque transmitted is 20000 Nm. Determine the minimum diameter of the shaft.*

Sol. Given :

Maximum shear stress, $\tau = 40$ N/mm²

Torque transmitted, $T = 20000$ Nm $= 20000 \times 10^3$ Nmm

Let D = Minimum diameter of the shaft in mm.

Using equation (5.4),

$$T = \frac{\pi}{16} \tau D^3$$

or

$$D = \left(\frac{16T}{\pi\tau}\right)^{1/3} = \left(\frac{16 \times 20000 \times 10^3}{\pi \times 40}\right)^{1/3} = \mathbf{136.2\ mm.}\quad \textbf{Ans.}$$

5.2.3. Torque Transmitted by a Hollow Circular Shafts. Torque transmitted by a hollow circular shaft is obtained in the same way as for a solid shaft. Consider a hollow shaft. Let it is subjected to a torque T as shown in Fig. 5.4. Take an elementary circular ring of thickness 'dr' at a distance r from the centre as shown in Fig. 5.4.

Let R_0 = Outer radius of the shaft

R_i = Inner radius of the shaft

r = Radius of elementary circular ring

dr = Thickness of the ring

τ = Maximum shear stress induced at outer surface of the shaft

q = Shear stress induced on the elementary ring

dA = Area of the elementary circular ring

$= 2\pi r \times dr$

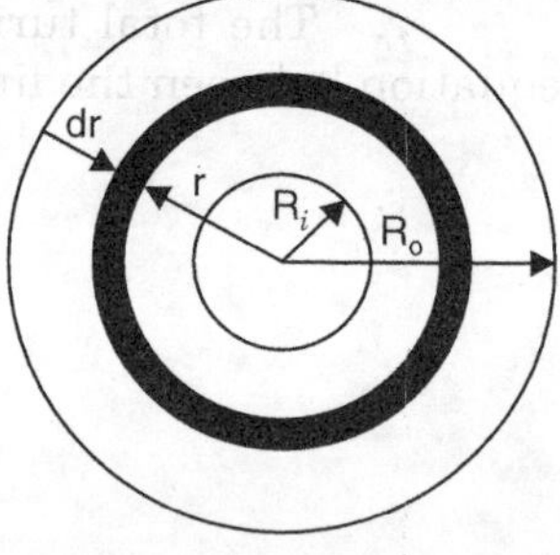

Fig. 5.4. Hollow shaft.

Shear stress at the elementary ring is obtained from equation (5.2) as

$$\frac{\tau}{R_0} = \frac{q}{r}$$

($\because$ Here outer radius $R = R_0$)

$$\therefore \quad q = \frac{\tau}{R_0} \times r$$

$\therefore$ Turning force on the ring = Stress × Area = $q \times dA$

$$= \frac{\tau}{R_0} r \times 2\pi r dr \qquad \left(\because \; q = \frac{\tau}{R_0} r\right)$$

$$= 2\pi \frac{\tau}{R_0} r^2 \, dr$$

Turning moment (dT) on the ring,

dT = Turning force × Distance of the ring from centre

$$= 2\pi \frac{\tau}{R_0} r^2 dr \times r = 2\pi \frac{\tau}{R_0} r^3 \, dr$$

The total turning moment (or total torque T) is obtained by integrating the above equation between the limits R_i and R_0.

$$\therefore \quad T = \int_{R_i}^{R_0} dT = \int_{R_i}^{R_0} 2\pi \frac{\tau}{R_0} r^3 \, dr$$

$$= 2\pi \frac{\tau}{R_0} \int_{R_i}^{R_0} r^3 \, dr$$

($\because$ τ and R_0 are constant and can be taken outside the integral)

$$= 2\pi \frac{\tau}{R_0} \left[\frac{r^4}{4}\right]_{R_i}^{R_0} = 2\pi \frac{\tau}{R_0} \left[\frac{R_0^4 - R_i^4}{4}\right]$$

$$= \frac{\pi}{2} \tau \left[\frac{R_0^4 - R_i^4}{R_0}\right] \qquad \text{...(5.5)}$$

Let D_0 = Outer diameter of the shaft

D_i = Inner diameter of the shaft.

Then $R_0 = \dfrac{D_0}{2}$ and $R_i = \dfrac{D_i}{2}$.

Substituting the values of R_0 and R_i in equation (5.6),

$$T = \frac{\pi}{2}\tau\left[\frac{\left(\frac{D_0}{2}\right)^4 - \left(\frac{D_i}{2}\right)^4}{\left(\frac{D_0}{2}\right)}\right] = \frac{\pi}{2}\tau\left[\frac{\frac{D_0^{\,4}}{16} - \frac{D_i^{\,4}}{16}}{\frac{D_0}{2}}\right]$$

$$= \frac{\pi}{2}\tau\left[\frac{D_0^{\,4} - D_i^{\,4}}{16} \times \frac{2}{D_0}\right]$$

$$= \frac{\pi}{16}\tau\left[\frac{D_0^{\,4} - D_i^{\,4}}{D_0}\right] \quad \text{...(5.6)}$$

5.3. POWER TRANSMITTED BY SHAFTS

Once the expression for torque (T) for a solid or a hollow shaft is obtained, power transmitted by the shafts can be determined.

Let N = r.p.m. of the shaft

T = Mean torque transmitted in Nm

ω = Angular speed of shaft.

Then Power $= \frac{2\pi NT^*}{60}$ watts ...(5.7)

$$= \omega \times T \qquad \left(\because \frac{2\pi N}{60} = \omega\right)$$

$$= T \times \omega \quad \text{...[5.7 (A)]}$$

Problem 5.3. *In a hollow circular shaft of outer and inner diameters of 20 cm and 10 cm respectively, the shear stress is not to exceed 40 N/mm². Find the maximum torque which the shaft can safely transmit.*

Sol. Given :

Outer diameter, D_0 = 20 cm = 200 mm

Inner diameter, D_i = 10 cm = 100 mm

Maximum shear stress, τ = 40 N/mm²

Let T = Maximum torque transmitted by the shaft.

Using equation (5.6),

$$T = \frac{\pi}{16}\tau\left[\frac{D_0^{\,4} - D_i^{\,4}}{D_0}\right] = \frac{\pi}{16} \times 40\left[\frac{200^4 - 100^4}{200}\right]$$

$$= \frac{\pi}{16} \times 40\left[\frac{16 \times 10^8 - 1 \times 10^8}{200}\right] = 58904860 \text{ Nmm}$$

= **58904.86 Nm. Ans.**

Problem 5.4. *Two shafts of the same material and of same lengths are subjected to the same torque, if the first shaft is of a solid circular section and the second shaft is of hollow*

*The torque T obtained by this formula is the average (or mean) torque.

circular section, whose internal diameter is 2/3 of the outside diameter and the maximum shear stress developed in each shaft is the same, compare the weights of the shafts.

(AMIE, Summer 1989)

Sol. Given :

Two shafts of the same material and same lengths (one is solid and other is hollow) transmit the same torque and develops the same maximum stress.

Let T = Torque transmitted by each shaft

τ = Max. shear stress developed in each shaft

D = Outer diameter of the solid shaft

D_0 = Outer diameter of the hollow shaft

D_i = Inner diameter of the hollow shaft = $\frac{2}{3} D_0$

W_s = Weight of the solid shaft

W_h = Weight of the hollow shaft

L = Length of each shaft

w = Weight density of the material of each shaft.

Torque transmitted by the solid shaft is given by equation (5.4)

$$T = \frac{\pi}{16} \tau D^3 \qquad \text{...}(i)$$

Torque transmitted by the hollow shaft is given by equation (5.6),

$$T = \frac{\pi}{16} \tau \left[\frac{D_0^4 - D_i^4}{D_0} \right] = \frac{\pi}{16} \tau \left[\frac{D_0^4 - (2/3\, D_0)^4}{D_0} \right]$$

$$= \frac{\pi}{16} \tau \left[\frac{D_0^4 - \frac{16}{81} D_0^4}{D_0} \right] = \frac{\pi}{16} \tau \times \frac{65\, D_0^4}{81 \times D_0}$$

$$= \frac{\pi}{16} \tau \times \frac{65\, D_0^3}{81} \qquad \text{...}(ii)$$

As torque transmitted by solid and hollow shafts are equal, hence equating equations (*i*) and (*ii*),

$$\frac{\pi}{16} \tau D^3 = \frac{\pi}{16} \tau \times \frac{65}{81} D_0^3$$

Cancelling $\frac{\pi}{16} \tau$ to both sides

or
$$D^3 = \frac{65}{81} D_0^3$$

$\therefore$
$$D = \left[\frac{65}{81} D_0^3 \right]^{1/3} = \left(\frac{65}{81} \right)^{1/3} D_0 = 0.929\, D_0 \qquad \text{...}(iii)$$

Now weight of solid shaft, W_s = Weight density × Volume of solid shaft

$= w \times$ Area of cross-section × Length

$$= w \times \frac{\pi}{4} D^2 \times L \qquad \text{...}(iv)$$

Weight of hollow shaft,

$$W_h = w \times \text{Area of cross-section of hollow shaft} \times \text{Length}$$

$$= w \times \frac{\pi}{4}[D_0^2 - D_i^2] \times L = w \times \frac{\pi}{4}[D_0^2 - (2/3\ D_0)^2] \times L$$

$$= w \times \frac{\pi}{4}\left[D_0^2 - \frac{4}{9}D_0^2\right] \times L = w \times \frac{\pi}{4} \times \frac{5}{9} D_0^2 \times L \quad ...(v)$$

Dividing equation (*iv*) by equation (*v*),

$$\frac{W_s}{W_h} = \frac{w \times \frac{\pi}{4} D^2 \times L}{w \times \frac{\pi}{4} \times \frac{5}{9} D_0^2 \times L} = \frac{9D^2}{5D_0^2}$$

$$= \frac{9}{5} \times \frac{(0.929 D_0)^2}{D_0^2} \quad [\because\ D = 0.929\ D_0 \text{ from equation } (iii)]$$

$$= \frac{9}{5} \times 0.929^2 \times \frac{D_0^2}{D_0^2} = \frac{1.55}{1}$$

$$\therefore \quad \frac{\text{Weight of solid shaft}}{\text{Weight of hollow shaft}} = \frac{\mathbf{1.55}}{\mathbf{1}}. \quad \textbf{Ans.}$$

Problem 5.5. *A solid circular shaft and a hollow circular shaft whose inside diameter is* $\left(\frac{3}{4}\right)$ *of the outside diameter, are of the same material, of equal lengths and are required to transmit a given torque. Compare the weights of these two shafts if the maximum shear stress developed in the two shafts are equal.* (AMIE, Winter 1988)

Sol. Given :

Dia. of hollow shaft, $D_i = \frac{3}{4}$ Dia. at outside

$$= \frac{3}{4} D_0 = 0.75\ D_0$$

Let L = Length of both shaft (equal length),

T = Torque transmitted by each shaft (equal torque),

τ = Maximum shear stress developed in each shaft (equal max. shear stress),

D = Dia. of solid shaft,

W_s = Weight of solid shaft, and

W_h = Weight of hollow shaft.

Torque transmitted by a solid shaft is given by equation (5.4) as

$$T = \frac{\pi}{16} \times \tau \times D^3 \quad ...(i)$$

Torque transmitted by a hollow shaft is given by equation (5.6) as

$$T = \frac{\pi}{16} \times \tau \times \left[\frac{D_0^4 - D_i^4}{D_0}\right] = \frac{\pi}{16} \times \tau \times \left[\frac{D_0^4 - (0.75\ D_0)^4}{D_0}\right]$$

$$= \frac{\pi}{16} \times \tau \times \frac{D_0^4}{D_0}[1 - (0.75)^4] = \frac{\pi}{16} \times \tau \times D_0^3 \times [1 - 0.3164]$$

$$= \frac{\pi}{16} \times \tau \times D_0^3 \times 0.6836 \quad ...(ii)$$

But torque transmitted by solid shaft

= Torque transmitted by hollow shaft.

Hence equating equations (*i*) and (*ii*), we get

$$\frac{\pi}{16} \times \tau \times D^3 = \frac{\pi}{16} \times \tau \times D_0^3 \times 0.6836 \quad \text{or} \quad D^3 = 0.6836\, D_0^3$$

$$\therefore \quad D = (0.6836)^{1/3} \times D_0 \quad \text{or} \quad D = 0.8809\, D_0 \quad ...(iii)$$

Now weight of solid shaft,

$$W_s = \rho \times g \times \text{Volume of solid shaft}$$

$$= \rho \times g \times \left(\frac{\pi}{4} D^2 \times L\right) \quad ...(iv)$$

Weight of hollow shaft,

$$W_h = \rho \times g \times \text{Volume of hollow shaft}$$

$$= \rho \times g \times \left[\frac{\pi}{4}(D_0^2 - D_i^2) \times L\right] = \rho \times g \times \frac{\pi}{4}\,[D_0^2 - (0.75D_0)^2] \times L$$

$$= \rho \times g \times \frac{\pi}{4}\,[D_0^2 - 0.5625D_0^2] \times L = \rho \times g \times \frac{\pi}{4} \times D_0^2(1 - 0.5625) \times L$$

$$= \rho \times g \times \frac{\pi}{4} \times D_0^2 \times 0.4375 \times L \quad ...(v)$$

Dividing equation (*iv*) by equation (*v*),

$$\frac{W_s}{W_h} = \frac{\rho \times g \times \frac{\pi}{4} \times D^2 \times L}{\rho \times g \times \frac{\pi}{4} D_0^2 \times 0.4375 \times L} = \frac{D^2}{0.4375\, D_0^2}$$

$$= \frac{(0.8809\, D_0)^2}{0.4375\, D_0^2} \quad [\because \text{ From } (iii),\ D = 0.8809\, D_0]$$

$$= \frac{0.776\, D_0^2}{0.4375\, D_0^2} = \mathbf{1.7737.} \quad \textbf{Ans.}$$

Problem 5.6. *A hollow shaft of external diameter 120 mm transmits 300 kW power at 200 r.p.m. Determine the maximum internal diameter if the maximum stress in the shaft is not to exceed 60 N/mm².* (AMIE, Summer 1990)

Sol. Given :

External dia., $D_0 = 120$ mm

Power, $P = 300$ kW $= 300{,}000$ W

Speed, $N = 200$ r.p.m.

Max. shear stress, $\tau = 60$ N/mm²

Let D_i = Internal dia. of shaft

Using equation (5.7),

$$P = \frac{2\pi NT}{60} \quad \text{or} \quad 300{,}000 = \frac{2\pi \times 200 \times T}{60}$$

$$\therefore \quad T = \frac{300{,}000 \times 60}{2\pi \times 200} = 14323.9 \text{ Nm}$$

$$= 14323.9 \times 1000 \text{ Nmm} = 14323900 \text{ Nmm}$$

Now using equation (5.6),

$$T = \frac{\pi}{16} \times \tau \times \frac{(D_0^4 - D_i^4)}{D_0}$$

or $$14323900 = \frac{\pi}{16} \times 60 \times \frac{(120^4 - D_i^4)}{120}$$

or $$\frac{14323900 \times 16 \times 120}{\pi \times 60} = 120^4 - D_i^4$$

$$145902000 = 207360000 - D_i^4$$

or $$D_i^4 = 207360000 - 145902000 = 61458000$$

$$\therefore \quad D_i = (61458000)^{1/4} = \mathbf{88.5 \text{ mm.}} \quad \textbf{Ans.}$$

Problem 5.7. *Find the maximum shear stress induced in a solid circular shaft of diameter 15 cm when the shaft transmits 150 kW power at 180 r.p.m.*

Sol. Given :

Diameter of shaft, $D = 15 \text{ cm} = 150 \text{ mm}$

Power transmitted, $P = 150 \text{ kW} = 150 \times 10^3 \text{ W}$

Speed of shaft, $N = 180$ r.p.m.

Let τ = Maximum shear stress induced in the shaft

Power transmitted is given by equation (5.7) as

$$P = \frac{2\pi NT}{60}$$

$$150 \times 10^3 = \frac{2\pi \times 180 \times T}{60}$$

$$\therefore \quad T = \frac{150 \times 10^3 \times 60}{2\pi \times 180} = 7957.7 \text{ Nm} = 7957700 \text{ Nmm}$$

Now using equation (5.4) as,

$$T = \frac{\pi}{16} \tau D^3$$

$$7957700 = \frac{\pi}{16} \times \tau \times 150^3$$

$$\therefore \quad \tau = \frac{16 \times 7957700}{\pi \times 150^3} = \mathbf{12 \text{ N/mm}^2.} \quad \textbf{Ans.}$$

Problem 5.8. *A solid cylindrical shaft is to transmit 300 kW power at 100 r.p.m.*

(a) If the shear stress is not to exceed 80 N/mm², find its diameter.

(b) What percent saving in weight would be obtained if this shaft is replaced by a hollow one whose internal diameter equals to 0.6 of the external diameter, the length, the material and maximum shear stress being the same ? (AMIE, Winter 1983)

Sol. Given :

Power, $P = 300 \text{ kW} = 300 \times 10^3 \text{ W}$

Speed, $N = 100$

Max. shear stress, $\tau = 80 \text{ N/mm}^2$

(*a*) Let D = Dia. of solid shaft

Using equation (5.7),

$$P = \frac{2\pi NT}{60}$$

$$300 \times 10^3 = \frac{2\pi \times 100 \times T}{60}$$

$$\therefore \quad T = \frac{300 \times 10^3 \times 60}{2\pi \times 100} = 28647.8 \text{ Nm} = 28647800 \text{ Nmm}$$

Now using equation (5.4),

$$T = \frac{\pi}{16} \times \tau \times D^3 \quad \text{or} \quad 28647800 = \frac{\pi}{16} \times 80 \times D^3$$

$$\therefore \quad D = \left(\frac{16 \times 28647800}{\pi \times 80}\right)^{1/3} = 121.8 \text{ mm}$$

$$= \textbf{say 122.0 mm.} \quad \textbf{Ans.}$$

(*b*) *Percent saving in weight*

Let D_0 = External dia. of hollow shaft

D_i = Internal dia. of hollow shaft

$= 0.6 \times D_0$. (given)

The length, material and maximum shear stress in solid and hollow shafts are given the same. Hence torque transmitted by solid shaft is equal to the torque transmitted by hollow shaft. But the torque transmitted by hollow shaft is given by equation (5.6).

$\therefore$ Using equation (5.6),

$$T = \frac{\pi}{16} \times \tau \times \frac{(D_0^4 - D_i^4)}{D_0}$$

$$= \frac{\pi}{16} \times 800 \times \frac{[D_0^4 - (0.6\, D_0)^4]}{D_0} \qquad (\because \; D_i = 0.6\, D_0)$$

$$= \pi \times 50 \times \frac{[D_0^4 - (0.6\, D_0)^4]}{D_0}$$

But torque transmitted by solid shaft

$= 28647800$ Nmm

$\therefore$ Equating the two torques, we get

$$28647800 = \pi \times 50 \times \left(\frac{0.8704\, D_0^4}{D_0}\right) = \pi \times 50 \times 0.8704\, D_0^3$$

$$\therefore \quad D_0 = \left(\frac{28647800}{\pi \times 50 \times 0.8704}\right)^{1/3} = 127.6 \text{ mm} = \text{say } 128 \text{ mm}$$

$\therefore$ Internal dia. $\quad D_i = 0.6 \times D_0 = 0.6 \times 128 = 76.8$ mm

Now let $\quad W_s$ = Weight of solid shaft,

and $\quad W_h$ = Weight of hollow shaft.

Then $\quad W_s$ = Weight density × Area of solid shaft × Length

$$= w \times \frac{\pi}{4} D^2 \times L \qquad \text{(where } w = \text{weight density)}$$

Similarly W_h = Weight density × Area of hollow shaft × Length

$$= w \times \frac{\pi}{4}[D_0^2 - D_i^2] \times L$$

($\because$ Both shafts are of same lengths and of same material)

Now percent saving in weight

$$= \frac{W_s - W_h}{W_s} \times 100$$

$$= \frac{w \times \frac{\pi}{4} D^2 \times L - w \times \frac{\pi}{4}[D_0^2 - D_i^2] \times L}{w \times \frac{\pi}{4} D^2 \times L} \times 100$$

$$= \frac{D^2 - (D_0^2 - D_i^2)}{D^2} \times 100 \quad \left(\text{Cancelling } w \times \frac{\pi}{4} \times L\right)$$

$$= \frac{122^2 - (128^2 - 75.8^2)}{122^2} \times 100 = \frac{14884 - (16364 - 5898)}{14884} \times 100$$

$$= \frac{14884 - 10486}{14884} \times 100 = \textbf{29.55\%.} \quad \textbf{Ans.}$$

Problem 5.9. *A solid steel shaft has to transmit 75 kW at 200 r.p.m. Taking allowable shear stress as 70 N/mm², find suitable diameter for the shaft, if the maximum torque transmitted at each revolution exceeds the mean by 30%.* (AMIE, Summer 1978)

Sol. Given :

Power transmitted, $P = 75 \text{ kW} = 75 \times 10^3 \text{ W}$

R.P.M. of the shaft, $N = 200$

Shear stress, $\tau = 70 \text{ N/mm}^2$

Let T = Mean torque transmitted

T_{max} = Maximum torque transmitted = $1.3\,T$

D = Suitable diameter of the shaft

Power is given by the relation,

$$= \frac{2\pi NT}{60}$$

or
$$75 \times 10^3 = \frac{2\pi \times 200 \times T}{60}$$

$\therefore$
$$T = \frac{75 \times 10^3 \times 60}{2\pi \times 200} = 3580.98 \text{ Nm} = 3580980 \text{ Nmm}$$

$\therefore$
$$T_{max} = 1.3\,T = 1.3 \times 3580980 = 4655274 \text{ Nmm.}$$

Maximum torque transmitted by a solid shaft is given by equation (5.4) as,

$$T_{max} = \frac{\pi}{16} \times \tau \times D^3$$

or
$$4655274 = \frac{\pi}{16} \times 70 \times D^3$$

$\therefore$
$$D = \left(\frac{16 \times 4655274}{\pi \times 70}\right)^{1/3} = \textbf{69.57 mm} \simeq \textbf{70 mm.} \quad \textbf{Ans.}$$

Problem 5.10. *A hollow shaft is to transmit 300 kW power at 80 r.p.m. If the shear stress is not to exceed 60 N/mm² and the internal diameter is 0.6 of the external diameter, find the external and internal diameters assuming that the maximum torque is 1.4 times the mean.*

Sol. Given :

Power transmitted, $P = 300 \text{ kW} = 300 \times 10^3 \text{ W}$

Speed of the shaft, $N = 80$ r.p.m.

Maximum shear stress, $\tau = 60 \text{ N/mm}^2$

Internal diameter, $D_i = 0.6 \times$ External diameter $= 0.6\, D_0$

Maximum torque, $T_{max} = 1.4$ times the mean torque $= 1.4 \times T$

Power is given by the relation,

$$P = \frac{2\pi NT}{60}$$

or
$$T = \frac{60 \times P}{2\pi N} = \frac{60 \times 300 \times 10^3}{2\pi \times 80} = 35809.8 \text{ Nm}$$

$\therefore$
$$T_{max} = 1.4\, T = 1.4 \times 35809.8 \text{ Nm} = 50133.7 \text{ Nm} = 50133700 \text{ Nmm.}$$

Now maximum torque transmitted by a hollow shaft is given by equation (5.6) as,

$$T_{max} = \frac{\pi}{16} \times \tau \times \left[\frac{D_0^4 - D_i^4}{D_0}\right]$$

or
$$50133700 = \frac{\pi}{16} \times 60 \times \left[\frac{D_0^4 - (0.6D_0)^4}{D_0}\right]$$

$$= \frac{\pi}{16} \times 60 \left[\frac{D_0^4 - .1296 D_0^4}{D_0}\right] = \frac{\pi}{16} \times 60 \times .8704\, D_0^3$$

$\therefore$
$$D_0 = \left(\frac{16 \times 50133700}{\pi \times 60 \times .8704}\right)^{1/3} = \mathbf{169.2 \simeq 170 \text{ mm.}} \quad \textbf{Ans.}$$

and
$$D_i = 0.6 \times D_0 = 0.6 \times 170 = \mathbf{102 \text{ mm.}} \quad \textbf{Ans.}$$

Problem 5.11. *A hollow shaft, having an inside diameter 60% of its outer diameter, is to replace a solid shaft transmitting the same power at the same speed. Calculate the percentage saving in material, if the material to be used is also the same.*

Sol. Given :

Let D_0 = Outer diameter of the hollow shaft

D_i = Inside diameter of the hollow shaft = 60% of D_0

$= \frac{60}{100} \times D_0 = 0.6\, D_0$

D = Diameter of the solid shaft

P = Power transmitted by hollow shaft *or* by solid shaft

N = Speed of each shaft

τ = Maximum shear stress induced in each shaft. Since material of both shafts is same and hence shear stress will be same.

Power by solid shaft or hollow shaft is given by

$$P = \frac{2\pi NT}{60}$$

$$\therefore \qquad T = \frac{60 \times P}{2\pi N} = \text{constant}$$

($\because$ P and N are same for solid and hollow shafts)

$\therefore$ Torque transmitted by solid shaft is the same as the torque transmitted by hollow shaft.

Torque transmitted by solid shaft is given by equation (5.4) as

$$T = \frac{\pi}{16}\tau D^3 \qquad \ldots(i)$$

Torque transmitted by hollow shaft is given by equation (5.6) as

$$T = \frac{\pi}{16}\tau\left[\frac{D_0^4 - D_i^4}{D_0}\right] = \frac{\pi}{16}\tau\left[\frac{D_0^4 - (.6D_0)^4}{D_0}\right]$$

$$= \frac{\pi}{16}\tau\left[\frac{D_0^4 - 0.1296D_0^4}{D_0}\right] = \frac{\pi}{16}\tau \times 0.8704D_0^3 \qquad \ldots(ii)$$

Since torque transmitted is the same and hence equating equations (*i*) and (*ii*)

$$\frac{\pi}{16}\tau D^3 = \frac{\pi}{16}\tau \times 0.8704\, D_0^3.$$

$$\therefore \qquad D = (0.8704)^{1/3}\, D_0 = 0.9548D_0.$$

$\therefore$ Area of solid shaft $= \frac{\pi}{4}D^2 = \frac{\pi}{4}(0.9548D_0)^2 = 0.716\, D_0^2$

Area of hollow shaft $= \frac{\pi}{4}[D_0^2 - D_i^2] = \frac{\pi}{4}[D_0^2 - (0.6D_0)^2]$

$$= \frac{\pi}{4}[D_0^2 - 0.36D_0^2] = \frac{\pi}{4} \times 0.64D_0^2 = 0.502D_0^2.$$

For the shafts of the same material, the weight of the shafts is proportional to the areas.

$\therefore$ Saving in material = Saving in area

$$= \frac{\text{Area of solid shaft} - \text{Area of hollow shaft}}{\text{Area of solid shaft}}$$

$$= \frac{0.716D_0^2 - 0.502D_0^2}{0.716D_0^2} = 0.2988.$$

$\therefore$ Percentage saving in material = 0.2988 × 100 = **29.88. Ans.**

5.3.1. Expression for Torque in terms of Polar Moment of Inertia. Polar moment of inertia of a plane area is defined as the moment of inertia of the area about an axis perpendicular to the plane of the figure and passing through the C.G. of the area. It is denoted by symbol J.

The torque in terms of polar moment of inertia (J) is obtained from equation [5.3 (*A*)] of Art. 5.3.

The moment (dT) on the circular ring is given by equation [5.3 (A)] as

$$dT = \frac{\tau}{R}\, 2\pi r^3 dr = \frac{\tau}{R}\, 2\pi r \times r^2 dr = \frac{\tau}{R}\, r^2 \times 2\pi r \times dr$$

$$= \frac{\tau}{R}\, r^2\, dA \qquad (\because\ dA = 2\pi r\, dr \text{ see Fig. 5.3})$$

$\therefore$ Total torque,

$$T = \int_0^R dT = \int_0^R \frac{\tau}{R}\, r^2\, dA = \frac{\tau}{R}\int_0^R r^2\, dA \qquad \text{...}(i)$$

But $r^2\, dA$ = Moment of inertia of the elementary ring about an axis perpendicular to the plane of Fig. 5.3 and passing through the centre of the circle.

$\therefore$ $\int_0^R r^2\, dA$ = Moment of inertia of the circle about an axis perpendicular to the plane of the circle and passing through the centre of the circle

$$= \text{Polar moment of inertia } (J) = \frac{\pi}{32}\, D^4.$$

Hence equation (i) becomes as

$$T = \frac{\tau}{R} \times J \qquad \left(\because\ J = \int_0^R r^2\, dA\right)$$

$$\therefore \quad \frac{T}{J} = \frac{\tau}{R} \qquad \text{...(5.8)}$$

But from equation (5.1), we have

$$\frac{\tau}{R} = \frac{C\theta}{L}$$

$$\therefore \quad \frac{T}{J} = \frac{\tau}{R} = \frac{C\theta}{L} \qquad \text{...(5.9)}$$

where C = Modulus of rigidity

θ = Angle of twist in radiation

L = Length of the shaft.

5.3.2. Polar Modulus. Polar modulus is defined as the ratio of the polar moment of inertia to the radius of the shaft. It is also called torsional section modulus. It is denoted by Z_p. Mathematically,

$$Z_p = \frac{J}{R}$$

(a) For a solid shaft, $J = \frac{\pi}{32} D^4$

$$\therefore \quad Z_p = \frac{\frac{\pi}{32} D^4}{R} = \frac{\frac{\pi}{32} D^4}{D/2} = \frac{\pi}{16}\, D^3 \qquad \text{...(5.10)}$$

(*b*) For a hollow shaft, $J = \frac{\pi}{32}(D_0^4 - D_i^4)$ ...(5.11)

$$\therefore \quad Z_p = \frac{\frac{\pi}{32}[D_0^4 - D_i^4]}{R}$$ (Here R is the outer radius)

$$\left(\because \ R = \frac{D_0}{2}\right)$$

$$= \frac{\frac{\pi}{32}[D_0^4 - D_i^4]}{D_0/2} = \frac{\pi}{16D_0} \times [D_0^4 - D_i^4] \quad \text{...(5.12)}$$

5.3.3. Strength of a Shaft and Torsional Rigidity. The strength of a shaft means the *maximum torque or maximum power* the shaft can transmit.

Torsional rigidity or stiffness of the shaft is defined as the product of modulus of rigidity (C) and polar moment of inertia of the shaft (J). Hence mathematically, the torsional rigidity is given as,

Torsional rigidity = $C \times J$.

Torsional rigidity is also defined as the torque required to produce a twist of one radian per unit length of the shaft.

Let a twisting moment T produces a twist of θ radians in a shaft of length L.

Using equation (5.9), we have

$$\frac{T}{J} = \frac{C\theta}{L} \quad \text{or} \quad C \times J = \frac{T \times L}{\theta}$$

But $C \times J$ = Torsional rigidity

$$\therefore \quad \text{Torsional rigidity} = \frac{T \times L}{\theta}$$

If L = one metre and θ = one radian

Then torsional rigidity = Torque.

Problem 5.12. *Determine the diameter of a solid steel shaft which will transmit 90 kW at 160 r.p.m. Also determine the length of the shaft if the twist must not exceed 1° over the entire length. The maximum shear stress is limited to 60 N/mm². Take the value of modulus of rigidity = 8 × 10⁴ N/mm².*

Sol. Given :

Power, $P = 90 \text{ kW} = 90 \times 10^3$ W

Speed, $N = 160$ r.p.m.

Angle of twist, $\theta = 1°$ or $\frac{\theta}{180}$ radian $\left(\because \ 1° = \frac{\pi}{180} \text{ radian}\right)$

Max. shear stress, $\tau = 60$ N/mm²

Modulus of rigidity, $C = 8 \times 10^4$ N/mm²

Let D = Diameter of the shaft and

L = Length of the shaft.

(*i*) *Diameter of the shaft*

Using equation (5.7),

$$P = \frac{2\pi NT}{60}$$

or $$90 \times 10^3 = \frac{2\pi \times 160 \times T}{60}$$

$$\therefore \quad T = \frac{90 \times 10^3 \times 60}{2\pi \times 160} = 5371.48 \text{ Nm} = 5371.48 \times 10^3 \text{ Nmm}$$

Now using equation (5.4),

$$T = \frac{\pi}{16} \tau D^3$$

or $$5371.48 \times 10^3 = \frac{\pi}{16} \times 60 \times D^3$$

$$\therefore \quad D^3 = \frac{5371.48 \times 10^3 \times 16}{\pi \times 60} = 455945$$

$$\therefore \quad D = (455945)^{1/3} = \mathbf{76.8 \text{ mm.}} \quad \textbf{Ans.}$$

(*ii*) *Length of the shaft*

Using equation (5.7),

$$\frac{\tau}{R} = \frac{C\theta}{L}$$

or $$\frac{60}{\left(\frac{76.8}{2}\right)} = \frac{8 \times 10^4 \times \pi}{L \times 180} \qquad \left(\because R = \frac{D}{2} = \frac{76.8}{2} \text{ mm}, \theta = \frac{\pi}{180} \text{ radian}\right)$$

or $$L = \frac{8 \times 10^4 \times \pi \times 76.8}{60 \times 180 \times 2} = \mathbf{893.6 \text{ mm.}} \quad \textbf{Ans.}$$

Problem 5.13. *Determine the diameter of a solid shaft which will transmit 300 kW at 250 r.p.m. The maximum shear stress should not exceed 30 N/mm² and twist should not be more than 1° in a shaft length of 2 m. Take modulus of rigidity = 1 × 10⁵ N/mm².*

Sol. Given :

Power transmitted, $P = 300 \text{ kW} = 300 \times 10^3 \text{ W}$

Speed of the shaft, $N = 250$ r.p.m.

Maximum shear stress, $\tau = 30 \text{ N/mm}^2$

Twist in shaft, $\theta = 1° = \frac{\pi}{180} = 0.01745$ radian

Length of shaft, $L = 2 \text{ m} = 2000 \text{ mm}$

Modulus of rigidity, $C = 1 \times 10^5 \text{ N/mm}^2$

Let D = Diameter of the shaft.

Power is given by the relation,

$$P = \frac{2\pi NT}{60}$$

or $$300 \times 10^3 = \frac{2\pi \times 250 \times T}{60}$$

$$\therefore \quad T = \frac{300 \times 10^3 \times 60}{2\pi \times 250} = 11459.1 \text{ Nm} = 11459.1 \times 10^3 \text{ Nmm}$$

(*i*) Diameter of the shaft when maximum shear stress,

$$\tau = 30 \text{ N/mm}^2$$

Maximum torque transmitted by a solid shaft is given by equation (5.4) as

$$T = \frac{\pi}{16} \times \tau \times D^3$$

$$\therefore \quad 11459100 = \frac{\pi}{16} \times 30 \times D^3$$

$$\therefore \quad D = \left(\frac{16 \times 11459100}{\pi \times 30}\right)^{1/3} = 124.5 \text{ mm} \quad \ldots(i)$$

(*ii*) Diameter of shaft when twist should not be more than 1°.

Using equation (5.9),

$$\frac{T}{J} = \frac{C\theta}{L}$$

where J = Polar moment of inertia of solid shaft

$$= \frac{\pi}{32} D^4$$

$$\therefore \quad \frac{11459100}{\frac{\pi}{32} D^4} = \frac{10^5 \times 0.01745}{2000}$$

$$\therefore \quad D^4 = \frac{32 \times 2000 \times 11459100}{10^5 \times \pi \times 0.01745} = 13377.81 \times 10^4$$

$$\therefore \quad D = (13377.81 \times 10^4)^{1/4} = 107.5 \text{ mm} \quad \ldots(ii)$$

The suitable diameter of the shaft is the greater* of the two values given by equations (*i*) and (*ii*).

∴ Diameter of the shaft = **124.5 mm say 125 mm. Ans.**

*(If diameter is taken smaller of the two values say 107.5 mm, then from equation $T = \frac{\pi}{16} \tau D^3$, the value of shear stress will be

$$11459100 = \frac{\pi}{16} \tau \times (107.5)^3$$

or $$11459100 = 243920 \tau$$

or $$\tau = \frac{11459100}{243920} = 46.978 \text{ N/mm}^2$$

which is more than the given value of 30 N/mm²).

Problem 5.14. *A hollow shaft of diameter ratio $\frac{3}{8}$ (internal dia. to outer dia.) is to transmit 375 kW power at 100 r.p.m. The maximum torque being 20% greater than the mean. The shear stress is not to exceed 60 N/mm² and twist in a length of 4 m not to exceed 2°. Calculate its external and internal diameters which would satisfy both the above conditions. Assume modulus of rigidity, C = 0.85 × 10⁵ N/mm².*

Sol. Given :

Diameter ratio, $\frac{D_i}{D_0} = \frac{3}{8}$

$\therefore \quad D_i = \frac{3}{8} D_0$

Power, $P = 375 \text{ kW} = 375000 \text{ W}$

Speed, $N = 100$ r.p.m.

Max. torque, $T_{max} = 1.2\, T_{mean}$

Length, $L = 4 \text{ m} = 4000 \text{ mm}$

Max. twist, $\theta = 2° = 2 \times \frac{\pi}{180}$ radians $= 0.0349$ radians

Modulus of rigidity, $C = 0.85 \times 10^5 \text{ N/mm}^2$

Power is given by, $P = \frac{2\pi NT}{60}$ Here torque is T_{mean}

or $$T = \frac{P \times 60}{2\pi N} = \frac{375000 \times 60}{2\pi \times 100} = 35810 \text{ Nm}$$

or $$T_{mean} = 35810 \text{ Nm}$$

$$\therefore \quad T_{max} = 1.2 \times T_{mean} = 1.2 \times 35810$$
$$= 42972 \text{ Nm} = 42972 \times 1000 \text{ Nmm.}$$

(i) Diameters of the shaft when shear stress is not to exceed 60 N/mm².

For the hollow shaft, the torque transmitted is given by

$$T_{max} = \frac{\pi}{16} \times \tau \times \frac{(D_0^{\,4} - D_i^{\,4})}{D_0}$$

or $$42972 \times 1000 = \frac{\pi}{16} \times 60 \times \frac{\left[D_0^{\,4} - \left(\frac{3}{8} D_0\right)^4\right]}{D_0}$$

or $$\frac{42972000 \times 16}{\pi \times 60} = \frac{D_0^{\,4}}{D_0}\left(1 - \frac{81}{4096}\right) = D_0^{\,3} \times \frac{4015}{4096}$$

or $$D_0^{\,3} = \frac{42972000 \times 16 \times 4096}{\pi \times 60 \times 4015}$$

$$\therefore \quad D_0 = \left(\frac{42972000 \times 16 \times 4096}{\pi \times 60 \times 4015}\right)^{1/3} = 154.97 \text{ mm say } 155 \text{ mm}$$

$$\therefore \quad D_i = \frac{3}{8} D_0 = \frac{3}{8} \times 155 \simeq 58.1 \text{ mm.}$$

(ii) Diameters of the shaft when the twist is not to exceed 2 degrees.

Using equation (5.9) in terms of torque and θ, we get

$$\frac{T}{J} = \frac{C \times \theta}{L}$$

or
$$\frac{42972000}{\frac{\pi}{32}[D_0^4 - D_i^4]} = \frac{(0.85 \times 10^5) \times 0.0349}{4000}$$

or
$$\frac{42972000 \times 4000 \times 32}{\pi \times 0.85 \times 10^5 \times 0.0349} = D_0^4 - D_i^4 = D_0^4 - \left(\frac{3}{8}D_0\right)^4 = D_0^4 - \frac{81}{4096}D_0^4$$

$$= D_0^4\left[1 - \frac{81}{4096}\right] = \frac{4015}{4096}D_0^4$$

$$\therefore \quad D_0^4 = \frac{42972000 \times 4000 \times 32 \times 4096}{\pi \times 0.85 \times 10^5 \times 0.0349 \times 4015}$$

$$\therefore \quad D_0 = 156.65 \text{ mm say } 157 \text{ mm}$$

and
$$D_i = \frac{3}{8} \times 156.65 = 58.74 \text{ mm say } 59 \text{ mm.}$$

The diameters of the shaft, which would satisfy both the conditions are the greater of the two values.

∴ External dia., D_0 = **157 mm. Ans.**

Internal dia., D_i = **59 mm. Ans.**

Problem 5.15. *A solid circular shaft transmits 75 kW power at 200 r.p.m. Calculate the shaft diameter, if the twist in the shaft is not to exceed 1° in 2 metres length of shaft, and shear stress is limited to 50 N/mm². Take C = 1 × 10⁵ N/mm².*

Sol. Given :

Power transmitted, $P = 75 \text{ kW} = 75 \times 10^3 \text{ W}$

Speed of the shaft, $N = 200$

Twist in the shaft, $\theta = 1°$

$$= \frac{\pi}{180} \text{ radians} = 0.01745 \text{ rad.}$$

Length of the shaft, $L = 2 \text{ m} = 2000 \text{ mm}$

Maximum shear stress, $\tau = 50 \text{ N/mm}^2$

Modulus of rigidity, $C = 1 \times 10^5 \text{ N/mm}^2$

$$P = \frac{2\pi NT}{60} \quad \text{or} \quad 75 \times 10^3 = \frac{2\pi \times 200 \times T}{60}$$

$$\therefore \quad T = \frac{75 \times 10^3 \times 60}{2\pi \times 200} = 3580.98 \text{ Nm} = 3580980 \text{ Nmm.}$$

(*i*) Diameter of the shaft when maximum shear stress is limited to 50 N/mm².

Using equation (5.4),

$$T = \frac{\pi}{16}\tau D^3 \quad \text{or} \quad 3580980 = \frac{\pi}{16} \times 50 \times D^3.$$

$$\therefore \quad D = \left(\frac{16 \times 3580980}{\pi \times 50}\right)^{1/3} = 71.3 \text{ mm} \qquad ...(i)$$

(*ii*) Diameter of the shaft when the twist in the shaft is not to exceed 1°.

Using equation (5.9),

$$\frac{T}{J} = \frac{C\theta}{L}$$

or

$$\frac{3580980}{\frac{\pi}{32}D^4} = \frac{10^5 \times 0.01745}{2000} \qquad \left(\because \; J = \frac{\pi}{32}D^4\right)$$

$$\therefore \qquad D = \left(\frac{32 \times 2000 \times 3580980}{\pi \times 10^5 \times 0.01745}\right)^{1/4} = 80.4 \text{ mm.} \qquad ...(ii)$$

The suitable diameter of the shaft is the greater value of the two diameters given by equations (*i*) and (*ii*) *i.e.*, 80.4 mm say 81 mm. **Ans.**

Problem 5.16. *A hollow shaft, having an internal diameter 40% of its external diameter, transmits 562.5 kW power at 100 r.p.m. Determine the external diameter of the shaft if the shear stress is not to exceed 60 N/mm² and the twist in a length of 2.5 m should not exceed 1.3 degrees. Assume maximum torque = 1.25 mean torque and modulus of rigidity = 9 × 10⁴ N/mm².*

Sol. Given :

Internal diameter, D_i = 40% of external diameter, $(D_0) = 0.40\, D_0$

Power transmitted, $P = 562.5 \text{ kW} = 562.5 \times 10^3$ W

Speed of the shaft, $N = 100$ r.p.m.

Maximum shear stress, $\tau = 60$ N/mm²

Twist in the shaft, $\theta = 1.3° = 1.3 \times \frac{\pi}{180}$ radians $= 0.02269$ rad

Length of shaft, $L = 2.5$ m $= 2500$ mm

Maximum torque, $T_{max} = 1.25 \times T_{mean}$

Modulus of rigidity, $C = 9 \times 10^4$ N/mm²

The horse power transmitted is given by

$$P = \frac{2\pi NT}{60}$$

or

$$562.5 \times 10^3 = \frac{2\pi \times 100 \times T_{mean}}{60} \qquad (\because \text{ Here } T = T_{mean})$$

or

$$T_{mean} = \frac{60 \times 562.5 \times 10^3}{2\pi \times 100} = 53714.7 \text{ Nm} = 53714700 \text{ Nmm}$$

$$\therefore \qquad T_{max} = 1.25 \times T_{mean} = 1.25 \times 53714700 = 67143375 \text{ Nmm.}$$

(*i*) Diameter of the hollow shaft when maximum shear stress is 60 N/mm².

Using equation (5.6) for torque in case of hollow shaft

$$T = \frac{\pi}{16}\tau\left[\frac{D_0^4 - D_i^4}{D_0}\right]$$

where $T = T_{max} = 67143375$ and $D_i = 0.4\, D_0$, $\tau = 60$ N/mm²

$$\therefore \qquad 67143375 = \frac{\pi}{16} \times 60\left[\frac{D_0^4 - (0.4D_0)^4}{D_0}\right]$$

$$= \frac{\pi}{16} \times 60 \left[\frac{D_0^4 - 0.0256 D_0^4}{D_0} \right]$$

$$= \frac{\pi}{16} \times 60 \times 0.9744 D_0^3 = 11.479 D_0^3$$

$$\therefore \quad D_0 = \left(\frac{67143375}{11.479} \right)^{1/3} = 179.6 \text{ mm} \quad \ldots(i)$$

(*ii*) Diameter of the shaft when the twist in the hollow shaft is not to exceed 1.3°.

Using equation (5.9), we have

$$\frac{T}{J} = \frac{C\theta}{L} \quad \ldots(ii)$$

where $T = T_{max} = 67143375$, $\theta = 1.3° = 0.02269$ rad., $L = 2500$ mm, $C = 9 \times 10^4$ N/mm²

and J = Polar moment of inertia for hollow shaft

$$= \frac{\pi}{32} [D_0^4 - D_i^4] \quad \text{[see equation (5.11)]}$$

$$= \frac{\pi}{32} [D_0^4 - (0.4 D_0)^4] \quad (\because \; D_i = 0.4 D_0)$$

$$= \frac{\pi}{32} \times 0.9744 D_0^4 = 0.09566 D_0^4.$$

Substituting these values in equation (*ii*), we get

$$\frac{67143375}{0.09566 D_0^4} = \frac{9 \times 10^4 \times 0.02269}{2500}$$

$$\therefore \quad D_0^4 = \frac{2500 \times 67143375}{0.09566 \times 9 \times 10^4 \times 0.02269} = 85928.215 \times 10^4$$

$$\therefore \quad D_0 = 85928.215 \times 10^4 = 171.2 \text{ mm} \quad \ldots(iii)$$

The external diameter of the shaft should be 179.6 mm say 180 mm (*i.e.*, greater of the two values given by equations (*i*) and (*iii*)). **Ans.**

5.3.4. Composite Shaft. A shaft made up of two or more different materials, and behaving as a single shaft is known as composite shaft. Hence in a composite shaft one type of shaft is rigidly seelved over another type of shaft. The total torque transmitted by a composite shaft is the sum of the torques transmitted by each individual shaft. But the angle of twist in each shaft will be equal.

Problem 5.17. *A composite shaft consists of a steel rod 60 mm diameter surrounded by a closely fitting tube of brass. Find the out-side diameter of the tube so that when a torque of 1000 Nm is applied to the composite shaft, it will be shared equally by the two materials. Take C for steel = 8.4 × 10⁴ N/mm² and C for brass = 4.2 × 10⁴ N/mm².*

Find also the maximum shear stress in each material and common angle of twist in a length of 4 m.

Sol. Given :

Dia. of steel rod, $d = 60$ mm

Torque, $T = 1000$ Nm $= 1000 \times 10^3$ Nmm

Value of C for steel, $C_s = 8.4 \times 10^4$ N/mm²

Value of C for brass, $C_b = 4.2 \times 10^4$ N/mm²

Length of composite shaft, $L = 4$ m $= 4000$ mm

Let D = Outside dia. of brass tube in mm,
τ_s = Shear stress in steel,
τ_b = Shear stress in brass.

The inner dia. of brass tube will be equal to dia. of steel rod.

∴ Inner dia. of brass tube,

$$d = 60 \text{ mm}$$

Polar moment of inertia for steel rod is given by,

$$J_s = \frac{\pi}{32} d^4 = \frac{\pi}{32} \times 60^4 \text{ mm}^4$$

Polar moment of inertia for brass tube is given by,

$$J_b = \frac{\pi}{32} [D^4 - 60^4] \text{ mm}^4$$

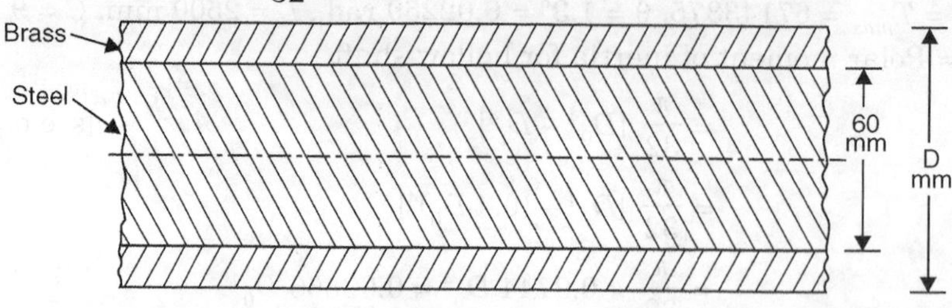

Fig. 5.5

Let T_s = Torque transmitted by steel rod, and
T_b = Torque transmitted by brass tube.

But total torque,

$$T = T_s + T_b = T_s + T_s \qquad (\because \; T_s = T_b \text{ given})$$

$$\therefore \qquad T_s = \frac{T}{2} = \frac{1000 \times 10^3}{2} \text{ Nmm} = 500 \times 10^3 \text{ Nmm}$$

and $$T_b = T_s = 500 \times 10^3 \text{ Nmm}$$

Using equation (5.9), we have

$$\frac{T}{J} = \frac{C \times \theta}{L} \quad \text{or} \quad T = \frac{C \times \theta \times J}{L}$$

For steel rod, we have

$$T_s = \frac{C_s \times \theta_s \times J_s}{L_s} \qquad \ldots(i)$$

For brass tube, we have

$$T_b = \frac{C_b \times \theta_b \times J_b}{L_b} \qquad \ldots(ii)$$

But $$T_s = T_b$$

Hence equating equations (*i*) and (*ii*), we get

$$\frac{C_s \times \theta_s \times J_s}{L_s} = \frac{C_b \times \theta_b \times J_b}{L_b}$$

But $$L_s = L_b = L$$

Hence above equation becomes as

$$C_s \times \theta_s \times J_s = C_b \times \theta_b \times J_b$$

But in a composite shaft, the angle of twist in each shaft is same.

$$\therefore \quad \theta_s = \theta_b$$

Now the above equation becomes as

$$C_s \times J_s = C_b \times J_b$$

or $$8.4 \times 10^4 \times \frac{\pi}{32} \times 60^4 = 4.2 \times 10^4 \times \frac{\pi}{32} [D^4 - 60^4]$$

or $$\frac{8.4}{4.2} \times 60^4 = (D^4 - 60^4) \text{ or } 2 \times 60^4 = D^4 - 60^4$$

or $$3 \times 60^4 = D^4 \quad \text{or} \quad (3)^{1/4} \times 60 = D$$

or $$D = \mathbf{78.98\ mm} \simeq \mathbf{79\ mm.} \quad \textbf{Ans.}$$

(*i*) *Shear stresses in each material*

Using the equation,

$$\frac{T}{J} = \frac{\tau}{R} \quad \text{or} \quad \tau = \frac{T \times R}{J}$$

(*a*) For a steel rod, we have

$$\tau_s = \frac{T_s \times \left(\frac{d}{2}\right)}{J_s} \qquad \left(\because R = \frac{d}{2}\right)$$

$$= \frac{500 \times 10^3 \times \left(\frac{60}{2}\right)}{\frac{\pi}{32} \times 60^4} = \mathbf{117.9\ N/mm^2.} \quad \textbf{Ans.}$$

(*b*) For a brass tube, we have

$$\tau_b = \frac{T_b \times \left(\frac{D}{2}\right)}{J_b} = \frac{500 \times 10^3 \times \left(\frac{79}{2}\right)}{\frac{\pi}{32}(D^4 - 60^4)}$$

$$= \frac{500 \times 10^3 \times 39.5}{\frac{\pi}{32}[79^4 - 60^4]} = \mathbf{77.6\ N/mm^2.} \quad \textbf{Ans.}$$

(*ii*) *Common angle of twist*

Using equation (*i*), we get

$$T_s = \frac{C_s \times \theta_s \times J_s}{L_s}$$

or $$\theta_s = \frac{T_s \times L_s}{C_s \times J_s} = \frac{500 \times 10^3 \times 4000}{8.4 \times 10^4 \times \frac{\pi}{32} \times 60^4} \qquad (\because L_s = L = 4000)$$

$$= 0.01871 \text{ radians}$$

$$= 0.01871 \times \frac{180}{\pi} \text{ degrees} = 1.072°.$$

But angle of twist in each shaft will be equal. The common angle to twist will be equal to the angle of twist in any shaft.

$\therefore$ Common angle of twist = **1.072°. Ans.**

Problem 5.18. *A composite shaft consists of copper rod of 30 mm diameter enclosed in a steel tube of external diameter 50 mm and 10 mm thick. The shaft is required to transmit a torque of 1000 Nm. Determine the shear stresses developed in copper and steel, if both the shafts have equal lengths and welded to a plate at each end, so that their twists are equal. Take modulus of rigidity for steel as twice that of copper.*

Sol. Given :

Dia. of copper rod, $d = 30$ mm

External dia. of steel, $D_0 = 50$ mm

Internal dia. (D_i) of steel shaft $= D - 2 \times$ Thickness

$= 50 - 2 \times 10 = 30$ mm

$\therefore \quad D_i = 30$ mm

Total torque, $T = 1000$ Nm $= 1000 \times 10^3$ Nmm

Let τ_s = Shear stress in steel shaft

τ_c = Shear stress in copper

C_s = Modulus of rigidity of steel

C_c = Modulus of rigidity of copper

L = Common length

θ = Common angle of twist $= \theta_c = \theta_s$

J_c = Polar moment of inertia of copper rod

$$= \frac{\pi}{32} \times d^4 = \frac{\pi}{32} \times 30^4 \text{ mm}^4$$

J_s = Polar moment of inertia of steel shaft

$$= \frac{\pi}{32} [D_0^4 - D_i^4] = \frac{\pi}{32} [50^4 - 30^4] \text{ mm}^4$$

$$= \frac{\pi}{32} [125 \times 10^4 - 81 \times 10^4] = \frac{\pi}{32} \times 44 \times 10^4 \text{ mm}^4$$

T_c = Torque transmitted by copper rod

T_s = Torque transmitted by steel shaft.

Now for a composite shaft,

Total torque,

$$T = T_c + T_s$$

or $$T_c + T_s = 1000 \times 10^3 \quad (\because \ T = 1000 \times 10^3) \quad \ldots(i)$$

Using the relation,

$$\frac{T}{J} = \frac{C \times \theta}{L} \quad \text{or} \quad \theta = \frac{T \times L}{J \times C}$$

For copper rod, $$\theta_c = \frac{T_c \times L_c}{T_c \times C_c}$$

For steel shaft, $$\theta_s = \frac{T_s \cdot L_s}{T_s \cdot C_s}$$

But $$\theta_c = \theta_s \quad (\because \text{ Angle of twist are same})$$

$$\therefore \quad \frac{T_c \, . \, L_c}{J_c \times C_c} = \frac{T_s \, . \, L_s}{J_s \, . \, C_s}$$

But $\quad L_c = L_s = L$

Hence the above equation becomes as

$$\frac{T_c}{J_c \times C_c} = \frac{T_s}{J_s \, . \, C_s}$$

or
$$T_c = \frac{J_c}{J_s} \cdot \frac{C_c}{C_s} \cdot T_s = \frac{\frac{\pi}{32} \times 30^4}{\frac{\pi}{32} \times 44 \times 10^4} \times \frac{C_c}{2C_c} \times T_s \qquad (\because \; C_s = 2C_c)$$

$$= \frac{81}{44} \times \frac{1}{2} \times T_s = \frac{81}{88} \times T_s$$

Substituting the value of T_c in equation (*i*), we get

$$\frac{81}{88} T_s + T_s = 1000{,}000$$

or
$$\frac{81T_s + 88T_s}{88} = 1000{,}000$$

or
$$169 \; T_s = 1000{,}000 \times 88$$

$$\therefore \quad T_s = \frac{1000{,}000 \times 88}{169} = 520710 \text{ Nmm}$$

Substituting this value in (*i*), we get

$$T_c + 520710 = 1000{,}000$$

$$\therefore \quad T_c = 1000{,}000 - 520710 = 479290 \text{ Nmm}$$

Now using the relation,

$$\frac{T}{J} = \frac{\tau}{R}$$

$$\therefore \quad \tau = \frac{T \times R}{J}$$

For a copper rod, we have

$$\tau_c = \frac{T_c \times \left(\frac{d}{2}\right)}{J_c} \qquad \left(\because \; R = \frac{d}{2}\right)$$

$$= \frac{479290 \times \frac{30}{2}}{\frac{\pi}{32} \times 30^4} = \frac{479290 \times 15 \times 32}{\pi \times 30^4} = \mathbf{90.407 \; N/mm^2}. \quad \textbf{Ans.}$$

And for a steel shaft, we have

$$\tau_s = \frac{T_s \times \left(\frac{D_0}{2}\right)}{J_s} = \frac{520710 \times \frac{50}{2}}{\frac{\pi}{32} \times 44 \times 10^4}$$

$$= \frac{520710 \times 25 \times 32}{\pi \times 44 \times 10^4} = \mathbf{301.358 \; N/mm^2}. \quad \textbf{Ans.}$$

5.4. COMBINED BENDING AND TORSION OF SOLID AND HOLLOW SHAFTS

When a shaft is transmitting torque or power, it is subjected to shear stresses. At the same time the shaft is also subjected to bending moments due to self weight of shafts. Due to bending moment, bending stresses are also set up in the shaft. Hence each particle in a shaft is subjected to shear stress and bending stress. For design purposes it is necessary to find the principal stresses, maximum shear stress and strain energy. The principal stresses and maximum shear stress when a shaft is subjected to bending and torsion, are obtained as :

Consider any point on the cross-section of a shaft.

Let T = Torque at the section

D = Diameter of the shaft

M = B.M. at the section.

The torque T will produce shear stress at a point whereas the B.M. will produce bending stress.

Let q = Shear stress at the point produced by torque T and

σ = Bending stress at the point produced by B.M. (M)

The shear stress at any point due to torque (T) is given by

$$\frac{q}{r} = \frac{T}{J} \qquad \left(\because \frac{q}{r} = \frac{\tau}{R} = \frac{T}{J}\right)$$

or

$$q = \frac{T}{J} \times r$$

The bending stress at any point due to bending moment (M) is given by

$$\frac{M}{I} = \frac{\sigma}{y} \text{ or } \sigma = \frac{M \times y}{I}$$

From Art. 1.4.4, we know that the angle θ made by the plane of maximum shear with the normal cross-section is given by,

$$\tan 2\theta = \frac{2\tau}{\sigma}$$

The bending stress and shear stress is maximum at a point on the surface of the shaft, where $r = R = \frac{D}{2}$ and $y = \frac{D}{2}$

Let σ_b = Maximum bending stress *i.e.*, on the surface of the shaft

$$= \frac{M}{I} \times \frac{D}{2} = \frac{M}{\frac{\pi}{64} D^4} \times \frac{D}{2} = \frac{32M}{\pi D^3}$$

τ = Maximum shear stress *i.e.*, on the surface of the shaft

$$= \frac{T}{J} \times R = \frac{T}{\frac{\pi}{32} D^4} \times \frac{D}{2} = \frac{16T}{\pi D^3}$$

The angle θ made by plane of max. shear with normal cross-section is given by

$$\tan 2\theta = \frac{2\tau}{\sigma} = \frac{2\tau}{\sigma_b} = \frac{2 \times \frac{16T}{\pi D^3}}{\frac{32M}{\pi D^3}} = \frac{T}{M} \qquad ...(5.13)$$

Major principal stress

$$= \frac{\sigma}{2} + \sqrt{\left(\frac{\sigma}{2}\right)^2 + \tau^2} = \frac{\sigma_b}{2} + \sqrt{\left(\frac{\sigma_b}{2}\right)^2 + \tau^2}$$

$$= \frac{32M}{2 \times \pi D^3} + \sqrt{\left(\frac{32M}{2 \times \pi D^3}\right)^2 + \left(\frac{16T}{\pi D^3}\right)^2}$$

$$= \frac{16}{\pi D^3}(M + \sqrt{M^2 + T^2}) \quad \text{...(5.14)}$$

and minor principal stress $= \frac{16}{\pi D^3}(M - \sqrt{M^2 + T^2})$ ...(5.15)

Maximum shear stress $= \frac{\text{Major principal stress} - \text{Minor principal stress}}{2}$

$$= \frac{16}{\pi D^3}(\sqrt{M^2 + T^2}) \quad \text{...(5.16)}$$

For a hollow shaft

Major principal stress $= \frac{16D_0}{\pi[D_0^4 - D_i^4]}(M + \sqrt{M^2 + T^2})$ ...(5.17)

Minor principal stress $= \frac{16D_0}{\pi[D_0^4 - D_i^4]}(M - \sqrt{M^2 + T^2})$ ...(5.18)

Maximum shear stress $= \frac{16D_0}{\pi[D_0^4 - D_i^4]}(\sqrt{M^2 + T^2})$ ...(5.19)

Problem 5.19. *A solid shaft of diameter 80 mm is subjected to a twisting moment of 8 MN mm and a bending moment of 5 MN mm at a point. Determine :*

(i) Principal stresses and

(ii) Position of the plane on which they act.

Sol. Given :

Diameter of shaft, $D = 80$ mm

Twisting moment, $T = 8$ MN mm $= 8 \times 10^6$ Nmm

Bending moment, $M = 5$ MN mm $= 5 \times 10^6$ Nmm.

The major principal stress is given by equation (5.14), as

Major principal stress $= \frac{16}{\pi D^3}(M + \sqrt{M^2 + T^2})$

$$= \frac{16}{\pi \times 80^3}\left(5 \times 10^6 + \sqrt{(5 \times 10^6)^2 + (8 \times 10^6)^2}\right)$$

$$= \frac{16 \times 10^6}{\pi \times 80^3}(5 + \sqrt{25 + 64}) = \mathbf{143.57\ N/mm^2}. \quad \textbf{Ans.}$$

Minor principal stress is given by equation (5.15).

∴ Minor principal stress

$$= \frac{16}{\pi D^3}(M - \sqrt{M^2 + T^2})$$

$$= \frac{16}{\pi \times 80^3}\left(5 \times 10^6 - \sqrt{(5 \times 10^6)^2 + (8 \times 10^6)^2}\right)$$

$$= \frac{16 \times 10^6}{\pi \times 80^3}(5 - \sqrt{25 + 64}) = -44.1 \text{ N/mm}^2$$

$$= \mathbf{44.1\ N/mm^2\ (tensile).} \quad \textbf{Ans.}$$

Position of plane is given by equation (5.13), as

$$\tan 2\theta = \frac{T}{M} = \frac{8 \times 10^6}{5 \times 10^6} = 1.6$$

$$\therefore \quad 2\theta = \tan^{-1} 1.6 = 57° \ 59.68', \text{ or } 237° \ 59.68'$$

or $$\theta = 28° \ 59.84' \quad \text{or} \quad \mathbf{118° \ 59.84'}. \quad \textbf{Ans.}$$

Problem 5.20. *The maximum allowable shear stress in a hollow shaft of external diameter equal to twice the internal diameter, is 80 N/mm². Determine the diameter of the shaft if it is subjected to a torque of 4 × 10⁶ Nmm and a bending moment of 3 × 10⁶ Nmm.*

Sol. Given :

Maximum shear stress $= 80$ N/mm²

Torque, $T = 4 \times 10^6$ Nmm

Bending moment, $M = 3 \times 10^6$ Nmm

Let D_0 = External diameter of shaft

D_i = Internal diameter of shaft

Then $D_0 = 2D_i$

Using equation (5.19) for a follow shaft, we get

$$\text{Maximum shear stress} = \frac{16 D_0}{\pi(D_0^4 - D_i^4)}(\sqrt{M^2 + T^2})$$

or $$80 = \frac{16 D_0}{\pi\left[D_0^4 - \left(\frac{D_0}{2}\right)^4\right]}[\sqrt{(3 \times 10^6)^2 + (4 \times 10^6)^2}]$$

$$= \frac{16 D_0 \times 10^6 [\sqrt{9 + 16}]}{\pi . D_0^4\left[1 - \frac{1}{16}\right]} = \frac{16 \times 10^6 \times 5}{\pi D_0^3 \times \frac{15}{16}}$$

$$D_0^3 = \frac{16 \times 10^6 \times 5 \times 16}{\pi \times 15 \times 80} = 0.3395 \times 10^6$$

$$\therefore \quad D_0 = (0.3395 \times 10^6)^{1/3} = 0.3395^{1/3} \times 10^2 = \mathbf{69.78 \ mm.} \quad \textbf{Ans.}$$

and $$D_i = \frac{D_0}{2} = \frac{69.78}{2} = \mathbf{34.89 \ mm.} \quad \textbf{Ans.}$$

5.4.1. Expression for Strain Energy Stored in a Body due to Torsion. Consider a solid shaft which is in torsion. Take an elementary ring of width dr at a radius r as shown in Fig. 5.6.

Let D = Diameter of shaft

l = Length of shaft

R = Radius of shaft $= \frac{D}{2}$

τ = Shear stress on the surface of the shaft *i.e.*, at radius R

C = Shear modulus or modulus of rigidity

U = Total shear strain energy in the shaft.

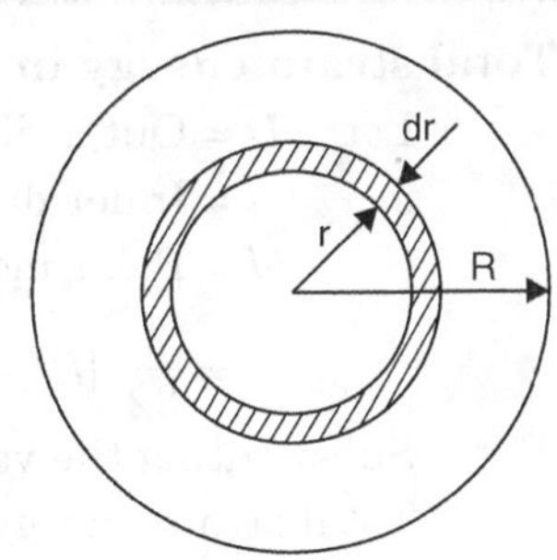

Fig. 5.6

Then shear stress due to torsion at a radius r from the centre is given by

$$\frac{\tau_r}{r} = \frac{\tau}{R} \quad \text{or} \quad \tau_r = \frac{r}{R} \times \tau$$

Area of elementary ring, $\quad dA = 2\pi r.dr$

$\therefore$ Volume of ring, $\quad V = dA.l = 2\pi r \times dr \times l$

The shear strain energy is given by equation (5.9) as,

$$\text{Shear strain energy} = \frac{(\text{Shear stress})^2}{2C} \times \text{Volume}$$

$\therefore$ Shear strain energy in the ring of radius r

$$= \frac{\left(\frac{r}{R} \times \tau\right)^2}{2C} \times 2\pi r \,.\, dr.l = \frac{\tau^2.l}{2CR^2} \times r^2 \times 2\pi r\, dr.$$

Total shear strain energy stored in the shaft is obtained by integrating the above equation between limits 0 to R.

$\therefore$ Total strain energy stored in the shaft,

$$U = \int_0^R \frac{\tau^2.l}{2CR^2} \,.\, r^2 \,.\, 2\pi r\, dr = \frac{\tau^2.l}{2CR^2} \int_0^R r^2 \,.\, 2\pi r\, dr$$

$$= \frac{\tau^2.l}{2CR^2} \int_0^R r^2 \,.\, dA. \qquad (\because \quad dA = 2\pi.dr)$$

$$= \frac{\tau^2.l}{2CR^2} J \qquad \left[\because \int_0^R r^2 \,.\, dA = J\right] \qquad ...(i)$$

where $\quad J$ = Polar moment of inertia of the shaft

$$= \frac{\pi}{32} \times D^4 \text{ for a solid shaft.}$$

$\therefore$ Total strain energy in the shaft due to torsion,

$$U = \frac{\tau^2.l}{2CR^2} \times \frac{\pi}{32} D^4$$

$$= \frac{\tau^2.l}{2CR^2} \times \frac{\pi}{32} \times (2R)^4 \qquad (\because \quad D = 2R)$$

$$= \frac{\tau^2.l}{2CR^2} \times \frac{\pi}{32} \times 16R^4 = \frac{\tau^2}{4C} \,.\, \pi R^2.l$$

$$= \frac{\tau^2}{4C} \,.\, V \qquad (\because \quad \text{Volume, } V = \pi R^2.l) \qquad ...(5.20)$$

Total strain energy in the hollow shaft due to torsion

Let D = Outer dia. of shaft

d = Inner dia. of shaft

J = Polar moment of inertia of hollow shaft

$$= \frac{\pi}{32}[D^4 - d^4]$$

Substituting the value of J in equation (i), we get

Total strain energy in the hollow shaft due to torsion,

$$U = \frac{\tau^2 . l}{2CR^2} \times \frac{\pi}{32}[D^4 - d^4]$$

$$= \frac{\tau^2 . l}{2C\left(\frac{D}{2}\right)^2} \times \frac{\pi}{32}[D^2 + d^2][D^2 - d^2] \qquad \left[\because R = \frac{D}{2}\right]$$

$$= \frac{\tau^2 . l \times 4}{2C . D^2} \times \frac{\pi}{32}[D^2 + d^2][D^2 - d^2]$$

$$= \frac{\tau^2}{4CD^2} \times \frac{\pi}{4} \times (D^2 - d^2) . l(D^2 + d^2)$$

$$= \frac{\tau^2}{4CD^2} \times V \times (D^2 + d^2) \qquad \left[\because V = \frac{\pi}{4}(D^2 - d^2) . l\right]$$

$$= \frac{\tau^2}{4CD^2}(D^2 + d^2) \times V \qquad \text{...(5.21)}$$

Problem 5.21. *Determine the maximum strain energy stored in a solid shaft of diameter 10 cm and of length 1.25 m, if the maximum allowable shear stress is 50 N/mm². Take $C = 8 \times 10^4$ N/mm².*

Sol. Given :

Dia. of shaft, $D = 10$ cm

∴ Area of shaft, $A = \frac{\pi}{4} \times 10^2 = 78.54 \text{ cm}^2 = 7854 \text{ mm}^2$

Length of shaft, $l = 1.25 \text{ m} = 125 \text{ cm}$

∴ Volume of shaft, $V = A \times l = 78.54 \times 125 = 9817.5 \text{ cm}^3 = 9817.5 \times 10^3 \text{ mm}^3$

Maximum allowable shear stress,

$\tau = 50 \text{ N/mm}^2$

(shear stress is maximum on the surface of the shaft)

Modulus of rigidity, $C = 8 \times 10^4 \text{ N/mm}^2$

Let U = Shear strain energy stored in the shaft.

Using equation (5.20), we get

$$U = \frac{\tau^2}{4C} \times V = \frac{50^2}{4 \times 8 \times 10^4} \times 9817.5 \times 10^3$$

= **76699 Nmm. Ans.**

Problem 5.22. *The external and internal diameters of a hollow shaft are 40 cm and 20 cm. Determine the maximum strain energy stored in the hollow shaft if the maximum allowable shear stress is 50 N/mm² and length of the shaft is 5 m. Take C = 8 × 10⁴ N/mm².*

Sol. Given :

External dia., $D = 40 \text{ cm} = 400 \text{ mm}$

Internal dia., $d = 20 \text{ cm} = 200 \text{ mm}$

∴ Area of cross-section, $A = \frac{\pi}{4}(40^2 - 20^2) = 942.47 \text{ cm}^2 = 94247 \text{ mm}^2$

Maximum allowable shear stress (*i.e.*, shear stress on the surface of the shaft),

$$\tau = 50 \text{ N/mm}^2$$

Length of shaft, $l = 5 \text{ m} = 500 \text{ cm}$

∴ Volume of hollow shaft,

$$V = A \times l = 942.47 \times 500 = 471235 \text{ cm}^3 = 471235 \times 10^3 \text{ mm}^3$$

Modulus of rigidity, $C = 8 \times 10^4 \text{ N/mm}^2$

Let U = Strain energy stored

Using equation (5.21), we have

$$U = \frac{\tau^2}{4CD^2}(D^2 + d^2) \times V$$

$$= \frac{50^2}{4 \times 8 \times 10^4 \times 400^2} \times (400^2 + 200^2) \times 471235 \times 10^3$$

$$= 4601900 \text{ Nmm} = \mathbf{4601.9 \text{ Nm}}. \quad \textbf{Ans.}$$

Problem 5.23. *Calculate the diameters of a hollow shaft of the same length and same cross-sectional area as a solid shaft of 15 cm diameter if the strain energy in the hollow shaft is 25% greater than that of solid shaft transmitting the same torque at the same maximum shear stress.*

Sol. Given :

Dia. of solid shaft, $D = 15 \text{ cm}$

∴ Area of solid shaft, $A = \frac{\pi}{4} \times 15^2 = 56.25\,\pi \text{ cm}^2$

Area of solid shaft = Area of hollow shaft

Length of solid shaft = Length of hollow shaft

∴ Volume of solid shaft = Volume of hollow shaft

Strain energy in solid shaft = 1.25 of strain energy in solid shaft

Maximum shear stress in solid shaft

= Maximum shear stress in hollow shaft

Let D_1 = External dia. of hollow shaft

D_2 = Internal dia. of hollow shaft

∴ Cross-sectional area of hollow shaft

$$= \frac{\pi}{4}[D_1^2 - D_2^2].$$

But cross-sectional area of hollow shaft

= Area of solid shaft (given)

$$\therefore \quad \frac{\pi}{4}(D_1^2 - D_2^2) = 56.25\pi$$

$$\therefore \quad D_1^2 - D_2^2 = 56.25 \times 4 = 225 \quad ...(i)$$

Let τ = Max. shear stress in solid as well as in hollow shafts,

U_s = Strain energy of solid shaft, and

U_h = Strain energy of hollow shaft.

Using equation (5.20) for the strain energy in solid shaft, we get

$$U_s = \frac{\tau^2}{4C} \times \text{Volume of solid shaft.}$$

Using equation (5.21) for the strain energy in hollow shaft,

$$U_h = \frac{\tau^2}{4CD_1^2}(D_1^2 - D_2^2) \times \text{Volume of hollow shaft}$$

(∵ External dia. = D_1, Internal dia. = D_2)

But

$$U_h = 1.25 \times U_s$$

or

$$\frac{\tau^2}{4CD_1^2}(D_1^2 + D_2^2) \times \text{Volume of hollow shaft}$$

$$= 1.25 \times \frac{\tau^2}{4C} \times \text{Volume of solid shaft}$$

or

$$\frac{\tau^2}{4CD_1^2}(D_1^2 + D_2^2) = 1.25 \frac{\tau^2}{4C}$$

(∵ Volume of hollow shaft = Volume of solid shaft)

or

$$\frac{D_1^2 + D_2^2}{D_1^2} = 1.25 \qquad \left(\text{Cancelling } \frac{\tau^2}{4C} \text{ to both sides}\right)$$

or

$$D_1^2 + D_2^2 = 1.25\, D_1^2$$

or

$$D_2^2 = 1.25\, D_1^2 - D_1^2 = 0.25\, D_1^2$$

$$\therefore \quad D_2 = \sqrt{0.25 D_1^2} = 0.5\, D_1 \quad ...(ii)$$

Substituting this value of D_2 in equation (i), we get

$$D_1^2 - 0.25\, D_1^2 = 225 \quad \text{or} \quad 0.75\, D_1^2 = 225$$

$$\therefore \quad D_1^2 = \frac{225}{0.75} = 300$$

$$\therefore \quad D_1 = \sqrt{300} = \mathbf{17.32\ cm.} \quad \textbf{Ans.}$$

Substituting this value of D_1 in equation (ii), we get

$$D_2 = 0.5 \times 17.32 = \mathbf{8.66\ cm.} \quad \textbf{Ans.}$$

Problem 5.24. *A solid circular shaft of 10 cm diameter of length 4 m is transmitting 112.5 kW power at 150 r.p.m. Determine : (i) the maximum shear stress induced in the shaft and (ii) Strain energy stored in the shaft. Take C = 8 × 10⁴ N/mm².*

Sol. Given :

Dia. of shaft, D = 10 cm = 100 mm

Length of shaft, l = 4 m = 4000 mm

Power, P = 112.5 kW = 112.5 × 10^3 W

Speed of shaft, $N = 150$ r.p.m.
Modulus of rigidity, $C = 8 \times 10^4$ N/mm^2
Let τ = Maximum shear stress induced in the shaft and
U = Strain energy stored in the shaft.

We know $$P = \frac{2\pi NT}{60}$$

or $$112.5 \times 10^3 = \frac{2\pi \times 150 \times T}{60}$$

$\therefore$ $$T = \frac{112.5 \times 10^3 \times 60}{2\pi \times 150} = 7159 \text{ Nm} = 7159000 \text{ Nmm}$$

But we know, $$T^* = \frac{\pi}{16} \times \tau \times D^3$$

or $$7159000 = \frac{\pi}{16} \times \tau \times 100^3$$

$\therefore$ $$\tau = \frac{7159000 \times 16}{\pi \times 10^6} = 36.5 \text{ N/mm}^2$$

Using equation (5.20) for strain energy,

$$U = \frac{\tau^2}{4C} \times \text{Volume of shaft}$$

$$= \frac{36.5^2}{4 \times 8 \times 10^4} \times \text{Volume of shaft}$$

$$= \frac{36.5^2}{4 \times 8 \times 10^4} \times \frac{\pi}{4} \times 100^2 \times 4000 \qquad \left(\because \text{ Volume} = \frac{\pi}{4} \times D^2 \times l\right)$$

$= \mathbf{130793}$ **Nmm. Ans.**

Problem 5.25. *A hollow shaft of internal diameter 10 cm, is subjected to pure torque and attains a maximum shear stress q on the outer surface of the shaft. If the strain energy stored in the hollow shaft is given by* $\frac{\tau^2}{3C} \times V$, *determine the external diameter of the shaft.*

Sol. Given :
Internal dia., $d = 10$ cm
Maximum shear stress $= \tau$

Strain energy stored, $U = \frac{\tau^2}{3C} \times V$, where V = Volume

Let D = External diameter of the hollow shaft.
Using equation (5.21) for the strain energy in hollow shaft

$$U = \frac{\tau^2}{4CD^2}(D^2 + d^2) \times V$$

*Please see equation (5.4) on page 291. Here $q = \tau$.

Equating the two values of strain energy, we get

$$\frac{\tau^2}{4CD^2}(D^2 + d^2) \times V = \frac{\tau^2}{3C} \times V$$

or $$\frac{D^2 + d^2}{4D^2} = \frac{1}{3} \qquad \left(\text{Cancelling } \frac{\tau^2}{C} \times V \text{ to both sides}\right)$$

or $$3D^2 + 3d^2 = 4D^2$$

or $$3d^2 = 4D^2 - 3D^2 = D^2$$

or $$\frac{D^2}{d^2} = 3 \quad \text{or} \quad \frac{D}{d} = \sqrt{3} = 1.732$$

$\therefore$ $$D = 1.732 \times d = 1.732 \times 10 = \mathbf{17.32 \text{ cm.}} \quad \textbf{Ans.}$$

5.5. TORSION OF THIN WALLED TUBES

Fig. 5.7 shows a thin circular tube

Let D = External diameter of the tube

t = Thickness of tube

T = Torsional resistance

θ = Angle of twist

L = Length of tube

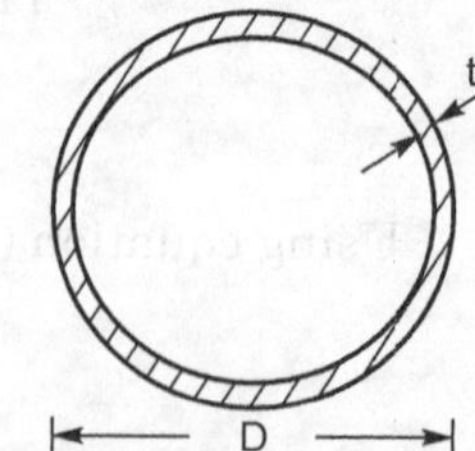

Fig. 5.7. Thin circular tube.

Thickness 't' is very small as compared to external diameter.

Using torsion equation, we get

$$\frac{T}{J} = \frac{\tau}{R}$$

where J = Polar moment of inertia of the section of tube

τ = Shear stress developed on the surface of tube

R = Radius of tube = $\frac{D}{2}$

T = Torque or torsional resistance or twisting moment.

Now the polar moment of inertia of the tube section (Tube is very thin) is given by,

$$J = \text{Area of section} \times \text{Radius square}$$

$$= \pi Dt \times R^2 \qquad [\because \text{ Area of section of thin tube} = \text{Circumference} \times \text{Thickness} = \pi D \times t]$$

$$= \pi Dt \times \left(\frac{D}{2}\right)^2 \qquad \left(\because R = \frac{D}{2}\right)$$

$$= \pi Dt \times \frac{D^2}{4}$$

$$= \pi \times t \times \frac{D^3}{4}$$

$\therefore$ Torque, $$T = \frac{\tau}{R} \times J$$

$$= \frac{\tau}{\left(\frac{D}{2}\right)} \times \pi t \times \frac{D^3}{4} = \tau \times \pi t \times \frac{D^2}{2}$$

$$= \frac{\tau \times \pi t \times D^2}{2} \quad \text{...(5.22)}$$

To find the angle of twist (θ), using equation

$$\frac{T}{J} = \frac{C \times \theta}{L}$$

or

$$\theta = \frac{T \times L}{J \times C}$$

$$= \frac{T \times L}{\left(\pi t \times \frac{D^3}{4}\right) \times C} = \frac{4TL}{\pi t \times D^3 \times C} \quad \left(\because \; J = \frac{\pi t \times D^3}{4}\right) \quad \text{...(5.23)}$$

Here angle of twist θ will be in radians.

Problem 5.26. *A thin tube is subjected to a twisting moment. Determine the safe twisting moment and angle of twist in a thin steel tube of external diameter 80 mm and 3 mm thick. The allowable shear stress is 80 N/mm² and modulus of rigidity is 75 kN/mm². The length of tube is 500 mm.*

Sol. Given :

External dia., $D = 80$ mm

Thickness, $t = 3$ mm, Length, $L = 500$ mm

Shear stress, $\tau = 80 \text{ N/mm}^2$

Modulus of rigidity, $C = 75 \text{ kN/mm}^2 = 75 \times 10^3 \text{ N/mm}^2$

(*i*) **Safe Twisting Moment (T)**

Using equation (5.22), we get

$$T = \frac{\tau \times \pi t \times D^2}{2}$$

[If τ is taken in N/mm², t in mm, and D also in mm, then torque T will be in Nmm]

$$= \frac{80 \times \pi \times 3 \times 80^2}{2} = 2413824 \text{ Nmm}$$

$$= \frac{2413824}{1000} = \mathbf{2413.824 \text{ Nm.}} \quad \textbf{Ans.}$$

(*ii*) **Angle of Twist (θ)**

Using equation (5.23), we get

$$\theta = \frac{4TL}{\pi t \times D^3 \times C}, \quad \text{where } L = 500 \text{ mm and } C = 75 \times 10^3 \text{ N/mm}^2$$

$$= \frac{4 \times 2413824 \times 500}{\pi \times 3 \times 80^3 \times 75 \times 10^3} \text{ radians}$$

[Here T is in Nmm, L, t and D are in mm and C is in N/mm² then θ will be in radians]

$$= 0.0133 \text{ radians}$$

$$= 0.0133 \times \frac{180}{\pi} \qquad \left[\begin{array}{r}\pi \text{ radians} = 180^\circ \\ \text{or} \quad 1 \text{ radian} = \dfrac{180^\circ}{\pi}\end{array}\right]$$

$$= \mathbf{0.7636^\circ.} \quad \textbf{Ans.}$$

HIGHLIGHTS

1. A shaft is in torsion, when equal and opposite torques are applied at the two ends of a shaft.
2. The relation of maximum shear stress induced in a shaft subjected to twisting moment is given by

$$\frac{\tau}{R} = \frac{C\theta}{L}$$

where τ = Maximum shear stress,
R = Radius of shaft,
C = Modulus of rigidity,
θ = Angle of twist in radian, and
L = Length of the shaft.

3. When a circular shaft is subjected to torsion, the shear stress at any point varies linearly from the axis to the surface *i.e.*,

$$\frac{\tau}{R} = \frac{q}{r}$$

where τ = Maximum shear stress on the surface of the shaft,
R = Radius of surface,
q = Shear stress at a point which is at a radius 'r'.

4. The shear stress is maximum on the surface of the shaft and is zero at the axis of the shaft.
5. The torque transmitted by a solid shaft is given by

$$T = \frac{\pi}{16} \tau D^3$$

where D = Dia. of solid shaft, and
τ = Max. shear stress.

6. The torque transmitted by a hollow circular shaft is given by

$$T = \frac{\pi}{16} \tau \left(\frac{D_0^4 - D_i^4}{D_0}\right)$$

where D_0 = External diameter, and
D_i = Internal diameter.

7. Relation between torque, polar moment of inertia and shear stress is given as

$$\frac{T}{J} = \frac{\tau}{R} = \frac{C\theta}{L}$$

where I = Polar moment of inertia

$$= \frac{\pi}{32} D^4 \qquad \text{... For a solid shaft}$$

$$= \frac{\pi}{32} [D_0^4 - D_1^4] \qquad \text{... For a hollow shaft.}$$

8. The ratio of the polar moment of inertia to the radius of the shaft is known as polar modulus or torsional section modulus. It is denoted by Z_p.

$$\therefore \quad Z_p = \frac{J}{R}$$

$$= \frac{\pi}{16} D^3 \qquad \text{... For a solid shaft}$$

$$= \frac{\pi}{16D_0} [D_0^4 - D_i^4] \qquad \text{... For a hollow shaft.}$$

9. Strength of a shaft means the maximum torque or maximum power the shaft can transmit.

10. The product of modulus of rigidity and polar moment of inertia of a shaft is known as torsional rigidity or stiffness of the shaft. Mathematically,

$$\text{Torsional rigidity} = C \times J$$

$$= \frac{T \times L}{\theta} \qquad \left(\because \frac{T}{J} = \frac{C\theta}{L} \quad \therefore \quad C \times J = \frac{T \times L}{\theta}\right)$$

$$= T \text{ if } L = 1 \text{ metre and } \theta = \text{one radian.}$$

11. The power transmitted by a shaft is given by

$$P = \frac{2\pi NT}{60}.$$

12. The strain energy stored in a shaft due to torsion is given by,

$$U = \frac{\tau^2}{4C} \times V \qquad \text{... For a solid shaft}$$

$$= \frac{\tau}{4C \, . \, D^2} (D^2 + d^2) \times V \qquad \text{... For a hollow shaft}$$

where D = External diameter of shaft,

d = Internal diameter of shaft,

T = Modulus of rigidity, and

τ = Shear stress on the surface of the shaft.

13. Polar moment of inertia (J) is given by

$$J = \frac{\pi}{32} D^4 \qquad \text{... For a solid shaft}$$

$$= \frac{\pi}{32} (D^4 - d^4) \qquad \text{... For hollow shaft.}$$

EXERCISE 5

(A) Theoretical Questions

1. Define the terms : Torsion, torsional rigidity and polar moment of inertia.
2. Derive an expression for the shear stress produced in a circular shaft which is subject to torsion. What are the assumptions made in the derivation ?
3. Prove that the torque transmitted by a solid shaft when subjected to torsion is given by

$$T = \frac{\pi}{16} \tau D^3$$

where D = Dia. of solid shaft and τ = Max. shear stress.

4. When a circular shaft is subjected to torsion show that the shear stress varies linearty from the axis to the surface. *(AMIE, Winter 1982)*

5. Derive the relation for a circular shaft when subjected to torsion as given below

$$\frac{T}{J} = \frac{\tau}{R} = \frac{C\theta}{L}$$

where T = Torque transmitted,
J = Polar moment of inertia,
τ = Max. shear stress,
R = Radius of the shaft,
C = Modulus rigidity,
θ = Angle of twist, and
L = Length of the shaft.

6. Find an expression for the torque transmitted by a hollow circular shaft of external diameter $= D_0$ and internal diameter $= D_i$.

7. Define the term 'Polar modulus'. Find the expressions for polar modulus for a solid shaft and for a hollow shaft.

8. What do you mean by 'strength of a shaft' ?

9. Define torsional rigidity of a shaft. Prove that the torsional rigidity is the torque required to produce a twist of one radian in a unit length of the shaft.

10. Prove that the strain energy stored in a body due to shear stress is given by,

$$U = \frac{\tau^2}{2C} \times V$$

where τ = Shear stress,
C = Modulus of rigidity, and
V = Volume of the body.

11. Find an expression for strain energy stored in a body which due to torsion or Prove that the strain energy stored in a body due to torsion is given by,

$$U = \frac{\tau^2}{4C} \times V$$

where τ = Shear stress on the surface of the shaft,
C = Modulus of rigidity, and V = Volume of the body.

12. A hollow shaft of external diameter D and internal diameter d is subjected to torsion, prove that the strain energy stored is given by,

$$U = \frac{\tau^2}{4CD^2}(D^2 + d^2) \times V$$

where V = Volume of the hollow shaft and
τ = Shear stress on the surface of the shaft.

(B) Numerical Problems

1. A solid shaft of 20 cm diameter is used to transmit torque. Find the maximum torque transmitted by the shaft if the maximum shear stress induced in the shaft is 50 N/mm^2.

[**Ans.** 78539.8 Nm]

2. The shearing stress in a solid shaft is not to exceed 45 N/mm^2 when the torque transmitted is 40000 Nm. Determine the minimum diameter of the shaft. [**Ans.** 16.49 mm]

3. Find the maximum torque transmitted by a hollow circular shaft of external diameter 30 cm and internal diameter 15 cm if the shear stress is not to exceed 40 N/mm^2. [**Ans.** 198.8 kN]

4. Two shafts of the same material and of same lengths are subjected to the same torque, if the first shaft is of a solid circular section and the second shaft is of hollow circular section, whose internal diameter is 0.7 times the outside diameter and the maximum shear stress developed in each shaft is the same, compare the weights of the shafts. $\left[\textbf{Ans. } \dfrac{1.633}{1}\right]$

5. Find the maximum shear stress induced in a solid circular shaft of diameter 20 cm when the shaft transmit 187.5 kW at 200 r.p.m. [**Ans.** 5.7 N/mm^2]

6. A solid circular shaft is to transmit 375 kW at 150 r.p.m.

   (*i*) Find the diameter of the shaft if the shear stress is not to exceed 65 N/mm^2.

   (*ii*) What per cent saving in weight would be obtained if this shaft is replaced by a hollow shaft whose internal diameter equal to 2/3rd of its external diameter, the length, the material and maximum shear stress being the same ? [**Ans.** (*i*) 12.29 cm ; (*ii*) 35.71%]

7. A solid shaft has to transmit 112.5 kW at 250 r.p.m. Taking allowable shear stress as 70 N/mm^2, find suitable diameter for the shaft, if the maximum torque transmitted at each revolution exceeds the mean by 20%. [**Ans.** 7.20 cm]

8. A hollow shaft is to transmit 337.5 kW at 100 r.p.m. If the shear stress is not to exceed 65 N/mm^2 and the internal diameter is 0.6 of the external diameter, find the external and internal diameters assuming that the maximum torque is 1.3 times the mean. [**Ans.** 15.52 cm, 9.312 cm]

9. Determine the diameter of a solid steel shaft which will transmit 112.5 kW at 200 r.p.m. Also determine the length of the shaft if the twist must not exceed 1.5° over the entire length. The maximum shear stress is limited to 55 N/mm^2. Take the value of modulus of rigidity $= 8 \times 10^4$ N/mm^2. [**Ans.** 7.9 cm, 150.4 cm]

10. Determine the diameter of a solid shaft which will transmit 337.5 kW at 300 r.p.m. The maximum shear stress should not exceed 35 N/mm^2 and twist should not be more than 1° in a shaft length of 2.5 m. Take modulus of rigidity $= 9 \times 10^4$ N/mm^2. [**Ans.** 11.57 cm]

11. A hollow shaft, having an internal diameter 50% of its external diameter transmits 600 kW at 150 r.p.m. Determine the external diameter of the shaft if the shear stress is not to exceed 65 N/mm^2 and the twist in a length of 3 m should not exceed 1.4 degrees. Assume maximum torque = 1.20 mean torque and modulus of rigidity $= 1 \times 10^5$ N/mm^2. [**Ans.** 15.70 cm]

UNIT III

Chapters

6

Helical and Leaf Springs

6.1. INTRODUCTION

Springs are the elastic bodies which absorb energy due to resilience. The absorbed energy may be released as and when required. A spring which is capable of absorbing the greatest amount of energy for the given stress, without getting permanently distorted, is known as the best spring. The important functions of the springs are :

1. To store energy.
2. To absorb shock or impact loading as in carriage springs.
3. To measure force as in spring balances.

6.1.1. Types of Springs. The important type of springs are :

(*a*) **Helical springs.** They are classified as :

(*i*) Close-coiled helical springs and

(*ii*) Open-coiled helical springs.

(*b*) **Leaf springs.** They are classified as :

(*i*) Semi-elliptical and

(*ii*) Quarter-elliptical.

6.2. HELICAL SPRINGS

Helical springs are the thick spring wires coiled into a helix. They are of two types :

1. Close-coiled helical springs and
2. Open-coiled helical springs.

6.2.1. Close-Coiled Helical Springs. Close-coiled helical springs are the springs in which helix angle is very small or in other words the pitch between two adjacent turns is small. A close-coiled helical spring carrying an axial load is shown in Fig. 6.1. As the helix angle in case of close-coiled helical springs is small, hence the bending effect on the spring is ignored and bending and shear stresses are neglected. Each section of the coils of a close-coiled helical springs are under torsion.

6.3. HELICAL SPRINGS UNDER AXIAL LOAD

Fig. 6.1 shows a close-coiled helical spring subjected to an axial load.

Let d = Diameter of spring wire = $2r$

p = Pitch of the helical spring

n = Number of coils
R = Mean radius of spring coil
W = Axial load on spring
C = Modulus of rigidity
τ = Max. shear stress induced in the wire
θ = Angle of twist in spring wire
δ = Deflection of spring due to axial load, and
l = Length of wire.

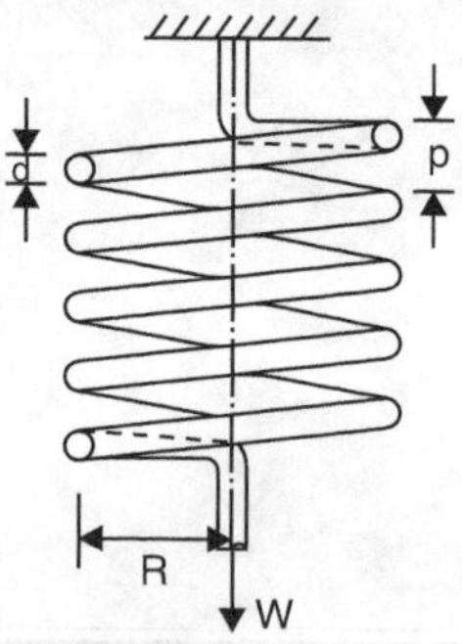

Fig. 6.1

Now twisting moment on the wire,

$$T = W \times R \quad \ldots(i)$$

But twisting moment is also given by torsion equation as

$$\frac{T}{J} = \frac{C \times \theta}{l} = \frac{\tau}{r}$$

or

$$T = \frac{J \times \tau}{r} = \frac{\left(\frac{\pi}{32} d^4\right) \times \tau}{(d/2)} \qquad \left[\because J = \frac{\pi}{32} d^4 \text{ and } r = \frac{d}{2}\right]$$

$$= \frac{\pi}{16} \times d^3 \times \tau$$

$$\therefore \quad T = \frac{\pi}{16} \tau d^3 \quad \ldots(ii)$$

Equating equations (*i*) and (*ii*), we get

$$W \times R = \frac{\pi}{16} \tau d^3 \quad \text{or} \quad \tau = \frac{16 W \times R}{\pi d^3} \quad \ldots(6.1)$$

Equation (6.1) gives the max. shear stress induced in the wire.

6.4. DEFLECTION OF SPRING BY ENERGY METHOD

Now length of one coil = πD or $2\pi R$

$\therefore$ Total length of the wire = Length of one coil × No. of coils or $l = 2\pi R \times n$.

As the every section of the wire is subjected to torsion, hence the strain energy stored by the spring due to torsion is given by :

$\therefore$ Strain energy stored in the spring,

$$U = \frac{1}{2} \times T \times \theta = \frac{1}{2} \times W \cdot R \times \theta$$

We know that

$$\frac{T}{J} = \frac{C \times \theta}{l}$$

$$\therefore \quad \theta = \frac{T \times l}{C \times J} = \frac{T \times l}{C \times \frac{\pi}{32} d^4}$$

$$= \frac{32 T \times l}{C \times \pi d^4}$$

$$= \frac{32 T \times 2\pi R \cdot n}{C \times \pi d^4} = \frac{64 TR \cdot n}{C d^4}$$

$$= \frac{64 \times W \,.\, R \times R \,.\, n}{Cd^4} = \frac{64 W \,.\, R^2 \,.\, n}{Cd^4}$$

Substituting the value of θ, we get

$$U = \frac{1}{2} \times W \,.\, R \times \frac{64 WR^2 \,.\, n}{Cd^4}$$

$$= \frac{32 W^2 \,.\, R^3 \,.\, n}{Cd^4} \quad \text{...(6.2)}$$

Work done on the spring $= \text{Average load} \times \text{Deflection}$

$$= \tfrac{1}{2} W \times \delta \quad (\because \text{ Deflection} = \delta)$$

Equating the work done on spring to the energy stored, we get

$$\tfrac{1}{2} W.\delta = \frac{32 W^2 R^3 \,.\, n}{Cd^4}$$

$\therefore$ Deflection,

$$\delta = \frac{64 WR^3 n}{Cd^4} \quad \text{...(6.3)}$$

Expression for stiffness of spring

The stiffness of spring,

$$s = \text{Load per unit deflection}$$

$$= \frac{W}{\delta} = \frac{W}{\dfrac{64 \,.\, WR^3 \,.\, n}{Cd^4}} = \frac{Cd^4}{64 \,.\, R^3 \,.\, n} \quad \text{...(6.4)}$$

Note. The solid length of the spring means the vertical distance between the coils when the coils are touching each other. There is no gap between the coils. The solid length is given by

Solid length = Number of coils × Dia. of wire = $n \times d$ ...(6.5)

Length of wire is different from solid length.

Length of wire = $2\pi R \times n$.

6.4.1. Solved Problems 6.1 to 6.9 for Close-Coiled Circular Wire Helical Springs

Problem 6.1. *A closely coiled helical spring is to carry a load of 500 N. Its mean coil diameter is to be 10 times that of the wire diameter. Calculate these diameters if the maximum shear stress in the material of the spring is to be 80 N/mm².*

Sol. Given :

Load on spring, $W = 500$ N

Max. shear stress, $\tau = 80$ N/mm^2

Let d = Diameter of wire

D = Mean diameter of coil

$\therefore$ $D = 10\,d$.

Using equation (6.1), $\tau = \dfrac{16 WR}{\pi d^3}$

or

$$80 = \frac{16 \times 500 \times \left(\dfrac{D}{2}\right)}{\pi d^3} \quad \left(\because R = \frac{D}{2}\right)$$

$$= \frac{8000 \times \left(\dfrac{10d}{2}\right)}{\pi d^3} \quad (\because D = 10d)$$

or $$80 \times \pi d^3 = 8000 \times 5d$$

or $$d^2 = \frac{8000 \times 5}{80 \times \pi} = 159.25$$

$\therefore$ $$d = \sqrt{159.2} = 12.6 \text{ mm} = \mathbf{1.26\ cm.}\quad \textbf{Ans.}$$

$\therefore$ $$D = 10 \times d = 10 \times 1.26 = \mathbf{12.6\ cm.}\quad \textbf{Ans.}$$

Problem 6.2. *In problem 6.1, if the stiffness of the spring is 20 N per mm deflection and modulus of rigidity = 8.4 × 10⁴ N/mm², find the number of coils in the closely coiled helical spring.*

Sol. Given :

Stiffness, $s = 20$ N/mm

Modulus of rigidity, $C = 8.4 \times 10^4$ N/mm^2

From problem 6.1,

$W = 500$ N, $\tau = 80$ N/mm^2

$d = 12.6$ mm and $D = 126$ mm

$\therefore$ $R = D/2 = 126/2 = 63$ mm

Let n = Number of coils in the spring

We know, stiffness $= \frac{\text{Load}}{\delta}$

or $$20 = \frac{500}{\delta}$$

$\therefore$ $$\delta = \frac{500}{20} = 25 \text{ mm}$$

Using equation (6.3),

$$\delta = \frac{64WR^3 . n}{C.d^4}$$

or $$25 = \frac{64 \times 500 \times (63)^3 \times n}{8.4 \times 10^4 \times 12.6^4} \quad (\because R = 63 \text{ mm})$$

$\therefore$ $$n = \frac{25 \times 8.4 \times 10^4 \times 12.6^4}{64 \times 500 \times (63)^3} = 6.6 \text{ say } 7.0$$

or $$n = \mathbf{7.}\quad \textbf{Ans.}$$

Problem 6.3. *A closely coiled helical spring of round steel wire 10 mm in diameter having 10 complete turns with a mean diameter of 120 mm is subjected to an axial load of 200 N. Determine : (i) the deflection of the spring (ii) maximum shear stress in the wire, (iii) stiffness of the spring. Take C = 8 × 10⁴ N/mm².*

Sol. Given :

Dia. of wire, $d = 10$ mm

No. of turns, $n = 10$

Mean dia. of coil, $D = 120$ mm

$\therefore$ Radius of coil, $R = D/2 = 60$ mm

Axial load, $W = 200$ N

Modulus of rigidity, $C = 8 \times 10^4$ N/mm^2

Let δ = Deflection of the spring
τ = Max. shear stress in the wire
s = Stiffness of the spring.

(*i*) Using equation (6.3),

$$\delta = \frac{64WR^3 \times n}{Cd^4} = \frac{64 \times 200 \times 60^3 \times 10}{8 \times 10^4 \times 10^4} = \textbf{34.5 mm.} \quad \textbf{Ans.}$$

(*ii*) Using equation (6.1),

$$\tau = \frac{16WR}{\pi d^3} = \frac{16 \times 200 \times 60}{\pi \times 10^3} = \textbf{61.1 N/mm}^2\textbf{.} \quad \textbf{Ans.}$$

(*iii*) Stiffness of the spring,

$$s = \frac{W}{\delta} = \frac{200}{34.5} = \textbf{5.8 N/mm.} \quad \textbf{Ans.}$$

Problem 6.4. *A close coiled helical spring of 100 mm mean diameter is made up of 10 mm diameter rod and has 20 turns. The spring carries an axial load of 200 N. Determine the shearing stress. Taking the value of modulus of rigidity = 8.4 × 10⁴ N/mm², determine the deflection when carrying this load. Also calculate the stiffness of the spring and the frequency of free vibration for a mass hanging from it.*

Sol. Given :

Mean diameter of coil, $D = 100$ mm

∴ Mean radius of coil, $R = 50$ mm

Diameter of rod, $d = 10$ mm

Number of turns, $n = 20$

Axial load, $W = 200$ N

Modulus of rigidity, $C = 8.4 \times 10^4$ N/mm^2

Let τ = Shear stress in the material of the spring
δ = Deflection of the spring due to axial load
s = Stiffness of spring
τ = Frequency of free vibration.

Using equation (6.1),

$$\tau = \frac{16WR}{\pi d^3} = \frac{16 \times 200 \times 50}{\pi \times 10^3} = \textbf{50.93 N/mm}^2\textbf{.} \quad \textbf{Ans.}$$

Deflection of the spring

Using equation (6.3),

$$\delta = \frac{64\,WR^3 \times n}{Cd^4} = \frac{64 \times 200 \times 50^3 \times 20}{8.4 \times 10^4 \times 10^4} = \textbf{38.095 mm.} \quad \textbf{Ans.}$$

Stiffness of the spring

$$\text{Stiffness} = \frac{\text{Load on spring}}{\text{Deflection of spring}} = \frac{200}{38.095} = \textbf{5.25 N/mm.} \quad \textbf{Ans.}$$

Frequency of free vibration

$$\delta = 38.095 \text{ mm} = 3.8096 \text{ cm}$$

Using the relation, $$\tau = \frac{1}{2\pi}\sqrt{\frac{g}{\delta}} = \frac{1}{2\pi}\sqrt{\frac{981}{3.8095}} = \textbf{2.55 cycles/sec.} \quad \textbf{Ans.}$$

Problem 6.5. *A closely coiled helical spring of mean diameter 200 mm is made of 30 mm diameter rod and has 16 turns. A weight of 3 kN is dropped on this spring. Find the height by which the weight should be dropped before striking the spring so that the spring may be compressed by 180 mm. Take C = 8 × 10⁴ N/mm².*

Sol. Given :

Mean dia. of coil, $D = 200$ mm

∴ Mean radius of coil, $R = \frac{200}{2} = 100$ mm

Dia. of spring rod, $d = 30$ mm

Number of turns, $n = 16$

Weight dropped, $W = 3$ kN $= 3000$ N

Compression of the spring, $\delta = 180$ mm

Modulus of rigidity, $C = 8 \times 10^4$ N/mm^2

Let h = Height through which the weight W is dropped

W = Gradually applied load which produces the compression of spring equal to 180 mm.

Now using equation (6.3),

$$\delta = \frac{64W.R^3.n}{Cd^4}$$

or
$$180 = \frac{64 \times W \times 100^3 \times 16}{8 \times 10^4 \times 30^4}$$

or
$$W = \frac{180 \times 8 \times 10^4 \times 30^4}{64 \times 100^3 \times 16} = 11390 \text{ N}$$

Work done by the falling weight on spring

= Weight falling $(h + \delta) = 3000\,(h + 180)$ N-mm

Energy stored in the spring $= \frac{1}{2}\, W \times \delta$

$= \frac{1}{2} \times 11390 \times 180 = 1025100$ N-mm.

Equating the work done by the falling weight on the spring to the energy stored in the spring, we get

$$3000(h + 180) = 1025100$$

or
$$h + 180 = \frac{1025100}{3000} = 341.7 \text{ mm}$$

∴ $h = 341.7 - 180 =$ **161.7 mm. Ans.**

Problem 6.6. *The stiffness of a close-coiled helical spring is 1.5 N/mm of compression under a maximum load of 60 N. The maximum shearing stress produced in the wire of the spring is 125 N/mm². The solid length of the spring (when the coils are touching) is given as 50 mm. Find : (i) diameter of wire, (ii) mean diameter of the coils and (iii) number of coils required. Take C = 4.5 × 10⁴ N/mm².*

Sol. Given :

Stiffness of spring, $s = 1.5$ N/mm

Load on spring, $W = 60$ N

Maximum shear stress, $\tau = 125$ N/mm^2

Solid length of spring, $= 50$ mm

Modulus of rigidity, $C = 4.5 \times 10^4$ N/mm^2.

Let d = Diameter of wire,

D = Mean dia. of coil, and

R = Mean radius of coil $= \frac{D}{2}$

n = Number of coils.

Using equation (6.4),

$$s = \frac{Cd^4}{64.R^3.n} \quad \text{or} \quad 1.5 = \frac{4.5 \times 10^4 \times d^4}{64 \times R^3 \times n}$$

$$\therefore \quad d^4 = \frac{1.5 \times 64 \times R^3 \times n}{4.5 \times 10^4} = 0.002133R^3 \times n \quad \text{...(i)}$$

Using equation (6.1),

$$\tau = \frac{16W \times R}{\pi d^3} \quad \text{or} \quad 125 = \frac{16 \times 60 \times R}{\pi d^3}$$

$$\therefore \quad R = \frac{125 \times \pi d^3}{16 \times 60} = 0.40906\, d^3 \quad \text{...(ii)}$$

Substituting the value of R in equation (*i*), we get

$$d^4 = 0.002133 \times (0.40906d^3)^3 \times n$$

$$= 0.002133 \times (0.40906)^3 \times d^9 \times n = 0.00014599 \times d^9 \times n$$

or
$$\frac{d^9\,.\,n}{d^4} = \frac{1}{0.00014599} \quad \text{or} \quad d^5\,.\,n = \frac{1}{0.00014599} \quad \text{...(iii)}$$

Now using equation (6.5),

Solid length $= n \times d$ or $50 = n \times d$

$$\therefore \quad n = \frac{50}{d} \quad \text{...(iv)}$$

Substituting this value of n in equation (*iii*), we get

$$d^5 \times \frac{50}{d} = \frac{1}{0.00014599}$$

or
$$d^4 = \frac{1}{0.00014599} \times \frac{1}{50} = 136.99$$

$\therefore$ $d = (136.99)^{1/4} =$ **3.42 mm. Ans.**

Substituting this value in equation (*iv*)

$$n = \frac{50}{d} = \frac{50}{3.42} = \mathbf{14.62 \text{ say } 15.} \quad \textbf{Ans.}$$

Also from equation (*ii*),

$$R = 0.40906\, d^3 = 0.40906 \times (3.42)^3 = 16.36 \text{ mm}$$

i.e., Mean dia. of coil, $D = 2R = 2 \times 16.36 =$ **32.72 mm. Ans.**

Problem 6.7. *A close-coiled helical spring has a stiffness of 10 N/mm. Its length when fully compressed, with adjacent coils touching each other is 40 cm. The modulus of rigidity of the material of the spring is 0.8 × 10^5 N/mm^2.*

(i) Determine the wire diameter and mean coil diameter if their ratio is $\frac{1}{10}$.

(ii) If the gap between any two adjacent coil is 0.2 cm, what maximum load can be applied before the spring becomes solid, i.e., adjacent coils touch ?

(iii) What is the corresponding maximum shear stress in the spring ? (AMIE, May 1976)

Sol. Given :

Stiffness of spring, $s = 10$ N/mm

Length of spring when fully compressed *i.e.*, solid length

$= 40$ cm $= 400$ mm

Modulus of rigidity, $C = 0.8 \times 10^5$ N/mm²

Let d = Diameter of wire of spring

D = Mean coil diameter

n = Number of turns

W = Maximum load applied when spring becomes solid

τ = Maximum shear stress induced in the wire.

Now $$\frac{d}{D} = \frac{1}{10}$$

Gap between any two adjacent coil = 0.2 cm = 2.0 mm

∴ Total gap in coils = Gap between two adjacent coil × Number of turns

$= 2 \times n$ mm.

When spring is fully compressed, there is no gap in the coils and hence maximum compression of the coil will be equal to the total gap in the coil.

∴ Maximum compression, $\delta = 2 \times n$ mm

Now using equation (6.4),

$$s = \frac{Cd^4}{64 \,.\, R^3 \,.\, n} \quad \text{or} \quad 10 = \frac{0.8 \times 10^5 \times d^4}{64 \,.\, R^3 \,.\, n}$$

$$\therefore \quad d^4 = \frac{10 \times 64 \times R^3 \times n}{0.8 \times 10^5} = \left(\frac{8}{10^3}\right) R^3 \times n \qquad ...(i)$$

But from equation (6.5),

Solid length $= n \times d$ or $400 = n \times d$.

$$\therefore \quad n = \frac{400}{d} \qquad ...(ii)$$

Substituting the value of n in equation (*i*),

$$d^4 = \left(\frac{8}{10^3}\right) \times R^3 \times \frac{400}{d} = 3.2 \times \frac{R^3}{d}$$

or $$d^5 = 3.2 \times R^3.$$

But mean coil radius,

$$R = \frac{D}{2}$$

$$\therefore \quad d^5 = 3.2 \times \left(\frac{D}{2}\right)^3 = \frac{3.2 \times D^3}{8} = 0.4\, D^3$$

or $$\frac{d^5}{D^3} = 0.4 \quad \text{or} \quad \frac{d^3}{D^3} \cdot d^2 = 0.4$$

or $$\left(\frac{1}{10}\right)^3 \cdot d^2 = 0.4 \qquad \left(\because \frac{d}{D} = \frac{1}{10}\right)$$

$\therefore$ $$d^2 = 0.4 \times 10^3 = 400$$

$\therefore$ $$d = \sqrt{400} = 20 \text{ mm} = \mathbf{2\ cm.} \quad \textbf{Ans.}$$

But $$\frac{d}{D} = \frac{1}{10}$$

$\therefore$ $$D = 10 \times d = 10 \times 2 = \mathbf{20.0\ cm.} \quad \textbf{Ans.}$$

Let us find first number of turns.

From equation (*ii*), we have

$$n = \frac{400}{d} = \frac{400}{20} = 20 \qquad (\because d = 20)$$

$\therefore$ $$\delta = 2 \times n = 2 \times 20 = 40 \text{ mm}$$

We know, stiffness of spring is given by

$$s = \frac{W}{\delta} \quad \text{or} \quad 10 = \frac{W}{40}$$

$\therefore$ $$W = 10 \times 40 = \mathbf{400\ N.} \quad \textbf{Ans.}$$

Using equation (6.1), we have

$$\tau = \frac{16 \cdot W \cdot R}{\pi d^3}$$

$$= \frac{16 \times 400 \times 100}{\pi \times 20^3} \qquad \left(\because R = \frac{D}{2} = \frac{20}{2} = 10 \text{ cm} = 100 \text{ mm}\right)$$

$$= \mathbf{25.465\ N/mm^2.} \quad \textbf{Ans.}$$

Problem 6.8. *Two close-coiled concentric helical springs of the same length, are wound out of the same wire, circular in cross-section and supports a compressive load 'P'. The inner spring consists of 20 turns of mean diameter 16 cm and the outer spring has 18 turns of mean diameter 20 cm. Calculate the maximum stress produced in each spring if the diameter of wire = 1 cm and P = 1000 N.*

Sol. Given :

Total load supported, $P = 1000$ N

Both the springs are of the same length of the same material and having same dia. of wire. Hence values of L, C and 'd' will be same.

For inner spring

No. of turns, $n_i = 20$

Mean dia., $D_i = 160$ mm $\therefore$ $R_i = \frac{160}{2} = 80$ mm

Dia. of wire, $d_i = 10$ mm

For outer spring

No. of turns, $n_0 = 18$

Mean dia., $D_0 = 200$ mm $\therefore$ $R_0 = \frac{200}{2} = 100$ mm

Dia. of wire, $d_0 = 10$ mm

Let W_i = Load carried by inner spring

W_0 = Load carried by outer spring

τ_i = Max. shear stress produced in inner spring

τ_0 = Max. shear stress produced in outer spring.

Now, $W_i + W_0$ = Total load carried = 1000 N ...(*i*)

Since there are two *close-coiled concentric* helical springs, hence deflection of both the springs will be same.

$$\therefore \quad \delta_0 = \delta_i \quad \text{where } \delta_0 = \text{Deflection of outer spring}$$
$$\delta_i = \text{Deflection of inner spring.}$$

The deflection of close-coiled spring is given by equation (6.3) as

$$\delta = \frac{64W \times R^3 \times n}{C \times d^4}$$

Hence for outer spring, we have

$$\delta_0 = \frac{64W_0 \times R_0^{\ 3} \times n_0}{C \times d_0^{\ 4}} = \frac{64W_0 \times 100^3 \times 18}{C \times 10^4} \quad (\because R_0 = 100, d_0 = 10)$$

Similarly for inner spring, we have

$$\delta_i = \frac{64W_i \times R_i^{\ 3} \times n_i}{C \times d_i^{\ 4}} = \frac{64W_i \times 80^3 \times 20}{C \times 10^4}$$

(Material of wires is same. Hence value of C will be same.)

But $\delta_0 = \delta_i$

$$\therefore \quad \frac{64W_0 \times 100^3 \times 18}{C \times 10^4} = \frac{64W_i \times 80^3 \times 20}{C \times 10^4}$$

or $$W_0 \times 100^3 \times 18 = W_i \times 80^3 \times 20$$

or $$W_0 = \frac{W_i \times 80^3 \times 20}{100^3 \times 18} = 0.569\ W_i.$$

Substituting the value of W_0 in equation (*i*), we get

$$W_i + 0.569\ W_i = 1000 \quad \text{or} \quad 1.569\ W_i = 1000$$

$$\therefore \quad W_i = \frac{1000}{1.569} = 637.3 \text{ N.}$$

But from equation (*i*), $W_i + W_0 = 1000$

$$\therefore \quad W_0 = 1000 - W_i = 1000 - 637.3 = 362.7 \text{ N.}$$

The maximum shear stress produced is given by equation (6.1) as

$$\tau = \frac{16\ WR}{\pi d^3}$$

For outer spring, the maximum shear stress produced is given by,

$$\tau_0 = \frac{64W_0 \times R_0}{\pi d_0^{\ 3}} = \frac{16 \times 362.7 \times 100}{\pi \times 10^3}$$

$$= \mathbf{184.72\ N/mm^2.} \quad \textbf{Ans.}$$

Similarly for inner spring, the maximum shear stress produced is given by,

$$\tau_i = \frac{16 \times W_i \times R_i}{\pi \times d_i^3} = \frac{16 \times 637.3 \times 80}{\pi \times 10^3}$$

$$= \mathbf{259.66\ N/mm^2.\quad Ans.}$$

Problem 6.9. *A closely coiled helical spring made of 10 mm diameter steel wire has 15 coils of 100 mm mean diameter. The spring is subjected to an axial load of 100 N. Calculate :*

(i) The maximum shear stress induced, *(ii) The deflection, and*

(iii) Stiffness of the spring.

Take modulus of rigidity, $C = 8.16 \times 10^4\ N/mm^2$.

(AMIE, Winter 1990 ; Converted to S.I. units)

Sol. Given :

Dia. of wire, $d = 10$ mm

Number of coils, $n = 15$

Mean dia. of coil, $D = 100$ mm

$\therefore$ Mean radius of coil, $R = \frac{100}{2} = 50$ mm

Axial load, $W = 100$ N

Modulus of rigidity, $C = 8.16 \times 10^4$ N/mm².

(i) Maximum shear stress induced

Using equation (6.1), $\tau = \frac{16WR}{\pi d^3} = \frac{16 \times 100 \times 50}{\pi \times 10^3} = \mathbf{24.46\ N/mm^2.\quad Ans.}$

(ii) The deflection (δ)

Using equation (6.3),

$$\delta = \frac{64W \times R^3 \times n}{C \times d^4} = \frac{64 \times 100 \times 50^3 \times 15}{8.16 \times 10^4 \times 10^4}$$

$$= \mathbf{14.7\ mm.\quad Ans.}$$

(iii) Stiffness of the spring

$$\text{Stiffness} = \frac{\text{Load on spring}}{\text{Deflection of spring}}$$

$$= \frac{100}{14.7} = \mathbf{6.802\ N/mm.\quad Ans.}$$

6.5. HELICAL SPRINGS UNDER AXIAL TWIST OR TORQUE

Fig. 6.2 shows a close-coiled spring subjected to axial torque.

Let T = Torque applied axially

d = Dia. of wire

D = Dia. of coil

l = Length of wire

$= 2\pi R \times n = \pi D \times n = \pi D \times n$

n = Number of turns

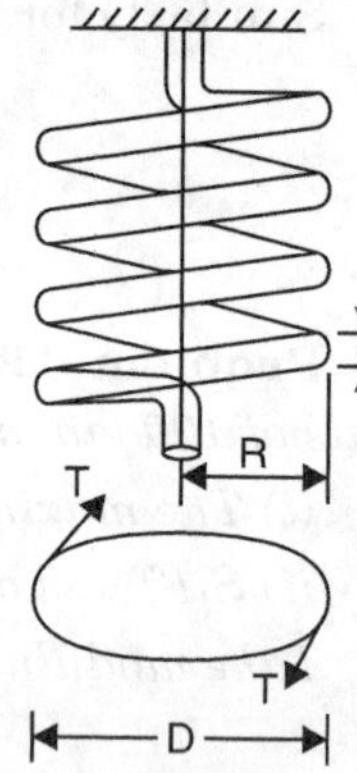

The axial torque T tends to wind up the springs by producing a pure bending moment at all cross-section. Hence all cross-section will be subjected to pure bending moment of magnitude T. The total strain energy stored in the wire due to bending moment M ($M = T$) is given as

$$U = \frac{M^2 \times l}{2EI}$$

$$= \frac{T^2 \times l}{2EI} \quad \text{...(i)} \ (\because \ M = T)$$

$$\left(\because \ \frac{M^2\,dx}{2EI} \text{ for length } dx \text{ and } \frac{M^2 \times l}{2EI} \text{ for length } l\right)$$

Fig. 6.2

Let ϕ = angle of rotation of one end of the spring relative to other end about the axis.

The strain energy due to torsion, is given as

$$U = \frac{1}{2} T \times \phi \quad \text{...(ii)}$$

Equating the two values given by equations (*i*) and (*ii*), we get

$$\frac{1}{2} T \times \phi = \frac{T^2 \times l}{2EI}$$

or

$$\phi = \frac{2T^2 \times l}{T \times 2EI} = \frac{T \times l}{EI} = \frac{T \times \pi D \times n}{E \times \frac{\pi d^4}{64}} \quad \left(\because \ l = 2\pi R \times n = \pi D \times n \text{ and } I = \frac{\pi d^4}{64}\right)$$

$$= \frac{64\ TD \times n}{Ed^4} \quad \text{...(6.6)}$$

Bending Stress (σ_b)

Using $\frac{M}{I} = \frac{\sigma_b}{y}$

Max. bending stress will be on the surface of the wire, where $y = \frac{d}{2}$.

$$\therefore \quad \sigma_b = \frac{M}{I} \times y = \frac{M}{\pi d^4/64} \times \frac{d}{2} = \frac{32M}{\pi d^3} \quad \text{...(6.7)}$$

6.5.1. Solved Problems 6.10 to 6.11 for Helical Spring under Axial Torque

Problem 6.10. *Determine : (i) axial twist, (ii) bending stress and (iii) the strain energy per unit volume of a close-coiled helical spring of round steel wire of 10 mm in diameter having 8 complete turns with a mean diameter of 150 mm, when an axial torque of 9 Nm is applied to the spring. Take E = 20.5 N/mm².*

Sol. Given :

Dia. of wire, d = 10 mm ; No. of turns, n = 8 ;

Mean dia. of coil, D = 150 mm ; Axial torque, T = 9 Nm = 9 × 1000 Nmm = 9000 Nmm

$E = 20.5 \times 10^4$ N/mm²

Now total length of wire, $l = \pi D \times n = \pi \times 150 \times 8$ mm

(i) Axial Twist (ϕ)

It is given by equation (6.6) as

$$\phi = \frac{64T \times D \times n}{Ed^4}$$

$$= \frac{64 \times 9000 \times 150 \times 8}{20.5 \times 10^4 \times (10)^4} = \mathbf{0.337\ rad.} \quad \textbf{Ans.}$$

(ii) Bending Stress (σ_b)

It is given by equation (6.7) as

$$\sigma_b = \frac{32M}{\pi d^3} \quad (M = T)$$

$$= \frac{32T}{\pi d^3} = \frac{32 \times 9000}{\pi \times 10^3} = \mathbf{91.7\ N/mm^2.} \quad \textbf{Ans.}$$

(iii) Strain Energy per unit Volume

Total strain energy due to axial torque

$$= \frac{1}{2} T \times \phi$$

$$= \frac{1}{2} \times 9000 \times 0.337$$

[$\because$ $T = 9000$ Nmm and $\phi = 0.337$ radians]

$$= 4500 \times 0.337 \text{ Nmm}$$

Strain energy per unit volume

$$= \frac{\text{Total strain Energy}}{\text{Volume of wire}} = \frac{4500 \times 0.337}{\left(\frac{\pi}{4} d^2\right) \times l}$$

$$= \frac{4 \times 4500 \times 0.337}{\pi \times 10^2 \times \pi \times 150 \times 8} = \mathbf{0.00512\ Nmm/mm^3.} \quad \textbf{Ans.}$$

Problem 6.11. *A close-coiled helical spring of circular section of mean coil diameter D extends 1 mm when subjected to an axial load W, and there is an angular rotation of 1 radian when a torque T is independently applied about the axis. Show that* $\frac{T}{W} = \frac{D^2(1+\mu)}{4}$ *where* μ *is Poisson's ratio.*

Also determine the Poisson's ratio if a load of 100 N extends the spring by 66.67 mm and torque of 400 Nmm produces an angular rotation of 72°. Take mean coil diameter of the spring as 25 mm.

Sol. Given :

Deflection, $\delta = 1$ mm when axial load = W

Angular rotation, $\phi = 1$ radian when torque applied = T

For axial load, the deflection δ is given by equation (6.3) as

$$\delta = \frac{64WR^3 \times n}{Cd^4}$$

$$= \frac{64W\left(\frac{D}{2}\right)^3 \times n}{Cd^4} \qquad \left(\because R = \frac{D}{2}\right)$$

$$= \frac{8WD^3 \times n}{Cd^4}$$

When δ = 1 mm, then above equation becomes as,

$$1 \text{ mm} = \frac{8WD^3 \times n}{Cd^4}$$

or $$W = \frac{Cd^4}{8D^3 \times n} \quad ...(i)$$

For axial torque, the angular rotation ϕ is given by equation (6.6) as

$$\phi = \frac{64T \times D \times n}{Ed^4}$$

For ϕ = 1 radian, the above equation becomes as

$$1 = \frac{64T \times D \times n}{Ed^4}$$

or $$T = \frac{Ed^4}{64 \times D \times n} \quad ...(ii)$$

Now equation (*ii*) divided by equation (*i*),

$$\frac{T}{W} = \left(\frac{Ed^4}{64D \times n}\right) \div \frac{Cd^4}{8D^3 \times n}$$

$$= \frac{Ed^4}{64D \times n} \times \frac{8D^3 \times n}{Cd^4}$$

$$= \frac{E \times D^2}{8C} \qquad \text{But } E = 2C(1 + \mu)$$

$$\therefore \quad \frac{T}{W} = \frac{2C(1+\mu)D^2}{8C} = \frac{D^2(1+\mu)}{4}. \quad \textbf{Ans. (Proved)}$$

In the above equation, W is axial load per unit deflection and T is torque per unit angular rotation.

Numerical Data :

Axial load = 100 N, δ = 66.67 mm,

Torque = 400 Nmm, $\phi = 72° = \frac{72 \times \pi}{180}$ radian = 0.4 π radian, D = 25 mm

Axial load per unit deflection,

$$W = \frac{\text{Axial load}}{\text{Deflection}} = \frac{100}{66.67} = 1.5 \text{ N/mm}$$

Torque per unit angular rotation,

$$T = \frac{\text{Torque}}{\phi} = \frac{400 \text{ Nmm}}{0.4\pi} = 318.3 \text{ Nmm/radian}$$

But $$\frac{T}{W} = \frac{D^2(1+\mu)}{4}$$

or $$\frac{318.3 \text{ Nmm/radian}}{1.5 \text{ N/mm}} = \frac{25^2(1+\mu)}{4} \text{ mm}^2$$

or $$\frac{318.3}{1.5} \text{ mm}^2 = \frac{625}{4}(1+\mu) \text{ mm}^2$$

or $$\frac{318.3}{1.5} \times \frac{4}{625} = 1 + \mu$$

or $$1.358 = 1 + \mu \text{ or } \mu = 1.358 - 1 = \mathbf{0.358.} \quad \textbf{Ans.}$$

6.6. OPEN-COILED HELICAL SPRING

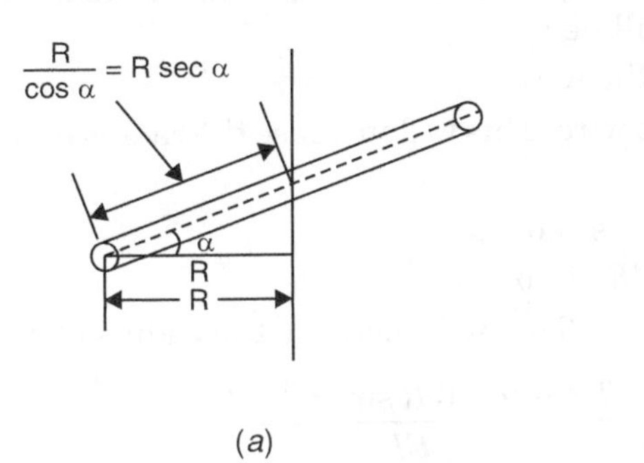

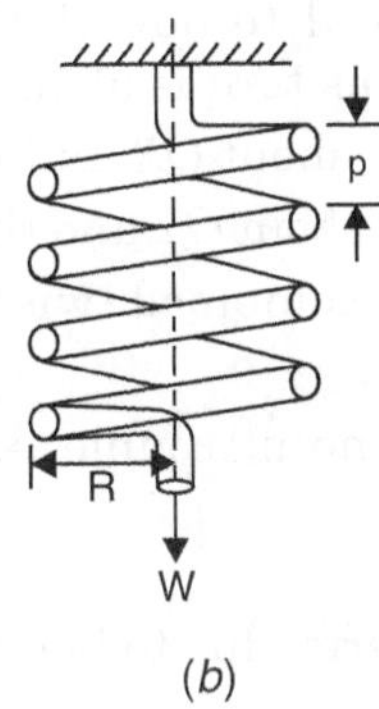

(a) (b)

Fig. 6.3

Fig. 6.3 shows an open-coiled-helical spring in which α is the angle of helix. The length of one coil will be $2\pi R \sec \alpha$ as shown in Fig. 6.3 (a). The total length of the wire will be

$$= 2\pi R \sec \alpha \times \text{No. of coils}$$

$$= 2\pi R \sec \alpha \times n, \text{ where } n = \text{number of coils,}$$

Refer to Fig. 6.4

Let OH = Horizontal axis

OV = Vertical axis

OX = Polar axis (or axis of twisting) at any normal cross-section and is inclined at an angle α to the horizontal axis OH

OY = Bending axis about which bending takes place and inclined at an angle α to the vertical axis OV

W = Axial load applied to the spring

T = Axial torque applied to the spring.

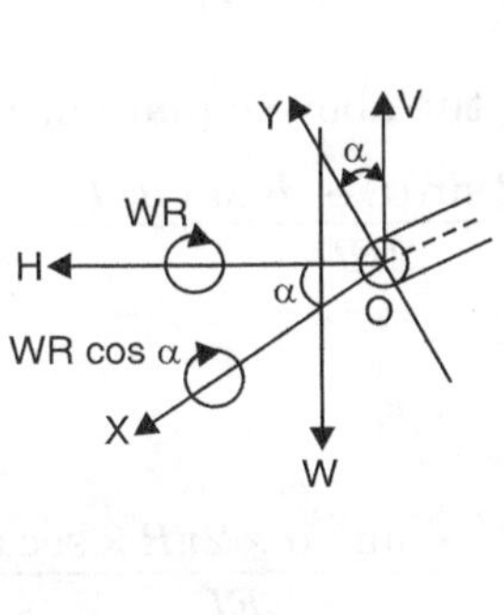

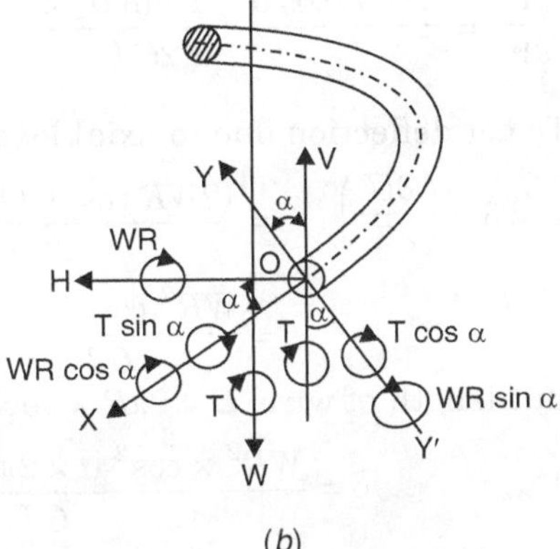

(a) (b)

Fig. 6.4

Due to axial load W, there will be a couple at O of magnitude $W \times R$ about OH. The components of this couple about OX and OY will be :

(i) $WR \cos \alpha$ about OX and this will twist the wire and

(ii) $WR \sin \alpha$ about OY and this will bend the wire. This component of couple is clockwise. This component about OY' will be—$WR \sin \alpha$ this component will be anti-clockwise. This will reduce the radius of curvature.

Due to axial torque T, there will be a torque at O of magnitude T about OV. The components of this torque about OX and OY will be :

(i) $T \sin \alpha$ about OX and this will twist the wire and

(ii) $T \cos \alpha$ about OY and this will bend the wire. This will increase the radius of curvature.

Hence the combined twisting couple

$$T_o = WR \cos \alpha + T \sin \alpha$$

and combined bending couple, $M_o = T \cos \alpha - WR \sin \alpha$

(This will increase the radius of curvature)

$$\text{Strain energy due to bending} = \frac{M_o{}^2 L}{2EI} = \frac{(T \cos \alpha - WR \sin \alpha)^2 \times L}{2EI}$$

$$\text{Strain energy due to twisting} = \frac{T_o{}^2 L}{2CJ} = \frac{(T \cos \alpha - WR \sin \alpha)^2 \times L}{2EI}$$

Total strain energy due to bending and twisting is

$$U = \frac{(WR \cos \alpha + T \sin \alpha)^2 \times L}{2CJ} + \frac{(T \cos \alpha - WR \sin \alpha)^2 \times L}{2EI} \qquad \text{...(6.8)}$$

From Castigliano's theorem, we have

$$\text{Axial deflection,} \quad \delta = \frac{\partial U}{\partial W} \text{ and}$$

$$\text{Axial rotation,} \quad \phi = \frac{\partial U}{\partial T}$$

Usually the loading is either W only or T only.

6.6.1. Open-Coiled Helical Spring Under Axial Load Only

(i) Deflection due to axial load W only,

Differentiating equation (6.8) w.r.t. W, we get

$$\frac{\partial U}{\partial W} = \frac{2\,(WR \cos \alpha + T \sin \alpha) \times R \cos \alpha \times L}{2CJ} + \frac{2\,(T \cos \alpha - WR \sin \alpha) \times (-\,R \sin \alpha \times L)}{2EI} \qquad \text{...(6.9)}$$

To get deflection due to axial load only, put $T = 0$, in the above equation,

$$\therefore \quad \delta = \left(\frac{\partial U}{\partial W}\right)_{T=0} = \frac{(2WR \cos \alpha)\,(R \cos \alpha)\,L}{2CJ} + \frac{2\,(-\,WR \sin \alpha)\,(-\,R \sin \alpha)\,L}{2EI}$$

$$= \frac{WR^2 \cos^2 \alpha \times L}{CJ} + \frac{WR^2 \sin^2 \alpha L}{EI} \qquad \text{...(6.10)}$$

But length of wire, $L = 2\pi R \times \sec \alpha \times n$

$$\therefore \quad \delta = \frac{WR^2 \times \cos^2 \alpha \times 2\pi R \times \sec \alpha \times n}{CJ} + \frac{WR^2 \times \sin^2 \alpha \times 2\pi R \times \sec \alpha \times n}{EI}$$

$$= 2WR^3 \times \pi \times n \times \sec \alpha \left[\frac{\cos^2 \alpha}{CJ} + \frac{\sin^2 \alpha}{EI}\right]$$

But $J = \frac{\pi}{32} d^4$ and $I = \frac{\pi}{64} d^4$

$$\therefore \quad \delta = 2\pi WR^3 \times n \times \sec\alpha \left[\frac{\cos^2\alpha}{C \times \frac{\pi}{32} d^4} + \frac{\sin^2\alpha}{E \times \frac{\pi}{64} \times d^4} \right]$$

$$= 2\pi WR^3 \times n \times \sec\alpha \left[\frac{32\cos^2\alpha}{C \times \pi d^4} + \frac{64\sin^2\alpha}{E \times \pi d^4} \right]$$

$$= 2\pi WR^3 \times n \times \sec\alpha \times \frac{32}{\pi d^4} \left[\frac{\cos^2\alpha}{C} + \frac{2\sin^2\alpha}{E} \right]$$

$$= \frac{64\, WR^3 \times n \times \sec\alpha}{d^4} \left[\frac{\cos^2\alpha}{C} + \frac{2\sin^2\alpha}{E} \right]$$

But $R = \frac{D}{2}$

$$\therefore \quad \delta = \frac{64\, W \times \left(\frac{D}{2}\right)^3 \times n \times \sec\alpha}{d^4} \left[\frac{\cos^2\alpha}{C} + \frac{2\sin^2\alpha}{E} \right]$$

$$= \frac{64\, W \times D^3 \times n \times \sec\alpha}{8 d^4} \left[\frac{\cos^2\alpha}{C} + \frac{2\sin^2\alpha}{E} \right]$$

$$= \frac{8\, W\, D^3 \times n \times \sec\alpha}{d^4} \left[\frac{\cos^2\alpha}{C} + \frac{2\sin^2\alpha}{E} \right]$$

$$= \frac{8 WD^3 \times n}{d^4 \times \cos\alpha} \left[\frac{\cos^2\alpha}{C} + \frac{2\sin^2\alpha}{E} \right] \quad \text{...(6.11)}$$

(*ii*) **Axial Rotation due to axial load W only** $\left(\textbf{axial rotation, } \phi = \frac{\partial U}{\partial T}\right)$.

Differentiating equation (6.8) w.r. to T, we get

$$\frac{\partial U}{\partial T} = \frac{2(WR\cos\alpha + T\sin\alpha) \times \sin\alpha \times L}{2CJ} + \frac{2(T\cos\alpha - WR\sin\alpha) \times \cos\alpha \times L}{2EI} \quad \text{...(6.12)}$$

To get rotation due to axial load only, put $T = 0$ in the above equation.

$$\therefore \quad \phi = \left(\frac{\partial U}{\partial T}\right)_{T=0} = \frac{2WR\cos\alpha \times \sin\alpha \times L}{2CJ} + \frac{2(-WR\sin\alpha) \times \cos\alpha \times L}{2EI}$$

$$= \frac{WRL \times \sin\alpha \times \cos\alpha}{CJ} - \frac{WR \times L \times \sin\alpha\cos\alpha}{EI}$$

where $\quad L = 2\pi R \times \sec\alpha \times n = \frac{2\pi R}{\cos\alpha} \times n$

$$\therefore \quad \phi = WR \times \frac{2\pi R \times n}{\cos\alpha} \times \frac{\sin\alpha \times \cos\alpha}{CJ} - WR \times \frac{2\pi R \times n}{\cos\alpha} \times \frac{\sin\alpha\cos\alpha}{EI}$$

$$= \frac{2\pi R^2 \times W \times n \times \sin\alpha}{CJ} - \frac{2\pi R^2 W \times n \times \sin\alpha}{EI}$$

$$= 2\pi R^2 \times W \times n \times \sin\alpha \left[\frac{1}{CJ} - \frac{1}{EI} \right]$$

Here $J = \frac{\pi}{32} d^4$ and $I = \frac{\pi}{64} d^4$

$$\phi = 2\pi R^2 \times W \times n \times \sin\alpha \left[\frac{1}{C \times \frac{\pi}{32} d^4} - \frac{1}{E \times \frac{\pi}{64} d^4} \right]$$

$$= 2\pi R^2 \times W \times n \times \sin\alpha \left[\frac{32}{C \times \pi d^4} - \frac{64}{E \times \pi d^4} \right]$$

$$= 2\pi R^2 \times W \times n \times \sin\alpha \times \frac{32}{\pi d^4} \left[\frac{1}{C} - \frac{2}{E} \right]$$

$$= \frac{64 R^2 \times W \times n \times \sin\alpha}{d^4} \left[\frac{1}{C} - \frac{2}{E} \right]$$

But $R = \frac{D}{2}$

$$\therefore \quad \phi = \frac{64 \times \left(\frac{D}{2} \right)^2 \times W \times n \times \sin\alpha}{d^4} \left[\frac{1}{C} - \frac{2}{E} \right]$$

$$= \frac{64 D^2 \times W \times n \times \sin\alpha}{4 d^4} \left[\frac{1}{C} - \frac{2}{E} \right]$$

$$= \frac{16 W D^2 \times n \times \sin\alpha}{d^4} \left[\frac{1}{C} - \frac{2}{E} \right] \quad ...(6.13)$$

(*iii*) **Stresses in the wire due to axial load only.** Due to axial load W, the torque $W \times R$ is about axis OH (Refer to Fig. 6.4). The components of this torque are $WR \cos\alpha$ about polar axis OX and $WR \sin\alpha$ about axis OY. The component $WR \cos\alpha$ about polar axis is the twisting moment and component $WR \sin\alpha$ about axis OY is the bending moment.

$\therefore \quad T = WR \cos\alpha$ and $M = WR \sin\alpha$

Twisting moment (T) produces shear stress (τ) in the wire whereas the bending moment (M) produces bending stress (σ_b) in the wire.

$\therefore$ The shear stress due to twisting moment is given by,

$$T = \frac{\pi}{16} \times \tau \times d^3 \quad \therefore \quad \tau = \frac{16T}{\pi d^3} \quad ...(6.14)$$

The bending stress due to bending moment is given by,

$$\frac{M}{I} = \frac{\sigma_b}{y} \quad \therefore \quad \sigma_b = \frac{M}{I} \times y = \frac{M}{\frac{\pi}{64} d^4} \times \frac{d}{2} = \frac{32M}{\pi d^3} \quad ...(6.15)$$

$$\text{Principal stresses} = \frac{\sigma_b}{2} \pm \sqrt{\left(\frac{\sigma_b}{2} \right)^2 + \tau^2} \quad ...(6.16)$$

$$\left(\because \text{ Principal stresses} = \frac{\sigma_x + \sigma_y}{2} \pm \sqrt{\left(\frac{\sigma_x - \sigma_y}{2} \right)^2 + \tau^2}. \text{ Here } \sigma_x = \sigma_b, \sigma_y = 0 \right)$$

$$\text{Maximum shear stress} = \sqrt{\left(\frac{\sigma_b}{2} \right)^2 + \tau^2} \quad ...(6.17)$$

6.6.2. Open-Coiled Helical Spring under Axial Torque only

(*i*) **Deflection** δ

As, $\delta = \left(\dfrac{\partial U}{\partial W}\right)_{W=0}$. Hence substitute $W = 0$ in equation (6.9), we get

$$\delta = \left(\frac{\partial U}{\partial W}\right)_{W=0} = \frac{2T \sin\alpha \times R\cos\alpha \times L}{2CJ} + \frac{2T\cos\alpha \times (-R\sin\alpha)\times L}{2EI}$$

$$= \frac{TRL \times \sin\alpha\cos\alpha}{CJ} - \frac{TRL\cos\alpha\sin\alpha}{EI}$$

$$= TRL \times \sin\alpha \times \cos\alpha\left[\frac{1}{CJ} - \frac{1}{EI}\right] \quad \text{...(6.18)}$$

$$= TR \times \frac{2\pi R \times n}{\cos\alpha} \times \sin\alpha \times \cos\alpha\left[\frac{1}{CJ} - \frac{1}{EI}\right] \qquad \left[\because\ L = \frac{2\pi R}{\cos\alpha}\times n\right]$$

$$= 2\pi T \times R^2 \times n \times \sin\alpha\left[\frac{1}{CJ} - \frac{1}{EI}\right] \quad \text{...(6.19)}$$

(*ii*) **Axial rotation** ϕ

$\phi = \left(\dfrac{\partial U}{\partial T}\right)_{W=0}$. Hence substitute $W = 0$ in equation (6.12), we get

$$\phi = \left(\frac{\partial U}{\partial T}\right)_{W=0} = \frac{2T\sin\alpha \times \sin\alpha \times L}{2CJ} + \frac{2T\cos\alpha \times \cos\alpha \times L}{2EI}$$

$$\therefore \quad \phi = \frac{T \times L \times \sin^2\alpha}{CJ} + \frac{T\times L \times \cos^2\alpha}{EI}$$

$$= T \times L\left[\frac{\sin^2\alpha}{CJ} + \frac{\cos^2\alpha}{EI}\right] \quad \text{...(6.20)}$$

(*iii*) **Stresses in the wire**

Due to axial torque T, the component of this torque about polar axis OX is the twisting moment and the component about axis OY is the bending moment

$\therefore$ Twisting moment $(T^*) = T\sin\alpha$

and bending moment $(M^*) = T\cos\alpha$

$\therefore$ Torsional shear stress due to twisting moment T^* is obtained as

$$T^* = \frac{\pi}{16}\times\tau\times d^3 \quad \therefore \quad \tau = \frac{16T^*}{\pi d^3} = \frac{16T\sin\alpha}{\pi d^3} \quad \text{...(6.21)}$$

and maximum stress due to bending moment (M^*) is obtained as

$$\frac{M^*}{I} = \frac{\sigma_b}{y} \quad \therefore \quad \sigma_b = \frac{M^*}{I}\times y = \frac{M^*}{\frac{\pi}{64}d^4}\times\frac{d}{2} = \frac{32M^*}{\pi d^3}$$

$$= \frac{32T\cos\alpha}{\pi d^3} \quad \text{...(6.22)}$$

$$\therefore \quad \text{Principal stresses} = \frac{\sigma_b}{2} \pm \sqrt{\left(\frac{\sigma_b}{2}\right)^2 + \tau^2} \quad \text{...(6.23)}$$

6.6.3. Solved Problems 6.12 to 6.13 for Open-Coiled Helical Spring

Problem 6.12. *Determine :*

(i) the axial deflection of an open-coiled helical spring and

(ii) the angle turned by the free end of the wire if free to rotate, when the coil is subjected with an axial load of 100 N only. The mean diameter of the coil is 125 mm whereas the diameter of the wire is 10 mm. The coil has 10 turns and the angle of the coils with a plane perpendicular to the axis of the coil is 30°. Take E = 205,000 N/mm² and C = 80,000 N/mm².

Sol. Given :

$W = 100$ N, $D = 125$ mm $\quad\therefore\quad R = \frac{125}{2} = 62.5$ mm, $d = 10$ mm

$n = 10$, $\alpha = 30°$, $E = 205{,}000$ N/mm² and $C = 80{,}000$ N/mm².

For axial load only.

(i) Axial Deflection

The deflection (δ) is given by equation (6.11) as

$$\delta = \frac{8WD^3 \times n}{d^4 \times \cos\alpha}\left[\frac{\cos^2\alpha}{C} + \frac{2\sin^2\alpha}{E}\right]$$

$$= \frac{8 \times 100 \times (125)^3 \times 10}{(10^4) \times \cos 30°}\left[\frac{\cos^2 30°}{80{,}000} + \frac{2\sin^2 30°}{205{,}000}\right]$$

[Take *W* in *N*, *D* and *d* in mm, *C* and *E* in N/mm², then δ will be in mm]

$$= \frac{8 \times 125^3}{10 \times 0.866}\left[\frac{0.866^2}{80{,}000} + \frac{2 \times 0.25}{205{,}000}\right]$$

$= 1804272.5\ [9.374 \times 10^{-6} + 2.44 \times 10^{-6}]$

$= 1804272.5 \times 10^{-6} \times 11.814$

$=$ **21.31 mm. Ans.**

(ii) The angle turned by free end of wire

The angle turned (or rotation) due to axial load only is given by equation (6.13) as

$$\phi = \frac{16WD^2 \times n \times \sin\alpha}{d^4}\left[\frac{1}{C} - \frac{2}{E}\right]$$

$$= \frac{16 \times 100 \times 125^2 \times 10 \times \sin 30°}{10^4}\left[\frac{1}{80{,}000} - \frac{2}{205{,}000}\right]$$

$= 12500\ [1.25 \times 10^{-5} - 0.4878 \times 10^{-5}]$

$= 12500 \times 10^{-5}\ [1.25 - 0.9756] = 12500 \times 10^{-5} \times 0.2744$

$= 0.0343$ rad. $= 0.0343 \times \frac{180}{\pi} =$ **1.965°. Ans.**

Problem 6.13. *An open-coiled helical spring of wire diameter 9 mm, mean coiled diameter 90 mm and helix angle 30° consists of 12 turns.*

Determine : (i) Load required to elongate the spring by 30 mm,

(ii) Bending and shear stresses produced by that load.

(iii) Axial torque that will produce a bending stress of 50 N/mm². Take E = 200,000 N/mm² and C = 82,000 N/mm².

Sol. Given :

Wire dia., $d = 9$ mm ; mean coil diameter, $D = 90$ mm ; helix angle, $\alpha = 30°$; no. of turns, $n = 12$; deflection of spring, $\delta = 30$ mm ; $E = 200,000$ N/mm² and $C = 80,000$ N/mm².

(i) Load required to elongate the spring by 30 mm

Let W = Load required

The deflection due to axial load for open-coiled spring is given by equation (6.11) as

$$\delta = \frac{8WD^3 \times n}{d^4 \times \cos\alpha}\left[\frac{\cos^2\alpha}{C} + \frac{2\sin^2\alpha}{E}\right]$$

$$\therefore \quad 30 = \frac{8W \times (90)^3 \times 12}{9^4 \times \cos 30°}\left[\frac{\cos^2 30°}{82000} + \frac{2\sin^2 30°}{200000}\right]$$

$$= \frac{8W \times 9^3 \times 10^3 \times 12}{9^4 \times 0.866}\left[\frac{0.75}{82000} + \frac{0.25}{100000}\right]$$

$$= \frac{8W \times 10^3 \times 12}{9 \times 0.866} \times \frac{1}{10^4}\left[\frac{0.75}{8.2} + \frac{0.25}{10}\right]$$

$$= \frac{8W \times 12}{9 \times 8.66}[0.0915 + 0.025]$$

$$= \frac{96W}{9 \times 8.66} \times 0.1165$$

$$\therefore \quad W = \frac{30 \times 9 \times 8.66}{96 \times 0.1165} = 209.06 \simeq \textbf{209 N.} \quad \textbf{Ans.}$$

(ii) Bending and shear stresses produced by axial load 209 N. Due to axial load W, the torque at a mean radius R about horizontal axis OH is $W \times R$. The component of this torque about polar axis OX is $W \times R \times \cos\alpha$. This component will twist the wire. The component about axis OY is $W \times R \times \sin\alpha$. This component will bend the wire (Refer to Fig. 6.5)

Fig. 6.5

$\therefore$ Twisting moment, $T = W \times R \times \cos\alpha$ and

Bending moment, $M = W \times R \times \sin\alpha$.

Twisting moment will produce shear stress whereas bending moment will produce bending stress. The shear stress (τ) due to twisting moment is obtained from

$$T = \frac{\pi}{16} \times \tau \times d^3$$

$$\therefore \quad \tau = \frac{16 \times T}{\pi d^3} = \frac{16 \times WR \times \cos\alpha}{\pi d^3} \qquad [\because \; T = W \times R \times \cos\alpha]$$

$$= \frac{16 \times 209 \times 45 \times \cos 30°}{\pi \times 9^3} \text{ N/mm}^2 = \textbf{56.9 N/mm}^2. \quad \textbf{Ans.}$$

The bending stress due to bending moment is obtained from

$$\frac{M}{I} = \frac{\sigma_b}{y} \quad \text{or} \quad \frac{WR\sin\alpha}{\frac{\pi}{64} \times d^4} = \frac{\sigma_b}{\left(\frac{d}{2}\right)} \qquad \left[\because \; y = \frac{d}{2}\right]$$

$$\therefore \qquad \sigma_b = (WR \sin \alpha) \times \frac{d}{2} \times \frac{64}{\pi d^4} = \frac{32\, WR \sin \alpha}{\pi d^3}$$

$$= \frac{32 \times 209 \times 45 \times \sin 30°}{\pi \times 9^3} = \mathbf{65.7\ N/mm^2.} \quad \textbf{Ans.}$$

(iii) Axial torque required to produce a bending stress of 50 N/mm²

Let T^* = Axial torque required.

The component of axial torque T^* about axis OY is $T^* \cos \alpha$. It will produce bending moment.

$\therefore$ Bending moment, $M = T^* \cos \alpha$

Using $\frac{M}{I} = \frac{\sigma_b}{y}$

or $$M = \frac{\sigma_b}{y} \times I = \frac{\sigma_b}{d/2} \times \frac{\pi}{64} \times d^4 = \frac{\sigma_b \times \pi \times d^3}{32}$$

or $$T^* \cos \alpha = \frac{\sigma_b \times \pi \times d^3}{32} \qquad (\because M = T^* \cos \alpha)$$

or $$T^* = \frac{50 \times \pi \times 9^3}{32 \times \cos 30°} = \mathbf{4132.18\ N/mm.} \quad \textbf{Ans.}$$

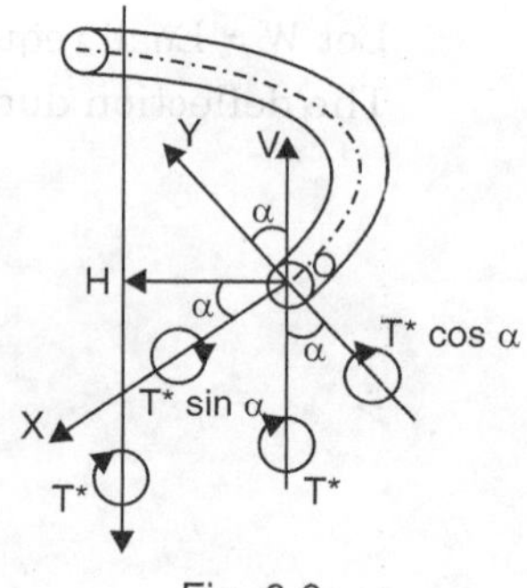

Fig. 6.6

6.7. LAMINATED OR LEAF SPRING

The laminated springs are used to absorb shocks in railway wagons, coaches and road vehicles (such as cars, lorries etc.).

Fig. 6.7 shows a laminated spring which consists of a number of parallel strips of a metal having different lengths and same width, placed one over the other. Initially all the plates are bent to the same radius and are free to slide one over the other. Fig. 6.7 shows the initial position of the spring, which is having some central deflection δ. The spring rests on the axis of the vehicle and its top plate is pinned at the ends to the chassis of the vehicle.

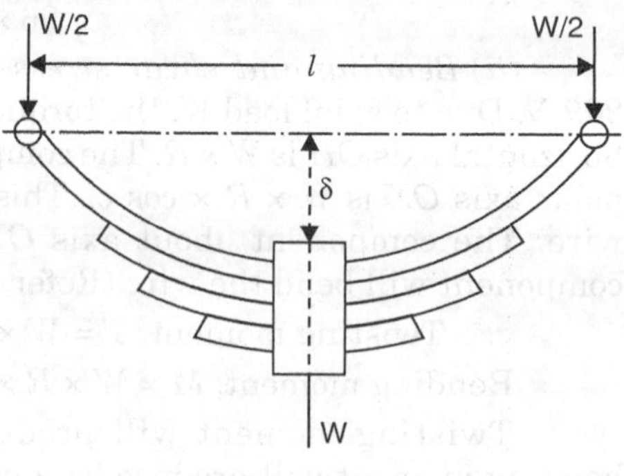

Fig. 6.7

When the spring is loaded to the designed load W, all the plates becomes flat and the central deflection (δ) disappears. This load is called **'Proof Load'.**

6.7.1 Semi-Elliptical Spring. Fig 6.7 shows a semi-elliptical spring.

Let b = Width of each plate

n = Number of plates

l = Span of spring

σ = Maximum bending stress developed in the plates

t = Thickness of each plate

W = Point load acting at the centre of the spring and

δ = Original deflection of the top spring.

(*a*) **Expression for maximum bending stress developed in the plate.** The load W acting at the centre of the lowermost plate, will be shared equally on the two ends of the top plate as shown in Fig. 6.7.

$\therefore$ B.M. at the centre = Load at one end $\times \frac{l}{2}$

or
$$M = \frac{W}{2} \times \frac{l}{2} = \frac{W.l}{4} \quad ...(i)$$

The moment of inertia of each plate,

$$I = \frac{bt^3}{12}$$

But the relation among bending stress (σ), bending moment (M) and moment of inertia (I) is given by

$$\frac{M}{I} = \frac{\sigma}{y} \quad \left(\text{Here } y = \frac{t}{2}\right)$$

or
$$M = \frac{\sigma}{y} \times I = \frac{\sigma \times \frac{bt^3}{12}}{\frac{t}{2}} = \frac{\sigma . bt^2}{6}$$

$\therefore$ Total resisting moment by n plates

$$= n \times M = \frac{n \times \sigma . bt^2}{6} \quad ...(ii)$$

As the maximum B.M. due to load is equal to the total resisting moment, therefore equating (*i*) and (*ii*),

$$\frac{W.l}{4} = \frac{n\sigma . bt^2}{6}$$

$\therefore$
$$\sigma = \frac{6W.l}{4.n.b.t^2} = \frac{3Wl}{2nbt^2} \quad ...(6.24)$$

Equation (6.24) gives the maximum stress developed in the plate of the spring.

(*b*) **Expression for central deflection of the leaf spring**

Now R = Radius of the plate to which they are bent.

From triangle ACO of Fig. 6.8, we have

$$AO^2 = AC^2 + CO^2$$

or
$$R^2 = \left(\frac{l}{2}\right)^2 + (R - \delta)^2$$

$$= \frac{l^2}{4} + R^2 + \delta^2 - 2R\delta$$

$$= \frac{l^2}{4} + R^2 - 2R\delta$$

(Neglecting δ^2 which is a small quantity)

$\therefore$
$$2R\delta = \frac{l^2}{4}$$

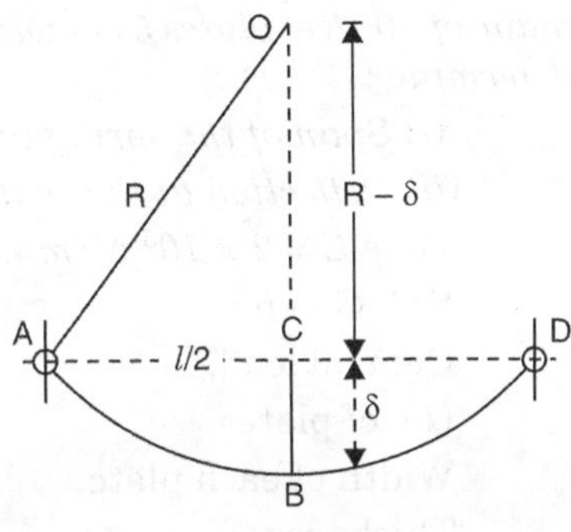

Fig. 6.8

$$\therefore \quad \delta = \frac{l^2}{4 \times 2R} = \frac{l^2}{8R} \qquad \ldots(iii)$$

But the relation between bending stress, modulus of elasticity and radius of curvature (R) is given by

$$\frac{\sigma}{y} = \frac{E}{R}$$

$$\therefore \quad R = \frac{E \times y}{\sigma} = \frac{E \times t}{2\sigma} \qquad \left(\text{Here } y = \frac{t}{2}\right)$$

Substituting this value of R in equation (iii), we get

$$\delta = \frac{l^2 \times 2\sigma}{8 \times E \times t} = \frac{\sigma . l^2}{4Et} = \frac{3Wl}{2nbt^2} \times \frac{l^2}{4Et} = \frac{3Wl^3}{8nEbt^3} \qquad \ldots(6.25)$$

$$\left[\because \quad \sigma = \frac{3Wl}{2n \times bt^2}\right]$$

Equation (6.25) gives the central deflection of the spring.

6.7.2. Quarter-Elliptical Spring. Fig. 6.9 shows a quarter-elliptical spring. This is also known as cantilever laminated spring.

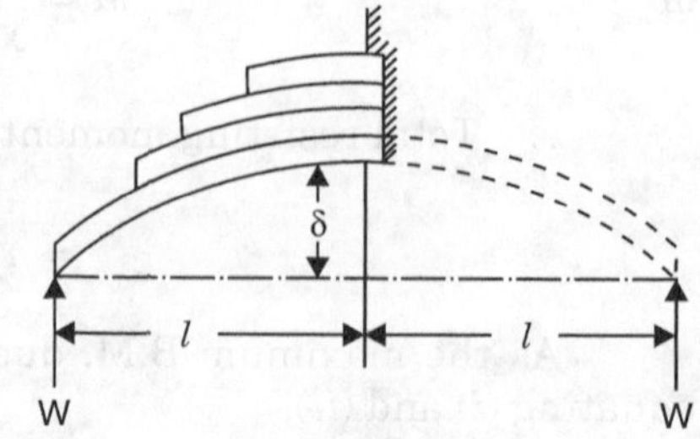

Fig. 6.9

This spring is equivalent to a laminated elliptical type spring of length $2l$ carrying a load $2W$ at the centre. Hence the stress and deflection in this spring can be obtained by replacing l by $2l$ and W by $2W$ for semi-elliptical spring expressions,

$$\therefore \quad \text{Deflection,} \quad \delta = \frac{3\,(2W)\,.\,(2l)^3}{8\,nEbt^3}$$

$$= \frac{6Wl^3}{nEbt^3} \qquad \ldots(6.26)$$

and

$$\text{stress, } \sigma = \frac{3(2W)\,(2l)}{2nbt^2} = \frac{6Wl}{nbt^2} \qquad \ldots(6.27)$$

6.7.3. Solved Problems for Leaf Spring

Problem 6.14. *A leaf spring carries a central load of 3000 N. The leaf spring is to be made of 10 steel plates 5 cm wide and 6 mm thick. If the bending stress is limited to 150 N/mm² determine :*

(i) Span of the spring and

(ii) Deflection at the centre of the spring.

Take $E = 2 \times 10^5$ N/mm².

Sol. Given :

Central load, $W = 3000$ N

No. of plates, $n = 10$

Width of each plate, $b = 5$ cm $= 50$ mm

Thickness, $t = 6$ mm

Bending stress, $\sigma = 150$ N/mm²

Modulus of elasticity, $E = 2 \times 10^5$ N/mm².

Let l = Span of spring

δ = Deflection at the centre of spring.

Using equation (6.24),

$$\sigma = \frac{3Wl}{2nbt^2}$$

or
$$150 = \frac{3 \times 3000 \times l}{2 \times 10 \times 50 \times 6^2}$$

$$\therefore \quad l = \frac{150 \times 2 \times 10 \times 50 \times 6^2}{3 \times 3000} = \mathbf{600\ mm.} \quad \textbf{Ans.}$$

Using equation (6.25) for deflection,

$$\delta = \frac{\sigma . l^2}{4Et} = \frac{150 \times 600^2}{4 \times 2 \times 10^5 \times 6} = \mathbf{11.25\ mm.} \quad \textbf{Ans.}$$

Problem 6.15. *A laminated spring 1 m long is made up of plates each 5 cm wide and 1 cm thick. If the bending stress in the plate is limited to 100 N/mm², how many plates would be required to enable the spring to carry a central point load of 2 kN ? If E = 2.1 × 10⁵ N/mm², what is the deflection under the load ?* (AMIE, Summer 1982)

Sol. Given :

Length of spring, $l = 1\text{ m} = 1000\text{ mm}$

Width of each plate, $b = 5\text{ cm} = 50\text{ mm}$

Thickness of each plate, $t = 1\text{ cm} = 10\text{ mm}$

Bending stress, $\sigma = 100\text{ N/mm}^2$

Central load on spring, $W = 2\text{ kN} = 2000\text{ N}$

Young's modulus, $E = 2.1 \times 10^5\text{ N/mm}^2$

Let n = Number of plates and

δ = Deflection under the load.

Using the equation (6.24),

$$\sigma = \frac{3Wl}{2nbt^2} \quad \text{or} \quad 100 = \frac{3 \times 2000 \times 1000}{2 \times n \times 50 \times 10^2}$$

or
$$n = \frac{3 \times 2000 \times 1000}{100 \times 2 \times 50 \times 100} = \mathbf{6.} \quad \textbf{Ans.}$$

Deflection under load

Using equation (6.25),

$$\delta = \frac{\sigma \times l^2}{4E \times t} = \frac{100 \times 1000^2}{4 \times 2.1 \times 10^5 \times 10} = \mathbf{11.9\ mm.} \quad \textbf{Ans.}$$

Problem 6.16. *For a quarter-elliptical spring, determine the least number of plates if the deflection under gradually applied load of 4.5 kN is not to exceed 100 mm. The span of the spring is 750 mm and width and thickness of each plate is 90 mm and 9 mm respectively. Take E = 2 × 10⁵ N/mm². Find also the maximum stress produced.*

Sol. Given :

Quarter-elliptical spring, $W = 4.5$ kN = 4500 N ; $\delta = 100$ mm ; $l = 750$ mm ; $b = 90$ mm ; $t = 9$ mm and $E = 2 \times 10^5$ N/mm^2.

Let n = least number of plates.

The deflection for quarter-elliptical spring is given by equation (6.26) as

$$\delta = \frac{6Wl^3}{nEbt^3}$$

[If W is in N ; l in mm, E in N/mm^2, b in mm and t in mm, then δ will be in mm.]

or
$$100 = \frac{6 \times 4500 \times (750)^3}{n \times 2 \times 10^5 \times 90 \times 9^3}$$

$$\therefore \quad n = \frac{6 \times 4500 \times (750)^3}{100 \times 2 \times 10^5 \times 90 \times 9^3} = \mathbf{8.68 \text{ say } 9.} \quad \textbf{Ans.}$$

$\therefore$ Least number of plates should be nine.

Maximum stress produced.

The stress for quarter-elliplical spring is given by equation (6.27) as

$$\sigma = \frac{6Wl}{nbt^2}$$

$$= \frac{6 \times 4500 \times 750}{9 \times 90 \times 9^2} = \mathbf{308.64 \text{ N/mm}^2.} \quad \textbf{Ans.}$$

HIGHLIGHTS

1. Springs are the elastic bodies which absorb energy due to resilience. Two important types of springs are :
   1. Laminated or leaf spring and
   2. Helical springs.
2. The maximum stress developed in the plates of a leaf spring is given by,

$$\tau = \frac{3 \,.\, W \,.\, l}{2n \,.\, b \,.\, t^2}$$

where W = Point load acting at the centre of leaf spring,
l = Span of leaf spring or length of leaf spring,
n = Number of plates,
b = Width of each plate, and
t = Thickness of each plate.

3. The central deflection (δ) of the laminated spring is given by,

$$\delta = \frac{\tau \,.\, l^2}{4E \,.\, t}$$

where τ = Maximum stress developed in the plates,
E = Modulus of elasticity,
l = Length of leaf spring, and
t = Thickness of each plate.

4. Helical springs are the thick spring wires coiled into a helix. They are of two types :
1. Close-coiled helical springs and
2. Open-coiled helical spring.

5. The maximum shear stress induced in the wire of a close-coiled helical spring which carries an axial load is given by,

$$\tau = \frac{16 . W . R}{\pi d^3}$$

where W = Axial load on the spring,
R = Mean radius of spring coil, and
d = Diameter of spring wire.

6. For a close-coiled helical spring which carries an axial load, we have

(*i*) strain energy stored,

$$U = \frac{32 . W^2 . R^3 . n}{Cd^4},$$

(*ii*) the deflection of the spring at the centre due to axial load is given as

$$\delta = \frac{64 . W . R^3 . n}{Cd^4} \text{ and}$$

(*iii*) the stiffness of the spring,

$$s = \frac{Cd^4}{64 . R^3 . n}$$

where W = Axial load on the spring,
n = Number of coils,
C = Modulus of rigidity, and
R = Mean radius of spring coil.

EXERCISE 6

(A) Theoretical Questions

1. What is a spring ? Name the two important types of spring.

2. Prove that the maximum stress developed in the plates of a leaf spring is given by

$$\sigma = \frac{3W . l}{2n . b . t^2}$$

where W = Point load acting at the centre of leaf spring,
l = Span of leaf spring or length of the leaf spring,
n = Number of plates,
b = Width of each plate, and
t = Thickness of each plate.

3. Prove that the central deflection of the laminated spring (or leaf spring) is given by

$$\delta = \frac{\sigma . l^2}{4Et}$$

where σ = Maximum stress developed in plates,
E = Modulus of elasticity,
l = Length of leaf spring, and
t = Thickness of each plate.

4. Define helical springs. Name the two important types of helical springs.
5. Prove that the maximum shear stress induced in the wire of a close-coiled helical spring is given by

$$\tau = \frac{16 . W . R}{\pi d^3}$$

where τ = Maximum shear stress induced in the wire,
W = Axial load on spring,
R = Mean radius of spring coil, and d = Diameter of spring wire.
6. Find an expression for the strain energy stored by the close-coiled helical spring when subjected to axial load W.
7. Prove that the deflection of a close-coiled helical spring at the centre due to axial load W is given by

$$\delta = \frac{64 . W . R^3 . n}{Cd^4}$$

where R = Mean radius of spring coil,
n = Number of coils,
C = Modulus of rigidity, and
d = Diameter of spring wire.

(B) Numerical Problems

1. A leaf spring carries a central load of 2.5 kN. The leaf spring is to be made of 10 steel plate 6 cm wide and 5 mm thick. If the bending stress is limited to 100 N/mm² determine : (*i*) length of the spring and deflection at the centre of the spring. Take $E = 2 \times 10^5$ N/mm². [**Ans.** 40 cm, 0.4 cm]
2. A laminated spring 0.9 m long is made up of plates each 5 cm wide and 1 cm thick. If the bending stress in the plate is limited to 120 N/mm², how many plates would be required to enable the spring to carry a central point load of 2.65 kN ? If $E = 2.0 \times 10^5$ N/mm², what is the deflection under the load ? [**Ans.** 6 plates, 1.215 cm]
3. A closely-coiled helical spring is to carry a load of 1 kN. Its mean coil diameter is to be 10 times that of wire diameter. Calculate these diameters if the maximum shear stress in the material of the spring is to be 90 N/mm². [**Ans.** 16.82 cm and 1.68 cm]
4. In question 3, if the stiffness of the spring is 20 N/mm deflection and modulus of rigidity = 8.4×10^4 N/mm², find the number of coils in the closely-coiled helical spring. [**Ans.** 9]
5. A closely-coiled helical spring of round steel wire 8 mm in diameter having 10 complete turns with a mean diameter of 10 cm is subjected to an axial load of 250 N. Determine : (*i*) the deflection of the spring, (*ii*) Maximum shear stress in the wire and (*iii*) stiffness of the spring. Take $C = 8 \times 10^4$ N/mm². [**Ans.** (*i*) 6.1 cm, (*ii*) 124.34 N/mm², (*iii*) 4.1 N/mm²]
6. A close-coiled helical spring of 10 cm mean diameter is made up of 1.0 cm diameter rod and has 20 turns. The spring carried an axial load of 300 N. Determine the shearing stress. Taking the value of modulus of rigidity = 8.2×10^4 N/mm², determine the deflection when carrying this load. Also calculate the stiffness of the spring and frequency of free vibration for a mass hanging from it. [**Ans.** 76.4 N/mm², 5.71 cm, 5.25 N/mm and 2.08 cycles/sec]

7

Columns and Struts

(A) COMBINED BENDING AND DIRECT STRESSES

7.1. BUCKLING AND STABILITY

If a long column fixed at both ends, is subjected to an axial compressive load, compressive stress will be produced. If the compressive load is gradually increased, a stage will reach when the column starts bending (also known buckling). This process is known as **buckling**. The load at which the column just buckles, is known as buckling load. Buckling takes about the axis having minimum radius of gyration or least moment of inertia.

The cement concrete columns are weak in tension. Hence a column will be stable when no tensile stress is produced anywhere in the section of the column. When an eccentric load is acting on a column, it produces direct stress as well as bending stress. The direct stress is compressive and is uniform along the cross-section of the column. The bending stress is compressive to one side of neutral axis whereas it is tensile on the other side of axis. The resultant stress at any point in the section is the algebraic sum of direct stress and bending stress.

The column will be **stable** when the load is applied on these columns in such a way that there is no tensile stress anywhere in the section. This process is known as stability.

7.2. SLENDERNESS RATIO

The slenderness ratio is defined as the ratio of the actual length of a column to its least radius of gyration.

$$\text{Mathematically, slenderness ratio} = \frac{\text{Actual length}}{\text{Least radius of gyration}} = \frac{L}{k}.$$

7.3. COMBINED BENDING AND DIRECT STRESSES

Direct stress alone is produced in a body when it is subjected to an axial tensile or compressive load. And bending stress is produced in the body, when it is subjected to a bending moment. But if a body is subjected to axial loads and also bending moments, then both the stresses (*i.e.*, direct and bending stresses) will be produced in the body. Both these stresses act normal to a cross-section, hence the two stresses may be algebraically added into a single resultant stress.

Consider the case of a column* subjected by a compressive load P acting along the axis of the column as shown in Fig. 7.1. This load will cause a direct compressive stress whose intensity will be uniform across the cross-section of the column.

---

* Column is a vertical member subjected to a compressive load.

Let σ_0 = Intensity of the stress

A = Area of cross-section

P = Load acting on the column.

Then stress, $\sigma_0 = \frac{\text{Load}}{\text{Area}} = \frac{P}{A}$

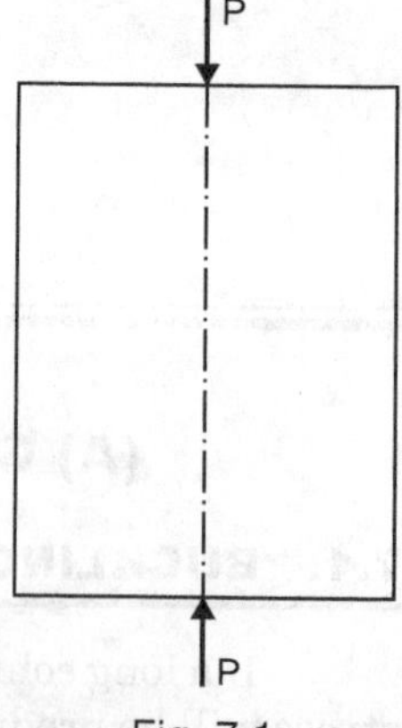

Fig. 7.1

Now consider the case of a column subjected by a compressive load P whose line of action is at a distance of 'e' from the axis of the column as shown in Fig. 7.2 (a). Here 'e' is known as eccentricity of the load. The eccentric load shown in Fig. 7.2 (a) will cause direct stress and bending stress. This is proved as discussed below :

1. In Fig. 7.2 (b), we have applied, along the axis of the column at point O, two equal and opposite forces P. The condition is not changed, because two equal and opposite forces at a point is equivalent to zero. Thus three forces are acting now on the column. One of the forces is shown in Fig. 7.2 (c) and the other two forces are shown in Fig. 7.2 (d).

2. The force shown in Fig. 7.2 (c) is acting along the axis of the column and hence this force will produce a direct stress.

3. The forces shown in Fig. 7.2 (d) will form a couple, whose moment will be $P \times e$. This couple will produce a bending stress.

Hence an eccentric* load will produce a direct stress as well as a bending stress. By adding these two stresses algebraically, a single resultant stress can be obtained.

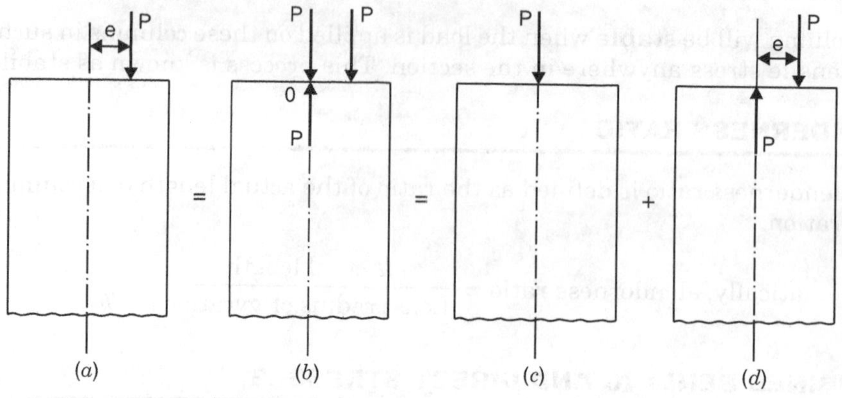

Fig. 7.2

7.4. RESULTANT STRESS WHEN A COLUMN OF RECTANGULAR SECTION IS SUBJECTED TO AN ECCENTRIC LOAD

A column of rectangular section subjected to an eccentric load is shown in Fig. 7.3. Let the load is eccentric with respect to the axis Y-Y as shown in Fig. 7.3 (b). It is acting at a distance 'e' from axis Y-Y. It is mentioned in Art. 7.3 that an eccentric load causes direct stress as well as bending stress. Let us calculate these stresses.

* Eccentric load is a load whose line of action does not coincide with the axis of the column.

Let P = Eccentric load on column

e = Eccentricity of the load

σ_0 = Direct stress

σ_b = Bending stress

b = Width of column

d = Depth of column

$\therefore$ Area of column section, $A = b \times d$

Now moment due to eccentric load P is given by,

$$M = \text{Load} \times \text{eccentricity}$$
$$= P \times e$$

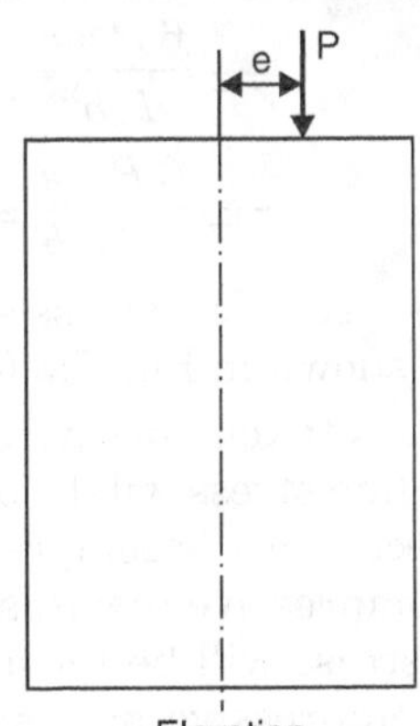

Fig. 7.3 (*a*)

The direct stress (σ_0) is given by,

$$\sigma_0 = \frac{\text{Load } (P)}{\text{Area}} = \frac{P}{A} \quad ...(i)$$

This stress is compressive and is uniform along the cross-section of the column as shown in Fig. 7.3 (*c*).

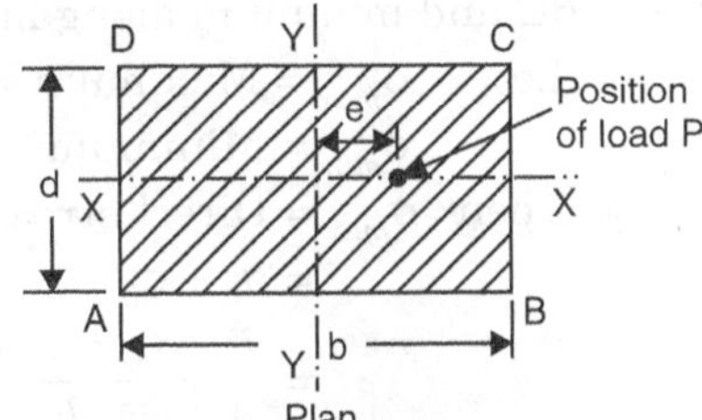

Fig. 7.3 (*b*)

The compressive stress is taken negative and hence it is acting downwards from base line.

The bending stress σ_b due to moment at any point of the column section at a distance y from the neutral axis Y-Y is given by

$$\frac{M}{I} = \frac{\sigma_b}{\pm y}$$

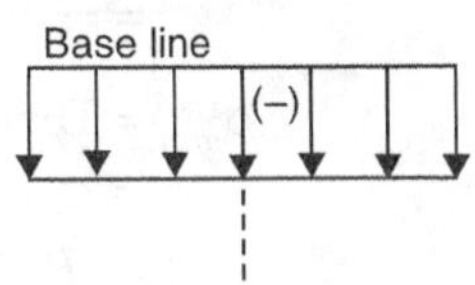

Fig. 7.3 (*c*) Direct stress across the section

Here the distance from the axis Y-Y is along axis X-X. Hence the above equation becomes

$$\frac{M}{I} = \frac{\sigma_b}{\pm x}$$

$$\therefore \quad \sigma_b = \pm \frac{M}{I} \times x \quad ...(ii)$$

where I = Moment of inertia of the column section about the neutral axis Y-Y = $\frac{d \, . \, b^3}{12}$

Substituting the value of I in equation (*ii*), we get

$$\sigma_b = \pm \frac{M}{\frac{d \, . \, b^3}{12}} \times x = \pm \frac{12 \, M}{d \, . \, b^3} \times x$$

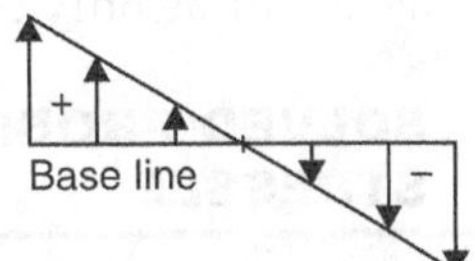

Fig. 7.3 (*d*) Bending stress across the section.

The bending stress depends upon the value of x from the axis Y-Y and its variation is linear as shown in Fig. 7.3 (*d*). On the axis Y-Y, the value of x is zero, hence bending stress is zero. On the lines BC and AD, x is maximum but of opposite sign. Hence bending stress will be maximum on lines BC and AD, but of opposite sign.

The bending stress at the extreme is obtained by substituting $x = \frac{b}{2}$ in the above equation.

$$\therefore \quad \sigma_b = \pm \frac{12 \, M}{d \, . \, b^3} \times \frac{b}{2} = \pm \frac{6 \, M}{d \, . \, b^2}$$

$$= \pm \frac{6\,P \times e}{d\,.\,b^2} \qquad (\because \quad M = P \times e)$$

$$= \pm \frac{6\,P \times e}{d\,.\,b\,.\,b} = \pm \frac{6\,P \times e}{A \times b} \qquad (\because \quad \text{Area} = b \times d = A)$$

The resultant stress at any point will be the algebraic sum of direct stress and bending stress as shown in Fig. 7.3 (*e*).

If x is taken positive on the same side of *Y-Y* as the load, then bending stress will be of the same type as the direct stress. Here direct stress is compressive and hence bending stress will also be compressive towards the right of the axis *Y-Y*. Similarly bending stress will be tensile towards the left of the axis *Y-Y*. Taking compressive stress as negative and tensile stress as positive we can find the maximum and minimum stress at the extremities of the section. The stress will be maximum along layer *BC* and minimum along layer *AD*.

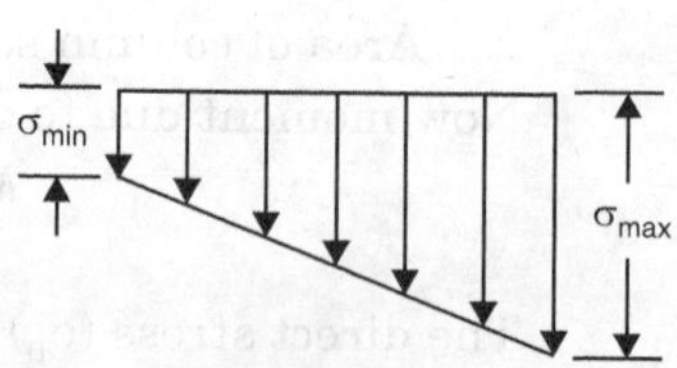

Fig. 7.3 (*e*) Resultant stress across the section.

Let σ_{max} = Maximum stress (*i.e.*, stress along *BC*)

σ_{min} = Minimum stress (*i.e.*, stress along *AD*)

Then σ_{max} = Direct stress + Bending stress

$$= \sigma_0 + \sigma_b$$

$$= \frac{P}{A} + \frac{6\,P\,.\,e}{A\,.\,b}$$

$$= \frac{P}{A}\left(1 + \frac{6 \times e}{b}\right) \qquad \text{...(7.1)}$$

and σ_{min} = Direct stress − Bending stress

$$= \sigma_0 - \sigma_b$$

$$= \frac{P}{A} - \frac{6\,P\,.\,e}{A\,.\,b} = \frac{P}{A}\left(1 - \frac{6 \times e}{b}\right) \qquad \text{...(7.2)}$$

These stresses are shown in Fig. 7.3 (*e*). The resultant stress along the width of the column will vary by a straight line law.

If in equation (7.2), σ_{min} is negative then the stress along the layer *AD* will be tensile. If σ_{min} is zero then there will be no tensile stress along the width of the column. If σ_{min} is positive then there will be only compressive stress along the width of the column.

7.5. SOLVED PROBLEMS 7.1 TO 7.7 FOR COMBINED BENDING AND DIRECT STRESSES

Problem 7.1. *A rectangular column of width 200 mm and of thickness 150 mm carries a point load of 240 kN at an eccentricity of 10 mm as shown in Fig. 7.4 (i). Determine the maximum and minimum stresses on the section.*

Sol. Given :

Width, b = 200 mm

Thickness, d = 150 mm

∴ Area, $A = b \times d$

$= 200 \times 150 = 30000$ mm^2

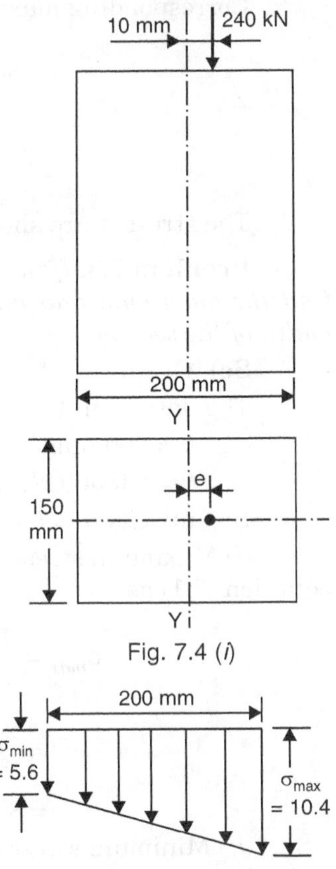

Fig. 7.4 (*i*)

Eccentric load,

$$P = 240 \text{ kN}$$
$$= 240000 \text{ N}$$

Eccentricity,

$$e = 10 \text{ mm}$$

Let σ_{max} = Maximum stress, and

σ_{min} = Minimum stress.

(*i*) Using equation (7.1), we get

$$\sigma_{max} = \frac{P}{A}\left(1 + \frac{6 \times e}{b}\right)$$

$$= \frac{240000}{30000}\left(1 + \frac{6 \times 10}{200}\right)$$

$$= 8(1 + 0.3) = \mathbf{10.4\ N/m^2}. \quad \textbf{Ans.}$$

(*ii*) Using equation (7.2), we get

$$\sigma_{min} = \frac{P}{A}\left(1 - \frac{6 \times e}{b}\right)$$

$$= \frac{240000}{30000}\left(1 - \frac{6 \times 10}{200}\right)$$

$$= 8(1 - 0.3) = \mathbf{5.6\ N/mm^2}. \quad \textbf{Ans.}$$

These stresses are shown in Fig. 7.4 (*ii*).

Fig. 7.4 (*ii*)

Problem 7.2. *If in problem 7.1, the minimum stress on the section is given zero then find the eccentricity of the point load of 240 kN acting on the rectangular column. Also calculate the corresponding maximum stress on the section.*

Sol. Given :

The data from Problem 7.1 is :

b = 200 mm, d = 150 mm, P = 240000 N, A = 30000 mm^2

Minimum stress,

$$\sigma_{min} = 0$$

Let e = Eccentricity

Using equation (7.2), we get

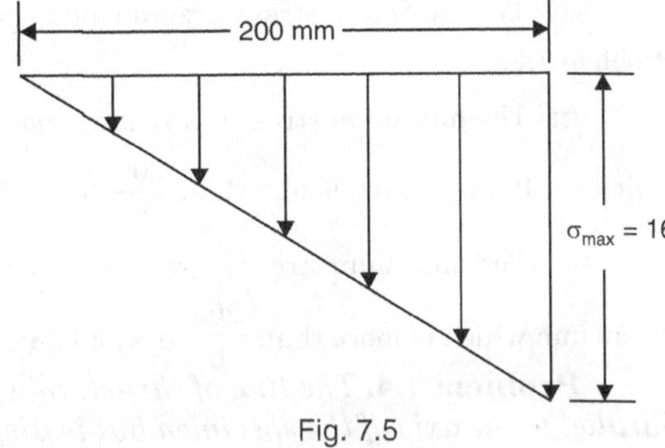

Fig. 7.5

$$\sigma_{min} = \frac{P}{A}\left(1 - \frac{6 \times e}{b}\right)$$

or
$$0 = \frac{240000}{30000}\left(1 - \frac{6 \times e}{200}\right)$$

or
$$1 - \frac{6 \times e}{200} = 0 \quad \text{or} \quad 1 = \frac{6 \times e}{200}$$

$$\therefore \quad e = \frac{200}{6} = \mathbf{33.33\ mm.} \quad \textbf{Ans.}$$

Corresponding maximum stress is obtained by using equation (7.1).

$$\therefore \quad \sigma_{max} = \frac{P}{A}\left(1+\frac{6\times e}{b}\right) = \frac{240000}{30000}\left(1+\frac{6\times 200}{\frac{6}{200}}\right)$$

$$= 8(1 + 1) = \mathbf{16\ N/mm^2}$$

The stresses are shown in Fig. 7.5.

Problem 7.3. *If in Problem 7.1, the eccentricity is given 50 mm instead of 10 mm then find the maximum and minimum stresses on the section. Also plot these stresses along the width of the section.*

Sol. Given :

The data from Problem 7.1 is :

$b = 200$ mm, $d = 150$ mm

$P = 240000$ N, $A = 30000$ mm^2

Eccentricity, $e = 50$ mm

(*i*) Maximum stress (σ_{max}) is given by equation (7.1) as

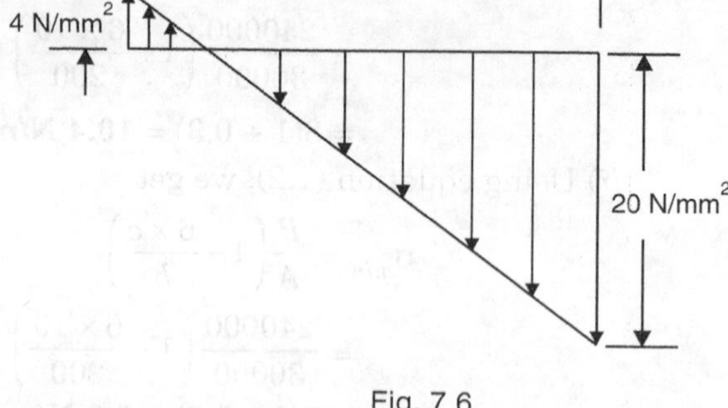

Fig. 7.6

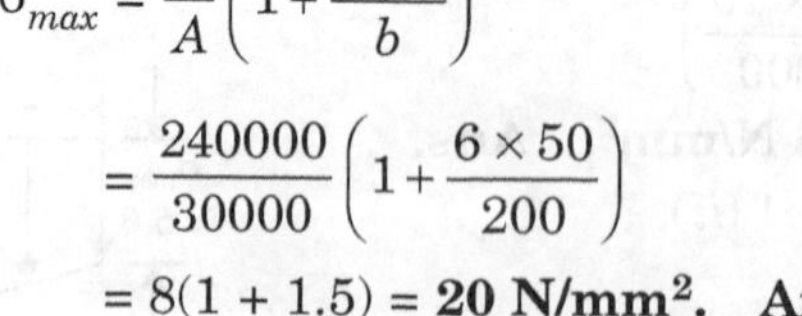

$$\sigma_{max} = \frac{P}{A}\left(1+\frac{6\times e}{b}\right)$$

$$= \frac{240000}{30000}\left(1+\frac{6\times 50}{200}\right)$$

$$= 8(1 + 1.5) = \mathbf{20\ N/mm^2}. \quad \textbf{Ans.}$$

(*ii*) Minimum stress (σ_{min}) is given by equation (7.2) as

$$\sigma_{min} = \frac{P}{A}\left(1-\frac{6\times e}{b}\right)$$

$$= \frac{240000}{30000}\left(1-\frac{6\times 50}{200}\right) = 8(1 - 1.5) = \mathbf{-\ 4\ N/mm^2}. \quad \textbf{Ans.}$$

Negative sign means tensile stress.

The stresses are plotted as shown in Fig. 7.6.

Note. From the above three problems, we have

(*i*) The minimum stress is zero when $e = \frac{200}{6}$ mm or $\frac{b}{6}$ mm (as $b = 200$). This is clear from Problem 7.2.

(*ii*) The minimum stress is +ve (*i.e.*, compressive) when $e < \frac{b}{6}$. This is clear from Problem 7.1 in which $e = 10$ mm which is less than $\frac{200}{6}$ (*i.e.*, 33.33).

(*iii*) The minimum stress is –ve (*i.e.*, tensile) when $e > \frac{b}{6}$. This is clear from Problem 7.3 in which $e = 50$ mm which is more than $\frac{200}{6}$ (*i.e.*, 33.33).

Problem 7.4. *The line of thrust, in a compression testing specimen 15 mm diameter, is parallel to the axis of the specimen but is displaced from it. Calculate the distance of the line of thrust from the axis when the maximum stress is 20% greater than the mean stress on a normal section.*

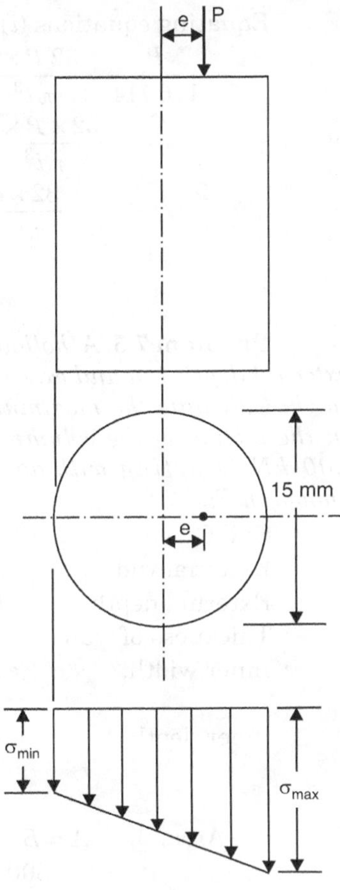

Fig. 7.7

Sol. Given :

Diameter, $d = 15$ mm

$\therefore$ Area, $A = \frac{\pi}{4} \times 15^2$

$= 176.714 \text{ mm}^2$

σ_{max} = 20% greater than mean

$= \frac{120}{100} \times$ mean stress

$= 1.2 \times$ mean stress.

Let P = Compressive load on specimen

e = Eccentricity

$$\text{Mean stress} = \frac{\text{Load}}{\text{Area}} = \frac{P}{176.714} \text{ N/mm}^2$$

We know that moment,

$$M = P \times e$$

Now bending stress is given by

$$\frac{M}{I} = \frac{\sigma_b}{x}$$

$$\therefore \quad \sigma_b = \frac{M}{I} \times x$$

$\therefore$ Maximum bending stress will be when $x = \pm \frac{d}{2}$.

Hence maximum bending stress is given by,

$$\sigma_b = \frac{M}{I} \times \left(\pm \frac{d}{2}\right)$$

$$= \pm \frac{M}{I} \times \frac{d}{2}$$

$$= \pm \frac{M}{\frac{\pi}{64} d^4} \times \frac{d}{2} \quad \left(\because \ I = \frac{\pi}{64} d^4\right)$$

$$= \pm \frac{32 M}{\pi d^3}$$

$$= \pm \frac{32 P \times e}{\pi d^3} \quad (\because \ M = P \times e)$$

Direct stress due to load is given by,

$$\sigma_0 = \frac{P}{A} = \frac{P}{176.714}$$

$\therefore$ Maximum stress = Direct stress × Bending stress

$= \sigma_0 + \sigma_b$

or $$\sigma_{max} = \frac{P}{176.714} + \frac{32 P \times e}{\pi d^3} \quad \text{...}(i)$$

But $\sigma_{max} = 1.2 \times$ Mean stress (given)

$$= 1.2 \times \frac{P}{176.714} \quad \text{...}(ii) \quad \left(\because \ \text{Mean stress} = \frac{P}{176.714}\right)$$

Equating equations (*i*) and (*ii*), we get

$$\frac{P}{176.714} + \frac{32\,P \times e}{\pi d^3} = 1.2 \times \frac{P}{176.714}$$

or
$$\frac{32 \times P \times e}{\pi d^3} = \frac{1.2\,P}{176.714} - \frac{P}{176.714} = \frac{0.2\,P}{176.714}$$

or
$$\frac{32 \times e}{\pi d^3} = \frac{0.2}{176.714} \quad \text{(Cancelling } P \text{ to both sides)}$$

$\therefore$
$$e = \frac{0.2 \times \pi \times d^3}{32 \times 176.714} = \frac{0.2 \times \pi \times 15^3}{32 \times 176.714} = \mathbf{0.375\ mm.} \quad \textbf{Ans.}$$

Problem 7.5. *A hollow rectangular column of external depth 1 m and external width 0.8 m is 10 cm thick. Calculate the maximum and minimum stress in the section of the column if a vertical load of 200 kN is acting with an eccentricity of 15 cm as shown in Fig. 7.8.*

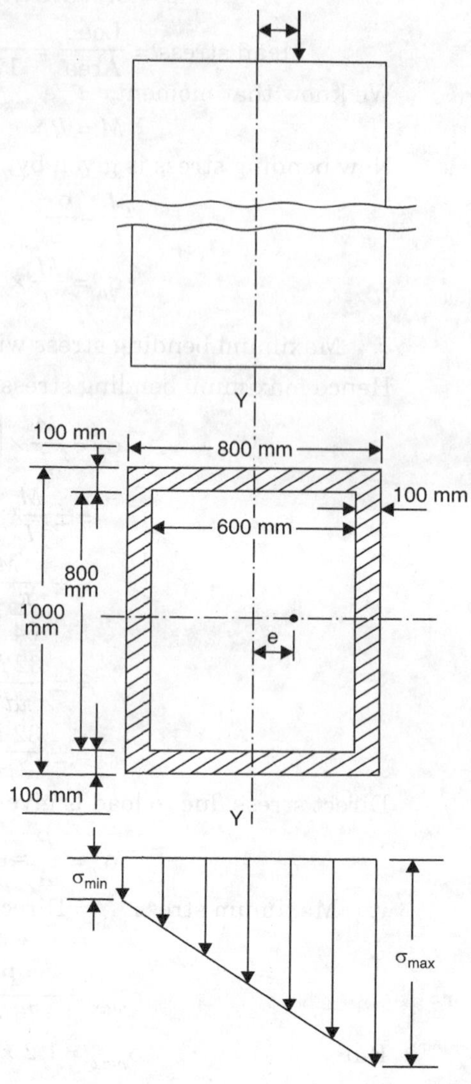

Fig. 7.8

Sol. Given :

External width, $B = 0.8$ m = 800 mm

External depth, $D = 1.0$ m = 1000 mm

Thickness of walls, $t = 10$ cm = 100 mm

Inner width, $b = B - 2 \times 100$

$= 800 - 200 = 600$ mm

Inner depth, $d = D - 2 \times t$

$= 1000 - 2 \times 100$

$= 800$ mm

$\therefore$ Area, $A = B \times D - b \times d$

$= 800 \times 1000 - 600 \times 800$

$= 800000 - 480000$

$= 320000\ \text{mm}^2$

M.O.I. about *Y-Y* axis is given by,

$$I = \frac{1000 \times 800^3}{12} - \frac{800 \times 600^3}{12}$$

$= 42.66 \times 10^9 - 14.4 \times 10^9$

$= 28.26 \times 10^9\ \text{mm}^4$

Eccentric load, $P = 200$ kN = 200,000 N

Eccentricity, $e = 15$ cm = 150 mm

We know that the moment,

$M = P \times e$

$= 200{,}000 \times 150$

$= 30{,}000{,}00$ Nmm

The bending stress is given by,

$$\frac{M}{I} = \frac{\sigma_b}{x}$$

$\therefore$
$$\sigma_b = \frac{M}{I} \times x$$

Maximum bending stress will be when

$$x = \pm 400$$

$$\therefore \quad \sigma_b = \frac{M}{I} \times (\pm 400)$$

$$= \pm \frac{30000000}{28.26 \times 10^9} \times 400 = \pm 0.4246 \text{ N/mm}^2$$

Direct stress is given by,

$$\sigma_0 = \frac{P}{A} = \frac{200000}{320000} = 0.625 \text{ N/mm}^2$$

$\therefore$ Maximum stress $= \sigma_0 + \sigma_b = 0.625 + 0.4246$

$= \mathbf{1.0496\ N/mm^2}$ **(Compressive). Ans.**

Minimum stress $= \sigma_0 - \sigma_b = 0.625 - 0.4246$

$= \mathbf{0.2004\ N/mm^2}$ **(Compressive). Ans.**

Problem 7.6. *A short column of external diameter 40 cm and internal diameter 20 cm carries an eccentric load of 80 kN. Find the greatest eccentricity which the load can have without producing tension on the cross-section.*

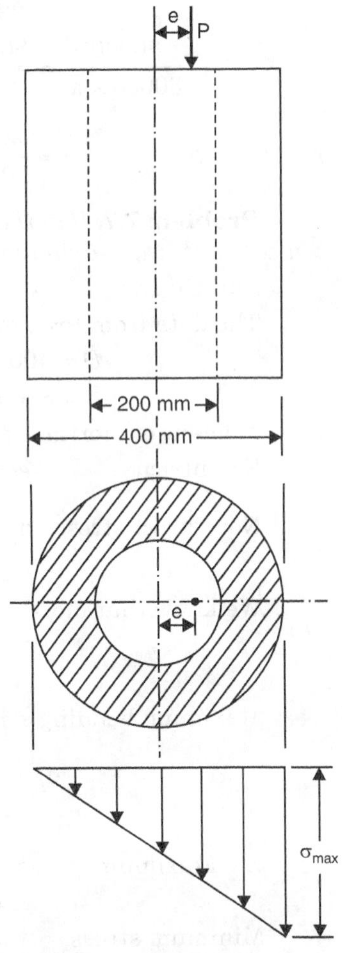

Fig. 7.9

Sol. Given :

External dia., $D = 40$ cm $= 400$ mm

Internal dia., $d = 20$ cm $= 200$ mm

$\therefore$ Area of cross-section,

$$A = \frac{\pi}{4}(D^2 - d^2)$$

$$= \frac{\pi}{4}(400^2 - 200^2) = 30000 \times \pi \text{ mm}^2$$

Moment of inertia

$$I = \frac{\pi}{64}(D^4 - d^4)$$

$$= \frac{\pi}{64}(400^4 - 200^4) = 3.75 \times 10^8 \times \pi \text{ mm}^4$$

Eccentric load, $P = 80$ kN $= 80000$ N

Let e = Eccentricity when there is no tension.

Now direct stress, $\sigma_0 = \dfrac{P}{A} = \dfrac{80000}{30000 \times \pi}$ ...(*i*)

We know that moment,

$$M = P \times e = 80000 \times e$$

Now bending stress (σ_b) is given by

$$\frac{M}{I} = \frac{\sigma_b}{x}$$

$$\therefore \quad \sigma_b = \frac{M \times x}{I}$$

The bending stress will be maximum when

$$x = \pm \frac{D}{2} = \pm \frac{400}{2} = \pm 200 \text{ mm}$$

∴ Maximum bending stress is given by,

$$\sigma_b = \frac{M \times (\pm 200)}{I} = \pm \frac{M \times 200}{I}$$

$$= \pm \frac{80000 \times e \times 200}{3.75 \times 10^8 \times \pi} \quad ...(ii)$$

Now minimum stress is given by,

$$\sigma_{min} = \sigma_0 - \sigma_b$$

$$= \frac{80000}{30000 \times \pi} - \frac{80000 \times e \times 200}{3.75 \times 10^8 \times \pi}$$

There will be no tension if $\sigma_{min} = 0$

∴ For no tension, we have

$$0 = \frac{80000}{30000 \times \pi} - \frac{80000 \times e \times 200}{3.75 \times 10^8 \times \pi}$$

or $$\frac{80000}{30000 \times \pi} = \frac{80000 \times e \times 200}{3.75 \times 10^8 \times \pi}$$

or $$e = \frac{3.75 \times 10^8 \times \pi \times 80000}{30000 \times \pi \times 80000 \times 200} = \textbf{62.5 mm.}$$ **Ans.**

Problem 7.7. *If in the Problem 7.6, the eccentricity of the point load is given as 150 mm, then calculate the maximum and minimum stress in the section.*

Sol. Given :

The data from Problem 7.6 is :

$D = 400$ mm, $d = 200$ mm

$P = 80000$ N, $A = 30000 \times \pi$ mm^2

Moment of inertia, $I = 3.75 \times 10^8 \times \pi$ mm^4

Eccentricity, $e = 150$ mm

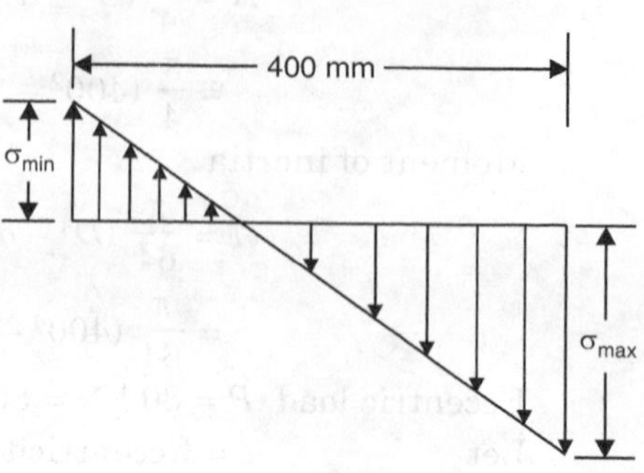

Fig. 7.10

Now direct stress, $\sigma_0 = \frac{P}{A} = \frac{80000}{30000 \times \pi}$

$= 0.8488$ N/mm^2

We know that moment,

$M = P \times e = 80000 \times 150$

$= 12000000$ Nmm

Maximum bending stress is given by,

$$\sigma_b = \frac{M \times y_{max}}{I} = \frac{12000000 \times (\pm 200)}{3.75 \times 10^8 \times \pi} \quad (\because \; y_{max} = \pm 200 \text{ mm})$$

$= \pm 2.037$ N/mm^2

∴ Maximum stress $= \sigma_0 + \sigma_b$

$= 0.8488 + 2.037 =$ **2.8858 N/mm² (Compressive). Ans.**

Minimum stress $= \sigma_0 - \sigma_b$

$= 0.8488 - 2.037 =$ **– 1.1882 N/mm² (Tensile). Ans.**

The stress distribution across the width is shown in Fig. 7.10.

7.6. RESULTANT STRESS WHEN A COLUMN OF RECTANGULAR SECTION IS SUBJECTED TO A LOAD WHICH IS ECCENTRIC TO BOTH AXES

A column of rectangular section *ABCD*, subjected to a load which is eccentric to both axes, is shown in Fig. 7.11.

Let P = Eccentric load on column

e_x = Eccentricity of load about *X-X* axis *i.e.*, distance from *X-X* axis

e_y = Eccentricity of load about *Y-Y* axis *i.e.*, distance from *Y-Y* axis

b = Width of column

d = Depth of column

σ_0 = Direct stress

$(\sigma_b)_{xx}$ = Bending stress due to eccentricity e_x

$(\sigma_b)_{yy}$ = Bending stress due to eccentricity e_y

M_x = Moment of load about *X-X* axis

$= P \times e_x$

M_y = Moment of load about *Y-Y* axis

$= P \times e_y$

I_{xx} = Moment of inertia about *X-X* axis

$= \dfrac{bd^3}{12}$

I_{yy} = Moment of inertia about *Y-Y* axis $= \dfrac{db^3}{12}$

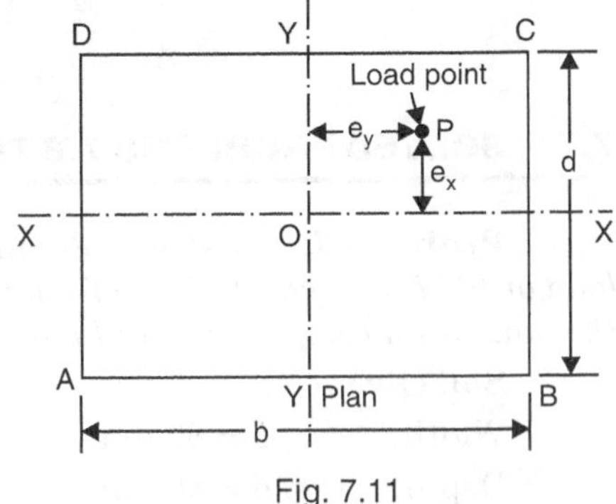

Fig. 7.11

Now, the eccentric load is equivalent to a central load P, together with a bending moment $P \times e_y$ about *Y-Y* and a bending moment $P \times e_x$ about *X-X*.

(*i*) The direct stress (σ_0) is given by,

$$\sigma_0 = \frac{P}{A} \qquad \text{...(}i\text{)}$$

(*ii*) The bending stress at a distance x from axis *Y-Y* due to eccentricity e_y is given by,

$$(\sigma_b)_{yy} = \frac{M_y \times x}{I_{yy}} = \frac{P \times e_y \times x}{I_{yy}} \qquad (\because \; M_y = P \times e_y) \quad \text{...(}ii\text{)}$$

In the above equation x varies from $-\dfrac{b}{2}$ to $+\dfrac{b}{2}$

(*iii*) The bending stress at a distance y from axis *X-X* due to eccentricity e_x is given by,

$$(\sigma_b)_{xx} = \frac{M_x \times y}{I_{xx}} = \frac{P \times e_x \times y}{I_{xx}}$$

In the above equation, y varies from $-\dfrac{d}{2}$ to $+\dfrac{d}{2}$

The resultant stress at any point on the section

$$= \sigma_0 \pm (\sigma_b)_{yy} \pm (\sigma_b)_{xx}$$

$$= \frac{P}{A} \pm \frac{M_y \times x}{I_{yy}} \pm \frac{M_x \cdot y}{I_{xx}} \qquad \text{...(7.3)}$$

(*i*) At the point *C*, the co-ordinates *x* and *y* are positive hence the resultant stress will be maximum.

(*ii*) At the point *A*, the co-ordinates *x* and *y* are negative and hence the resultant stress will be minimum.

(*iii*) At the point *B*, *x* is +ve and *y* is –ve and hence resultant stress

$$= \frac{P}{A} + \frac{M_y \cdot x}{I_{yy}} - \frac{M_x \cdot y}{I_{xx}}$$

(*iv*) At the point *D*, *x* is –ve and *y* is +ve and hence resultant stress

$$= \frac{P}{A} - \frac{M_y \cdot x}{I_{yy}} + \frac{M_x \cdot y}{I_{xx}}.$$

7.7. SOLVED PROBLEMS 7.8 TO 7.10 WHEN LOAD IS ECCENTRIC TO BOTH AXES

Problem 7.8. *A short column of rectangular cross-section 80 mm by 60 mm carries a load of 40 kN at a point 20 mm from the longer side and 35 mm from the shorter side. Determine the maximum compressive and tensile stresses in the section.*

Sol. Given :

Width, $b = 80$ mm

Depth, $d = 60$ mm

∴ Area, $A = 80 \times 60 = 4800$ mm^2

Point load, $P = 40$ kN $= 40000$ N

Eccentricity of load about *X-X* axis,

$e_x = 10$ mm

Eccentricity of load about *Y-Y* axis,

$e_y = 5$ mm

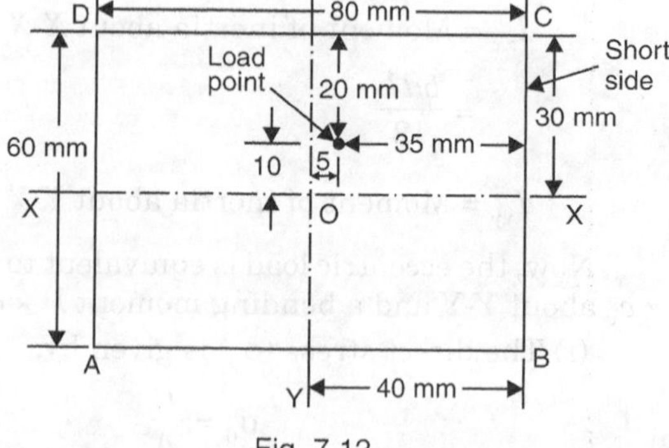

Fig. 7.12

Moment of load about *X-X* axis,

$M_x = P \times e_x = 40000 \times 10$

$= 400000$ Nmm

Moment of load about *Y-Y* axis,

$M_y = P \times e_y = 40000 \times 5 = 200000$ Nmm

Moment of inertia about *X-X* axis,

$$I_{xx} = \frac{1}{12} \times 80 \times 60^3 = 1440000 \text{ mm}^4$$

Similarly, $I_{yy} = \frac{1}{12} \times 60 \times 80^3 = 2560000$ mm^4

(*i*) The maximum compressive stress will be at point *C* where *x* and *y* are positive. The value of $x = 40$ mm and $y = 30$ mm at *C*.

Hence maximum compressive stress is given by equation (7.3)

$$= \frac{P}{A} + \frac{M_y \times x}{I_{yy}} + \frac{M_x \times y}{I_{xx}} \quad \text{(Taking +ve sign)}$$

$$= \frac{40000}{4800} + \frac{200000 \times 40}{2560000} + \frac{400000 \times 30}{1440000}$$

$= 8.33 + 3.125 + 8.33 =$ **19.785 N/mm². Ans.**

(*ii*) The maximum tensile stress will be at point A where $x = -40$ mm and $y = -30$ mm. Hence, using equation (7.3), we get

$$\text{Resultant stress at } A = \frac{P}{A} + \frac{M_y \times x}{I_{yy}} + \frac{M_x \times y}{I_{xx}} = \frac{P}{A} + \frac{M_y \times (-40)}{I_{yy}} + \frac{M_x \times (-30)}{I_{xx}}$$

$$= \frac{40000}{4800} - \frac{200000 \times 40}{2560000} - \frac{400000 \times 30}{1440000}$$

$$= 8.33 - 3.125 - 8.33 = \mathbf{-3.125\ N/mm^2}. \quad \textbf{Ans.}$$

Problem 7.9. *A column is rectangular in cross-section of 300 mm × 400 mm in dimensions. The column carries an eccentric point load of 360 kN on one diagonal at a distance of quarter diagonal length from a corner. Calculate the stresses at all four corners. Draw stress distribution diagrams for any two adjacent sides.* (Bhavnagar University, Feb. 1992)

Sol. Given :

Width, $b = 300$ mm

Depth, $d = 400$ mm

∴ Area, $A = b \times d = 300 \times 400$

$= 12 \times 10^4\ \text{mm}^2$

Eccentric load, $P = 360\ \text{kN} = 360000\ \text{N}$

The eccentric load is acting at point E, where distance EC = one quarter of diagonal AC.

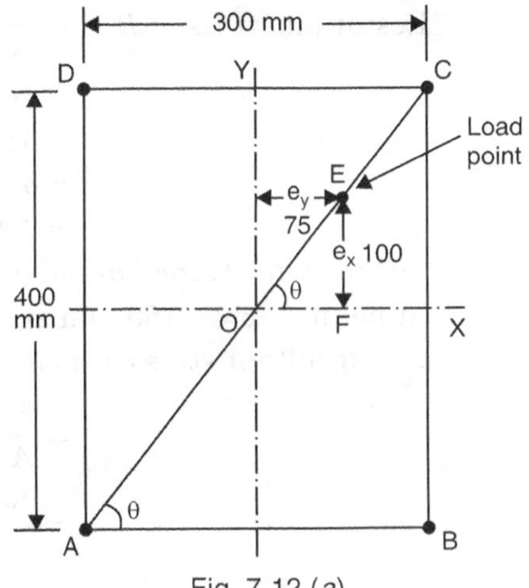

Fig. 7.12 (*a*)

Now diagonal $AC = \sqrt{300^2 + 400^2} = 500$

∴ In ΔCAB, $\tan\theta = \frac{4}{3}$

$\cos\theta = \frac{3}{5}$ and $\sin\theta = \frac{4}{5}$

Also, $OE = EC = \frac{1}{4}$ of AC

$= \frac{1}{4} \times 500 = 125$ mm

And $e_x = EF = OE \sin\theta = 125 \times \frac{4}{5} = 100$ mm

$e_y = OF = OE \cos\theta = 125 \times \frac{3}{5} = 75$ mm

Moment of load about X-X axis,

$$M_x = P \times e_x = 360000 \times 100 = 36000000\ \text{Nmm}$$

Moment of load about Y-Y axis,

$$M_y = P \times e_y = 360000 \times 75 = 27000000\ \text{Nmm}$$

Also, $I_{xx} = \frac{1}{12} \times 300 \times 400^3 = 16 \times 10^8\ \text{mm}^4$

$I_{yy} = \frac{1}{12} \times 400 \times 300^3 = 9 \times 10^8\ \text{mm}^4$

The resultant stress at any point is given by equation (7.3) as

$$\text{Resultant stress} = \frac{P}{A} + \frac{M_y \times x}{I_{yy}} + \frac{M_x \times y}{I_{xx}}$$

(i) Resultant stress at point C

At point C, x = 150 mm and y = 200 m

∴ Resultant stress at C

$$= \frac{P}{A} + \frac{M_y \times 150}{I_{yy}} + \frac{M_x \times 200}{I_{xx}}$$

$$= \frac{360000}{12 \times 10^4} + \frac{27000000 \times 150}{9 \times 10^8} + \frac{36000000 \times 200}{16 \times 10^8}$$

$= 3 + 4.5 + 4.5$ N/mm^2 = **12 N/mm^2 (compressive). Ans.**

(ii) Resultant stress at point B

At point B, x = 150 mm and y = – 200 mm

Resultant stress at B

$$= \frac{P}{A} + \frac{M_y \times 150}{I_{yy}} + \frac{M_x \times (-200)}{16 \times 10^8}$$

$$= \frac{360,000}{12 \times 10^4} + \frac{27000000 \times 150}{9 \times 10^8} - \frac{36000000 \times 200}{16 \times 10^8}$$

$= 3 + 4.5 - 4.5$

= **3 N/mm^2 (compressive). Ans.**

(iii) Resultant stress at point A

At point A, x = – 150 mm and y = – 200 mm

∴ Resultant stress at point A

$$= \frac{P}{A} + \frac{M_y \times (-150)}{I_{yy}} + \frac{M_x \times (-200)}{I_{xx}}$$

$$= \frac{360000}{12 \times 10^4} - \frac{27000000 \times 150}{9 \times 10^8} - \frac{36000000 \times 200}{16 \times 10^8}$$

$= 3 - 4.5 - 4.5$

= **– 6 N/mm^2 (Tensile). Ans.**

(iv) Resultant stress at point D

At point D, x = – 150 mm and y = 200 mm

∴ Resultant stress at point D

$$= \frac{P}{A} + \frac{M_y(-150)}{I_{yy}} + \frac{M_x \times 200}{I_{xx}}$$

$$= \frac{360000}{12 \times 10^4} - \frac{27000000 \times 150}{9 \times 10^8} + \frac{36000000 \times 200}{16 \times 10^8}$$

$= 3 - 4.5 + 4.5$

= **3 N/mm^2 (compressive). Ans.**

Stress distribution for AB and BC (i.e., two adjacent sides)

Fig. 7.12 (*b*) shows the stress distribution along two adjacent sides (*i.e.*, *AB* and *BC*). At point *A*, resultant stress is 6 N/mm^2 (tensile) whereas at point *B*, the resultant stress is 3 N/mm^2 (compressive). Take AE = 6 N/mm^2 and BF = 3 N/mm^2. Join *E* to *F*.

For side *BC*, the resultant stress at *B* is 3 N/mm^2 (compressive) whereas at point *C* the resultant stress is 12 N/mm^2 (compressive).

Take BH = 3 N/mm^2 (compressive) and CG = 12 N/mm^2 (compressive).

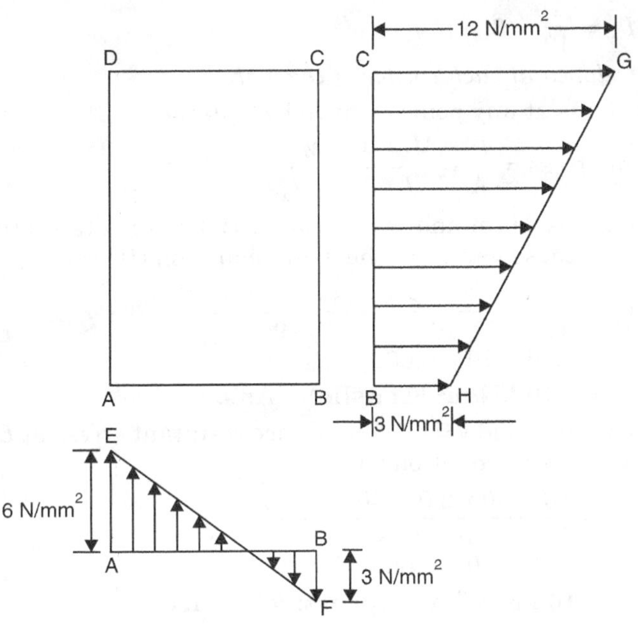

Fig. 7.12 (*b*)

Problem 7.10. *A masonry pier of 3 m × 4 m supports a vertical load of 80 kN as shown in Fig. 7.13.*

(*a*) *Find the stresses developed at each corner of the pier.*

(*b*) *What additional load should be placed at the centre of the pier, so that there is no tension anywhere in the pier section ?*

(*c*) *What are the stresses at the corners with the additional load in the centre ?*

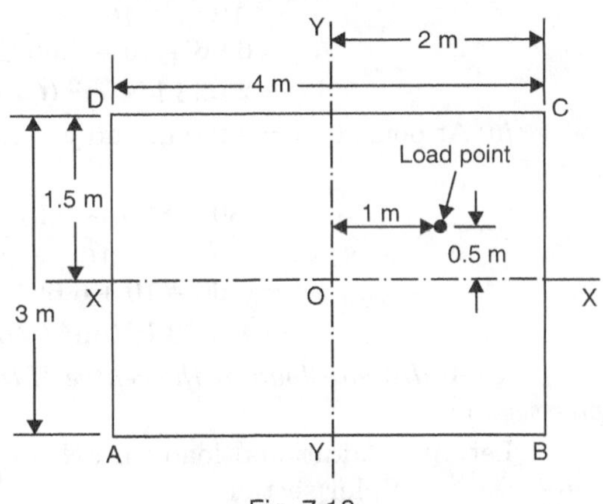

Fig. 7.13

Sol. Given :

Width $b = 4$ m

Depth, $d = 3$ m

∴ Area, $A = 4 \times 3 = 12 \text{ mm}^2$

Point load, $P = 80$ kN

Eccentricity of load about *X-X* axis,

$$e_x = 0.5 \text{ m}$$

Eccentricity of load about *Y-Y* axis,

$$e_y = 1.0 \text{ m}$$

Moment of load about *X-X* axis,

$$M_x = P \times e_x = 80 \times 0.5 = 40 \text{ kNm}$$

Similarly, $M_y = P \times e_y = 80 \times 1.0 = 80 \text{ kNm}$

Moment of inertia about *X-X* axis,

$$I_{xx} = \frac{1}{12} \times 4 \times 3^3 = 9 \text{ mm}^4$$

Similarly, $I_{yy} = \frac{1}{12} \times 3 \times 4^3 = 16 \text{ m}^4$.

(a) Stresses developed at each corner of the pier

The resultant stress at any point is given by equation (7.3).

$$\text{Hence resultant stress} = \frac{P}{A} + \frac{M_y \times x}{I_{yy}} + \frac{M_x \times y}{I_{xx}} \quad ...(i)$$

(*i*) At point A, $x = -2.0$ m and $y = -1.5$ m. Hence resultant stress at A (*i.e.*, σ_A) is obtained by substituting these values in the above equation (*i*).

$$\therefore \quad \sigma_A = \frac{80}{12} + \frac{80 \times (-2.0)}{16} + \frac{40 \times (-1.5)}{9}$$
$$= 6.66 - 10 - 6.66$$
$$= \mathbf{-10 \text{ kN/m}^2 \text{ (Tensile)}}. \quad \textbf{Ans.}$$

(*ii*) A point B, $x = 2.0$ m and $y = -1.5$ m. Hence resultant stress at B (*i.e.*, σ_B) is obtained by substituting these values in equation (*i*).

$$\therefore \quad \sigma_B = \frac{80}{12} + \frac{80 \times 2.0}{16} + \frac{40 \times (-1.5)}{9}$$
$$= 6.66 + 10 - 6.66$$
$$= \mathbf{10 \text{ kN/m}^2 \text{ (Compressive)}}. \quad \textbf{Ans.}$$

(*iii*) At point C, $x = 2.0$ m and $y = 1.5$ m. Hence resultant stress at C (*i.e.*, σ_C) is given by,

$$\therefore \quad \sigma_C = \frac{80}{12} + \frac{80 \times 2.0}{16} + \frac{40 \times 1.5}{9}$$
$$= 6.66 + 10 + 6.66$$
$$= \mathbf{23.33 \text{ kN/m}^2 \text{ (Compressive)}}. \quad \textbf{Ans.}$$

(*iv*) At point D, $x = -2.0$ m and $y = 1.5$ m. Hence resultant stress at D (*i.e.*, σ_D) is given by,

$$\therefore \quad \sigma_D = \frac{80}{12} + \frac{80 \times (-2.0)}{16} + \frac{40 \times 1.5}{9}$$
$$= 6.66 - 10 + 6.66$$
$$= \mathbf{+3.33 \text{ kN/m}^2 \text{ (Compressive)}}. \quad \textbf{Ans.}$$

(b) Additional load at the centre of the pier, so that there is no tension anywhere in the pier section.

Let W = Additional load (in kN) placed at the centre for no tension anywhere in the pier section.

The above load is compressive and will cause a compressive stress

$$= \frac{W}{A} = \frac{W}{12} \text{ kN/m}^2 \quad (\because \; A = 12 \text{ m}^2)$$

As this load is placed at the centre, it will produce a uniform compressive stress across the section of the pier. But we know that there is tensile stress at point A having magnitude = 10 kN/m². Hence the compressive stress due to load W should be equal to tensile stress at A.

$$\therefore \quad \frac{W}{12} = 10$$

or $W = 10 \times 12 =$ **120 kN. Ans.**

(c) *Stresses at the corners with the additional load at the centre*

Stress due to additional load = $\frac{W}{A} = \frac{120}{12}$ = 10 kN/m^2 (Compressive)

This stress is uniform across the cross-section of the pier. Hence to find the stresses at the corners with this additional load, we must add the stress 10 kN/m^2 in each value of the stresses already existing in the corners.

$\therefore$ Stress at A, $\sigma_A = -10 + 10 = \mathbf{0}$. **Ans.**

Similarly, $\sigma_B = 10 + 10 = \mathbf{20\ kN/m^2}$. **Ans.**

$\sigma_C = 23.33 + 10 = \mathbf{33.33\ kN/m^2}$. **Ans.**

and $\sigma_D = 3.33 + 10 = \mathbf{13.33\ kN/m^2}$. **Ans.**

7.8. MIDDLE THIRD RULE FOR RECTANGULAR SECTIONS [*i.e., KERNEL OF SECTION*]

The cement concrete columns are weak in tension. Hence the load must be applied on these columns in such a way that there is no tensile stress anywhere in the section. But when an eccentric load is acting on a column, it produces direct stress as well as bending stress. The resultant stress at any point in the section is the algebraic sum of the direct stress and bending stress.

Consider a rectangular section of width 'b' and depth 'd' as shown in Fig. 7.14. Let this section is subjected to a load which is eccentric to the axis Y-Y.

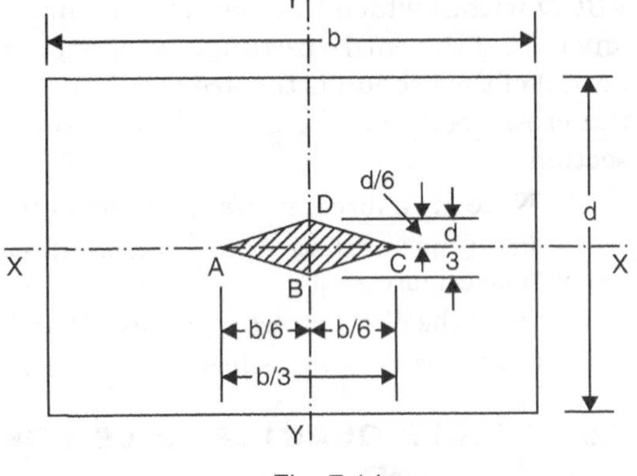

Fig. 7.14

Let P = Eccentric load acting on the column

e = Eccentricity of the load

A = Area of the section.

Then from equation (7.2), we have the minimum stress as

$$\sigma_{min} = \frac{P}{A}\left(1 - \frac{6 \times e}{b}\right) \qquad ...(i)$$

If σ_{min} is –ve, then stress will be tensile. But if σ_{min} is zero (or positive) then there will be no tensile stress along the width of the column.

Hence for no tensile stress along the width of the column,

$$\sigma_{min} \geq 0$$

or $$\frac{P}{A}\left(1 - \frac{6 \times e}{b}\right) \geq 0 \quad \text{or} \quad \left(1 - \frac{6 \times e}{b}\right) \geq 0$$

or $$1 \geq \frac{6 \times e}{b} \quad \text{or} \quad \frac{b}{6} \geq e$$

or $$e \leq \frac{b}{6} \qquad ...(7.4)$$

The above result shows that the eccentricity '*e*' must be less than or equal to $\frac{b}{6}$. Hence the greatest eccentricity of the load is $\frac{b}{6}$ from the axis *Y-Y*. Hence if the load is applied at any distance less than $\frac{b}{6}$ from the axis, on any side of the axis *Y-Y*, the stresses are wholly compressive. Hence the range within which the load can be applied so as not to produce any tensile stress, is $\frac{b}{6}+\frac{b}{6}=\frac{2b}{6}=\frac{b}{3}$ or *within the middle third of the base.*

Similarly, if the load had been eccentric with respect to the axis *X-X*, the condition that tensile stress will not occur is when the eccentricity of the load with respect to this axis *X-X* does not exceed $\frac{d}{6}$. Hence the range within which the load may be applied is $\frac{d}{6}+\frac{d}{6}=\frac{2d}{6}=\frac{d}{3}$ or within the middle third of the depth.

If it is possible that the load is likely to be eccentric about both the axes *X-X* and *Y-Y*, the condition that tensile stress will not occur is when the load is applied anywhere within the rhombus *ABCD* whose diagonals are $AC = \frac{b}{3}$ and $BD = \frac{d}{3}$ as shown in Fig. 7.14. This figure *ABCD* within which the load may be applied anywhere so as not to produce tensile stress in any part of the entire rectangular section, is called the *Core or Kernel* of the section. Hence the kernel of the section is the area within which the line of action of the eccentric load *P* must cut the cross-section if the stress is not to become tensile in any part of the entire rectangular section.

Note. (*i*) If direct stress (σ_0) is equal to bending stress (σ_b), then the tensile stress will be zero.

(*ii*) If the direct stress (σ_0) is more than bending stress (σ_b), then the stress throughout the section will be compressive.

(*iii*) If the direct stress (σ_0) is less than bending stress (σ_b), then there will be tensile stress.

(*iv*) Hence for no tensile stress, $\sigma_0 \geq \sigma_b$.

7.9. MIDDLE QUARTER RULE FOR CIRCULAR SECTIONS [*i.e., KERNEL OF SECTION*]

Consider a circular section of diameter '*d*' as shown in Fig. 7.15. Let this section is subjected to a load which is eccentric to the axis *Y-Y*.

Let P = Eccentric load

e = Eccentricity of the load

A = Area of the section $= \frac{\pi}{4}d^2$

Now direct stress,

$$\sigma_0 = \frac{P}{A} = \frac{P}{\frac{\pi}{4}d^2} = \frac{4P}{\pi d^2}$$

Moment, $M = P \times e$

Bending stress (σ_b) is given by,

$$\frac{M}{I} = \frac{\sigma_b}{y} \quad \text{or} \quad \sigma_b = \frac{M \times y}{I}$$

Maximum bending stress will be when

$$y = \pm \frac{d}{2}$$

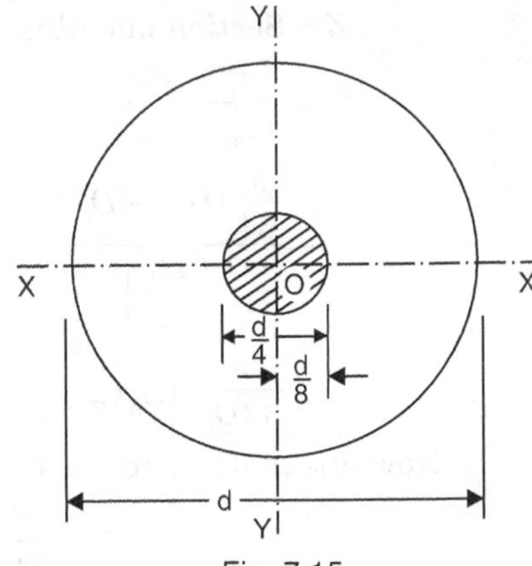

Fig. 7.15

$\therefore$ Maximum bending stress is given by,

$$\sigma_b = \frac{M}{I} \times \left(\pm \frac{d}{2}\right) = \pm \frac{P \times e \times \frac{d}{2}}{\frac{\pi}{64} d^4}$$

$$= \pm \frac{32 \times P \times e}{\pi d^3}$$

Now minimum stress is given by,

$$\sigma_{min} = \sigma_0 - \sigma_b$$

$$= \frac{4P}{\pi d^2} - \frac{32P \times e}{\pi d^3}$$

For no tensile stress, $\sigma_{min} \geq 0$

or $$\frac{4P}{\pi d^2} - \frac{32P \times e}{\pi d^3} \geq 0 \quad \text{or} \quad \frac{4P}{\pi d^2}\left(1 - \frac{8d}{d}\right) \geq 0$$

or $$1 - \frac{8 \times e}{d} \geq 0 \quad \text{or} \quad 1 \geq \frac{8 \times e}{d} \quad \text{or} \quad e \leq \frac{d}{8} \qquad \text{...(7.5)}$$

The above result shows that the eccentricity 'e' must be less than or equal to $\frac{d}{8}$. It means that the load can be eccentric, on any side of the centre of the circle, by an amount equal to $\frac{d}{8}$. Thus, if the line of action of the load is within a circle of radius equal to $\frac{d}{8}$ or of diameter equal to one-fourth of the main circle as shown in Fig. 7.15, then the stress will be compressive throughout the circular section. This is known as *Middle Quarter Rule.*

7.9.1. Kernel of Hollow Circular Section *(Or Value of Eccentricity for Hollow Circular Section)*

Let D_0 = External diameter, and

D_i = Internal diameter

P = Eccentric load

e = Eccentricity of the load

A = Area of section

$$= \frac{\pi}{4} [D_0^2 - D_i^2]$$

M = Moment due to eccentric load $P = P \times e$

Z = Section modulus

$$= \frac{I}{y_{max}}$$

$$= \frac{\frac{\pi}{64}[D_0^4 - D_i^4]}{\left(\frac{D_0}{2}\right)} \qquad \left(\because \; y_{max} = \frac{D_0}{2}\right)$$

$$= \frac{\pi}{32D_0}[D_0^4 - D_i^4]$$

Now direct stress (σ_0) is given by

$$\sigma_0 = \frac{P}{A} \qquad ...(i)$$

The direct stress is compressive and uniform throughout the section.

Bending stress (σ_b) is given by

$$\frac{M}{I} = \frac{\sigma_b}{y}$$

$$\therefore \qquad \sigma_b = \frac{M}{I} \times y = \frac{M}{\left(\frac{I}{y}\right)}$$

$$= \frac{M}{Z} \qquad \left(\because \; \frac{I}{y} = Z\right) \qquad ...(ii)$$

The bending stress may be tensile or compressive. The resultant stress at any point is the algebraic sum of direct stress and bending stress. There will be no tensile stress at any point if the bending stress is less than or equal to direct stress at that point.

Hence for no tensile stress,

Bending stress ≤ Direct stress

or $\qquad \sigma_b \le \sigma_0$

Substituting the values of σ_0 and σ_b from equations (i) and (ii), we get

$$\frac{M}{Z} \le \frac{P}{A}$$

$$\frac{P \times e}{Z} \le \frac{P}{A} \qquad (\because \; M = P \times e)$$

or $$\frac{e}{Z} \le \frac{1}{A} \qquad \text{(cancelling } P \text{ to both sides)}$$

or $$e \le \frac{Z}{A} \qquad ...(7.6)$$

$$\le \frac{\frac{\pi}{32D_0}[D_0^4 - D_i^4]}{\frac{\pi}{4}[D_0^2 - D_i^2]}$$

$$\leq \frac{4\pi}{32\pi D_0} \frac{(D_0^2 + D_i^2)(D_0^2 - D_i^2)}{(D_0^2 - D_i^2)}$$

$$\leq \frac{1}{8D_0} (D_0^2 + D_i^2) \quad \text{...(7.7)}$$

The above result shows that the eccentricity 'e' must be less than or equal to $(D_0^2 + D_i^2)/(8D_0)$. It means that the load can be eccentric, on any side of the centre of the circle, by an amount equal to $(D_0^2 + D_i^2)/(8D_0)$. Thus, if the line of action of the load is within a circle of diameter equal to $(D_0^2 + D_i^2)/(4D_0)$, then the stress will be compressive throughout.

$\therefore$ Diameter of kernel $= \dfrac{D_0^2 + D_i^2}{4D_0}$.

7.9.2. Kernel of Hollow Rectangular Section (*Or Value of Eccentricity for Hollow Rectangular Section*)

Refer to Fig. 7.16

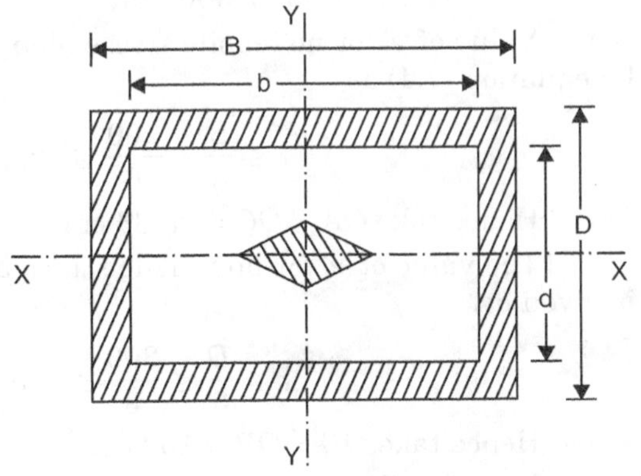

Fig. 7.16

Let B = Outer width of rectangular section

D = Outer depth

b = Internal width

d = Internal depth

A = Area of section

$= B \times D - b \times d$

$$I_{xx} = \frac{BD^3}{12} - \frac{bd^3}{12}$$

$$y_{max} = \frac{D}{2}$$

$$\therefore \quad Z_{xx} = \frac{I_{xx}}{y_{max}}$$

$$= \frac{\left(\frac{BD^3}{12} - \frac{bd^3}{12}\right)}{D/2} = \frac{BD^3 - bd^3}{6D}$$

Similarly,

$$Z_{yy} = \frac{I_{yy}}{x_{max}} = \frac{\frac{DB^3}{12} - \frac{db^3}{12}}{B/2} = \frac{DB^3 - db^3}{6B}$$

For no tensile stress at any section, the value of e is given by equation (7.6).

$$\therefore \quad e \leq \frac{Z}{A} \quad \text{or} \quad e_x \leq \frac{Z_{xx}}{A} \quad \text{and} \quad e_y \leq \frac{Z_{yy}}{A}$$

or

$$e_x \leq \frac{\left[\frac{(BD^3 - bd^3)}{6D}\right]}{(BD - bd)} \leq \frac{(BD^3 - bd^3)}{6D(BD - bd)} \quad \text{...(7.8)}$$

and

$$e_y = \frac{\left(\frac{DB^3 - db^3}{6B}\right)}{(BD - bd)} = \frac{DB^3 - db^3}{6B(BD - bd)} \quad \text{...[7.8 (A)]}$$

It means that the load can be eccentric on either side of the geometrical axis by an amount equal to $\frac{(BD^3 - bd^3)}{6D(BD - bd)}$ and $\frac{DB^3 - db^3}{6B(BD - bd)}$ along x-axis and y-axis respectively.

7.9.3. Solved Problems 7.11 to 7.12 Based on Middle Third and Middle Quarter Rules

Problem 7.11. *Draw neat sketches of kernel of the following cross-sections :*

(*i*) *Rectangular section 200 mm × 300 mm*

(*ii*) *Hollow circular cylinder with external dia = 300 mm and thickness = 50 mm*

(*iii*) *Square with 400 cm² area.* (Bhavnagar University, Feb. 1992)

Sol. Given :

(*i*) *Rectangular Section*

$$B = 200 \text{ mm}$$
$$D = 300 \text{ mm}$$

Value of 'e' for no tensile stress along width is given by equation (7.4) as

$$e \le \frac{B}{6} \le \frac{200}{6} \le 33.33 \text{ cm}$$

Hence take $OA = OC = 33.33$ cm

The value of 'e' for no tensile stress along the depth is given by,

$$e \le \frac{D}{3} \le \frac{300}{3} \le 100 \text{ cm}$$

Hence take $OD = OB = 100$ cm

Now, join A to B, B to C, C to D and D to A. The figure $ABCD$ represents the kernel of the given rectangular section as shown in Fig. 7.17.

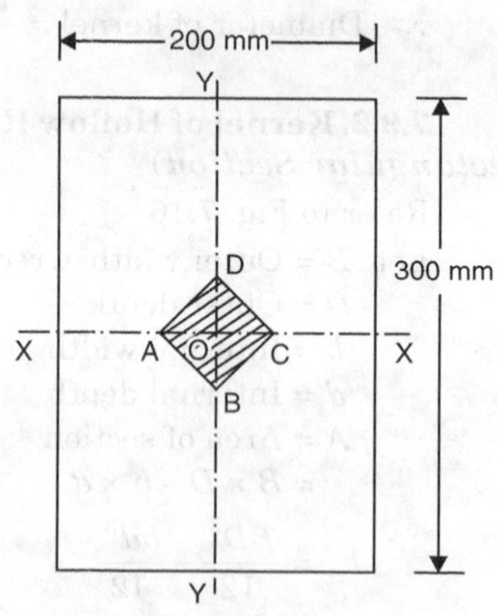

Fig. 7.17

(*ii*) *Kernel for Hollow Circular Section*

Given :

External dia., $D_0 = 300$ m

Thickness, $t = 50$ mm

∴ Internal dia., $D_i = D_0 - 2 \times t$

$= 300 - 2 \times 50 = 200$ mm

For hollow cylindrical section, for no tensile stress, the value of e *is* given by equation (7.7) as

$$e \le \frac{1}{8D_0} = (D_0^2 + D_i^2)$$

$$\le \frac{1}{8 \times 300} (300^2 + 200^2)$$

$$\le \frac{1}{2400} (90000 + 40000)$$

$$\le \frac{130000}{2400} \le 54.16 \text{ mm}$$

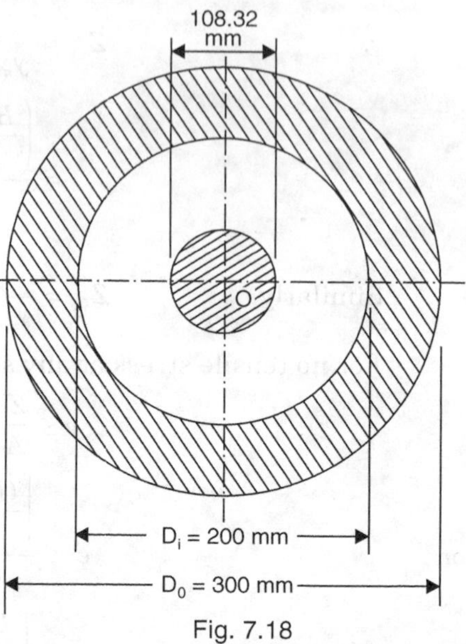

Fig. 7.18

Taking O as centre and radius equal to 54.16 mm (or dia. = 2 × 54.16 = 108.32 mm) draw a circle. This circle is the kernel of the hollow circular section of external dia. = 300 mm and internal dia. = 200 mm, as shown in Fig. 7.18.

(iii) Kernel for Square Section

Given :

Area = 400 cm^2

∴ One side of square = $\sqrt{400}$ = 20 cm

For no tensile stress, the value of 'e' for the square section is given by equation (7.4) as

$$e \le \frac{\text{Side}}{6}\ [B \text{ or } D] \le \frac{20}{6} \le 3.33 \text{ cm}$$

Hence take $OA = OC = OB = OD = 3.33$ cm

Join $ABCDA$ as shown in Fig. 7.19 (a). Then $ABCD$ is the kernel of given square section.

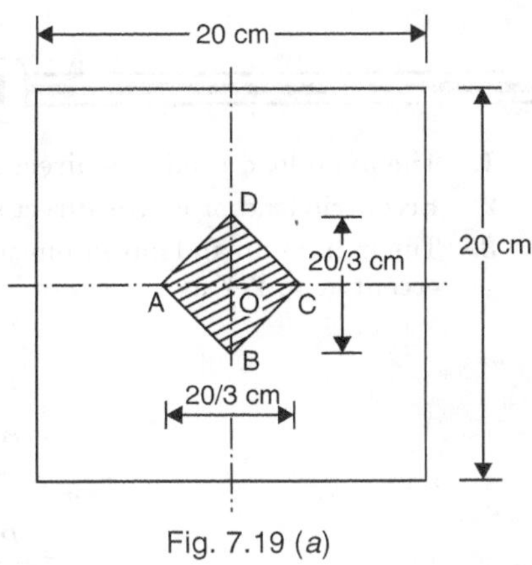

Fig. 7.19 (a)

Problem 7.12. *Draw neat sketch of kernel of a hollow rectangular section of outer cross-section 300 mm × 200 mm and inner cross-section 150 mm × 100 mm.*

Sol. Given :

Outer rectangular section, B = 300 mm, D = 200 mm

Inner rectangular section, b = 150 mm and d = 100 mm.

For no tensile stress the value of 'e' along x-axis and along y-axis are given by equations (7.8) and (7.8 A) respectively.

Using equation (7.8), we get

$$e_x \le \frac{(BD^3 - bd^3)}{6D(BD - bd)}$$

$$= \frac{(300 \times 200^3 - 150 \times 100^3)}{6 \times 200(300 \times 200 - 150 \times 100)}$$

$$= \frac{100^3 \times (2400 - 150)}{12 \times 100^3 \times (6 - 1.5)} = \frac{2250}{12 \times 4.5} = 41.67 \text{ mm}$$

Fig. 7.19 (b)

Hence take $OA = OC = 41.67$ mm in Fig. 7.19 (b).

Using equation (7.8 A), we get

$$e_y \le \frac{(DB^3 - db^3)}{6B(BD - bd)}$$

$$= \frac{(200 \times 300^3 - 100 \times 150^3)}{6 \times 300(300 \times 200 - 150 \times 100)} = \frac{100^3(5400 - 337.5)}{6 \times 3 \times 10^3(6 - 1.5)}$$

$$= \frac{5062.5}{18 \times 4.5} = 62.5 \text{ mm}$$

Hence, take $OD = OB = 62.5$ mm in Fig. 7.19 (*b*).

Now, join *A* to *B*, *B* to *C*, *C* to *D* and *D* to *A*. The figure *ABCD* represents the kernel of the given hollow rectangular section.

HIGHLIGHTS (7A)

1. The axial load produces direct stress (σ_0).
2. Eccentric load produces direct stress as well as bending stress (σ_b).
3. The maximum and minimum stress at any point in a section which is subjected to a load which is eccentric to *Y-Y* axis is given by,

$$\sigma_{max} = \text{Direct stress} + \text{Bending stress}$$

$$= \frac{P}{A}\left(1 + \frac{6 \times e}{b}\right) \quad \text{...For a rectangular section}$$

and

$$\sigma_{min} = \text{Direct stress} - \text{Bending stress}$$

$$= \frac{P}{A}\left(1 - \frac{6 \times e}{b}\right) \quad \text{...For a rectangular section}$$

where P = Eccentric load

A = Area of section

e = Eccentricity

b = Width of the section.

4. If $\sigma_0 = \sigma_b$, the tensile stress will be zero across the section.
5. If $\sigma_0 > \sigma_b$, there will be no tensile stress across the section.
6. If $\sigma_0 < \sigma_b$, there will be tensile stress across the section.
7. The resultant stress at any point when a symmetrical column section is subjected to a load which is eccentric to both the axis, is given by,

$$\sigma = \frac{P}{A} \pm \frac{M_y \times x}{I_{yy}} \pm \frac{M_x \cdot y}{I_{xx}}$$

where P = Eccentric load

A = Area of the section

M_y = Moment of load about *Y-Y* axis

I_{yy} = Moment of inertia about *Y-Y* axis

M_x and I_{xx} = Moment and moment of inertia about *X-X* axis respectively.

The values of *x* and *y* are positive on the same side on which load is acting.

8. For unsymmetrical sections, subjected to eccentric load, first of all the C.G. of the section is determined. Then moment of inertia of the section about an axis passing through the C.G. is obtained. After that stresses are obtained.
9. For a rectangular section, there will be no tensile stress if the load is on either axis within the middle third of the section.
10. For a circular section of diameter '*d*', there will be no tensile stress if the load lies in a circle of diameter $\frac{d}{4}$ with centre *O* of the main circular section. This is known as 'middle quarter rule for circular sections'.

11. For no tensile stress, the value of eccentricity e is given by

$$e \leq \frac{d}{8}$$ ...For circular section

$$\leq \frac{1}{8D_0}(D_0^2 + D_i^2)$$ ...For hollow circular section with D_0 as external dia. and D_i as internal dia.

$$\leq \frac{b}{6} \text{ and } \frac{d}{6}$$ ...For rectangular section

$$\leq \frac{\text{One side of square}}{6}$$ ...For square section

$$e_x \leq \frac{(BD^3 - bd^3)}{6D(BD - bd)}$$ ...For hollow rectangular section with B and D as outer width and depth and b and d as inner width and depth

$$e_y \leq \frac{(DB^3 - db^3)}{6B(BD - bd)}.$$

EXERCISE (7A)

(A) Theoretical Questions

1. What do you mean by direct stress and bending stress ?
2. Prove that an eccentric load causes a direct stress as well as bending stress.
3. Find an expression for the maximum and minimum stresses when a rectangular column is subjected to a load which is eccentric to Y-Y axis.
4. Prove that for rectangular section subjected to eccentric load, the maximum and minimum stresses are given by :

$$\sigma_{max} = \frac{P}{A}\left(1 + \frac{6e}{b}\right) \quad \text{and} \quad \sigma_{min} = \frac{P}{A}\left(1 - \frac{6e}{b}\right)$$

where P = Eccentric load,
A = Area of the system,
b = Width of section,
and e = Eccentricity.

5. How will you find the maximum and minimum stresses at the base of a symmetrical column, when it is subjected to load which is eccentric to both axis ?
6. Find and expression for the maximum and minimum stresses at the base of an unsymmetrical column which is subjected to an eccentric load.
7. What do you mean by the following terms :
   (*i*) Middle third rule for rectangular sections, and
   (*ii*) Middle quarter rule for circular sections.
8. Prove that for no tension at the base of a short column :
   (*i*) of rectangular section, the line of action of the load should be within the middle third, and
   (*ii*) of circular section, the line of action of the load should be within the middle quarter.
9. Draw a neat sketches of Kernel of the following cross-sections :
   (*i*) Rectangular 200 mm × 300 mm.
   (*ii*) Hollow circular cylinder with external dia. = 300 mm, thickness = 50 mm.
   (*iii*) Square with 400 cm^2 area. *(Bhavnagar University, Feb. 1992)*

(B) Numerical Problems

1. A rectangular column of width 120 mm and of thickness 100 mm carries a point load of 120 kN at an eccentricity of 10 mm. Determine the maximum and minimum stresses at the base of the column. **[Ans.** 15 N/mm^2, 5 N/mm^2]
2. If in the above problem, the minimum stress at the base of the section is given as zero then find the eccentricity of the point load of 120 kN acting on the rectangular section. Also calculate the corresponding maximum stress on the section. **[Ans.** 20 mm, 20 N/mm^2]
3. If in Q. 1, the eccentricity is given as 30 mm, then find the maximum and minimum stress on the section. Also plot these stress along the width of the section. **[Ans.** – 5 N/mm^2, 25 N/mm^2]
4. In a tension specimen 13 mm in a diameter the line of pull is parallel to the axis of the specimen but is displaced from it. Determine the distance of the line of pull from the axis, when the maximum stress is 15% greater than the mean stress on a section normal to the axis. **[Ans.** 0.25 mm]
5. A hollow rectangular column is having external and internal dimensions as 120 cm deep × 80 cm wide and 90 cm deep × 50 cm wide respectively. A vertical load of 200 kN is transmitted in the vertical plane bisecting 120 cm side and at an eccentricity of 10 cm from the geometric axis of the section. Calculate the maximum and minimum stresses in the section. **[Ans.** 0.61 N/mm^2 and 0.17 N/mm^2]
6. A short column of diameter 40 cm carries an eccentric load of 80 kN. Find the greatest eccentricity which the load can have without producing tension on the cross-section. **[Ans.** 5 cm]
7. A short column of external diameter 50 cm and internal diameter 30 cm carries an eccentric load of 100 kN. Find the greatest eccentricity which the load can have without producing tension on the cross-section. **[Ans.** 8.5 cm]
8. A hollow circular column of 25 cm external and 20 cm internal diameter respectively carries an axial load of 200 kN. It also carries a load of 100 kN on a bracket whose line of action is 20 cm from the axis of the column. Determine the maximum and minimum stress at the base section. **[Ans.** 39 N/mm^2 (comp.), 5.13 N/mm^2 (tension)]
9. A column section 30 cm external diameter and 15 cm internal diameter supports an axial load of 2.6 MN and an eccentric load of *PN* at an eccentricity of 40 cm. If the compressive and tensile stresses are not to exceed 140 N/mm^2 and 60 N/mm^2 respectively, find the magnitude of load *P*. **[Ans.** 766.8 kN]
10. A rectangular pier of 1.5 m × 1.0 m is subjected to a compressive load of 450 kN as shown in Fig. 7.20. Find the stresses on all four corners of the pier.
[Ans. σ_A = 0.45 N/mm^2, σ_B = + 0.15 N/mm^2, σ_C = 1.05 N/mm^2, σ_D = 0.45 N/mm^2]

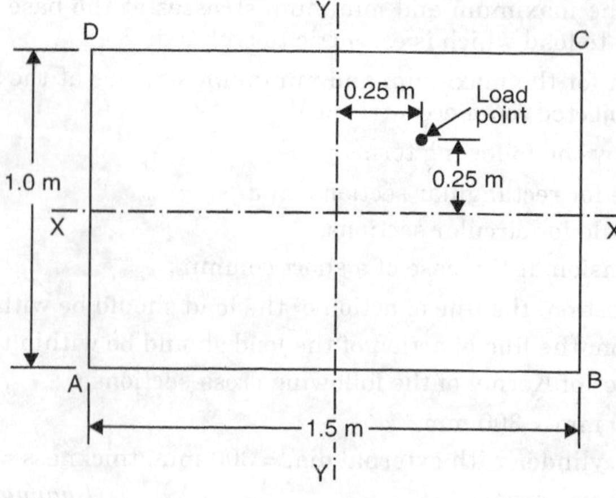

Fig. 7.20

(B) COLUMNS AND STRUTS

7.10. COLUMNS AND STRUTS

Column or strut is defined as a member of a structure, which is subjected to axial compressive load. If the member of the structure is vertical and both of its ends are fixed rigidly while subjected to axial compressive load, the member is known as *column*, for example a vertical pillar between the roof and floor. If the member of the structure is not vertical and one or both of its ends are hinged or pin joined, the bar is known as *strut*. Examples of struts are : connecting rods, piston rods etc.

7.10.1. Failure of a Column. The failure of a column takes place due to the anyone of the following stresses set up in the columns :

(*i*) Direct compressive stresses,

(*ii*) Buckling stresses, and

(*iii*) Combined of direct compressive and buckling stresses.

7.10.2. Failure of a Short Column. A short column of uniform cross-sectional area A, subjected to an axial compressive load P, is shown in Fig. 7.21. The compressive stress induced is given by

$$p = \frac{P}{A}$$

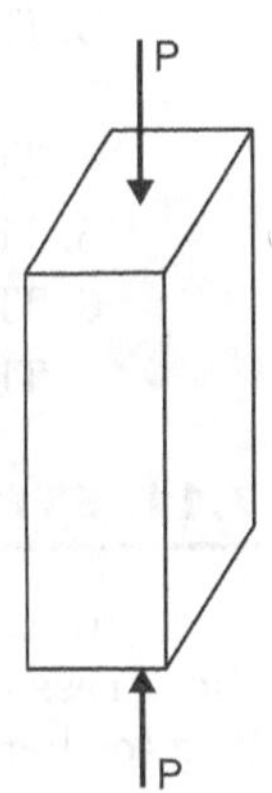

Fig. 7.21

If the compressive load on the short column is gradually increased, a stage will reach when the column will be on the point of failure by crushing. The stress induced in the column corresponding to this load is known as crushing stress and the load is called crushing load.

Let P_c = Crushing load,

σ_c = Crushing stress, and

A = Area of cross-section.

Then $$\sigma_c = \frac{P_c}{A}.$$

All short columns fail due to crushing.

7.10.3. Failure of a Long Column. A long column of uniform cross-sectional area A and of length l, subjected to an axial compressive load P, is shown in Fig. 7.22. A column is known as long column if the length of the column in comparison to its lateral dimensions, is very large. Such columns do not fail by crushing alone, but also by bending (also known buckling) as shown in Fig. 7.22. The load at which the column just buckles, is known as *buckling load or critical load or crippling load*. The buckling load is less than the crushing load for a long column. Actually the value of buckling load for long columns is low whereas for short columns the value of buckling load is relatively high.

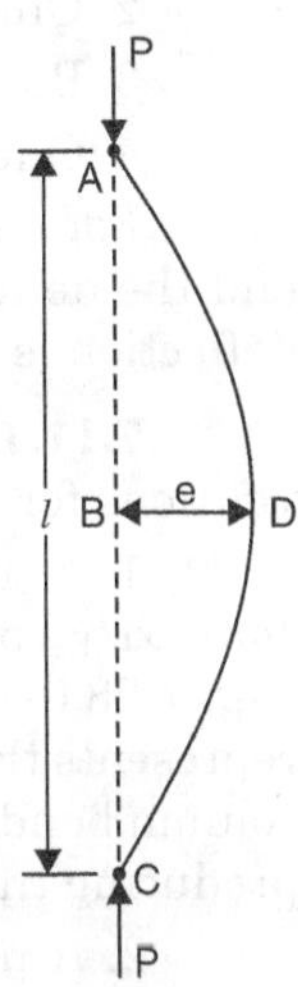

Fig. 7.22

Refer to Fig. 7.22.

Let l = Length of a long column

P = Load (compressive) at which the column has just buckled

A = Cross-sectional area of the column

e = Maximum bending of the column at the centre

σ_0 = Stress due to direct load = $\frac{P}{A}$

σ_b = Stress due to bending at the centre of the column = $\frac{P \times e}{Z}$

where Z = Section modulus about the axis of bending.

The extreme stresses on the mid-section are given by

Maximum stress = $\sigma_0 + \sigma_b$

and Minimum stress = $\sigma_0 - \sigma_b$.

The column will fail when maximum stress (*i.e.*, $\sigma_0 + \sigma_b$) is more than the crushing stress σ_c. But in case of long columns, the direct compressive stresses are negligible as compared to buckling stresses. Hence very long columns are subjected to buckling stresses only.

7.10.4. Assumptions Made in the Euler's Column Theory

The following assumptions are made in the Euler's column theory :

1. The column is initially perfectly straight and the load is applied axially.
2. The cross-section of the column is uniform throughout its length.
3. The column material is perfectly elastic, homogeneous and isotropic and obeys Hooke's law.
4. The length of the column is very large as compared to its lateral dimensions.
5. The direct stress is very small as compared to the bending stress.
6. The column will fail by buckling alone.
7. The self-weight of column is negligible.

7.11. STRUTS WITH DIFFERENT END CONDITION

In case of long columns, the stress due to direct load is very small in comparison with the stress due to buckling. Hence the failure of long columns take place entirely due to buckling (or bending). The following four types of end conditions of the columns are important :

1. Both the ends of the column are hinged (or pinned).
2. One end is fixed and the other end is free.
3. Both the ends of the column are fixed.
4. One end is fixed and the other is pinned.

For a hinged end, the deflection is zero. For a fixed end the deflection and slope are zero. For a free end the deflection is not zero.

7.11.1. Sign Conventions. The following sign conventions for the bending of the columns will be used :

1. A moment which will bend the column with its *convexity* towards its initial central line as shown in Fig. 7.23 (a) is taken as positive. In Fig. 7.23 (a), AB represents the initial centre line of a column. Whether the column bends taking the shape AB' or AB'', the moment producing this type of curvature is positive.

2. A moment which will tend to bend the column with its *concavity* towards its initial centre line as shown in Fig. 7.23 (b) is taken as negative.

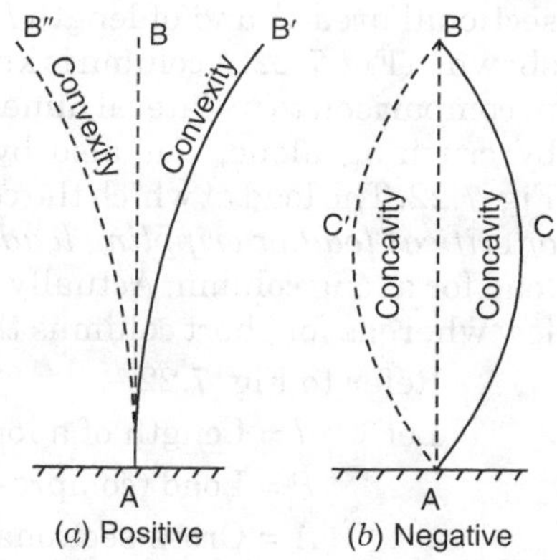

Fig. 7.23

7.12. EULER'S THEORY FOR PIN-ENDED COLUMNS

From Euler's theory and experimental results, the crippling load for different end conditions, is obtained as explained in the next articles.

7.12.1. Crippling Load When Both the Ends of the Column are Hinged. The load at which the column just buckles (or bends) is called crippling load. Consider a column AB of length l and uniform cross-sectional area, hinged at both of its ends A and B. Let P be the crippling load at which the column has just buckled. Due to the crippling load, the column will deflect into a curved form ACB as shown in Fig. 7.24.

Consider any section at a distance x from the end A.

Let y = Deflection (lateral displacement) at the section.

The moment due to the crippling load at the section $= -P \cdot y$

(–ve sign is taken due to sign convention given in Art. 7.11.1)

But moment $= EI \dfrac{d^2y}{dx^2}$.

Equating the two moments, we have

$$EI \frac{d^2y}{dx^2} = -P \cdot y \quad \text{or} \quad EI \frac{d^2y}{dx^2} + P \cdot y = 0$$

or
$$\frac{d^2y}{dx^2} + \frac{P}{EI} \cdot y = 0$$

Fig. 7.24

The solution* of the above differential equation is

$$y = C_1 \cdot \cos\left(x\sqrt{\frac{P}{EI}}\right) + C_2 \cdot \sin\left(x\sqrt{\frac{P}{EI}}\right) \quad \text{...}(i)$$

where C_1 and C_2 are the constants of integration. The values of C_1 and C_2 are obtained as given below :

(*i*) At A, $x = 0$ and $y = 0$ (See Fig. 7.24)

Substituting these values in equation (*i*), we get

$$0 = C_1 \cdot \cos 0^\circ + C_2 \sin 0$$
$$= C_1 \times 1 + C_2 \times 0 \qquad (\because \cos 0 = 1 \text{ and } \sin 0 = 0)$$
$$= C_1$$
$$\therefore \quad C_1 = 0. \quad \text{...}(ii)$$

(*ii*) At B, $x = l$ and $y = 0$ (See Fig. 7.24).

Substituting these values in equation (*i*), we get

$$0 = C_1 \cdot \cos\left(l \times \sqrt{\frac{P}{EI}}\right) + C_2 \cdot \sin\left(l \times \sqrt{\frac{P}{EI}}\right)$$

---

*The equation $\dfrac{d^2y}{dx^2} + \dfrac{P}{EI} \times y = 0$ can be written as $\dfrac{d^2y}{dx^2} + \alpha^2 y = 0$, where $\alpha^2 = \dfrac{P}{EI}$ or $\alpha = \sqrt{\dfrac{P}{EI}}$

The solution of the equation is $y = C_1 \cos(\alpha x) + C_2 \sin(\alpha x)$

$$= C_1 \cos\left(\sqrt{\frac{P}{EI}} \times x\right) + C_2 \sin\left(\sqrt{\frac{P}{EI}} \times x\right) \text{ as } \alpha = \sqrt{\frac{P}{EI}}$$

$$= 0 + C_2 \,.\, \sin\left(l \times \sqrt{\frac{P}{EI}}\right) \qquad [\because C_1 = 0 \text{ from equation } (ii)]$$

$$= C_2 \sin\left(l\sqrt{\frac{P}{EI}}\right) \qquad ...(iii)$$

From equation (*iii*), it is clear that either $C_2 = 0$

or $$\sin\left(l\sqrt{\frac{P}{EI}}\right) = 0.$$

As $C_1 = 0$, then if C_2 is also equal to zero, then from equation (*i*) we will get $y = 0$. This means that the bending of the column will be zero or the column will not bend at all. Which is not true.

$\therefore$ $$\sin\left(l\sqrt{\frac{P}{EI}}\right) = 0$$

$$= \sin 0 \text{ or } \sin \pi \text{ or } \sin 2\pi \text{ or } \sin 3\pi \text{ or} ...$$

or $$l\sqrt{\frac{P}{EI}} = 0 \text{ or } \pi \text{ or } 2\pi \text{ or } 3\pi \text{ or} ...$$

Taking the least practical value,

$$l\sqrt{\frac{P}{EI}} = \pi$$

or $$P = \frac{\pi^2 EI}{l^2}. \qquad ...(7.9)$$

7.12.2. Crippling Load When One end of the Column is Fixed and the Other end is Free. Consider a column *AB*, of length *l* and uniform cross-sectional area, fixed at the end *A* and free at the end *B*. The free end will sway sideways when load is applied at free end and curvature in the length *l* will be similar to that of upper half of the column whose both ends are hinged. Let *P* is the crippling load at which the column has just buckled. Due to the crippling load *P*, the column will deflect as shown in Fig. 7.25 in which *AB* is the original position of the column and *AB′*, is the deflected position due to crippling load *P*.

Consider any section at a distance *x* from the fixed end *A*.

Let y = Deflection (or lateral displacement) at the section

a = Deflection at the free end *B*.

Then moment at the section due to the crippling load $= P(a - y)$

(+ve sign is taken due to sign convention given in Art. 7.11.1)

But moment is also $= EI \dfrac{d^2y}{dx^2}$

$\therefore$ Equating the two moments, we get

$$EI\frac{d^2y}{dx^2} = P(a - y) = P \,.\, a - P \,.\, y$$

or $$EI\frac{d^2y}{dx^2} + P \,.\, y = P \,.\, a$$

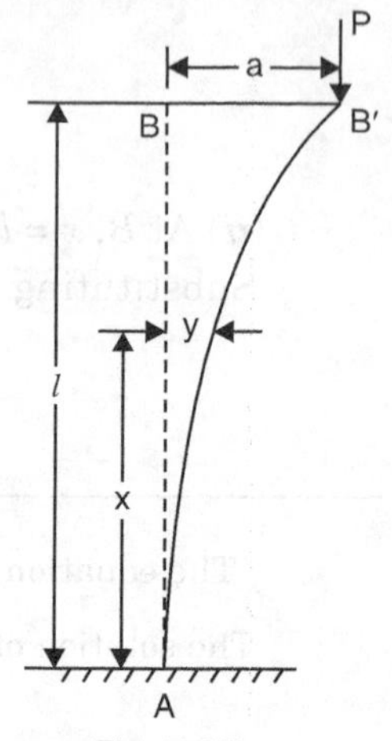

Fig. 7.25

or $$\frac{d^2y}{dx^2} + \frac{P}{EI} \cdot y = \frac{P}{EI} \cdot a. \qquad ...(A)$$

The solution* of the differential equation is

$$y = C_1 \cdot \cos\left(x\sqrt{\frac{P}{EI}}\right) + C_2 \cdot \sin\left(x\sqrt{\frac{P}{EI}}\right) + a \qquad ...(i)$$

where C_1 and C_2 are constant of integration. The values of C_1 and C_2 are obtained from boundary conditions. The boundary conditions are :

(*i*) For a fixed end, the deflection as well as slope is zero.

Hence at end A (which is fixed), the deflection $y = 0$ and also slope $\frac{dy}{dx} = 0$.

Hence at A, $x = 0$ and $y = 0$

Substituting these values in equation (*i*), we get

$$\begin{aligned} 0 &= C_1 \cdot \cos 0 + C_2 \sin 0 + a \\ &= C_1 \times 1 + C_2 \times 0 + a \qquad (\because \cos 0 = 1, \sin 0 = 0) \\ &= C_1 + a \end{aligned}$$

$$\therefore \quad C_1 = -a \qquad ...(ii)$$

At A, $x = 0$ and $\frac{dy}{dx} = 0$.

Differentiating the equation (*i*) w.r.t. x, we get

$$\frac{dy}{dx} = C_1 \cdot (-1)\sin\left(x\sqrt{\frac{P}{EI}}\right) \cdot \sqrt{\frac{P}{EI}} + C_2 \cos\left(x\sqrt{\frac{P}{EI}}\right) \cdot \sqrt{\frac{P}{EI}} + 0$$

$$= -C_1 \cdot \sqrt{\frac{P}{EI}} \sin\left(x\sqrt{\frac{P}{EI}}\right) + C_2 \cdot \sqrt{\frac{P}{EI}} \cos\left(x\sqrt{\frac{P}{EI}}\right)$$

But at A, $x = 0$ and $\frac{dy}{dx} = 0$.

$\therefore$ The above equation becomes as

$$\begin{aligned} 0 &= -C_1 \cdot \sqrt{\frac{P}{EI}} \sin 0 + C_2 \sqrt{\frac{P}{EI}} \cos 0 \\ &= -C_1\sqrt{\frac{P}{EI}} \times 0 + C_2 \cdot \sqrt{\frac{P}{EI}} \times 1 = C_2\sqrt{\frac{P}{EI}} \,. \end{aligned}$$

---

*The equation (*A*) can be written as

$$\frac{d^2y}{dx^2} + \alpha^2 \times y = \alpha^2 \times a, \text{ where } \alpha^2 = \frac{P}{EI} \text{ or } \alpha = \sqrt{\frac{P}{EI}}$$

The complete solution of this equation is $y = C_1 \cos(\alpha \,.\, x) + C_2 \sin(\alpha \,.\, x) + \frac{\alpha^2 \times a}{\alpha^2}$

$$= C_1 \cos\left(\sqrt{\frac{P}{EI}} \times x\right) + C_2 \sin\left(\sqrt{\frac{P}{EI}} \times x\right) + a$$

From the above equation it is clear that either $C_2 = 0$.

or $$\sqrt{\frac{P}{EI}} = 0.$$

But for the crippling load P, the value of $\sqrt{\frac{P}{EI}}$ cannot be equal to zero.

$\therefore \quad C_2 = 0.$

Substituting the values of $C_1 = -a$ and $C_2 = 0$ in equation (*i*), we get

$$y = -a \cdot \cos\left(x\sqrt{\frac{P}{EI}}\right) + a. \qquad \text{...}(iii)$$

But at the free end of the column, $x = l$ and $y = a$.

Substituting these values in equation (*iii*), we get

$$a = -a \cdot \cos\left(l \cdot \sqrt{\frac{P}{EI}}\right) + a$$

or $$0 = -a \cdot \cos\left(l \cdot \sqrt{\frac{P}{EI}}\right) \text{ or } a \cos\left(l \cdot \sqrt{\frac{P}{EI}}\right) = 0$$

But 'a' cannot be equal to zero

$$\therefore \quad \cos\left(l \cdot \sqrt{\frac{P}{EI}}\right) = 0 = \cos\frac{\pi}{2} \text{ or } \cos\frac{3\pi}{2} \text{ or } \cos\frac{5\pi}{2} \text{ or ......}$$

$$\therefore \quad l \cdot \sqrt{\frac{P}{EI}} = \frac{\pi}{2} \text{ or } \frac{3\pi}{2} \text{ or } \frac{5\pi}{2} \text{ ......}$$

Taking the least practical value,

$$l \cdot \sqrt{\frac{P}{EI}} = \frac{\pi}{2} \quad \text{or} \quad \sqrt{\frac{P}{EI}} = \frac{\pi}{2l}$$

or $$P = \frac{\pi^2 EI}{4l^2} \qquad \text{...}(7.10)$$

7.12.3. Crippling Load When Both the Ends of the Column are Fixed. Consider a column AB of length l and uniform cross-sectional area fixed at both of its ends A and B as shown in Fig. 7.26. Let P is the crippling load at which the column has buckled. Due to the crippling load P, the column will deflect as shown in Fig. 7.26. Due to fixed ends, there will be fixed end moments (say M_0) at the ends A and B. The fixed end moments will be accting in such direction so that slope at the fixed ends becomes zero.

Consider a section at a distance x from the end A. Let the deflection of the column at the section is y. As both the ends of the column are fixed and the column carries a crippling load, there will be some fixed end moments at A and B.

Let M_0 = Fixed end moments at A and B.

Then moment at the section = $M_0 - P.y$

But moment at the section is also = $EI \dfrac{d^2y}{dx^2}$

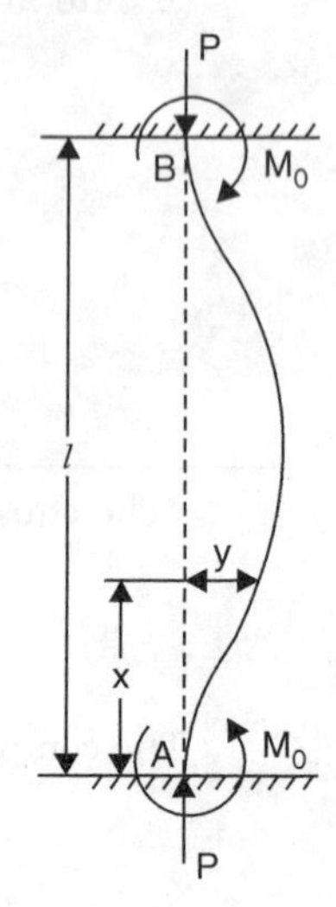

Fig. 7.26

∴ Equating the two moments, we get

$$EI\frac{d^2y}{dx^2} = M_0 - P.y$$

or
$$EI\frac{d^2y}{dx^2} + P \,.\, y = M_0$$

or
$$\frac{d^2y}{dx^2} + \frac{P}{EI} \,.\, y = \frac{M_0}{EI} \qquad ...(A)$$

$$= \frac{M_0}{EI} \times \frac{P}{P} = \frac{P}{EI} \,.\, \frac{M_0}{P}$$

The solution* of the above differential equation is

$$y = C_1 \,.\, \cos\left(x \,.\, \sqrt{\frac{P}{EI}}\right) + C_2 \,.\, \sin\left(x \,.\, \sqrt{\frac{P}{EI}}\right) + \frac{M_0}{P} \qquad ...(i)$$

where C_1 and C_2 are constant of integration and their values are obtained from boundary conditions. Boundary conditions are :

(*i*) At A, $x = 0$, $y = 0$ and also $\frac{dy}{dx} = 0$ as A is a fixed end.

(*ii*) At B, $x = l$, $y = 0$ and also $\frac{dy}{dx} = 0$ as B is also a fixed end.

Substituting the value $x = 0$ and $y = 0$ in equation (*i*), we get

$$0 = C_1 \times 1 + C_2 \times 0 + \frac{M_0}{P} \qquad (\because \ \cos 0 = 1)$$

$$= C_1 + \frac{M_0}{P}$$

∴
$$C_1 = -\frac{M_0}{P} \qquad ...(ii)$$

Differentiating the equation (*i*), with respect to x, we get

$$\frac{dy}{dx} = C_1 \,.\, (-1)\sin\left(x \,.\, \sqrt{\frac{P}{EI}}\right) . \sqrt{\frac{P}{EI}} + C_2 \cos\left(x \,.\, \sqrt{\frac{P}{EI}}\right) . \sqrt{\frac{P}{EI}} + 0$$

$$= -C_1 \sin\left(x \,.\, \sqrt{\frac{P}{EI}}\right) . \sqrt{\frac{P}{EI}} + C_2 \cos\left(x \,.\, \sqrt{\frac{P}{EI}}\right) . \sqrt{\frac{P}{EI}}$$

---

*The equation (*A*) can be written as

$$\frac{d^2y}{dx^2} + \alpha^2 \times y = \frac{M_0}{EI}, \text{ where } \alpha^2 = \frac{P}{EI} \quad \text{or} \quad \alpha = \sqrt{\frac{P}{EI}}$$

The complete solution of this equation is

$$y = C_1 \cos(\alpha \,.\, x) + C_2 \sin(\alpha \,.\, x) + \frac{M_0}{EI \times \alpha^2}$$

$$= C_1 \cos\left(\sqrt{\frac{P}{EI}} \times x\right) + C_2 \sin\left(\sqrt{\frac{P}{EI}} \times x\right) + \frac{M_0}{EI \times \frac{P}{EI}}$$

$$= C_1 \cos\left(x \times \sqrt{\frac{P}{EI}}\right) + C_2 \sin\left(x \times \frac{P}{\sqrt{EI}}\right) + \frac{M_0}{P}$$

Substituting the value $x = 0$ and $\frac{dy}{dx} = 0$, the above equation becomes

$$0 = -C_1 \times 0 + C_2 \times 1 \times \sqrt{\frac{P}{EI}} \qquad (\because \quad \sin 0 = 0 \text{ and } \cos 0 = 1)$$

$$= C_2 \sqrt{\frac{P}{EI}}\,.$$

From the above equation, it is clear that either $C_2 = 0$ or $\sqrt{\frac{P}{EI}} = 0$. But for a given crippling load P, the value of $\sqrt{\frac{P}{EI}}$ cannot be equal to zero.

$\therefore \qquad C_2 = 0.$

Now substituting the values of $C_1 = -\frac{M_0}{P}$ and $C_2 = 0$ in equation (*i*), we get

$$y = -\frac{M_0}{P} \cos\left(x\sqrt{\frac{P}{EI}}\right) + 0 + \frac{M_0}{P}$$

$$= -\frac{M_0}{P} \cos\left(x\sqrt{\frac{P}{EI}}\right) + \frac{M_0}{P} \qquad \text{...(iii)}$$

At the end B of the column, $x = l$ and $y = 0$.

Substituting these values in equation (*iii*), we get

$$0 = -\frac{M_0}{P} \cos\left(l\,.\sqrt{\frac{P}{EI}}\right) + \frac{M_0}{P}$$

or
$$\frac{M_0}{P} \cos\left(l\,.\sqrt{\frac{P}{EI}}\right) = \frac{M_0}{P}$$

or
$$\cos\left(l\,.\sqrt{\frac{P}{EI}}\right) = \frac{M_0}{P} \times \frac{P}{M_0} = 1 = \cos 0,\ \cos 2\pi,\ \cos 4\pi,\ \cos 6\pi,\ \ldots\ldots$$

$$\therefore \qquad l\,.\sqrt{\frac{P}{EI}} = 0,\ 2\pi,\ 4\pi,\ 6\pi,\ \ldots\ldots$$

Taking the least practical value,

$$l\,.\sqrt{\frac{P}{EI}} = 2\pi \quad \text{or} \quad P = \frac{\pi^2 EI}{l^2}\,. \qquad \text{...(7.11)}$$

7.12.4. Crippling Load When One End of the Column is Fixed and the Other End is Hinged (or Pinned). Consider a column AB of length l and uniform cross-sectional area fixed at the end A and hinged at the end B as shown in Fig. 7.27. Let P is the crippling load at which the column has buckled. Due to the crippling load P, the column will deflect as shown in Fig. 7.27.

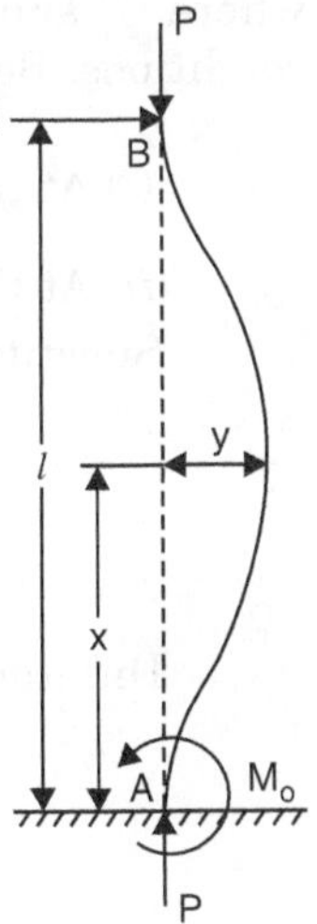

Fig. 7.27

There will be fixed end moment (M_0) at the fixed end A. This will try to bring back the slope of deflected column zero at A. Hence it will be acting anticlock wise at A. The fixed end moment M_0 at A is to be balanced. This will be balanced by a horizontal reaction (H) at the top end B as shown in Fig. 7.27.

Consider a section at a distance x from the end A.

Let y = Deflection of the column at the section,

M_0 = Fixed end moment at A, and

H = Horizontal reaction at B.

The moment at the section = Moment due to crippling load at B

+ Moment due to horizontal reaction at B

$$= -P \,.\, y + H \,.\, (l - x)$$

But the moment at the section is also

$$= EI \frac{d^2y}{dx^2}$$

Equating the two moments, we get

$$EI \frac{d^2y}{dx^2} = -P \,.\, y + H\,(l - x)$$

or

$$EI \frac{d^2y}{dx^2} + P \,.\, y = H\,(l - x)$$

or

$$\frac{d^2y}{dx^2} + \frac{P}{EI} \,.\, y = \frac{H}{EI}\,(l - x) \qquad \text{(Dividing by } EI\text{)} \quad ...(A)$$

$$= \frac{H}{EI}\,(l - x) \times \frac{P}{P} = \frac{P}{EI} \,.\, \frac{H(l - x)}{P}$$

The solution* of the above differential equation is

$$y = C_1 \cos\left(x\sqrt{\frac{P}{EI}}\right) + C_2 \sin\left(x\sqrt{\frac{P}{EI}}\right) + \frac{H}{P}\,(l - x) \qquad ...(i)$$

---

*The equation (A) can be written as

$$\frac{d^2y}{dx^2} + \alpha^2 \,.\, y = \frac{H}{EI}\,(l - x) \text{ where } \alpha^2 = \frac{P}{EI} \quad \text{or} \quad \alpha = \sqrt{\frac{P}{EI}}$$

The complete solution of this equation is

$$y = C_1 \cos(\alpha \,.\, x) + C_2 \sin(\alpha \,.\, x) + \frac{H\,(l - x)}{EI \times \alpha^2}$$

$$= C_1 \cos\left(\sqrt{\frac{P}{EI}} \times x\right) + C_2 \sin\left(\sqrt{\frac{P}{EI}} \times x\right) + \frac{H}{EI} \times \frac{(l - x)}{\left(\frac{P}{EI}\right)}$$

$$= C_1 \cos\left(x \times \sqrt{\frac{P}{EI}}\right) + C_2 \sin\left(x \times \sqrt{\frac{P}{EI}}\right) + \frac{H}{P}\,(l - x)$$

where C_1 and C_2 are constants of integration and their values are obtained from boundary conditions. Boundary conditions are :

(*i*) At the fixed end A, $x = 0$, $y = 0$ and also $\frac{dy}{dx} = 0$

(*ii*) At the hinged end B, $x = l$ and $y = 0$.

Substituting the value $x = 0$ and $y = 0$ in equation (*i*), we get

$$0 = C_1 \times 1 + C_2 \times 0 + \frac{H}{P}(l - 0) = C_1 + \frac{H.l}{P}$$

$$\therefore \quad C_1 = -\frac{H}{P}.l \qquad ...(ii)$$

Differentiating the equation (*i*) w.r.t. x, we get

$$\frac{dy}{dx} = C_1(-1)\sin\left(x.\sqrt{\frac{P}{EI}}\right).\sqrt{\frac{P}{EI}} + C_2\cos\left(x.\sqrt{\frac{P}{EI}}\right).\sqrt{\frac{P}{EI}} - \frac{H}{P}$$

$$= -C_1\sin\left(x.\sqrt{\frac{P}{EI}}\right).\sqrt{\frac{P}{EI}} + C_2\cos\left(x.\sqrt{\frac{P}{EI}}\right).\sqrt{\frac{P}{EI}} - \frac{H}{P}$$

At A, $x = 0$ and $\frac{dy}{dx} = 0$.

$$\therefore \quad 0 = -C_1 \times 0 + C_2 . 1 . \sqrt{\frac{P}{EI}} - \frac{H}{P} \qquad (\because \sin 0 = 0, \cos 0 = 1)$$

$$= C_2\sqrt{\frac{P}{EI}} - \frac{H}{P} \quad \text{or} \quad C_2 = \frac{H}{P}\sqrt{\frac{EI}{P}}.$$

Substituting the values of $C_1 = -\frac{H}{P}, l$ and $C_2 = \frac{H}{P}\sqrt{\frac{EI}{P}}$ in equation (*i*), we get

$$y = -\frac{H}{P}.l\cos\left(x\sqrt{\frac{P}{EI}}\right) + \frac{H}{P}\sqrt{\frac{EI}{P}}\sin\left(x\sqrt{\frac{P}{EI}}\right) + \frac{H}{P}(l - x)$$

At the end B, $x = l$ and $y = 0$.

Hence the above equation becomes as

$$0 = -\frac{H}{P}l\cos\left(l\sqrt{\frac{P}{EI}}\right) + \frac{H}{P}\sqrt{\frac{EI}{P}}\sin\left(l\sqrt{\frac{P}{EI}}\right) + \frac{H}{P}(l - l)$$

$$= -\frac{H}{P}l\cos\left(l\sqrt{\frac{P}{EI}}\right) + \frac{H}{P}\sqrt{\frac{EI}{P}}\sin\left(l\sqrt{\frac{P}{EI}}\right) + 0$$

or
$$\frac{H}{P}\sqrt{\frac{EI}{P}}\sin\left(l\sqrt{\frac{P}{EI}}\right) = \frac{H}{P}l\cos\left(l\sqrt{\frac{P}{EI}}\right)$$

or
$$\sin\left(l\sqrt{\frac{P}{EI}}\right) = \frac{H}{P}.l \times \frac{P}{H} \times \sqrt{\frac{P}{EI}}.\cos\left(l\sqrt{\frac{P}{EI}}\right)$$

$$= l.\sqrt{\frac{P}{EI}}.\cos\left(l.\sqrt{\frac{P}{EI}}\right)$$

or $$\tan\left(l\sqrt{\frac{P}{EI}}\right) = l \, . \sqrt{\frac{P}{EI}} \, .$$

The solution to the above equation is, $l \, . \sqrt{\frac{P}{EI}}$ = 4.5 radians

Squaring both sides, we get

$$l^2 \, . \frac{P}{EI} = 4.5^2 = 20.25$$

$$\therefore \quad P = 20.25 \frac{EI}{l^2}$$

But approximately $20.25 = 2\pi^2$

$$\therefore \quad P = \frac{2\pi^2 EI}{l^2} \quad ...(7.12)$$

7.13. EFFECTIVE LENGTH (OR EQUIVALENT LENGTH) OF A COLUMN

The effective length of a given column with given end conditions is the length of an equivalent column of the same material and cross-section with hinged ends, and having the value of the crippling load equal to that of the given column. Effective length is also called equivalent length.

Let L_e = Effective length of a column,

l = Actual length of the column, and

P = Crippling load for the column.

Then the crippling load for any type of end condition is given by

$$P = \frac{\pi^2 EI}{L_e^2} \, . \quad ...(7.13)$$

The crippling load (P) in terms of actual length and effective length and also the relation between effective length and actual length are given in Table 7.1.

TABLE 7.1

S.No.	*End conditions of column*	*Crippling load in terms of*		*Relation between effective length and actual length*
		Actual length	*Effective length*	
1.	Both ends hinged	$\frac{\pi^2 EI}{l^2}$	$\frac{\pi^2 EI}{L_e^2}$	$L_e = l$
2.	One end is fixed and other is free	$\frac{\pi^2 EI}{4l^2}$	$\frac{\pi^2 EI}{L_e^2}$	$L_e = 2l$
3.	Both ends fixed	$\frac{4\pi^2 EI}{l^2}$	$\frac{\pi^2 EI}{L_e^2}$	$L_e = \frac{l}{2}$
4.	One end fixed and other is hinged	$\frac{4\pi^2 EI}{l^2}$	$\frac{\pi^2 EI}{L_e^2}$	$L_e = \frac{l}{\sqrt{2}}$

There are two values of moment of inertia *i.e.,* I_{xx} and I_{yy}.

The value of I (moment of inertia) in the above expressions should be taken as the least value of the two moments of inertia as the column will tend to bend in the direction of least moment of inertia.

7.13.1. Crippling Stress in Terms of Effective Length and Radius of Gyration. The moment of inertia (I) can be expressed in terms of radius of gyration (k) as

$$I = Ak^2 \text{ where } A = \text{Area of cross-section.}$$

As I is the least value of moment of inertia, then

$$k = \text{Least radius of gyration of the column section.}$$

Now crippling load P in terms of effective length is given by

$$P = \frac{\pi^2 EI}{L_e^2} = \frac{\pi^2 E \times Ak^2}{L_e^2} \qquad (\because \quad I = Ak^2)$$

$$= \frac{\pi^2 E \times A}{\frac{L_e^2}{k^2}} = \frac{\pi^2 E \times A}{\left(\frac{L_e}{k}\right)^2} \qquad \text{...(7.14)}$$

And the stress corresponding to crippling load is given by

$$\text{Crippling stress} = \frac{\text{Crippling load}}{\text{Area}} = \frac{P}{A}$$

$$= \frac{\pi^2 E \times A}{A\left(\frac{L_e}{k}\right)^2} \qquad \text{(Substituting the value of } P\text{)}$$

$$= \frac{\pi^2 E}{\left(\frac{L_e}{k}\right)^2} \qquad \text{...(7.15)}$$

7.13.2. Slenderness Ratio. The ratio of the actual length of a column to the least radius of gyration of the column, is known as slenderness ratio.

Mathematically, slenderness ratio is given by

$$\text{Slenderness ratio} = \frac{\text{Actual length}}{\text{Least radius of gyration}} = \frac{l}{k} \qquad \text{...(7.16)}$$

7.13.3. Limitation of Euler's Formula. From equation (7.14), we have

$$\text{Crippling stress} = \frac{\pi^2 E}{\left(\frac{L_e}{k}\right)^2}$$

For a column with both ends hinged, $L_e = l$. Hence Crippling stress becomes as $= \dfrac{\pi^2 E}{\left(\frac{l}{k}\right)^2}$,

where $\dfrac{l}{k}$ is slenderness ratio.

If the slenderness ratio $\left(i.e., \frac{l}{k}\right)$ is small, the crippling stress (or the stress at failure) will be high. But for the column material, the crippling stress cannot be greater than the crushing stress. Hence when the slenderness ratio is less than a certain limit, Euler's formula gives a value of crippling stress greater than the crushing stress. In the limiting case, we can find the value of l/k for which crippling stress is equal to crushing stress.

For example, for a mid steel column with both ends hinged,

Crushing stress = 330 N/mm^2

Young's modulus, $E = 2.1 \times 10^5$ N/mm^2.

Equating the crippling stress to the crushing stress corresponding to the minimum value of slenderness ratio, we get

Crippling stress = Crushing stress

or
$$\frac{\pi^2 E}{\left(\frac{l}{k}\right)^2} = 330 \quad \text{or} \quad \frac{\pi^2 \times 2.1 \times 10^5}{\left(\frac{l}{k}\right)^2} = 330$$

$$\therefore \quad \left(\frac{l}{k}\right)^2 = \frac{\pi^2 \times 2.1 \times 10^5}{330} = 6282$$

$$\therefore \quad \frac{l}{k} = \sqrt{6282} = 79.27, \text{ say } 80.$$

Hence, if the slenderness ratio is less than 80 for mild steel column with both ends hinged, then Euler's formula will not be valid.

7.14. SOLVED PROBLEMS 7.13 TO 7.25 BASED ON EULER'S THEORY

Problem 7.13. *A solid round bar 3 m long and 5 cm in diameter is used as a strut with both ends hinged. Determine the crippling (or collapsing) load. Take E = 2.0 × 10^5 N/mm^2.*

Sol. Given :

Length of bar, $l = 3$ m = 3000 mm

Diameter of bar, $d = 5$ cm = 50 mm

Young's modulus, $E = 2.0 \times 10^5$ N/mm^2

Moment of inertia, $I = \frac{\pi}{64} \times 5^4 = 30.68 \text{ cm}^4 = 30.68 \times 10^4 \text{ mm}^4$

Let P = Crippling load.

As both the ends of the bar are hinged, hence the crippling load is given by equation (7.1).

$$\therefore \quad P = \frac{\pi^2 EI}{l^2} = \frac{\pi^2 \times 2 \times 10^5 \times 30.68 \times 10^4}{3000^2}$$

$$= 67288 \text{ N} = \mathbf{67.288 \text{ kN}}. \quad \textbf{Ans.}$$

Problem 7.14. *For the problem 7.13 determine the crippling load, when the given strut is used with the following conditions :*

(i) One end of the strut is fixed and the other end is free

(ii) Both the ends of strut are fixed

(iii) One end is fixed and other is hinged.

Sol. Given :

The data from Problem 7.13, is l = 3000 mm, diameter = 50 mm, $E = 2.0 \times 10^5$ N/mm^2 and $I = 30.68 \times 10^4$ mm^4.

Let P = Crippling load.

(i) Crippling load when one end is fixed and other is free

Using equation (7.10), $P = \dfrac{\pi^2 EI}{4l^2} = \dfrac{\pi^2 \times 2 \times 10^5 \times 30.68 \times 10^4}{4 \times 3000^2}$ = **16822 N. Ans.**

Alternate Method

The crippling load for any type of end condition is given by equation (7.13) as,

$$P = \frac{\pi^2 EI}{L_e^2} \quad ...(i)$$

where L_e = Effective length.

The effective length (L_e) when one end is fixed and other end is free from Table 7.1 on page 415 is given as

$$L_e = 2l = 2 \times 3000 = 6000 \text{ mm}$$

Substituting the value of L in equation (i), we get

$$P = \frac{\pi^2 \times 2 \times 10^5 \times 30.68 \times 10^4}{6000^2} = \textbf{16822 N. Ans.}$$

(ii) Crippling load when both the ends are fixed

Using equation (7.12), $P = \dfrac{4\pi^2 EI}{l^2} = \dfrac{4\pi^2 \times 2 \times 10^5 \times 30.68 \times 10^4}{3000^2}$

= 269152 N = **269.152 kN. Ans.**

Alternate Method

Using equation (7.13), $P = \dfrac{\pi^2 EI}{L_e^2}$

where L_e = Effective length

$= \dfrac{l}{2}$ (when both the ends are fixed)

$= \dfrac{3000}{2}$ ($\because$ l = 3000)

= 1500 mm

$$\therefore \quad P = \frac{\pi^2 \times 2.0 \times 10^5 \times 30.68 \times 10^4}{1500^2} = \textbf{269152 N. Ans.}$$

(iii) Crippling load when one end is fixed and the other is hinged

Using equation (7.12), $P = \dfrac{2\pi^2 EI}{l^2} = \dfrac{2 \times \pi^2 \times 2.0 \times 10^5 \times 30.68 \times 10^4}{3000^2}$ = **134576 N. Ans.**

Alternate Method

Using equation (7.13), $P = \dfrac{\pi^2 EI}{L_e^2}$

where L_e = Effective length

$$= \frac{l}{\sqrt{2}} \text{ (when one end is fixed and the other is hinged)}$$

$$= \frac{3000}{\sqrt{2}}$$

$$\therefore \quad P = \frac{\pi^2 \times 2.0 \times 10^5 \times 30.68 \times 10^4}{\left(\frac{3000}{\sqrt{2}}\right)^2} = \textbf{134576 N. Ans.}$$

Problem 7.15. *A column of timber section 15 cm × 20 cm is 6 metre long both ends being fixed. If the Young's modulus for timber = 17.5 kN/mm², determine :*

(i) Crippling load and

(ii) Safe load for the column if factor of safety = 3.

Sol. Given :

Dimension of section = 15 cm × 20 cm

Actual length, l = 6 m = 6000 mm

Young's modulus, E = 17.5 kN/mm²

(i) Let P = Crippling load

Using equation (7.13), we get

$$P = \frac{\pi^2 EI}{L_e^2} \quad \text{...}(i)$$

where L_e = Effective length

$$= \frac{l}{2} \quad \text{(when both the ends are fixed)}$$

$$= \frac{6000}{2} = 3000 \text{ mm} \quad (\because \ l = 6000 \text{ mm})$$

I = Least value of moment of inertia

Moment of inertia of the section about X-X axis,

$$I_{XX} = \frac{15 \times 20^3}{12} = 10000 \text{ cm}^4$$

$$= 10000 \times 10^4 \text{ mm}^4$$

And moment of inertia of the section about Y-Y axis,

$$I_{YY} = \frac{20 \times 15^3}{12} = 5625 \text{ cm}^4$$

$$= 5625 \times 10^4 \text{ mm}^4.$$

Since I_{YY} is less than I_{XX}, therefore the column will tend to buckle in Y-Y direction.

And the value of I will be the least value of the two moment of inertia.

$$\therefore \quad I = 5625 \text{ cm}^4 = 5625 \times 10^4 \text{ mm}^4$$

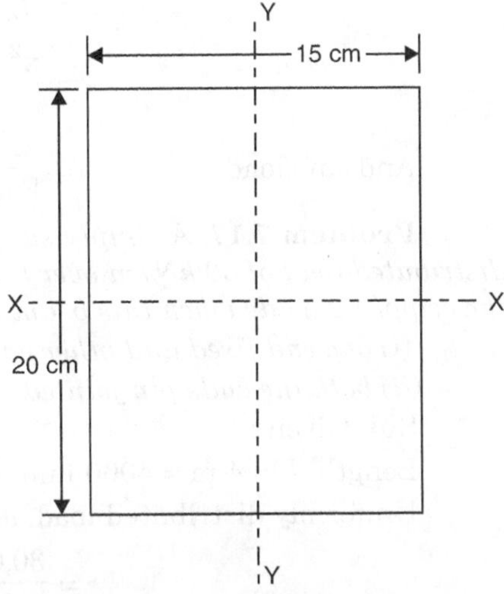

Fig. 7.28

Substituting the values of $I = 5625 \times 10^4$ mm^4 and $L = 3000$ mm in equation (*i*), we get

$$P = \frac{\pi^2 \times 17.5 \times 5625 \times 10^4}{3000} = \mathbf{1079.48\ kN.}\quad \textbf{Ans.}$$

(*ii*) *Safe load for the column*

Factor of safety = 3.0 (given)

$\therefore$ Safe load $= \dfrac{\text{Crippling load}}{\text{Factor of safety}} = \dfrac{1079.48}{3}$ = **359.8 say 360 kN. Ans.**

Problem 7.16. *A hollow mild steel tube 6 m long 4 cm internal diameter and 6 mm thick is used as a strut with both ends hinged. Find the crippling load and safe load taking factor of safety as 3. Take $E = 2 \times 10^5$ N/mm^2.*

Sol. Given :

Length of tube, l = 6 m = 600 cm

Internal dia., d = 4 cm

Thickness, t = 5 mm = 0.5 cm

$\therefore$ External dia., $D = d + 2t = 4 + 2 \times 0.5 = 4 + 1 = 5$ cm

Young's modulus, $E = 2 \times 10^5$ N/mm^2

Factor of safety = 3.0

Moment of inertia of section, $I = \dfrac{\pi}{64}(D^4 - d^4) = \dfrac{\pi}{64}[5^4 - 4^4]$ cm^4

$$= \frac{\pi}{64}(625 - 256) = 18.11 \text{ cm}^4 = 18.11 \times 10^4 \text{ mm}^4$$

Since both ends of the strut are hinged,

$\therefore$ Effective length, $L_e = l$ = 600 cm = 6000 mm

Let P = Crippling load

Using equation (7.13), we get

$$P = \frac{\pi^2 EI}{L_e^{\,2}}$$

$$= \frac{\pi^2 \times 2.0 \times 10^5 \times 18.11 \times 10^4}{6000^2} = \mathbf{9929.9\ say\ 9930\ N.}\quad \textbf{Ans.}$$

And safe load $= \dfrac{\text{Crippling load}}{\text{Factor of safety}} = \dfrac{9930}{3.0}$ = **3310 N. Ans.**

Problem 7.17. *A simply supported beam of length 4 metre is subjected to a uniformly distributed load of 30 kN/m over the whole span and deflects 15 mm at the centre. Determine the crippling loads when this beam is used as a column with the following conditions :*

(*i*) *one end fixed and other end hinged*

(*ii*) *both the ends pin jointed.* (Annamalai University, 1990)

Sol. Given :

Length, L = 4 m = 4000 mm

Uniformly distributed load, w = 30 kN/m = 30,000 N/m

$$= \frac{30{,}000}{1000} \text{ N/mm} = 30 \text{ N/mm}$$

Deflection at the centre, δ = 15 mm.

For a simply supported beam, carrying U.D.L. over the whole span, the deflection at the centre is given by,

$$\delta = \frac{5}{384} \times \frac{w \times L^4}{EI}$$

or $$15 = \frac{5}{384} \times \frac{30 \times 4000^4}{EI}$$

$\therefore$ $$EI = \frac{5}{384} \times \frac{30 \times 4000^4}{15}$$

$$= \frac{5}{384} \times \frac{3 \times 256}{15} \times 10^{13} = \frac{2}{3} \times 10^{13} \text{ N mm}^2.$$

(*i*) *Crippling load when the beam is used as a column with one end fixed and other end hinged.*

The crippling load P for this case in terms of actual length is given by equation (7.12) as

$$P = \frac{2\pi^2 \times EI}{L_e^2}, \text{ where } l = \text{actual length} = 4000 \text{ mm}$$

$$= \frac{2 \times \pi^2 \times \frac{2}{3} \times 10^{13}}{4000^2} = \textbf{8224.5 kN.} \quad \textbf{Ans.}$$

(*ii*) *Crippling load when both the ends are pin-jointed*

This is given by equation (7.9) in terms of actual length as

$$P = \frac{2\pi^2 \times EI}{l^2}, \quad \text{where } l = \text{actual length} = 4000 \text{ mm}$$

$$= \frac{\pi^2 \times \frac{2}{3} \times 10^{13}}{4000^2} = \textbf{4112.25 kN.} \quad \textbf{Ans.}$$

Problem 7.18. *A solid round bar 4 m long and 5 cm in diameter was found to extend 4.6 mm under a tensile load of 50 kN. This bar is used as a strut with both ends hinged. Determine the buckling load for the bar and also the safe load taking factor of safety as 4.0.*

Sol. Given :

Actual length of bar, L = 4 m = 4000 mm

Dia. of bar, d = 5 cm

$\therefore$ Area of bar, $A = \frac{\pi}{4} \times 5^2 = 6.25\pi \text{ cm}^2 = 6.25\pi \times 10^2 \text{ mm}^2 = 625\pi \text{ mm}^2$

Extension of bar, δL = 4.6 mm

Tensile load, W = 50 kN = 50000 N.

In this problem, the values of Young's modulus (E) is not given. But it can be calculated from the given data.

$$\therefore \text{Young's modulus,} \quad E = \frac{\text{Tensile stress}}{\text{Tensile strain}} = \frac{\left(\frac{\text{Tensile load}}{\text{Area}}\right)}{\left(\frac{\text{Extension of bar}}{\text{Length of bar}}\right)}$$

$$\left(\because \text{ Stress} = \frac{\text{Load}}{\text{Area}} \text{ and strain} = \frac{\delta L}{L}\right)$$

$$= \frac{\left(\frac{W}{A}\right)}{\frac{\delta L}{L}} = \frac{W}{A} \times \frac{L}{\delta L} = \frac{50000}{625\,\pi} \times \frac{4000}{4.6} = 2.214 \times 10^4 \text{ N/mm}^2.$$

Since the strut is hinged at its both ends,

$\therefore$ Effective length, L_e = Actual length = 4000 mm

Let P = Crippling or buckling load.

Using equation (7.13), we get

$$P = \frac{\pi^2 EI}{L_e^2}$$

$$= \frac{\pi^2 \times 2.214 \times 10^4 \times \frac{\pi}{64} \times 5^4 \times 10^4}{4000 \times 4000} \qquad \left(\because \; I = \frac{\pi}{64} \times 5^4 \times 10^4 \text{ mm}^4\right)$$

= 4189.99 say 4190 N. Ans.

And safe load $= \frac{\text{Crippling load}}{\text{Factor of safety}} = \frac{4190}{4}$ = **1047.5 N. Ans.**

Problem 7.19. *A hollow alloy tube 5 m long with external and internal diameters 40 mm and 25 mm respectively was found to extend 6.4 mm under a tensile load of 60 kN. Find the buckling load for the tube when used as a column with both ends pinned. Also find the safe load for the tube, taking a factor of safety = 4.* (AMIE, Summer 1989)

Sol. Given :

Actual length, L = 5 m = 5000 mm

External dia., D = 40 mm

Internal dia., d = 25 mm

Extension, δL = 6.4 mm

Tensile load, W = 60 kN = 60,000 N

Safety factor = 4

Cross-section area, $A = \frac{\pi}{4}(D^2 - d^2) = \frac{\pi}{4}(40^2 - 25^2) = 766 \text{ mm}^2$

Moment of inertia, $I = \frac{\pi}{64}(D^4 - d^4) = \frac{\pi}{64}(40^4 - 25^4)$

$$= \frac{\pi}{64}(2560000 - 390625) = \frac{\pi}{64} \times 2169375 = 106500 \text{ mm}^4$$

The value of Young's modulus is obtained as given below

$$E = \frac{\text{Tensile stress}}{\text{Tensile strain}} = \frac{\left(\frac{W}{A}\right)}{\left(\frac{\delta L}{L}\right)}$$

$$= \frac{\left(\frac{60,000}{766}\right)}{\left(\frac{6.4}{5000}\right)} = \frac{60,000}{766} \times \frac{5000}{64} = 6.11945 \times 10^4 \text{ N/mm}^2$$

Since the column is pinned at both the ends.

$\therefore$ Effective length, L_e = Actual length = 5000 mm

Let P = Buckling load

Using equation (7.13), we get

$$P = \frac{\pi^2 EI}{L_e^2} = \frac{\pi^2 \times 6.11945 \times 10^4 \times 106500}{5000^2} = \mathbf{2573\ N.}\quad \textbf{Ans.}$$

And safe load $= \dfrac{\text{Buckling load}}{\text{Factor of safety}} = \dfrac{2573}{4} = \mathbf{643.2\ N.}$ **Ans.**

Problem 7.20. *Calculate the safe compressive load on a hollow cast iron column (one end rigidly fixed and other hinged) of 15 cm external diameter, 10 cm internal diameter and 10 m in length. Use Euler's formula with a factor of safety of 5 and E = 95 kN/mm².*

(AMIE, Winter 1981)

Sol. Given :

External dia., D = 15 cm

Internal dia., d = 10 cm

Actual length of column, l = 10 m = 10000 mm

Factor of safety = 5.0

Young's modulus, E = 95 kN/mm² = 95000 N/mm²

Moment of inertia of hollow column,

$$I = \frac{\pi}{64}(D^4 - d^4) = \frac{\pi}{64}(15^4 - 10^4) = \frac{\pi}{64}(50625 - 10000)$$

$$= 1994.175 \text{ cm}^4 = 1994.175 \times 10^4 \text{ mm}^4$$

Since one end of the column is fixed and other end is hinged.

$\therefore$ Effective length, $L_e = \dfrac{l}{\sqrt{2}} = \dfrac{10000}{\sqrt{2}}$

Let P = Crippling load.

Using equation (7.13), we get

$$P = \frac{\pi^2 EI}{L_e^2}$$

$$= \frac{\pi^2 \times 95000 \times 1994.175 \times 10^4}{\left(\frac{10000}{\sqrt{2}}\right)^2} \qquad \left(\because\ L = \frac{10000}{\sqrt{2}}\right)$$

$$= \frac{\pi^2 \times 95000 \times 1994.175 \times 10^4 \times 2}{10000 \times 10000} = 373950 \text{ N} = 373.95 \text{ kN}$$

But safe load $= \dfrac{\text{Crippling load}}{\text{Factor of safety}} = \dfrac{373.95}{5} = \mathbf{74.79\ kN.}$ **Ans.**

Problem 7.21. *Determine Euler's crippling load for an I-section joist 40 cm × 20 cm × 1 cm and 5 m long which is used as a strut with both ends fixed. Take Young's modulus for the joist as 2.1 × 10⁵ N/mm².*

Sol. Given :

Dimensions of I-section = 40 cm × 20 cm × 1 cm

Length actual, l = 5 m = 5000 mm

Young's modulus, $E = 2.1 \times 10^5$ N/mm².

Moment of inertia of the section about X-X axis,

I_{XX} = M.O.I. of rectangle of dimension 20 cm × 40 cm – M.O.I. of rectangle of dimension [(20 – 1), (40 – 1 – 1)]

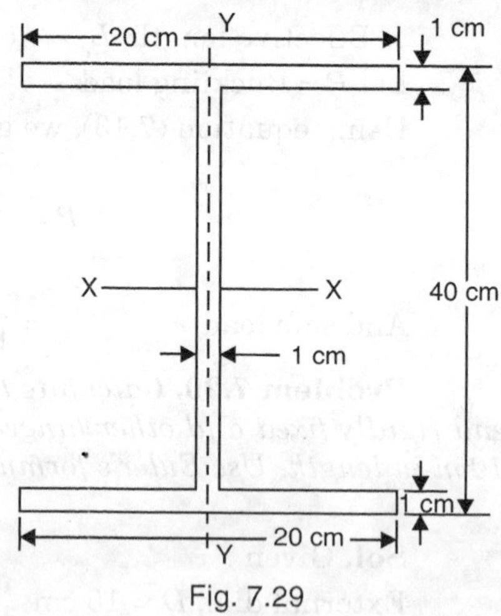

Fig. 7.29

$$= \frac{1}{12} bd^3 - \frac{1}{12} b_1 d_1^{\,3}$$

$$= \frac{1}{12} \times 20 \times 40^3 - \frac{1}{12} \times 19 \times 38^3$$

$$(\because \; b_1 = 19, d_1 = 38)$$

$$= 19786 \text{ cm}^4.$$

Similarly the moment of inertia of the section about Y-Y axis

I_{YY} = M.O.I. of rectangle of dimension (38 × 1) + M.O.I. of two rectangles of dimension (1 × 20)

$$= \frac{1}{12} \times 38 \times 1^3 + 2 \times \frac{1}{12} \times 1 \times 20^3$$

$$= 3.166 \times 1333.33 = 1336.5 \text{ cm}^4.$$

∴ Least value of the moment of inertia is about Y-Y axis.

∴ $I = 1336.5 \text{ cm}^4 = 1336.5 \times 10^4 \text{ mm}^4$

As both the ends of the strut are fixed

∴ Effective length, $L_e = \frac{l}{2} = \frac{5000}{2} = 2500$ mm

Let P = Crippling load.

Using equation (7.13), we get

$$P = \frac{\pi^2 EI}{L_e^{\,2}} = \frac{\pi^2 \times 2.1 \times 10^5 \times 1336.5 \times 10^4}{2500^3}$$

$$= 4432080 \text{ N} = \mathbf{4.432 \text{ MN}}. \quad \textbf{Ans.}$$

Problem 7.22. *Determine the crippling load for a T-section of dimensions 10 cm × 10 cm × 2 cm and of length 5 m when it is used as strut with both of its ends hinged. Take Young's modulus, $E = 2.0 \times 10^5$ N/mm².*

Sol. Given :

Dimensions of T-section = 10 cm × 10 cm × 2 cm

Length actual, l = 5 m = 5000 mm

Young's modulus, $E = 2.0 \times 10^5$ N/mm².

First of all, calculate the C.G. of the section. The given section is symmetrical about the axis Y-Y, hence the C.G. of the section will lie on Y-Y axis.

Let $\bar{y}$ = Distance of C.G. of the section from bottom end.

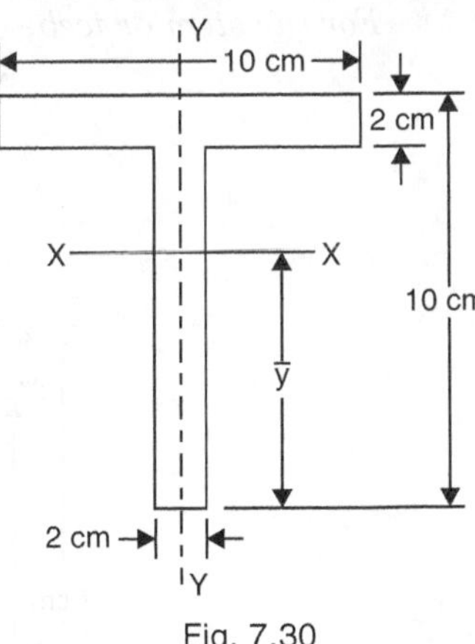

Fig. 7.30

For the flange, we have $a_1 = 10 \times 2 = 20$ cm²

y_1 = Distance of C.G. of area a_1 from the bottom end

$= 8 + 1 = 9$ cm

For the web, we have $a_2 = 8 \times 2 = 16$ cm²

y_2 = Distance of C.G. of area a_2 from bottom end

$= \frac{8}{2} = 4$ cm

Using the relation, $\bar{y} = \frac{a_1y_1 + a_2y_2}{a_1 + a_2}$

$= \frac{20 \times 9 + 16 \times 4}{20 + 16} = \frac{180 + 64}{36}$

$= 6.777$ cm

Moment of inertia of the section about the axis X-X,

$$I_{XX} = \left(\frac{10 \times 8^3}{12} + 20 \times 2.223^2\right) + \left(\frac{2 \times 8^3}{12} + 16 \times 2.777^2\right)$$

$= (6.667 + 98.834) + (85.333 + 123.387) = 314.221$ cm⁴.

Moment of inertia of the section about the axis Y-Y,

$$I_{YY} = \frac{2 \times 10^3}{12} + \frac{8 \times 2^3}{12} = 166.67 + 5.33 = 172 \text{ cm}^4.$$

Least value of moment of inertia is about Y-Y axis

$\therefore$ $I = 172 \text{ cm}^4 = 172 \times 10^4 \text{ mm}^4$

Since the strut is hinged at both of its end

$\therefore$ Effect length, $L_e = l = 5000$ mm

Let P = Crippling load

Using equation (7.13), we get

$$P = \frac{\pi^2 EI}{L_e^2} = \frac{\pi^2 \times 2.0 \times 10^5 \times 172 \times 10^4}{5000^3} = \textbf{135805.7 N.} \quad \textbf{Ans.}$$

Problem 7.23. *Calculate the Euler's critical load for a strut of T-section, the flange width being 10 cm, overall depth 8 cm and both flange and stem 1 cm thick. The strut is 3 m long and is built in at both ends. Take E = 2 × 10⁵ N/mm⁵.*

Sol. Given :

Actual length, $l = 3$ m $= 3000$ mm

Value of $E = 2 \times 10^5$ N/mm²

The dimensions of T-section are shown in Fig. 7.30 (a). First of all, calculate the C.G. of the section. The given section is symmetrical about the Y-Y axis.

Hence the C.G. will lie on Y-Y axis.

Let $\bar{y}$ = Distance of C.G. of the section from the bottom end.

For the flange a_1 = Area of flange = $10 \times 1 = 10$ cm²

y_1 = Distance of C.G. of area a_1 from the bottom end

$= 7 + \frac{1}{2} = 7.5$ cm.

For the stem or web, a_2 = Area of stem = 7 × 1 = 7 cm^2

y_2 = Distance of C.G. of area a_2 from the bottom end

$= \frac{7}{2} = 3.5$ cm

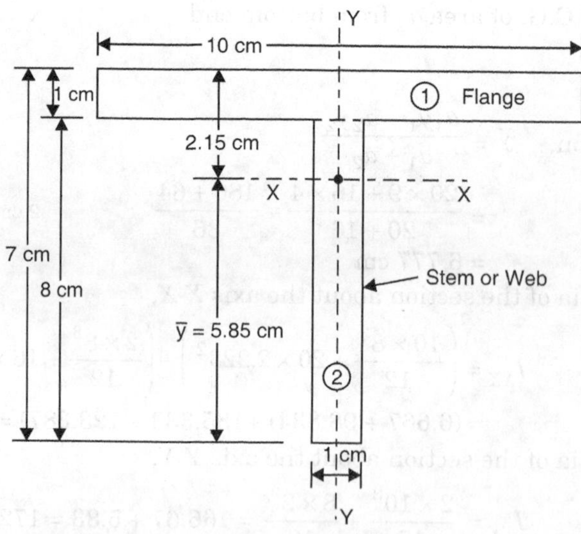

Fig. 7.30 (*a*)

Now using the relation, $\bar{y} = \frac{a_1 \times y_1 + a_2 \times y_2}{a_1 + a_2} = \frac{10 \times 7.5 + 7 \times 3.5}{(10 + 7)}$

$= \frac{75 + 24.5}{17} = \frac{99.5}{17} = 5.85$ cm.

Now calculate the moment of inertia about *X-X* axis and *Y-Y* axis.

$$I_{XX} = \left[\frac{10 \times 1^3}{12} + a_1 \times (2.15 - 0.5)^2\right] + \left[\frac{1 \times 7^3}{12} + a_a \times (5.85 - 3.5)^2\right]$$

$$= \left[\frac{10}{12} + 10 \times 1.65^2\right] + \left[\frac{343}{12} + 7 \times 2.35^2\right] = 95.298 \text{ cm}^4$$

and $$I_{YY} = \frac{1 \times 10^3}{12} + \frac{7 \times 1^3}{12} = 83.33 + 0.583 = 83.916 \text{ cm}^4.$$

The least value of moment of inertia is about *Y*-axis.

$\therefore$ $$I = I_{YY} = 83.916 \text{ cm}^4 = 839160 \text{ mm}^4.$$

As the strut is fixed at both ends, hence its effective length (L_e) will be half of its actual length.

$\therefore$ $$L_e = \frac{l}{2} = \frac{3000}{2} = 1500 \text{ mm}$$

Let P = Euler's critical load.

Using equation (7.13), $P = \frac{\pi^2 EI}{L_e^2} = \frac{\pi^2 \times 2 \times 10^5 \times 839160}{1500^3}$

= 736190 N = **736.19 kN. Ans.**

Problem 7.24. *Determine the ratio of buckling strengths of two columns one hollow and the other solid. Both are made of the same material and have the same length, cross-sectional area and end conditions. The internal diameter of hollow column is half of its external diameter.*

Sol. Given :

Let d = Dia. of solid column

D = External dia. of hollow column

$\therefore$ Internal dia. of hollow shaft $= \dfrac{D}{2}$

Both the shafts are made of the same material and have the same length, cross-sectional area and end conditions. As areas of solid and hollow shafts are equal.

$\therefore$ Area of solid shaft = Area of hollow shaft

or
$$\frac{\pi}{2} d^2 = \frac{\pi}{2}\left[D^2 - \left(\frac{D}{2}\right)^2\right] \text{ or } d^2 = \left[D^2 - \frac{D^2}{4}\right] = \frac{3D^2}{4}$$

$$\therefore \quad d = \frac{\sqrt{3}D}{2} \qquad \ldots(i)$$

Buckling load of a column as given by Euler's formula is

$$P = \frac{\pi^2 EI}{L^2} \qquad \text{[see equation (7.13)]}$$

As the shafts are having the same length, same material, same cross-sectional areas and same end conditions. Hence the value of A, E and L for both shafts are same.

$$\therefore \quad P \propto I \qquad (\because \; A, E \text{ and } L \text{ are same}) \ldots(ii)$$

Let P_1 = Buckling load of the hollow column

P_2 = Buckling load of solid column

I_1 = Least moment of inertia of the hollow column

$$= \frac{\pi}{64}\left[D^4 - \left(\frac{D}{2}\right)^2\right] = \frac{\pi}{64}\left[D^4 - \frac{D^4}{16}\right] = \frac{\pi}{64} \times \frac{15}{16} D^4$$

I_2 = Least moment of inertia of the solid column

$$= \frac{\pi}{64} d^4.$$

From equation (ii), we have

$$\frac{P_1}{I_1} = \frac{P_2}{I_2} \quad \text{or} \quad \frac{P_1}{P_2} = \frac{I_1}{I_2}$$

$$= \frac{\frac{\pi}{64} \times \frac{15}{16} D^4}{\frac{\pi}{64} d^4} \qquad \left(\because \; I_1 = \frac{\pi}{64} \times \frac{15}{16} D^4,\ I_2 = \frac{\pi}{64} d^4\right)$$

$$= \frac{15}{16} \frac{D^4}{d^4} = \frac{15}{16} \frac{D^4}{\left(\frac{\sqrt{3}D}{2}\right)^4} \qquad \left(\because \; d = \frac{\sqrt{3}D}{2}\right)$$

$$= \frac{15}{16} \times \frac{D^4}{\frac{9}{16} D^4} = \frac{15}{9} = \frac{5}{3} = 1.667$$

$$\therefore \quad \frac{\text{Buckling load of hollow column}}{\text{Buckling load of solid column}} = \mathbf{1.667.} \quad \textbf{Ans.}$$

Problem 7.25. *Using Euler's formula, calculate the critical stresses for a series of struts having slenderness ratio of 40, 80, 120, 160 and 200 under the following conditions :*

(i) both ends hinged and

(ii) both ends fixed.

Take $E = 2.05 \times 10^5$ N/mm^2. (Annamalai University, 1990)

Sol. Given :

(i) Critical stresses when both ends hinged

Slenderness ratios, $\frac{l}{k} = 40, 80, 120, 160$ and 200

The critical stress or crippling stress is given by equation (7.15).

$$\therefore \quad \text{Critical stress} = \frac{\pi^2 \times E}{\left(\frac{L_e}{k}\right)^2}$$

where L_e = Effective length

But for both ends hinged, $L_e = l$

where l = actual length.

$$\therefore \quad \text{Critical stress} = \frac{\pi^2 E}{\left(\frac{l}{k}\right)^2} \qquad (\because \quad L = l) \text{ ...(ii)}$$

When $\frac{l}{k} = 40$, the critical stress becomes as

$$= \frac{\pi^2 \times E}{40^2} = \frac{\pi^2 \times 2.05 \times 10^5}{1600} = \mathbf{1264.54\ N/mm^2.} \quad \textbf{Ans.}$$

When $\frac{l}{k} = 80$, the critical stress becomes as

$$= \frac{\pi^2 \times E}{80^2} = \frac{\pi^2 \times 2.05 \times 10^5}{6400} = \mathbf{316.135\ N/mm^2.} \quad \textbf{Ans.}$$

When $\frac{l}{k} = 120$, the critical stress becomes as

$$= \frac{\pi^2 \times E}{120^2} = \frac{\pi^2 \times 2.05 \times 10^5}{14400} = \mathbf{140.5\ N/mm^2.} \quad \textbf{Ans.}$$

When $\frac{l}{k} = 160$, the critical stress

$$= \frac{\pi^2 \times E}{160^2} = \frac{\pi^2 \times 2.05 \times 10^5}{25600} = \mathbf{79.03\ N/mm^3.} \quad \textbf{Ans.}$$

When $\frac{l}{k} = 200$, the critical stress

$$= \frac{\pi^2 \times E}{200^2} = \frac{\pi^2 \times 2.05 \times 10^5}{400,00} = \mathbf{50.58\ N/mm^2.}\quad \textbf{Ans.}$$

(ii) Critical stresses when both ends fixed

Using equation (7.15),

$$\text{Crippling or critical stress} = \frac{\pi^2 \times E}{\left(\frac{L_e}{k}\right)^2}$$

When both ends fixed, $L_e = \frac{l}{2}$

$$\therefore \quad \text{Critical stress} = \frac{\pi^2 \times E}{\left(\frac{l}{2k}\right)^2} = \frac{4\pi^2 \times E}{\left(\frac{l}{k}\right)^2}$$

When $\frac{l}{k} = 40$, the critical stress becomes as

$$= \frac{4\pi^2 \times E}{\left(\frac{l}{k}\right)^2} = \frac{4 \times \pi^2 \times 2.05 \times 10^5}{40^2}$$

$$= \frac{4 \times \pi^2 \times 2.05 \times 10^5}{1600} = \mathbf{5058.16\ N/mm^2.}\quad \textbf{Ans.}$$

When $\frac{l}{k} = 80$, the critical stress becomes as

$$= \frac{4\pi^2 \times E}{\left(\frac{l}{k}\right)^2} = \frac{4 \times \pi^2 \times 2.05 \times 10^5}{80^2} = \mathbf{1264.54\ N/mm^2.}\quad \textbf{Ans.}$$

When $\frac{l}{k} = 120$, the critical stress becomes as

$$= \frac{4\pi^2 \times E}{120^2} = \frac{4 \times \pi^2 \times 2.05 \times 10^5}{120^2} = \mathbf{562.02\ N/mm^2.\ Ans.}$$

When $\frac{l}{k} = 160$, the critical stress becomes as

$$= \frac{4\pi^2 \times 2.05 \times 10^5}{160^2} = \mathbf{316.135\ N/mm^2.\ Ans.}$$

When $\frac{l}{k} = 200$, the critical stress becomes as

$$= \frac{4\pi^2 \times 2.05 \times 10^5}{200^2} = \mathbf{202.32\ N/mm^2.\ Ans.}$$

7.15. RANKINE GOURDAN FORMULA

In Art. 7.12, we have learnt that Euler's formula gives correct results only for very long columns. But what happens when the column is a short or the column is not a very long. On the basis of results of experiments performed by Rankine Gourdan, they established an empirical formula which is applicable to all columns whether they are short or long. The empirical formula given by Rankine Gourdan is known as Rankine Gourdan formula, which is given as

$$P = \frac{\sigma_c \, . \, A}{1 + a \, . \left(\frac{L_e}{k}\right)^2} \qquad ...(7.17)$$

where $a = \dfrac{\sigma_c}{\pi^2 E}$ and is known as Rankine's constant

and P = crippling load by Rankine formula.

The equation (7.17) gives crippling load by Rankine's formula. As the Rankine formula is empirical formula, the value of 'a' is taken from the results of the experiments and is not calculated from the values of σ_c and E.

The values of σ_c and a for different columns material are given below in Table 7.2.

TABLE 7.2

S. No.	*Material*	*σ_c in N/mm^2*	*a*
1.	Wrought Iron	250	$\frac{1}{9000}$
2.	Cast Iron	550	$\frac{1}{1600}$
3.	Mild Steel	320	$\frac{1}{7500}$
4.	Timber	50	$\frac{1}{750}$

7.16. SOLVED PROBLEMS 7.26 TO 7.32 BASED ON RANKINE'S FORMULA

Problem 7.26. *The external and internal diameter of a hollow cast iron column are 5 cm and 4 cm respectively. If the length of this column is 3 m and both of its ends are fixed, determine the crippling load using Rankine's formula. Take the values of* $\sigma_c = 550$ *N/mm*^{2} *and* $a = \dfrac{1}{1600}$ *in Rankine's formula.*

Sol. Given :

External dia., D = 5 cm

Internal dia., d = 4 cm

$\therefore$ Area, $A = \dfrac{\pi}{4}(5^2 - 4^2) = 2.25\pi \text{ cm}^2 = 2.25\pi \times 10^2 \text{ mm}^2 = 225\pi \text{ mm}^2$

Moment of Inertia, $I = \frac{\pi}{64} [5^4 - 4^4] = 5.7656\ \pi\ \text{cm}^4$

$= 5.7656\pi \times 10^4\ \text{mm}^4 = 57656\pi\ \text{mm}^4$

$\therefore$ Least radius of gyration,

$$k = \sqrt{\frac{I}{A}} = \sqrt{\frac{57656\pi}{225\pi}} = 25.625\ \text{mm}$$

Length of column, $l = 3\ \text{m} = 3000\ \text{mm}$

As both the ends are fixed,

$\therefore$ Effective length, $L_e = \frac{l}{2} = \frac{3000}{2} = 1500\ \text{mm}$

Crushing stress, $\sigma_c = 550\ \text{N/mm}^2$

Rankine's constant, $a = \frac{1}{1600}$

Let P = Crippling load by Rankine's formula

Using equation (7.17), we have

$$P = \frac{\sigma_c \,.\, A}{1 + \left(\frac{L_e}{k}\right)^2} = \frac{550 \times 225\pi}{1 + \frac{1}{1600} \times \left(\frac{1500}{25.625}\right)^2}$$

$$= \frac{550 \times 225\pi}{3.1415} = \mathbf{123750\ N.}\quad \textbf{Ans.}$$

Problem 7.27. *A hollow cylindrical cast iron column is 4 m long with both ends fixed. Determine the minimum diameter of the column if it has to carry a safe load of 250 kN with a factor of safety of 5. Take the internal diameter as 0.8 times the external diameter. Take* $\sigma_c = 550\ N/mm^2$ *and* $a = \frac{1}{1600}$ *in Rankine's formula.* (AMIE, Winter 1983)

Sol. Given :

Length of column, $l = 4\ \text{m} = 4000\ \text{mm}$

End conditions = Both ends fixed

$\therefore$ Effective length, $L_e = \frac{l}{2} = \frac{4000}{2} = 2000\ \text{mm}$

Safe load, = 250 kN

Factor of safety, = 5

Let External dia., $= D$

Internal dia. $= 0.8 \times D$

Crushing stress, $\sigma_c = 550\ \text{N/mm}^2$

Value of 'a' $= \frac{1}{1600}$ in Rankine's formula

Now factor of safety $= \frac{\text{Cirippling load}}{\text{Safe load}}$ or $5 = \frac{\text{Cirippling load}}{250}$

$\therefore$ Crippling load, $P = 5 \times 250 = 1250\ \text{kN} = 1250000\ \text{N}$

Area of column, $A = \frac{\pi}{4} [D^2 - (0.8D)^2]$

$= \frac{\pi}{4} [D^2 - 0.64D^2] = \frac{\pi}{4} \times 0.36D^2 = \pi \times 0.09D^2$

Moment of Inertia, $I = \frac{\pi}{64}[D^4 - (0.8D^4] = \frac{\pi}{64}[D^4 - 0.4096D^4]$

$$= \frac{\pi}{64} \times 0.5904 \times D^4 = 0.009225 \times \pi \times D^4$$

But $I = A \times k^2$, where k is radius of gyration

$$\therefore \quad k = \sqrt{\frac{I}{A}} = \sqrt{\frac{0.009225 \times \pi D^4}{\pi \times 0.09 \times D^2}} = 0.32D$$

Now using equation (7.17), $P = \frac{\sigma_c \, . \, A}{1 + a\left(\frac{L_e}{k}\right)^2}$

$$\text{or} \quad 1250000 = \frac{550 \times \pi \times 0.09\, D^2}{1 + \frac{1}{1600} \times \left(\frac{2000}{0.32D}\right)^2} \qquad (\because \; A = \pi \times 0.09D^2)$$

$$\frac{1250000}{550 \times \pi \times 0.09} = \frac{D^2}{1 + \frac{24414}{D^2}} \quad \text{or} \quad 8038 = \frac{D^2 \times D^2}{D^2 + 24414}$$

or $\quad 8038D^2 + 8038 \times 24414 = D^4$ or $D^4 - 8038D^2 - 8038 \times 24414 = 0$

or $\quad D^4 - 8038\, D^2 - 196239700 = 0.$

The above equations is a quadratic equation in D^2. The solution is

$$\therefore \quad D^2 = \frac{8038 \pm \sqrt{8038^2 + 4 \times 1 \times 196239700}}{2}$$

$$\left(\text{Roots} = \frac{-b \pm \sqrt{b^2 - 4ac}}{2a}\right)$$

$$= \frac{8038 \pm \sqrt{646094 + 784958800}}{2} = \frac{8038 \pm 29147}{2}$$

$$= \frac{8038 + 29147}{2} \text{ (The other root is not possible)}$$

$$= 18592.5 \text{ mm}^2$$

$$\therefore \quad D = \sqrt{18592.5} = 136.3 \text{ mm}$$

∴ External diameter = **136.3 mm. Ans.**

Internal diameter = 0.8 × 1363 = **109 mm. Ans.**

Problem 7.28. *A 1.5 m long column has a circular cross-section of 5 cm diameter. One of the ends of the column is fixed in direction and position and other end is free. Taking factor of safety as 3, calculate the safe load using :*

(a) *Rankine's formula, take yield stress, $\sigma_c = 560\ N/mm^2$ and $a = \frac{1}{1600}$ for pinned ends.*

(b) *Euler's formula, Young's modulus for C.I. = $1.2 \times 10^5\ N/mm^2$.*

(AMIE, Summer 1976)

Sol. Given :

Length, $l = 1.5 \text{ m} = 1500 \text{ mm}$

Diameter, $d = 5 \text{ cm}$

$\therefore$ Area, $A = \frac{\pi}{4} \times 5^2 = 19.635 \text{ cm}^2 = 19.635 \times 10^2 \text{ mm}^2$

Moment of inertia, $I = \frac{\pi}{64} \times 5^4 = 30.7 \text{ cm}^4 = 30.7 \times 10^4 \text{ mm}^4$

and least radius of gyration, $k = \sqrt{\frac{I}{A}} = \sqrt{\frac{30.7 \times 10^4}{19.635 \times 10^2}} = 12.5 \text{ mm.}$

End conditions = One end is fixed and other end is free.

$\therefore$ Effective length, $L_e = 2l = 2 \times 1500 = 3000 \text{ mm}$

Factor of safety = 3.

(a) Safe load by Rankine's formula

Yield stress, $\sigma_c = 560 \text{ N/mm}^2$

Rankine's constant, $a = \frac{1}{1600}$

Let P = Crippling load by Rankine's formula

Using equation (7.17), we have

$$P = \frac{\sigma_c \times A}{1 + a\left(\frac{L_e}{k}\right)^2}$$

$$= \frac{560 \times 1963.5}{1 + \frac{1}{1600} \times \left(\frac{3000}{12.5}\right)^2} \qquad (\because \; L = 3000 \text{ mm and } k = 12.5)$$

$$= 29708.1 \text{ N}$$

$$\therefore \text{ Safe load} = \frac{\text{Crippling load}}{\text{Factor of safety}}$$

$$= \frac{29708.1}{3} = \mathbf{9902.7 \text{ N.}} \quad \textbf{Ans.}$$

(b) Safe load by Euler's formula

Young's Modulus, $E = 1.2 \times 10^5 \text{ N/mm}^2$

Let P = Crippling load by Euler's formula

Using equation (7.13), $P = \frac{\pi^2 EI}{L_e^2}$

$$= \frac{\pi^2 \times 1.2\ 10^5 \times 30.7 \times 10^4}{3000^2} \qquad (\because \; L = 3000 \text{ mm})$$

$$= 40200 \text{ N}$$

$$\therefore \text{ Safe load} = \frac{\text{Crippling load}}{\text{Factor of safety}} = \frac{420200}{3} = \mathbf{1340 \text{ N.}} \quad \textbf{Ans.}$$

Problem 7.29. *A short length of tube, 4 cm internal diameter and 5 cm external diameter, failed in compression at a load of 240 kN. When a 2 metre length of the same tube was* *tested as a strut with fixed ends, the load at failure was 158 kN. Assuming that σ_c in Rankine's formula is given by the first test, find the value of the constant a in the same formula. What will be the crippling load of this tube if it is used as a strut 3 m long with one end fixed and the other hinged ?*

(AMIE, Summer 1983)

Sol. Given :

External diameter, $D = 5$ cm

Internal diameter, $d = 4$ cm

$\therefore$ Area, $$A = \frac{\pi}{4}(5^2 - 4^2) = \frac{9\pi}{4} = 2.25\ \pi\ \text{cm}^2 = 225\ \pi\ \text{mm}^2$$

Moment of inertia, $$I = \frac{\pi}{64}[5^4 - 4^4] = \frac{\pi}{64}(625 - 256)$$
$$= 5.7656 \times \pi\ \text{cm}^4 = 57656\ \pi\ \text{mm}^4$$

$\therefore$ Least radius of gyration, $k = \sqrt{\dfrac{I}{A}} = \sqrt{\dfrac{57656\ \pi}{225\ \pi}} = 16$ mm

Crushing load $= 240$ kN.

The value of σ_c in Rankine's formula is given by the crushing load of 240 kN.

$\therefore$ The value of $$\sigma_c = \frac{\text{Crushing load of 240 kN}}{\text{Area}}$$
$$= \frac{240}{225\ \pi} = 0.3395\ \text{kN/mm}^2$$

Length of the strut, $l = 2$ m $= 2000$ mm

End condition = Both the ends are fixed

$\therefore$ Effective length, $L_e = \dfrac{l}{2} = \dfrac{2000}{2} = 1000$ mm

Crushing load of strut, $P = 158$ kN.

(i) Value of constant 'a'

Let a = Value of Rankine's constant

Using equation (7.17), we have

$$P = \frac{\sigma_c \, . \, A}{1 + a \, . \left(\dfrac{L_e}{k}\right)^2}$$

or $$158 = \frac{0.33953 \times 225\ \pi}{1 + a \, . \left(\dfrac{1000}{16}\right)^2} = \frac{239.99}{1 + 3906.25 \times a}$$

$$1 + 3906.25 \times a = \frac{239.99}{158} = 1.5189.$$

$\therefore$ $$a = \frac{1.5189 - 1.0}{3906.25} = 0.0001328 = \mathbf{\frac{1}{7530}}. \quad \textbf{Ans.}$$

(ii) Crippling load for the strut of length 3 m when one end is fixed and other is hinged

Actual length, $l = 3$ m $= 3000$ mm

End conditions = One end fixed and other is hinged

$\therefore$ Effective length, $L_e = \dfrac{l}{\sqrt{2}} = \dfrac{3000}{\sqrt{2}}$

Let P = Crippling load.

Using equation (7.17),

$$P = \frac{\sigma_c \,.\, A}{1 + a.\left(\dfrac{L_e}{k}\right)^2}$$

$$= \frac{0.33953 \times 225\pi}{1 + \dfrac{1}{7530} \times \left(\dfrac{3000}{\sqrt{2} \times 16}\right)^2} \qquad \left(\because\ k = 16, a = \frac{1}{7530}\right)$$

$$= \frac{0.33953 \times 225 \times \pi}{1 + 2.3344} = \textbf{71.97 kN.} \quad \textbf{Ans.}$$

Problem 7.30. *Find the Euler crushing load for a hollow cylindrical cast iron column 20 cm external diameter and 25 mm thick if it is 6 m long and is hinged at both ends. Take $E = 1.2 \times 10^6\ N/mm^2$.*

Compare the load with the crushing load as given by the Rankine's formula, taking $\sigma_c = 550\ N/mm^2$ and $a = \dfrac{1}{1600}$; for what length of the column would these two formulae give the same crushing load ?

Sol. Given :

External dia., D = 20 cm

Thickness, t = 25 mm = 2.5 cm

$\therefore$ Internal dia., $d = (D - 2 \times t) = 20 - 2 \times 2.5 = 15$ cm.

Area, $A = \dfrac{\pi}{4}(20^2 - 15^2) = \dfrac{175\,\pi}{4} = 137.44\ \text{cm}^2 = 13744\ \text{mm}^2$

Moment of inertia, $I = \dfrac{\pi}{64}[20^4 - 15^4] = \dfrac{\pi}{64}(160000 - 50625)$

$= 5368.93\ \text{cm}^4 = 53689300\ \text{mm}^4$

$\therefore$ Least radius of gyration,

$$k = \sqrt{\frac{I}{A}} = \sqrt{\frac{53689300}{15744}} = 62.5\ \text{mm}$$

Length of column, l = 6 m = 6000 mm

End conditions = Both ends are hinged

$\therefore$ Effective length, $L_e = l = 6000$ mm

Value of $E = 1.2 \times 10^5\ \text{N/mm}^2$.

Euler's crushing load is given by equation (7.13),

$$P = \frac{\pi^2 EI}{{L_e}^2}$$

$$= \frac{\pi^2 \times 1.2 \times 10^5 \times 53689300}{6000^2} = \textbf{1766307 N.} \quad \textbf{Ans.}$$

Crushing load by Rankine's formula

The value of $\sigma_c = 550 \text{ N/mm}^2$

Value of $a = \dfrac{1}{1600}$

Let P = Crushing load by Rankine's formula

Using equation (7.17),

$$P = \frac{\sigma_c \, . \, A}{1 + a \, . \left(\dfrac{L_e}{k}\right)^2} = \frac{550 \times 13744}{1 + \dfrac{1}{1600} \times \left(\dfrac{6000}{62.5}\right)^2} = \textbf{1118224.8 N. Ans.}$$

The length of the column for which the above two formulae gives the same crushing load

Let L = Length of the column

Crushing load by Euler's formula $= \dfrac{\pi^2 EI}{L^2}$

Crushing load by Rankine's formula $= \dfrac{\sigma_c \, . \, A}{1 + a\left(\dfrac{L}{k}\right)^2}$

Equating the crushing loads given by the above two formulae, we get

$$\frac{\pi^2 EI}{L^2} = \frac{\sigma_c \, . \, a}{1 + a\left(\dfrac{L}{k}\right)^2}$$

Substituting the all known values except 'L', we get

$$\frac{\pi^2 \times 1.2 \times 10^5 \times 53689300}{L^2} = \frac{550 \times 13744}{1 + \dfrac{1}{1600} \times \left(\dfrac{L}{62.5}\right)^2}$$

or $$\frac{\pi^2 \times 1.2 \times 10^5 \times 53689300}{550 \times 13744} = \frac{L^2}{1 + \dfrac{L^2}{6250000}}$$

or $$8411800 = \frac{L^2}{1 + \dfrac{L^2}{6250000}}$$

or $$8411800\left(1 + \frac{L^2}{6250000}\right) = L^2$$

or $$8411800 + \frac{8411800}{6250000} = L^2 = L^2$$

or $$8411800 + 1.346 \, L^2 = L^2 \text{ or } 1.346 \, L^2 - L^2 = -\,8411800$$

or $$0.346 \, L^2 = -\,8411800 \quad \text{or} \quad L = \sqrt{\frac{-\,8411800}{0.346}}$$

The above equation gives the imaginary value of length L.

Hence it is not possible to have the same length of the columns, which have the same crushing load for the two given formulae. **Ans.**

Problem 7.31. *A hollow cast iron column 200 mm outside diameter and 150 mm inside diameter, 8 m long has both ends fixed. It is subjected to an axial compressive load. Taking a factor of safety as 6, $\sigma_c = 560\ N/mm^2$, $a = \frac{1}{1600}$, determine the safe Rankine load.*

Sol. Given :

External dia., $D = 200$ mm

Internal dia., $d = 150$ mm

Length, $l = 8$ m $= 8000$ mm

End conditions = Both the ends are fixed

Crushing stress, $\sigma_c = 560$ N/mm²

Rankine's constant, $a = \frac{1}{1600}$

Safety factor = 6

$$\text{Area of cross-section, } A = \frac{\pi}{4}(D^2 - d^2) = \frac{\pi}{4}(200^2 - 150^2)$$

$$= \frac{\pi}{4}(40000 - 22500) = 13744 \text{ mm}^2$$

$$\text{Moment of inertia, } I = \frac{\pi}{64}(D^4 - d^4) = \frac{\pi}{64}(200^4 - 150^4)$$

$$= \frac{\pi}{64}(1600000000 - 506250000) = 53689000 \text{ mm}^4$$

$$\text{Least radius of gyration, } k = \sqrt{\frac{I}{A}} = \sqrt{\frac{53689000}{13744}} = 62.5 \text{ mm}$$

Let P = Crippling load by Rankine formula.

$$\text{Using equation (7.17), } P = \frac{\sigma_c \times A}{1 + a\left(\frac{L_e}{k}\right)^2}$$

$$\text{where } L_e = \text{Effective length} = \frac{l}{2} = \frac{8000}{2} = 4000 \text{ mm}$$

$$\therefore \quad P = \frac{560 \times 13744}{1 + \frac{1}{1600} \times \left(\frac{4000}{62.5}\right)^2} = \frac{7696640}{1 + 2.56} = \frac{7696640}{3.56}$$

$$= 2161977 \text{ N} = 2161.977 \text{ kN}$$

$$\therefore \text{Safe load} = \frac{\text{Crippling load}}{\text{Factor of safety}} = \frac{2161.977}{6} = \mathbf{360.3295\ kN.} \quad \textbf{Ans.}$$

Problem 7.32. *A hollow C.I. column whose outside diameter is 200 mm has a thickness of 20 mm. It is 4.5 m long and is fixed at both ends. Calculate the safe load by Rankine's formula using a factor of safety of 4. Calculate the slenderness ratio and the ratio of Euler's and Rankine's critical loads. Take $\sigma_c = 550\ N/mm^2$, $a = \frac{1}{1600}$ in Rankine's formula and $E = 9.4 \times 10^4\ N/mm^2$.*

(AMIE, Winter 1979 ; Annamalai University, 1991)

Sol. Given :

Outside diameter, $D = 200$ mm

Thickness, $t = 20$ mm

$\therefore$ Inside diameter, $d = D - 2 \times t = 200 - 2 \times 20 = 160$ mm

Area, $A = \frac{\pi}{4}(D^2 - d^2) = \frac{\pi}{4}(200^2 - 160^2) = 11310 \text{ mm}^2$

Moment of inertia, $I = \frac{\pi}{64}(D^4 - d^4) = \frac{\pi}{64}(200^4 - 160^4) = 46370000 \text{ mm}^4$

And the least radius of gyration,

$$k = \sqrt{\frac{I}{A}} = \sqrt{\frac{46370000}{11310}} = 64 \text{ mm}$$

Length of column, $l = 4.5$ m $= 4500$ mm

End condition = Both the ends are fixed

$\therefore$ Effective length, $L_e = \frac{l}{2} = \frac{4500}{2} = 2250$ mm

Factor of safety = 4

Value of $\sigma_c = 550 \text{ N/mm}^2$

Value of $a = \frac{1}{1600}$

Value of $E = 9.4 \times 10^4 \text{ N/mm}^2$.

(i) Slenderness ratio

Using equation (7.16), we get

Slenderness ratio $= \frac{l}{k} = \frac{4500}{64} =$ **70.30. Ans.**

(ii) Safe load by Rankine's formula

Let P = Crippling load by Rankine's formula

Using equation (7.17), $P = \dfrac{\sigma_c \times A}{1 + a\left(\dfrac{L_e}{k}\right)^2} = \dfrac{550 \times 11310}{1 + \dfrac{1}{1600}\left(\dfrac{2250}{64}\right)^2} = 351100$ N

$\therefore$ Safe load $= \dfrac{\text{Crippling load}}{\text{Factor of safety}} = \dfrac{3511000}{4} =$ **8777 N. Ans.**

(iii) Ratio of Euler's and Rankine's critical loads

Let P_E = Euler's critical load

Euler's critical load is given by equation (7.13)

$$P_E = \frac{\pi^2 EI}{L_e^2} = \frac{\pi^2 \times 9.4 \times 10^4 \times 46370000}{2250^2} = 849770 \text{ N}$$

$$\therefore \frac{\text{Euler's critical load}}{\text{Rankine's critical load}} = \frac{P_E}{P}$$

$$= \frac{8497700}{3511000} \qquad (\because P = 3511000 \text{ N})$$

$=$ **2.42. Ans.**

HIGHLIGHTS (7B)

1. A vertical member of a structure, which is subjected to axial compressive load and is fixed at both of its ends, is known as a column.
2. Strut is a member of a structure which is not vertical or whose one or both of its ends are hinged or pin joined.
3. All short columns fail due to crushing whereas long columns fail due to buckling and crushing.
4. The load at which the column just buckles is known as buckling load or critical load or crippling load.
5. The crippling load for a column by Euler's formula for different end conditions is given by

$$P = \frac{\pi^2 EI}{l^2} \quad \text{when both ends are hinged}$$

$$= \frac{\pi^2 EI}{4l^2} \quad \text{when one end is fixed and other is free}$$

$$= \frac{4\pi^2 EI}{l^2} \quad \text{when both ends are fixed}$$

$$= \frac{2\pi^2 EI}{l^2} \quad \text{when one end is fixed and other is hinged}$$

where l = Actual length of column
E = Young's modulus of the material of the column
I = Least moment of inertia of the column.

6. The effective length of a given column with given end conditions is the length of an equivalent column of the same material and cross-section with hinged ends, and having the value of crippling load equal to that of the given column.
7. The crippling load for any type of end condition is given by

$$P = \frac{\pi^2 EI}{L_e^2}$$

where L_e = Effective length
$= l$ when both ends hinged
$= 2l$ when one end is fixed and other is free
$= \frac{l}{2}$ when both ends are fixed
$= \frac{l}{\sqrt{2}}$ when one end fixed and other is hinged

where l = Actual length of the column.

8. Crippling load and crippling stress in terms of effective length and radius of gyration are given by

Crippling loads, $$P = \frac{\pi^2 E.A}{\left(\frac{L_e}{k}\right)^2}$$

and Crippling stress $= \dfrac{\pi^2 E}{\left(\dfrac{L_e}{k}\right)^2}$

where L_e = Effective length and k = Least radius of gyration $= \sqrt{\dfrac{I}{A}}$

where I = Least moment of Inertia.

9. Slenderness ratio is the ratio of the effective length of the column to the least radius of gyration. Mathematically,

Slenderness ratio $= \dfrac{L_e}{k}$.

10. The crippling load by Rankine's formula is given by

$$P = \frac{\sigma_c \times A}{1 + a\left(\dfrac{L_e}{k}\right)^2}$$

where σ_c = Ultimate crushing stress
A = Area of cross-section of column
a = Rankine's constant
L_e = Effective length
k = Least radius of gyration.

EXERCISE (7B)

(A) Theoretical Questions

1. Explain the assumptions made in Euler's column theory. How far are the assumptions valid in practice ? *(AMIE, Summer 1982)*
2. Define the terms : column, strut and crippling load.
3. Explain how the failure of a short and of a long column takes place ?
4. What do you mean by end conditions of a column ? What are the important end conditions for a column ? Explain them.
5. What is 'equivalent length of a column' ? How is the concept used in the column theory ? *(AMIE, Winter 1982)*
6. What is 'equivalent length of a column' ? Give the ratios of equivalent length and actual length of columns with various end conditions. *(AMIE, Summer 1985)*
7. Derive an expression for the Euler's crippling load for a long column with following end conditions.
   (*a*) Both ends are hinged (*b*) Both ends are fixed.
8. Explain how Rankine-Gordon formula is used to calculate the intensity of stress in short, intermediate and long columns. *(AMIE, Winter 1981)*
9. Prove that the crippling load by Euler's formula for a column having one end fixed and other end free is given by

$$P = \frac{\pi^2 EI}{4l^2}$$

where l = Actual length of the column,
E = Young's modulus, and
l = Least moment of inertia.

10. Find an expression for crippling load for a long column when one end of the column is fixed and other end is hinged.

11. Prove that crippling stress by Euler's formula is given by $f_c = \dfrac{\pi^2 E}{\left(\dfrac{L_e}{k}\right)^2}$.

12. Define slenderness ratio. State the limitations of Euler's formula.

13. How will you justify that Rankine's formula is applicable for all lengths of columns, ranging from short to long columns.

14. What is a Rankine's constant ? What is the approximate value of Rankine's constant for cast iron column ?

15. Deduce an expression for the Euler's crippling load of an 'ideal column' pin-joined at each end. Explain the limitations, if any, in using the formula.

16. Derive the equation for the Euler's crippling load for a column with one end fixed and the other is free. *(Annamalai University, 1991)*

17. Derive expression for Euler's buckling load for a long column of length L with both ends fixed, from first principle. Mention the assumptions made in the derivation.

(Bangalore University, 1991)

(B) Numerical Problems

1. A solid round bar 4 m long and 6 cm in diameter is used as a strut with both ends hinged. Determine the crippling load. Take $E = 2 \times 10^5$ N/mm^2. [**Ans.** 78.486 kN]

2. For the problem 1, determine the crippling loads when the given strut is used with the following conditions :

(*i*) One end is fixed and other end is free (*ii*) Both the ends are fixed and

(*iii*) One end is fixed and other is hinged. [**Ans.** (*i*) 19.621 kN ; (*ii*) 313.94 kN ; (*iii*) 156.96 kN]

3. A column of timber section 10 cm × 15 cm is 5 m long both ends being fixed. If the Young's modulus for timber = 17.5 kN/mm^2, determine :

(*i*) Crippling load, and (*ii*) Safe load for the column if factor of safety = 3.

[**Ans.** (*i*) 45.4 kN ; (*ii*) 115.1 kN]

4. A hollow mild steel tube 5 m long, 4 cm internal diameter and 5 mm thick is used as a strut with both ends hinged. Find the crippling load and safe load taking factor of safety as 3. Taking $E = 2 \times 10^5$ N/mm^2. [**Ans.** 14.99 N and 4766 N]

5. A solid circular bar 5 m long and 4 cm in diameter was found to extend 4.5 mm under a tensile load of 48 kN. The bar is used as a strut with both ends hinged. Determine the buckling load for the bar and also the safe load taking factor of safety as 3.0. [**Ans.** 2105.5 N and 701.8 N]

6. Calculate the safe compressive load on a hollow cast iron column (one end rigidly fixed and other hinged) of 10 cm external diameter, 7 cm internal diameter and 8 m in length. Use Euler's formula with a factor of safety of 4 and $E = 95$ kN/mm^2. [**Ans.** 27.3 kN]

7. Determine Euler's crippling load for an I-section joist 30 cm × 15 cm × 2 cm and 5 m long which is used as a strut with both ends fixed. Take Young's modulus for the joist as 2×10^5 N/mm^2.

[**Ans.** 3.6 MN]

8. Determine the crippling load for a T-section of dimensions 12 cm × 12 cm × 2 cm and of length 6 cm when it is used as a strut with both of its ends hinged. Take $E = 2 \times 10^5$ N/mm^2.

[**Ans.** 161.57 kN]

9. Determine the ratio of buckling strengths of two columns one hollow and the other solid. Both are made of the same material and have the same length, cross-sectional area and end conditions. The internal diameter of hollow column is 2/3rd of its external diameter. [**Ans.** 2.6 : 1]

10. The external and internal diameter of a hollow cast iron column are 5 cm and 3 cm respectively. If the length of this column is 4 m and both of its ends are fixed, determine the crippling load using Rankine's formula. Take the value of $f_c = 550$ N/mm^2 and $a = \frac{1}{1600}$ in Rankine's formula.

[**Ans.** 540.09 kN]

11. A hollow cylindrical cast iron column is 6 m long with both ends fixed. Determine the minimum diameter of the column if it has to carry a safe load of 300 kN with a factor of safety of 4. Take the internal diameter as 0.7 times the external diameter. Take $f_c = 550$ N/mm^2 and $a = \frac{1}{1600}$ in Rankine's formula. [**Ans.** $D = 9.53$ cm, $d = 6.67$ cm]

12. A 2.0 m long column has a circular cross-section of 6 cm diameter. One of the ends of the column is fixed in direction and position and other end is free. Taking factor of safety as 3, calculate the safe load using

(*i*) Rankine's formula take yield stress $f_c = 550$ N/mm^2 and $a = \frac{1}{1600}$ for pinned ends,

(*ii*) Euler's formula, Young's modulus for *C.I.* = 1.3×10^5 N/mm^2.

[**Ans.** (*i*) 11.4 kN and (*ii*) 170 kN]

13. A short length of tube, 5 cm internal diameter and 6 cm external diameter, failed in compression at a load of 250 kN. When 2.5 m length of the same tube was tested as a strut with fixed ends, the load at failure was 150 kN. Assuming that f_c in Rankine's formula is given by the first test, find the value of constant '*a*' in the same formula. What will be the crippling load of this tube if it is used as a strut 3.2 m long with one end fixed and the other hinged. $\left[\textbf{Ans.}\ \frac{1}{6148}, 78.5\text{ kN}\right]$

14. Find the Euler crushing load for a hollow cylindrical cast iron column 15 cm external diameter and 25 mm thick if it is 6 m long and is hinged at both ends. Take $E = 8 \times 10^4$ N/mm^2. Compare the load with the crushing load as given by the Rankine's formula, taking $f_c = 550$ N/mm^2 and $a = \frac{1}{1600}$; for what length of the column would these two formulae give the same crushing load ?

[**Ans.** 665 kN, 260.77 kN, 507 cm]

15. A mild steel tube is 7.5 cm in diameter and 0.25 cm thick. A short length of this tube is tested in compression and is found to yield at 500 N/mm^2. The modulus of elasticity of the material of the tube is 2×10^5 N/mm^2. A length of 2 m when tested as a strut with free ends failed with a load of 180 kN force. Assuming the failing stress in Rankine formula to be the yield stress of the material, find the value of Rankine's constant '*a*'. Find also the crushing load as by the Euler's formula.

$\left[\textbf{Ans.}\ a - \frac{1}{8000} \text{ and } P\ 225\text{ kN}\right]$

16. An alloy steel tube is 7.5 cm external diameter and 0.25 cm wall thickness. A very short length of this tube was tested in compression and found to yield at a load of 320 kN. A length of 2 metres when tested as a strut with hinged ends buckled at a load of 170 kN. Assuming the failing stress in Rankine formula to be the yield stress of the material, find the value of Rankine's constant '*a*'.

$\left[\textbf{Ans.}\ a = \frac{1}{6980}\right]$

17. A 1.5 m long column has a circular cross-section of 0.5 cm diameter. One of the ends of the column is fixed in direction and position and the other end is free. Taking factor of safety as 3, calculate the safe load using ; (*i*) Rankine formula with $f_c = 560$ N/mm^2 and $a = \frac{1}{1600}$ for pinned ends and (*ii*) Euler's formula with E for C.I. = 1.2×10^5 N/mm^2. (*AMIE, Summer 1976*)

[**Ans.** (*i*) 9.9 kN and (*ii*) 13.45 kN]

18. Determine the external diameter and internal diameter of a hollow circular cast iron column, which carries a load of 1000 kN. The length of the column is 6 metre. The internal diameter is to be one half that of outer diameter. Use Rankine's formula with $f = 560$ N/mm^2 and $a = \frac{1}{1600}$. Take a factor of safety 4. One end is fixed and the other end is free.

[**Ans.** $D = 35.4$ cm, $d = 17.7$ cm]

UNIT IV

Chapters

8

Thin Cylinders and Spheres

8.1. INTRODUCTION AND DIFFERENCE BETWEEN THIN WALLED AND THICK WALLED PRESSURE VESSELS

The vessels such as boilers, compressed air receivers etc., are of cylindrical and spherical forms. These vessels are generally used for storing fluids (liquid or gas) under pressure. The walls of such vessels are thin as compared to their diameters. If the thickness of the wall of the pressure vessel is less than $\frac{1}{15}$ to $\frac{1}{20}$ of its internal diameter, the pressure vessel is known as a *thin*. In case of thin pressure, the stress distribution is assumed uniform over the thickness of the wall.

If the ratio of thickness to internal diameter is more than $\frac{1}{20}$, then the pressure vessel is known as thick. The hoop stress in case of thick pressure vessel will not be uniform across the thickness.

8.2. THIN CYLINDRICAL VESSEL SUBJECTED TO INTERNAL PRESSURE

Fig. 8.1 shows a thin cylindrical vessel in which a fluid under pressure is stored.

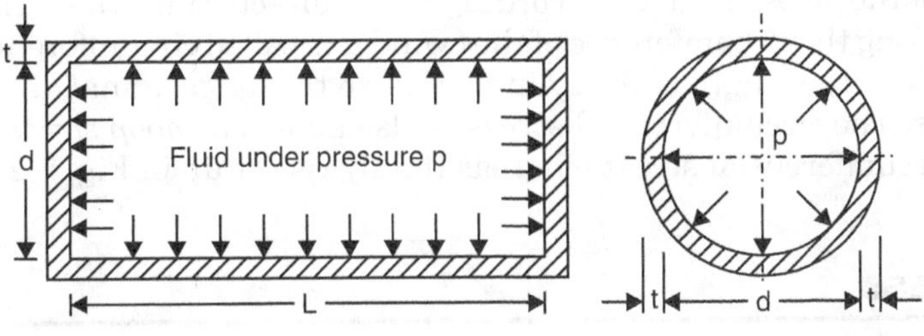

Fig. 8.1

Let d = Internal diameter of the thin cylinder

t = Thickness of the wall of the cylinder

p = Internal pressure of the fluid

L = Length of the cylinder.

On account of the internal pressure p, the cylindrical vessel may fail by splitting up in any one of the two ways as shown in Fig. 8.2 (a) and 8.2 (b).

The forces, due to pressure of the fluid acting vertically upwards and downwards on the thin cylinder, tend to burst the cylinder as shown in Fig. 8.2 (*a*).

The forces, due to pressure of the fluid, acting at the ends of the thin cylinder, tend to burst the thin cylinder as shown in Fig. 8.2 (*b*).

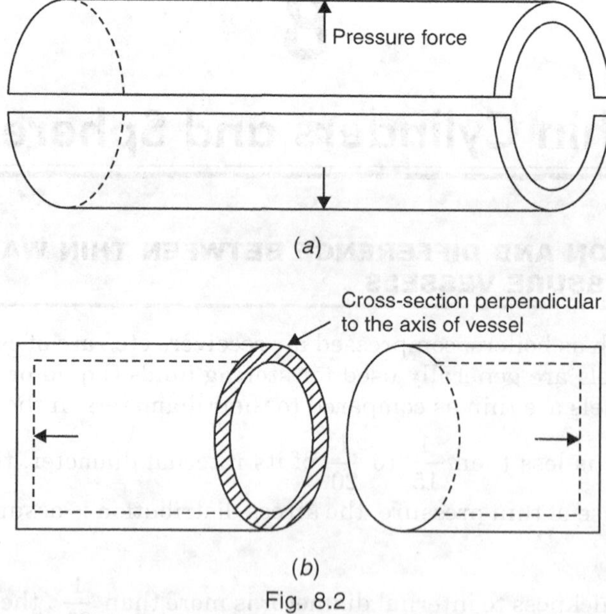

Fig. 8.2

8.2.1. Stresses in a Thin Cylindrical Vessel Subjected to Internal Pressure. When a thin cylindrical vessel is subjected to internal fluid pressure, the stresses in the wall of the cylinder on the cross-section along the axis and on the cross-section perpendicular to the axis are set up. These stresses are tensile and are known as :

1. Circumferential stress (or hoop stress) and
2. Longitudinal stress.

The name of the stress is given according to the direction in which the stress is acting. The stress acting along the circumference of the cylinder is called circumferential stress whereas the stress acting along the length of the cylinder (*i.e.*, in the longitudinal direction) is known as longitudinal stress. The circumferential stress is also known as *hoop stress*. The stress set up in Fig. 8.2 (*a*) is circumferential stress whereas the stress set up in Fig. 8.2 (*b*) is longitudinal stress.

8.3. HOOP STRESS

Consider a thin cylindrical vessel subjected to an internal fluid pressure. The circumferential stress will be set up in the material of the cylinder, if the bursting of the cylinder takes place as shown in Fig. 8.3 (*a*).

The expression for hoop stress or circumferential stress (σ_1) is obtained as given below.

Let p = Internal pressure of fluid

d = Internal diameter of the cylinder

t = Thickness of the wall of the cylinder

σ_1 = Circumferential or hoop stress in the material.

The bursting will take place if the force due to fluid pressure is more than the resisting force due to circumferential stress set up in the material. In the limiting case, the two force should be equal.

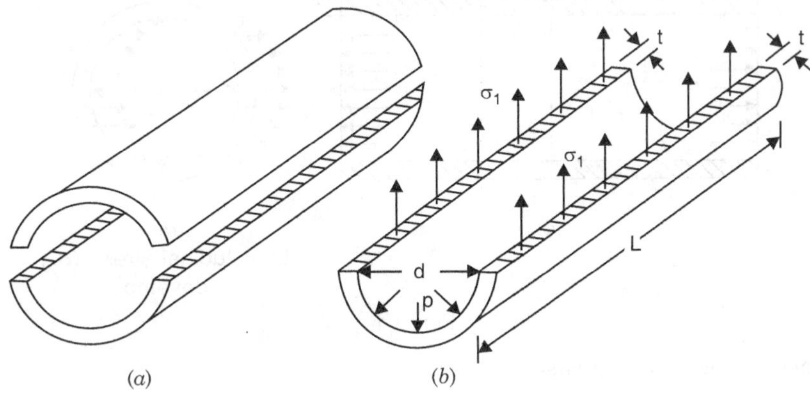

Fig. 8.3

Force due to fluid pressure $= p \times$ Area on which p is acting

$= p \times (d \times L)$ ...(i)

($\because$ p is acting on projected area $d \times L$)

Force due to circumferential stress

$= \sigma_1 \times$ Area on which σ_1 is acting

$= \sigma_1 \times (L \times t + L \times t)$

$= \sigma_1 \times 2Lt = 2\sigma_1 \times L \times t$ ...(ii)

Equating (i) and (ii), we get

$$p \times d \times L = 2\sigma_1 \times L \times t$$

$$\therefore \quad \sigma_1 = \frac{pd}{2t} \text{ (cancelling } L) \quad ...(8.1)$$

This stress is tensile as shown in Fig. 8.3 (b).

8.4. AXIAL STRESS OR LONGITUDINAL STRESS

Consider a thin cylindrical vessel subjected to internal fluid pressure. The longitudinal stress will be set up in the material of the cylinder, if the bursting of the cylinder takes place along the section AB of Fig. 8.4 (a).

The longitudinal stress (σ_2) developed in the material is obtained as :

Let p = Internal pressure of fluid stored in thin cylinder

d = Internal diameter of cylinder

t = Thickness of the cylinder

σ_2 = Longitudinal stress in the material.

The bursting will take place if the force due to fluid pressure acting on the ends of the cylinder is more than the resisting force due to longitudinal stress (σ_2) developed in the material as shown in Fig. 8.4 (b). In the limiting case, both the forces should be equal.

Force due to fluid pressure $= p \times$ Area on which p is acting

$= p \times \frac{\pi}{4} d^2$

Resisting force $= \sigma_2 \times$ Area on which σ_2 is acting

$= \sigma_2 \times \pi d \times t$

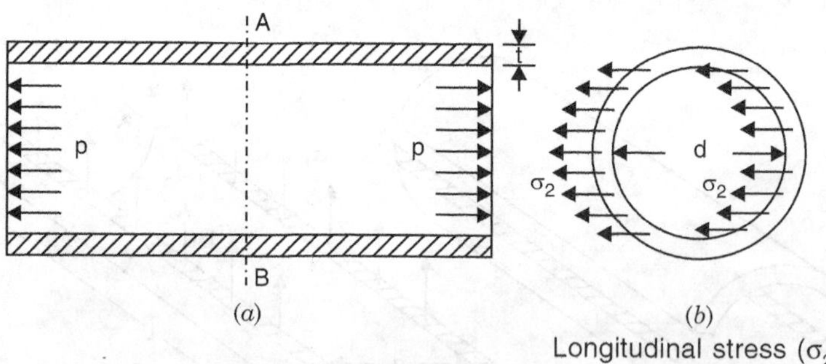

(a)

(b) Longitudinal stress (σ_2) develop

Fig. 8.4

∴ Hence in the limiting case

Force due to fluid pressure = Resisting force

$$p \times \frac{\pi}{4} d^2 = \sigma_2 \times \pi d \times t$$

$$\therefore \quad \sigma_2 = \frac{p \times \frac{\pi}{4} d^2}{\pi d \times t} = \frac{pd}{4t} \qquad \ldots(8.2)$$

The stress σ_2 is also tensile.

Equation (8.2) can be written as

$$\sigma_2 = \frac{pd}{2 \times 2t} = \frac{1}{2} \times \sigma_1 \qquad \left(\because \ \sigma_1 = \frac{pd}{2t}\right)$$

or Longitudinal stress = Half of circumferential stress.

This also means that circumferential stress (σ_1) is two times the longitudinal stress (σ_2). Hence in the material of the cylinder the permissible stress should be less than the circumferential stress. Or in other words, the circumferential stress should not be greater than the permissible stress.

8.4.1. Maximum Shear Stress. At any point in the material of the cylindrical shell, there are two principal stresses, namely a circumferential stress of magnitude $\sigma_1 = \frac{pd}{2t}$ acting circumferentially and a longitudinal stress of magnitude $\sigma_2 = \frac{pd}{4t}$ acting parallel to the axis of the shell. These two stresses are tensile and perpendicular to each other.

$$\therefore \text{ Maximum shear stress} \quad \tau_{max} = \frac{\sigma_1 - \sigma_2}{2} = \frac{\frac{pd}{2t} - \frac{pd}{4t}}{2} = \frac{pd}{8t} \qquad \ldots[8.2\,(A)]$$

Note. (*i*) If the thickness of the thin cylinder is to be determined then equation (8.1) should be used.

(*ii*) If maximum permissible stress in the material is given. This stress should be taken circumferential stress (σ_1).

(*iii*) While using equations (8.1) and (8.2), the units of p, σ_1 and σ_2 should be same. They should be expressed either in N/mm^2 or N/m^2. Also the units of d and t should be same. They may be in metre (m) or millimetre (mm).

8.5. SOLVED PROBLEMS 8.1 TO 8.8 BASED ON HOOP AND AXIAL (LONGITUDINAL) STRESSES

Problem 8.1. *A cylindrical pipe of diameter 1.5 m and thickness 1.5 cm is subjected to an internal fluid pressure of 1.2 N/mm². Determine :*

(i) Longitudinal stress developed in the pipe, and

(ii) Circumferential stress developed in the pipe.

Sol. Given :

Dia. of pipe, $d = 1.5$ m

Thickness, $t = 1.5 \text{ cm} = 1.5 \times 10^{-2}$ m

Internal fluid pressure, $p = 1.2$ N/mm²

As the ratio $\frac{t}{d} = \frac{1.5 \times 10^{-2}}{1.5} = \frac{1}{100}$, which is less than $\frac{1}{20}$, hence this is a case of thin cylinder.

Here unit of pressure (p) is in N/mm². Hence the unit of σ_1 and σ_2 will also be in N/mm².

(*i*) The longitudinal stress (σ_2) is given by equation (8.2) as,

$$\sigma_2 = \frac{p \times d}{4t}$$

$$= \frac{1.2 \times 1.5}{4 \times 1.5 \times 10^{-2}} = \mathbf{30\ N/mm^2.} \quad \textbf{Ans.}$$

(*ii*) The circumferential stress (σ_1) is given by equation (8.1) as

$$\sigma_1 = \frac{pd}{2t}$$

$$= \frac{1.2 \times 1.5}{2 \times 1.5 \times 10^{-2}} = \mathbf{60\ N/mm^2.} \quad \textbf{Ans.}$$

Problem 8.2. *A cylinder of internal diameter 2.5 m and of thickness 5 cm contains a gas. If the tensile stress in the material is not to exceed 80 N/mm², determine the internal pressure of the gas.*

Sol. Given :

Internal dia. of cylinder, $d = 2.5$ m

Thickness of cylinder, $t = 5 \text{ cm} = 5 \times 10^{-2}$ m

Maximum permissible stress = 80 N/mm²

As maximum permissible stress is given. Hence this should be equal to circumferential stress (σ_1).

We know that the circumferential stress should not be greater than the maximum permissible stress. Hence take circumferential stress equal to maximum permissible stress.

$$\therefore \quad \sigma_1 = 80 \text{ N/mm}^2$$

Let p = Internal pressure of the gas

Using equation (8.1),

$$\sigma_1 = \frac{pd}{2t}$$

or $$p = \frac{2t \times \sigma_1}{d} = \frac{2 \times 5 \times 10^{-2} \times 80}{2.5}$$ (Here unit of σ_1 is in N/mm², hence unit of p will also be in N/mm²)

$$= \mathbf{3.2\ N/mm^2.} \quad \textbf{Ans.}$$

Problem 8.3. *A cylinder of internal diameter 0.50 m contains air at a pressure of 7 N/mm² (gauge). If the maximum permissible stress induced in the material in 80 N/mm², find the thickness of the cylinder.*

Sol. Given :

Internal dia. of cylinder, $d = 0.50$ m

Internal pressure of air, $p = 7$ N/mm²

Maximum permissible stress in the material means the circumferential stress (σ_1).

As stated earlier that the circumferential stress should not be greater than the maximum permissible stress. Hence take circumferential stress equal to maximum permissible stress.

$\therefore$ Circumferential stress, $\sigma_1 = 80$ N/mm²

Let t = Thickness of the cylinder

Using equation (8.1),

$$\sigma_1 = \frac{pd}{2t}$$

$$\therefore \quad t = \frac{pd}{2 \times \sigma_1} = \frac{7 \times 0.50}{2 \times 80} = 0.021875 \text{ m}$$

(Here p and σ_1 are in N/mm², d is in m hence t will be in m)

= **2.188 cm. Ans.**

If the value of t is taken 2.1875 cm, the stress induced will be 80 N/mm². If the value of t is less than 2.1875 cm, the stress induced will be more than 80 N/mm². But the stress induced should not be more than 80 N/mm². If the value of t is taken more than 2.1875 cm (say t = 2.188 cm), the stress induced will be less than 80 N/mm².

Hence take t = **2.188 cm or say 2.2 cm. Ans.**

Problem 8.4. *A thin cylinder of internal diameter 1.25 m contains a fluid at an internal pressure of 2 N/mm². Determine the maximum thickness of the cylinder if :*

(*i*) *The longitudinal stress is not to exceed 30 N/mm².*

(*ii*) *The circumferential stress is not to exceed 45 N/mm².*

Sol. Given :

Internal dia. of cylinder, $d = 1.25$ m

Internal pressure of fluid, $p = 2$ N/mm²

Longitudinal stress, $\sigma_2 = 30$ N/mm²

Circumferential stress, $\sigma_1 = 45$ N/mm²

Using equation (8.1),

$$\sigma_1 = \frac{pd}{2t}$$

$$\therefore \quad t = \frac{p \times d}{2 \times \sigma_1} = \frac{2 \times 1.25}{2 \times 45} = 0.0277 \text{ m}$$

$$= 2.77 \text{ cm.} \quad \ldots(i)$$

Using equation (8.2),

$$\sigma_2 = \frac{pd}{4t}$$

$$\therefore \quad t = \frac{pd}{4 \times \sigma_2} = \frac{2 \times 1.25}{4 \times 30} = 0.0208 \text{ m}$$

$$= 2.08 \text{ cm.} \quad \ldots(ii)$$

The longitudinal or circumferential stresses induced in the material are inversely proportional to the thickness (t) of the cylinder. Hence the stress induced will be less if the value of 't' is more. Hence take the maximum value of 't' calculated in equations (i) and (ii)

From the equations (i) and (ii) it is clear that t should not be less than 2.77 cm.

Take t = **2.80 cm. Ans.**

Problem 8.5. *A water main 80 cm diameter contains water at a pressure head of 100 m. If the weight density of water is 9810 N/m^3, find the thickness of the metal required for the water main. Given the permissible stress as 20 N/mm^2.* (AMIE, Summer 1974)

Sol. Given :

Dia. of main, $d = 80$ cm

Pressure head of water, $h = 100$ m

Weight density of water, $w = \rho \times g = 1{,}000 \times 9.81 = 9810$ N/m^3

Permissible stress = 20 N/mm^2

Permissible stress is equal to circumferential stress (σ_1)

or $\sigma_1 = 20$ N/mm^2

Pressure of water inside the water main,

$$p = \rho \times g \times h = wh = 9810 \times 100 \text{ N/m}^2$$

Here σ_1 is in N/mm^2, hence pressure (p) should also be N/mm^2. The value of p in N/mm^2 is given as

$$p = \frac{9810 \times 100}{1000^2 \text{ mm}^2} \text{ N/mm}^2 \qquad (\because \ 1 \text{ m} = 1000 \text{ mm})$$

$$= 0.981 \text{ N/mm}^2$$

Let t = Thickness of the metal required.

Using equation (8.1),

$$\sigma_1 = \frac{p \times d}{2 \times t} \qquad \text{(Here '}d\text{' is in cm hence '}t\text{' will also be in cm)}$$

$$\therefore \quad t = \frac{p \times d}{2 \times \sigma_1} = \frac{0.981 \times 80}{2 \times 20} = \textbf{2 cm. Ans.}$$

8.5.1. Efficiency of a Joint. The cylindrical shells such as boilers are having two types of joints namely longitudinal joint and circumferential joint. In case of a joint, holes are made in the material of the shell for the rivets. Due to the holes, the area offering resistance decreases. Due to the decrease in area, the stress (which is the equal to the force divided by the area) developed in the material of the shell will be more.

Hence in case of rivetted shell the circumferential and longitudinal stresses are greater than what are given by equations (8.1) and (8.2). If the efficiency of a longitudinal joint and circumferential joint are given then the circumferential and longitudinal stresses are obtained as :

Let η_l = Efficiency of a longitudinal joint, and

η_c = Efficiency of the circumferential joint.

Then the circumferential stress (σ_1) is given as

$$\sigma_1 = \frac{p \times d}{2t \times \eta_l} \qquad \text{...(8.3)}$$

and the longitudinal stress (σ_2) is given as

$$\sigma_2 = \frac{p \times d}{4t \times \eta_c} \qquad ...(8.4)$$

Note. (*i*) In longitudinal joint, the circumferential stress is developed whereas in circumferential joint, the longitudinal stress is developed.

(*ii*) Efficiency of a joint means the efficiency of a longitudinal joint.

(*iii*) If efficiencies of a joint are given, the thickness of the thin shell is determined from equation (8.3).

Problem 8.6. *A boiler is subjected to an internal steam pressure of 2 N/mm². The thickness of boiler plate is 2.6 cm and permissible tensile stress is 120 N/mm². Find out the maximum diameter, when efficiency of longitudinal joint is 90% and that of circumferential joint is 40%.* (AMIE, 1976)

Sol. Given :

Internal steam pressure, $p = 2 \text{ N/mm}^2$

Thickness of boiler plates, $t = 2.0$ cm

Permissible tensile stress $= 120 \text{ N/mm}^2$

In case of a joints, the permissible stress may be circumferential stress or longitudinal stress.

Efficiency of longitudinal joint, $\eta_l = 90\% = 0.90$

Efficiency of circumferential joint, $\eta_c = 40\% = 0.40$.

Max. diameter for circumferential stress is given by equation (8.3).

$\therefore$ Using equation (8.3),

$$\sigma_1 = \frac{p \times d}{2 \times \eta_l \times t}$$

where σ_1 = Given permissible stress = 120 N/mm²

$$\therefore \quad 120 = \frac{2 \times d}{2 \times 0.90 \times 2.0}$$

(Here p and σ_1 are in same units. Thickness is in cm hence 'd' will be in cm)

$$\therefore \quad d = \frac{120 \times 2 \times 0.9 \times 2.0}{2} = 216.0 \text{ cm.} \qquad ...(i)$$

Max. diameter for longitudinal stress is given by equation (8.4).

$\therefore$ Using equation (8.4),

$$\sigma_2 = \frac{p \times d}{4 \times \eta_c \times t}$$

where σ_2 = Given permissible stress
= 120 N/mm².

$$\therefore \quad 120 = \frac{2 \times d}{4 \times 0.4 \times 2.0}$$

$$\therefore \quad d = \frac{120 \times 4 \times 0.4 \times 2.0}{2} = 192 \text{ cm.} \qquad ...(ii)$$

The longitudinal or circumferential stresses induced in the material are directly proportional to diameter (d). Hence the stress induced will be less if the value of 'd' is less. Hence take the minimum value of 'd' calculated from equations (*i*) and (*ii*).

Maximum diameter of the boiler is equal to the minimum* value of diameter given by equation (*i*) and (*ii*).

Hence maximum diameter, d = **192 cm. Ans.**

(*Please note that if d is taken as equal to 216.0 cm, the longitudinal stress (σ_2) will be more than the given permissible value as shown below :

$$\sigma_2 = \frac{p \times d}{4 \times \eta_c \times t} = \frac{2 \times 216}{4 \times 0.4 \times 2.0} = 135 \text{ N/mm}^2.)$$

Problem 8.7. *A boiler shell is to be made of 15 mm thick plate having a limiting tensile stress of 120 N/mm². If the efficiencies of the longitudinal and circumferential joints are 70% and 30% respectively determine :*

(*i*) *The maximum permissible diameter of the shell for an internal pressure of 2 N/mm², and*

(*ii*) *Permissible intensity of internal pressure when the shell diameter is 1.5 m.*

Sol. Given :

Thickness of boiler shell, t = 15 mm

Limiting tensile stress = 120 N/mm²

Limiting tensile stress may be circumferential stress or langitudinal stress.

Efficiency of longitudinal joint, η_l = 70% = 0.70

Efficiency of circumferential joint, η_c = 30% = 0.30.

(*i*) *Maximum permissible diameter for an internal pressure,*

$$p = \mathbf{2\ N/mm^2.}$$

The boiler shell should be designed for the limiting tensile stress of 120 N/mm². First consider the limiting tensile stress as circumferential stress and then as longitudinal stress. The minimum diameter of the two case will satisfy the condition.

(*a*) Taking limiting tensile stress = Circumferential stress

$$\sigma_1 = 120 \text{ N/mm}^2.$$

$$\therefore \quad \sigma_1 = 120 \text{ N/mm}^2$$

But σ_1 is also given by equation (8.3) as

$$\sigma_1 = \frac{p \times d}{2 \times \eta_l \times t}$$

$$\therefore \quad 120 = \frac{2 \times d}{2 \times 0.7 \times 15} \quad \text{(Here } t \text{ is in mm hence '}d\text{' will be in mm)}$$

$$\therefore \quad d = \frac{120 \times 2 \times 0.7 \times 15}{2} = 1260 \text{ mm} \quad \ldots(i)$$

(*b*) Taking limiting tensile stress = Longitudinal stress

$$(\sigma_2) = 120 \text{ N/mm}^2.$$

$$\therefore \quad \sigma_2 = 120 \text{ N/mm}^2$$

Using equation (8.4),

$$\sigma_2 = \frac{p \times d}{4 \times \eta_c \times t}$$

$$\text{or} \quad 120 = \frac{2 \times d}{4 \times 0.30 \times 15}$$

$$\therefore \quad d = \frac{120 \times 4 \times 0.3 \times 15}{2} = 1080 \text{ mm} \quad ...(ii)$$

Thus the maximum diameter of the shell, in order both the conditions may be satisfied = **1080 mm. Ans.**

(*ii*) *Permissible intensity of internal pressure when the shell diameter is 1.5 m.*

or $$d = 1.5 \text{ m} = 1500 \text{ mm}.$$

(*a*) Taking limiting tensile stress = Circumferential stress (σ_1)

$$= 120 \text{ N/mm}^2$$

Using equation (8.3),

$$\sigma_1 = \frac{p \times d}{2 \times \eta_l \times t}$$

or $$120 = \frac{p \times 1500}{2 \times 0.7 \times 15} \quad \text{(Here } \sigma_1 \text{ is in N/mm}^2 \text{ hence '}p\text{' will be in N/mm}^2\text{)}$$

$$\therefore \quad p = \frac{120 \times 2 \times 0.7 \times 15}{1500} = 1.68 \text{ N/mm}^2 \quad ...(i)$$

(*b*) Taking limiting tensile stress = Longitudinal stress (σ_2)

$$= 120 \text{ N/mm}^2$$

Using equation (8.4),

$$\sigma_2 = \frac{p \times d}{4 \times \eta_c \times t}$$

$$120 = \frac{p \times 1500}{4 \times 0.30 \times 15}$$

$$\therefore \quad p = \frac{120 \times 4 \times 0.30 \times 15}{1500} = 1.44 \text{ N/mm}^2 \quad ...(ii)$$

Hence in order both the conditions may be satisfied the maximum permissible internal pressure is equal to the minimum* value of pressure given by (*i*) and (*ii*).

∴ Maximum permissible internal pressure = 1.44 N/mm^2

(*If p is taken equal to 1.68 N/mm^2, then longitudinal stress (σ_2) will be,

$$\sigma_2 = \frac{p \times d}{4 \times \eta_c \times t} = \frac{1.68 \times 1500}{4 \times 0.30 \times 15} = 140 \text{ N/mm}^2.$$

This value is more than the given limiting tensile stress.

Problem 8.8. *A cylinder of thickness 1.5 cm, has to withstand maximum internal pressure of 1.5 N/mm^2. If the ultimate tensile stress in the material of the cylinder is 300 N/mm^2, factor of safety 3.0 and joint efficiency 80%, determine the diameter of the cylinder.*

Sol. Given :

Thickness of cylinder, t = 1.5 cm

Internal pressure, p = 1.5 N/mm^2

Ultimate tensile stress = 300 N/mm^2

Factor of safety = 3.0

∴ Working stress or σ_1

$$= \frac{\text{Ultimate tensile stress}}{\text{Factor of safety}} = \frac{300}{3} = 100 \text{ N/mm}^2$$

Joint efficiency, η = 80% = 0.80

Joint efficiency means the efficiency of longitudinal joint (or η_l),

$\therefore$ $\eta_l = 0.80.$

The stress corresponding to longitudinal joint is given by equation (8.3).

Using equation (8.3),

$$\sigma_1 = \frac{p \times d}{2 \times \eta_l \times t}$$

$$100 = \frac{1.5 \times d}{2 \times 0.80 \times 1.5}$$

$$\therefore \quad d = \frac{100 \times 2 \times 0.80 \times 1.5}{1.5} = \textbf{160 cm = 1.6 m. Ans.}$$

8.6. STRAINS IN THIN CYLINDERS DUE TO INTERNAL PRESSURE

When a fluid having internal pressure (p) is stored in a thin cylindrical shell, due to internal pressure of the fluid the stresses set up at any point of the material of the shell are :

(*i*) Hoop or circumferential stress (σ_1), acting on longitudinal section.

(*ii*) Longitudinal stress (σ_2) acting on the circumferential section.

These stresses are principal stresses, as they are acting on principal planes. The stress in the third principal plane is zero as the thickness (t) of the cylinder is very small. Actually the stress in the third principal plane is radial stress which is very small for thin cylinders and can be neglected.

Let p = Internal pressure of fluid
L = Length of cylindrical shell
d = Diameter of the cylindrical shell
t = Thickness of the cylindrical shell
E = Modulus of Elasticity for the material of the shell
σ_1 = Hoop stress in the material
σ_2 = Longitudinal stress in the material
μ = Poisson's ratio
δd = Change in diameter due to stresses set up in the material
δL = Change in length
δV = Change in volume.

The values of σ_1 and σ_2 are given by equations (8.1) and (8.2) as

$$\sigma_1 = \frac{pd}{2t}$$

$$\sigma_2 = \frac{p \times d}{4t}$$

Let e_1 = Circumferential strain,
e_2 = Longitudinal strain.

Then **circumferential strain,**

$$e_1 = \frac{\sigma_1}{E} - \frac{\mu\sigma_2}{E} \quad \text{...(8.5)}$$

$$= \frac{pd}{2tE} - \frac{\mu pd}{4tE} \quad \left(\because \ \sigma_1 = \frac{pd}{2t} \text{ and } \sigma_2 = \frac{pd}{4t}\right)$$

$$= \frac{pd}{2tE}\left[1 - \frac{\mu}{2}\right] \quad \text{...(8.6)}$$

and **longitudinal strain,**

$$e_2 = \frac{\sigma_2}{E} - \frac{\mu\sigma_1}{E} \quad \text{...(8.7)}$$

$$= \frac{pd}{4tE} - \frac{\mu pd}{2tE} \quad \text{(substituting values of } \sigma_1 \text{ and } \sigma_2\text{)}$$

$$= \frac{pd}{2tE}\left(\frac{1}{2} - \mu\right) \quad \text{...(8.8)}$$

But circumferential strain is also given as,

$$e_1 = \frac{\text{Change in circumference due to pressure}}{\text{Original circumference}}$$

$$= \frac{\text{Final circumference} - \text{Original circumference}}{\text{Original circumference}}$$

$$= \frac{\pi(d + \delta d) - \pi d}{\pi d}$$

$$= \frac{\pi d + \pi\delta d - \pi d}{\pi d} = \frac{\pi\delta d}{\pi d}$$

$$= \frac{\delta d}{d}\left(\text{or} = \frac{\text{Change in diameter}}{\text{Original diameter}}\right) \quad \text{...(8.9)}$$

Equating the two values of e_1 given by equations (8.6) and (8.9), we get

$$\frac{\delta d}{d} = \frac{pd}{2tE}\left[1 - \frac{\mu}{2}\right]$$

8.6.1. Circumferential Strain. It is given by

$$\frac{\delta d}{d} = \frac{pd}{2tE}\left(1 - \frac{\mu}{2}\right) \quad \text{...(8.10)}$$

$\therefore$ Change in diameter,

$$\delta d = \frac{pd^2}{2tE}\left(1 - \frac{\mu}{2}\right) \quad \text{...(8.11)}$$

Similarly longitudinal strain is also given as,

$$e_2 = \frac{\text{Change in length due to pressure}}{\text{Original length}}$$

$$= \frac{\delta L}{L} \quad \text{...(8.12)}$$

Equating the two values of e_2 given by equations (8.8) and (8.12).

$$\therefore \quad \frac{\delta L}{L} = \frac{pd}{2tE}\left(\frac{1}{2} - \mu\right)$$

8.6.2. Longitudinal Strain. It is given by

$$\frac{\delta L}{L} = \frac{pd}{2tE}\left(\frac{1}{2} - \mu\right) \quad \text{...(8.13)}$$

$\therefore$ Change in length,

$$\delta L = \frac{p \times d \times L}{2tE}\left(\frac{1}{2} - \mu\right) \quad ...(8.14)$$

8.6.3. Volumetric Strains. It is defined as change in volume divided by original volume.

$\therefore$ Volumetric strain $= \dfrac{\delta V}{V}$

But change in volume (δV) = Find volume – Original volume

Original volume (V) = Area of cylindrical shell × Length

$$= \frac{\pi}{4} d^2 \times L$$

Final volume = (Final area of cross-section) × Final length

$$= \frac{\pi}{4}[d + \delta d]^2 \times [L + \delta L]$$

$$= \frac{\pi}{4}[d^2 + (\delta d)^2 + 2d\,\delta d] \times [L + \delta L]$$

$$= \frac{\pi}{4}[d^2 L + (\delta d)^2 L + 2d\,L\delta d + \delta L d^2 + \delta L\,(\delta d)^2 + 2d\,\delta d \delta L]$$

Neglecting the smaller quantities such as $(\delta d)^2 L$, $\delta L(\delta d)^2$ and $2d\,\delta d\delta L$, we get

Final volume $= \dfrac{\pi}{4}[d^2 L + 2d\,L\delta d + \delta L\,d^2]$

$\therefore$ Change in volume (δV) $= \dfrac{\pi}{4}[d^2 L + 2\,dL\delta d + \delta L d^2] - \dfrac{\pi}{4} d^2 \times L$

$$= \frac{\pi}{4}[2d\,L\delta d + \delta L d^2]$$

$\therefore$ Volumetric strains $= \dfrac{\delta V}{V} = \dfrac{\frac{\pi}{4}[2d\,L\delta d + \delta L d^2]}{\frac{\pi}{4} d^2 \times L}$

$$= \frac{2\delta d}{d} + \frac{\delta L}{L} \quad ...(8.15)$$

$$= 2\,e_1 + e_2 \qquad \left(\because \frac{\delta d}{d} = e_1, \frac{\delta L}{L} = e_2\right) \quad ...(8.16)$$

$$= 2 \times \frac{pd}{2Et}\left[1 - \frac{\mu}{2}\right] + \frac{pd}{2Et}\left(\frac{1}{2} - \mu\right)$$

(Substituting the values of e_1 and e_2)

$$= \frac{pd}{2Et}\left(2 - \frac{2\mu}{2} + \frac{1}{2} - \mu\right)$$

$$= \frac{pd}{2Et}\left(2 + \frac{1}{2} - \mu - \mu\right)$$

$$= \frac{pd}{2Et}\left(\frac{5}{2} - 2\mu\right) \quad ...(8.17)$$

Also change in volume (δV) $= V(2e_1 + e_2)$. ...(8.18)

8.7. SOLVED PROBLEMS 8.9 TO 8.19 BASED ON STRAINS PRODUCED IN THIN CYLINDERS

Problem 8.9. *Calculate : (i) the change in diameter, (ii) change in length and (iii) change in volume of a thin cylindrical shell 100 cm diameter, 1 cm thick and 5 m long when subjected to internal pressure of 3 N/mm². Take the value of E = 2 × 10⁵ N/mm² and Poisson's ratio, μ = 0.3.*

Sol. Given :

Diameter of shell, $d = 100$ cm

Thickness of shell, $t = 1$ cm

Length of shell, $L = 5 \text{ m} = 5 \times 100 = 500$ cm

Internal pressure, $p = 3 \text{ N/mm}^2$

Young's modulus, $E = 2 \times 10^5 \text{ N/mm}^2$

Poisson's ratio, $\mu = 0.30$

(*i*) Change in diameter (δd) is given by equation (8.11) as

$$\delta d = \frac{pd^2}{2tE}\left[1 - \frac{\mu}{2}\right]$$

$$= \frac{3 \times 100^2}{2 \times 1 \times 2 \times 10^5}\left[1 - \frac{1}{2} \times 0.30\right]$$

$$= \frac{3}{40}[1 - 0.15] = \mathbf{0.06375 \text{ cm}}. \quad \textbf{Ans.}$$

(*ii*) Change in length (δL) is given by equation (8.14) as

$$\delta L = \frac{pdL}{2tE}\left[\frac{1}{2} - \mu\right]$$

$$= \frac{3 \times 100 \times 500}{2 \times 1 \times 2 \times 10^5}\left[\frac{1}{2} - 0.30\right]$$

$$= \frac{15}{40} \times 0.20 = \mathbf{0.075 \text{ cm}}. \quad \textbf{Ans.}$$

(*iii*) Change in volume (δV) is given by equation (8.18) as

$$\delta V = V[2e_1 + e_1]$$

$$= V\left[2\frac{\delta d}{d} + \frac{\delta L}{L}\right] \qquad \left(\because e_1 = \frac{\delta d}{d}, e_2 = \frac{\delta L}{L}\right)$$

Substituting the values of δd, δL, d and L, we get

$$\delta V = V\left[2 \times \frac{0.06375}{100} + \frac{0.075}{500}\right]$$

$$= V[0.001275 + 0.00015] = 0.001425\, V.$$

But $V = \text{Original volume} = \frac{\pi}{4} d^2 L$

$$= \frac{\pi}{4} \times 100^2 \times 500 \text{ cm}^3 = 3926990.817 \text{ cm}^3$$

$\therefore$ $\delta V = 0.001425 \times 3926990.817 = \mathbf{5595.96 \text{ cm}^3}$. **Ans.**

Problem 8.10. *A cylindrical thin drum 80 cm in diameter and 3 m long has a shell thickness of 1 cm. If the drum is subjected to an internal pressure of 2.5 N/mm², determine (i) change in diameter, (ii) change in length and (iii) change in volume.*

Take E = 2 × 10⁵ N/mm² : Poisson's ratio = 0.25. (Annamalai University 1990)

Sol. Given :

Diameter of drum, $d = 80$ cm

Length of drum, $L = 3 \text{ m} = 3 \times 100 = 300$ cm

Thickness of drum, $t = 1$ cm

Internal pressure, $p = 2.5 \text{ N/mm}^2$

Young's modulus, $E = 2 \times 10^5 \text{ N/mm}^2$

Poisson's ratio, $\mu = 0.25$

(*i*) Change in diameter (δd) is given by equation (8.11) as

$$\delta d = \frac{pd^2}{2tE}\left(1 - \frac{1}{2} \times \mu\right)$$

$$= \frac{2.5 \times 80^2}{2 \times 1 \times 2 \times 10^5}\left[1 - \frac{1}{2} \times 0.25\right]$$

$$= 0.04\,[1 - 0.125] = \mathbf{0.035\ cm.}\quad \textbf{Ans.}$$

(*ii*) Change in length (δL) is given by equation (8.14) as

$$\delta L = \frac{pdL}{2tE}\left[\frac{1}{2} - \mu\right]$$

$$= \frac{2.5 \times 80 \times 300}{2 \times 1 \times 2 \times 10^5}\left[\frac{1}{2} - 0.25\right] = \mathbf{0.0357\ cm.}\quad \textbf{Ans.}$$

(*iii*) Using equation (8.15) for volumetric strain $\left(\frac{\delta V}{V}\right)$, we have

$$\frac{\delta V}{V} = 2\frac{\delta d}{d} + \frac{\delta L}{L}$$

$$= 2 \times \frac{0.035}{80} + \frac{0.0375}{300} \qquad \left(\because \begin{matrix} \delta d = 0.035, & \delta L = 0.0375 \\ d = 80, & L = 300 \end{matrix}\right)$$

$$= 0.000875 + 0.000125 = 0.001$$

$$\therefore \quad \delta V = 0.001 \times V$$

where volume $V = \frac{\pi}{4} d^2 \times L = \frac{\pi}{4} \times 80^2 \times 300 = 1507964.473 \text{ cm}^3$

$\therefore$ Change in volume, $\delta V = 0.001 \times 1507964.473 =$ **1507.96 cm³. Ans.**

Problem 8.11. *A cylindrical shell 90 cm long 20 cm internal diameter having thickness of metal as 8 mm is filled with fluid at atmospheric pressure. If an additional 20 cm³ of fluid is pumped into the cylinder, find (i) the pressure exerted by the fluid on the cylinder and (ii) the hoop stress induced. Take E = 2 × 10⁵ N/mm² and μ = 0.3.*

Sol. Given :

Length of cylinder, $L = 90$ cm

Diameter of cylinder, $d = 20$ cm

Thickness of cylinder, $t = 8 \text{ mm} = 0.8 \text{ cm}$

Volume of additional fluid $= 20 \text{ cm}^3$

Volume of cylinder, $V = \frac{\pi}{4} d^2 \times L = \frac{\pi}{4} \times 20^2 \times 90$

$= 28274.33 \text{ cm}^3$

Increase in volume, $\delta V =$ Volume of additional fluid

$= 20 \text{ cm}^3$

(*i*) Let p = Pressure exerted by fluid on the cylinder

$E = 2 \times 10^5 \text{ N/mm}^2$

$\mu = 0.3.$

Now using equation (8.16), volumetric strain is given as

$$\frac{\delta V}{V} = 2 e_1 + e_2$$

or
$$\frac{20}{28274.33} = 2 e_1 + e_2 \quad ..(i)$$

But e_1 and e_2 are circumferential and longitudinal strains and are given by equations (8.6) and (8.8) respectively as

$$e_1 = \frac{pd}{2Et}\left[1 - \frac{1}{2} \times \mu\right]$$

and
$$e_2 = \frac{pd}{2tE}\left(\frac{1}{2} - \mu\right).$$

Substituting these values in equation (*i*), we get

$$\frac{20}{28274.33} = \frac{2pd}{2Et}\left[1 - \frac{1}{2} \times \mu\right] + \frac{pd}{2tE}\left[\frac{1}{2} - \mu\right]$$

$$= \frac{2p \times 20}{2 \times 2 \times 10^5 \times 0.8}\left[1 - \frac{1}{2} \times 0.3\right] + \frac{p \times 20}{0.8 \times 2 \times 10^5}\left[\frac{1}{2} - 0.3\right]$$

or
$$0.000707 = \frac{p}{8000} \times 0.85 + \frac{p}{8000} \times 0.20 = \frac{1.05p}{8000}$$

$\therefore$
$$p = \frac{0.000707 \times 8000}{1.05} = \mathbf{5.386 \ N/mm^2}. \quad \textbf{Ans.}$$

(*ii*) Hoop stress (σ_1) is given by equation (8.1) as

$$\sigma_1 = \frac{pd}{2t} = \frac{5.386 \times 20}{2 \times 0.8} = \mathbf{67.33 \ N/mm^2}. \quad \textbf{Ans.}$$

Problem 8.12. *A cylindrical vessel whose ends are closed by means of rigid flange plates, is made of steel plate 3 mm thick. The length and the internal diameter of the vessel are 50 cm and 25 cm respectively. Determine the longitudinal and hoop stresses in the cylindrical shell due to an internal fluid pressure of 3 N/mm². Also calculate the increase in length, diameter and volume of the vessel. Take E = 2 × 10⁵ N/mm² and μ = 0.3.* (AMIE, Winter 1984)

Sol. Given :

Thickness, $t = 3 \text{ mm} = 0.3 \text{ cm}$

Length of the cylindrical vessel, $L = 50 \text{ cm}$

Internal diameter, $d = 25 \text{ cm}$

Internal fluid pressure, $p = 3 \text{ N/mm}^2$

Young's modulus, $E = 2 \times 10^5$ N/mm²

Poisson's ratio, $\mu = 0.3$

Let σ_1 = Hoop stress and

σ_2 = Longitudinal stress.

Using equation (8.1) for hoop stress,

$$\sigma_1 = \frac{p \times d}{2t} = \frac{3 \times 25}{2 \times 0.3} = \mathbf{125\ N/mm^2.} \quad \textbf{Ans.}$$

Using equation (8.2) for longitudinal stress,

$$\sigma_2 = \frac{p \times d}{4t} = \frac{3 \times 25}{4 \times 0.3} = \mathbf{62.5\ N/mm^2.} \quad \textbf{Ans.}$$

Using equation (8.5) for circumferential strain,

$$e_1 = \frac{\sigma_1}{E} - \frac{\mu \times \sigma_2}{E}$$

$$= \frac{1}{E}[\sigma_1 - \mu \times \sigma_2]$$

$$= \frac{1}{2 \times 10^5}[125 - 62.5 \times 0.3] \quad (\because \mu = 0.3, \sigma_1 = 125 \text{ and } \sigma_2 = 62.5)$$

$$= \frac{1}{2 \times 10^5}(125 - 18.75) = \frac{106.25}{2 \times 10^5}$$

$$= 53.125 \times 10^{-5}.$$

But circumferential strain is also given by equation (8.9) as

$$e_1 = \frac{\delta d}{d}$$

Equating the two values of circumferential strain e_1, we get

$$\frac{\delta d}{d} = 53.125 \times 10^{-5}$$

$$\therefore \quad \delta d = 53.125 \times 10^{-5} \times d = 53.125 \times 10^{-5} \times 25 = 0.0133 \text{ cm}$$

$\therefore$ Increase in diameter, δd = **0.0133 cm. Ans.**

Longitudinal strain is given by equation (8.7) as

$$e_2 = \frac{\delta L}{L} = \frac{\sigma_2}{E} - \frac{\mu \times \sigma_1}{E}$$

$$= \frac{1}{E}[\sigma_2 - \mu \times \sigma_1]$$

$$= \frac{1}{2 \times 10^5}[62.5 - 125 \times 0.3] = \frac{1}{2 \times 10^5}[62.5 - 37.5]$$

$$= \frac{2.5}{2 \times 10^5} = 12.5 \times 10^{-5}$$

$\therefore$ Increase in length, $\delta L = 12.5 \times 10^{-5} \times L$

$= 12.5 \times 10^{-5} \times 50 =$ **0.00625 cm. Ans.**

Volumetric strain is given by equation (8.16), as

$$\frac{\delta V}{V} = 2\frac{\delta d}{d} + \frac{\delta l}{l}$$

$$= 2e_1 + e_2 = 2 \times 53.125 \times 10^{-5} + 12.5 \times 10^{-5}$$
$$= 106.25 \times 10^{-5} + 12.5 \times 10^{-5} = 118.75 \times 10^{-5}$$

$\therefore$ Increase in volume,

$$\delta V = 118.75 \times 10^{-5} \times V$$

$$= 118.75 \times 10^{-5} \times \frac{\pi}{4} \times 25^2 \times 50 \qquad \left(\because \text{ volume} = \frac{\pi}{4} d^2 \times L\right)$$

$$= \mathbf{29.13\ cm^2.}\quad \textbf{Ans.}$$

Problem 8.13. *A cylindrical vessel is 1.5 m diameter and 4 m long is closed at ends by rigid plates. It is subjected to an internal pressure of 3 N/mm². If the maximum principal stress is not to exceed 150 N/mm², find the thickness of the shell. Assume E = 2 × 10⁵ N/mm² and Poisson's ratio = 0.25. Find the changes in diameter, length and volume of the shell.*

Sol. Given :

Dia., $d = 1.5 \text{ m} = 1500 \text{ mm}$

Length, $L = 4 \text{ m} = 4000 \text{ mm}$

Internal pressure, $p = 3 \text{ N/mm}^2$

Max. principal stress $= 150 \text{ N/mm}^2$

Max. principal stress means the circumferential stress

$\therefore$ Circumferential stress, $\sigma_1 = 150 \text{ N/mm}^2$

Value of $E = 2 \times 10^5 \text{ N/mm}^2$.

Poisson's ratio, $\mu = 0.25$

Let t = thickness of the shell,

δd = change in diameter,

δL = change in length, and

δV = change in volume.

(*i*) Using equation (8.1),

$$\sigma_1 = \frac{p \times d}{2t}$$

$$\therefore \quad t = \frac{p \times d}{2 \times \sigma_1} = \frac{3 \times 1500}{2 \times 150}$$

(Here p and σ_1 are in same units, 'd' is in mm hence 't' will be in mm)

$$= \mathbf{15\ mm.}\quad \textbf{Ans.}$$

(*ii*) Using equation (8.11),

$$\delta d = \frac{pd^2}{2t \times E}\left(1 - \frac{1}{2} \times \mu\right)$$

$$= \frac{3 \times 1500^2}{2 \times 15 \times 2 \times 10^5}\left(1 - \frac{1}{2} \times 0.25\right) = \mathbf{0.984\ mm.}\quad \textbf{Ans.}$$

(*iii*) Using equation (8.14),

$$\delta L = \frac{p \times d \times L}{2t \times E}\left(\frac{1}{2} - \mu\right)$$

$$= \frac{3 \times 1500 \times 4000}{2 \times 15 \times 2 \times 10^5}\left(\frac{1}{2} - 0.25\right)$$

$$= \mathbf{0.75\ mm.}\quad \textbf{Ans.}$$

(*iv*) Using equation (8.17),

$$\frac{\delta V}{V} = \frac{p \times d}{2E \times t}\left(\frac{5}{2} - 2 \times \mu\right)$$

$$= \frac{3 \times 1500}{2 \times 2 \times 10^5 \times 15}\left(\frac{5}{2} - 2 \times 0.25\right) = \frac{3 \times 1500 \times 2}{4 \times 10^5 \times 15}$$

$$\therefore \quad \delta V = \frac{3}{2000} \times V = \frac{3}{2000} \times \left(\frac{\pi}{4} \times d^2 \times L\right)$$

$$= \frac{3}{2000} \times \left(\frac{\pi}{4} \times 1500^2 \times 4000\right) = \mathbf{10602875\ mm^3.} \quad \textbf{Ans.}$$

Problem 8.14. *A closed cylindrical vessel made of steel plates 4 mm thick with plane ends, carries fluid under a pressure of 3 N/mm². The dia. of cylinder is 25 cm and length is 75 cm, calculate the longitudinal and hoop stresses in the cylinder wall and determine the change in diameter, length and volume of the cylinder. Take E = 2.1 × 10⁵ N/mm² and μ = 0.286.*

Sol. Given :

Thickness, $t = 4$ mm

Fluid pressure, $p = 3$ N/mm^2

Diameter, $d = 25$ cm $= 250$ mm

Length, $L = 75$ cm $= 750$ mm

Value of $E = 2.1 \times 10^5$ N/mm^2

Poisson's ratio, $\mu = 0.286$

Let σ_1 = Hoop stress,

σ_2 = Longitudinal stress,

δd = Change in diameter,

δL = Change in length, and

δV = Change in volume.

(*i*) Longitudinal stress is given by equation (8.2) as

$$\sigma_2 = \frac{p \times d}{4 \times t}$$

$$= \frac{3 \times 250}{4 \times 4} = \mathbf{46.875\ N/mm^2.} \quad \textbf{Ans.}$$

(*ii*) Hoop stress is given by equation (8.1) as

$$\sigma_1 = \frac{p \times d}{2 \times t}$$

$$= \frac{3 \times 250}{2 \times 4} = \mathbf{93.75\ N/mm^2.} \quad \textbf{Ans.}$$

(*iii*) The change in diameter is given by equation (8.11) as

$$\delta d = \frac{p \times d^2}{2t \times E}\left(1 - \frac{1}{2} \times \mu\right)$$

$$= \frac{3 \times 250^2}{2 \times 4 \times 2.1 \times 10^5}\left(1 - \frac{1}{2} \times 0.286\right) = \mathbf{0.0956\ mm.} \quad \textbf{Ans.}$$

(*iv*) The change in length is given by equation (8.14) as

$$\delta L = \frac{p \times d \times L}{2E \times t}\left(\frac{1}{2} - \mu\right)$$

$$= \frac{3 \times 250 \times 750}{2 \times 2.1 \times 10^5 \times 4}\left(\frac{1}{2} - 0.286\right) = \mathbf{0.0716\ mm.} \quad \textbf{Ans.}$$

(*v*) The change in volume is given by equation (8.17) as

$$\frac{\delta V}{V} = \frac{p \times d}{2t \times E}\left(\frac{5}{2} - 2 \times \mu\right)$$

$$= \frac{3 \times 250}{2 \times 4 \times 2.1 \times 10^5}\left(\frac{5}{2} - 2 \times 0.286\right)$$

$$= \frac{750}{16.8 \times 10^5}(2.5 - 0.572) = \frac{750 \times 1.928}{16.8 \times 10^5}$$

$$\therefore \quad \delta V = \frac{750 \times 1.928}{16.8 \times 10^5} \times V$$

$$= \frac{750 \times 1.928}{16.8 \times 10^5} \times \left(\frac{\pi}{4} \times 250^2 \times 750\right)$$

$$\left(\because \ V = \frac{\pi}{4} d^2 \times L = \frac{\pi}{4} \times 250^2 \times 750\right)$$

$$= \mathbf{31680\ mm^3.} \quad \textbf{Ans.}$$

Problem 8.15. *A cylindrical shell 3 metres long which is closed as the ends has an internal diameter of 1 m and a wall thickness of 15 mm. Calculate the circumferential and longitudinal stresses induced and also changes in the dimensions of the shell, if it is subjected to an internal pressure of 1.5 N/mm². Take E = 2 × 10⁵ N/mm² and μ = 0.3.*

(Annamalai University 1991)

Sol. Given :

Length of shell, $L = 3$ m $= 300$ cm

Internal diameter, $d = 1$ m $= 100$ cm

Wall thickness, $t = 15$ mm $= 1.5$ cm

Internal pressure, $p = 1.5$ N/mm^2

Young's modulus, $E = 2 \times 10^5$ N/mm^2

Poisson's ratio, $\mu = 0.3$

Let σ_1 = Circumferential (or Hoop) stress, and

σ_2 = Longitudinal stress.

Using equation (8.1) for hoop stress,

$$\sigma_1 = \frac{pd}{2t}$$

$$= \frac{1.5 \times 100}{2 \times 1.5} = \mathbf{50\ N/mm^2.} \quad \textbf{Ans.}$$

Using equation (8.2) for longitudinal stress,

$$\sigma_2 = \frac{p \times d}{4t}$$

$$= \frac{1.5 \times 100}{4 \times 1.5} = \mathbf{25\ N/mm^2}. \quad \textbf{Ans.}$$

Changes in the dimensions

Using equation (8.11) for the change in diameter (δd),

$$\delta d = \frac{pd^2}{2tE}\left(1 - \frac{1}{2} \times \mu\right)$$

$$= \frac{1.5 \times 100^2}{2 \times 1.5 \times 2 \times 10^5}\left(1 - \frac{1}{2} \times 0.3\right) \qquad (\because \mu = 0.3)$$

$$= \frac{1}{4 \times 10^3}(1 - 0.15) = \frac{0.85}{4 \times 10^3}$$

$$= \mathbf{0.2125 \times 10^{-3}\ cm}. \quad \textbf{Ans.}$$

Using equation (8.14) for change in length, we get

$$\delta L = \frac{p \times d \times L}{2tE}\left(\frac{1}{2} - \mu\right)$$

$$= \frac{1.5 \times 100 \times 300}{2 \times 1.5 \times 2 \times 10^8}\left(\frac{1}{2} - 0.3\right)$$

$$= \frac{10 \times 100 \times 300}{4 \times 10^6} \times 0.2 = \frac{0.06}{4} = \mathbf{0.015\ cm}. \quad \textbf{Ans.}$$

Using equation (8.17) for volumetric strain, we get

$$\frac{\delta V}{V} = \frac{p \times d}{2Et}\left[\frac{5}{2} - 2\mu\right]$$

$$= \frac{5 \times 100}{2 \times 2 \times 10^5 \times 1.5}[2.5 - 2 \times 0.3] \qquad (\because \mu = 0.3)$$

$$= 0.25 \times 10^{-3} \times [2.5 - 0.6]$$

$$= 0.25 \times 10^{-3} \times 1.9 = 0.475 \times 10^{-3}$$

$\therefore$ Change in volume, $\delta V = 0.475 \times 10^{-3} \times V$

where V = Original volume

$$= \frac{\pi}{4} d^2 \times L = \frac{\pi}{4} \times 100^2 \times 300 = 2356194.49\ \text{cm}^3.$$

$$\therefore \qquad \delta V = 0.475 \times 10^{-3} \times 2356194.49 = \mathbf{1119.19\ cm^3}. \quad \textbf{Ans.}$$

Problem 8.16. *A thin cylindrical shell with following dimensions is filled with a liquid at atmospheric pressure : Length = 1.2 m, external diameter = 20 cm, thickness of metal = 8 mm.*

Find the value of the pressure exerted by the liquid on the walls of the cylinder and the hoop stress induced if an additional volume of 25 cm³ of liquid is pumped into the cylinder. Take E = 2.1 × 10⁵ N/mm² and Poisson's ratio = 0.33.

Sol. Given :

Length, L = 1.2 m = 1200 mm

External dia. D = 20 cm = 200 mm

Thickness, $t = 8$ mm

$\therefore$ Internal dia., $d = D - 2 \times t = 200 - 2 \times 8 = 184$ mm

Additional volume, $\delta V = 25 \text{ cm}^3 = 25 \times 10^3 \text{ mm}^3 = 25000 \text{ mm}^3$

Value of $E = 2.1 \times 10^5 \text{ N/mm}^2$

Poisson's ratio, $\mu = 0.33$

Let p = Pressure exerted, and

σ_1 = hoop stress produced.

Volume of liquid or inside volume of cylinder,

$$V = \frac{\pi}{4} d^2 \times L$$

$$= \frac{\pi}{4} \times 184^2 \times 1200 = 31908528 \text{ mm}^3$$

(*i*) Using equation (8.17),

$$\frac{\delta V}{V} = \frac{p \times d}{2E \times t}\left(\frac{5}{2} - 2\mu\right)$$

or

$$\frac{25000}{31908528} = \frac{p \times 184}{2 \times 2.1 \times 10^5 \times 8}\left(\frac{5}{2} - 2 \times 0.33\right)$$

$\therefore$

$$p = \frac{25000 \times 2 \times 2.1 \times 10^5 \times 8}{31908528 \times 184 \times (2.5 - 0.66)} = \mathbf{7.77 \ N/mm^2}. \quad \textbf{Ans.}$$

(*ii*) Using equation (8.1),

$$\sigma_1 = \frac{p \times d}{2t} = \frac{7.77 \times 184}{2 \times 8} = \mathbf{89.42 \ N/mm^2}. \quad \textbf{Ans.}$$

Problem 8.17. *A hollow cylindrical drum 600 mm in diameter and 3 m long, has a shell thickness of 10 mm. If the drum is subjected to an internal air pressure of 3 N/mm², determine the increase in its volume. Take E = 2 × 10⁵ N/mm² and Poisson's ratio = 0.3 for the material.*

Sol. Given :

External diameter, $D = 600$ mm

Length of drum, $L = 3 \text{ m} = 3000$ mm

Thickness of drum, $t = 10$ mm

Internal pressure, $p = 3 \text{ N/mm}^2$

Young's modulus, $E = 2 \times 10^5 \text{ N/mm}^2$

Poisson's ratio, $\mu = 0.3$

Internal dia., $d = D - 2 \times t = 600 - 2 \times 10 = 580$ mm

Using equation (8.17),

$$\frac{\delta V}{V} = \frac{p \times d}{2E \times t}\left(\frac{5}{2} - 2\mu\right)$$

$$= \frac{3 \times 580}{2 \times 2 \times 10^5} \times 10\left(\frac{5}{2} - 2 \times 0.3\right) = 0.000435 \times 1.9 = 0.0008265$$

$\therefore$

$$\delta V = 0.0008265 \times V$$

$$= 0.0008265 \times \left(\frac{\pi}{4} \times d^2 \times L\right) = 0.0008265 \times \left(\frac{\pi}{4} \times 580^2 \times 3000\right)$$

$$= \mathbf{792623000 \ mm^3}. \quad \textbf{Ans.}$$

8.7.1. A Thin Cylindrical Vessel Subjected to Internal Fluid Perssure and a Torque. When a thin cylindrical vessel is subjected to internal fluid pressure (p), the stresses set up in the material of the vessel are circumferential stress σ_1 and longitudinal stress σ_2. These two stresses are tensile and are acting perpendicular to each other. If the cylindrical vessel is subjected to a torque, shear stresses will also be set up in the material of the vessel.

Hence at any point in the material of the cylindrical vessel, there will be two tensile stresses mutually perpendicular to each other accompanied by a shear stress. The major principal stress, the minor principal stress and maximum shear stress will be obtained as given in Art. 1.6.

Let σ_1 = Circumferential stress (tensile)

σ_2 = Longitudinal stress (tensile)

τ = Shear stress due to torque.

The major principal stress $$= \frac{\sigma_1 + \sigma_2}{2} + \sqrt{\left(\frac{\sigma_1 - \sigma_2}{2}\right)^2 + \tau^2} \qquad ...(8.19)$$

Minor principal stress $$= \frac{\sigma_1 + \sigma_2}{2} - \sqrt{\left(\frac{\sigma_1 - \sigma_2}{2}\right)^2 + \tau^2} \qquad ...(8.20)$$

and maximum shear stress $= \frac{1}{2}$ [Major principal stress – Minor principal stress] ...(8.21)

Problem 8.18. *A thin cylindrical tube 80 mm internal diameter and 5 mm thick, is closed at the ends and is subjected to an internal pressure of 6 N/mm². A torque of 2009600 Nmm is also applied to the tube. Find the hoop stress, longitudinal stress, maximum and minimum principal stresses and the maximum shear stress.*

Sol. Given :

Internal diameter, $d = 80$ mm

Thickness of tube, $t = 5$ mm

Internal pressure, $p = 6$ N/mm^2

Torque applied, $T = 2009600$ Nmm

Let σ_1 = Hoop stress and

σ_2 = Longitudinal stress.

Using equation (8.1) for hoop stress,

$$\sigma_1 = \frac{p \times d}{2t} = \frac{6 \times 80}{2 \times 5} = \mathbf{48\ N/mm^2}. \quad \textbf{Ans.}$$

Using equation (8.2) for longitudinal stress,

$$\sigma_2 = \frac{p \times d}{4t} = \frac{6 \times 80}{4 \times 5} = \mathbf{24\ N/mm^2}. \quad \textbf{Ans.}$$

Maximum and minimum principal stresses

The stresses σ_1 and σ_2 are tensile stresses. But the cylindrical tube is also subjected to torque. Due to torque, shear stress will be produced in the tube. As the thickness of the wall of the tube is small, the shear stress is assumed to be uniform throughout the thickness.

Let τ = Shear stress in the wall of the tube

$\therefore$ Shear force = Shear stress × Area

$= \tau \times (\pi d \times t)$ ($\because$ Area = Circumference × Thickness = $\pi d \times t$)

$$= \tau \times \pi \times 80 \times 5$$
$$= 400\pi\tau$$

and torque, $T = \text{Shear force} \times \frac{d}{2} = 400\pi \times \tau \times \frac{d}{2}$

$$= 400\pi \times \tau \times \frac{80}{2} = 16000\pi \times \tau \text{ Nmm}$$

But torque applied $= 2009600$ Nmm (given)

$\therefore$ Equating the two values of the torque, we get

$$16000\pi \times \tau = 2009600$$

$$\therefore \quad \tau = \frac{2009600}{16000\pi} = 40 \text{ N/mm}^2.$$

Hence the material of the tube is subjected to two tensile stresses (*i.e.*, $\sigma_1 = 48$ N/mm^2 and $\sigma_2 = 24$ N/mm^2) accompanied by a shear stress ($\tau = 40$ N/mm^2).

$\therefore$ Maximum principal stress is given by equation (8.19) as

$$\text{Max. principal stress} = \frac{\sigma_1 + \sigma_2}{2} + \sqrt{\left(\frac{\sigma_1 - \sigma_2}{2}\right)^2 + \tau^2}$$

$$= \frac{48 + 24}{2} + \sqrt{\left(\frac{48 - 24}{2}\right)^2 + 40^2}$$

$$= 36 + \sqrt{12^2 + 1600}$$

$$= 36 + 41.76 = \mathbf{77.7 \text{ N/mm}^2 \text{ (tensile)}}. \quad \textbf{Ans.}$$

Minimum principal stress is given by equation (8.20) as

$$\text{Min. principal stress} = \frac{\sigma_1 + \sigma_2}{2} - \sqrt{\left(\frac{\sigma_1 - \sigma_2}{2}\right)^2 + \tau^2}$$

$$= \frac{48 + 24}{2} - \sqrt{\left(\frac{48 - 24}{2}\right)^2 + 40^2}$$

$$= 36 - 41.76$$

$$= \mathbf{-5.76 \text{ N/mm}^2 \text{ (compressive)}}. \quad \textbf{Ans.}$$

(– ve sign shows compressive stress)

Max. shear stress is given by equation (8.21) as

$$\therefore \quad \text{Maximum shear stress} = \frac{\text{Max. principal stress} - \text{Min. principal stress}}{2}$$

$$= \frac{77.76 - (-5.76)}{2} = \frac{77.76 + 5.76}{2} = \frac{83.52}{2}$$

$$= \mathbf{41.76 \text{ N/mm}^2}. \quad \textbf{Ans.}$$

Problem 8.19. *A copper cylinder, 90 cm long, 40 cm external diameter and wall thickness 6 mm has its both ends closed by rigid blank flanges. It is initially full of oil at atmospheric pressure. Calculate the additional volume of oil which must be pumped into it in order to raise the oil pressure to 5 N/mm² above atmospheric pressure. For copper assume E = 1.0 × 10⁵ N/mm² and Poisson's ratio = 1/3. Take bulk modulus of oil as 2.6 × 10³ N/mm².* (AMIE)

Sol. Given :

Length of cylinder, $L = 90$ cm

External diameter, $D = 40$ cm

Wall thickness, $t = 6$ mm $= 0.6$ cm.

$\therefore$ Internal diameter, d = External dia. – 2 × Wall thickness

$= 40 - 2 \times 0.6 = 38.8$ cm.

Initial volume of oil, V = Internal volume of cylinder

$$= \frac{\pi}{4} \times d^2 \times L$$

$$= \frac{\pi}{4} \times 38.8 \times 90 = 106413 \text{ cm}^3$$

Increase in oil pressure, $p = 5$ N/mm^2.

Young's modulus for copper, $E = 1.0 \times 10^5$ N/mm^2

Poisson's ratio, $\mu = \frac{1}{3}$.

*Bulk modulus of oil, $k = 2.6 \times 10^5$ N/mm^2

Due to internal pressure of fluid inside the cylinder, there will be a change in the dimensions of the cylinder. Due to this, there will be a increase in the volume of the cylinder. Let us first calculate this increase in volume of the cylinder.

Let δV_1 = Increase in volume of cylinder.

Then volumetric strain $= \frac{\delta V_1}{V}$.

But volumetric strain due to fluid pressure is given by equation (8.17), as

$$\frac{\delta V_1}{V} = \frac{pd}{2Et}\left(\frac{5}{2} - 2\mu\right) \quad \text{(Here } \delta V_1 = \text{Increase of volume instead of } \delta V)$$

$$= \frac{5 \times 38.8}{2 \times 1.0 \times 10^5 \times 0.6}\left(\frac{5}{2} - 2 \times \frac{1}{3}\right) \qquad \left(\because \mu = \frac{1}{3}\right)$$

$$= \frac{5 \times 38.8}{2 \times 10^5 \times 0.6}(2.5 - 0.667) = 0.00296.$$

$\therefore$ $\delta V_1 = 0.00296 \times V$

$= 0.00296 \times 106413$ $\qquad (\because V = 106413 \text{ cm}^3)$

$= 314.98$ cm^3.

As bulk modulus of oil is given, then due to increase of fluid pressure on the oil, the original volume of oil will decrease. Let us find this decrease in volume of the oil.

Let δV_2 = Decrease in volume of oil due to increase of pressure.

$$\text{* Bulk modulus, } k = \frac{\text{Increase of pressure}}{\left(\frac{\text{Decrease in volume}}{\text{Volume}}\right)} = \frac{dp}{\left(\frac{dV}{V}\right)}.$$

Bulk modulus is given as

$$k = \frac{\text{Increase in pressure of oil}}{\left(\dfrac{\text{Decrease in volume of oil}}{\text{Original volume of oil}}\right)} = \frac{p}{\left(\dfrac{\delta V_2}{V}\right)}$$

$$= \frac{p \times V}{\delta V_2}$$

$$\therefore \qquad \frac{\delta V_2}{V} = \frac{p}{k}$$

or

$$\delta V_2 = \frac{p}{k} \times V$$

$$= \frac{5}{2.6 \times 10^3} \times 106413 = 204.64 \text{ cm}^3.$$

$\therefore$ Resultant additional space created in the cylinder

= Increase in volume of cylinder + Decrease in volume of oil

$= \delta V_1 + \delta V_2$

$= 314.98 + 204.64 = 519.62 \text{ cm}^3$.

$\therefore$ **Additional quantity of oil which must be pumped in order to raise the oil pressure to 5 N/mm²**

$= \mathbf{519.62 \text{ cm}^3}$. **Ans.**

8.8. THIN WALLED SPHERICAL SHELLS

Fig. 8.5 shows a thin spherical shell of internal diameter 'd' and thickness 't' and subjected to an internal fluid pressure 'p'. The fluid inside the shell has a tendency to split the shell into two hemispheres along x-x axis.

The force (P) which has a tendency to split the shell

$$= p \times \frac{\pi}{4} d^2$$

The area resisting this force $= \pi.d.t$

$\therefore$ Hoop or circumferential stress (σ_1) induced in the material of the shell is given by,

$$\sigma_1 = \frac{\text{Force } P}{\text{Area resisting the force } P}$$

$$= \frac{p \times \frac{\pi}{4} d^2}{\pi.d.t} = \frac{p.d}{4t}$$

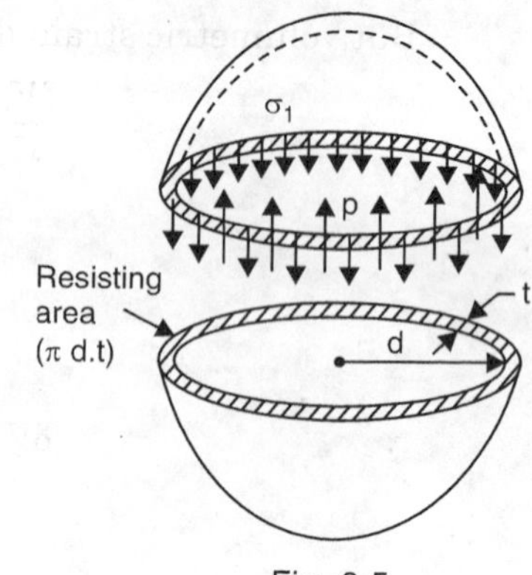

Fig. 8.5

The stress σ_1 is tensile in nature.

The fluid inside the shell is also having tendency to split the shell into two hemispheres along Y-Y axis. Then it can be shown that the tensile hoop stress will also be equal to $\frac{p.d}{4t}$. Let this stress is σ_2.

$$\therefore \qquad \sigma_2 = \frac{p \times d}{4t}.$$

The stress σ_2 will be at right angles to σ_1.

8.9. SOLVED PROBLEMS 8.20 TO 8.21 FOR THIN SPHERICAL SHELLS

Problem 8.20. *A vessel in the shape of a spherical shell of 1.20 m internal diameter and 12 mm shell thickness is subjected to pressure of 1.6 N/mm². Determine the stress induced in the material of the vessel.*

Sol. Given :

Internal diameter, $d = 1.2 \text{ m} = 1.2 \times 10^3 \text{ mm}$

Shell thickness, $t = 12 \text{ mm}$

Fluid pressure, $p = 1.6 \text{ N/mm}^2$

The stress induced in the material of spherical shell is given by,

$$\sigma_1 = \frac{p.d}{4t} = \frac{1.6 \times 1.2 \times 10^3}{4 \times 12} = \mathbf{40 \text{ N/mm}^2}. \quad \textbf{Ans.}$$

Problem 8.21. *A spherical vessel 1.5 m diameter is subjected to an internal pressure of 2 N/mm². Find the thickness of the plate required if maximum stress is not to exceed 150 N/mm² and joint efficiency is 75%.*

Sol. Given :

Diameter of shell, $d = 1.5 \text{ m} = 1.5 \times 10^3 \text{ mm}$

Fluid pressure, $p = 2 \text{ N/mm}^2$

Stress in material, $\sigma_1 = 150 \text{ N/mm}^2$

Joint efficiency, $\eta = 75\% \text{ or } 0.75$

Let t = thickness of the plate.

The stress induced is given by,

$$\sigma_1 = \frac{p.d}{4t.\eta}$$

or

$$t = \frac{p.d}{4.\eta.\sigma_1} = \frac{2 \times 1.5 \times 10^3}{4 \times 0.75 \times 150} = \mathbf{6.67 \text{ mm.}} \quad \textbf{Ans.}$$

8.10. VOLUMETRIC STRAIN IN A THIN SPHERICAL SHELL DUE TO AN INTERNAL PRESSURE

In previous article, we have seen that the stresses σ_1 and σ_2 at any point are equal (each equal to $p.d/4t$) and like. There is no shear stress at any point in the shell. $\left(\text{Maximum shear stress} = \frac{\sigma_1 - \sigma_2}{2} = \frac{p \times d}{4t} - \frac{p \times d}{4t} = 0\right)$. The stresses σ_1 and σ_2 are acting at right angles to each other.

$\therefore$ Strain in any one direction is given by,

$$e = \frac{\sigma_1}{E} - \frac{\mu \times \sigma_1}{E}$$

$$= \frac{\sigma_1}{E} - \frac{\mu \times \sigma_1}{E} \qquad \left(\because \ \sigma_1 = \sigma_2 = \frac{p \times d}{4t}\right)$$

$$= \frac{\sigma_1}{E}(1-\mu) = \frac{p \times d}{4tE}(1-\mu)$$

We know that strain in any direction is also

$$= \frac{\delta d}{d}$$

$$\therefore \qquad \frac{\delta d}{d} = \frac{p \times d}{4tE}(1-\mu) \qquad ...(i)$$

8.10.1. Volumetric strain $\left(\frac{\mathbf{dV}}{\mathbf{V}}\right)$. The ratio of change of volume to the original volume is known as volumetric strain. If V = original volume and dV = change in volume. Then $\frac{dV}{V}$ = volumetric strain.

Let V = Original volume

$$= \frac{\pi}{6} d^3 \qquad \left(\because \text{ For a sphere, } V = \frac{4}{3}\pi r^3 = \frac{\pi}{6} d^3\right)$$

Taking the differential of the above equation, we get

$$dV = \frac{\pi}{6} \times 3d^2 \times d(d)$$

Hence

$$\frac{dV}{V} = \frac{\frac{\pi}{6} \times 3d^2 \times d(d)}{\frac{\pi}{6} \times d^3}$$

$$= 3\frac{d(d)}{d} \qquad ...(ii)$$

But from equation (*i*), we have

$$\frac{\delta d}{d} \text{ or } \frac{d(d)}{d} = \frac{p \times d}{4tE}(1-\mu)$$

Substituting this value in equation (*ii*), we get

$$\frac{dV}{V} = \frac{3 \times p \times d}{4tE}(1-\mu).$$

Problem 8.22. *A spherical shell of internal diameter 0.9 m and of thickness 10 mm is subjected to an internal pressure of 1.4 N/mm². Determine the increase in diameter and increase in volume. Take $E = 2 \times 10^5$ N/mm² and $\mu = \frac{1}{3}$.*

Sol. Given :

Internal diameter, $d = 0.9 \text{ m} = 0.9 \times 10^3 \text{ mm}$

Thickness of shell, $t = 10$ mm

Fluid pressure, $p = 1.4 \text{ N/mm}^2$

Value of $E = 2 \times 10^5 \text{ N/mm}^2$

Value of $\mu = \frac{1}{3}$

Using the relation,

$$\frac{\delta d}{d} = \frac{p \times d}{4tE}(1 - \mu)$$

or

$$\frac{\delta d}{d} = \frac{p \times d}{4tE}(1 - \mu)$$

$$= \frac{1.4 \times (0.9 \times 10^3)}{4 \times 10 \times 2 \times 10^5}\left(1 - \frac{1}{3}\right) = 105 \times 10^{-6}$$

∴ Increase in diameter,

$$\delta d = 105 \times 10^{-6} \times 0.9 \times 10^3$$

$$= 94.5 \times 10^{-3} \text{ mm} = \mathbf{0.0945 \text{ mm.}} \quad \textbf{Ans.}$$

Now, volumetric strain $= 3 \times \frac{\delta d}{d}$

$$= 3 \times 105 \times 10^{-6} = 315 \times 10^{-6}$$

or

$$\frac{\delta V}{V} = 315 \times 10^{-6}$$

∴ Increase in volume,

$$\delta V = 315 \times 10^{-6} \times V$$

$$= 315 \times 10^{-6} \times \left(\frac{\pi}{6} d^3\right) \qquad \left(\because V = \frac{\pi}{6} d^3 \text{ for a sphere}\right)$$

$$= 315 \times 10^{-6} \times \frac{\pi}{6} \times (0.9 \times 10^3)^3$$

$$= \mathbf{12028.5 \text{ mm}^3.} \quad \textbf{Ans.}$$

HIGHLIGHTS

1. If the thickness of the wall of the cylindrical vessel is less than 1/20 of its internal diameter, the cylindrical vessel is known as a thin cylinder. Or if $\frac{t}{d} < \frac{1}{20}$, the cylinder is a thin cylinder.
2. In case of thin cylinders, the stress distribution is assumed uniform over the thickness of wall.
3. When a thin cylindrical vessel is subjected to internal fluid pressure, the stresses set up in the material of a thin cylinder are :
   (*i*) Circumferential or hoop stress and
   (*ii*) Longitudinal stress.
4. The circumferential stress (σ_1) is given by

   $$\sigma_1 = \frac{p \times d}{2t}$$

   where p = Internal fluid pressure,
   d = Internal diameter of the thin cylinder, and
   t = Thickness of the wall of the cylinder.
5. The longitudinal stress (σ_2) is given by

   $$\sigma_2 = \frac{p \times d}{4t}.$$

6. The circumferential and longitudinal stresses are tensile stresses.
7. The circumferential stress is twice the longitudinal stress.
8. If the thickness of the thin cylinder is to be determined then circumferential stress is used.
9. Maximum shear stress at any point in a thin cylinder, subjected to internal fluid pressure is given by,

$$\text{Max. shear stress} = \frac{\sigma_1 - \sigma_2}{2} = \frac{pd}{8t}.$$

10. If η_l = Efficiency of longitudinal joint, and
    η_c = Efficiency of a circumferential joint
    then the circumferential stress (σ_1) and longitudinal stress (σ_2) are given by

$$\sigma_1 = \frac{p \times d}{2t \times \eta_l} \text{ and } \sigma_2 = \frac{p \times d}{4t \times \eta_c}.$$

11. In longitudinal joint, the circumferential stress is developed whereas in circumferential joint the longitudinal stress is developed.
12. The circumferential strain (e_1) and longitudinal strain (e_2) are given by

$$e_1 = \frac{pd}{2tE}\left(1 - \frac{1}{2} \times \mu\right) \quad \text{and} \quad e_2 = \frac{pd}{2tE}\left(\frac{1}{2} - \mu\right).$$

13. The circumferential strain (e_1) and longitudinal strain (e_2) are also equal to

$$e_1 = \frac{\delta d}{d} \quad \text{and} \quad e_2 = \frac{\delta L}{L}.$$

14. The change in diameter, of a cylindrical shell subjected to internal fluid pressure p, is given by

$$\delta d = \frac{pd^2}{2tE}\left(1 - \frac{1}{2} \times \mu\right)$$

where μ = Poisson's ratio.

15. The change in length, of a cylindrical shell subjected to internal fluid pressure p, is given by

$$\delta L = \frac{p \times d \times L}{2tE}\left(\frac{1}{2} - \mu\right).$$

16. The change in volume, of a cylindrical shell subjected to internal fluid pressure p, is given by

$$\delta V = \frac{pd}{2Et}\left(\frac{5}{2} - 2 \times \mu\right) \times \text{volume.}$$

17. When a thin cylindrical shell is subjected to internal fluid pressure p and a torque, then the principal stresses are given by

$$\text{Major principal stress} = \frac{\sigma_1 + \sigma_2}{2} + \sqrt{\left(\frac{\sigma_1 - \sigma_2}{2}\right)^2 + \tau^2}$$

$$\text{Minor principal stress} = \frac{\sigma_1 + \sigma_2}{2} - \sqrt{\left(\frac{\sigma_1 - \sigma_2}{2}\right)^2 + \tau^2} \text{ and}$$

$$\text{Maximum shear stress} = \frac{1}{2}\,[\text{Major principal stress} - \text{Minor principal stress}]$$

where σ_1 = Circumferential stress,
σ_2 = Longitudinal stress, and
τ = Shear stress due to torque.

18. The hoop stress in the thin spherical shell is given by

$$\sigma_1 = \frac{p \times d}{4 \times t}.$$

EXERCISE 8

(A) Theoretical Questions

1. Define thin cylinders. Name the stresses set up in a thin cylinder subjected to internal fluid pressure.
2. (*a*) Prove that the circumference stress (σ_1) and longitudinal stress (σ_2) are given by,

$$\sigma_1 = \frac{p \times d}{2t}$$

$$\sigma_2 = \frac{p \times d}{4t}$$

where p = Internal fluid pressure
d = Internal diameter of thin cylinder
t = Thickness of wall of the thin cylinder.

(*b*) Derive an expression for circumferential stress and longitudinal stress for a thin shell subjected to an internal pressure. *(Bhavnagar University, Feb. 1992)*

3. (*a*) Derive the expressions for hoop stress and longitudinal stress in a thin cylinder with ends closed by rigid flanges and subjected to an internal fluid pressure p. Take the internal diameter and shell thickness of the cylinder to be 'd' and 't' respectively. *(AMIE, Summer 1983, 85)*

(*b*) Derive from the first principles the expressions for circumferential and longitudinal stresses in a thin cylinder closed at both ends and subjected to internal fluid pressure. *(AMIE, Winter 1988)*

4. Show that in thin cylinder shells subjected to internal fluid pressure, the circumferential stress is twice the longitudinal stress. *(AMIE, Summer 1982)*
5. While resigning a cylindrical vessel, which stress should be used for calculating the thickness of the cylindrical vessel ?
6. Prove that maximum shear stress at any point in a thin cylinder, subjected to internal fluid pressure is given by,

$$\text{Max. shear stress} = \frac{p \times d}{8t}$$

where p = Internal fluid pressure,
d = Internal dia. of thin cylinder, and
t = Wall thickness of cylinder.

7. Find the expression for circumferential stress and longitudinal stress for a longitudinal joint and circumferential joint.
8. Prove that the circumferential strain (e_1) and longitudinal strain (e_2), produced in a thin cylinder when subjected to internal fluid pressure (p), are given by

$$d_1 = \frac{pd}{2tE}\left(1 - \frac{1}{2} \times \mu\right) \text{ and}$$

$$e_2 = \frac{pd}{2tE}\left(\frac{1}{2} - \mu\right)$$

where p = Internal fluid pressure,
d = Internal dia. of thin cylinder,
t = Thickness of wall of thin cylinder, and
μ = Poisson's ratio.

9. A cylindrical shell is subjected to internal fluid pressure, find an expression for change in diameter and change in length of the cylinder.

10. (*a*) Prove that volumetric strain in case of a thin cylinder, subjected to internal fluid pressure is equal to two times the circumferential strain plus longitudinal strain.

(*b*) Show that when a thin walled cylindrical vessel of dia. D, length L and thickness t is subjected to an internal pressure p, the change in volume is given by

$$\frac{\pi \times p \times D^3 \times L}{16tE}(5 - 4 \times \mu).$$ (*AMIE, Summer 1990*)

11. Find an expression for the change in volume of a thin cylindrical shell subjected to internal fluid pressure.

12. Write down the expression for major principal stress, minor principal stress and maximum shear stress when a thin cylindrical shell is subjected to internal fluid pressure and a torque.

13. Show that when a thin walled spherical vessel of dia. 'd' and thickness 't' is subjected to internal fluid pressure 'p', the increase in volume equal to

$$\frac{\pi}{8} \times \frac{pd^4}{tE}\left(1 - \frac{1}{\mu}\right)$$

where E = Elastic modulus and
μ = Poisson's ratio. (*AMIE, Summer 1989*)

(B) Numerical Problems

1. A cylindrical pipe of diameter 2.0 m and thickness 2.0 cm is subjected to an internal fluid pressure of 1.5 N/mm². Determine :
   (*i*) Longitudinal stress, and
   (*ii*) Circumferential stress developed in the pipe material. [**Ans.** 37.5 N/mm², 75 N/mm²]
2. A cylinder of internal diameter 3.0 m and of thickness 6 cm contains a gas. If the tensile stress in the material is not to exceed 70 N/mm², determine the internal pressure of the gas. [**Ans.** 2.8 N/mm²]
3. A cylinder of internal diameter 0.60 m contains air at a pressure of 7.5 N/mm² (gauge). If the maximum permissible stress induced in the material is 75 N/mm², find the thickness of the cylinder. [**Ans.** 3 cm]
4. A thin cylinder of internal diameter 2.0 m contains a fluid at an internal pressure of 3 N/mm². Determine the maximum thickness of the cylinder if (*i*) the longitudinal stress is not to exceed 30 N/mm² and (*ii*) the circumferential stress is not to exceed 40 N/mm². [**Ans.** 7.5 cm]
5. A water main 90 cm diameter contains water at a pressure head of 110 m. If the weight density of water is 9810 N/mm³, find the thickness of the metal required for the water main. Given the permissible stress as 22 N/mm². [**Ans.** 2.25 cm]
6. A boiler is subjected to an internal steam pressure of 3 N/mm². The thickness of the boiler plate is 2.5 cm and the permissible tensile stress is 125 N/mm². Find out the maximum diameter, when efficiency of longitudinal joint is 90% and that of circumferential joint is 35%. [**Ans.** 145.83 cm]
7. A boiler shell is to be made of 20 mm thick plate having a limiting tensile stress of 125 N/mm². If the efficiencies of the longitudinal and circumferential joints are 80% and 30% respectively, determine :
   (*i*) The maximum permissible diameter of the shell for an internal pressure of 2.5 N/mm², and
   (*ii*) Permissible intensity of internal pressure when the shell diameter is 1.6 m.
   [**Ans.** (*i*) 120 cm, (*ii*) 1.875 N/mm²]
8. A cylinder of thickness 2.0 cm has to withstand maximum internal pressure of 2 N/mm². If the ultimate tensile stress in the material of the cylinder is 292 N/mm², factor of safety 4 and joint efficiency 80%, determine the diameter of the cylinder. [**Ans.** 116.8 cm]

9. A thin cylindrical shell of 120 cm diameter, 1.5 cm thick and 6 m long is subjected to internal fluid pressure of 2.5 N/mm^2. If the value of $E = 2 \times 10^5$ N/mm^2 and Poisson's ratio = 0.3, calculate : (*i*) change in diameter, (*ii*) change in length, and (*iii*) change in volume.

[**Ans.** (*i*) 0.051 m ; (*ii*) 0.06 cm ; (*iii*) 6449.7 cm^3]

10. A cylindrical shell 100 cm long 20 cm internal diameter having thickness of metal as 10 mm is filled with fluid at atmospheric pressure. If an additional 20 cm^3 of fluid is pumped into cylinder find (*i*) the pressure exerted by the fluid on the cylinder and (*ii*) the hoop stress induced. Take $E = 2 \times 10^5$ N/mm^2 and $\mu = 0.3$. [**Ans.** 10.05 N/mm^2, 100.52 N/mm^2]

11. A cylindrical vessel whose ends are closed by means of rigid flange plates, is made of steel plate 4 mm thick. The length and the internal diameter of the vessel are 100 cm and 30 cm respectively. Determine the longitudinal and hoop stresses in the cylindrical shell due to an internal fluid pressure of 2 N/mm^2. Also calculate the increase in length, diameter and volume of the vessel. Take $E = 2 \times 10^5$ N/mm^2 and $\mu = 0.3$.

[**Ans.** 37.5 N/mm^2, 75 N/mm^2, 0.075 cm, 0.0095 cm, 50.36 cm^3]

12. A thin cylindrical tube 100 mm internal diameter and 5 mm thick is closed at the ends and is subjected to an internal pressure of 5 N/mm^2. A torque of 22000 Nm is also applied to the tube. Find the hoop stress, longitudinal stress, maximum and minimum principal stresses and the maximum shear stress.

[**Ans.** 50 N/mm^2, 25 N/mm^2, 28 N/mm^2, 68.16 N/mm, 6.84 N/mm^2, 30.66 N/mm^2]

13. A copper cylinder, 100 cm long, 50 cm external diameter and wall thickness 5 mm has its both ends closed by rigid blank flanges. It is initially full of oil at atmospheric pressure. Calculate the additional volume of oil which must be pumped into it in order to raise the oil pressure to 4 N/mm^2 above atmospheric pressure. For copper assume, $E = 1.0 \times 10^5$ N/mm^2 and Poisson's ratio = 0.3. Take bulk modulus of oil as 2.5×10^3 N/mm^2. [**Ans.** 486.3 cm^3]

14. A vessel in the shape of a spherical shell of 1.4 m internal diameter and 4.5 mm thickness is subjected to a pressure of 1.8 N/mm^2. Determine the stress induced in the material of the vessel.

[**Ans.** 140 N/mm^2]

15. A thin spherical of 1.20 m internal diameter is subjected to an internal pressure of 1.6 N/mm^2. If the permissible stress in the plate material is 80 N/mm^2 and joint efficiency is 75%, find the minimum thickness. [**Ans.** 8 mm]

16. A thin spherical shell of internal diameter 1.5 m and of thickness 8 mm is subjected to an internal pressure of 1.5 N/mm^2. Determine the increase in diameter and increase in volume. Take $E = 2 \times 10^5$ N/mm^2 and $\mu = 0.3$. [**Ans.** 0.369 mm, 1304×10^3 mm^3]

9

Thick Cylinders

9.1. INTRODUCTION

In the last chapter, we have mentioned that if the ratio of thickness to internal diameter of a cylindrical shell is less than about 1/20, the cylindrical shell is known as thin cylinders. For them it may be assumed with reasonable accuracy that the hoop and longitudinal stresses are constant over the thickness and the radial stress is small and can be neglected. If the ratio of thickness to internal diameter is more than 1/20, then cylindrical shell is known as thick cylinders.

The hoop stress in case of a thick cylinder will not be uniform across the thickness. Actually the hoop stress will vary from a maximum value at the inner circumference to a minimum value at the outer circumference.

9.2. RADIAL, AXIAL AND CIRCUMFERENTIAL STRESSES IN A THICK CYLINDER

Fig. 9.1 (*a*) shows a thick cylinder subjected to a internal fluid pressure.

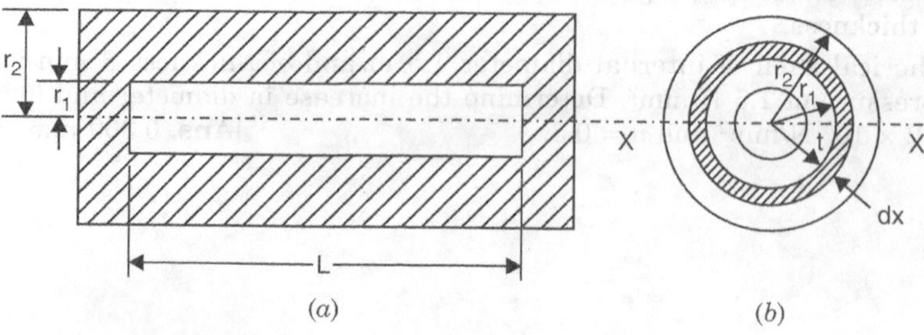

Fig. 9.1

Let r_2 = External radius of the cylinder,

r_1 = Internal radius of the cylinder, and

L = Length of cylinder.

Consider an elementary ring of the cylinder of radius x and thickness dx as shown in Fig. 9.1 (*b*) and 9.2.

Let p_x = Radial pressure on the inner surface of the ring

$p_x + dp_x$ = Radial pressure on the outer surface of the ring

σ_x = Hoop stress induced in the ring.

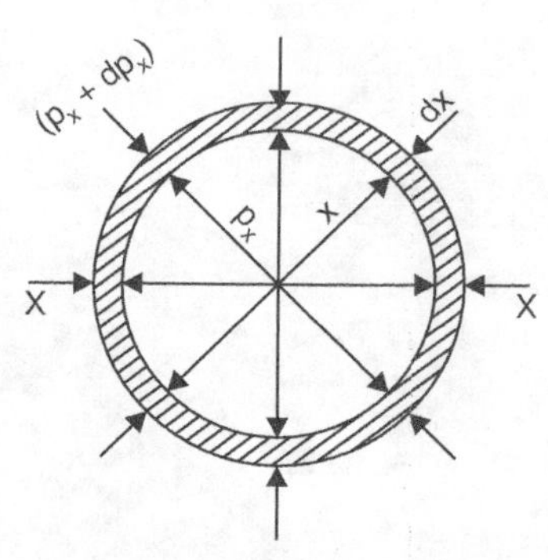

Fig. 9.2

Take a longitudinal section *x-x* and consider the equilibrium of half of the ring of Fig. 9.2 as shown in Fig. 9.2 (*a*) or in Fig. 9.2 (*b*).

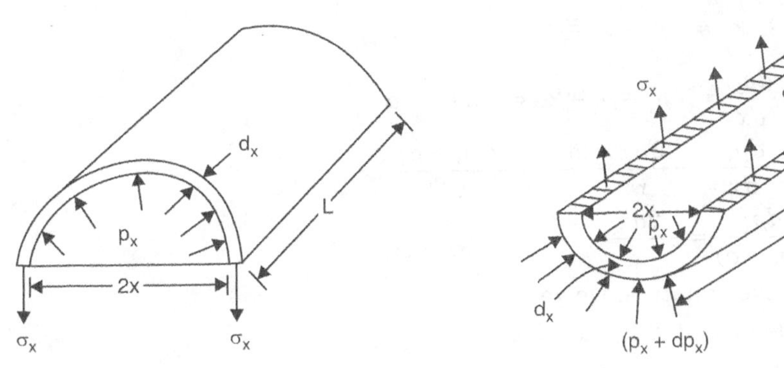

Fig. 9.2 (*a*) Fig. 9.2 (*b*)

Bursting force

$= p_x(2xL) - (p_x + dp_x) \times 2(x + dx) \, . \, L$

$= 2L \, [p_x \, . \, x - (p_x \, . \, x + p_x \, . \, dx + x dp_x + dp_x \, . \, dx)]$

$= 2L \, [- p_x \, . \, dx - x \, . \, dp_x]$ (Neglecting $dp_x \, . \, dx$ which is a small quantity)

$= - 2L \, (p_x dx + x \, . \, dp_x)$ ...(*i*)

Resisting force = Hoop stress × Area on which it acts $= \sigma_x \times 2dx. \, L$ ...(*ii*)

Equating the resisting force to the bursting force, we get

$$\sigma_x \times 2dx \, . \, L = - 2L \, (p_x \, . \, dx + x \, . \, dp_x)$$

or
$$\sigma_x = - p_x - x \frac{dp_x}{dx} \quad \text{...}(iii)$$

In case of thick cylinder it may be assumed that longitudinal strain at any point in the section is constant. This means that cross-sections remain plane after straining and this is true for sections, remote from any end fixing. As longitudinal strain is constant, hence longitudinal stress will also be constant.

Let σ_2 = Longitudinal stress.

Hence at any point at a distance *x* from the centre, three principal stresses are acting : They are :

(*i*) the radial compressive stress, p_x

(*ii*) the hoop (or circumferential) tensile stress, σ_x

(*iii*) the longitudinal tensile stress σ_2.

The longitudinal strain (e_2) at this point is given by,

$$e_2 = \frac{\sigma_2}{E} - \frac{\mu \sigma_x}{E} - \left(\frac{- \mu p_x}{E} \right) \frac{\mu p_x}{E} = \frac{\sigma_2}{E} - \frac{\mu \sigma_x}{E} + \frac{\mu p_x}{E}$$

But longitudinal strain is constant.

$$\therefore \quad \frac{\sigma_2}{E} - \frac{\mu \sigma_x}{E} + \frac{\mu p_x}{E} = \text{constant}$$

But σ_2 is also constant, and for the material of the cylinder E and μ are also constant.

$\therefore \quad \sigma_x - p_x$ = constant

$= 2a$, where a is constant

$\therefore \qquad \sigma_x = p_x + 2a \qquad ...(iv)$

Equating the two values of σ_x given by equations (*iii*) and (*iv*), we get

$$p_x + 2a = -p_x - x\frac{dp_x}{dx}$$

or
$$x \cdot \frac{dp_x}{dx} = -p_x - p_x - 2a = -2p_x - 2a$$

or
$$\frac{dp_x}{dx} = -\frac{2p_x}{x} - \frac{2a}{x} = \frac{-2(p_x + a)}{x}$$

or
$$\frac{dp_x}{(p_x + a)} = -\frac{2dx}{x}$$

Integrating the above equation, we get

$$\log_e (p_x + a) = -2\log_e x + \log_e b$$

where $\log_e b$ is a constant of integration.

The above equation can also be written as

$$\log_e (p_x + a) = -\log_e x^2 + \log_e b$$
$$= \log_e \frac{b}{x^2}$$

$\therefore$
$$p_x + a = \frac{b}{x^2}$$

or
$$p_x = \frac{b}{x^2} - a \qquad ...(9.1)$$

Substituting the values of p_x in equation (*iv*), we get

$$\sigma_x = \frac{b}{x^2} - a + 2a = \frac{b}{x^2} + a \qquad ...(9.2)$$

The equation (9.1) gives the radial pressure p_x and equation (9.2) gives the hoop stress at any radius x. These two equations are called *Lame's equations*. The constants 'a' and 'b' are obtained from boundary conditions, which are :

(*i*) at $x = r_1$, $p_x = p_0$ or the pressure of fluid inside the cylinder, and

(*ii*) at $x = r_2$, $p_x = 0$ or atmosphere pressure.

After knowing the values of 'a' and 'b', the hoop stress can be calculated at any radius.

9.2.1. Alternate Method for finding stresses in a thick-cylinder. In case of thick cylinders, due to internal fluid pressure, the three principal stresses acting at a point are :

(*i*) radial pressure (p) which is compressive,

(*ii*) circumferential stress or hoop stress (σ_1) which is tensile, and

(*iii*) longitudinal stress (σ_2), which is also tensile.

Let e_1 = circumferential strain,

e_2 = longitudinal strain, and

e_r = radial strain.

In case of thick cylinder, it may be assumed that longitudinal strain (e_2) is constant, which means that cross-section remain plane after straining.

Consider a circular ring of radius r and thickness 'dr'. Due to internal fluid pressure, let the radius r increases to $(r + u)$ and increase in the thickness *dr be du.*

$\therefore$ Initial radius = r whereas final radius = $(r + u)$

$$\text{Now circumferential strain} = \frac{\text{Final circumference} - \text{Original circumference}}{\text{Original circumference}}$$

$$= \frac{2\pi(r+u) - 2\pi r}{2\pi r} = \frac{2\pi u}{2\pi r} = \frac{u}{r}$$

or $$e_1 = \frac{u}{r} \qquad (\because \text{ Circumferential strain} = e_1) \qquad ...(i)$$

Now original thickness of ring $= dr$

Final thickness of ring $= dr + du$

$$\therefore \text{ Radial strain} = \frac{\text{Final thickness of ring} - \text{Original thickness}}{\text{Original thickness}}$$

$$= \frac{(dr + du) - dr}{dr} = \frac{du}{dr}$$

or $$e_r = \frac{du}{dr} \qquad (\because \text{ radial strain} = e_r) \qquad ...(ii)$$

The circumferential strain, longitudinal strain and radial strains in terms of stresses and Poisson's ratio are also given by,

$$e_1 = \frac{\sigma_1}{E} - \frac{\mu\sigma_2}{E} - \left(\frac{-\mu p}{E}\right) = \frac{\sigma_1}{E} - \frac{\mu\sigma_2}{E} + \frac{\mu p}{E} \qquad (\because \ p \text{ is compressive}) \qquad ...(iii)$$

$$e_2 = \frac{\sigma_2}{E} - \frac{\mu\sigma_1}{E} + \frac{\mu p}{E} \qquad ...(iv)$$

$$e_r = \frac{-p}{E} - \frac{\mu\sigma_1}{E} - \frac{\mu\sigma_2}{E} \qquad ...(v)$$

Equating the two values of circumferential strain (e_1) given by equations (*i*) and (*iii*). Also equate the two values of radial strain (e_r) given by equations (*ii*) and (*v*), we get

$$\frac{u}{r} = \frac{\sigma_1}{E} - \frac{\mu\sigma_2}{E} + \frac{\mu p}{E} \qquad [\text{Equations } (i) = (iii)] \qquad ...(vi)$$

$$\frac{du}{dr} = \frac{-p}{E} - \frac{\mu\sigma_1}{E} - \frac{\mu\sigma_2}{E} \qquad [\text{Equations } (ii) = (v)] \qquad ...(vii)$$

In equations (*vi*) and (*vii*), u and du are unknowns.

Let us first eliminate 'u' from equations (*vi*) and (*vii*).

From equation (*vi*),

$$u = \left(\frac{\sigma_1}{E} - \frac{\mu\sigma_2}{E} + \frac{\mu p}{E}\right) \times r$$

Differentiating the above equation with respect to r, we get

$$\frac{du}{dr} = \left(\frac{\sigma_1}{E} - \frac{\mu\sigma_2}{E} + \frac{\mu p}{E}\right) + \frac{r}{E}\left(\frac{d\sigma_1}{dr} - \frac{\mu d\sigma_2}{dr} + \frac{\mu dp}{dr}\right) \qquad ...(viii)$$

($\because$ σ_1, σ_2 and p are function of r)

Equating the two values of $\frac{du}{dr}$ given by equations (*vii*) and (*viii*),

$$\frac{p}{E} - \frac{\mu\sigma_1}{E} - \frac{\mu\sigma_2}{E} = \left(\frac{\sigma_1}{E} - \frac{\mu\sigma_2}{E} + \frac{\mu p}{E}\right) + \frac{r}{E}\left(\frac{d\sigma_1}{dr} - \frac{\mu d\sigma_2}{dr} + \frac{\mu dp}{dr}\right)$$

or $$-p-\mu\sigma_1-\mu\sigma_2=\sigma_1-\mu\sigma_2+\mu p+r\left(\frac{d\sigma_1}{dr}-\mu\frac{d\sigma_2}{dr}+\mu\frac{dp}{dr}\right)$$

(Cancelling E to both sides)

or $$0=p+\mu\sigma_1+\mu\sigma_2+\sigma_1-\mu\sigma_2+\mu p+r\left(\frac{d\sigma_1}{dr}-\mu\frac{d\sigma_2}{dr}+\mu\frac{dp}{dr}\right)$$

$$=p(1+\mu)+\sigma_1(1+\mu)+r\left(\frac{d\sigma_1}{dr}-\mu\frac{d\sigma_2}{dr}+\mu\frac{dp}{dr}\right)$$

$$=(p+\sigma_1)(1+\mu)+r\left(\frac{d\sigma_1}{dr}-\mu\frac{d\sigma_2}{dr}+\mu\frac{dp}{dr}\right) \quad \text{....(ix)}$$

To find the value of σ_1 in terms of p, the value of $\frac{d\sigma_2}{dr}$ must be substituted in the above equation (*ix*).

Longitudinal strain (e_2) is given by equation (*iv*) as

$$e_2=\frac{\sigma_2}{E}-\frac{\mu\sigma_1}{E}+\frac{\mu p}{E}$$

Since longitudinal strain (e_2) is assumed constant. Its differentiation with respect to r will be zero. Differentiating the above equation with respect to 'r', we get

$$0=\frac{1}{E}\left(\frac{d\sigma_2}{dr}-\mu\frac{d\sigma_1}{dr}+\frac{\mu dp}{dr}\right) \qquad \left(\because \frac{de_2}{dr}=0\right)$$

or $$0=\frac{d\sigma_2}{dr}-\mu\frac{d\sigma_1}{dr}+\mu\frac{dp}{dr}$$

or $$\frac{d\sigma_2}{dr}=\mu\frac{d\sigma_1}{dr}-\mu\frac{dp}{dr}$$

Substituting the value of $\frac{d\sigma_2}{dr}$ in equation (*ix*), we get

$$0=(p+\sigma_1)(1+\mu)+r\left[\frac{d\sigma_1}{dr}-\mu\left(\mu\frac{d\sigma_1}{dr}-\mu\frac{dp}{dr}\right)+\mu\frac{dp}{dr}\right]$$

$$=(p+\sigma_1)(1+\mu)+r\left[\frac{d\sigma_1}{dr}(1-\mu^2)+\mu\frac{dp}{dr}(1+\mu)\right]$$

$$=(p+\sigma_1)(1+\mu)+r(1-\mu^2)\frac{d\sigma_1}{dr}+\mu r(1+\mu)\frac{dp}{dr}$$

$$=(p+\sigma_1)(1+\mu)+r(1+\mu)(1-\mu)\frac{d\sigma_1}{dr}+\mu r(1+\mu)\frac{dp}{dr}$$

$$=(1+\mu)\left[(p+\sigma_1)+r(1-\mu)\frac{d\sigma_1}{dr}+\mu r\frac{dp}{dr}\right] \qquad [\because (1+\mu)\neq 0]$$

or $$0=(p+\sigma_1)+r(1-\mu)\frac{d\sigma_1}{dr}+\mu r\frac{dp}{dr} \quad \text{...(x)}$$

For the equilibrium of the element, resolving the forces acting on the element in the radial direction we get [Refer to Fig. 9.2 (*c*)],

$$(p+dp)(r+dr)\,\delta\theta-p\times r\times\delta\theta+2\sigma_1\times dr\times\sin\frac{\delta\theta}{2}=0$$

As $\delta\theta$ is small, hence $\sin\frac{\delta\theta}{2}\simeq\frac{\delta\theta}{2}$. Now the above equation becomes

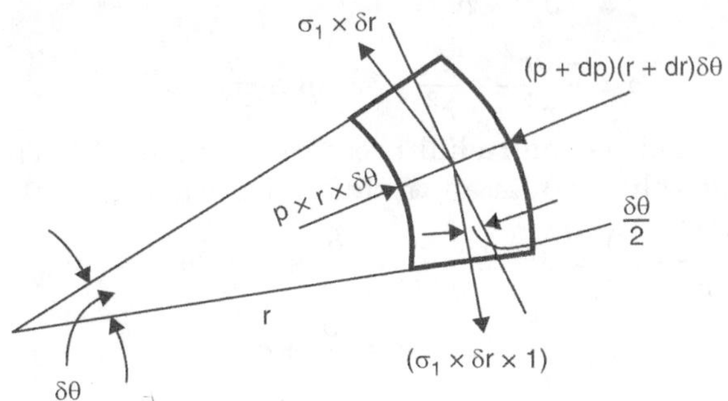

Fig. 9.2 (*c*)

$$(p + dp)(r + dr)\ \delta\theta - pr\delta\theta + 2\sigma_1 \times dr \times \frac{\delta\theta}{2} = 0$$

or $(p + dp)(r + dr) - pr + \sigma_1 \times dr = 0$ [Cancelling δθ to both sides]

$$pr + pdr + rdp + dp \times dr - pr + \sigma_1 \times dr = 0$$

or $pdr + rdp + \sigma_1\, dr = 0$

[$dp \times dr$ is the product of two small quantities and can be neglected]

or $$p + r\frac{dp}{dr} + \sigma_1 = 0 \quad \text{or} \quad p + \sigma_1 = -r\frac{dp}{dr} \qquad ...(xi)$$

Substituting the value of $(p + \sigma_1)$ in equation (*x*), we get

$$0 = -r\frac{dp}{dr} + r(1-\mu)\frac{d\sigma_1}{dr} + \mu r\frac{dp}{dr}$$

$$0 = -r\frac{dp}{dr}(1-\mu) + r(1-\mu)\frac{d\sigma_1}{dr}$$

or $$0 = -\frac{dp}{dr} + \frac{d\sigma_1}{dr}$$ [Cancelling $r(1-\mu)$ to both sides]

or $$\frac{d\sigma_1}{dr} - \frac{dp}{dr} = 0$$

Integrating, we get

$$\sigma_1 - p = \text{constant}$$

$$= 2a \qquad \text{(let the constant} = 2a) \qquad ...(xii)$$

In equations (*xii*) and (*xi*), the two unknown are σ_1 and p. Let us find the value of σ_1 from equation (*xii*) and substitute this value in equation (*xi*).

From equation (*xii*), $\sigma_1 = 2a + p$

Substituting this value of σ_1 in equation (*xi*), we get

$$p + (2a + p) = -r\frac{dp}{dr}$$

or $$2p + 2a = -r\frac{dp}{dr} \quad \text{or} \quad 2p + r\frac{dp}{dr} = -2a$$

or $$\frac{1}{r}\frac{d(pr^2)}{dr} = -2a \quad \text{or} \quad \frac{d}{dr}(pr^2) = -2ar$$

Integrating, we get

$$pr^2 = \frac{-2ar^2}{2} + b \text{ where } b \text{ is another constant.}$$

$$= -ar^2 + b$$

or $$p = \frac{-ar^2}{r^2} + \frac{b}{r^2} \quad \text{or} \quad p = \frac{b}{r^2} - a \qquad ...(xiii)$$

The above equation gives the radial pressure at radius 'r'. This equation is same as equation (9.1). To find the value of σ_1 at any radius, substitute this value of p in equation (*xiii*).

$$\therefore \quad \sigma_1 - \left(\frac{b}{r^2} - a\right) = 2a \quad \text{or} \quad \sigma_1 - \frac{b}{r^2} + a = 2a$$

or $$\sigma_1 = \frac{b}{r^2} + 2a - a = \frac{b}{r^2} + a \qquad ...(xiv)$$

The above equation gives the circumferential stress (or hoop stress) at any radius r. This equation is same as equation (9.2).

9.3. SOLVED PROBLEMS 9.1 TO 9.2 FOR THICK CYLINDERS

Problem 9.1. *Determine the maximum and minimum hoop stress across the section of a pipe of 400 mm internal diameter and 100 mm thick, when the pipe contains a fluid at a pressure of 8 N/mm². Also sketch the radial pressure distribution and hoop stress distribution across the section.* (Annamalai University, 1991)

Sol. Given :

Internal dia. $= 400$ mm

$\therefore$ Internal radius, $r_1 = \frac{400}{2} = 200$ mm

Thickness $= 100$ mm

$\therefore$ External dia. $= 400 + 2 \times 100 = 600$ mm

$\therefore$ External radius, $r_2 = \frac{600}{2} = 300$ mm

Fluid pressure, $p_0 = 8$ N/mm²

or at $x = r_1$, $p_x = p_0 = 8$ N/mm²

The radial pressure (p_x) is given by equation (9.1) as

$$p_x = \frac{b}{x^2} - a \qquad ...(i)$$

Now apply the boundary conditions to the above equations. The boundary conditions are :

1. At $x = r_1 = 200$ mm, $p_x = 8$ N/mm²
2. At $x = r_2 = 300$ mm, $p_x = 0$

Substituting these boundary conditions in equation (*i*), we get

$$8 = \frac{b}{200^2} - a = \frac{b}{40000} - a \qquad ...(ii)$$

and $$0 = \frac{b}{300^2} - a = \frac{b}{90000} - a \qquad ...(iii)$$

Subtracting equation (*iii*) from equation (*ii*), we get

$$8 = \frac{b}{40000} - \frac{b}{90000} = \frac{9b - 4b}{360000} = \frac{5b}{360000}$$

$$\therefore \quad b = \frac{360000 \times 8}{5} = 576000$$

Substituting this value in equation (*iii*), we get

$$0 = \frac{576000}{90000} - a \quad \text{or} \quad a = \frac{576000}{90000} = 6.4$$

The values of 'a' and 'b' are substituted in the hoop stress.

Now hoop stress at any radius x is given by equation (9.2) as

$$\sigma_x = \frac{b}{x^2} + a = \frac{576000}{x^2} + 6.4$$

At $x = 200$ mm, $\sigma_{200} = \frac{576000}{200^2} + 6.4 = 14.4 + 6.4 = \mathbf{20.8\ N/mm^2}$. **Ans.**

At $x = 300$ mm, $\sigma_{300} = \frac{576000}{300^2} + 6.4 = 6.4 + 6.4 = \mathbf{12.8\ N/mm^2}$. **Ans.**

Fig. 9.3 shows the radial pressure distribution and hoop stress distribution across the section. AB is taken a horizontal line. $AC = 8$ N/mm². The variation between B and C is parabolic. The curve BC shows the variation of radial pressure across AB.

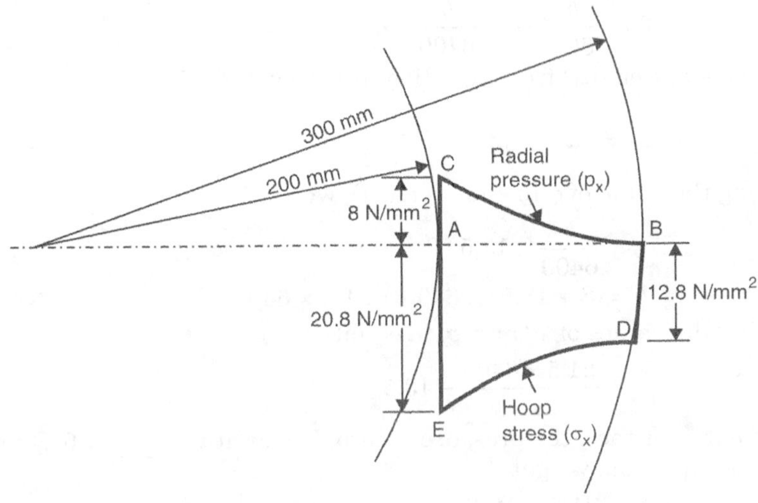

Fig. 9.3

The curve DE which is also parabolic, shows the variation of hoop stress across AB. Values $BD = 12.8$ N/mm² and $AE = 20.8$ N/mm². The radial pressure is compressive whereas the hoop stress is tensile.

Problem 9.2. *Find the thickness of metal necessary for a cylindrical shell of internal diameter 160 mm to withstand an internal pressure of 8 N/mm². The maximum hoop stress in the section is not to exceed 35 N/mm².*

Sol. Given :

Internal dia. $= 160$ mm

$\therefore$ Internal radius, $r_1 = \frac{160}{2} = 80$ mm

Internal pressure $= 8$ N/mm²

This means at $x = 80$ mm, $p_x = 8$ N/mm²

Maximum hoop stress, $\sigma_x = 35$ N/mm²

The maximum hoop stress is at the inner radius of the shell.

Let r_2 = External radius.

The radial pressure and hoop stress at any radius x are given by equations (9.1) and (9.2) is

$$p_x = \frac{b}{x^2} - a \qquad ...(i)$$

and

$$\sigma_x = \frac{b}{x^2} + a \qquad ...(ii)$$

Let us now apply the boundary conditions. The boundary conditions are :

at $x = 80$ mm, $p_x = 8$ N/mm^2 and $\sigma_x = 35$ N/mm^2

Substituting $x = 80$ mm and $p_x = 8$ N/mm^2 in equation (*i*), we get

$$8 = \frac{b}{80^2} - a = \frac{b}{6400} - a \qquad ...(iii)$$

Substituting $x = 80$ mm and $\sigma_x = 35$ N/mm^2 in equation (*ii*), we get

$$35 = \frac{b}{80^2} + a = \frac{b}{6400} + a \qquad ...(iv)$$

Subtracting equation (*iii*) from equation (*iv*), we get

$$27 = 2a \quad \text{or} \quad a = \frac{27}{2} = 13.5$$

Substituting the value of a in equation (*iii*), we get

$$8 = \frac{b}{6400} - 13.5$$

or

$$b = (8 + 13.5) \times 6400 = 21.5 \times 6400$$

Substituting the values of 'a' and 'b' in equation (*i*),

$$p_x = \frac{21.5 \times 6400}{x^2} - 13.5$$

But at the outer surface, the pressure is zero. Hence at $x = r_2$, $p_x = 0$. Substituting these values in the above equation, we get

$$0 = \frac{21.5 \times 6400}{r_1^2} - 13.5$$

or

$$r_2^2 = \frac{21.5 \times 6400}{13.5} \quad \text{or} \quad r_2 = \sqrt{\frac{21.5 \times 6400}{13.5}} = 100.96 \text{ mm}$$

$\therefore$ Thickness of the shell, $t = r_2 - r_1$

$= 100.96 - 80 =$ **20.96 mm. Ans.**

9.4. COMPOUND CYLINDERS

From problem 9.1, we find that the hoop stress is maximum at the inner radius and it decreases towards the outer radius. The hoop stress is tensile in nature and it is caused by the internal fluid pressure inside the cylinder. This maximum hoop stress at the inner radius is always greater than the internal fluid pressure. Hence the maximum fluid pressure inside the cylinder is limited corresponding to the condition that the hoop stress at the inner radius reaches the permissible value. In case of cylinders which have to carry high internal fluid pressures, some methods of reducing the hoop stress have to be devised.

One method is to wind* strong steel wire under tension on the cylinder. The effect of the wire is to put the cylinder wall under an initial compressive stress.

Another method is to shrink one cylinder over the other. Due to this, the inner cylinder will be put into initial compression whereas the outer cylinder will be put into initial tension. If now the compound cylinder is subjected to internal fluid pressure, both the inner and outer cylinders will be subjected to hoop tensile stress. The net effect of the initial stresses due to shrinking and those due to internal fluid pressure is to make the resultant stresses more or less uniform.

Fig. 9.4 shows a compound thick cylinder made up of two cylinders.

Let r_2 = Outer radius of compound cylinder

r_1 = Inner radius of compound cylinder

r^* = Radius at the junction of the two cylinders (*i.e.*, outer radius of inner cylinder or inner radius of outer cylinder)

p^* = Radial pressure at the junction of the two cylinders.

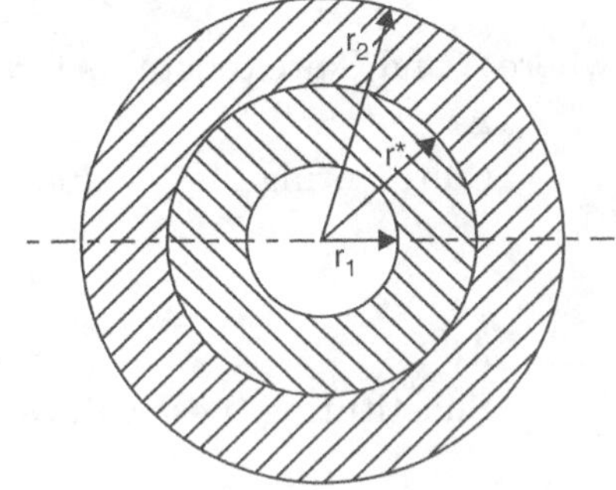

Fig. 9.4

Let us now apply Lame's equation *i.e.*, equations (9.1) and (9.2) for the initial conditions (*i.e.*, after shrinking the outer cylinder over the inner cylinder and fluid under pressure is not admitted into the inner cylinder).

(i) For outer cylinder

The Lame's equations at a radius x for outer cylinder are given by

$$p_x = \frac{b_1}{x^2} - a_1 \quad \text{...}(i) \qquad \sigma_x = \frac{b_1}{x^2} + a_1 \quad \text{...}(ii)$$

where a_1, b_1 are constants for outer cylinder.

At $x = r_2$, $p_x = 0$. And at $x = r^*$, $p_x = p^*$

Substituting these conditions in equation (*i*), we get

$$0 = \frac{b_1}{{r_2}^2} - a_1 \quad \text{...}(iii) \qquad p^* = \frac{b_1}{r^{*2}} - a_1 \quad \text{...}(iv)$$

From equations (*iii*) and (*iv*), the constants a_1 and b_1 can be determined. These values are substituted in equation (*ii*). And then hoop stresses in the outer cylinder due to shrinking can be obtained.

(ii) For inner cylinder

The Lame's equations for inner cylinder at a radius x are given by

$$p_x = \frac{b_2}{x^2} - a_2, \qquad \sigma_x = \frac{b_2}{x^2} + a_2$$

where a_2, b_2 are constants for inner cylinder.

At $x = r_1$, $p_x = 0$ as fluid under pressure is not admitted into the inner cylinder. And at $x = r^*$, $p_x = p^*$.

Substituting these values in the above value of p_x, we get

$$0 = \frac{b_2}{{r_2}^2} - a_2 \quad \text{...}(v) \quad \text{and} \quad p^* = \frac{b_2}{r^{*2}} - a_2 \quad \text{...}(vi)$$

From equations (v) and (vi), the constants a_2 and b_2 can be determined. These values are substituted in σ_x. And then hoop stresses are obtained.

Hoop stresses in compound cylinder due to internal fluid pressure alone

When the fluid under pressure is admitted into the compound cylinder, the hoop stresses are set in the compound cylinder. To find these stresses, the inner cylinder and outer cylinder will together will be considered as one thick shell, Let p = internal fluid pressure. Now the Lame's equations are applied, which are given by

$$p_x = \frac{B}{x^2} - A \quad \text{...(vii)} \quad \text{and} \quad \sigma_x = \frac{B}{x^2} + A \quad \text{...(viii)}$$

where A and B are constants for single thick shell due to internal fluid pressure.

At $x = r_2, p_x = 0.$

Substituting these values in equation (vii), we get

$$0 = \frac{B}{r_2^2} - A \quad \text{...(ix)}$$

At $x = r_1, p_x = p.$

Substituting these values in equation (vii), we get

$$p = \frac{B}{r_1^2} - A \quad \text{...(x)}$$

From equations (ix) and (x), the constants A and B can be determined. These values are substituted in equation ($viii$). And then hoop stresses across the section can be obtained.

The resultant hoop stresses will be the algebraic sum of the hoop stresses caused due to shrinking and those due to internal fluid pressure.

9.5. SOLVED PROBLEM 9.3 FOR COMPOUND THICK CYLINDER

Problem 9.3. *A compound cylinder is made by shrinking a cylinder of external diameter 300 mm and internal diameter of 250 mm over another cylinder of external diameter 250 mm and internal diameter 200 mm. The radial pressure at the junction after shrinking is 8 N/mm². Find the final stresses set up across the section, when the compound cylinder is subjected to an internal fluid pressure of 84.5 N/mm².*

Sol. Given :

For outer cylinder :

External diameter = 300 mm

∴ External radius,

$$r_2 = \frac{300}{2} = 150 \text{ mm}$$

Internal diameter = 250 mm

∴ Radius at the junction,

$$r^* = \frac{250}{2} = 125 \text{ mm.}$$

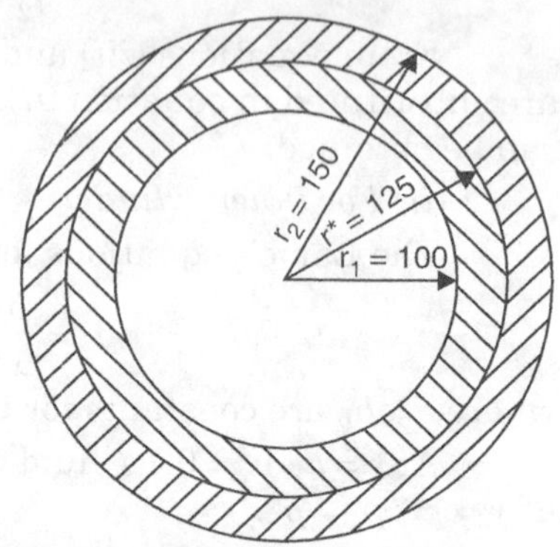

Fig. 9.5

For inner cylinder

Internal diameter = 200 mm

∴ Internal radius

$$r_1 = \frac{200}{2} = 100 \text{ mm}$$

Radical pressure due to shrinking at the junction,

$$p^* = 8 \text{ N/mm}^2$$

Fluid pressure in the compound cylinder,

$$p = 84.5 \text{ N/mm}^2.$$

(*i*) ***Stresses due to shrinking in the outer and inner cylinders before the fluid pressure is admitted.***

(*a*) Lame's equations for outer cylinders are

$$p_x = \frac{b_1}{x^2} - a_1 \quad ...(i) \qquad \text{and} \qquad \sigma_x = \frac{b_1}{x^2} + a_1 \quad ...(ii)$$

At $x = 150$ mm, $p_x = 0$.

Substituting these values in equation (*i*),

$$0 = \frac{b_1}{150^2} - a_1 = \frac{b_1}{22500} - a_1 \quad ...(iii)$$

At $x = r^* = 125$ mm, $p_x = p^* = 8$ N/mm^2.

Substituting these values in equation (*i*), we get

$$8 = \frac{b_1}{125^2} - a_1 = \frac{b_1}{15625} - a_1 \quad ...(iv)$$

Subtracting equation (*iii*) from equation (*iv*), we get

$$8 = -\frac{b_1}{22500} + \frac{b_1}{15625} = \frac{(-15625 + 22500)\, b_1}{22500 \times 15625}$$

$$\therefore \qquad b_1 = \frac{8 \times 22500 \times 15625}{(-15625 + 22500)} = 409090.9$$

Substituting the value of b_1 in equation (*iii*), we get

$$0 = \frac{409090.9}{22500} - a_1 \quad \text{or} \quad a_1 = \frac{409090.9}{22500} = 18.18$$

Substituting the values of a_1 and b_1 in equation (*ii*), we get

$$\sigma_x = \frac{409090.9}{x^2} + 18.18$$

The above equation gives the hoop stress in the outer cylinder due to shrinking. The hoop stress at the outer and inner surface of the outer cylinder is obtained by substituting $x = 150$ mm and $x = 125$ mm respectively in the above equation.

$$\therefore \qquad \sigma_{150} = \frac{409090.9}{150^2} + 18.18 = 36.36 \text{ N/mm}^2 \text{ (tensile)}$$

and

$$\sigma_{125} = \frac{409090.9}{125^2} + 18.18 = 44.36 \text{ N/mm}^2 \text{ (tensile)}.$$

(*b*) Lame's equations for the inner cylinder are :

$$p_x = \frac{b_2}{x^2} - a_2 \quad ...(v) \qquad \text{and} \qquad \sigma_x = \frac{b_2}{x^2} + a_2 \quad ...(vi)$$

At $x = r_1 = 100$ mm, $p_x = 0$ (There is no fluid under pressure.)

Substituting these values in equation (*v*), we get

$$0 = \frac{b_2}{100^2} - a_2 = \frac{b_2}{10000} - a_2 \quad ...(vii)$$

At $x = r^* = 125$ mm, $p_x = p^* = 8$ N/mm². Substituting these values in equation (v), we get

$$8 = \frac{b_2}{125^2} - a_2 = \frac{b_2}{15625} - a_2 \qquad ...(viii)$$

Subtracting equation (vii) from equation ($viii$), we get

$$8 = \frac{b_2}{15625} - \frac{b_2}{10000}$$

$$= \frac{b_2(10000 - 15625)}{15625 \times 10000} = \frac{-5625\, b_2}{15625 \times 10000}$$

$$\therefore \quad b_2 = \frac{8 \times 15625 \times 10000}{5625} = -222222.2$$

Substituting the value b_2 in equation (vii), we get

$$0 = -\frac{222222.2}{10000} - a_2$$

or $$a_2 = -22.22$$

Substituting the values of a_2 and b_2 in equation (vi), we get

$$\sigma_x = -\frac{222222.2}{x^2} - 22.22$$

Hence the hoop stress for the inner cylinder is obtained by substituting $x = 125$ mm respectively in the above equation.

$$\therefore \quad \sigma_{125} = -\frac{222222.2}{125^2} - 22.22$$

$$= -14.22 - 22.22 = -36.22 \text{ N/mm}^2 \text{ (compressive)}$$

and $$\sigma_{100} = -\frac{222222.2}{100^2} - 22.22$$

$$= -22.22 - 22.22 = -44.44 \text{ N/mm}^2 \text{ (compressive)}$$

(*ii*) *Stresses due to fluid pressure alone*

When the fluid under pressure is admitted inside the compound cylinder, the two cylinders together will be considered as one single unit. The hoop stresses are calculated by Lame's equations, which are

$$p_x = \frac{B}{x^2} - A \quad ...(ix) \qquad \text{and} \qquad \sigma_x = \frac{B}{x^2} + A \quad ...(x)$$

where A and B are constants.

At $x = 100$ mm, $p_x = p = 84.5$ N/mm². Substituting the values in equation (ix), we get

$$84.5 = \frac{B}{100^2} - A = \frac{B}{10000} - A \qquad ...(xi)$$

At $x = 150$ mm, $p_x = 0$. Substituting these values in equation (ix), we get

$$0 = \frac{B}{150^2} - A = \frac{B}{22500} - A \qquad ...(xii)$$

Subtracting equation (xii) from equation (xi), we get

$$84.5 = \frac{B}{10000} - \frac{B}{22500}$$

$$= \frac{B(22500 - 10000)}{10000 \times 22500} = \frac{12500 \times B}{10000 \times 22500}$$

$$\therefore \quad B = \frac{84.5 \times 10000 \times 22500}{12500} = 1521000$$

Substituting this values in equation (*xii*), we get

$$0 = \frac{1521000}{22500} - A \quad \text{or} \quad A = \frac{1521000}{22500} = 67.6$$

Substituting the values of A and B in equation (x), we get

$$\sigma_x = \frac{1521000}{x^2} + 67.6$$

Hence the hoop stresses due to internal fluid pressure alone are given by,

$$\sigma_{100} = \frac{1521000}{100^2} + 67.6 = 219.7 \text{ N/mm}^2 \text{ (tensile)}$$

$$\sigma_{125} = \frac{1521000}{125^2} + 67.6 = 97.344 + 67.6$$

$$= 164.94 \text{ N/mm}^2$$

$$\sigma_{150} = \frac{1521000}{150^2} + 67.6 = 67.6 + 67.6 = 135.2 \text{ N/mm}^2.$$

The resultant stresses will be the algebraic sum of the initial stresses due to shrinking and those due to internal fluid pressure.

Inner cylinder

$F_{100} = \sigma_{100}$ due to shrinkage + σ_{100} due to internal fluid pressure

$= - 44.44 + 219.7 =$ **175.26 N/mm² (tensile). Ans.**

$F_{125} = \sigma_{125}$ due to shrinkage + σ_{125} due to internal fluid pressure

$= - 36.22 + 164.94 =$ **128.72 N/mm² (tensile). Ans.**

Outer cylinder

$F_{125} = \sigma_{125}$ due to shrinkage + σ_{125} due to internal fluid pressure

$= 44.36 + 164.94 =$ **209.3 N/mm² (tensile). Ans.**

$F_{150} = \sigma_{150}$ due to shrinkage + σ_{150} due to internal fluid pressure

$= 36.36 + 135.2 =$ **171.56 N/mm² (tensile). Ans.**

9.6. STRESSES IN ROTATING SHAFT AND CYLINDERS

This article deals with the study of stresses developed due to rotation in circular discs and cylinders. The machine members of the rotating type and bodies like circular discs, cylinders, flywheels etc., invariably rotate at high speeds. Due to rotation, centrifugal stresses are developed in these members. These centrifugal stresses are equivalent to outward radial pressure on vessels. Due to outward centrifugal stresses, the radial and circumferential stresses (or hoop stresses) are developed as in the case of cylinders and shells subjected to internal fluid pressure. Therefore the analysis of rotating members will be almost similar to thin or thick cylinder subjected to fluid pressure.

Let us first derive the expression for the stress developed due to rotation in a thin cylinder.

9.6.1. Rotational Stresses in Thin Cylinders. Fig. 9.6 shows a thin cylinder of mean radius r and thickness 't', rotating at an angular velocity ω about its axis.

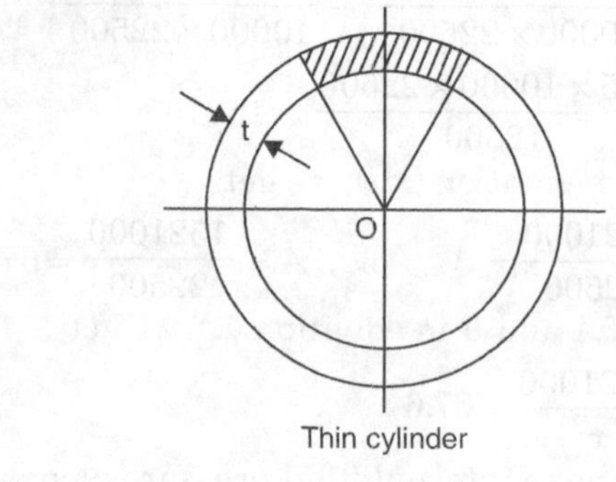

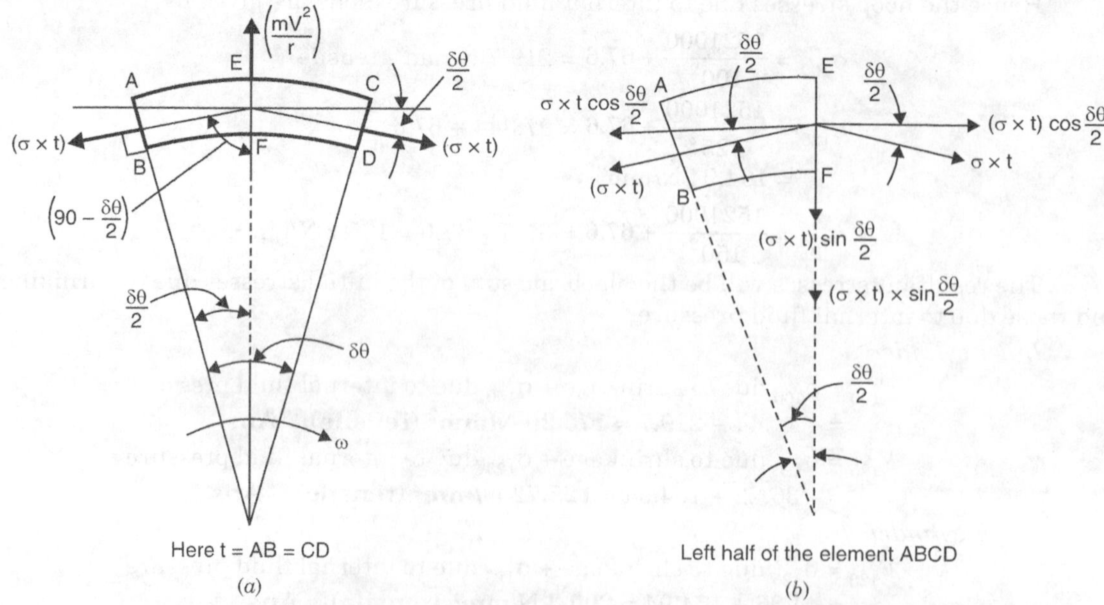

Fig. 9.6

Let r = mean radius of the cylinder

t = thickness of the cylinder

ω = angular speed of the cylinder

ρ = density of the material of cylinder

Due to rotation of cylinder, centrifugal force will be acting radially outwards on the walls of the cylinder. This centrifugal force will produce a circumferential (or hoop) stress σ. For a thin cylinder, this hoop stress σ may be assumed constant.

Consider a small element $ABCD$ of the rotating cylinder. Let this element makes an angle $\delta\theta$ at the centre as shown in Fig. 9.6 (a). Consider unit length of this element perpendicular to the plane of paper.

The forces acting on the element are :

(i) Centrifugal force $\left(\dfrac{mV^2}{r} \text{ or } m\omega^2 r\right)$ acting radially outwards. Here m is the mass of the element per unit length.

$\therefore \qquad m$ = mass of element

$= \rho \times$ volume of element

$= \rho \times$ (area of element) $\times$ unit length

$= \rho \times [(r \times \delta\theta) \times t] \times 1$

$= \rho \times r \times \delta\theta \times t$

$\therefore$ Centrifugal force $= m\omega^2 r = (\rho r \delta\theta t)\, \omega^2 r$

$= \rho r^2 \times \omega^2 \times \delta\theta \times t$

(*ii*) Tensile force due to hoop stress (σ) on the face *AB*. This force is equal to ($\sigma \times t \times 1$) and acts perpendicular to face *AB*.

(*iii*) Tensile force due to hoop stress (σ) on the face *CD*. This force is equal to ($\sigma \times t \times 1$) and acts perpendicular to face *CD*.

The horizontal component $\sigma \times t \times \cos \frac{\delta\theta}{2}$ on the face *AB* and *CD* are equal and opposite.

The radial components $\sigma \times t \times \sin \frac{\delta\theta}{2}$ are acting towards centre and they will be added.

Resolving the forces radially for equilibrium, we get

Centrifugal force $= \sigma t \times \sin \frac{\delta\theta}{2} + \sigma t \times \sin \frac{\delta\theta}{2}$

or $\qquad \rho r^2 \times \omega^2 \times \delta\theta \times t = 2 \times \sigma \times t \times \sin \frac{\delta\theta}{2}$

$= 2 \times \sigma \times t \times \frac{\delta\theta}{2} \qquad \left(\text{As } \delta\theta \text{ is very small, hence } \sin \frac{\delta\theta}{2} \approx \frac{\delta\theta}{2}\right)$

or $\qquad \rho r^2 \times \omega^2 = \sigma$ (Concelling $\delta\theta \times t$ to both sides)

or $\qquad \sigma = \rho r^2 \times \omega^2$ ...(9.3)

The equation (9.3) gives the expression for hoop stress produced in a thin cylinder due to rotation. The above equation can also be used approximately for rim-type flywheels.

9.6.2. Solved Problems 9.4 to 9.6 for Rotating Thin Cylinders

Problem 9.4. *A rim-type flywheel is rotating at a speed of 2400 r.p.m. If the mean diameter of the flywheel is 750 mm and density of the material of the wheel is 8000 kg/m³, then find the hoop stress produced in the rim due to rotation.*

Sol. Given :

Speed, $N = 2400$ r.p.m. $\therefore \quad \omega = \frac{2\pi N}{60} = \frac{2\pi \times 2400}{60} = 80\pi$ rad/s

Mean dia, $d = 750$ mm $\therefore \quad r = \frac{750}{2} = 375$ mm $= 0.375$ m

Density, $\rho = 8000$ kg/m^3

Let σ = hoop stress produced.

Using equation (9.3), we get

$$\sigma = \rho \times r^2 \times \omega^2$$

$$= 8000 \times 0.375^2 \times (80\pi)^2 \text{ N/m}^2$$

$$= 71.0485 \times 10^6 \text{ N/m}^2 = \mathbf{71.0485\ MN/m^2}. \quad \textbf{Ans.}$$

Problem 9.5. *If in problem 9.4 E = 200 GN/m² then what will be the change in diameter of the flywheel due to rotation.*

Sol. Given :

From problem 9.4,

Original dia, $d = 750$ mm, $\sigma = 71.0485$ MN/m^2 $= 71.0485 \times 10^6$ N/m^2

$E = 200$ GN/m^2 $= 200 \times 10^9$ N/m^2

Due to hoop stress, circumferential strain $= \dfrac{\sigma}{E}$

$$= \frac{71.0485 \times 10^6}{200 \times 10^9} = 0.3552 \times 10^{-3}$$

But circumferential strain is also equal to $\dfrac{\delta d}{d}$

where δd = change in diameter

$$\therefore \quad \frac{\delta d}{d} = 0.3552 \times 10^{-3}$$

or $$\delta d = 0.3552 \times 10^{-3} \times d = 0.3552 \times 10^{-3} \times 750$$

$$= \mathbf{0.2664\ mm.} \quad \textbf{Ans.}$$

Problem 9.6. *Find the speed of rotation of a wheel of diameter 750 mm, if the hoop stress is not to exceed 120 MN/m². The wheel has a thin rim and density of the wheel is 7200 kg/ m³.*

Sol. Given :

Dia, $d = 750$ mm $\quad \therefore \quad r = \dfrac{750}{2} = 375$ mm $= 0.375$ m

Max. hoop stress, $\sigma = 120$ MN/m^2 $= 120 \times 10^6$ N/m^2

Density, $\rho = 7200$ kg/m^3.

Using equation (9.3), we get

$$\sigma = \rho \times r^2 \times \omega^2$$

or $$120 \times 10^6 = 7200 \times 0.375^2 \times \omega^2$$

or $$\omega = \sqrt{\frac{120 \times 10^6}{7200 \times 0.375^2}} = 344.26 \text{ rad/s}$$

But $$\omega = \frac{2\pi N}{60}$$

$$\therefore \quad N = \frac{60 \times \omega}{2\pi} = \frac{60 \times 344.26}{2\pi} = \mathbf{3287.4\ r.p.m.} \quad \textbf{Ans.}$$

9.7. STRESSES IN A ROTATING THIN DISC

In case of thin disc, the stress in axial direction is zero. But in case of thick disc (or long cylinder) the stress in axial direction (*i.e.*, longitudinal direction) will not be zero.

In case of thin disc, only two stresses namely circumferential and radial stresses are existing. The longitudinal (or axial) stress is zero.

Thin disc is different than thin cylinder. In case of thin cylinder the difference of radii $(R_2 - R_1)$ is small. But in case of thin disc the thickness of disc perpendicular to plane of paper is small. The difference of radii $(R_2 - R_1)$ is not small.

A thin disc of inner radius r_1 and outer radius r_2 rotating about its axis is shown in Fig. 9.7. Due to rotation the radial and circumferential stresses are developed in the disc.

Consider an element ABCD of the disc at a radius 'r' of radial width dr and thickness 't'. Let the element subtends an angle $d\theta$ at the centre.

Let σ_r = Radial stress on the face AB,

$\sigma_r + d\sigma_r$ = Radial stress on the face CD,

σ_c = Circumferential stress on faces BC and AD.

The forces due to these stresses along with centrifugal force acting on the element are shown on the enlarged view of the element.

Now consider the equilibrium of the element $ABCD$ of radius 'r' radial width 'dr' and thickness 't' subtending an angle $d\theta$ at the centre as shown in Fig. 9.7.

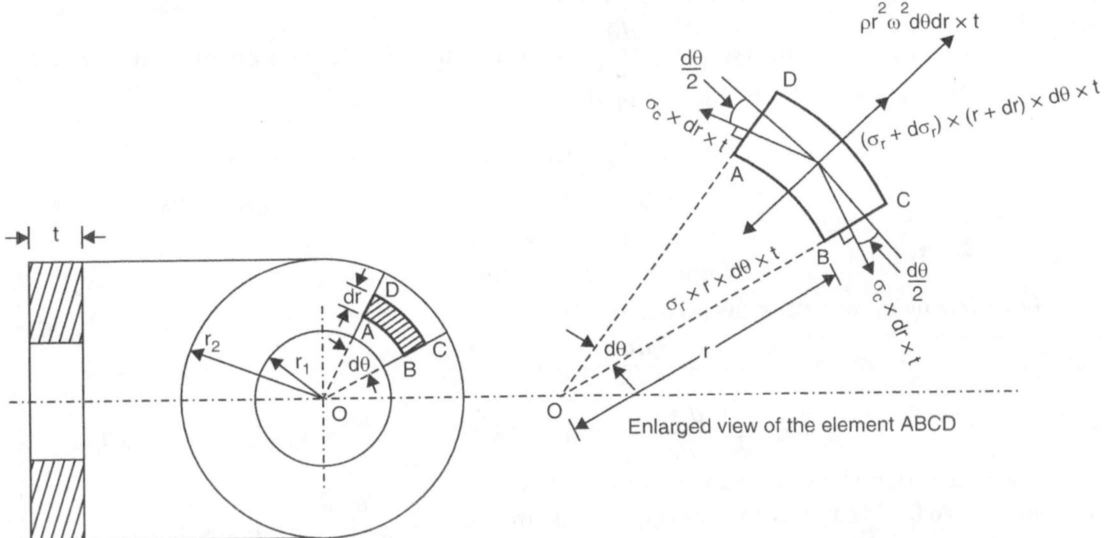

Fig. 9.7

The forces acting on the element are :

(*i*) Radial force on face AB and equal to $\sigma_r \times r\, d\theta \times t$,

(*ii*) Radial force on face CD and equal to $(\sigma_r + d\sigma_r)(r + dr) \times d\theta \times t$,

(*iii*) Circumferential force on face $BC = \sigma_c \times dr \times t$

(*iv*) Circumferential force on face $AD = \sigma_c \times dr \times t$

(*v*) Centrifugal force $= m \times \omega^2 \times r$ where

m = mass of element = ρ × volume of element

$= \rho \times (r \times d\theta \times dr \times t)$

$\therefore$ Centrifugal force $= (\rho \times r \times d\theta \times dr \times t) \times \omega^2 \times r$

$= \rho r^2 \omega^2\, d\theta\, dr \times t$

The forces in circumferential direction will be equal and opposite.

Resolving the force in radial direction, we get

$$\sigma_r \times r\,d\theta \times t + 2\sigma_c \times dr \times t \times \sin\frac{d\theta}{2}$$

$$= (\sigma_r + d\sigma_r)\,(r + dr)\,d\theta \times t + \text{centrifugal force} \qquad ...(A)$$

where centrifugal force $= m \times \omega^2 \times r$ where m = mass of element

$$= \rho \times \text{volume of element}$$

$$= \rho \times (rd\theta \times dr \times t)$$

$$= (\rho \times rd\theta \times dr \times t) \times \omega^2 \times r$$

$$= \rho r^2 \times \omega^2 \times d\theta \times dr \times t$$

and $\sin\dfrac{d\theta}{2} \simeq \dfrac{d\theta}{2}$ as $d\theta$ is very small

Hence the above equation (A) becomes as

$$\sigma_r \times rd\theta \times t + 2\sigma_c \times dr \times t \times \frac{d\theta}{2} = (\sigma_r + d\sigma_r)\,(r + dr)\,d\theta \times t + \rho r^2\,\omega^2 \times d\theta \times dr \times t$$

Cancelling $d\theta \times t$ to both sides, we get

$$\sigma_r \times r + 2\sigma_c \times dr \times \frac{1}{2} = (\sigma_r + d\sigma_r)(r + dr) + \rho r^2\,\omega^2 \times dr$$

or $$\sigma_r \times r + \sigma_c \times dr = [\sigma_r \times r + \sigma_r \times dr + rd\sigma_r + (d\sigma_r) \times dr] + \rho r^2\,\omega^2 \times dr$$

or $$\sigma_c \times dr = \sigma_r \times dr + r\,d\sigma_r + \rho r^2\,\omega^2 \times dr$$

[Neglecting the product of two small quantities *i.e.*, $(d\sigma_r) \times dr$]

Dividing by dr to both sides, we get

$$\sigma_c = \sigma_r + r\frac{d\sigma_r}{dr} + \rho r^2\,\omega^2$$

or $$(\sigma_c - \sigma_r) = r\frac{d\sigma_r}{dr} + \rho r^2\,\omega^2 \qquad ...(9.4) \quad ...(i)$$

There are two unknowns in equation (i). They are σ_c and σ_r. To find their values, we require one more equation. The second equation is obtained from stress strain equations. Hence first find circumferential and radial strain.

When the disc is rotating at high speed, the radius increases due to centrifugal force. Also the radial thickness dr increases, let the radius r becomes $(r + u)$ after straining and dr becomes $(dr + du)$ as shown in Fig. 9.7 (a).

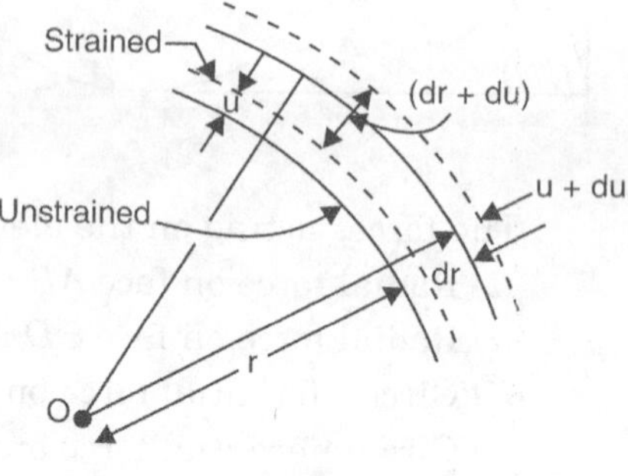

Fig. 9.7 (a)

∴ Circumferential strain,

$$e_c = \frac{\text{Final circumference} - \text{Initial circumference}}{\text{Initial circumference}}$$

$$= \frac{2\pi(r + u) - 2\pi r}{2\pi r} = \frac{u}{r}$$

and radial strain, $$e_r = \frac{\text{Final radial width} - \text{Initial radial width}}{\text{Initial radial width}}$$

$$= \frac{(dr + du) - (dr)}{dr} = \frac{du}{dr}$$

From stress strain equations the radial and circumferential strains are also given by,

$$e_c = \frac{\sigma_c}{E} - \frac{\mu\sigma_r}{E}$$

and
$$e_r = \frac{\sigma_r}{E} - \frac{\mu\sigma_c}{E}$$

Equating the two values of e_c and e_r, we get

Circumferential strain, $$e_c = \frac{\sigma_c}{E} - \frac{\mu\sigma_r}{E} = \frac{u}{r} \quad ...(ii)$$

and radial strain, $$e_r = \frac{\sigma_r}{E} - \frac{\mu\sigma_c}{E} = \frac{du}{dr} \quad ...(iii)$$

In equations (*ii*) and (*iii*), u is the increase in radius r due to rotation and du is the increase in radial thickness dr. These two values u and du are unknowns. They can be eliminated from equations (*ii*) and (*iii*) as given below :

From equation (*ii*), we have

$$u = \frac{r}{E}(\sigma_c - \mu\sigma_r)$$

or
$$E \times u = r(\sigma_c - \mu\sigma_r)$$

Differentiating the above equation w.r.t. r, we get

$$E\frac{du}{dr} = (\sigma_c - \mu\sigma_r) + r\left(\frac{d\sigma_c}{dr} - \mu\frac{d\sigma_r}{dr}\right) \quad ...(iv)$$

But from equation (*iii*), we have

$$\frac{du}{dr} = \frac{\sigma_r}{E} - \frac{\mu\sigma_c}{E}$$

or
$$E\frac{du}{dr} = \sigma_r - \mu\sigma_c \quad ...(v)$$

Equating the two values of $E\frac{du}{dr}$ given by equations (*iv*) and (*v*), we get

$$(\sigma_c - \mu\sigma_r) + r\left(\frac{d\sigma_c}{dr} - \mu\frac{d\sigma_r}{dr}\right) = \sigma_r - \mu\sigma_c$$

or
$$(\sigma_c - \mu\sigma_r) + r\left(\frac{d\sigma_c}{dr} - \mu\frac{d\sigma_r}{dr}\right) - \sigma_r + \mu\sigma_c = 0$$

or
$$\sigma_c(1+\mu) - \sigma_r(1+\mu) + r\left(\frac{d\sigma_c}{dr} - \mu\frac{d\sigma_r}{dr}\right) = 0$$

or
$$(\sigma_c - \sigma_r)(1+\mu) + r\left(\frac{d\sigma_c}{dr} - \mu\frac{d\sigma_r}{dr}\right) = 0 \quad ...(vi)$$

Substituting the value of $(\sigma_c - \sigma_r)$ from equation (*i*), into above equation (*vi*), we get

$$\left[\frac{rd\sigma_r}{dr} + \rho r^2 \times \omega^2\right][1+\mu] + r\left(\frac{d\sigma_c}{dr} - \mu\frac{d\sigma_r}{dr}\right) = 0$$

or
$$\left[r\frac{d\sigma_r}{dr} + \rho r^2 \times \omega^2\right](1+\mu) + r\frac{d\sigma_c}{dr} - \mu \times r \times \frac{d\sigma_r}{dr} = 0$$

or $$\left(r\frac{d\sigma_r}{dr}\right)(1+\mu)+\rho r^2\,\omega^2\,(1+\mu)+r\,\frac{d\sigma_c}{dr}-\mu\times r\times\frac{d\sigma_r}{dr}=0$$

or $$r\,\frac{d\sigma_r}{dr}+\mu\times r\,\frac{d\sigma_r}{dr}+\rho r^2\times\omega^2\,(1+\mu)+r\,\frac{d\sigma_c}{dr}-\mu\times r\,\frac{d\sigma_r}{dr}=0$$

or $$r\,\frac{d\sigma_r}{dr}+\rho r^2\times\omega^2\,(1+\mu)+r\,\frac{d\sigma_c}{dr}=0$$

Dividing by 'r' to both sides, we get

$$\frac{d\sigma_r}{dr}+\rho\times r\times\omega^2\,(1+\mu)+\frac{d\sigma_c}{dr}=0$$

or $$\frac{d\sigma_r}{dr}+\frac{d\sigma_c}{dr}=-\rho\times r\times\omega^2\,(1+\mu)$$

Integrating, we get

$$\int d\sigma_r+\int d\sigma_c=\int -\rho r\omega^2\,(1+\mu)\,dr$$

$$\sigma_r+\sigma_c=-\rho\times\frac{r^2}{2}\times\omega^2\times(1+\mu)+C_1 \qquad \text{...}(vii)$$

where C_1 is a constant of integration and it is independent of r.

Subtracting equation (i) from equation (vii), we get

$$2\sigma_r=-\frac{\rho\times r^2\times\omega^2\times(1+\mu)}{2}+C_1-r\,\frac{d\sigma_r}{dr}-\rho r^2\times\omega^2$$

$$=-\frac{\rho\times r^2\times\omega^2}{2}\,[(1+\mu)+2]+C_1-r\,\frac{d\sigma_r}{dr}$$

or $$2\sigma_r+r\,\frac{d\sigma_r}{dr}=-\frac{\rho\times\omega^2\times r^2}{2}\,(3+\mu)+C_1$$

Multiplying by r to both sides, we get

$$2\times r\times\sigma_r+r^2\,\frac{d\sigma_r}{dr}=-\frac{\rho\times\omega^2\times r^3}{2}\,(3+\mu)+C_1\times r$$

or $$\frac{d}{dr}\,(r^2\times\sigma_r)=-\frac{\rho\times\omega^2\times r^3}{2}\,(3+\mu)+C_1\times r$$

Integrating both sides, we get

$$r^2\times\sigma_r=-\frac{\rho\times\omega^2}{2}\times\frac{r^4}{4}\times(3+\mu)+\frac{C_1\times r^2}{2}+C_2$$

(where C_2 is another constant of integration and is also independent of r)

or $$\sigma_r=-\frac{\rho\times\omega^2\times r^2}{8}\,(3+\mu)+\frac{C_1}{2}+\frac{C_2}{r^2}$$

$$=\frac{C_1}{2}+\frac{C_2}{r^2}-\frac{\rho\times\omega^2\times r^2}{8}\,(3+\mu) \qquad \text{...(9.5)}$$

Equation (9.5) gives the expression for radial stress in terms of constants, C_1 and C_2. Now let us find the value of σ_c. To get the value of σ_c, substitute equation (9.5) into equation (vii).

$$\therefore \quad \frac{C_1}{2}+\frac{C_2}{r^2}-\frac{\rho\times\omega^2\times r^2}{8}\,(3+\mu)+\sigma_c=-\frac{\rho\times r^2\times\omega^2}{2}\,(1+\mu)+C_1$$

or $$\sigma_c = -\frac{C_1}{2} - \frac{C_2}{r^2} + \frac{\rho \times \omega^2 \times r^2}{8}(3 + \mu) - \frac{\rho \times r^2 \times \omega^2}{2}(1 + \mu) + C_1$$

$$= \frac{C_1}{2} - \frac{C_2}{r^2} + \frac{\rho \times \omega^2 \times r^2}{8}(3 + \mu - 4 - 4\mu)$$

$$= \frac{C_1}{2} - \frac{C_2}{r^2} + \frac{\rho \times \omega^2 \times r^2}{8}(-1 - 3\mu)$$

$$= \frac{C_1}{2} - \frac{C_2}{r^2} - \frac{\rho \times \omega^2 \times r^2}{8}(1 + 3\mu) \qquad ...(9.6)$$

The constants C_1 and C_2 are obtained from boundary conditions. In equations (9.5) and (9.6), L.H.S. is in N/m^2, hence every term on R.H.S. should be in N/m^2. Hence C_1 and C_2 will be in N/m^2.

1st Case : Solid Disc. From equations (9.5) and (9.6), it is clear that the stresses set up in a rotating disc will become infinite at the centre of the disc where $r = 0$ $\left(\because \frac{C_2}{r^2} = \frac{C_2}{0} = \infty\right)$. But the stresses at the centre, cannot be infinite. Hence C_2 should be zero. Substituting $C_2 = 0$ in these equations, we get

$$\sigma_r = \frac{C_1}{2} - \frac{\rho \times \omega^2 \times r^2}{8}(3 + \mu) \qquad ...(A)$$

and $$\sigma_c = \frac{C_1}{2} - \frac{\rho \times \omega^2 \times r^2}{8}(1 + 3\mu) \qquad ...(B)$$

The value of C_1 will be obtained from boundary condition.

At the outer radius, $\sigma_r = 0$. This means at $r = r_2$, $\sigma_r = 0$

Substituting $\sigma_r = 0$ at $r = r_2$ in equation (A), we get

$$0 = \frac{C_1}{2} - \frac{\rho \times \omega^2 \times r_2^2}{8}(3 + \mu)$$

or $$\frac{C_1}{2} = \frac{\rho \times \omega^2 \times r_2^2}{8}(3 + \mu)$$

or $$C_1 = \frac{\rho \times \omega^2 \times r_2^2}{4}(3 + \mu)$$

Substituting the value of C_1 in equations (A) and (B), we get

$$\sigma_r = \frac{\rho \times \omega^2 \times r_2^2}{4 \times 2}(3 + \mu) - \frac{\rho \times \omega^2 \times r^2}{8}(3 + \mu)$$

$$= \frac{\rho \times \omega^2}{8}(3 + \mu)\,[r_2^2 - r^2] \qquad ...(9.7)$$

and $$\sigma_c = \frac{\rho \times \omega^2 \times r_2^2}{4 \times 2}(3 + \mu) - \frac{\rho \times \omega^2 \times r^2}{8}(1 + 3\mu)$$

$$= \frac{\rho \times \omega^2}{8}\,[(3 + \mu)r_2^2 - (1 + 3\mu)r^2] \qquad ...(9.8)$$

At the centre, $r = 0$ [Substituting $r = 0$ in equations (9.7) and (9.8)]

$$\therefore \quad \sigma_r = \frac{\rho \times \omega^2}{8}(3 + \mu)r_2^{\,2}$$

and
$$\sigma_c = \frac{\rho \times \omega^2}{8}(3 + \mu)r_2^{\,2}$$

Hence maximum stresses are at the centre and they are equal.

$$\therefore \quad (\sigma_r)_{max} = (\sigma_c)_{max} = \frac{\rho \times \omega^2}{8}(3 + \mu)\, r_2^{\,2} \qquad ...(9.9)$$

At the outer radius, radial stress (σ_r) is zero, but circumferential stress is not zero. By substituting $r = r_2$ in equation (9.8), we get circumferential stress at outer radius. This is equal to

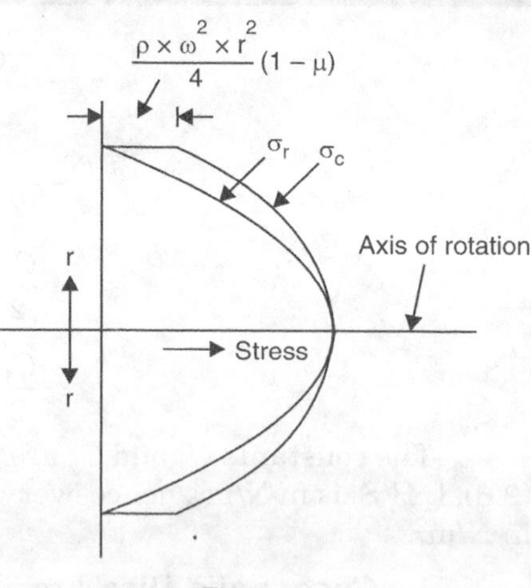

Fig. 9.8

$$\sigma_c \text{ at outer radius} = \frac{\rho \times \omega^2}{8}[(3 + \mu)r_2^{\,2} - (1 + 3\mu)r_2^{\,2}]$$

$$= \frac{\rho \times \omega^2}{8} \times r_2^{\,2}\,[3 + \mu - 1 - 3\mu]$$

$$= \frac{\rho \times \omega^2 \times r_2^2}{8}(2 - 2\mu) = \frac{\rho \times \omega^2 \times r_2^2}{4}(1 - \mu) \qquad ...(9.10)$$

The variation of circumferential and radial stresses with radius are shown in Fig. 9.8.

2nd Case : Hollow Disc. The equations (9.5) and (9.6) gives the radial and circumferential stresses in terms of constants as

$$\sigma_r = \frac{C_1}{2} + \frac{C_2}{r^2} - \frac{\rho \times \omega^2 \times r^2}{8}(3 + \mu)$$

and
$$\sigma_c = \frac{C_1}{2} - \frac{C_2}{r^2} - \frac{\rho \times \omega^2 \times r^2}{8}(1 + 3\mu)$$

To find C_1 and C_2, the boundary conditions are used. If r_1 and r_2 are the inner and outer radii, then radial stress is zero at these radii. This means $\sigma_r = 0$ at $r = r_1$ and at $r = r_2$.

Substituting these values, we get two equations as

$$0 = \frac{C_1}{2} + \frac{C_2}{r_1^2} - \frac{\rho \times \omega^2 \times r_1^{\,2}}{8}(3 + \mu) \qquad [\text{Here } \sigma_r = 0 \text{ at } r = r_1] \quad ...(i)$$

$$0 = \frac{C_1}{2} + \frac{C_2}{r_2^2} - \frac{\rho \times \omega^2 \times r_2^2}{8}(3 + \mu) \qquad [\text{Here } \sigma_r = 0 \text{ at } r = r_2] \quad ...(ii)$$

Subtracting equation (*ii*) from equation (*i*), we get

$$0 = \frac{C_2}{r_1^2} - \frac{C_2}{r_2^2} - \frac{\rho \times \omega^2 \times r_1^2}{8}(3 + \mu) + \frac{\rho \times \omega^2 \times r_2^2}{8}(3 + \mu)$$

$$= C_2\left[\frac{1}{r_1^2} - \frac{1}{r_2^2}\right] - \frac{\rho \times \omega^2}{8}(3 + \mu)\,[r_1^{\,2} - r_2^{\,2}]$$

or $$C_2\left[\frac{r_2^{\,2}-r_1^{\,2}}{r_1^{\,2}r_2^{\,2}}\right]=\frac{\rho\times\omega^2}{8}(3+\mu)\,[r_1^{\,2}-r_2^{\,2}]$$

$$=-\frac{\rho\times\omega^2}{8}(3+\mu)(r_2^{\,2}-r_1^{\,2})$$

or $$\frac{C_2}{r_1^2r_2^2}=-\frac{\rho\times\omega^2}{8}(3+\mu)\qquad\text{[Cancelling }(r_2^{\,2}-r_1^{\,2})\text{ to both sides]}$$

$$\therefore\qquad C_2=-\frac{\rho\times\omega^2}{8}(3+\mu)\,r_1^{\,2}r_2^{\,2}\qquad\ldots(iii)$$

Substituting the value of C_2 in equation (i), we get

$$0=\frac{C_1}{2}+\left[-\frac{\rho\times\omega^2}{8}(3+\mu)\,r_1^2\,r_2^2\right]\times\frac{1}{r_1^{\,2}}-\frac{\rho\times\omega^2\times r_1^2}{8}\times(3+\mu)$$

$$=\frac{C_1}{2}-\frac{\rho\times\omega^2\times r_2^2}{8}(3+\mu)-\frac{\rho\times\omega^2\times r_1^2}{8}(3+\mu)$$

or $$-\frac{C_1}{2}=-\frac{\rho\times\omega^2}{8}(3+\mu)\,[r_2^{\,2}+r_1^{\,2}]$$

or $$C_1=\frac{\rho\times\omega^2}{4}\times(3+\mu)\,(r_2^{\,2}+r_1^{\,2})\qquad\ldots(iv)$$

Substituting the values of C_1 and C_2 is equations (9.5) and (9.6), we get

$$\sigma_r=\frac{\rho\times\omega^2}{4\times2}(3+\mu)\,(r_2^{\,2}+r_1^{\,2})+\left[-\frac{\rho\times\omega^2}{8}\,\frac{(3+\mu)r_1^2r_2^2}{r^2}\right]-\frac{\rho\times\omega^2\times r^2}{8}(3+\mu)$$

$$=\frac{\rho\times\omega^2}{8}\times(3+\mu)\left[(r_2^2+r_1^2)-\frac{r_1^2r_2^2}{r^2}-r^2\right]\qquad\ldots(9.11)$$

and $$\sigma_c=\frac{\rho\times\omega^2}{4\times2}(3+\mu)\times(r_2^{\,2}+r_1^{\,2})-\left[-\frac{\rho\times\omega^2}{8}\,\frac{(3+\mu)r_1^2r_2^2}{r^2}\right]-\frac{\rho\times\omega^2\times r^2}{8}(1+3\mu)$$

$$=\frac{\rho\times\omega^2}{8}(3+\mu)\left[(r_2^2+r_1^2)+\frac{r_1^2r_2^2}{r^2}-\frac{r^2(1+3\mu)}{(3+\mu)}\right]\qquad\ldots(9.12)$$

Equations (9.11) and (9.12) give the expression for radial stress and circumferential stress at any radius 'r' in a disc with a central hole.

Maximum value of σ_c. From equation (9.12), it is clear that circumferential stress goes on decreasing with the increase of radius. Hence σ_c is maximum when radius is minimum *i.e.*, when $r = r_1$, σ_c is maximum. Substituting $r = r_1$ in equation (9.12), we get maximum value of σ_c.

$$\therefore\qquad(\sigma_c)_{max}=\frac{\rho\times\omega^2}{8}(3+\mu)\left[(r_2^2+r_1^2)+\frac{r_1^2r_2^2}{r_1^2}-\frac{r_1^2(1+3\mu)}{(3+\mu)}\right]$$

$$=\frac{\rho\times\omega^2}{8}(3+\mu)\left[r_2^2+r_1^2+r_2^2-\frac{r_1^2(1+3\mu)}{(3+\mu)}\right]$$

$$= \frac{\rho \times \omega^2}{8}(3+\mu)\left[2r_2^2 + \frac{r_1^2(3+\mu) - r_1^2(1+3\mu)}{(3+\mu)}\right]$$

$$= \frac{\rho \times \omega^2}{8}(3+\mu)\left[2r_2^2 + \frac{r_1^2}{(3+\mu)}(3+\mu-1-3\mu)\right]$$

$$= \frac{\rho \times \omega^2}{8}(3+\mu)\left[2r_2^2 + \frac{r_1^2}{(3+\mu)}(2-2\mu)\right]$$

$$= \frac{\rho \times \omega^2}{8}(3+\mu)\left[2r_2^2 + 2r_1^2\frac{(1-\mu)}{(3+\mu)}\right]$$

$$= \frac{\rho \times \omega^2}{8}(3+\mu) \times 2\left[r_2^2 + \frac{r_1^2(1-\mu)}{3+\mu}\right]$$

$$= \frac{\rho \times \omega^2}{4}(3+\mu)\left[r_2^2 + \frac{(1-\mu)}{3+\mu}r_1^2\right] \qquad ...(9.13)$$

Equation (9.13) gives the maximum value of σ_c.

If $r_1 \to r_2 = r$, then

$$(\sigma_c)_{max} = \frac{\rho \times \omega^2}{4}(3+\mu)\left[r^2 + \frac{(1-\mu)}{3+\mu}r^2\right]$$

$$= \frac{\rho \times \omega^2}{4}(3+\mu)\left[\frac{r^2(3+\mu)+(1-\mu)r^2}{(3+\mu)}\right]$$

$$= \frac{\rho \times \omega^2}{4} \times r^2(3+\mu+1-\mu) = \rho \times \omega^2 \times r^2$$

which is same as of a thin rotating cylinder.

Maximum value of σ_r. The value of σ_r will be maximum, if $\frac{d\sigma_r}{dr} = 0$. Hence differentiating equation (9.11) w.r.t. r and patting $\frac{d\sigma_r}{dr} = 0$, we get

$$\frac{d}{dr}\left[\frac{\rho\omega^2}{8}(3+\mu)\left(r_2^2 + r_1^2 - \frac{r_1^2 r_2^2}{r^2} - r^2\right)\right] = 0$$

or $$\frac{\rho \times \omega^2}{8}(3+\mu)\left(0 + 0 - \frac{r_1^2\, r_2^2(-2)}{r^3} - 2r\right) = 0$$

or $$\frac{2r_1^2 r_2^2}{r^3} - 2r = 0$$

or $$r_1^2\, r_2^2 - r^4 = 0 \quad \text{or} \quad r^4 = r_1^2\, r_2^2$$

or $$r^2 = r_1 r_2 \quad \text{or} \quad r = \sqrt{r_1 \times r_2} \qquad ...(9.14)$$

Equation (9.14) gives the condition for σ_r to be maximum.

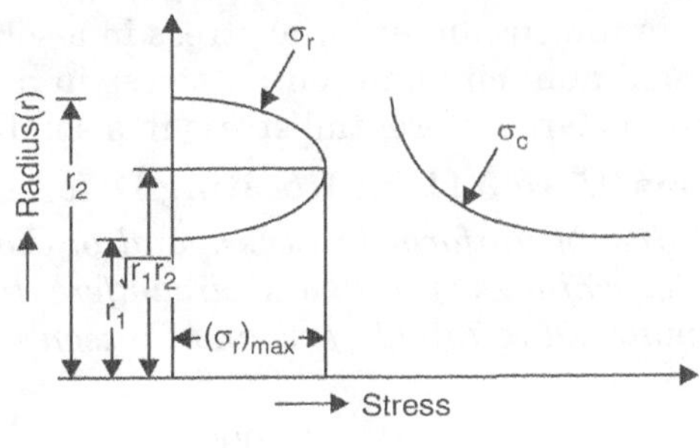

Fig. 9.9

Substituting $r = \sqrt{r_1 \times r_2}$ in equation (9.11), we get maximum value of σ_r.

$$\therefore \quad (\sigma_r)_{max} = \frac{\rho \times \omega^2}{8} (3 + \mu) \left[r_2^2 + r_1^2 - \frac{r_1^2 r_2^2}{r_1 r_2} - r_1 r_2 \right]$$

$$= \frac{\rho \times \omega^2}{8} (3 + \mu) [r_2^2 + r_1^2 - 2r_1 r_2]$$

$$= \frac{\rho \times \omega^2}{8} (3 + \mu) (r_2 - r_1)^2 \quad ...(9.15)$$

Value of σ_c at the outer surface of a disc with central hole

Substituting $r = r_2$ in equation (9.12), we get the value of σ_c at outer radius.

$$\therefore \quad \sigma_c = \frac{\rho \times \omega^2}{8} (3 + \mu) \left[r_2^2 + r_1^2 + \frac{r_1^2 r_2^2}{r_2^2} - r_2^2 \frac{(1 + 3\mu)}{3 + \mu} \right]$$

$$= \frac{\rho \times \omega^2}{8} (3 + \mu) \left[r_2^2 + 2r_1^2 - r_2^2 \frac{(1 + 3\mu)}{3 + \mu} \right]$$

$$= \frac{\rho \times \omega^2}{8} (3 + \mu) \left[2r_1^2 + r_2^2 \frac{(3 + \mu - 1 - 3\mu)}{3 + \mu} \right]$$

$$= \frac{\rho \times \omega^2}{8} (3 + \mu) \left[2r_1^2 + \frac{r_2^2}{(3 + \mu)} (2 - 2\mu) \right]$$

$$= \frac{\rho \times \omega^2}{4} (3 + \mu) \left[r_1^2 + \left(\frac{1 - \mu}{3 + \mu} \right) r_2^2 \right] \quad ...(9.16)$$

3rd Case : Disc With a Very small r_1. For a disc with a pin hole at the centre, $r_1 \to 0$. Hence substituting $r_1 = 0$ in the equations (9.13) and (9.15), we get the expression for σ_r and σ_c for a disc with pin hole at the centre.

$$\therefore \quad (\sigma_r)_{max} = \frac{\rho \times \omega^2}{8} (3 + \mu) r_2^2 = \frac{\rho \times \omega^2 \times r_2^2}{8} (3 + \mu) \quad ...(9.17)$$

and

$$(\sigma_c)_{max} = \frac{\rho \times \omega^2}{4} (3 + \mu) r_2^2 = \frac{\rho \times \omega^2 \times r_2^2}{4} (3 + \mu) \quad ...(9.18)$$

The maximum circumferential stress in a solid disc is given by equation (9.9), as

$$(\sigma_c)_{max} \text{ for solid disc} = \frac{\rho \times \omega^2}{8} (3 + \mu) r_2^2.$$

If we compare the maximum circumferential stress in a solid disc and in a disc with a pin hole at the centre, the maximum circumferential stress in a disc with a pin hole at the centre is two times the maximum circumferential stress in a solid disc.

9.7.1. Solved Problems 9.7 to 9.11 for Rotating Disc

Problem 9.7. *A steel disc of uniform thickness and of diameter 900 mm is rotating about its axis at 3000 r.p.m. Determine the radial and circumferential stresses at the centre and outer radius. The density of material is 7800 kg/m² and Poisson's ratio = 0.3.*

Sol. Given :

Diameter = 900 mm $\therefore$ Radius of disc, $r = \dfrac{900}{2} = 450$ mm = 0.45 m

Speed, $N = 3000$ r.p.m.

$\therefore$ Angular speed, $\omega = \dfrac{2\pi N}{60} = \dfrac{2\pi \times 3000}{60} = 100\,\pi$ rad/s

Density, $\rho = 7800$ kg/m³

Poisson's ratio, $\mu = 0.3$

The radial and circumferential stresses are given by equations (9.5) and (9.6) as

$$\sigma_r = \frac{C_1}{2} + \frac{C_2}{r^2} - \frac{\rho \times \omega^2 \times r^2}{8}(3+\mu) \quad ...(i)$$

and

$$\sigma_c = \frac{C_1}{2} - \frac{C_2}{r^2} - \frac{\rho \times \omega^2 \times r^2}{8}(1+3\mu) \quad ...(ii)$$

As the stresses cannot have infinite value at $r = 0$, hence C_2 should be zero. Hence equations (*i*) and (*ii*) becomes as

$$\sigma_r = \frac{C_1}{2} - \frac{\rho \times \omega^2 \times r^2}{8}(3+\mu) \quad ...(iii)$$

and

$$\sigma_c = \frac{C_1}{2} - \frac{\rho \times \omega^2 \times r^2}{8}(1+3\mu) \quad ...(iv)$$

Also at the outer radius, $\sigma_r = 0$. This means at $r = 0.45$ m,

$\sigma_r = 0$. Substituting these values in equation (*iii*), we get

$$0 = \frac{C_1}{2} - \frac{\rho \times \omega^2 \times (0.45)^2}{8}(3+\mu)$$

or

$$\frac{C_1}{2} = \frac{\rho \times \omega^2 \times (0.45)^2}{8}(3+\mu) \quad \text{or} \quad C_1 = \frac{\rho \times \omega^2 \times (0.45)^2}{4}(3+\mu)$$

$$= \frac{7800 \times (100\,\pi)^2 \times (0.45)^2 (3 + 0.3)}{4} = 128.6 \times 10^6 \text{ N/m}^2$$

Substituting the value of C_1 in equations (*iii*) and (*iv*), we get

$$\sigma_r = \frac{128.6 \times 10^6}{2} - \frac{\rho \times \omega^2 \times r^2}{8}(3+\mu)$$

and

$$\sigma_c = \frac{128.6 \times 10^6}{2} - \frac{\rho \times \omega^2 \times r^2}{8}(1+3\mu)$$

At $r = 0$,

$$\sigma_r = \sigma_c = \frac{128.6 \times 10^6}{2} = 64.3 \times 10^6 \text{ N/m}^2 = \mathbf{64.3\ MN/m^2}. \quad \textbf{Ans.}$$

At $r = 0.45$, $\sigma_r = 0$

$$\sigma_c = \frac{128.6 \times 10^6}{2} - \frac{7800 \times (100\,\pi)^2 \times (0.45)^2}{8}(1 + 3 \times 0.3)$$

$$= 64.3 \times 10^6 - 37.024 \times 10^6 \text{ N/m}^2$$

$$= 27.276 \times 10^6 \text{ N/m}^2 = \mathbf{27.276 \text{ MN/m}^2}. \quad \textbf{Ans.}$$

Alternate Method

(i) Stresses at the centre. The radial stress and circumferential stress at the centre of a solid disc are maximum and are equal. They are given by equation (9.9) as

$$(\sigma_r)_{max} = (\sigma_c)_{max} = \frac{\rho \times \omega^2}{8}(3 + \mu)\, r_2^2$$

where r_2 = Outer radius = 0.45 m

$$\therefore \quad (\sigma_r)_{max} = (\sigma_c)_{max} = \frac{7800 \times (100\,\pi)^2}{8}(3 + 0.3) \times (0.45)^2$$

$$= 64.3 \times 10^6 \text{ N/m}^2 = \mathbf{64.3 \text{ MN/m}^2}. \quad \textbf{Ans.}$$

(ii) Stresses at the outer radius

The radial stress at the outer surface is zero.

$\therefore \quad \sigma_r = 0$

The circumferential stress at outer radius is given by equation (9.10) as

$$\sigma_c = \frac{\rho \times \omega^2 \times r_2^2}{4}(1 - \mu) = \frac{7800 \times (100\,\pi)^2 \times 0.45^2}{4}(1 - 0.3)$$

$$= 27.28 \times 10^6 \text{ N/m}^2 = \mathbf{27.28 \text{ MN/m}^2}. \quad \textbf{Ans.}$$

At any section, there are radial and circumferential stresses which are pendicular to each other and tensile in nature. There is no shear stress. Hence these stresses are also known as **principal stresses.** Hence the above stresses are also principal stresses.

Problem 9.8. *If for the problem 9.7, the disc is having a central hole of 150 mm diameter, then determine :*

(i) circumferential stress at inner radius and outer radius,

(ii) radius at which radial stress is maximum, and

(iii) maximum radial stress.

Sol. Given :

From Problem 9.7,

Outer radius, $r_2 = 0.45$ m,

Angular speed, $\omega = (100\,\pi)$ rad/s

$\rho = 7800$ kg/m^3, $\mu = 0.3$

Inner dia = 150 mm, $\therefore$ Inner radius, $r_1 = \frac{150}{2} = 75$ mm $= 0.075$ mm

(i) Circumferential stress (σ_c) at inner and outer radii

The circumferential stress in a disc with central hole is maximum at the inner radius. This stress is given by equation (9.13) as

$$(\sigma_c)_{max} = \frac{\rho \times \omega^2}{4}(3 + \mu)\left[r_2^2 + \left(\frac{1 - \mu}{3 + \mu}\right) r_1^2\right]$$

$$= \frac{7800 \times (100\,\pi)^2}{4}(3 + 0.3)\left[0.45^2 + \left(\frac{(1 - 0.3)}{3 + 0.3}\right)(0.075)^2\right]$$

$$= 635.109 \times 10^6\ [0.2025 + 1.193 \times 10^{-3}]$$

$$= 129.367 \times 10^6\ \text{N/m}^2 = \mathbf{129.367\ MN/m^2}. \quad \textbf{Ans.}$$

The circumferential stress at the outer radius is given by equation (9.16) as

$$\sigma_c = \frac{\rho \times \omega^2}{4}(3+\mu)\left[r_1^2 + \left(\frac{1-\mu}{3+\mu}\right)r_2^2\right]$$

$$= \frac{7800 \times (100\,\pi)^2 \times (3+0.3)}{4}\left[0.075^2 + \left(\frac{1-0.3}{3+0.3}\right) \times 0.45^2\right]$$

$$= 635.109 \times 10^6\ [5.625 \times 10^{-3} + 0.04295]$$

$$= 30.85 \times 10^6\ \text{N/m}^2 = \mathbf{30.85\ MN/m^2}. \quad \textbf{Ans.}$$

(ii) Radius at which radial stress is maximum

The radius at which radial stress is maximum is given by equation (9.14) as

$$r = \sqrt{r_1 \times r_2} = \sqrt{0.075 \times 0.45} = \mathbf{0.1837\ m}. \quad \textbf{Ans.}$$

(iii) Maximum radial stress

The maximum radial stress is given by equation (9.15) as

$$(\sigma_r)_{max} = \frac{\rho \times \omega^2}{8}(3+\mu)(r_2 - r_1)^2$$

$$= \frac{7800 \times (100\,\pi)^2}{8}(3+0.3)(0.45-0.075)^2$$

$$= 317.5545 \times 10^6\ (0.375)^2$$

$$= 44.656 \times 10^6\ \text{N/m}^2 = \mathbf{44.656\ MN/m^2}. \quad \textbf{Ans.}$$

Problem 9.9. *For the data given in Problem 9.7, plot the variations of circumferential and radial stresses along the radius of the solid disc. Take radius along y-axis and stresses along x-axis. Also plot these variations taking 'r' along x-axis and stresses along y-axis.*

Sol. Given :

Data from Problem 9.7

$$N = 3000 \text{ r.p.m.}, \ \omega = (100\,\pi) \text{ rad/s}$$

$$\rho = 7800 \text{ kg/m}^3, \ \mu = 0.3, \text{ outer radius, } r_2 = 0.45 \text{ m}$$

(i) Circumferential stress (σ_c)

The circumferential stress for a solid disc at any radius 'r' is given by equation (9.8) as

$$\sigma_c = \frac{\rho \times \omega^2}{8}[(3+\mu)r_2^2 - (1+3\mu)r^2]$$

$$= \frac{7800 \times (100\,\pi)^2}{8}[(3+0.3) \times 0.45^2 - (1 + 3 \times 0.3)\,r^2]$$

$$= 96.228 \times 10^6\ [0.66825 - 1.9\,r^2]$$

By substituting different values of r in the above equation, different values of σ_c are obtained as given below :

At $r = 0$, $\quad \sigma_c = 96.228 \times 10 \times 0.66825 = 64.3 \times 10^6\ \text{N/m}^2 = 64.3\ \text{MN/m}^2$

$= 0.05$ m, $\sigma_c = 96.228 \times 10^6\ [0.66825 - 1.9 \times 0.05^2] = 63.84 \times 10^6\ \text{N/m}^2 = 63.84\ \text{MN/m}^2$

$= 0.1$ m, $\sigma_c = 96.228 \times 10^6\ [0.66825 - 1.9 \times 0.1^2] = 62.47 \times 10^6\ \text{N/m}^2 = 62.47\ \text{MN/m}^2$

$= 0.2$ m, $\sigma_c = 96.228 \times 10^6\ [0.66825 - 1.9 \times 0.2^2] = 56.99 \times 10^6\ \text{N/m}^2 = 56.99\ \text{MN/m}^2$

$= 0.3$ m, $\sigma_c = 96.228 \times 10^6\ [0.66825 - 1.9 \times 0.3^2] = 47.85 \times 10^6\ \text{N/m}^2 = 47.85\ \text{MN/m}^2$

$= 0.4$ m, $\sigma_c = 96.228 \times 10^6\ [0.66825 - 1.9 \times 0.4^2] = 35.05 \times 10^6\ \text{N/m}^2 = 35.05\ \text{MN/m}^2$

$= 0.45$ m, $\sigma_c = 96.228 \times 10^6\ [0.66825 - 1.9 \times 0.45^2] = 27.28 \times 10^6\ \text{N/m}^2 = 27.28\ \text{MN/m}^2$

(ii) Radial stress (σ_r)

The radial stress for a solid disc at any radius 'r' is given by equation (9.7) as

$$\sigma_r = \frac{\rho \times \omega^2}{8}(3 + \mu)(r_2^2 - r^2)$$

$$= \frac{7800 \times (100\,\pi)^2}{8}(3 + 0.3)(0.45^2 - r^2)\ \text{N/m}^2$$

$$= 317.5545 \times 10^6\ (0.2025 - r^2)\ \text{N/m}^2$$

$$= 317.5545\ (0.2025 - r^2)\ \text{MN/m}^2$$

By substituting different values of 'r' in the above equation, different values of σ_r are obtained as given below :

At $r = 0$, $\sigma_r = 317.5545\ (0.2025 - 0) = 64.3\ \text{MN/m}^2$

$r = 0.05$ m, $\sigma_r = 317.5545\ (0.2025 - 0.05^2) = 63.5\ \text{MN/m}^2$

$r = 0.1$ m, $\sigma_r = 317.5545\ (0.2025 - 0.1^2) = 61.13\ \text{MN/m}^2$

$r = 0.2$ m, $\sigma_r = 317.5545\ (0.2025 - 0.2^2) = 51.6\ \text{MN/m}^2$

$r = 0.3$ m, $\sigma_r = 317.5545\ (0.2025 - 0.3^2) = 35.7\ \text{MN/m}^2$

$r = 0.4$ m, $\sigma_r = 317.5545\ (0.2025 - 0.4^2) = 13.5\ \text{MN/m}^2$

$r = 0.45$ m, $\sigma_r = 317.5545\ (0.2025 - 0.45^2) = 0$

The variation of σ_c and σ_r along the radius (taking radius along y-axis and stress along x-axis) are plotted as shown in Fig. 9.10.

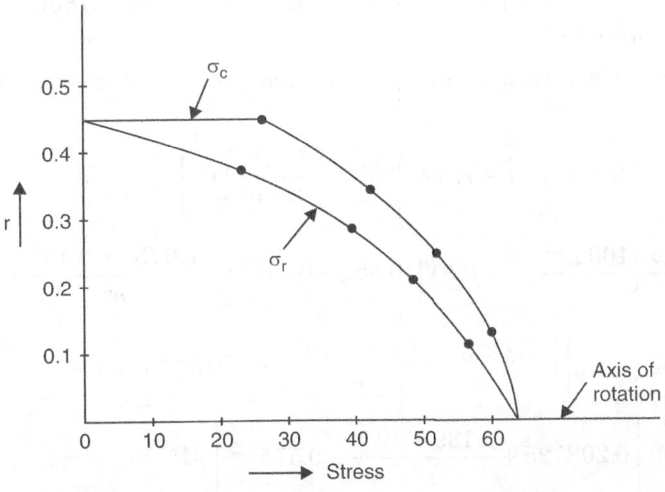

Fig. 9.10

If radius is taken along x-axis and stresses along y-axis, then variation of σ_c and σ_r will be as shown in Fig. 9.11.

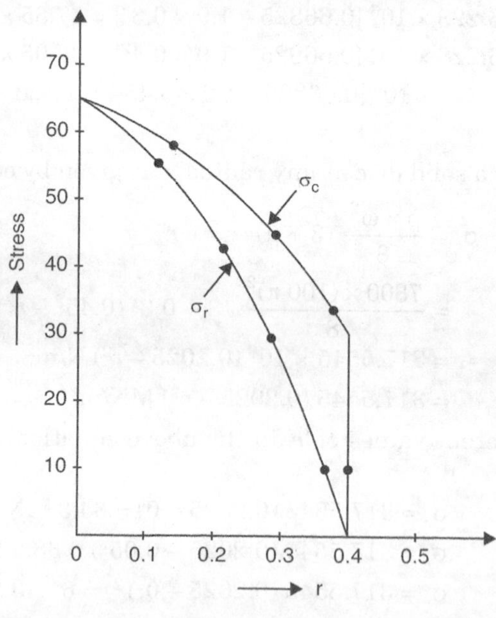

Fig. 9.11

Problem 9.10. *For the problem 9.8, plot the variation of circumferential and radial stresses along the radius of disc with a central hole.*

Sol. Given :

Data from Problem 9.8,

Outer radius, $r_2 = 0.45$ m ; inner radius, $r_1 = 0.075$ m

$N = 3000$ r.p.m. ; $\omega = 100\ \pi$ rad/s ; $\mu = 0.3$, $\rho = 7800$ kg/m^3

(i) Circumferential stress (σ_c)

The circumferential stress for a disc with a hole at the centre is given by equation (9.12), as

$$\sigma_c = \frac{\rho \times \omega^2}{8}(3+\mu)\left[(r_2^2 + r_1^2) + \frac{r_1^2 r_2^2}{r^2} - \left(\frac{1+3\mu}{3+\mu}\right)r^2\right]$$

$$= \frac{7800 \times (100\ \pi)^2}{8}(3+0.3)\left[(0.45^2 + 0.075^2) + \frac{0.075^2 \times 0.45^2}{r^2} - \left(\frac{1+3\times 0.3}{3+0.3}\right)r^2\right]$$

$$= 317.55 \times 10^6\left[0.208125 + \frac{1.139 \times 10^{-3}}{r^2} - 0.575\, r^2\right] \text{N/m}^2$$

$$= 317.55\left[0.208125 + \frac{1.139 \times 10^{-3}}{r^2} - 0.575\, r^2\right] \text{MN/m}^2$$

By substituting different values of r in the above equation, different values of σ_c are obtained. Let us start from inner radius *i.e.,* from $r = 0.075$ m

At $r = 0.075$ m, $\sigma_c = 317.55\left(0.208125 + \frac{1.139 \times 10^{-3}}{0.075^2} - 0.575 \times 0.075^2\right)$

$$= 317.55\ [0.208125 + 0.2025 - 0.00323] = 129.368\ \text{MN/m}^2$$

At $r = 0.1$ m, $\sigma_c = 317.55\left(0.208125 + \frac{1.139 \times 10^{-3}}{0.1^2} - 0.575 \times 0.1^2\right)$

$$= 317.55\ (0.208125 + 0.1139 - 0.00575) = 100.43\ \text{MN/m}^2$$

At $r = 0.2$ m, $\sigma_c = 317.55\left(0.208125 + \frac{1.139 \times 10^{-3}}{0.2^2} - 0.575 \times 0.2^2\right)$

$$= 317.55\ (0.208125 + 0.0285 - 0.023) = 67.83\ \text{MN/m}^2$$

At $r = 0.3$ m, $\sigma_c = 317.55\left(0.208125 + \frac{1.139 \times 10^{-3}}{0.3^2} - 0.575 \times 0.3^2\right)$

$$= 317.55\ (0.208125 + 0.01265 - 0.05175) = 50.05\ \text{MN/m}^2$$

At $r = 0.4$ m, $\sigma_c = 317.55\left(0.208125 + \frac{1.139 \times 10^{-3}}{0.4^2} - 0.575 \times 0.4^2\right)$

$$= 317.55\ (0.208125 + 0.007118 - 0.092) = 39.13\ \text{MN/m}^2$$

At $r = 0.45$ m, $\sigma_c = 317.55\left(0.208125 + \frac{1.139 \times 10^{-3}}{0.45^2} - 0.575 \times 0.45^2\right)$

$$= 317.55\ (0.208125 + 0.005625 - 0.1164) = 30.9\ \text{MN/m}^2.$$

(ii) Radial stress (σ_r)

The radial stress for a disc with a hole at the centre is given by equation (9.11) as

$$\sigma_r = \frac{\rho \times \omega^2}{8}(3 + \mu)\left[(r_2^2 + r_1^2) - \frac{r_1^2 r_2^2}{r^2} - r^2\right]$$

$$= \frac{7800 \times (100\,\pi)^2}{8}(3 + 0.3)\left[(0.45^2 + 0.075^2) - \frac{0.075^2 \times 0.45^2}{r^2} - r^2\right]$$

$$= 317.55\left[0.208125 - \frac{1.139 \times 10^{-3}}{r^2} - r^2\right] \times 10^6\ \text{N/m}^2$$

$$= 317.55\left[0.208125 - \frac{1.139 \times 10^{-3}}{r^2} - r^2\right]\ \text{MN/m}^2$$

By substituting the different values of r in the above equation, different values of σ_r are obtained. Also $\sigma_r = 0$ at $r = 0.075$ m and at $r = 0.45$ m. Also we have got in problem 9.9, that σ_r is maximum at $r = \sqrt{r_1 r_2} = \sqrt{0.075 \times 0.45} = 0.1837$ m. Let us start the calculations from inner radius (= 0.075 m).

At $r = 0.075$ m, $\sigma_r = 317.55\left[0.208125 - \frac{1.139 \times 10^{-3}}{0.075^2} - 0.075^2\right] = 0$

At $r = 0.1$ m, $\sigma_r = 317.55\left[0.208125 - \dfrac{1.139 \times 10^{-3}}{0.1^2} - 0.1^2\right]$

$= 317.55\ [0.208125 - 0.1139 - 0.01] = 26.74\ \text{MN/m}^2$

At $r = 0.1837$ m, $\sigma_r = 317.55\left[0.208125 - \dfrac{1.139 \times 10^{-3}}{0.1837^2} - 0.1837^2\right]$

$= 317.55\ [0.208125 - 0.03375 - 0.03375] = 44.65\ \text{MN/m}^2$

$r = 0.2$ m, $\sigma_r = 317.55\left[0.208125 - \dfrac{1.139 \times 10^{-3}}{0.2^2} - 0.2^2\right]$

$= 317.55\ [0.208125 - 0.02847 - 0.04] = 44.34\ \text{MN/m}^2$

$r = 0.3$ m, $\sigma_r = 317.55\left[0.208125 - \dfrac{1.139 \times 10^{-3}}{0.3^2} - 0.3^2\right]$

$= 317.55\ [0.208125 - 0.01265 - 0.09] = 33.49\ \text{MN/m}^2$

$r = 0.4$ m, $\sigma_r = 317.55\left[0.208125 - \dfrac{1.139 \times 10^{-3}}{0.4^2} - 0.4^2\right]$

$= 317.55\ [0.208125 - 0.007118 - 0.16] = 13.02\ \text{MN/m}^2$

$r = 0.45$ m, $\sigma_r = 317.55\left[0.208125 - \dfrac{1.139 \times 10^{-3}}{0.45^2} - 0.45^2\right]$

$= 317.55\ [0.208125 - 0.005625 - 0.2025] = 0\ \text{MN/m}^2$.

The variation of σ_c and σ_r along the radius is shown in Fig. 9.12 taking radius along y-axis and stresses along x-axis.

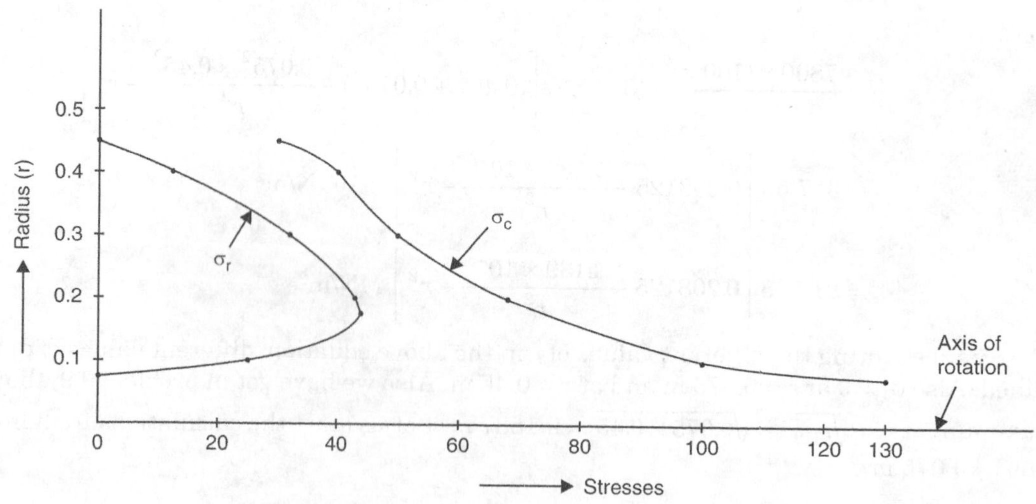

Fig. 9.12

Fig. 9.13 shows the variation of σ_c and σ_r along the radius, when radius is taken along x-axis and stresses along y-axis.

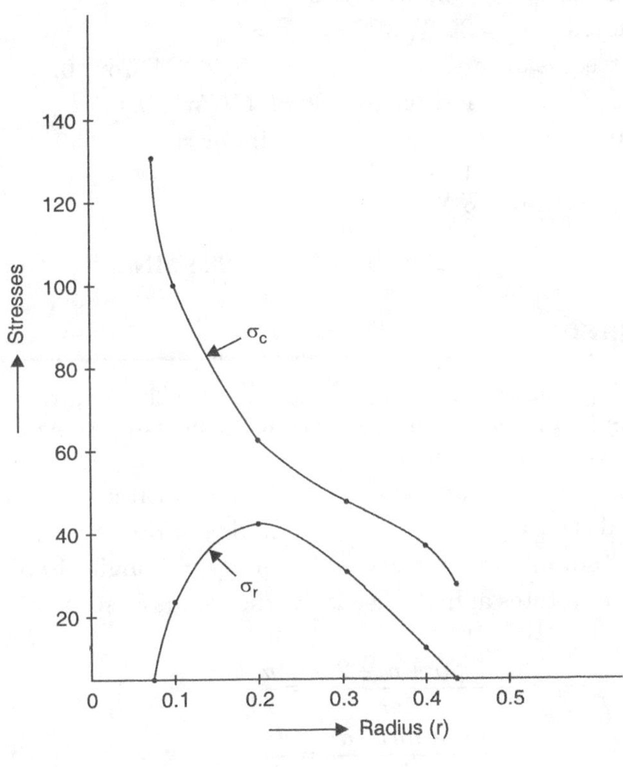

Fig. 9.13

Problem 9.11. *For the problems 9.7 and 9.8, find the maximum shear stress in the solid disc and in the disc with a central hole.*

Sol. The maximum shear stress at any radius is given by, $\tau_{max} = \frac{1}{2}(\sigma_c - \sigma_r)$

(i) Solid disc

In case of solid disc the stresses calculated in problem 9.1 at the centre are $\sigma_r = \sigma_c =$ 64.3 MN/m^2, whereas the stresses at outer radius are $\sigma_r = 0$ and $\sigma_c = 27.276$ MN/m^2.

Hence principal stresses at the centre are : 64.3 MN/m^2, 64.3 MN/m^2, 0

$\therefore$ Shear stress at centre $= \frac{1}{2}(64.3 - 64.3) = 0$

The principal stresses at the outer radius are : 27.276 MN/m^2, 0, 0

$\therefore$ Shear stress at outer radius $= \frac{1}{2}(27.276 - 0) = 13.638$ MN/m^2

$\therefore$ Maximum shear stress is at the outer radius and equal to 13.638 MN/m^2

$\therefore$ τ_{max} = **13.638 MN/m^2. Ans.**

(ii) Disc with a central hole

From Problem 9.8, the calculated values of stresses are :

At the inner radius, $\sigma_c = 129.367$ MN/m^2, $\sigma_r = 0$

At the outer radius, $\sigma_c = 30.85$ MN/m^2, $\sigma_r = 0$

$\therefore$ Principal stresses at inner radius are : 129.367 MN/m^2, 0, 0

Principal stresses at outer radius are : 30.85 MN/m^2, 0, 0.

Hence maximum shear stress will be at the inner radius.

$$\therefore \qquad \sigma_{max} = \frac{1}{2}(\sigma_c - \sigma_r)$$

$$= \frac{1}{2}(129.367 - 0) = \mathbf{64.683\ MN/m^2}. \quad \textbf{Ans.}$$

9.8. LONG CYLINDERS

In case of rotating long cylinders, it is assumed that the longitudinal strain (e_l) even at high speeds of rotation is constant (this means that the cross-sections remain plane, which is true for the sections away from the ends.)

Let σ_c = Circumferential (or hoop stress) and e_c = Circumferential strain

σ_r = Radial stress and e_r = Radial strain,

σ_l = Longitudinal (or axial) stress and e_l = Longitudinal strain.

When the cylinder rotates at high speeds, let the radius r becomes $(r + u)$ and 'dr' becomes $(dr + du)$. Then circumferential strain,

$$e_c = \frac{2\pi(r+u) - 2\pi r}{2\pi r} = \frac{u}{r}$$

Radial strain,
$$e_r = \frac{(dr + du) - dr}{dr} = \frac{du}{dr}$$

The stresses σ_c, σ_r and σ_l are acting on any element of a section of the cylinder. The strains produced by these stresses in circumferential direction, radial direction and longitudinal direction are circumferential strain, radial strain and longitudinal strains respectively. These strains in terms of stresses are given by,

Circumferential strain,
$$e_c = \frac{\sigma_c}{E} - \frac{\mu}{E}(\sigma_r + \sigma_l) = \frac{u}{r} \qquad ...(i)$$

Radial strain,
$$e_r = \frac{\sigma_r}{E} - \frac{\mu}{E}(\sigma_c + \sigma_l) = \frac{du}{dr} \qquad ...(ii)$$

and longitudinal strain,
$$e_l = \frac{\sigma_l}{E} - \frac{\mu}{E}(\sigma_r + \sigma_c) \qquad ...(iii)$$

where μ = Poisson's ratio, and

E = Young's modulus

In equations (*i*) and (*ii*), u is the increase of radius r due to rotation and du is the increase in thickness dr. These two values u and du are unknowns. They can be eliminated from equations (*i*) and (*ii*) as given below :

From equation (*i*), we have

$$u = \frac{r}{E}[\sigma_c - \mu(\sigma_r + \sigma_l)]$$

or
$$Eu = r[\sigma_c - \mu(\sigma_r + \sigma_l)]$$

Differentiating the above equation w.r.t. r, we get

$$E\frac{du}{dr} = [\sigma_c - \mu(\sigma_r + \sigma_l)] + r\left[\frac{d\sigma_c}{dr} - \mu\left(\frac{d\sigma_r}{dr} + \frac{d\sigma_l}{dr}\right)\right] \quad ...(iv)$$

But from equation (*iii*),

$$\frac{du}{dr} = \frac{\sigma_r}{E} - \frac{\mu}{E}(\sigma_c + \sigma_l)$$

or
$$E\frac{du}{dr} = \sigma_r - \mu(\sigma_c + \sigma_l) \quad ...(v)$$

Equating the two values of $E\frac{du}{dr}$ given by equations (*iv*) and (*v*), we get

$$[\sigma_c - \mu(\sigma_r + \sigma_l)] + r\left[\frac{d\sigma_c}{dr} - \mu\left(\frac{d\sigma_r}{dr} + \frac{d\sigma_l}{dr}\right)\right] = \sigma_r - \mu(\sigma_c + \sigma_l)$$

or
$$\sigma_c - \mu\sigma_r - \mu\sigma_l + r\left[\frac{d\sigma_c}{dr} - \mu\left(\frac{d\sigma_r}{dr} + \frac{d\sigma_l}{dr}\right)\right] = \sigma_r - \mu\sigma_c - \mu\sigma_l$$

$$\sigma_c(1+\mu) - \sigma_r(1+\mu) + r\left[\frac{d\sigma_c}{dr} - \mu\left(\frac{d\sigma_r}{dr} + \frac{d\sigma_l}{dr}\right)\right] = 0$$

or
$$(1+\mu)(\sigma_c - \sigma_r) + r\left[\frac{d\sigma_c}{dr} - \mu\left(\frac{d\sigma_r}{dr} + \frac{d\sigma_l}{dr}\right)\right] = 0 \quad ...(vi)$$

In equation (*vi*), the value of $\frac{d\sigma_l}{dr}$ should be substituted in terms of σ_c and σ_r. Now from equation (*iii*), we have

$$e_l = \frac{1}{E}[\sigma_l - \mu(\sigma_c + \sigma_r)]$$

or $\sigma_l - \mu(\sigma_c + \sigma_r) = E \times e_l$

where e_l = longitudinal strain which is assumed constant.

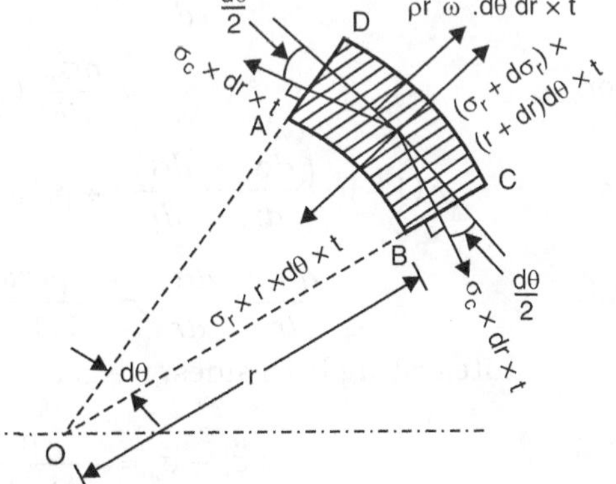

Fig. 9.14

To get the value of $\frac{d\sigma_l}{dr}$, differentiate the above equation w.r.t. r.

$$\therefore \quad \frac{d\sigma_l}{dr} - \mu\left(\frac{d\sigma_c}{dr} + \frac{d\sigma_r}{dr}\right) = 0$$

[$\because$ Differentiation of a constant ($E \times e_l$ is constant) is zero]

or
$$\frac{d\sigma_l}{dr} = \mu\left(\frac{d\sigma_c}{dr} + \frac{d\sigma_r}{dr}\right)$$

Substituting this value of $\frac{d\sigma_l}{dr}$ in equation (*vi*), we get

$$(1+\mu)(\sigma_c-\sigma_r)+r\left[\frac{d\sigma_c}{dr}-\mu\left\{\frac{d\sigma_r}{dr}+\mu\left(\frac{d\sigma_c}{dr}+\frac{d\sigma_r}{dr}\right)\right\}\right]=0$$

or
$$(1+\mu)(\sigma_c-\sigma_r)+r\left[\frac{d\sigma_c}{dr}-\mu\frac{d\sigma_r}{dr}-\mu^2\frac{d\sigma_c}{dr}-\mu^2\frac{d\sigma_r}{dr}\right]=0$$

or
$$(1+\mu)(\sigma_c-\sigma_r)+r\left[(1-\mu^2)\frac{d\sigma_c}{dr}-\mu\frac{d\sigma_r}{dr}(1+\mu)\right]=0$$

or
$$(1+\mu)(\sigma_c-\sigma_r)+r(1-\mu^2)\frac{d\sigma_c}{dr}-\mu\times r\frac{d\sigma_r}{dr}(1+\mu)=0$$

or
$$(1+\mu)(\sigma_c-\sigma_r)+r(1+\mu)(1-\mu)\frac{d\sigma_c}{dr}-\mu\times r\times(1+\mu)\frac{d\sigma_r}{dr}=0$$

or
$$(\sigma_c-\sigma_r)+r(1-\mu)\frac{d\sigma_c}{dr}-\mu\times r\times\frac{d\sigma_r}{dr}=0 \quad ...(vii)$$

[Cancelling $(1+\mu)$]

Now considering the equilibrium of an element *ABCD* of radius '*r*', radial width *dr* and thickness '*t*' subtending an angle $d\theta$ at the centre as shown in Fig. 9.14, we get the same equation as for the case of rotating disc *i.e.*, equation (9.4),

or
$$(\sigma_c-\sigma_r)=r\frac{d\sigma_r}{dr}+\rho r^2\omega^2 \quad ...(viii)$$

Substituting this value of $(\sigma_c-\sigma_r)$ in equation (*vii*), we get

$$r\frac{d\sigma_r}{dr}+\rho r^2\omega^2+r(1-\mu)\frac{d\sigma_c}{dr}-\mu\times r\times\frac{d\sigma_r}{dr}=0$$

or
$$r\frac{d\sigma_r}{dr}(1-\mu)+\rho r^2\omega^2+r(1-\mu)\frac{d\sigma_c}{dr}=0$$

or
$$r(1-\mu)\left(\frac{d\sigma_r}{dr}+\frac{d\sigma_c}{dr}\right)+\rho r^2\omega^2=0$$

or
$$\frac{d\sigma_r}{dr}+\frac{d\sigma_c}{dr}=-\frac{\rho r^2\omega^2}{r(1-\mu)}=-\frac{\rho\times r\times\omega^2}{(1-\mu)}$$

Integrating both sides, we get

$$\sigma_r+\sigma_c=-\frac{\rho}{(1-\mu)}\times\omega^2\times\frac{r^2}{2}+C_1 \quad ...(ix)$$

where C_1 is a constant of integration.

To get the values of σ_r and σ_c, the two equations (*ix*) and (*viii*) are to be solved. As equation (*viii*) contains $\frac{d\sigma_r}{dr}$, we should eliminate σ_c from these equations. This can be done by subtracting equation (*viii*) from equation (*ix*), then we get

$$2\sigma_r=\left(-\frac{\rho}{(1-\mu)}\times\omega^2\times\frac{r^2}{2}+C_1\right)-r\frac{d\sigma_r}{dr}-\rho r^2\omega^2$$

or $$2\sigma_r + r\frac{d\sigma_r}{dr} = -\frac{\rho}{(1-\mu)} \times \omega^2 \times \frac{r^2}{2} - \rho r^2 \times \omega^2 + C_1$$

$$= -\frac{\rho \times \omega^2 \times r^2}{2}\left(\frac{1}{(1-\mu)} + 2\right) + C_1$$

$$= -\frac{\rho \times \omega^2 \times r^2}{2}\left(\frac{1+2-2\mu}{(1-\mu)}\right) + C_1$$

$$= -\frac{\rho \times \omega^2 \times r^2}{2}\left(\frac{3-2\mu}{1-\mu}\right) + C_1$$

Multiplying by r to both sides, we get

$$2r\sigma_r + r^2\frac{d\sigma_r}{dr} = -\frac{\rho \times \omega^2 \times r^3}{2}\left(\frac{3-2\mu}{1-\mu}\right) + C_1 r$$

or $$\frac{d}{dr}(r^2 \times \sigma_r) = -\frac{\rho \times \omega^2 \times r^3}{2}\frac{(3-2\mu)}{1-\mu} + C_1 r$$

Integrating the above equation, we get

$$r^2 \times \sigma_r = -\frac{\rho \times \omega^2 \times r^4}{2 \times 4}\frac{(3-2\mu)}{1-\mu} + \frac{C_1 \times r^2}{2} + C_2$$

where C_2 is another constant of integration.

$$\therefore \quad \sigma_r = -\frac{\rho \times \omega^2 \times r^2}{8}\left(\frac{3-2\mu}{1-\mu}\right) + \frac{C_1}{2} + \frac{C_2}{r^2}$$

or $$\sigma_r = \frac{C_1}{2} + \frac{C_2}{r^2} - \frac{\rho \times \omega^2 \times r^2}{8}\left(\frac{3-2\mu}{1-\mu}\right) \qquad ...(9.19)$$

The equation (9.19) gives the expression for radial stress set up in a rotating long cylinder. To get the expression for circumferential stress (*i.e.*, σ_c), substitute the above value in equation (*ix*).

$$\therefore \quad \left[\frac{C_1}{2} + \frac{C_2}{r^2} - \frac{\rho \times \omega^2 \times r^2}{8}\frac{(3-2\mu)}{1-\mu}\right] + \sigma_c = -\frac{\rho}{(1-\mu)} \times \frac{\omega^2 \times r^2}{2} + C_1$$

$$\therefore \quad \sigma_c = -\frac{\rho}{(1-\mu)} \times \frac{\omega^2 \times r^2}{2} + C_1 - \frac{C_1}{2} - \frac{C_2}{r^2} + \frac{\rho \times \omega^2 \times r^2}{8}\frac{(3-2\mu)}{1-\mu}$$

$$= \frac{C_1}{2} - \frac{C_2}{r^2} - \frac{\rho \times \omega^2 \times r^2}{8(1-\mu)}[4 - (3-2\mu)]$$

$$= \frac{C_1}{2} - \frac{C_2}{r^2} - \frac{\rho \times \omega^2 \times r^2}{8(1-\mu)}(1 + 2\mu)$$

or $$\sigma_c = \frac{C_1}{2} - \frac{C_2}{r^2} - \frac{\rho \times \omega^2 \times r^2}{8}\left(\frac{1+2\mu}{1-\mu}\right) \qquad ...(9.20)$$

The equation (9.20), gives the expression for circumferential stress set up in a rotating long cylinder.

1st Case : Solid Cylinder Stresses in a Solid Cylinder. For a solid cylinder, at the centre $r = 0$, the stresses will become infinite. $\left(\text{As } \frac{C_2}{r^2} = \frac{C_2}{0} = \infty\right)$ But the stresses cannot be infinite at the centre. Hence in equations (9.19) and (9.20), the constant C_2 should be zero *i.e.*, $C_2 = 0$. Hence the stresses for solid cylinder will be as :

$$\sigma_r = \frac{C_1}{2} - \frac{\rho \times \omega^2 \times r^2}{8}\left(\frac{3-2\mu}{1-\mu}\right) \qquad ...(x)$$

and

$$\sigma_c = \frac{C_1}{2} - \frac{\rho \times \omega^2 \times r^2}{8}\left(\frac{1+2\mu}{1-\mu}\right) \qquad ...(xi)$$

The constant C_1 from these equations will be obtained from boundary condition.

Now for a solid cylinder on the free surface, $\sigma_r = 0$. This means at $r = r_2$ $\sigma_r = 0$ where r_2 = radius of the surface of the shaft.

From equation (*x*) in which $\sigma_r = 0$ at $r = r_2$, we get

$$0 = \frac{C_1}{2} - \frac{\rho \times \omega^2 \times r_2^2}{8}\left(\frac{3-2\mu}{1-\mu}\right)$$

or

$$\frac{C_1}{2} = \frac{\rho \times \omega^2 \times r_2^2}{8}\left(\frac{3-2\mu}{1-\mu}\right)$$

Substituting the above value in equations (*x*) and (*xi*), we get

$$\sigma_r = \frac{\rho \times \omega^2 \times r_2^2}{8}\left(\frac{3-2\mu}{1-\mu}\right) - \frac{\rho \times \omega^2 \times r^2}{8}\left(\frac{3-2\mu}{1-\mu}\right)$$

$$= \frac{\rho \times \omega^2}{8}\left(\frac{3-2\mu}{1-\mu}\right)(r_2^2 - r^2) \qquad ...(9.21)$$

The stress is maximum at the centre where $r = 0$

$$\therefore \qquad (\sigma_r)_{max} = \frac{\rho \times \omega^2}{8}\left(\frac{3-2\mu}{1-\mu}\right)r_2^2$$

and

$$\sigma_c = \frac{\rho \times \omega^2 \times r_2^2}{8}\left(\frac{3-2\mu}{1-\mu}\right) - \frac{\rho \times \omega^2 \times r^2}{8}\left(\frac{1+2\mu}{1-\mu}\right)$$

$$= \frac{\rho \times \omega^2 \times r_2^2}{8}\left(\frac{3-2\mu}{1-\mu}\right) - \frac{\rho \times \omega^2 \times r^2}{8}\left(\frac{1+2\mu}{1-\mu}\right) \times \frac{(3-2\mu)}{(3-2\mu)}$$

[Multiply and divide the last term by $(3 - 2\mu)$]

$$= \frac{\rho \times \omega^2}{8} \times \left(\frac{3-2\mu}{1-\mu}\right)\left[r_2^2 - r^2\left(\frac{1+2\mu}{3-2\mu}\right)\right]$$

$$= \frac{\rho \times \omega^2}{8} \times \left(\frac{3-2\mu}{1-\mu}\right)\left[r_2^2 - \left(\frac{1+2\mu}{3-2\mu}\right)r^2\right]$$

The stress is maximum at the centre where $r = 0$

$$\therefore \quad (\sigma_c)_{max} = \frac{\rho \times \omega^2}{8} \left(\frac{3-2\mu}{1-\mu}\right) \frac{r^2}{2} \quad ...(9.22)$$

Hence in case of solid cylinder, the circumferential and radial stresses are maximum at the centre and they are equal

$$\therefore \quad (\sigma_r)_{max} = (\sigma_c)_{max} = \frac{\rho \times \omega^2}{8} \left(\frac{3-2\mu}{1-\mu}\right) \times r_2^2 \quad ...(9.23)$$

2nd Case : Hollow Cylinder. The circumferential and radial stresses in a rotating long cylinder are given by equations as

$$\sigma_c = \frac{C_1}{2} - \frac{C_2}{r^2} - \frac{\rho \times \omega^2 \times r^2}{8} \left(\frac{1+2\mu}{1-\mu}\right) \quad ...(i)$$

and

$$\sigma_r = \frac{C_1}{2} + \frac{C_2}{r^2} - \frac{\rho \times \omega^2 \times r^2}{8} \left(\frac{3-2\mu}{1-\mu}\right) \quad ...(ii)$$

The values of C_1 and C_2 are obtained from boundary conditions.

For hollow cylinders, at $r = r_1$, $\sigma_r = 0$ and also at $r = r_2$, $\sigma_r = 0$

Substituting these values, we get two equations as

$$0 = \frac{C_1}{2} + \frac{C_2}{r^2} - \frac{\rho \times \omega^2 \times r_1^2}{8} \left(\frac{3-2\mu}{1-\mu}\right) \quad \text{[Here } \sigma_r = 0 \text{ at } r = r_1] \quad ...(iii)$$

$$0 = \frac{C_1}{2} + \frac{C_2}{r^2} - \frac{\rho \times \omega^2 \times r_2^2}{8} \left[\frac{3-2\mu}{1-\mu}\right] \quad \text{[Here } \sigma_r = 0 \text{ at } r = r_2] \quad ...(iv)$$

To eliminate C_1 from these equation, subtract equation (*iv*) from equation (*iii*),

$$\therefore \quad 0 = \frac{C_2}{r_1^2} - \frac{\rho \times \omega^2}{8} \times r_1^2 \times \frac{(3-2\mu)}{(1-\mu)} - \frac{C_2}{r_2^2} + \frac{\rho \times \omega^2}{8} \times r_2^2 \times \left(\frac{3-2\mu}{1-\mu}\right)$$

$$= C_2 \left[\frac{1}{r_1^2} - \frac{1}{r_2^2}\right] - \frac{\rho \times \omega^2}{8} \times \left(\frac{3-2\mu}{1-\mu}\right) [r_1^2 - r_2^2]$$

$$= C_2 \frac{(r_2^2 - r_1^2)}{r_1^2 r_2^2} + \frac{\rho \times \omega^2}{8} \left(\frac{3-2\mu}{1-\mu}\right) (r_2^2 - r_1^2)$$

$$= \frac{C_2}{r_1^2 r_2^2} + \frac{\rho \times \omega^2}{8} \left(\frac{3-2\mu}{1-\mu}\right) \quad \text{[Cancelling } (r_2^2 - r_1^2)]$$

$$\therefore \quad C_2 = -\frac{\rho \times \omega^2}{8} \left(\frac{3-2\mu}{1-\mu}\right) r_1^2 r_2^2$$

Substituting this value of C_2 in equation (*iii*), we get the value of C_1 as

$$0 = \frac{C_1}{2} + \frac{1}{r_1^2} \left[-\frac{\rho \times \omega^2}{8} \left(\frac{3-2\mu}{1-\mu}\right) r_1^2 r_2^2\right] - \frac{\rho \times \omega^2}{8} \times r_1^2 \times \left(\frac{3-2\mu}{1-\mu}\right)$$

$$= \frac{C_1}{2} - \frac{\rho \times \omega^2}{8} \left(\frac{3-2\mu}{1-\mu}\right) r_2^2 - \frac{\rho \times \omega^2}{8} \times r_1^2 \times \left(\frac{3-2\mu}{1-\mu}\right)$$

or $$\frac{C_1}{2} = \frac{\rho \times \omega^2}{8}\left(\frac{3-2\mu}{1-\mu}\right) r_2^{\,2} + \frac{\rho \times \omega^2}{8} \times \left(\frac{3-2\mu}{1-\mu}\right) \times r_1^{\,2}$$

$$= \frac{\rho \times \omega^2}{8}\left(\frac{3-2\mu}{1-\mu}\right)[r_2^{\,2} + r_1^{\,2}]$$

or $$C_1 = \frac{\rho \times \omega^2}{4}\left(\frac{3-2\mu}{1-\mu}\right)[r_1^{\,2} + r_2^{\,2}]$$

Substituting the values of C_1 and C_2 in equations (*i*) and (*ii*), we get

$$\sigma_c = \frac{\rho \times \omega^2}{2 \times 4}\left(\frac{3-2\mu}{1-\mu}\right)(r_1^{\,2} + r_2^{\,2}) - \left[-\frac{\rho \times \omega^2}{8}\left(\frac{3-2\mu}{1-\mu}\right)\frac{r_1^{\,2} r_2^{\,2}}{r^2}\right] - \frac{\rho \times \omega^2 \times r^2}{8}\left(\frac{1+2\mu}{1-\mu}\right)$$

$$= \frac{\rho \times \omega^2}{8}\left(\frac{3-2\mu}{1-\mu}\right)[r_1^{\,2} + r_2^{\,2}] + \frac{\rho \times \omega^2}{8}\left(\frac{3-2\mu}{1-\mu}\right) \times \frac{r_1^2 \times r_2^2}{r^2} - \frac{\rho \times \omega^2 \times r^2}{8}\left(\frac{1+2\mu}{1-\mu}\right)$$

$$= \frac{\rho \times \omega^2}{8}\left(\frac{3-2\mu}{1-\mu}\right)(r_1^{\,2} + r_2^{\,2}) + \frac{\rho \times \omega^2}{8}\left(\frac{3-2\mu}{1-\mu}\right)\left[\frac{r_1^2 - r_2^2}{r^2} - \frac{1+2\mu}{3-2\mu}r^2\right]$$

$$= \frac{\rho \times \omega^2}{8}\left(\frac{3-2\mu}{1-\mu}\right)\left[r_1^2 + r_2^2 + \frac{r_1^2 r_2^2}{r^2} - \left(\frac{1+2\mu}{3-2\mu}\right)r^2\right] \quad \text{...(9.24)}$$

and $$\sigma_r = \frac{\rho \times \omega^2}{2 \times 4}\left(\frac{3-2\mu}{1-\mu}\right)(r_1^{\,2} + r_2^{\,2}) + \left[-\frac{\rho \times \omega^2}{8} \times \left(\frac{3-2\mu}{1-\mu}\right)\frac{r_1^2 \times r_2^2}{r^2}\right] - \frac{\rho \times \omega^2 \times r^2}{8} \times \left(\frac{3-2\mu}{1-\mu}\right)$$

$$= \frac{\rho \times \omega^2}{8}\left(\frac{3-2\mu}{1-\mu}\right)\left[r_1^2 + r_2^2 - \frac{r_1^2 \times r_2^2}{r^2} - r^2\right] \quad \text{...(9.25)}$$

Equation (9.24) gives the expression for circumferential stress and equation (9.25) gives the expression for radial stress in a hollow long rotating cylinder.

Maximum circumferential stress $(\sigma_c)_{max}$

From equation (9.24), it is clear that circumferential stress will be maximum when '*r*' is minimum.

Hence circumferential stress will be maximum at the inner radius where '*r*' is minimum (*i.e.*, $r = r_1$)

Hence substituting $r = r_1$ in equation (9.24), we get max. value of σ_c

$$\therefore \quad (\sigma_c)_{max} = \frac{\rho \times \omega^2}{8}\left(\frac{3-2\mu}{1-\mu}\right)\left[r_1^2 + r_2^2 + \frac{r_1^2 \times r_2^2}{r_1^2} - \left(\frac{1+2\mu}{3-2\mu}\right)r_1^{\,2}\right]$$

$$= \frac{\rho \times \omega^2}{8} \times \left(\frac{3-2\mu}{1-\mu}\right)\left[r_1^2 + 2r_2^2 - \left(\frac{1+2\mu}{3-2\mu}\right)r_1^{\,2}\right] \quad \text{...(9.26)}$$

Maximum radial stress $(\sigma_r)_{max}$

From equation (9.25), we cannot say that σ_r will be maximum when r is minimum.

Hence for σ_r to be maximum $\frac{d(\sigma_r)}{dr}$ should be zero.

or $$\frac{d}{dr}\left[\frac{\rho\times\omega^2}{8}\left(\frac{3-2\mu}{1-\mu}\right)\left[r_1^2+r_2^2-\frac{r_1^2\times r_2^2}{r^2}-r^2\right]\right]=0$$

or $$\frac{\rho\times\omega^2}{8}\left(\frac{3-2\mu}{1-\mu}\right)\left[0+0-\frac{r_1^2\times r_2^2\times(-2)}{r^3}-2r\right]=0$$

or $$+\frac{2r_1^2\times r_2^2}{r^3}-2r=0$$

or $$2r_1^{\,2}\times r_2^{\,2}-2r^4=0$$

or $$r^4=r_1^{\,2}\times r_2^{\,2} \quad \text{or} \quad r^2=r_1\times r_2$$

or $$r=\sqrt{r_1\times r_2} \qquad \text{...(9.27)}$$

Equation (9.27) gives the value of r at which radial stress will be maximum. Substituting this value of 'r' in equation (9.25), we get maximum value of radial stress.

$\therefore$ $$(\sigma_r)_{max}=\frac{\rho\times\omega^2}{8}\left(\frac{3-2\mu}{1-\mu}\right)\left[r_1^2+r_2^2-\frac{r_1^2\times r_2^2}{r_1\times r_2}-r_1r_2\right]$$

$$=\frac{\rho\times\omega^2}{8}\left(\frac{3-2\mu}{1-\mu}\right)[r_1^{\,2}+r_2^{\,2}-2r_1r_2]=\frac{\rho\times\omega^2}{8}\left(\frac{3-2\mu}{1-\mu}\right)[r_2-r_1]^2 \qquad \text{...(9.28)}$$

Equation (9.28) gives the value of maximum radial stress.

9.9. SOLVED PROBLEMS 9.12 TO 9.13 FOR ROTATING LONG CYLINDERS

Problem 9.12. *A long cylinder is rotating at a speed of 3600 r.p.m. The diameter of the cylinder is 500 mm and density of its material is 8000 kg/m^3. If Poisson's ratio = 0.3, then :*

(i) Calculate the maximum stress in the cylinder and

(ii) Plot the circumferential stress and radial stress along the radius.

Sol. Given :

Diameter, $d_2 = 500$ mm

$\therefore$ Radius, $r_2=\frac{d_2}{2}=\frac{500}{2}=250\text{ mm}=0.25\text{ m}$

Speed of rotation, $N = 3600$ r.p.m.

$\therefore$ Angular speed, $\omega=\frac{2\pi N}{60}=\frac{2\pi\times3600}{60}=120\,\pi\text{ rad/s}$

Density, $\rho = 8000$ kg/m^3, Poisson's ratio, $\mu = 0.3$

(*i*) *Maximum stress in the cylinder.* Maximum circumferential and radial stresses occur at the centre, where $r = 0$. It is given by equation (9.23) as

$$(\sigma_c)_{max}=(\sigma_r)_{max}=\frac{\rho\times\omega^2\times r_2^2}{8}\left(\frac{3-2\mu}{1-\mu}\right)$$

$$=\frac{8000\times(120\,\pi)^2\times(0.25)^2}{8}\left(\frac{3-2\times0.3}{1-0.3}\right)$$

$$= \frac{8000 \times 14400 \times \pi^2 \times 0.0625}{8} \times \frac{2.4}{0.7} \text{ N/m}^2$$

$= 30.45 \times 10^6$ N/m^2 = **30.45 MN/m^2. Ans.**

(ii) Plot of the stresses

Circumferential stress (σ_c) : Circumferential stress is given by equation (9.22), as

$$\sigma_c = \frac{\rho \times \omega^2}{8}\left(\frac{3-2\mu}{1-\mu}\right)\left[r_2^2 - \left(\frac{1+2\mu}{3-2\mu}\right)r^2\right]$$

$$= \frac{8000 \times (120\,\pi)^2}{8}\left(\frac{3-2\times 0.3}{1-0.3}\right)\left[0.25^2 - \left(\frac{1+2\times 0.3}{3-2\times 0.3}\right)r^2\right]$$

$$= \frac{8000 \times 14400 \times \pi^2}{8} \times \frac{2.4}{0.7}\left[0.0625 - \frac{1.6}{2.4}r^2\right]$$

$= 4.872 \times 10^8\,[0.0625 - 0.667\,r^2]$ N/m^2

$= 487.2 \times 10^6\,[0.0625 - 0.667\,r^2]$ N/m^2

$= 487.2\,(0.0625 - 0.667\,r^2)$ MN/m^2

When $r = 0$, $\sigma_c = 487.2 \times 0.0625$ MN/m^2 = 30.45 MN/m^2

$r = 0.05$ m, $\sigma_c = 487.2\,(0.0625 - 0.667 \times 0.05^2) = 29.63$ MN/m^2

$r = 0.10$ m, $\sigma_c = 487.2\,(0.0625 - 0.667 \times 0.1^2) = 27.2$ MN/m^2

$r = 0.15$ m, $\sigma_c = 487.2\,(0.0625 - 0.667 \times 0.15^2) = 23.14$ MN/m^2

$r = 0.20$ m, $\sigma_c = 487.2\,(0.0625 - 0.667 \times 0.2^2) = 17.45$ MN/m^2

$r = 0.25$ m, $\sigma_c = 487.2\,(0.0625 - 0.667 \times 0.25^2) = 10.14$ MN/m^2

Radial stress

Radial stress is given by equation (9.21),

$$\sigma_r = \frac{\rho \times \omega^2\,(r_2^2 - r^2)}{8}\left[\frac{3-2\mu}{1-\mu}\right]$$

$$= \frac{8000 \times (120\,\pi)^2}{8}(0.25^2 - r^2)\left(\frac{3-2\times 0.3}{1-0.3}\right)$$

$$= 1000 \times 14400 \times \pi^2\,(0.0625 - r^2)\left(\frac{2.4}{0.7}\right)$$

$$= 1000 \times 14400 \times \pi^2 \times \frac{2.4}{0.7}\,(0.0625 - r^2)$$

$= 487.276 \times 10^6\,(0.0625 - r^2)$

$= 487.276\,(0.0625 - r^2)$ MN/m^2

When $r = 0$, $\sigma_r = 487.276\,(0.0625) = 30.45$ MN/m^2

$r = 0.05$ m, $\sigma_r = 487.276\,(0.0625 - 0.05^2) = 29.88$

$r = 0.1$ m, $\sigma_r = 487.276\,(0.0625 - 0.1^2) = 25.58$

$r = 0.15$ m, $\sigma_r = 487.276\,(0.0625 - 0.15^2) = 19.49$

$r = 0.20$ m, $\sigma_r = 487.276\,(0.0625 - 0.2^2) = 10.96$

$r = 0.25$ m, $\sigma_r = 487.276\,(0.0625 - 0.25^2) = 0$

The variation of circumferential and radial stresses are shown in Fig. 9.15, in which stresses are along x-axis and radius along y-axis.

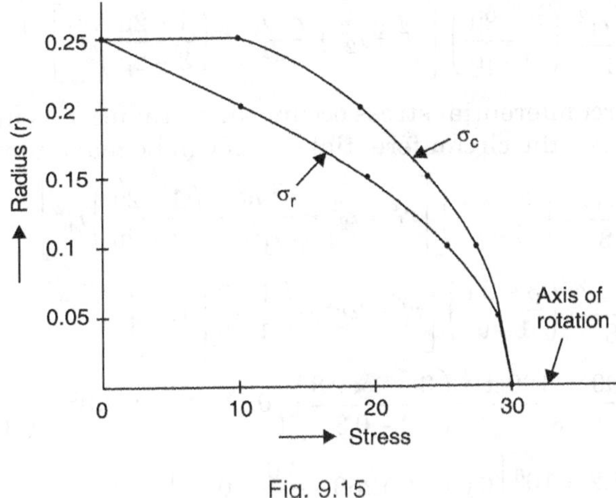

Fig. 9.15

Fig. 9.16 shows the variation of these stresses when radius is taken along x-axis and stresses along y-axis.

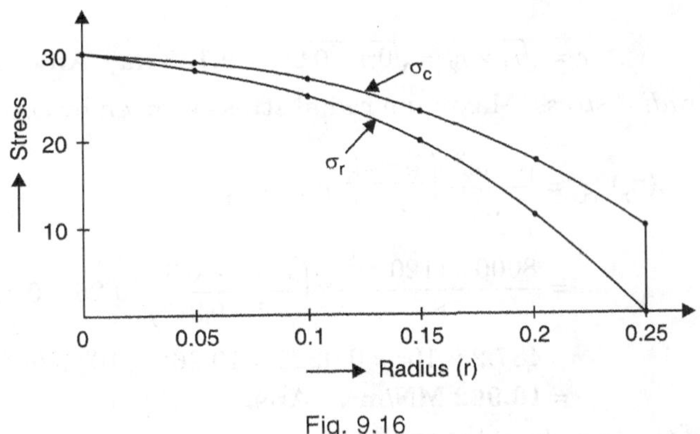

Fig. 9.16

Problem 9.13. *If in problem 9.12, the long rotating cylinder is having a central hole of diameter 200 mm, then determine :*

(i) Maximum circumferential stress

(ii) Radius at which radial stress is maximum,

(iii) Maximum radial stress,

(iv) The variation of circumferential and radial stresses along the radius of the cylinder. Take all other data from problem 9.12.

Sol. Given :

Outer diameter = 500 mm $\quad\therefore\quad$ Outer radius, $r_2 = 250$ mm = 0.25 m

Inner diameter = 200 mm $\quad\therefore\quad$ Inner radius, $r_1 = 100$ mm = 0.1 m

Speed, $N = 3600$ r.p.m ; $\omega = 120\,\pi$ rad/s ; $\rho = 8000$ kg/m^3, $\mu = 0.3$

(i) Maximum circumferential stress $(\sigma_c)_{max}$. The circumferential stress at any radius is given by equation (9.24) as

$$\sigma_c = \frac{\rho \times \omega^2}{8}\left(\frac{3-2\mu}{1-\mu}\right)\left[r_1^2 + r_2^2 + \frac{r_1^2 r_2^2}{r^2} - \left(\frac{1+2\mu}{3-2\mu}\right)r^2\right]$$

The maximum circumferential stress occurs where radius 'r' is minimum. Hence at the inner radius (*i.e.*, at $r = r_1$), the circumferential stress will be maximum,

$$\therefore \quad (\sigma_c)_{max} = \frac{\rho \times \omega^2}{8}\left(\frac{3-2\mu}{1-\mu}\right)\left[r_1^2 + r_2^2 + \frac{r_1^2 r_2^2}{r_1^2} - \left(\frac{1+2\mu}{3-2\mu}\right)r_1^2\right]$$

$$= \frac{\rho \times \omega^2}{8}\left(\frac{3-2\mu}{1-\mu}\right)\left[r_1^2 + 2r_2^2 - \left(\frac{1+2\mu}{3-2\mu}\right)r_1^2\right]$$

$$= \frac{8000 \times (120\,\pi)^2}{8}\left(\frac{3-2\times 0.3}{1-0.3}\right)\left[0.1^2 + 2\times(0.25)^2 - \left(\frac{1+2\times 0.3}{3-2\times 0.3}\right)\times 0.1^2\right]$$

$$= 487.2 \times 10^6\left[0.01 + 0.1250 - \frac{1.6}{2.4}\times 0.1^2\right]$$

$$= 62.5 \times 10^6 \text{ N/m}^2 = \mathbf{62.5\ MN/m^2}. \quad \textbf{Ans.}$$

(ii) Radius at which radial stress is maximum. The radial stress is maximum at radius given by,

$$r = \sqrt{r_1 \times r_2} = \sqrt{0.1 \times 0.25} = \mathbf{0.158\ m.} \quad \textbf{Ans.}$$

(iii) Maximum radial stress. Maximum radial stress is given by equation (9.28) as

$$(\sigma_r)_{max} = \frac{\rho \times \omega^2}{8}\left(\frac{3-2\mu}{1-\mu}\right)(r_2 - r_1)^2$$

$$= \frac{8000 \times (120\,\pi)^2}{8} \times \left(\frac{3-2\times 0.3}{1-0.3}\right)(0.25 - 0.1)^2$$

$$= 487.2 \times 10^6 \times 0.0225 = 10.962 \times 10^6 \text{ N/m}^2$$

$$= \mathbf{10.962\ MN/m^2}. \quad \textbf{Ans.}$$

(iv) Variation of stresses along the radius

(a) Circumferential stress (σ_c). The circumferential stress at any radius is given by equation (9.24) as

$$\sigma_c = \frac{\rho \times \omega^2}{8}\left(\frac{3-2\mu}{1-\mu}\right)\left[r_1^2 + r_2^2 + \frac{r_1^2 \times r_2^2}{r^2} - \left(\frac{1+2\mu}{3-2\mu}\right)r^2\right]$$

$$= \frac{8000 \times (120\,\pi)^2}{8}\left[\frac{3-2\times 0.3}{1-0.3}\right]\left[0.1^2 + 0.25^2 + \frac{0.1^2 \times 0.25^2}{r^2} - \left(\frac{1+2\times 0.3}{3-2\times 0.3}\right)r^2\right]$$

$$= 487.2 \times 10^6 \times \left[0.01 + 0.0625 + \frac{0.000625}{r^2} - 0.667 \times r^2\right] \text{N/m}^2$$

$$= 487.2 \times \left(0.0725 + \frac{0.000625}{r^2} - 0.667 \times r^2\right) \text{MN/m}^2$$

By substituting the different values of 'r' in the above equation, we get different values of σ_c. Let us start the calculation from inner radius (*i.e.*, $r = r_1 = 0.1$)

At $r = 0.1$, $\sigma_c = 487.2\left[0.0725 + \frac{0.000625}{0.1^2} - 0.667 \times 0.1^2\right] = 62.5 \text{ MN/m}^2$

At $r = 0.125$, $\sigma_c = 487.2\left[0.0725 + \frac{0.000625}{0.125^2} - 0.667 \times 0.125^2\right]$

$= 487.2\ [0.0725 + 0.04 - 0.0104] = 49.73 \text{ MN/m}^2$

At $r = 0.15$, $\sigma_c = 487.2\left[0.0725 + \frac{0.000625}{0.15^2} - 0.667 \times 0.15^2\right]$

$= 487.2\ [0.0725 + 0.0277 - 0.015] = 41.5 \text{ MN/m}^2$

At $r = 0.175$, $\sigma_c = 487.2\left[0.0725 + \frac{0.000625}{0.175^2} - 0.667 \times 0.175^2\right]$

$= 487.2\ [0.0725 + 0.0204 - 0.0204] = 35.32 \text{ MN/m}^2$

At $r = 0.2$, $\sigma_c = 487.2\left[0.0725 + \frac{0.000625}{0.2^2} - 0.667 \times 0.2^2\right]$

$= 487.2\ [0.0725 + 0.0156 - 0.02668] = 29.92 \text{ MN/m}^2$

At $r = 0.225$, $\sigma_c = 487.2\left[0.0725 + \frac{0.000625}{0.225^2} - 0.667 \times 0.225^2\right]$

$= 487.2\ [0.0725 + 0.0123 - 0.0337] = 24.86 \text{ MN/m}^2$

At $r = 0.25$, $\sigma_c = 487.2\left[0.0725 + \frac{0.000625}{0.25^2} - 0.667 \times 0.25^2\right]$

$= 487.2\ [0.0725 + 0.01 - 0.04168] = 19.88 \text{ MN/m}^2$

(*b*) *Radial Stress* (σ_r). The radial stress at any radius 'r' is given by equation (9.25) as

$$\sigma_r = \frac{\rho \times \omega^2}{8} \times \left(\frac{3 - 2\mu}{1 - \mu}\right)\left[r_1^2 + r_2^2 - \frac{r_1^2 r_2^2}{r^2} - r^2\right]$$

$$= \frac{8000 \times (120\,\pi)^2}{8} \times \left(\frac{3 - 2 \times 0.3}{1 - 0.3}\right)\left[0.1^2 + 0.25^2 - \frac{0.1^2 \times 0.25^2}{r^2} - r^2\right]$$

$$= 487.2 \times 10^6 \times \left[0.01 + 0.0625 - \frac{0.000625}{r^2} - r^2\right] \text{ N/m}^2$$

$$= 487.2 \times \left[0.0725 - \frac{0.000625}{r^2} - r^2\right] \text{ MN/m}^2$$

By substituting the different values of 'r' in the above equation, we get different values of σ_r. Also σ_r is zero at $r = 0.1$ m and $r = 0.25$ m.

At $r = 0.1$ m, $\sigma_r = 0$

At $r = 0.125$ m, $\sigma_r = 487.2 \times \left[0.0725 - \frac{0.000625}{0.125^2} - 0.125^2\right]$

$= 487.2 \times [0.0725 - 0.04 - 0.0156] = 8.23 \text{ MN/m}^2$

At $r = 0.15$ m, $\sigma_r = 487.2 \times \left[0.0725 - \frac{0.000625}{0.15^2} - 0.15^2\right]$

$= 487.2\ [0.0725 - 0.0277 - 0.0225] = 10.86\ \text{MN/m}^2$

We have already calculated that σ_r is maximum at $r = 0.158$ m and maximum value of σ_r is 10.962 MN/m²

$\therefore$ At $r = 0.158$ m, $\sigma_r = 10.962\ \text{MN/m}^2$

At $r = 0.175$ m, $\sigma_r = 487.2 \left[0.0725 - \frac{0.000625}{0.175^2} - 0.175^2\right]$

$= 487.2 \times [0.0725 - 0.0204 - 0.0306] = 10.47\ \text{MN/m}^2$

At $r = 0.2$ m, $\sigma_r = 487.2 \left[0.0725 - \frac{0.000625}{0.2^2} - 0.2^2\right]$

$= 487.2\ [0.0725 - 0.0156 - 0.04] = 8.23\ \text{MN/m}^2$

At $r = 0.225$ m, $\sigma_r = 487.2 \left[0.0725 - \frac{0.000625}{0.225^2} - 0.225^2\right]$

$= 487.2\ [0.0725 - 0.0123 - 0.0506] = 4.66\ \text{MN/m}^2$

At $r = 0.25$ m, $\sigma_r = 0$

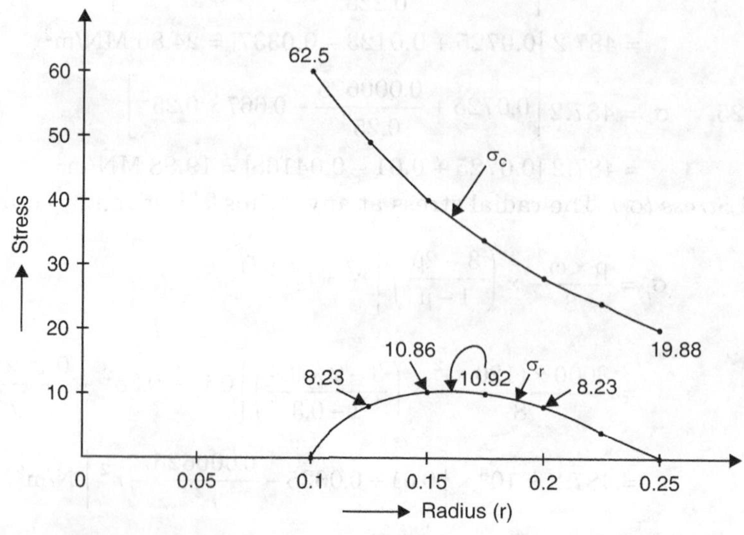

Fig. 9.17

The variation of circumferential and radial stresses are shown in Fig. 9.17.

9.10. STRESSES AND INITIAL DIFFERENCE IN RADII AT THE JUNCTION OF A COMPOUND CYLINDER DUE TO INTERFERENCE FITS

By shrinking the outer cylinder over the inner cylinder, some compressive stresses are produced in the inner cylinder. In order to shrink the outer cylinder over the inner cylinder, the inner diameter of the outer cylinder should be slightly less than the outer diameter of the inner cylinder. Now the outer cylinder is heated and inner cylinder is inserted into it. After cooling, the outer cylinder shrinks over the inner cylinder. Thus inner cylinder is put into compression and outer cylinder is put into tension. After shrinking the outer radius of inner cylinder decreases whereas the inner radius of outer cylinder increases from the initial values.

Let r_2 = Outer radius of the outer cylinder

r_1 = Inner radius of the inner cylinder

r^* = Radius of junction after shrinking or it is common radius after shrinking

p^* = Radial pressure at the junction after shrinking.

Before shrinking the outer radius of the inner cylinder is slightly more than r^* and inner radius of the outer cylinder is slightly less than r^*.

For the outer and inner cylinder Lame's equations are used. These equations are

$$p_x = \frac{b}{x^2} - a \quad \text{and} \quad \sigma_x = \frac{b}{x^2} + a$$

The values of constants a and b will be different for each cylinder.

Let the constants for inner cylinder be a_2, b_2 and for outer cylinder a_1, b_1.

The radial pressure at the junction (*i.e.*, p^*) is same for outer cylinder and inner cylinder.

At the junction, $x = r^*$ and $p_x = p^*$. Hence radial pressure at the junction.

$$p^* = \frac{b_1}{r^{*2}} - a_1 = \frac{b_2}{r^{*2}} - a_2 \quad \text{...(A)}$$

or

$$\frac{b_1 - b_2}{r^{*2}} = (a_1 - a_2) \quad \text{...(B)}$$

or

$$(b_1 - b_2) = r^{*2}(a_1 - a_2)$$

Now the hoop strain (or circumferential strain) in the cylinders at any point

$$= \frac{\sigma_x}{E} + \frac{p_x}{mE} \quad \text{...(C)}$$

But circumferential strain

$$= \frac{\text{Increase in circumference}}{\text{Original circumference}}$$

$$= \frac{2\pi(r + dr) - 2\pi r}{2\pi r} = \frac{dr}{r} \quad \text{...(D)}$$

= Radial strain

Hence equating the two values of circumferential strain given by equations (*C*) and (*D*), we get

$$\frac{dr}{r} = \frac{\sigma_x}{E} + \frac{p_x}{mE} \quad \text{...(i)}$$

On shrinking, at the junction there is extension in the inner radius of the outer cylinder and compression in the outer radius of the inner cylinder.

At the junction where $x = r^*$, increase in the inner radius of outer cylinder

$$= r^*\left(\frac{\sigma_x}{E} + \frac{p_x}{mE}\right) \quad \text{...(ii)}$$

But for outer cylinder at the junction, we have

$$\sigma_x = \frac{b_1}{r^{*2}} + a_1 \quad \text{and} \quad p_x = \frac{b_1}{r^{*2}} - a_1$$

where a_1 and b_1 are constants for outer cylinders.

Substituting the values of σ_x and p_x in equation (*ii*), we get

Increase in the inner radius of outer cylinder

$$= r^*\left[\frac{1}{E}\left(\sigma_x + \frac{p_x}{m}\right)\right] = r^*\left[\frac{1}{E}\left(\frac{b_1}{r^{*2}} + a_1\right) + \frac{1}{mE}\left(\frac{b_1}{r^{*2}} - a_1\right)\right]$$

Similarly, decrease in the outer radius of the inner cylinder is obtained from equation (*i*) as

$$= -\,r^*\left[\frac{\sigma_x}{E} + \frac{p_x}{mE}\right] \quad \text{(– ve sign is due to decrease)} \quad \text{...(iii)}$$

But for inner cylinder at the function, we have

$$\sigma_x = \frac{b_2}{r^{*2}} + a_2 \quad \text{and} \quad p_x = \frac{b_2}{r^{*2}} - a_2$$

Substituting these values in equation (*iii*), we get

Decrease in the outer radius of inner cylinder

$$= -\,r^*\left[\frac{1}{E}\left(\frac{b_2}{r^{*2}} + a_2\right) + \frac{1}{mE}\left(\frac{b_2}{r^{*2}} - a_2\right)\right] \quad \text{...(iv)}$$

But the original difference in the outer radius of the inner cylinder and inner radius of the outer cylinder.

= Increase in inner radius of outer cylinder

+ Decrease in outer radius of the inner cylinder

$$= r^*\left[\frac{1}{E}\left(\frac{b_1}{r^{*2}} + a_1\right) + \frac{1}{mE}\left(\frac{b_1}{r^{*2}} - a_1\right)\right] - r^*\left[\frac{1}{E}\left(\frac{b_2}{r^{*2}} + a_2\right) + \frac{1}{mE}\left(\frac{b_2}{r^{*2}} - a_2\right)\right]$$

$$= \frac{r^*}{E}\left[\left(\frac{b_1}{r^{*2}} + a_1\right) - \left(\frac{b_2}{r^{*2}} + a_2\right)\right] + \frac{r^*}{mE}\left[\left(\frac{b_1}{r^{*2}} - a_1\right) - \left(\frac{b_2}{r^{*2}} - a_2\right)\right]$$

But from equation (*A*),

$$\frac{b_1}{r^{*2}} - a_1 = \frac{b_2}{r^{*2}} - a_2.$$

Hence second part of the above equation is zero. Hence above equation becomes as

Original difference of radii at the junction

$$= \frac{r^*}{E}\left[\left(\frac{b_1}{r^{*2}} + a_1\right) - \left(\frac{b_2}{r^{*2}} + a_2\right)\right]$$

$$= \frac{r^*}{E}\left[\frac{(b_1 - b_2)}{r^{*2}} + (a_1 - a_2)\right]$$

$$= \frac{r^*}{E}\,[(a_1 - a_2) + (a_1 - a_2)] \qquad \left[\because \text{ From equation } (B), \frac{b_1 - b_2}{r^{*2}} = a_1 - a_2\right]$$

$$= \frac{2r^*}{E}(a_1 - a_2) \qquad \ldots(9.29)$$

The values of a_1 and a_2 are obtained from the given conditions. The value of a_1 is for outer cylinder whereas of a_2 is for inner cylinder.

9.11. SOLVED PROBLEMS 9.14 TO 9.15 FOR STRESSES DUE TO INTERFERENCE FITS

Problem 9.14. *A steel cylinder of 300 mm external diameter is to be shrunk to another steel cylinder of 150 mm internal diameter. After shrinking the diameter at the junction is 250 mm and radial pressure at the common junction is 28 N/mm². Find the original difference in radii at the junction. Take E = 2 × 10⁵ N/mm².*

Sol. Given :

External dia. of outer cylinder = 300 mm

∴ Radius, $r_2 = 150$ mm

Internal dia. of inner cylinder = 150 mm

∴ Radius, $r_1 = 75$ mm

Diameter at the junction = 250 mm

∴ Radius, $r^* = 125$ mm

Radial pressure at the junction, $p^* = 28$ N/mm²

Value of $E = 2 \times 10^5$ N/mm²

Using equation (9.29), we get

Original difference of radii at the junction

$$= \frac{2r^*}{E}(a_1 - a_2) \qquad \ldots(i)$$

First find the values of a_1 and a_2 from the given conditions. These are the constants for outer cylinder and inner cylinder respectively. They are obtained by using Lame's equations.

For outer cylinder

$$p_x = \frac{b_1}{x^2} - a_1$$

(*i*) At junction, $x = r^* = 125$ mm and $p_x = p^* = 28$ N/mm²

(*ii*) At $x = 150$ mm, $p_x = 0$.

Substituting these two conditions in the above equation, we get

$$28 = \frac{b_1}{125^2} - a_1 = \frac{b_1}{15625} - a_1 \qquad \ldots(ii)$$

and

$$0 = \frac{b_1}{150^2} - a_1 = \frac{b_1}{22500} - a_1 \qquad \ldots(iii)$$

Solving equations (*ii*) and (*iii*), we get

$$b_1 = 1432000 \quad \text{and} \quad a_1 = 63.6.$$

For inner cylinder

$$p_x = \frac{b_2}{x^2} - a_2$$

(*i*) At junction, $x = r^* = 125$ mm and $p_x = p^* = 28$ N/mm^2

(*ii*) At $x = 75$ mm, $p_x = 0$.

Substituting these two conditions in the above equation, we get

$$28 = \frac{b_2}{125^2} - a_2 = \frac{b_2}{15625} - a_2 \quad \text{....}(iv)$$

and
$$0 = \frac{b_2}{75^2} - a_2 = \frac{b_2}{5625} - a_2 \quad \text{...}(v)$$

Solving equations (*iv*) and (*v*), we get

$$b_2 = -246100$$

and
$$a_2 = -43.75$$

Now substituting the values of a_2 and a_1 in equation (*i*), we get

Difference of radii at the junction

$$= \frac{2 \times 125}{2 \times 10^5}[63.6 - (-43.75)]$$

$$= \frac{125}{10^5} \times 107.35 = \mathbf{0.13\ mm.} \quad \textbf{Ans.}$$

Problem 9.15. *A steel tube of 200 mm external diameter is to be shrunk on to another steel tube of 60 mm internal diameter. The diameter at the junction after shrinking is 120 mm. Before shrinking on, the difference of diameters at the junction is 0.08 mm. Calculate the radial pressure at the junction and the hoop stresses developed in the two tubes after shrinking on. Take E as 2 × 10^5 N/mm^2.*

Sol. Given :

External dia. of outer tube = 200 mm

∴ Radius, $r_2 = 100$ mm

Internal dia. of inner tube = 60 mm

∴ Radius, $r_1 = 30$ mm

The diameter at the junction after shrinking = 120 mm

∴ Radius, $r^* = 60$ mm

Before shrinking on the difference of dia. at the junction = 0.08 mm

∴ Difference of original radii = 0.04 mm

Value of $E = 2 \times 10^5$ N/mm^2

Let p^* = Radial pressure at the junction

Using equation (9.29),

Original difference of radii at junction

$$= \frac{2r^*}{E}(a_1 - a_2)$$

or
$$0.04 = \frac{2 \times 60}{2 \times 10^5}(a_1 - a_2) \quad \text{or} \quad \frac{0.04 \times 2 \times 10^5}{2 \times 60} = (a_1 - a_2)$$

or
$$(a_1 - a_2) = \frac{200}{3} \quad \text{...}(i)$$

Now using Lame's equation for outer tube

$$p_x = \frac{b_1}{x^2} - a_1 \quad ...(ii) \qquad \text{and} \qquad \sigma_x = \frac{b_1}{x^2} + a_1 \quad ...(iii)$$

At $x = 100$ mm, $p_x = 0$.

Substituting these values in equation (*ii*),

$$0 = \frac{b_1}{100^2} - a_1 = \frac{b_1}{10000} - a_1 \quad ...(iv)$$

At $x = 60$ mm, $p_x = p^*$.

Substituting these values in equation (*ii*),

$$p^* = \frac{b_1}{60^2} - a_1 = \frac{b_1}{3600} - a_1 \quad ...(v)$$

Now applying Lame's equation for inner tube

$$p_x = \frac{b_2}{x^2} - a_2 \quad ...(vi) \qquad \text{and} \qquad \sigma_x = \frac{b_2}{x^2} + a_2 \quad ...(vii)$$

At $x = 30$ mm, $p_x = 0$.

Substituting these values in equation (*vi*),

$$0 = \frac{b_2}{30^2} - a_2 = \frac{b_2}{900} - a_2 \quad ...(viii)$$

At $x = 60$ mm, $p_x = p^*$.

Substituting these values in equation (*vi*),

$$p^* = \frac{b_2}{60^2} - a_2 = \frac{b_2}{3600} - a_2 \quad ....(ix)$$

Equating the two values of p^*, given by equations (*v*) and (*ix*), we get

$$\frac{b_2}{3600} - a_2 = \frac{b_1}{3600} - a_1$$

or

$$\frac{b_2 - b_1}{3600} = a_2 - a_1 \quad ...(x)$$

But from equation (*iv*), $b_1 = 10000\, a_1$

and, from equation (*viii*), $b_2 = 900\, a_2$

Substituting these values in equation (*x*), we get

$$\frac{900\, a_2 - 10000\, a_1}{3600} = a_2 - a_1$$

or

$$900\, a_2 - 10000\, a_1 = 3600\, a_2 - 3600\, a_1$$

or

$$900\, a_2 - 3600\, a_2 = -3600\, a_1 + 10000\, a_1$$

or

$$-2700\, a_2 = 6400\, a_1$$

or

$$a_1 = -\frac{2700}{6400} a_2 = -\frac{27}{64} a_2 \quad ...(xi)$$

Substituting the value of a_1 in equation (*i*), we get

$$-\frac{27}{64} a_2 - a_2 = \frac{200}{3}$$

or

$$-\frac{(27\, a_2 + 64\, a_2)}{64} = \frac{200}{3} \quad \text{or} \quad a_2 = -\frac{200 \times 64}{3 \times 91} = -46.88$$

Substituting this value in equation (*xi*), we get

$$a_1 = +\frac{27}{64} \times 46.88 = +19.77$$

$$b_1 = 10000 \times a_1 = 10000 \times (19.77) = 197700$$

and
$$b_2 = 900 \times a_2 = -900 \times 46.88 = -42192$$

(*i*) *Radial pressure at the junction* (p^*)

Substituting the values of a_2 and b_2 in equation (*ix*), we get

$$p^* = \frac{b_2}{3600} - a_2 = -\frac{42192}{3600} + 46.88$$

$$= \mathbf{+\ 35.16\ N/mm^2.} \quad \textbf{Ans.}$$

The value of p^* can also be obtained by substituting a_1 and b_1 in equation (*v*)

$$p^* = \frac{b_1}{3600} - a_1$$

$$= \frac{197700}{3600} - 19.77 = 54.916 - 19.77$$

$$= \mathbf{35.146\ N/mm^2.} \quad \textbf{Ans.}$$

(*ii*) *Hoop stresses in the two tubes after shrinking on*

The hoop stresses can be calculated from equations (*iii*) and (*vii*).

(*a*) *For outer tube*

$$\sigma_x = \frac{b_1}{x^2} + a_1 = \frac{197700}{x^2} + 19.77 \qquad (\because\ b_1 = 197700,\ a_1 = 19.77)$$

$\therefore$
$$\sigma_{100} = \frac{197700}{100^2} + 19.77$$

$$= \mathbf{39.54\ N/mm^2\ (tensile).} \quad \textbf{Ans.}$$

and
$$\sigma_{60} = \frac{197700}{60^2} + 19.77 = \mathbf{74.68\ N/mm^2\ (tensile).} \quad \textbf{Ans.}$$

(*b*) *For inner tube*

$$\sigma_x = \frac{b_2}{x^2} + a_2$$

$$= -\frac{42192}{x^2} - 46.88 \qquad (\because\ b_2 = -42192,\ a_2 = -46.88)$$

$\therefore$
$$\sigma_{60} = -\frac{42192}{60^2} - 46.88$$

$$= \mathbf{-\ 58.6\ N/mm^2\ (compressive).} \quad \textbf{Ans.}$$

and
$$\sigma_{30} = -\frac{42192}{30^2} - 46.88$$

$$= \mathbf{-\ 93.76\ N/mm^2\ (compressive).} \quad \textbf{Ans.}$$

HIGHLIGHTS

1. The hoop stress is maximum at the inner circumference and minimum at the outer circumference of a thick cylinder.
2. The radial pressure (p_x) and hoop stress (σ_x) at any radius x in case of a thick cylinder are given by

$$p_x = \frac{b}{x^2} - a \quad \text{and} \quad \sigma_x = \frac{b}{x^2} + a .$$

The above equations are known as Lame's equations. In these equations a and b are constants whose values are obtained from boundary conditions.

3. The hoop stresses in a compound thick cylinder is the algebraic sum of the hoop stresses caused due to shrinking and those due to internal fluid pressure.
4. The hoop stress in case of thin cylinders are reduced by wire winding on the cylinders.
5. The hoop stress in case of thick cylinders are reduced by shrinking one cylinder over another cylinder.
6. The original difference of radii at the junction of a compound cylinder for shrinkage is given by,

$$dr = \frac{2r^*}{E}(a_1 - a_2)$$

where r^* = Radius of junction after shrinking

E = Young's modulus

a_1 and a_2 = Constants whose values are obtained from boundary conditions.

7. The hoop stress (or circumferential stress) in a rotating thin cylinder is given by

$$\sigma = \rho \times \omega^2 \times r^2 \quad \text{or} \quad \rho \times v^2.$$

8. For a rotating solid disc, the stresses at any radius r are :

$$\sigma_c = \frac{\rho \times \omega^2}{8} [(3 + \mu) r_2^2 - (1 + 3\mu) r^2]$$

and

$$\sigma_r = \frac{\rho \times \omega^2}{8} (3 + \mu) [r_2^2 - r^2]$$

where σ_c = Circumferential stress,

σ_r = Radial stress,

ω = Angular velocity = $\frac{2\pi N}{60}$.

μ = Poisson's ratio.

9. At the centre of rotating solid disc, the radial and circumferential stress are maximum and are equal.

$$(\sigma_r)_{max} = (\sigma_c)_{max} = \frac{\rho \times \omega^2}{8} (3 + \mu) r_2^2.$$

10. The circumferential stress at outer radius of a solid disc is given by

$$\sigma_c = \frac{\rho \times \omega^2 \times r_2^2}{4} (1 - \mu).$$

11. For a rotating disc with a central hole :

(*i*) The stresses at any radius are :

$$\sigma_r = \frac{\rho \times \omega^2}{8}(3 + \mu)\left[(r_1^2 + r_2^2) - \frac{r_1^2 r_2^2}{r^2} - r^2\right]$$

and
$$\sigma_c = \frac{\rho \times \omega^2}{8}(3 + \mu)\left[(r_1^2 + r_2^2) + \frac{r_1^2 r_2^2}{r^2} - \left(\frac{1 + 3\mu}{3 + \mu}\right) r^2\right]$$

(*ii*) The circumferential stress is maximum at inner radius and is given by

$$(\sigma_c)_{max} = \frac{\rho \times \omega^2}{4}(3 + \mu)\left[r_2^2 + \left(\frac{1 - \mu}{3 + \mu}\right) r_1^2\right]$$

(*iii*) The radial stress is maximum at the radius given by

$$r = \sqrt{r_1 \times r_2}$$

(*iv*) The maximum radial stress is given by

$$(\sigma_r)_{max} = \frac{\rho \times \omega^2}{8}(3 + \mu)\,[(r_2 - r_1)^2]$$

(*v*) The value of σ_c at the outer radius is

$$(\sigma_c)_{at\ r = r_2} = \frac{\rho \times \omega^2}{4}(3 + \mu)\left[r_1^2 + \left(\frac{1 + \mu}{3 + \mu}\right) r_2^2\right].$$

12. The maximum radial and circumferential stresses in a disc with pin hole at the centre are :

$$(\sigma_r)_{max} = \frac{\rho \times \omega^2 \times r_2^2}{8}(3 + \mu)$$

and
$$(\sigma_c)_{max} = \frac{\rho \times \omega^2 \times r_2^2}{4}(3 + \mu).$$

13. The maximum circumferential stress in a disc with a pin hole at the centre is two times the maximum circumferential stress in a solid disc.

14. For a rotating solid long cylinder :

(*i*) stresses at any radius are

$$\sigma_c = \frac{\rho \times \omega^2}{8}\left(\frac{3 - 2\mu}{1 - \mu}\right)\left[r_2^2 - \left(\frac{1 + 2\mu}{3 - 2\mu}\right) r^2\right]$$

and
$$\sigma_r = \frac{\rho \times \omega^2}{8}\left(\frac{3 - 2\mu}{1 - \mu}\right)[r_2^2 - r^2]$$

(*ii*) σ_r is maximum at the centre and is given by

$$(\sigma_r)_{max} = \frac{\rho \times \omega^2}{8}\left(\frac{3 - 2\mu}{1 - \mu}\right) \times r_2^2$$

(*iii*) σ_c is maximum at the centre and is given by

$$(\sigma_c)_{max} = \frac{\rho \times \omega^2}{8}\left(\frac{3 - 2\mu}{1 - \mu}\right) \times r_2^2$$

(*iv*)
$$(\sigma_r)_{max} = (\sigma_c)_{max} = \frac{\rho \times \omega^2}{8}\left(\frac{3 - 2\mu}{1 - \mu}\right) \times r_2^2.$$

15. For a rotating hollow cylinder :

(*i*) The stresses at any radius are :

$$\sigma_c = \frac{\rho \times \omega^2}{8}\left(\frac{3-2\mu}{1-\mu}\right)\left[r_1^2 + r_2^2 + \frac{r_1^2\, r_2^2}{r^2} - \left(\frac{1+2\mu}{3-2\mu}\right)r^2\right]$$

$$\sigma_r = \frac{\rho \times \omega^2}{8}\left(\frac{3-2\mu}{1-\mu}\right)\left[r_1^2 + r_2^2 - \frac{r_1^2\, r_2^2}{r^2} - r^2\right]$$

(*ii*) σ_c will be maximum at inner radius and is equal to

$$(\sigma_c)_{\max} = \frac{\rho \times \omega^2}{8}\left(\frac{3-2\mu}{1-\mu}\right)\left[r_1^2 + 2r_2^2 - \left(\frac{1+2\mu}{3+2\mu}\right)r_1^2\right]$$

(*iii*) σ_r is maximum at $r = \sqrt{r_1 \times r_2}$ and is given by

$$(\sigma_r)_{\max} = \frac{\rho \times \omega^2}{8}\left(\frac{3-2\mu}{1-\mu}\right)[(r_2 - r_1)^2].$$

EXERCISE 9

(A) Theoretical Questions

1. Differentiate between a thin cylinder and a thick cylinder. Find an expression for the radial pressure and hoop stress at any point in case of a thick cylinder.
2. What do you mean by Lame's equations ? How will you derive these equations ?
3. "The hoop stress is minimum at the outer surface and is maximum at the inner surface of a thick cylinder", prove this statement. Sketch the radial pressure distribution and hoop stress distribution across the section of a thick cylinder.
4. What do you mean by a thick compound cylinder ? How will you determine the hoop stresses in a thick compound cylinder ?
5. What are the different methods of reducing hoop stresses ? Explain the terms : Wire winding of thin cylinders and shrinking one cylinder over another cylinder.
6. Prove that the original difference in radii at the junction of a compound cylinder for shrinkage is given by

$$dr = \frac{2r^*}{E}(a_1 - a_2)$$

where r^* = Common radius after shrinking

E = Young's modulus

a_1 and a_2 = Constants whose values are obtained from the boundary conditions.

7. Find an expression for the circumferential stress developed in a thin rotating cylinder.
8. Find an expression for the circumferential and radial stresses developed in a rotating solid disc.
9. Prove that the circumferential and radial stresses are maximum and are equal at the centre of a rotating solid disc.
10. Prove that the circumferential stress at the outer radius of a rotating solid disc is given by

$$\sigma_c = \frac{\rho \times \omega^2 \times r_2^2}{4}(1-\mu).$$

11. Prove that the radius at which radial stress is maximum in a rotating disc with a central hole is given by $r = \sqrt{r_1 \times r_2}$.

Also find an expression for the maximum radial stress.

12. Prove that the circumferential stress in a rotating disc with a pin hole at the centre is two times the maximum circumferential stress in a rotating solid disc.
13. Prove that in case of a long solid rotating cylinder the maximum radial and circumferential stresses at the centre are equal and they are $\frac{\rho \times \omega^2}{8}\left(\frac{3-2\mu}{1-\mu}\right) \times r_2^2$ in magnitude.
14. Prove that in case of a rotating hollow cylinder, the radial stress is maximum at radius = $\sqrt{r_1 \times r_2}$ and is given by

$$(\sigma_r)_{max} = \frac{\rho \times \omega^2}{8}\left(\frac{3-2\mu}{1-\mu}\right)[(r_2 - r_1)^2].$$

(B) Numerical Problems

1. Determine the maximum hoop stress across the section of a pipe of external diameter 600 mm and internal diameter 440 mm, when the pipe is subjected to an internal fluid pressure of 0 N/mm^2.
[Ans. 99.9 N/mm^2]
2. Find the thickness of metal necessary for a cylindrical shell of internal diameter 150 mm to withstand an internal pressure of 50 N/mm^2. The maximum hoop stress in the section is not to exceed 150 N/mm^2.
[Ans. 31 mm] *(Annamalai University, 1990)*

[Hint. $p_x = \frac{b}{x^2} - a$. And $\sigma_x = \frac{b}{x^2} + a$.

Let r_2 = External radius in mm, r_1 = Internal radius.

And x is any radius.

At x = 75 mm, p_x = 50 N/mm^2 and σ_x = 150 N/mm^2

$$\therefore \quad 50 = \frac{b}{5625} - a \quad ...(i) \qquad 150 = \frac{b}{5625} + a \quad ...(ii)$$

Solve equations (*i*) and (*ii*). The values of a and b are a = 50 and b = 562500. Now at $x = r_2$, $p_x = 0$

$\therefore \quad 0 = \frac{b}{r_2^2} - a$. Find r_2. Then r_2 = 106 mm. Now t = 106 – 75 = 31 mm.]

3. A compound cylinder is made by shrinking a cylinder of external diameter 200 mm and internal diameter 160 mm over another cylinder of external diameter 160 mm and internal diameter 120 mm. The radial pressure at the junction after shrinking is 8 N/mm^2. Find the final stresses set up across the section, when the compound cylinder is subjected to an internal fluid pressure of 60 N/mm^2.
[Ans. Inner F_{60} = 90.9 and F_{80} = 57.9 N/mm^2 ; Outer F_{80} = 122.9 and F_{100} = 25.9 N/mm^2]
4. Find the hoop stress developed in the thin rim of a wheel of radius 400 mm, when it is rotating at a speed of 3000 r.p.m. The density of the material of wheel is 8000 kg/m^3.
[Ans. 126.33 MN/m^2]
5. A steel disc of uniform thickness and diameter 900 mm is rotating about its axis at 2400 r.p.m. Determine :
(*i*) radial and circumferential stresses at the centre, and
(*ii*) circumferential stress at the outer radius.
Take ρ = 7800 kg/m^2 and Poisson's ratios = 0.3 **[Ans.** (*i*) 41.152 MN/m^2, (*ii*) 17.46 MN/m^2]
6. If for the question 2, the disc is having a central hole of 200 mm diameter, then determine :
(*i*) circumferential stress at inner radius and outer radius.
(*ii*) radius at which radial stress is maximum and

(*iii*) maximum radial stress

(*iv*) maximum shear stress.

[**Ans.** (*i*) 82.16 MN/m^2, 21.5 MN/m^2, (*ii*) 0.212 m, (*iii*) 24.89 MN/m^2, (*iv*) 41.58 MN/m^2]

7. For the data given in question 2, plot the variation of circumferential and radial stresses along the radius.

8. For the data given in question 3, plot the variation of circumferential and radial stresses along the radius.

9. A long cylinder is rotating at a speed of 4200 r.p.m. The diameter of the cylinder is 400 mm and density of its material is 7800 kg/m^3. If Poisson's ratio = 0.3, then

(*i*) calculate the maximum stress in the cylinder and

(*ii*) plot the variation of circumferential and radial stresses along the radius.

[**Ans.** (*i*) 25.866 MN/m^2]

10. If in question 9 above, the long rotating cylinder is having a central hole of diameter 100 mm, then determine :

(*i*) maximum circumferential stress,

(*ii*) radius at which radial stress is maximum,

(*iii*) maximum radial stress,

(*iv*) variation of circumferential and radial stresses along the radius of the cylinder. Take all data from question 9. [**Ans.** (*i*) 52.27 MN/m^2, (*ii*) 0.1 m, (*iii*) 14.55 MN/m^2]

11. Calculate : (*i*) maximum and minimum circumferential stress, (*ii*) maximum and minimum radial stress in a thin uniform disc of inner diameter 50 mm and outer diameter 250 mm when rotating at 8000 r.p.m. Take μ = 0.33 and ρ = 8000 kg/m^3.

[**Ans.** (*i*) 74.9 MN/m^2, 17.9 MN/m^2, (*ii*) 23.8 MN/m^2, 0]

UNIT V

Chapters

10

Curved Beams

10.1. INTRODUCTION

For a straight beam, the bending equation $\frac{M}{I} = \frac{\sigma}{y} = \frac{E}{R}$ has been derived in chapter 3. This equation can be applied, with sufficient accuracy, to the beams or bars having small initial curvature. However the machine members such as crane hooks, chain links and rings etc., which are having large initial curvature, the simple bending equation cannot be used. This chapter deals with the bending of curved bars, which are having large initial curvature. Actually curved bars means the bars of large initial curvature.

Generally, if the radius of curvature is more than 5 times the depth of the beam, the beam is said to be having small initial curvature. But if radius of curvature is less than 5 times the depth, the beam is said to be having large initial curvature. Hence for large initial curvature, the radius of curvature is small. Also for curved beams, the **neutral and centroidal axes** donot coincide.

10.2. ASSUMPTIONS MADE IN THE DERIVATION OF STRESSES IN A CURVED BAR

Before deriving the expression for the stresses developed in a curved bar when it is subjected to some bending moments, the following assumption are made :

1. Transverse sections which are plane before bending remain plane after bending.

2. Hook's law is applicable. This means the working stresses are below the limit of proportionality.

3. The longitudinal fibres of the bar, parallel to centroidal axis exert no pressure on each other. This means the distance between any longitudinal fibre from centroidal axis is same before and after bending.

4. Each layer of the beam is free to expand or contract, independently of the layer, above or below it.

10.3. BENDING OF BEAMS WITH LARGE INITIAL CURVATURE

Fig. 10.1 shows the two positions of a curved bar, one position [*i.e.*, Fig. 10.1 (*a*)] is before bending whereas the second position [*i.e.*, Fig. 10.1 (*b*)] is after bending when some moment M is applied at the ends of the bar. The centroidal axis is shown by line LM and LM' in the two positions respectively.

Consider any fibre EF at a distance 'y' from the centroidal axis LM. This fibre takes the position of EF' when the moment M is applied. Then the fibre EF' will be at a distance of 'y'' from the centroidal axis LM'.

Refer to Fig. 10.1 (*a*). Here :

R = radius of curvature of centroidal axis LM,

θ = angle subtended by centroidal axis LM at the centre of curvature O,

y = distance of fibre EF from centroidal axis LM.

Refer to Fig. 10.1 (*b*). Here :

M = uniform bending moment applied to the bar,

R' = radius of curvature of centroidal axis LM',

θ' = angle subtended by centroid axis LM' at the centre of curvature O',

y' = distance of fibre EF' from centroidal axis LM'.

Let σ = stress in the strained fibre EF' due to bending moment M, and

e = strain in the same fibre EF',

e_0 = strain in centroidal layer LM.

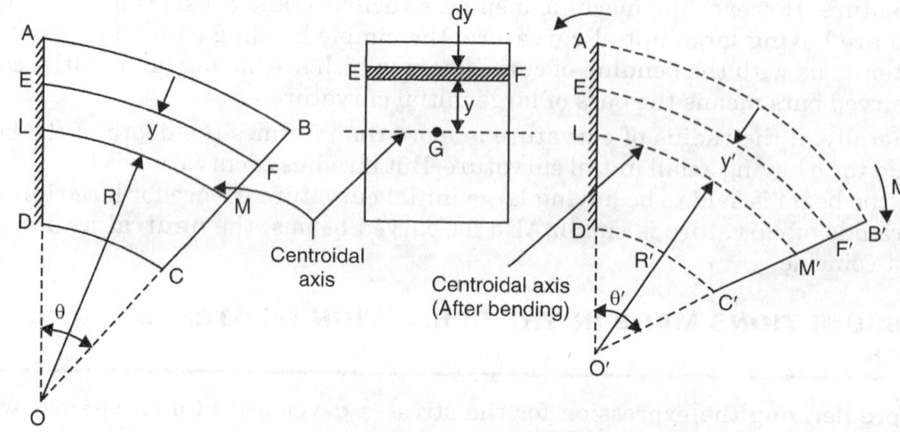

(*a*) Curved bar ABCD before bending (*b*) Curved bar AB'C'D after bending

Fig. 10.1

Now, from Figs. 10.1 (*a*) and 10.1 (*b*), we have

$$EF = (R + y) \times \theta \quad \text{and} \quad EF' = (R' + y') \times \theta'$$

Also $$LM = R \times \theta \quad \text{and} \quad LM' = R' \times \theta'$$

$\therefore$ Strain,
$$e = \frac{EF' - EF}{EF} = \frac{(R' + y') \times \theta' - (R + y) \times \theta}{(R + y) \times \theta}$$

$$= \frac{(R' + y') \times \theta'}{(R + y) \times \theta} - \frac{(R + y) \times \theta}{(R + y) \times \theta} = \frac{(R' + y')}{(R + y)} \times \frac{\theta'}{\theta} - 1$$

or
$$1 + e = \frac{(R' + y')}{(R + y)} \times \frac{\theta'}{\theta} \qquad \text{...(i)}$$

Also, strain,
$$e_0 = \frac{LM' - LM}{LM} = \frac{R' \times \theta' - R \times \theta}{R \times \theta} = \frac{R'}{R} \times \frac{\theta'}{\theta} - 1$$

or
$$1 + e_0 = \frac{R'}{R} \times \frac{\theta'}{\theta} \qquad \text{...(ii)}$$

Dividing equation (*i*) by equation (*ii*), we get

$$\frac{1+e}{1+e_0} = \frac{(R'+y')}{(R+y)} \times \frac{\theta'}{\theta} \Big/ \left(\frac{R'}{R} \times \frac{\theta'}{\theta}\right)$$

$$= \frac{(R'+y')}{(R+y)} \times \frac{R}{R'} \qquad \left(\text{Cancelling } \frac{\theta'}{\theta} \text{ from numerator and denominator}\right)$$

$$= \frac{\left(\dfrac{R'+y'}{R'}\right)}{\left(\dfrac{R+y}{R}\right)} = \frac{1+\dfrac{y'}{R'}}{1+\dfrac{y}{R}}$$

$$\therefore \qquad 1 + e = (1 + e_0)\left[\frac{1+\dfrac{y'}{R'}}{1+\dfrac{y}{R}}\right]$$

or

$$e = (1 + e_0)\left[\frac{1+\dfrac{y'}{R'}}{1+\dfrac{y}{R}}\right] - 1$$

$$= \frac{(1+e_0)\left(1+\dfrac{y'}{R'}\right) - \left(1+\dfrac{y}{R}\right)}{\left(1+\dfrac{y}{R}\right)} = \frac{1+\dfrac{y'}{R'}+e_0+e_0\times\dfrac{y'}{R'}-1-\dfrac{y}{R}}{\left(1+\dfrac{y}{R}\right)}$$

$$= \frac{\dfrac{y'}{R'}+e_0+e_0\dfrac{y'}{R'}-\dfrac{y}{R}}{\left(1+\dfrac{y}{R}\right)} = \frac{\dfrac{y'}{R'}+e_0\dfrac{y'}{R'}+e_0-\dfrac{y}{R}}{\left(1+\dfrac{y}{R}\right)}$$

But according to assumption (3), $y = y'$

$$\therefore \qquad e = \frac{\dfrac{y}{R'}+e_0\dfrac{y}{R'}+e_0-\dfrac{y}{R}}{\left(1+\dfrac{y}{R}\right)}$$

$$= \frac{\dfrac{y}{R'}+e_0\dfrac{y}{R'}+e_0-\dfrac{y}{R}+e_0\dfrac{y}{R}-e_0\dfrac{y}{R}}{\left(1+\dfrac{y}{R}\right)}$$

$$\left[\text{Adding and subtracting the term } \left(e_0\frac{y}{R}\right)\right]$$

$$= \frac{e_0+e_0\dfrac{y}{R}+\dfrac{y}{R'}-\dfrac{y}{R}+e_0\dfrac{y}{R'}-e_0\dfrac{y}{R}}{\left(1+\dfrac{y}{R}\right)}$$

$$= \frac{e_0\left(1+\dfrac{y}{R}\right)+y\left(\dfrac{1}{R'}-\dfrac{1}{R}\right)+e_0 y\left(\dfrac{1}{R'}-\dfrac{1}{R}\right)}{\left(1+\dfrac{y}{R}\right)}$$

$$= \frac{e_0\left(1+\frac{y}{R}\right)+\left(\frac{1}{R'}-\frac{1}{R}\right)(y+e_0 y)}{\left(1+\frac{y}{R}\right)}$$

$$= e_0 + \frac{\left(\frac{1}{R'}-\frac{1}{R}\right)(y+e_0 y)}{\left(1+\frac{y}{R}\right)} = e_0 + \frac{y(1+e_0)\left(\frac{1}{R'}-\frac{1}{R}\right)}{\left(1+\frac{y}{R}\right)} \quad ...(iii)$$

The equation (*iii*) gives the expression for the strain produced in fibre *EF* under the action of moment *M*.

Now the expression for stress produced in the fibre *EF* can be obtained.

We know that $\frac{\text{Stress}}{\text{Strain}} = E$

or $\text{stress} = E \times \text{strain}$

or $\sigma = E \times e$

$$= E \times \left[e_0 + \frac{y(1+e_0)\left(\frac{1}{R'}-\frac{1}{R}\right)}{\left(1+\frac{y}{R}\right)}\right] \quad ...(iv)$$

The above stress is produced due to bending moment *M*. Hence this stress is known as bending stress.

Consider a small strip of area *dA* at a distance *y* from centroidal axis *LM*.

The force on the strip $= \text{stress} \times \text{area of strip}$

$= \sigma \times dA$

Moment of this force about centroidal axis $= \text{Force} \times y$

$= (\sigma \times dA) \times y$

The above expression gives the resisting moment offered by the strip. Total resisting moment is obtained by integrating the above equation.

$\therefore$ Total resisting moment $= \int (\sigma \times dA \times y)$

For equilibrium, the total resisting moment must be equal to applied moment *M*.

$$\therefore \quad M = \int \sigma \times dA \times y \quad \text{or} \quad \int \sigma \times y \times dA$$

$$= \int E \times \left[e_0 + \frac{y(1+e_0)\left(\frac{1}{R'}-\frac{1}{R}\right)}{\left(1+\frac{y}{R}\right)}\right] \times y \times dA$$

$$\left[\because \text{ From equation } (iv), \quad \sigma = E \times \left[e_0 + \frac{y(1+e_0)\left(\frac{1}{R'}-\frac{1}{R}\right)}{\left(1+\frac{y}{R}\right)}\right]\right]$$

$$= E \int e_0 \, y \, dA + E \int \frac{y^2 (1 + e_0) \left(\frac{1}{R'} - \frac{1}{R} \right)}{\left(1 + \frac{y}{R} \right)} dA$$

$$= 0 + E(1 + e_0) \left(\frac{1}{R'} - \frac{1}{R} \right) \int \frac{y^2 \, dA}{\left(1 + \frac{y}{R} \right)}$$

$$\left[\because \ (i) \int y \, dA = 0 \quad \text{and} \quad (ii) \, e_0, R' \text{ and } R \text{ are constants} \right]$$

$$= E(1 + e_0) \left(\frac{1}{R'} - \frac{1}{R} \right) \int \frac{y^2 \, dA}{\left(1 + \frac{y}{R} \right)}$$

Let $\int \frac{y^2 \, dA}{\left(1 + \frac{y}{R} \right)} = Ah^2$ ...(v)

where h^2 = a constant for the cross-section of the bar

$\therefore$ $$M = E(1 + e_0) \left(\frac{1}{R'} - \frac{1}{R} \right) \times A \times h^2$$ ...(vi)

In the above equation e_0 and R' are unknown.

Let us find the value of e_0 and R'

We have seen that force on the strip of area $dA = \sigma \times dA$

$\therefore$ Total force on the cross-section will be obtained by integrating the above equation.

$\therefore$ Total force, $F = \int \sigma \times dA = \int E \times \left[e_0 + \frac{y(1 + e_0) \left(\frac{1}{R'} - \frac{1}{R} \right)}{\left(1 + \frac{y}{R} \right)} \right] dA$

$$\left[\text{From equation } (iv), \ \sigma = E \times \left[e_0 + \frac{y(1 + e_0) \left(\frac{1}{R'} - \frac{1}{R} \right)}{\left(1 + \frac{y}{R} \right)} \right] \right]$$

$$= E \int e_0 \, dA + E \int \frac{y(1 + e_0) \left(\frac{1}{R'} - \frac{1}{R} \right)}{\left(1 + \frac{y}{R} \right)} dA$$

$$= E \times e_0 \times A + E(1 + e_0) \left(\frac{1}{R'} - \frac{1}{R} \right) \int \frac{y}{\left(1 + \frac{y}{R} \right)} dA.$$

where A = Area of cross-section of the beam.

But for pure bending, there is no force on the section of the beam.

$\therefore$ $F = 0$

or $$0 = E \times e_0 \times A + E(1 + e_0)\left(\frac{1}{R'} - \frac{1}{R}\right)\int \frac{y}{\left(1+\frac{y}{R}\right)} dA \quad \text{...(vii)}$$

Let us now find the value of $\int \frac{y}{\left(1+\frac{y}{R}\right)} dA$.

We know that $\int \frac{y^2}{\left(1+\frac{y}{R}\right)} dA = Ah^2$ from equation (v). Hence let us use this value to find the value of $\int \frac{y}{\left(1+\frac{y}{R}\right)} dA$

$$\int \frac{y}{\left(1+\frac{y}{R}\right)} dA = \int \frac{\left(y + \frac{y^2}{R} - \frac{y^2}{R}\right)}{\left(1+\frac{y}{R}\right)} dA$$

$$\left(\text{Adding and subtracting } \frac{y^2}{R} \text{ to the numerator}\right)$$

$$= \int \frac{\left[y\left(1+\frac{y}{R}\right) - \frac{y^2}{R}\right]}{\left(1+\frac{y}{R}\right)} dA = \int \left[y - \frac{1}{R} \times \frac{y^2}{\left(1+\frac{y}{R}\right)}\right] dA$$

$$= \int y\, dA - \int \frac{1}{R} \times \frac{y^2}{\left(1+\frac{y}{R}\right)} dA$$

$$= 0 - \frac{1}{R}\int \frac{y^2}{\left(1+\frac{y}{R}\right)} dA \qquad \left[\because \int y\, dA = 0\right]$$

$$= -\frac{1}{R} \times Ah^2 \qquad \left[\because \int \frac{y^2}{\left(1+\frac{y}{R}\right)} dA = Ah^2\right]$$

Substituting the value of $\int \frac{y\,.\,dA}{\left(1+\frac{y}{R}\right)}$ in equation (vii), we get

$$0 = E \times e_0 \times A + E(1 + e_0)\left(\frac{1}{R'} - \frac{1}{R}\right) \times \left(-\frac{1}{R} \times Ah^2\right)$$

or $$E(1 + e_0)\left(\frac{1}{R'} - \frac{1}{R}\right) \times \frac{1}{R} \times Ah^2 = E \times e_0 \times A$$

or $$(1+e_0)\left(\frac{1}{R'}-\frac{1}{R}\right)\times\frac{h^2}{R}=e_0$$ [Cancelling $E \times A$ to both sides]

or $$(1+e_0)\left(\frac{1}{R'}-\frac{1}{R}\right)=\frac{e_0\times R}{h^2} \quad \text{...(viii)}$$

Substituting the value of $(1+e_0)\left(\frac{1}{R'}-\frac{1}{R}\right)$ in equation (*vi*),

We get $$M = E\times\frac{e_0\times R}{h^2}\times A\times h^2$$
$$= E\times e_0\times R\times A$$

or $$e_0=\frac{M}{E\times R\times A} \quad \text{...(ix)}$$

From equation (*ix*) it is clear that e_0 is constant for a given value of M, A and R.

Now the expression for stress (σ) can be obtained by substituting the value of $(1+e_0)$ $\left(\frac{1}{R'}-\frac{1}{R}\right)$ from (*viii*) in equation (*iv*).

$$\therefore \quad \sigma = E\times\left[e_0+\frac{y\times\left(\frac{e_0\times R}{h^2}\right)}{\left(1+\frac{y}{R}\right)}\right]$$

$$\left[\because\ (1+e_0)\left(\frac{1}{R'}-\frac{1}{R}\right)=\frac{e_0\times R}{h^2} \text{ from equation (viii)}\right]$$

$$= E\times e_0\left[1+\frac{\frac{y\times R}{h^2}}{\left(1+\frac{y}{R}\right)}\right]$$

$$= E\times\frac{M}{ERA}\left[1+\frac{\left(\frac{y\times R}{h^2}\right)}{\left(\frac{R+y}{R}\right)}\right] \qquad \left[\because \text{ From equation (ix) } e_0=\frac{M}{ERA}\right]$$

$$= \frac{M}{R\times A}\left[1+\frac{R^2}{h^2}\left(\frac{y}{R+y}\right)\right] \quad \text{...(10.1)}$$

The equation (10.1), gives the expression for tensile stress. The value of '*y*' will be –ve, on the downward side of centroidal axis *LM*. Hence stress will be opposite to the stress given by equation (10.1).

$$\therefore \quad \sigma\text{ (compressive)} = \frac{M}{RA}\left[1+\frac{R^2}{h^2}\left(\frac{-y}{R-y}\right)\right]$$

$$= \frac{M}{RA}\left[1-\frac{R^2}{h^2}\times\left(\frac{y}{R-y}\right)\right] \quad \text{...(10.2)}$$

The expression for the stress given by equation (10.1) or (10.2) is known as **Winkler-Bach Formula.** The distribution of the stress given by equation (10.1) or (10.2) is hyperbolic (and not linear as in the case of straight beams). The stress distribution is shown in Fig. 10.2.

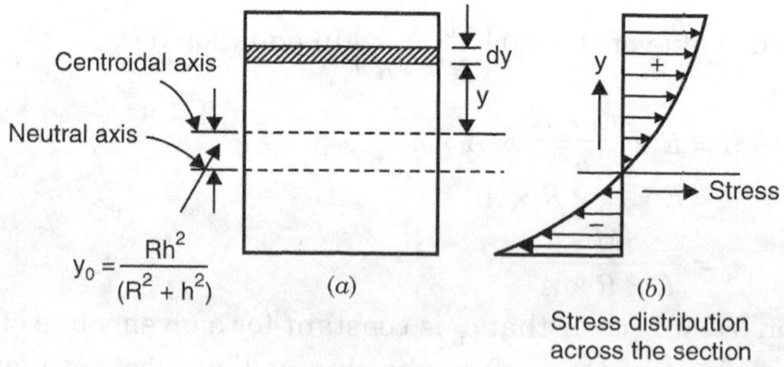

Fig. 10.2

10.4. POSITION OF NEUTRAL AXIS

Let y_0 = distance of neutral axis from centroidal axis.

The stress (σ) is zero at the neutral axis. Hence position of neutral axis is obtained by substituting $\sigma = 0$ and $y = y_0$ in equation (10.1).

$$\therefore \quad 0 = \frac{M}{R \times A}\left[1 + \frac{R^2}{h^2}\left(\frac{y_0}{R + y_0}\right)\right] = 0 \quad \text{or} \quad 1 + \frac{R^2}{h^2}\left(\frac{y_0}{R + y_0}\right) = 0$$

$$\text{or} \quad \frac{R^2}{h^2}\left(\frac{y_0}{R + y_0}\right) = -1 \quad \text{or} \quad R^2 y_0 = -h^2 \times R - h^2 \times y_0$$

$$\text{or} \quad y_0(R^2 + h^2) = -h^2 \times R$$

$$\text{or} \quad y_0 = -\frac{R \times h^2}{R^2 + h^2} \qquad \text{...(10.3)}$$

But 'y_0' is the distance from centroidal axis . As y_0 is –ve, this means the neutral axis will be at a distance of 'y_0' below the centroidal axis. Fig. 10.2 (*a*) shows the position of neutral axis and Fig. 10.2 (*b*) shows the stress distribution.

10.4.1. Sign Convention. The stress across the cross-section will be tensile or compressive according to the following conventions :

1. The bending moment M will be positive if it increases the curvature of the beam and negative if it decreases the curvature. (Increase of curvature of beam will decrease the radius of curvature).

2. The distance 'y' is positive if it is measured from the centroidal axis towards the convex side of beam, and negative when measured towards the concave side (or towards the centre of curvature).

3. The positive stress means tensile stress, whereas negative stress means compressive stress.

10.4.2. Determination of Factor h² for Various Sections

From equation (v), we have

$$\int \frac{y^2\, dA}{\left(1+\frac{y}{R}\right)} = Ah^2$$

or

$$h^2 = \frac{1}{A}\int \frac{y^2}{\left(1+\frac{y}{R}\right)}\, dA = \frac{1}{A}\int \frac{Ry^2}{R+y}\, dA = \frac{R}{A}\int \frac{y^2}{(R+y)}\, dA$$

There is some difficulty in integrating the term $\frac{y^2}{(R+y)}$, as y is existing in numerator and denominator. Hence y^2 should be illiminated from numerator so that integration can be done easily. By adding and subtracting 'R^2' to y^2, integration becomes easy.

$$\therefore \quad h^2 = \frac{R}{A}\int \left(\frac{y^2+R^2-R^2}{R+y}\right) dA = \frac{R}{A}\int \left(\frac{y^2-R^2}{(R+y)} + \frac{R^2}{R+y}\right) dA$$

$$= \frac{R}{A}\int \left[\frac{(y-R)(y+R)}{(R+y)} + \frac{R^2}{R+y}\right] dA = \frac{R}{A}\int \left[(y-R) + \frac{R^2}{R+y}\right] dA$$

$$= \frac{R}{A}\left[\int y\, dA - \int R\, dA + \int \left(\frac{R^2}{R+y}\right) dA\right]$$

$$= \frac{R}{A}\left[0 - R \times A + \int \frac{R^2}{(R+y)}\, dA\right]$$

$$\left(\because \int y\, dA = 0, \int R\, dA = R\int dA = R \times A\right)$$

$$= -R^2 + \frac{R}{A}\int \left(\frac{R^2}{R+y}\right) dA$$

$$= \frac{R^3}{A}\int \left(\frac{1}{R+y}\right) dA - R^2 \qquad \text{...(10.4)}$$

By substituting the values of 'y' and dA for various section in equation (10.4), the value of h^2 is obtained. After knowing the value of h^2, the position of neutral axis is obtained from equation

$$y_0 = -\frac{R \times h^2}{R^2 + h^2}$$

10.5. POSITION OF NEUTRAL AXIS FOR RECTANGULAR SECTION

A rectangular section of a curved beam of width 'b' and depth 'd' is shown in Fig. 10.3. The centre of curvature of the curved beam is at O. The radius of curvature from centroidal axis is R. Consider an elementary strip of width 'b' and depth 'dy' at a distance 'y' from the centroidal axis.

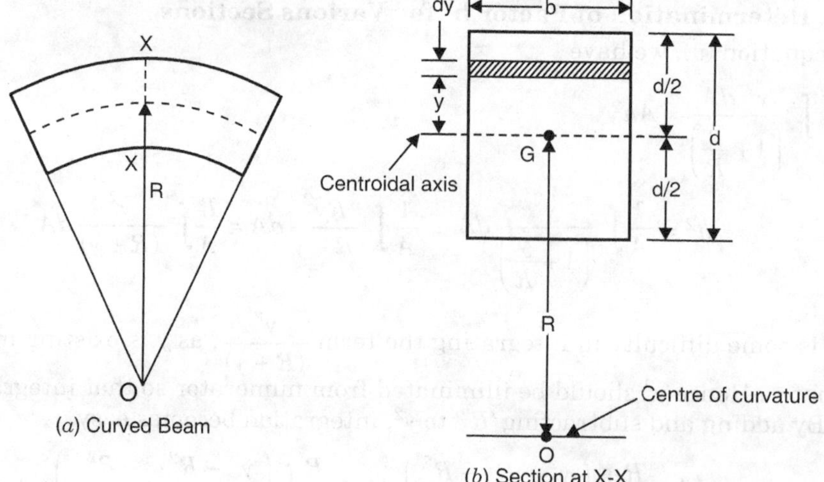

(*a*) Curved Beam

(*b*) Section at X-X

Fig. 10.3

Now area of strip, $dA = b \times dy$

Area of section, $A = b \times d$

From equation (10.4), we have

$$h^2 = \frac{R^3}{A} \int \left(\frac{1}{R+y} \right) dA - R^2 = \frac{R^3}{b \times d} \int_{-d/2}^{+d/2} \frac{1 \times b \times dy}{(R+y)} - R^2$$

$$(\because \quad dA = b \times dy)$$

$$= \frac{R^3 \times b}{b \times d} \int_{-d/2}^{+d/2} \left(\frac{1}{R+y} \right) dy - R^2 = \frac{R^3}{d} \Big[\log_e (R+y) \Big]_{-d/2}^{d/2} - R^2$$

$$= \frac{R^3}{d} \left[\log_e \left(R + \frac{d}{2} \right) - \log_e \left(R - \frac{d}{2} \right) \right] - R^2$$

$$= \frac{R^3}{d} \left[\log_e \frac{(R + d/2)}{(R - d/2)} \right] - R^2 = \frac{R^3}{d} \left[\log_e \frac{(2R+d)}{(2R-d)} \right] - R^2 \qquad \text{...(10.5)}$$

Also $\log_e \left(\frac{2R+d}{2R-d} \right) = \log_e \left(\frac{R+d/2}{R-d/2} \right) = \frac{d}{R} \left[1 + \frac{1}{3} \left(\frac{d}{2R} \right)^2 + \frac{1}{5} \left(\frac{d}{2R} \right)^4 + \ldots\ldots \right]$

$$\therefore \quad h^2 = \frac{R^3}{d} \left[\frac{d}{R} \left\{ 1 + \frac{1}{3} \left(\frac{d}{2R} \right)^2 + \frac{1}{5} \left(\frac{d}{2R} \right)^4 + \ldots\ldots \right\} \right] - R^2$$

$$= R^2 \left[1 + \frac{1}{3} \left(\frac{d}{2R} \right)^2 + \frac{1}{5} \left(\frac{d}{2R} \right)^4 + \ldots\ldots \right] - R^2$$

$$= R^2 \left[\frac{1}{3} \left(\frac{d}{2R} \right)^2 + \frac{1}{5} \left(\frac{d}{2R} \right)^4 + \ldots\ldots \right] \qquad \text{...[10.5 (A)]}$$

The position of N.A. is given by

$$y_0 = \frac{-Rh^2}{R^2 + h^2}$$

10.6. POSITION OF NEUTRAL AXIS FOR TRIANGULAR SECTION

A triangular section of a curved beam of width 'b' and height 'd' is shown in Fig. 10.4. The centre of curvature of the curved beam is at O. The radius of curvature from the centroidal axis is R. Consider a strip of width b^* and depth 'dy' at a distance 'y' from the centroidal axis and also at a distance 'r' from the axis of curvature.

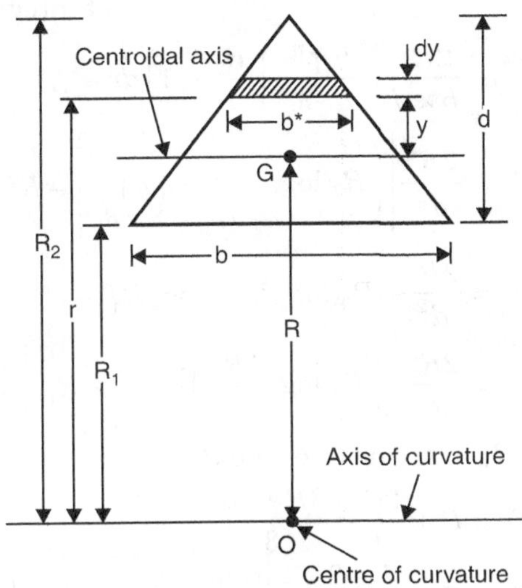

Fig. 10.4

$$\therefore \qquad r = R + y$$

and
$$dr = 0 + dy = dy$$

Let R_1 = distance between base of triangle and axis of curvature.

R_2 = distance between apex of triangle and axis of curvature.

Area of triangle, $A = \dfrac{b \times d}{2}$

Area of strip, $dA = b^* \times dy$

$$= \frac{b}{d}(R_2 - r) \times dy = \frac{b}{d}(R_2 - r) \times dr \qquad (\because \; dy = dr)$$

where b^* is obtained from

$$\frac{b^*}{b} = \frac{(R_2 - r)}{d}$$

or
$$b^* = \frac{b}{d}(R_2 - r)$$

Now from equation (10.4), we have

$$h^2 = \frac{R^3}{A}\int \frac{dA}{(R + y)} - R^2$$

$$= \frac{R^3}{\left(\frac{b \times d}{2}\right)} \int \frac{\frac{b}{d}(R_2 - r)\,dr}{r} - R^2 \qquad (\because \quad R + y = r)$$

(limit of integration will be according to dr in which r is the distance from axis of curvature. Hence limits of integration will be from R_1 to R_2)

$$= \frac{2R^3}{b \times d} \times \frac{b}{d} \int_{R_1}^{R_2} \left(\frac{R_2}{r} - 1\right) dr - R^2$$

$$= \frac{2R^3}{d^2} \left[\left(R_2 \log_e r \right)_{R_1}^{R_2} - \left(r \right)_{R_1}^{R_2} \right] - R^2$$

$$= \frac{2R^3}{d^2} [R_2(\log_e R_2 - \log_e R_1) - (R_2 - R_1)] - R^2$$

$$= \frac{2R^3}{d^2} \cdot \left[R_2 \log_e \left(\frac{R_2}{R_1}\right) - (R_2 - R_1) \right] - R^2 \qquad ...(10.6)$$

Let us convert R_2 and R_1 in terms of R and d.

Now $\qquad R_2 = R + \frac{2d}{3} = \frac{3R + 2d}{3}$

$$R_1 = R - \frac{d}{3} = \frac{3R - d}{3}$$

and $\qquad R_2 - R_1 = d$

Hence the above equation (10.6) becomes as

$$h^2 = \frac{2R^3}{d^2} \left[\frac{(3R + 2d)}{3} \left\{ \log_e \left(\frac{3R + 2d}{3R - d} \right) \right\} - d \right] - R^2 \qquad ...[10.6\,(A)]$$

Now the position of Neutral Axis (N.A.) is given by,

$$y_0 = \frac{-Rh^2}{R^2 + h^2}.$$

10.7. POSITION OF NEUTRAL AXIS FOR TRAPEZOIDAL SECTION

A trapezoidal section of a curved beam of widths b_1 at the bottom, b_2 at the top and of height d is shown in Fig. 10.5. The centre of curvature of the curved beam is at O. The radius of curvature from the centroidal axis is R. Consider a strip of width b and depth 'dy' at a distance of y from the centroid axis and at a distance r from the axis of curvature.

$\therefore \qquad r = R + y$

and $\qquad dr = 0 + dy = dy$

Let R_1 = distance of bottom layer of trapezoidal from axis of curvature

R_2 = distance of top layer of trapezoidal from axis of curvature

Area of trapezoidal $\qquad = \frac{(b_1 + b_2)}{2} \times d$

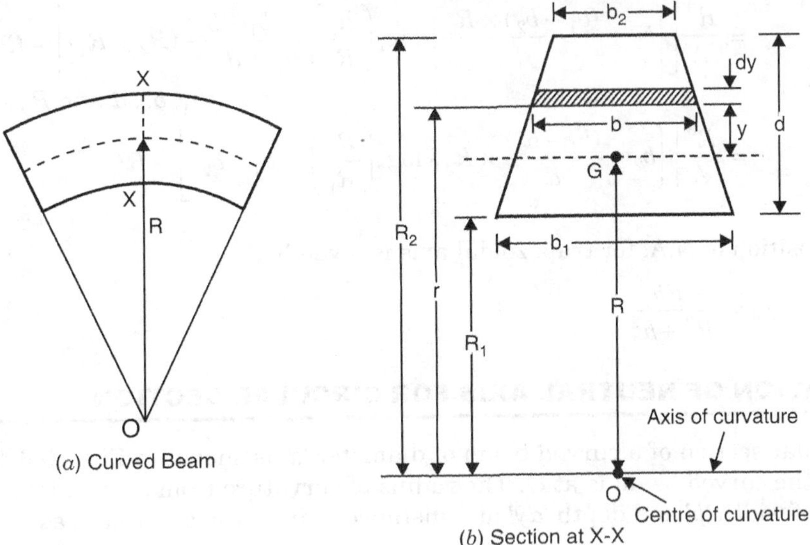

(a) Curved Beam

(b) Section at X-X

Fig. 10.5

Area of strip, $dA = b \times dy$ where $b = b_2 + \left(\frac{b_1 - b_2}{d}\right)(R_2 - r)$

$$= \left[b_2 + \left(\frac{b_1 - b_2}{d}\right)(R_2 - r)\right] dy = \left[b_2 + \left(\frac{b_1 - b_2}{d}\right)(R_2 - r)\right] dr \qquad (\because \quad dy = dr)$$

Now from equation (10.4), we have

$$h^2 = \frac{R^3}{A}\int \frac{dA}{(R + y)} - R^2$$

$$= \frac{R^3}{A}\int \frac{\left[b_2 + \left(\frac{b_1 - b_2}{d}\right)(R_2 - r)\right]dr}{(R + y)} - R^2 \qquad \left[\because \quad dA = b_2 + \left(\frac{b_1 - b_2}{d}\right)(R_2 - r)\right]$$

$$= \frac{R^3}{A}\int \frac{\left[b_2 + \left(\frac{b_1 - b_2}{d}\right)(R_2 - r)\right]dr}{r} - R^2$$

$$= \frac{R^3}{A}\int \left[\frac{b_2}{r} + \frac{(b_1 - b_2) \times R_2}{d \times r} - \frac{(b_1 - b_2) \times r}{d \times r}\right] dr - R^2$$

[Here the limits of integration will be according to '*dr*' in which *r* is the distance from the axis of curvature. Hence limits will be from R_1 to R_2 (*i.e.*, distances of bottom and top layers of trapezoidal section from axis of curvature.)]

$$\therefore \qquad h^2 = \frac{R^3}{A}\left[\int_{R_1}^{R_2} \left\{b_2 + \left(\frac{b_1 - b_2}{d}\right) \times R_2\right\} \times \frac{1}{r}\, dr - \int_{R_1}^{R_2} \left(\frac{b_1 - b_2}{d}\right) dr\right] - R^2$$

$$= \frac{R^3}{A}\left[\left\{b_2 + \left(\frac{b_1 - b_2}{d}\right) \times R_2\right\}\left(\log_e r\right)_{R_1}^{R_2} - \frac{(b_1 - b_2)}{d}\left(r\right)_{R_1}^{R_2}\right] - R^2$$

$$= \frac{R^3}{A}\left[\left\{b_2 + \frac{(b_1 - b_2) \times R_2}{d}\right\} \log_e\left(\frac{R_2}{R_1}\right) - \frac{(b_1 - b_2)}{d}(R_2 - R_1)\right] - R^2$$

(b_1, b_2, d and R_2 are constants)

$$= \frac{R^3}{A}\left[\left\{b_2 + \frac{(b_1 - b_2)}{d} \times R_2\right\} \log_e\left(\frac{R_2}{R_1}\right) - (b_1 - b_2)\right] - R^2 \quad ...(10.7)$$

$(\because \; R_2 - R_1 = d)$

Now position of N.A. for trapezoidal axis is given by,

$$y_0 = \frac{-Rh^2}{R^2 + h^2}$$

10.8. POSITION OF NEUTRAL AXIS FOR CIRCULAR SECTION

A circular section of a curved beam of diameter 'd' is shown in Fig. 10.6. The centre of curvature of the curved beam is at O. The radius of curvature from centroidal axis is R. Consider a strip of width 'b' and depth 'dy' at a distance 'y' from the centroidal axis.

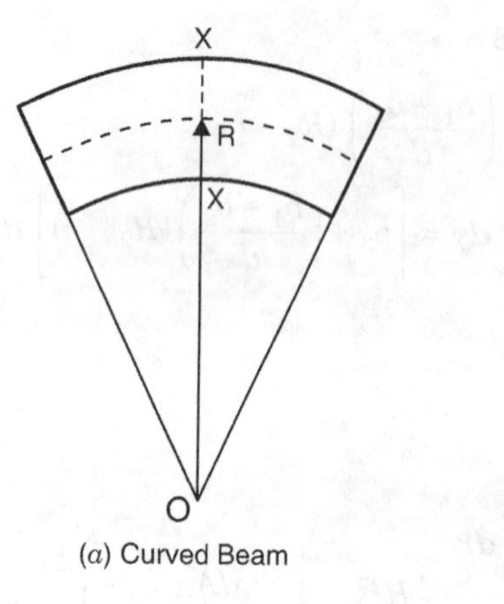

(a) Curved Beam

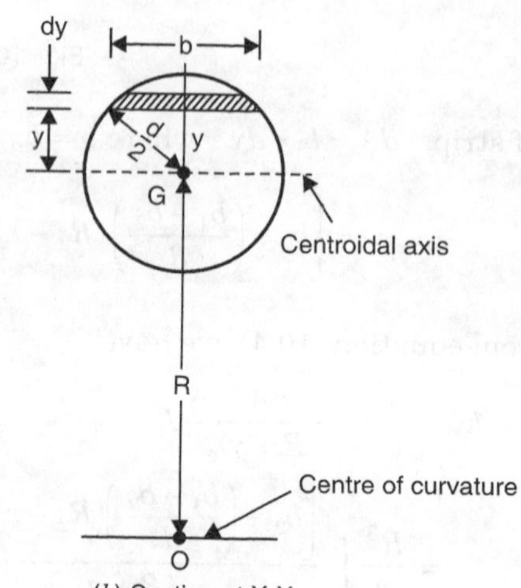

(b) Section at X-X

Fig. 10.6

Area of strip, $dA = b \times dy$ where $b = \left[\sqrt{(d/2)^2 - y^2}\right] \times 2$

$$= \left[2\sqrt{\left(\frac{d}{2}\right)^2 - y^2}\right] \times dy$$

Total area of section, $A = \frac{\pi}{4} d^2$

From equation (10.4), we have

$$h^2 = \frac{R^3}{A}\int_{-d/2}^{+d/2} \frac{dA}{(R+y)} - R^2 = \frac{R^3}{\frac{\pi}{4}d^2}\int_{-d/2}^{+d/2} \frac{\left(2\sqrt{\left(\frac{d}{2}\right)^2 - y^2}\right) dy}{(R+y)} - R^2$$

$$= \frac{8R^3}{\pi d^2} \int_{-d/2}^{+d/2} \frac{\sqrt{\frac{d^2}{4} - y^2}}{(R+y)} dy - R^2$$

The above integral can be expanded by binomial expression and then integrated, we will get

$$h^2 = \frac{d^2}{16} + \frac{1}{128} \frac{d^4}{R^2} + \ldots\ldots . \quad \ldots(10.8)$$

Now the position of N.A. for circular section is given by,

$$y_0 = \frac{-Rh^2}{R^2 + h^2}$$

10.9. SOLVED PROBLEMS 10.1 TO 10.3 FOR CURVED BEAMS

Problem 10.1. *Determine : (i) location of neutral axis, (ii) maximum and minimum stress, and (iii) ratio of maximum and minimum stress, when a curved beam of rectangular cross-section of width 20 mm and of depth 40 mm is subjected to pure bending of moment + 600 Nm. The beam is curved in a plane parallel to depth. The mean redius of curvature is 50 mm. Also plot the variation of the stresses across the section.*

Sol. Given :

Curved beam of rectangular section,

Width, b = 20 mm, depth, d = 40 mm, ∴ Area, $A = b \times d = 20 \times 40 = 800$ mm²

Radius of curvature, R = 50 mm.

Bending moment, M = + 600 Nm = 600 × 1000 = 600,000 Nmm

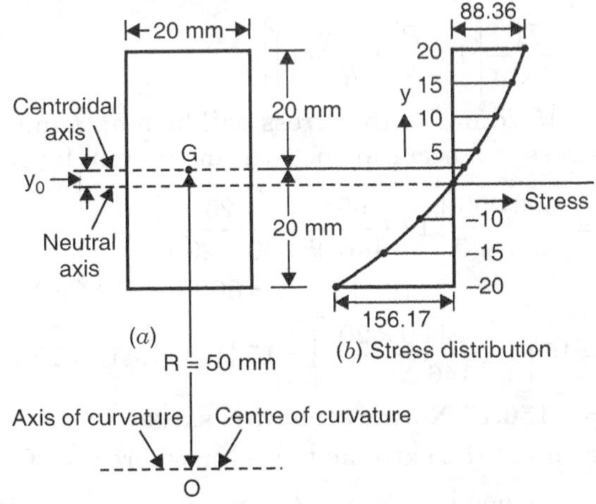

Fig. 10.7

(i) Location of neutral axis. Let y_0 = distance of neutral axis from centroidal axis using equation (10.3), we get

$$y_0 = -\frac{R \times h^2}{R^2 + h^2} \quad \ldots(i)$$

Let us find now the value of h^2 for rectangular section

The value of h^2 for rectangular section is given by equation (10.5) as

$$h^2 = \frac{R^3}{d}\left[\log_e\left(\frac{2R+d}{2R-d}\right)\right] - R^2 = \frac{50^3}{40}\left[\log_e\left(\frac{2\times 50+40}{2\times 50-40}\right)\right] - 50^2$$

$$= 3125\left[\log_e\left(\frac{140}{60}\right)\right] - 2500 = 3125\,[\,\log_e(2.333)] - 2500$$

$$= 3125 \times 0.847 - 2500 = 2646.8 - 2500 = 146.8$$

$$\therefore \quad y_0 = -\frac{50\times(146.8)}{50^2+(146.8)} = -\,\mathbf{2.773\ mm.\quad Ans.}$$

–ve sign means the neutral axis will be below central axis at a distance of 2.773 mm.

The value of h^2 can also be obtained from equation [10.5 (A)] as

$$h^2 = R^2\left[\frac{1}{3}\left(\frac{d}{2R}\right)^2 + \frac{1}{5}\left(\frac{d}{2R}\right)^4 + \ldots\ldots\right]$$

$$= 50^2\left[\frac{1}{3}\times\left(\frac{40}{2\times 50}\right)^2 + \frac{1}{5}\left(\frac{40}{2\times 50}\right)^4 + \ldots\ldots\right]$$

$$= 2500\left[\frac{1}{3}\times 0.16 + \frac{1}{5}\times 0.0256\right]$$

$$= 2500\,[0.0533 + 0.00512] = 2500 \times .05842 = 146.05$$

$$\therefore \quad y_0 = \frac{-50\times 146.05}{50^2+146.05} = \frac{-50\times 146.05}{2646.05} = -2.759 = -\,\mathbf{2.76\ mm.\quad Ans.}$$

(*ii*) *Maximum and minimum stresses.* The stress at any layer in the cross-section at a distance 'y' from centroidal axis is given by equation (10.1) as

$$\sigma = \frac{M}{R\times A}\left[1 + \frac{R^2}{h^2}\left(\frac{y}{R+y}\right)\right]$$

For a given value of M, R and A, the stress will be maximum when y is –ve and maximum. Hence maximum stress will occur at the extreme bottom layer where $y = -20$ mm.

$$\therefore \quad \sigma_{\max} = \frac{600{,}000}{50\times 800}\left[1 + \frac{50^2}{146.05}\left(\frac{-20}{50-20}\right)\right]$$

[$\because$ $A = 800$ mm², $h^2 = 146.05$, $M = 600{,}000$ Nmm]

$$= 15\left[1 - \frac{2500\times 20}{146.05\times 30}\right] = 15\,[1 - 11.41] = 15 \times (-10.41)$$

$= -$ **156.17 N/mm². (compressive) Ans.**

Minimum stress occurs at the extreme top layer where $y = 20$ mm

$$\therefore \quad \sigma_{\min} = \frac{600{,}000}{50\times 800}\left[1 + \frac{50^2}{146.05}\left(\frac{20}{50+20}\right)\right] = 15\left[1 + \frac{2500\times 20}{146.05\times 70}\right] = 15\,[1 + 4.89]$$

$= 15 \times 5.89 =$ **88.36 N/mm². (tensile) Ans.**

(*iii*) *Ratio of* σ_{max} *and* σ_{min}

$$\therefore \quad \frac{\sigma_{\max}}{\sigma_{\min}} = \frac{156.17}{88.36} \qquad \text{(Here take magnitude only)}$$

$=$ **1.767. Ans.**

(*iv*) *Plotting the stresses across the cross-section.* Stress at any distance '*y*' from centroidal axis is given by equation (10.1) as

$$\sigma = \frac{M}{R \times A}\left[1 + \frac{R^2}{h^2}\left(\frac{y}{R+y}\right)\right] = \frac{600{,}000}{50 \times 800}\left[1 + \frac{50^2}{146.05}\left(\frac{y}{50+y}\right)\right]$$

$$= 15\left[1 + 17.11\left(\frac{y}{50+y}\right)\right] \qquad ...(ii)$$

By substituting the various values of y in equation (*ii*), different values of stress are obtained.

At $y = 0$, $\sigma = 15\left[1 + \frac{17.11 \times 0}{50+0}\right] = 15 \times 1 = 15 \text{ N/mm}^2$

At $y = 5$ mm, $\sigma = 15\left[1 + 17.11 \times \left(\frac{5}{50+5}\right)\right] = 38.33 \text{ N/mm}^2$

At $y = 10$ mm, $\sigma = 15\left[1 + 17.11 \times \left(\frac{10}{50+10}\right)\right] = 57.77 \text{ N/mm}^2$

At $y = 15$ mm, $\sigma = 15\left[1 + 17.11 \times \left(\frac{15}{50+15}\right)\right] = 74.22 \text{ N/mm}^2$

At $y = 20$ mm, $\sigma = 15\left[1 + 17.11 \times \left(\frac{20}{50+20}\right)\right] = 88.36 \text{ N/mm}^2$

At $y = y_0 = -2.76$ mm, $\sigma = 0$ at neutral axis

At $y = -5$ mm, $\sigma = 15\left[1 + 17.11 \times \left(\frac{-5}{50-5}\right)\right] = -13.5 \text{ N/mm}^2$

At $y = -10$ mm, $\sigma = 15\left[1 + 17.11 \times \frac{(-10)}{(50-10)}\right] = -49.16 \text{ N/mm}^2$

At $y = -15$ mm, $\sigma = 15\left[1 + 17.11 \times \left(\frac{-15}{50-15}\right)\right] = -95 \text{ N/mm}^2$

At $y = -20$ mm, $\sigma = -156.17 \text{ N/mm}^2$

The variation of the stress across the section is plotted in Fig. 10.7 (*b*).

Problem 10.2. *Determine* : (*i*) *Location of neutral axis,*

(*ii*) *Maximum and minimum stresses.*

When a curved beam of trapezoidal section of bottom width 30 mm, top width 20 mm and height 40 mm is subjected to pure bending moment of + 600 Nm. The bottom width is towards the centre of curvature. The radius of curvature is 50 mm and beam is curved in a plane parallel to depth. Also plot the variation of stresses across the section.

Sol. Given :

$b_1 = 30$ mm, $b_2 = 20$ mm, $d = 40$ mm, $M = +600$ Nm $= 600 \times 1000$ Nmm

$R = 50$ mm.

Let us first find the location of centroidal axis. The distance 'd_1' of centroidal axis from the side b_1 is given by,

$$d_1 = \left(\frac{b_1 + 2b_2}{b_1 + b_2}\right) \times \frac{d}{3} = \left(\frac{30 + 2 \times 20}{30 + 20}\right) \times \frac{40}{3} = \frac{70}{50} \times \frac{40}{3} = 18.67 \text{ mm}$$

and $d_2 = 40 - 18.67 = 21.33$ mm

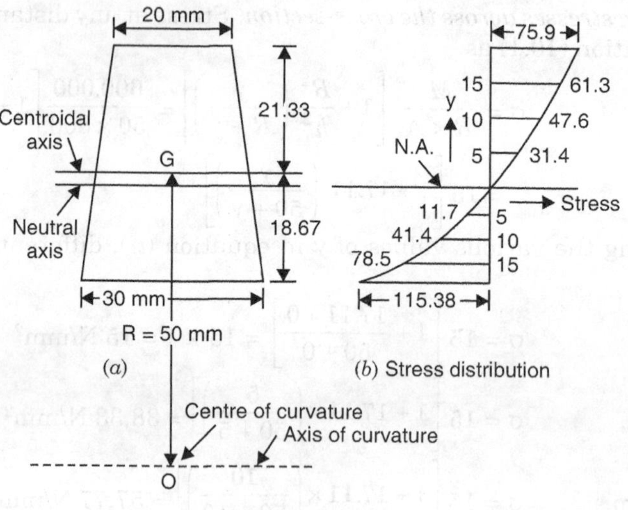

Fig. 10.8

Let us now find R_1, R_2 and area A.

$$R_1 = R - d_1 = 50 - 18.67 = 31.33 \text{ mm}$$
$$R_2 = R + d_2 = 50 + 21.33 = 71.33 \text{ mm}$$
$$A = \text{Area of trapezoidal section}$$
$$= \left(\frac{b_1 + b_2}{2}\right) \times d = \left(\frac{30 + 20}{2}\right) \times 40 = 1000 \text{ mm}^2$$

Let us now find h^2 for trapezoidal section. Using equation (10.7), we get

$$h^2 = \frac{R^3}{A}\left[\left\{b_2 + \left(\frac{b_1 - b_2}{d}\right) \times R_2\right\} \log_e\left(\frac{R_2}{R_1}\right) - (b_1 - b_2)\right] - R^2$$

$$= \frac{50^3}{1000}\left[\left\{20 + \frac{30 - 20}{40} \times 71.33\right\} \log_e\left(\frac{71.33}{31.33}\right) - (30 - 20)\right] - 50^2$$

$$= 125\ [\{20 + 17.83\} \times 0.8227 - 10] - 2500 = 140.34$$

(*i*) *Location of neutral axis.* Let y_0 = distance of neutral axis from centroidal axis. The location of neutral axis is given by equation (10.3) as

$$y_0 = -\frac{R \times h^2}{R^2 + h^2} = -\frac{50 \times 140.34}{50^2 + 140.34}$$

$$= -\frac{50 \times 140.34}{2640.34} = \mathbf{-2.65 \text{ mm.} \quad Ans.}$$

–ve sign means the neutral axis will be below the centroidal axis at a distance of 2.65 mm.

(*ii*) *Maximum and minimum stresses.* The stress at any layer at a distance '*y*' from centroidal axis is given by equation (10.1) as

$$\sigma = \frac{M}{R \times A}\left[1 + \frac{R^2}{h^2}\left(\frac{y}{R + y}\right)\right]$$

The stress will be maximum at the extreme bottom layer where

$$y = -18.67 \text{ mm}$$

$$\therefore \qquad \sigma_{max} = \frac{600{,}000}{50 \times 1000}\left[1 + \frac{50^2}{140.34}\left(\frac{-18.67}{50 - 18.67}\right)\right]$$

$$= 12\,[1 + (-10.61)] = 12 \times (-9.61) = \mathbf{-115.38\ N/mm^2}. \quad \textbf{Ans.}$$

$$= 115.38 \text{ N/mm}^2 \text{ (compressive)}$$

Minimum stress occurs at the extreme top layer where $y = 21.33$ mm

$$\therefore \qquad \sigma_{min} = \frac{600{,}000}{50 \times 1000}\left[1 + \frac{50^2}{140.34} \times \left(\frac{21.33}{50 + 21.33}\right)\right]$$

$$= 12\,(1 + 5.326) = \mathbf{75.9\ N/mm^2.\ (tensile)} \quad \textbf{Ans.}$$

(*iii*) *Plotting the stresses across the section*

$$\sigma = \frac{M}{R \times A}\left[1 + \frac{R^2}{h^2}\left(\frac{y}{R + y}\right)\right] = \frac{60000}{50 \times 1000}\left[1 + \frac{50^2}{140.34}\left(\frac{y}{50 + y}\right)\right]$$

$$= 12\left[1 + 17.8 \times \left(\frac{y}{50 + y}\right)\right]$$

At $y = 21.33$ mm, $\sigma = 75.9$ N/mm² (Already calculated)

At $y = 15$ mm, $\sigma = 12\left[1 + \dfrac{17.8 \times 15}{(50 + 15)}\right] = 61.3$ N/mm²

At $y = 10$ mm, $\sigma = 12\left(1 + \dfrac{17.8 \times 10}{(50 + 10)}\right) = 47.6$ N/mm²

At $y = 5$ mm, $\sigma = 12\left(1 + \dfrac{17.8 \times 5}{(50 + 5)}\right) = 31.4$ N/mm²

At $y = 0$, $\sigma = 12\,(1 + 0) = 12$ N/mm²

At $y = -5$ mm, $\sigma = 12\left(1 + \left(\dfrac{17.8 \times (-5)}{(50 - 5)}\right)\right) = -11.7$ N/mm²

At $y = -10$ mm, $\sigma = 12\left[1 + \dfrac{17.8 \times (-10)}{(50 - 10)}\right] = -41.4$ N/mm²

At $y = -15$ mm, $\sigma = 12\left[1 + \dfrac{17.8 \times (-15)}{(50 - 15)}\right] = -78.5$ N/mm²

At $y = -18.67$ mm, $\sigma = -115.38$ N/mm² (Already calculated)

Fig. 10.8(*b*) shows the stress distribution across the section.

Problem 10.3. *Determine : (i) position of neutral axis, and (ii) maximum and minimum stresses when a curved beam of circular section of diameter 100 mm is subjected to pure bending moment of + 11.5 kNm. The radius of curvature is 100 mm.*

Sol. Given :

$d = 100$ mm, B.M., $M = +11.5$ kNm $= 11.5 \times 10^3$ Nm

$= (11.5 \times 10^3) \times 10^3$ Nmm $= 11.5 \times 10^6$ Nmm, radius of curvature, $R = 100$ mm.

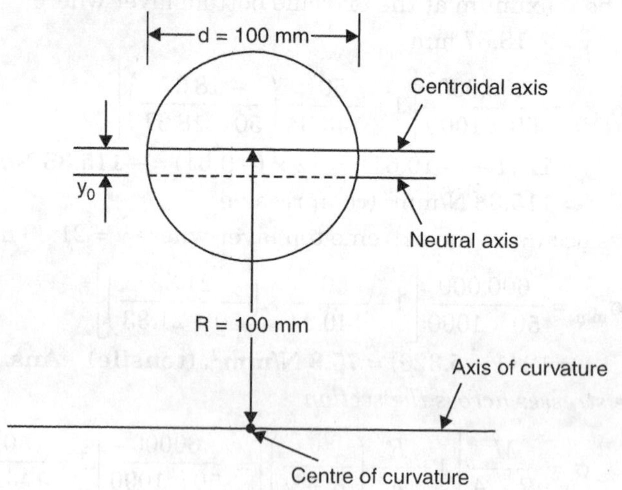

Fig. 10.9

(*i*) *Position of neutral axis.* Let y_0 = distance of neutral axis from centroidal axis the position of neutral axis is given by equation (10.3) as

$$y_0 = \frac{R \times h^2}{R^2 + h^2} \quad \text{...}(i)$$

The value of h^2 for circular section is given by equation (10.8) as,

$$h^2 = \frac{d^2}{16} + \frac{1}{128} \times \frac{d^4}{R^2} + ... = \frac{100^2}{16} + \frac{1}{128} \times \frac{100^4}{100^2}$$

(Here R = 100 mm and d also = 100 mm)

$$= \frac{10000}{16} + \frac{10000}{128} = \frac{90000}{128} = 703.125$$

Substituting this value of h^2 in equation (*i*), we get

$$y_0 = -\frac{100 \times 703.125}{100^2 + 703.125} = -\mathbf{6.57\ mm.} \quad \textbf{Ans.}$$

–ve sign means the neutral axis will be below the centroidal axis.

(*ii*) *Maximum and minimum stresses.* The stress at any layer at a distance of *y* from centroidal axis is given by equation (10.1) as

$$\sigma = \frac{M}{R \times A}\left[1 + \frac{R^2}{h^2}\left(\frac{y}{R + y}\right)\right]$$

where A = Area of cross-section = $\frac{\pi}{4} \times d^2 = \frac{\pi}{4} \times 100^2 = 2500\,\pi\ \text{mm}^2$

$M = 11.5 \times 10^6$ Nmm

The stress will be maximum at the extreme bottom layer where $y = -50$ mm

$$\therefore \quad \sigma_{max} = \frac{11.5 \times 10^6}{100 \times 2500\,\pi}\left[1 + \frac{100^2}{703.125}\left(\frac{-50}{100 - 50}\right)\right]$$

$$= 14.636\,[1 - 14.222] = -193.52\ \text{N/mm}^2$$

$$= \mathbf{193.52\ N/mm^2\ (compressive).} \quad \textbf{Ans.}$$

The stress will be minimum at the extreme top layer where $y = + 50$ mm.

$$\therefore \quad \sigma_{min} = \frac{11.5 \times 10^6}{100 \times 2500\,\pi}\left[1 + \frac{100^2}{703.125}\left(\frac{50}{100+50}\right)\right]$$

$= 14.636\,[1 + 4.74] =$ **84.02 N/mm². (tensile) Ans.**

10.10. STRESS IN A CURVED BAR SUBJECTED TO DIRECT STRESSES AND BENDING STRESSES

In the previous article, we have considered the bending of the curved bars which are subjected to pure bending. Hence the stress across the cross-section is only bending stress which is due to B.M. only. But in case of curved members like hooks, rings and chains which carry vertical loads, there is bending stress as well as direct stress.

Hence the stress at any point in the cross-section of these members will be equal to algebraic sum of direct stress (σ_0) due to vertical load (P) and bending stress (σ_b). It is assumed that vertical load P acts through the centroid of the section and the stress due to this load is equal to $\frac{P}{A}$ at each point on the cross-section. Hence the resultant stress at any point will be

$$\sigma_R = \sigma_0 + \sigma_b \text{ where } \sigma_b \text{ from equation (10.1)}$$

$$= \sigma = \frac{M}{A \times R}\left[1 + \frac{R^2}{h^2}\left(\frac{y}{R+y}\right)\right] \quad \text{...(10.9)}$$

$$\therefore \quad \sigma_R = \frac{P}{A} + \frac{M}{A \times R}\left[1 + \frac{R^2}{h^2}\left(\frac{y}{R+y}\right)\right] \quad \text{...[10.9 (A)]}$$

If P is tensile load, then, $\frac{P}{A}$ will be +ve and if P is compressive load, then $\frac{P}{A}$ will be –ve.

10.11. STRESS IN A CRANE HOOK

Fig. 10.10 shows a hook, which supports a vertical load P. The centre of curvature of the hook is at O. Consider a horizontal section AC passing through the centre of curvature of the hook. This section carries the maximum stress. This horizontal section may be trapezoidal or any other shape. The centroidal axis of the section is represented by BB.

Let P = Load supported by hook.

e = Eccentricity of the load *i.e.,* the distance between the load line and centroidal axis.

R = Radius of curvature of the curved hook *i.e.,* distance between axis of curvature and centroidal axis.

The other dimensions such as b_1, b_2, d, R_1 and R_2 are as usual. The bending moment caused by load P is given by,

$$M = P \times e$$

This bending moment is –ve as it decreases the curvature of the hook. Hence $M = -P \times e$ the resultant stress at any point across the section is given by equation (10.9) as

$$\sigma_R = \sigma_0 + \sigma_b$$

$$= \frac{P}{A} + \frac{M}{A \times R}\left[1 + \frac{R^2}{h^2}\left(\frac{y}{R+y}\right)\right]$$

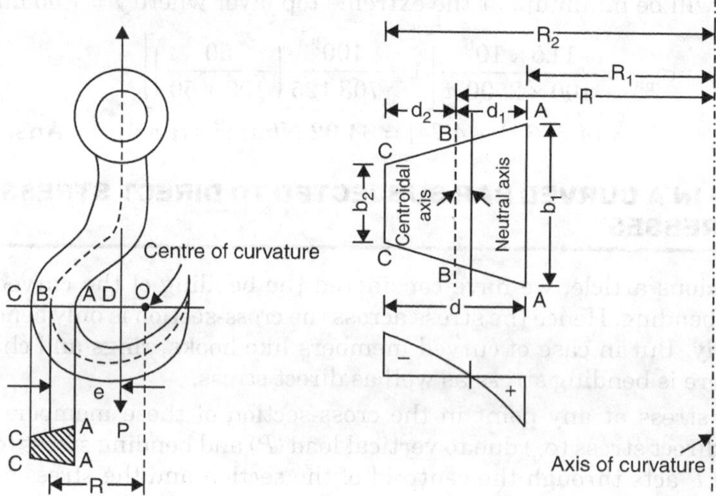

Fig. 10.10

Here load P is tensile, hence direct stress $\left(\sigma_0 = \frac{P}{A}\right)$ will be tensile stress and hence it will be +ve. Bending moment $M = -P \times e$.

Hence above equation becomes as

$$\sigma_R = \frac{P}{A} - \frac{P \times e}{A \times R}\left[1 + \frac{R^2}{h^2}\left(\frac{y}{R+y}\right)\right]$$

The bending moment here produces tensile stress at AA and compressive stress at CC. The vertical load P produces tensile stress over the whole section.

10.12. SOLVED PROBLEMS 10.4 TO 10.6 FOR STRESS IN A CRANE HOOK

Problem 10.4. *A hook carries a load of 7.5 kN and the load line is at a distance of 20 mm from the inner edge of the section which is trapezoidal. The load line also passes through the centre of curvature of the hook. The dimensions of the central horizontal trapezoidal section are : inner width = 30 mm ; outer width = 15 mm ; depth = 30 mm. Calculate the maximum and minimum stresses. Also plot the variation of stress across the section.*

Sol. Given :

Load, $P = 7.5 \text{ kN} = 7.5 \times 1000 \text{ N}.$

Distance of load line from inner edge of the section = 20 mm. Centre of curvature lies on load line. This means the radius of curvature and eccentricity will be equal.

$\therefore \quad R = e$

Dimensions of trapezoidal section.

Inner width, $b_1 = 30$ mm, outer width, $b_2 = 15$ mm and depth, $d = 30$ mm.

Refer to Fig. 10.11.

Let us first find the location of centroidal axis.

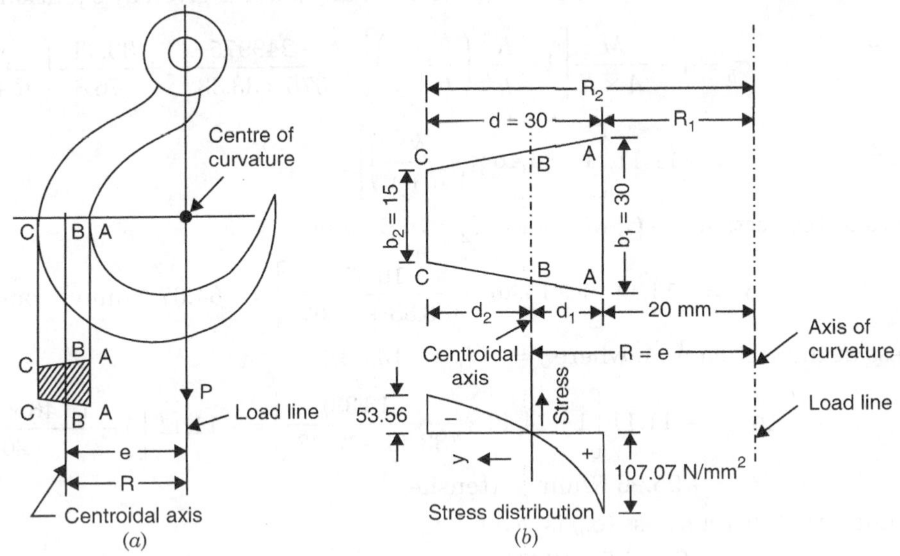

Fig. 10.11

The distance d_1 of the centroidal axis from the side b_1 is given by

$$d_1 = \left(\frac{b_1 + 2b_2}{b_1 + b_2}\right) \times \frac{d}{3} = \left(\frac{30 + 2 \times 15}{30 + 15}\right) \times \frac{30}{3} = \frac{60}{45} \times 10 = 13.33 \text{ mm}$$

$\therefore$ $$d_2 = d - d_1 = 30 - 13.33 = 16.67 \text{ mm}$$

From Fig. 10.11, it is clear that

$$R_1 = 20 \text{ mm}, R = 20 + d_1 = 20 + 13.33 = 33.33 \text{ mm},$$

$$R_2 = 20 + d = 20 + 30 = 50 \text{ mm}$$

$$e = R = 33.33 \text{ mm}$$

Area of section, $$A = \left(\frac{b_1 + b_2}{2}\right) \times d = \frac{30 + 15}{2} \times 30 = 675 \text{ mm}^2$$

Bending moment, $M = -P \times e$ (–ve sign is used as the applied load decreases the curvature of the hook)

$$= -(7.5 \times 1000) \times 33.33 = -249975 \text{ Nmm}$$

Let us now find the value of h^2 for trapezoidal section. It is given by equation (10.7) as

$$h^2 = \frac{R^3}{A}\left[\left\{b_2 + \frac{(b_1 - b_2) \times R_2}{d}\right\} \log_e\left(\frac{R_2}{R_1}\right) - (b_1 - b_2)\right] - R^2$$

$$= \frac{33.33^3}{675}\left[\left\{15 + \frac{30 - 15}{30} \times 50\right\} \log_e\left(\frac{50}{20}\right) - (30 - 15)\right] - 33.33^2$$

$$= 54.85\ [40 \log_e 2.5 - 15] - 1110.8$$

$$= 1187.6 - 1110.8 = 76.8 \text{ mm}^2.$$

(*i*) *Max. and min. stresses.* The bending stress at any point is given by equation (10.1) as

$$\sigma_b = \sigma = \frac{M}{A \times R}\left[1 + \frac{R^2}{h^2}\left(\frac{y}{R+y}\right)\right] = \frac{-249975}{675 \times 33.33}\left[1 + \frac{33.33^2}{76.8}\left(\frac{y}{R+y}\right)\right]$$

$$= -11.11\left[1 + 14.46 \times \left(\frac{y}{R+y}\right)\right]$$

Stress at the section $C - C$, where $y = d_2 = 16.67$ mm,

$$\sigma_{cc} = -11.11\left[1 + 14.46 \times \frac{16.67}{(33.33 + 16.67)}\right] = -64.67 \text{ N/mm}^2 \text{ (compressive)}$$

Stress at the section A–A, where $y = -d_1 = -13.33$,

$$\sigma_{AA} = -11.11\left[1 + 14.46 \times \frac{(-13.33)}{(33.33 - 13.33)}\right] = -11.11\left[1 - \frac{14.46 \times 13.33}{20}\right]$$

$$= +95.96 \text{ N/mm}^2 \text{ (tensile)}$$

We know that direct stress (σ_0) is given by

$$\sigma_0 = \frac{P}{A} = \frac{7.5 \times 1000}{675} = 11.11 \text{ N/mm}^2 \text{ (tensile)}$$

The direct stress is uniform at all points in the cross-section

∴ Resultant stress at $AA = \sigma_0 + \sigma_{AA}$

$= 11.11 + 95.96 =$ **107.07 N/mm² (tensile). Ans.**

Resultant stress at $CC = \sigma_0 + \sigma_{cc}$

$= 11.11 - 64.67$

= – 53.56 N/mm² (compressive). Ans.

(*ii*) *Plotting of stresses.* The resultant stress at any distance *y* from the centroidal axis is given by,

$$\sigma_r = \sigma_0 + \sigma_b \ (= \sigma)$$

$$= 11.11 + \frac{M}{A \times R}\left[1 + \frac{R^2}{h^2} \times \left(\frac{y}{R+y}\right)\right]$$

$$= 11.11 + \frac{-249975}{675 \times 33.33}\left[1 + \frac{33.33^2}{76.8} \times \left(\frac{y}{33.33 + y}\right)\right]$$

$$= 11.11 - 11.11\left[1 + 14.46 \times \left(\frac{y}{33.33 + y}\right)\right]$$

At $y = 0$, $\sigma_r = 11.11 - 11.11 = 0$

At $y = 5$ mm, $\sigma_r = 11.11 - 11.11\left[1 + \frac{14.46 \times 5}{33.33 + 5}\right] = -20.95 \text{ N/mm}^2$

At $y = 10$ mm, $\sigma_r = 11.11 - 11.11\left[1 + \frac{14.46 \times 10}{33.33 + 10}\right] = -37.07 \text{ N/mm}^2$

At $y = 15$ mm, $\sigma_r = 11.11 - 11.11\left[1 + \frac{14.46 \times 15}{33.33 + 15}\right] = -49.86 \text{ N/mm}^2$

At $y = 16.67 \text{ mm} = d_2, \sigma_r = -53.56 \text{ N/mm}^2$ (Already calculated)

At $y = -5 \text{ mm}, \sigma_r = 11.11 - 11.11\left[1 + \frac{14.46 \times (-5)}{33.33 - 5}\right] = 28.35 \text{ N/mm}^2$

At $y = -10 \text{ mm}, \sigma_r = 11.11 - 11.11\left[1 + \frac{14.46 \times (-10)}{33.33 - 10}\right] = 68.86 \text{ N/mm}^2$

At $y = -13.33 \text{ mm}, \sigma_r = 107.07 \text{ N/mm}^2$

The variation of stress across the cross-section is shown in Fig. 10.11 (*b*).

Problem 10.5. *A central horizontal section of a hook is a trapezoidal section shown in Fig. 10.12. Calculate the maximum compressive and tensile stresses in the section of the hook when it carries a load of 37.5 kN, the load line passes through the centre of curvature.*

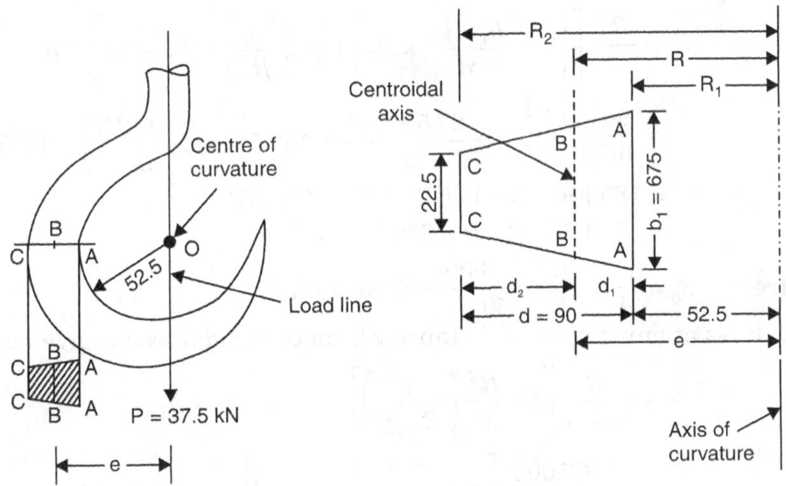

Fig. 10.12

Sol. Given : $P = 37.5 \text{ kN} = 37.5 \times 1000 \text{ N}$

Load line passes through centre of curvature

∴ Eccentricity, $e = R$ (radius of curvature)

where e = distance between load line and centroidal axis

and R = distance between centre of curvature and centroidal axis.

Dimensions of trapezoidal section are :

$b_1 = 67.5 \text{ mm}, b_2 = 22.5 \text{ mm}, d = 90 \text{ mm}$

Refer to Fig. 10.12.

Let us first find the location of centroidal axis.

The distance d_1 of the centroidal axis from the side b_1 is given by

$$d_1 = \left(\frac{b_1 + 2b_2}{b_1 + b_2}\right) \times \frac{d}{3} = \left(\frac{67.5 + 2 \times 22.5}{67.5 + 22.5}\right) \times \frac{90}{3} = \frac{112.5}{90} \times \frac{90}{3} = 37.5 \text{ mm}$$

∴ $d_2 = d - d_1 = 90 - 37.5 = 52.5 \text{ mm}$

From Fig. 10.12, it is clear that

$$R_1 = 52.5 \text{ mm. Then } R = R_1 + d_1 = 52.5 + 37.5 = 90 \text{ mm}$$
$$R_2 = 52.5 + d = 52.5 + 90 = 142.5 \text{ mm}$$
$$e = R = 90 \text{ mm}$$

Area of the section,

$$A = \left(\frac{b_1 + b_2}{2}\right) \times d = \frac{67.5 + 22.5}{2} \times 90 = 4050 \text{ mm}^2$$

Bending moment, $M = -P \times e$ (–ve sign is used as the applied load decreases the curvature of the hook)

$$= -(37.5 \times 1000) \times 90 \text{ Nmm} = -3375000 \text{ Nmm}$$

Let us now find the value of h^2 for trapezoidal axis. It is given by equation (10.7) as

$$h^2 = \frac{R^3}{A}\left[\left\{b_2 + \frac{b_1 - b_2}{d} \times R_2\right\} \log_e\left(\frac{R_2}{R_1}\right) - (b_1 - b_2)\right] - R^2$$
$$= \frac{90^3}{4050}\left[\left\{22.5 + \frac{67.5 - 22.5}{90} \times 142.5\right\} \log_e\left(\frac{142.5}{52.5}\right) - (67.5 - 22.5)\right] - 90^2$$
$$= 180\,[(93.75 \times \log_e 2.714^3) - 45] - 8100$$
$$= 8750.26 - 8100 = 650.26 \text{ mm}^2$$

Direct stress, $\sigma_0 = \dfrac{P}{A} = \dfrac{37.5 \times 1000}{4050} = 9.26 \text{ N/mm}^2$

Bending stress at any layer at a distance y from centroidal axis is given by,

$$\sigma = \frac{M}{A \times R}\left[1 + \frac{R^2}{h^2}\left(\frac{y}{R + y}\right)\right]$$
$$= -\frac{3375000}{4050 \times 90}\left[1 + \frac{90^2}{650.26}\left(\frac{y}{90 + y}\right)\right] \qquad [\because \; M = -3375000 \text{ N mm}]$$
$$= -9.26\left[1 + \frac{12.456y}{90 + y}\right] \qquad \ldots(i)$$

Bending stress at section C–C where $y = d_2 = 52.5$ mm becomes as

$$\sigma_{cc} = -9.26\left[1 + \frac{12.456 \times 52.5}{90 + 52.5}\right] = -51.75 \text{ N/mm}^2$$

∴ Resultant stress at section $C - C$ becomes

$$(\sigma_r)_{cc} = \sigma_0 + \sigma_{cc}$$
$$= 9.26 - 51.75 = \mathbf{-42.49 \text{ N/mm}^2 \text{ (Compressive)}}. \quad \textbf{Ans.}$$

Bending stress at section $A - A$ where $y = -d_1 = -37.5$ mm becomes as

$$\sigma_{AA} = -9.26\left[1 + \frac{12.456 \times (-37.5)}{90 - 37.5}\right] \qquad [\text{Substitute } y = -37.5 \text{ in equation } (i)]$$
$$= 73.12 \text{ N/mm}^2$$

∴ Resultant stress at section $A - A$ becomes as

$$(\sigma_r)_{AA} = \sigma_0 + \sigma_{AA} = 9.26 + 73.12 = \mathbf{82.38 \text{ N/mm}^2 \text{ (Tensile)}}. \quad \textbf{Ans.}$$

Problem 10.6. *The curved member shown in Fig. 10.13 has a solid circular cross-section 0.10 m in diameter. If the maximum tensile and compressive stresses in the member are not to exceed 150 MPa and 200 MPa respectively, determine the value of load P that can safely be carried by the member.*

(AMIE, Summer 1996)

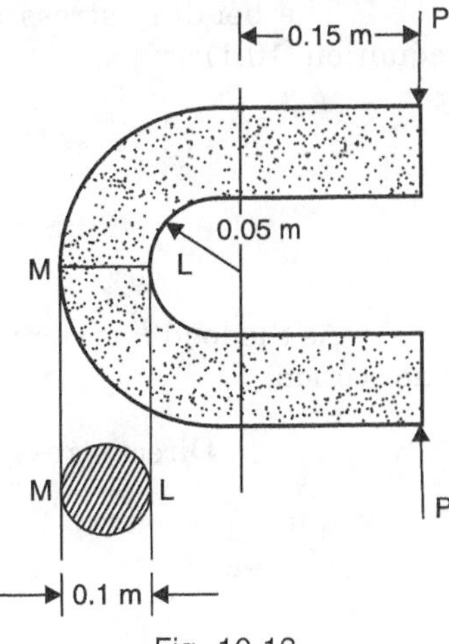

Fig. 10.13

Sol. Given :

Dia of section, $d = 0.10$ m

Max. tensile stress,

$$(\sigma_r)_t = 150 \text{ MPa} = 150 \times 10^6 \text{ N/m}^2$$

Max. compressive stress,

$$(\sigma_r)_c = 200 \text{ MPa} = 200 \times 10^6 \text{ N/m}^2$$

Refer to Fig. 10.13 (*a*)

Area of cross-section,

$$A = \frac{\pi}{4} d^2 = \frac{\pi}{4} (0.1^2) = 7.854 \times 10^{-3} \text{ m}^2$$

e = Eccentricity = distance between load line and centroidal axis
$= 0.15 + 0.05 + 0.05 = 0.25$ m

R = Radius of curvature
= distance between centre of curvature and centroidal axis
$= 0.05 + 0.05 = 0.1$ m

Let us first find the value of h^2 for circular section. It is given by equation (10.8) as

$$h^2 = \frac{d^2}{16} + \frac{1}{128} \times \frac{d^4}{R^2} + ...$$

$$= \frac{0.1^2}{16} + \frac{1}{128} \times \frac{0.1^4}{0.1^2} \qquad (\because \; R = 0.1 \text{ m also } d = 0.1 \text{ m})$$

$$= 7.031 \times 10^{-4} \text{ m}^2$$

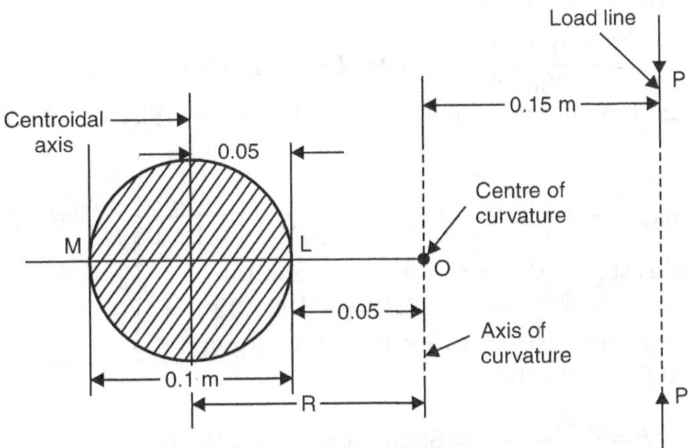

Fig. 10.13 (*a*)

Bending moment, $M = P \times e$ (+ve sign is used as load P increases the curvature)

$$= P \times 0.25 = 0.25 \, P \text{ Nm}$$

The bending stress at any section at a distance y from centroidal axis is given by equation (10.1) as

$$\sigma_b = \frac{M}{A \times R}\left[1 + \frac{R^2}{h^2} \times \left(\frac{y}{R+y}\right)\right] \quad \text{[Here } \sigma_b = \sigma]$$

$$= \frac{0.25\,P}{(7.854 \times 10^{-3}) \times 0.1}\left[1 + \frac{0.1^2}{7.031 \times 10^{-4}} \times \left(\frac{y}{0.1+y}\right)\right] \quad ...(i)$$

As the load P is compressive, hence direct stress (σ_0) will be compressive at all points of the section.

$\therefore$ Direct stress, $\sigma_0 = \dfrac{P}{A} = \dfrac{P}{7.854 \times 10^{-3}}$ N/m² (compressive)

$$= -\frac{P}{7.854 \times 10^{-3}} \quad \text{(–ve sign due to compressive stress)}$$

$$= -127.32\,P$$

As the load is compressive, the bending stress will be compressive at the inner point L of the cross-section where $y = -0.05$ m and it will be tensile at the outer point M of the cross-section where $y = 0.05$ m.

1st Case. Max. compressive bending stress will be obtained by substituting $y = -0.05$ m in equation (*i*).

$$\therefore \text{ Compressive } (\sigma_b)_{max} = \frac{0.25\,P}{(7.854 \times 10^{-3} \times 0.1)}\left[1 + \frac{0.1^2}{7.031 \times 10^{-4}} \times \left(\frac{-0.05}{0.1 - 0.05}\right)\right]$$

$$= 318.3\,P\,[1 - 14.22] = -4208.8\,P$$

$\therefore$ Max. compressive stress, $(\sigma_r)_c = \sigma_0 + (\sigma_b)_{max}$ (compressive)

$$= -127.32\,P - 4208.8\,P = -4336.12\,P.$$

But maximum compressive stress is given as 200×10^6 N/m²

$\therefore$ Maximum compressive stress, $(\sigma_r)_c = -200 \times 10^6$ N/m²

(–ve sign is due to compressive stress)

$\therefore$ $-200 \times 10^6 = -4336.12\,P$

$$\therefore \quad P = \frac{200 \times 10^6}{4336.12} = 46124 \text{ N} = 46.124 \text{ kN} \quad ...(ii)$$

2nd Case. Max. tensile bending stress will be obtained by substituting $y = +0.05$ m in equation (*i*).

$$\therefore \quad (\sigma_b)_{max} \text{ (tensile)} = 318.3\,P\left[1 + \frac{0.1^2}{7.031 \times 10^{-4}} \times \left(\frac{0.05}{0.1 + .05}\right)\right] = 1827.32\,P$$

$\therefore$ Max. tensile stress, $(\sigma_r)_t = \sigma_0 + (\sigma_b)_{max}$ (tensile)

$$= -127.32\,P + 1827.32\,P = 1700\,P$$

But max tensile stress, $(\sigma_r)_t = 150 \times 10^6$ N/m² (given)

$\therefore$ $1700\,P = 150 \times 10^6$

$$\therefore \quad P = \frac{150 \times 10^6}{1700} = 88235.3 \text{ N} = 88.235 \text{ kN} \quad ...(iii)$$

Comparing equations (*ii*) and (*iii*), the safe load will be lesser of the two.

$\therefore$ $P =$ **46.124 kN. Ans.**

10.13. STRESSES IN CIRCULAR RING SUBJECTED TO TENSION OR COMPRESSION

Fig. 10.14 (*a*) shows a circular ring subjected to a tensile load *P*. Cut the ring in two parts through *A* and *B* as shown in Fig. 10.14 (*b*). Fix the ends *A* and *B*. Now each half of the ring carries loads of $\frac{1}{2}$ *P* and fixing moments M_0 at the ends *A* and *B*. It is required to find the stresses in the ring. Due to symmetry, the stress distribution in the two halves of the ring will be the same.

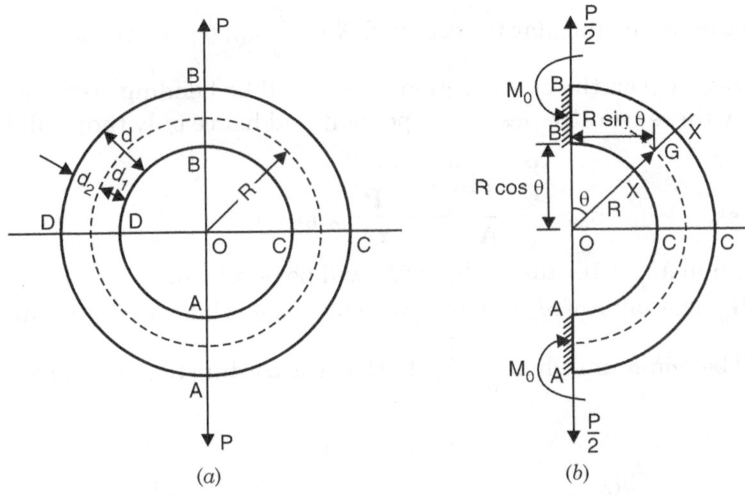

Fig. 10.14

Consider a section *X* – *X* at an angle θ with the vertical.

Let *M* = bending moment with which the section *X* – *X* is subjected.

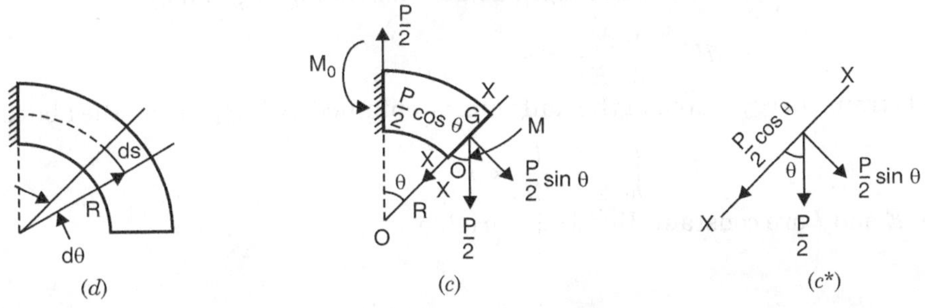

Fig. 10.14

If we know the value of *M*, then stresses can be calculated across the section *X–X*. But the determining the moment *M* at this section is statically indeterminate.

In case of straight beams which is subjected to transverse loading, there is bending moment and shear force at any section. The strain energy at any point is mainly due to bending. But a section of the curved beam has strain energy due to bending and also due to thrust. Here we assume the curved beam as the straight beam for determining M_0 (and hence *M*). So we neglect the effect of thrust in the calculation of strain energy.

Now, the bending moment (M) with which the section $X–X$ is subjected is given as

$$M = M_0 - \frac{P}{2} \times (R \sin \theta) \quad ...(i)$$

where R = radius of the centre line of the ring.

Refer to free-body diagram shown in Fig. 10.14 (*c*) of the ring upto section $X–X$, the vertical downward force $\frac{P}{2}$ has been resolved in two components, one along the radial direction $\left(=\frac{P}{2}\cos\theta\right)$ and other perpendicular to section $X–X$ $\left(=\frac{P}{2}\sin\theta\right)$. The component $\frac{P}{2} \times \sin\theta$ will produce direct stress (σ_0) on the cross-section. There will be bending stress also acting on the cross-section. Only these two stresses are important and hence only they will be considered.

Direct stress, $$\sigma_0 = \frac{\frac{\mathbf{P}}{\mathbf{2}} \times \sin\theta}{\mathbf{A}} = \frac{\mathbf{P}}{\mathbf{2A}} \times \sin\theta$$

Let us first find M_0. After that value of M will be obtained.

Value of $\mathbf{M_0}$. The value of M_0 is determined by using principle of minimum strain energy.

Strain energy will be minimum when $\frac{\partial U}{\partial M_0} = 0$. Hence according to principle of minimum strain energy

$$\therefore \quad \frac{\partial U}{\partial M_0} = 0 \quad ...(ii)$$

Refer to Fig. 10.14 (*d*).

Let a small element of length ds of the ring subtends an angle $d\theta$ at the centre.

Then strain energy* stored in length ds due to bending is given by

$$dU = \frac{1}{2}\frac{M^2}{EI} ds$$

Total strain energy stored in the half ring, is obtained by integrating the above equation.

$$\therefore \quad U = \int_0^{\pi} \frac{1}{2} \times \frac{M^2}{EI} \times ds$$

Here E and I are constant. But M is variable.

---

*dU = Work done by B.M. = Average B.M. × Change of angle = $\frac{1}{2} M \times d\theta$.

Length, $ds = R \times d\theta \quad \therefore \quad d\theta = \frac{ds}{R}$. Also $\frac{M}{I} = \frac{E}{R} \quad \therefore \quad R = \frac{EI}{M}$. Now

$$d\theta = \frac{ds}{R} = \frac{ds}{(EI/M)} = \frac{M\,ds}{EI}$$

$$\therefore \quad dU = \frac{1}{2} M \times d\theta = \frac{1}{2} M \times \frac{M\,ds}{EI} = \frac{1}{2}\frac{M^2}{EI} ds.$$

Differentiating partially the above equation w.r.t. M_0, we get

$$\frac{\partial U}{\partial M_0} = \int_0^{\pi} \frac{1}{2EI} \times 2M \times \frac{\partial M}{\partial M_0}\, ds \quad \text{...(iii)}$$

Let us now find the value of $\frac{\partial M}{\partial M_0}$.

We know from equation (*i*) that

$$M = M_0 - \frac{1}{2} P \times (R \sin \theta)$$

Differentiating partially with respect to M_0, we get

$$\therefore \quad \frac{\partial M}{\partial M_0} = 1 \qquad \text{(For partial differentiation, only } M_0 \text{ is variable.)}$$

Substituting this value in equation (*iii*), we get

$$\frac{\partial U}{\partial M_0} = \int_0^{\pi} \frac{1}{2EI} \times 2\,M \times 1\, ds = \int_0^{\pi} \frac{M}{EI} \times R d\theta \qquad (\because \quad ds = Rd\theta)$$

$$= \int_0^{\pi} \frac{\left(M_0 - \frac{1}{2} \times P \times R \sin \theta\right) \times R d\theta}{EI} \qquad \left(\because \; M = M_0 - \frac{P}{2} \times R \sin \theta\right)$$

But according to the principle of minimum strain energy,

$$\frac{\partial U}{\partial M_0} = 0$$

$$\therefore \quad \int_0^{\pi} \frac{\left(M_0 - \frac{1}{2} \times P \times R \sin \theta\right) R \times d\theta}{EI} = 0 \text{ or } \int_0^{\pi} \left(M_0 - \frac{1}{2} \times P \times R \sin \theta\right) d\theta = 0$$

or $$\left[M_0 \times \theta + \frac{1}{2} P \times R \times \cos \theta\right]_0^{\pi} = 0$$

or $$M_0 (\pi - 0) + \frac{1}{2} \times P \times R \times [\cos \pi - \cos 0] = 0$$

or $$M_0 \times \pi + \frac{1}{2} \times P \times R \times [-1 - 1] = 0 \quad \text{or} \quad M_0 \times \pi + \frac{1}{2} \times P \times R \times (-2) = 0$$

$$\therefore \quad M_0 = \frac{P \times R}{\pi} = 0.318\, P \times R$$

Now the value of bending moment M can be obtained by substituting M_0 in equation (*i*), as

$$M = M_0 - \frac{1}{2} \times P \times R \times \sin \theta = \frac{P \times R}{\pi} - \frac{1}{2} \times P \times R \times \sin \theta \qquad \left[\because \; M_0 = \frac{P \times R}{\pi}\right]$$

$$= P \times R \left(\frac{1}{\pi} - \frac{1}{2} \sin \theta\right) \quad \text{...(10.10)}$$

Value of M at section BB where $\theta = 0$, becomes

$$M_B = P \times R \left(\frac{1}{\pi}\right) = 0.318 \times P \times R$$

Value of M at section C-C is obtained by substituting

$$\theta = \frac{\pi}{2} \text{ in above equation,}$$

$$\therefore \quad M \text{ at } C\text{-}C, M_C = P \times R\left(\frac{1}{\pi} - \frac{1}{2}\sin\frac{\pi}{2}\right) = P \times R\left(\frac{1}{\pi} - \frac{1}{2} \times 1\right) \qquad \left[\because \sin\frac{\pi}{2} = 1\right]$$

$$= P \times R\,[0.318 - 0.5] = -0.182 \times P \times R$$

The bending moment at $\theta = 0$ is +ve whereas bending moment at $\theta = \frac{\pi}{2}$ is –ve. This means M is zero between $\theta = 0$ to $\theta = 90°$. The value of θ, where M is zero is obtained by substituting $M = 0$ in equation (10.10),

$$\therefore \quad 0 = P \times R \times \left(\frac{1}{\pi} - \frac{1}{2}\sin\theta\right)$$

or $$\frac{1}{2}\sin\theta = \frac{1}{\pi} \quad \text{or} \quad \sin\theta = \frac{2}{\pi} = 0.6366$$

$$\theta = 39.5°$$

The bending moment (M) can be obtained at any section of the ring. Now the bending stress can be determined from equation (10.1) as

$$\sigma_b = \frac{M}{A \times R}\left[1 + \frac{R^2}{h^2}\left(\frac{y}{R+y}\right)\right]$$

The bending stress will be maximum at the outer edge and inner edge of the cross-section where $y = +d_2$ and $y = -d_1$

where d_2 = distance of the extreme outer edge from centre line

d_1 = distance of the extreme inner edge from centre line [Refer to Fig. 10.14 (*a*)]

The direct stress (σ_0) will be due to pull $\frac{P}{2} \times \sin\theta$

$$\therefore \quad \sigma_0 = \frac{P}{2A} \times \sin\theta$$

The resultant stress (σ_r) at any point will be equal to algebraic sum of direct stress and bending stress.

$$\therefore \quad \sigma_r = \sigma_0 + \sigma_b$$

(*i*) **Stresses when θ = 0**

Direct stress, $\sigma_0 = \frac{P}{2A} \times \sin\theta = \frac{P}{2A} \times \sin 0° = 0$

Bending stress, $\sigma_b = \frac{M}{A \times R}\left[1 + \frac{R^2}{h^2}\left(\frac{y}{R+y}\right)\right]$

where M = bending moment and given by equation (10.10) as

$$M = P \times R\left(\frac{1}{\pi} - \frac{1}{2}\sin\theta\right) = P \times R \times \left(\frac{1}{\pi} - 0\right) = \frac{P \times R}{\pi}$$

$\therefore$ Resultant stress, $\sigma_r = \sigma_0 + \sigma_b$

(*ii*) **Stresses when** $\theta = \frac{\pi}{2} = 90°$

Direct stress, $$\sigma_0 = \frac{P}{2A} \sin 90° = \frac{P}{2A}$$

Bending stress, $$\sigma_b = \frac{M}{A \times R}\left[1 + \frac{R^2}{h^2}\left(\frac{y}{R+y}\right)\right]$$

where $$M = P \times R\left(\frac{1}{\pi} - \frac{1}{2}\sin\theta\right) = P \times R\left(\frac{1}{\pi} - \frac{1}{2}\sin 90°\right) = P \times R\left(\frac{1}{\pi} - 0.5\right)$$

Hence resultant stress,

$$\sigma_r = \sigma_0 + \sigma_b.$$

10.14. SOLVED PROBLEM 10.7 FOR STRESS IN CIRCULAR RINGS

Problem 10.7. *A closed ring of mean radius of curvature 90 mm is subjected to a pull of 3 kN. The line of action of the load passes through the centre of the ring. Calculate the maximum tensile and compressive stresses in the material of the ring if the ring is circular in cross-section with diameter equal to 15 mm.*

Sol. Given :

Mean radius of curvature of closed ring, $R = 90$ mm

Pull, $P = 3$ kN $= 3 \times 1000$ N $= 3000$ N

Cross-section of ring is circular with $d = 15$ mm

∴ $$r = \frac{d}{2} = \frac{15}{2} = 7.5 \text{ mm}$$

∴ Area of cross-section, $$A = \frac{\pi}{4} d^2 = \frac{\pi}{4} \times 15^2 = 176.70 \text{ mm}^2$$

Direct stress is given by, $$\sigma_0 = \frac{P}{2A} \times \sin\theta \quad \text{...}(i)$$

The bending stress (σ_b) is given by,

$$\sigma_b = \frac{M}{A \times R}\left[1 + \frac{R^2}{h^2} \times \left(\frac{y}{R+y}\right)\right] \quad \text{...}(ii)$$

where M = bending moment and is given by equation (10.10)

$$= P \times R\left(\frac{1}{\pi} - \frac{1}{2}\sin\theta\right) \quad \text{...}(iii)$$

Let us find first the value of h^2 for circular section. It is given by equation (10.8) as

$$h^2 = \frac{d^2}{16} + \frac{1}{128}\frac{d^4}{R^2} + \ldots$$

where d = Dia. of circular cross-section

= 15 mm here

R = Radius of curvature

= 90 mm here

$$= \frac{15^2}{16} + \frac{1}{128} \times \frac{15^4}{90^2}$$
$$= 14.0625 + 0.0488 = 14.1113 \simeq 14.11$$

(*i*) *Stresses when* $\theta = 0°$. From equation, (*i*), the direct stress,

$$\sigma_0 = \frac{P}{2A} \times \sin 0° = 0$$

Bending moment from equation (*iii*),

$$M = P \times R\left(\frac{1}{\pi} - \frac{1}{2} \times \sin 0°\right) = P \times R \times \left(\frac{1}{\pi} - 0\right)$$

$$= \frac{P \times R}{\pi} = \frac{3000 \times 90}{\pi} \text{ Nmm} = 85943 \text{ Nmm}$$

Bending stress (σ_b) from equation (*ii*),

$$\sigma_b = \frac{M}{A \times R}\left[1 + \frac{R^2}{h^2} \times \left(\frac{y}{R + y}\right)\right]$$

$$= \frac{85943}{176.7 \times 90}\left[1 + \frac{90^2}{14.11} \times \frac{y}{(90 + y)}\right]$$

$[\because \; A = 176.7 \text{ mm}^2 \text{ and } h^2 = 14.11]$

The bending stress at the outer most edge of the cross-section where $y = \frac{d}{2} = \frac{15}{2} = 7.5$ mm is given by

$$(\sigma_b)_0 = \frac{85943}{176.7 \times 90}\left[1 + \frac{90^2}{14.11}\left(\frac{7.5}{90 + 7.5}\right)\right] = 244 \text{ N/mm}^2 \quad \text{(tensile)}$$

The bending stress at the inner most edge of the cross-section where $y = -\frac{d}{2} = -7.5$ mm.

$$(\sigma_b)_i = \frac{85943}{176.7 \times 90}\left[1 + \frac{90^2}{14.11}\left(\frac{-7.5}{90 - 7.5}\right)\right]$$

$$= -276.6 \text{ N/mm}^2 \text{ (compressive)}$$

$\therefore$ Resultant stress at the outer-most edge,

$$(\sigma_r)_0 = \sigma_0 + (\sigma_b)_0 = 0 + 244 = 244 \text{ N/mm}^2 \text{ (tensile)}$$

Similarly resultant stress at the inner most edge,

$$(\sigma_r)_i = \sigma_0 + (\sigma_b)_i = 0 - 276.6 = \mathbf{-276.6 \text{ N/mm}^2} \quad \textbf{(compressive)}$$

Ans

(*ii*) *Stresses when* $\theta = 90° = \frac{\pi}{2}$

Direct stress, $\sigma_0 = \frac{P}{2A} \times \sin 90° = \frac{P}{2A}$

$$= \frac{3000}{2 \times 176.7} = 8.49 \text{ N/mm}^2 \text{ (tensile)}$$

Bending moment, $M = P \times R\left(\frac{1}{\pi} - \frac{1}{2} \times \sin 90°\right)$

$$= 3000 \times 90\left(\frac{1}{\pi} - \frac{1}{2}\right) = 270000\,(0.3183 - 0.5) = -49059 \text{ Nmm}$$

$$\therefore \quad \sigma_b = \frac{M}{A \times R}\left[1 + \frac{R^2}{h^2} \times \left(\frac{y}{R+y}\right)\right] = \frac{-49059}{176.7 \times 90}\left[1 + \frac{90^2}{14.11} \times \left(\frac{y}{90+y}\right)\right]$$

$$= -3.085\left[1 + 574 \times \left(\frac{y}{90+y}\right)\right]$$

The bending stress at outer-most edge where $y = 7.5$ mm,

$$(\sigma_b)_0 = -3.085\left[1 + 574 \times \left(\frac{7.5}{90+7.5}\right)\right] = -139.3 \text{ N/mm}^2 \text{ (compressive)}$$

The bending stress at inner most edge where $y = -7.5$ mm,

$$(\sigma_b)_i = -3.085\left[1 + 574 \times \left(\frac{-7.5}{90-7.5}\right)\right] = 157.9 \text{ N/mm}^2 \quad \text{(tensile)}$$

$\therefore$ Resultant stress at outer most edge,

$$(\sigma_r)_0 = \sigma_0 + (\sigma_b)_0$$

$$= 8.49 + (-139.3) = -130.81 \text{ N/mm}^2 \text{ (compressive)}$$

Resultant stress at inner most edge,

$$(\sigma_r)_i = \sigma_0 + (\sigma_b)_i = 8.49 + 157.9 = 166.39 \text{ N/mm}^2 \text{ (tensile)}$$

Comparing the resultant stresses when $\theta = 0$ and when $\theta = \frac{\pi}{2}$, we get

Maximum tensile stress $=$ **244 N/mm². Ans.**

Maximum compressive stress $=$ **276.6 N/mm². Ans.**

10.14.1. Stresses in a Chain Link. A chain link consists of two parts. One part at the middle is a straight portion of length L, whereas the other part at the ends is semi-circular as shown in Fig. 10.15.

The stresses in the link can be determined by the same method as applied to circular ring.

Let L = Length of the straight portion of the link

R = Mean radius of curvature of the circular portion

P = Pull on the chain link

M = B.M. at any section in the circular portion

M_1 = B.M. at any section in the straight portion

Consider half portion of the link, subjected to pull $P/2$ as shown in Fig. 10.15 (*b*). Let M_0 = fixing end moments at section *BB*. Consider any section *X–X* making an angle θ with vertical section *BB*.

The bending moment (M) at the section *X–X* in the circular portion is given by

$$M = M_0 - \frac{P}{2} \times R \sin\theta \quad \text{...(i)}$$

At section *A–A* where $\theta = \frac{\pi}{2} = 90°$, the above bending moment becomes,

$$M = M_0 - \frac{P}{2} \times R \times \sin 90° = M_0 - \frac{P}{2} \times R \quad \text{...(ii)}$$

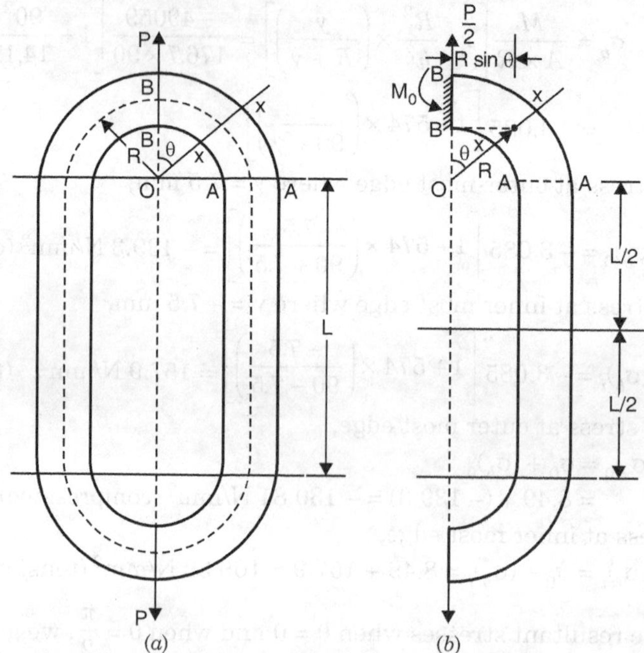

Fig. 10.15

As the section *AA* is common to circular portion and straight portion, hence B.M. for straight portion $= M_0 - \frac{P}{2} \times R$.

Value of M_0. Now the total strain energy stored in half portion of link is equal to the strain energy stored in circular portion plus strain energy stored in straight portion.

or $$U = 2\int_0^{\pi/2} (dU)_{\text{circular}} + 2\int_0^{L/2} (dU)_{\text{straight}}$$

$\left[\right.$ In half portion of link, there are two circular ends $\left(\text{from angle 0 to } \frac{\pi}{2}\right)$ and one straight portion of length L $\left.\left(\text{or two straight portions of lengths 0 to } \frac{L}{2}\right)\right]$

or $$U = 2\int_0^{\pi/2} \frac{M^2}{2EI}\, ds + 2\int_0^{L/2} \frac{M^2}{2EI}\, dx$$

Differentiating partially the above equation w.r.t., M_0, we get

$$\frac{\partial U}{\partial M_0} = \frac{2}{2EI}\left[\int_0^{\pi/2} (2M) \times \frac{\partial M}{\partial M_0} \times ds + \int_0^{L/2} (2M) \times \frac{\partial M}{\partial M_0}\, dx\right]$$

The strain energy will be minimum if $\frac{\partial U}{\partial M_0} = 0$

Hence for minimum strain energy,

$$\frac{2}{2EI}\left[\int_0^{\pi/2}(2M)\times\frac{\partial M}{\partial M_0}\times ds+\int_0^{L/2}(2M)\frac{\partial M}{\partial M_0}\,dx\right]=0$$

or
$$\int_0^{\pi/2}(2M)\times\frac{\partial M}{\partial M_0}\,ds+\int_0^{L/2}(2M)\frac{\partial M}{\partial M_0}\,dx=0$$

or
$$\int_0^{\pi/2}M\frac{\partial M}{\partial M_0}\,ds+\int_0^{L/2}M\frac{\partial M}{\partial M_0}\,dx=0 \quad ...(iii)$$

For circular portion, $M=M_0-\dfrac{P\times R}{2}\sin\theta$

$\therefore$ $\dfrac{\partial M}{\partial M_0}=1$ (For partial differentiation, only M_0 is variable)

and also $ds=R\times d\theta$

$$\therefore \int_0^{\pi/2}M\frac{\partial M}{\partial M_0}\times ds=\int_0^{\pi/2}\left(M_0-\frac{P\times R}{2}\sin\theta\right)\times 1\times Rd\theta$$

$$=R\left(M_0\times\theta+\frac{PR}{2}\cos\theta\right)_0^{\pi/2}$$

$$=R\left[M_0\left(\frac{\pi}{2}-0\right)+\frac{PR}{2}\left(\cos\frac{\pi}{2}-\cos 0^\circ\right)\right]$$

$$=R\left[M_0\times\frac{\pi}{2}+\frac{PR}{2}(0-1)\right]=R\left[M_0\times\frac{\pi}{2}-\frac{P\times R}{2}\right] \quad ...(iv)$$

For straight portion,

$$M=M_0-\frac{P\times R}{2}$$

$$\therefore \frac{\partial M}{\partial M_0}=1$$

$$\therefore \int_0^{L/2}M\frac{\partial M}{\partial M_0}\,dx=\int_0^{L/2}\left(M_0-\frac{P\times R}{2}\right)\times 1\times dx$$

$$=\int_0^{L/2}\left(M_0-\frac{PR}{2}\right)dx=\left[\left(M_0-\frac{PR}{2}\right)\times x\right]_0^{L/2}$$

$$=\left(M_0-\frac{PR}{2}\right)\times\frac{L}{2} \quad ...(v)$$

Substituting equations (*iv*) and (*v*) in equation (*iii*), we get

$$R\left[M_0\times\frac{\pi}{2}-\frac{P\times R}{2}\right]+\left(M_0-\frac{P\times R}{2}\right)\times\frac{L}{2}=0$$

$$M_0\times R\times\frac{\pi}{2}-\frac{P\times R^2}{2}+M_0\times\frac{L}{2}-\frac{P\times R}{2}\times\frac{L}{2}=0$$

or
$$M_0\left(R\times\frac{\pi}{2}+\frac{L}{2}\right)-\frac{PR}{2}\left(R+\frac{L}{2}\right)=0$$

or $$M_0\left(R\times\frac{\pi}{2}+\frac{L}{2}\right)=\frac{PR}{2}\left(R+\frac{L}{2}\right)$$

or $$\frac{M_0}{2}(R\times\pi+L)=\frac{P\times R}{2}\left(R+\frac{L}{2}\right)$$

or $$M_0(\pi R+L)=P\times R\left(R+\frac{L}{2}\right) \qquad \left[\text{Cancelling } \frac{1}{2}\right]$$

or $$M_0=\frac{P\times R(2R+L)}{2(\pi R+L)}$$

$$=\frac{P\times R}{2}\left(\frac{2R+L}{\pi R+L}\right) \qquad ...(10.11)$$

Substituting this value of M_0 in equation (*i*), we get the moment at any section in the curved portion as

$$M=\frac{P\times R}{2}\left(\frac{2R+L}{\pi R+L}\right)-\frac{P}{2}\times R\sin\theta$$

$$=\frac{P\times R}{2}\left[\frac{2R+L}{\pi R+L}-\sin\theta\right] \qquad ...(10.12)$$

When $\theta=\frac{\pi}{2}=90°$ (*i.e.*, at section *AA*) this moment becomes as

$$M=\frac{P\times R}{2}\left(\frac{2R+L}{\pi R+L}-1\right)=\frac{P\times R}{2}\left(\frac{2R+L-\pi R-L}{\pi R+L}\right)$$

$$=\frac{P\times R}{2}\left(\frac{2R-\pi R}{\pi R+L}\right) \qquad ...(10.13)$$

As πR is more than $2R$, hence the term $(2R-\pi R)$ is –ve. Hence the bending moment at section *AA* will be negative. The equation (10.13) also gives the bending moment in the straight portion. Once the bending moments in the curved portion and in the straight portion are known, the bending stresses can easily be calculated. These portions will also be subjected to direct stresses due to pull of $\frac{P}{2}\sin\theta$ in the curved portion and a pull of $\frac{P}{2}$ in the straight portion.

Stresses in the Curved Portion

The stress at any section is the algebraic sum of direct stress and bending stress.

∴ $$\sigma_r=\sigma_0+\sigma_b$$

where $$\sigma_0=\frac{P}{2A}\times\sin\theta \qquad ...(10.14)$$

and $$\sigma_b=\frac{M}{A\times R}\left[1+\frac{R^2}{h^2}\times\left(\frac{y}{R+y}\right)\right]$$

where $$M=\frac{P\times R}{2}\left[\frac{2R+L}{\pi R+L}-\sin\theta\right] \qquad ...(10.15)$$

At any section, there will two maximum values of (σ_b).

One value will be at $y=+d_2$ and other value will be at

$y=-d_1$.

Here d_2 = distance of extreme outer edge of the section from centre line of curved portion and d_1 = distance of extreme inner edge of the section from centre line.

Stresses in Straight Portion

Here σ_0 = direct stress

$$= \frac{P}{2A} \quad \text{...(10.16)}$$

σ_b = bending stress

$$= \frac{M}{I} \times y \quad \text{where } M = \frac{P \times R}{2}\left(\frac{2R - \pi R}{\pi R + L}\right) \quad \text{...(10.17)}$$

Here also $y = +d_2$ and $y = -d_1$

σ_r = resultant stress

$= \sigma_0 + \sigma_b$.

10.14.2. Solved Problem 10.8 for Stresses in a Chain Link

Problem 10.8. *A pull of 25 kN is applied to a simple chain link shown in Fig. 10.16. The mean radius of curvature of the semi-circular ends is 40 mm. The link is circular in cross-section with radius = 20 mm. If the length of straight portion is 40 mm, determine the stresses in the link.*

Sol. Given :

$P = 25 \text{ kN} = 25 \times 1000 \text{ N}$

$R = 40 \text{ mm}$

Cross-section of link is circular with $r = 20$ mm

$\therefore$ $d = 2r = 2 \times 20 = 40$ mm

$L = 40$ mm

Area of cross-section, $A = \pi r^2$

$= \pi \times 20^2 \text{ mm}^2$

$= 400\pi \text{ mm}^2 = 1256.6 \text{ mm}^2$

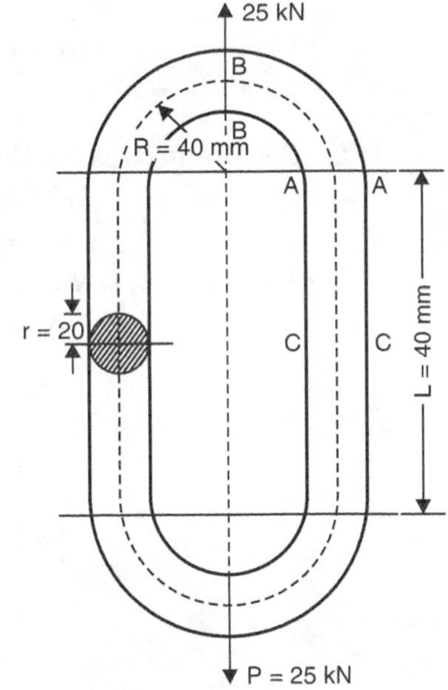

Fig. 10.16

Let us find the value of h^2 for circular section. It is given by equation (10.8) as

$$h^2 = \frac{d^2}{16} + \frac{1}{128}\frac{d^4}{R^2} + \ldots\ldots$$

$$= \frac{40^2}{16} + \frac{1}{128} \times \frac{40^4}{40^2}$$

$$= 100 + 12.5 = 112.5$$

(i) Stresses in curved portion

(a) At θ = 0 i.e., section BB

Direct stress, $\sigma_0 = \frac{P}{2A} \times \sin\theta = 0$

Bending stress, $\sigma_b = \frac{M}{A \times R}\left[1 + \frac{R^2}{h^2}\left(\frac{y}{R+y}\right)\right]$

where M = bending moment in curved portion and is given by equation (10.12) as

$$M = \frac{P \times R}{2}\left(\frac{2R + L}{\pi R + L} - \sin\theta\right)$$

$$= \frac{P \times R}{2}\left(\frac{2R+L}{\pi R+L} - \sin 0°\right) \quad (\text{at } \theta = 0°)$$

$$= \frac{P \times R}{2}\left(\frac{2R+L}{\pi R+L}\right) = \frac{25000 \times 40}{2}\left(\frac{2 \times 40 + 40}{\pi \times 40 + 40}\right)$$

$$= 362187.6 \text{ Nmm}$$

$$\therefore \quad \sigma_b = \frac{362187.6}{1256.6 \times 40}\left[1 + \frac{40^2}{112.5}\left(\frac{y}{40+y}\right)\right]$$

$$= 7.205\left[1 + 14.22\left(\frac{y}{40+y}\right)\right]$$

Bending stress at the extreme outer edge where $y = r = +\,20$ mm

$$\therefore \quad (\sigma_b)_0 = 7.205\left[1 + 14.22\left(\frac{20}{40+20}\right)\right] = 41.35 \text{ N/mm}^2$$

$\therefore$ Resultant stress at the extreme outer edge,

$$(\sigma_r)_0 = \sigma_0 + (\sigma_b)_0$$

$$= 0 + 41.35 = \mathbf{41.35\ N/mm^2\ (tensile).} \quad \textbf{Ans.}$$

Bending stress at the extreme inner edge where $y = r = -\,20$ mm,

$$(\sigma_b)_i = 7.205\left[1 + 14.22\,\frac{(-20)}{(40-20)}\right]$$

$$= -\,95.25 \text{ N/mm}^2 \text{ (compressive)}$$

$\therefore$ Resultant stress at the extreme inner edge,

$$(\sigma_r)_i = \sigma_0 + (\sigma_b)_i$$

$$= 0 + (-\,95.25) = \mathbf{-\,95.25\ N/mm^2\ (compressive).} \quad \textbf{Ans.}$$

(*b*) *at* $\theta = 90°$ *i.e., section AA*

$$\sigma_0 = \text{direct stress}$$

$$= \frac{P}{2A} \times \sin\theta = \frac{P}{2A} \times \sin 90° = \frac{P}{2A}$$

$$= \frac{25000}{2 \times 1256.6} = 9.95 \text{ N/mm}^2 \text{ (tensile)}$$

$$\sigma_b = \text{bending stress}$$

$$= \frac{M}{A \times R}\left[1 + \frac{R^2}{h^2}\left(\frac{y}{R+y}\right)\right]$$

where

$$M = \frac{P \times R}{2}\left(\frac{2R+L}{\pi R+L} - \sin 90°\right)$$

$$= \frac{P \times R}{2}\left(\frac{2R+L}{\pi R+L} - 1\right)$$

$$= \frac{25000 \times 40}{2}\left(\frac{2 \times 40 + 40}{\pi \times 40 + 40} - 1\right)$$

$$= 500000\,(0.724 - 1) = -\,137812 \text{ Nmm.}$$

$$= \frac{-137812}{1256.6 \times 40}\left[1 + \frac{40^2}{112.5}\left(\frac{y}{40+y}\right)\right]$$

$$= -2.74\left[1 + 14.22\left(\frac{y}{40+y}\right)\right]$$

Bending stress at outer edge where $y = 20$ mm

$$(\sigma_b)_0 = -2.74\left[1 + 14.22 \times \left(\frac{20}{40+20}\right)\right]$$

$$= -15.73 \text{ N/mm}^2 \text{ (compressive)}$$

Resultant stress at outer edge

$\therefore$
$$(\sigma_r)_0 = \sigma_0 + (\sigma_b)_0$$

$$= 9.95 + (-15.73) = \mathbf{-5.78 \text{ N/mm}^2 \text{ (compressive)}}. \quad \textbf{Ans.}$$

Bending stress at the inner edge where $y = -20$ mm,

$$(\sigma_b)_i = -2.74\left[1 + 14.22\frac{(-20)}{(40-20)}\right]$$

$$= 36.22 \text{ N/mm}^2 \text{ (tensile)}$$

$\therefore$ Resultant stress at the inner edge,

$$(\sigma_r)_i = \sigma_0 + (\pi_b)_i$$

$$= 9.95 + 36.22 = \mathbf{46.17 \text{ N/mm}^2 \text{ (tensile)}}. \quad \textbf{Ans.}$$

(ii) Stresses in straight portion

Here
$$\sigma_0 = \text{direct stress}$$

$$= \frac{P}{2A} = \frac{25000}{2 \times 1256.6} = 9.95 \text{ N/mm}^2 \text{ (tensile)}$$

$$\sigma_b = \text{bending stress}$$

$$= \frac{M}{I} \times y$$

where M = bending moment

$$= \frac{P \times R}{2}\left(\frac{2R+L}{\pi R+L} - 1\right) = -137812 \text{ Nmm (Already calculated)}$$

$$I = \text{M.O.I.} = \frac{\pi}{64} d^4 = \frac{\pi}{64} \times 40^4 \text{ mm}^2$$

$\therefore$
$$\sigma_b = \frac{-137812}{\frac{\pi}{64} \times 40^4} \times y$$

The bending stress at any point on outer edge of straight portion where $y = +20$ mm,

$$(\sigma_b)_0 = \frac{-137812 \times 64}{\pi \times 40^4} \times 20 = -21.93 \text{ N/mm}^2 \text{ (compressive)}$$

The bending stress at any point on inner edge of straight portion where $y = -20$ mm,

$$(\sigma_b)_i = \frac{-137812 \times 64}{\pi \times 40^4} \times (-20) = +21.93 \text{ N/mm}^2 \text{ (tensile)}$$

$\therefore$ Total stress at any point on outer edge of straight portion is,

$$(\sigma_r)_0 = \sigma_0 + (\sigma_b)_0 = 9.95 + (-21.93)$$

$$= \mathbf{-11.98 \text{ N/mm}^2 \text{ (compressive)}}. \quad \textbf{Ans.}$$

And total stress at any point on inner edge of the straight portion is,

$$(\sigma_r)_i = \sigma_0 + (\sigma_b)_i = 9.95 + 21.93 = \mathbf{31.88\ N/mm^2\ (tensile).}\quad \textbf{Ans.}$$

The maximum compressive stress in the link is 95.25 N/mm^2 at the inner edge of curved portion where load is applied and maximum tensile stress is 46.17 N/mm^2 just at the junction of curved and straight portion at the inner edge.

HIGHLIGHTS

1. The stress at any point in case of curved bar subjected to a bending moment is given by,

$$\sigma = \frac{M}{R \times A}\left[1 + \frac{R^2}{h^2}\left(\frac{y}{R+y}\right)\right]$$

where σ = Bending stress (*i.e.*, σ_b)

M = Bending moment with which the bar is subjected

R = Radius of curvature of curved bar, or it is the distance of axis of curvature from centroidal axis

A = Area of cross-section

y = Distance of any point from centroidal axis

h^2 = is a constant for a cross-section

$$= \frac{1}{A}\int \frac{y^2 dA}{\left(1 + \frac{y}{R}\right)}.$$

2. The distance of neutral axis from centroidal axis is given by,

$$y_0 = -\frac{R \times h^2}{(R^2 + h^2)}$$

– ve sign shows that neutral axis is below the centroidal axis.

3. The value of h^2 for different cross-sections is given as :

(*i*) For rectangular section :

$$h^2 = \frac{R^3}{d}\left[\log_e\left(\frac{2R+d}{2R-d}\right)\right] - R^2 = R^2\left[\frac{1}{3}\left(\frac{d}{2R}\right)^2 + \frac{1}{5}\left(\frac{d}{2R}\right)^4 + \ldots\ldots\right]$$

where d = Depth of section,

R = Radius of curvature or distance of axis of curvature from centroidal axis.

The value of 'h^2' is independent of width 'b'

(*ii*) For triangular section.

(*a*) $h^2 = \dfrac{2R^3}{d^2}\left[R_2 \log_e\left(\dfrac{R_2}{R_1}\right) - (R_2 - R_1)\right] - R^2$

where R_2 = Distance of apex of triangle from axis of curvature

R_1 = Distance of base of triangle from axis of curvature

d = Depth or height of triangle

(*b*) The value of h^2 for triangular section, in terms of R and d is also given as :

$$h^2 = \frac{2R^3}{d^2}\left[\left(\frac{3R+2d}{3}\right)\left\{\log_e\left(\frac{3R+2d}{3R-d}\right)\right\} - d\right] - R^2$$

(*iii*) For trapezoidal section.

$$h^2 = \frac{R^3}{A}\left[\left\{b_2 + \left(\frac{b_1 - b_2}{d}\right) \times R_2\right\} \log_e \left(\frac{R_2}{R_1}\right) - (b_1 - b_2)\right] - R^2$$

where b_1 = Width of section at base,
b_2 = Width of section at top,
d = Depth of trapezoidal section
R_2 = Distance of top layer of section from axis of curvature
R_1 = Distance of bottom layer of section from axis of curvature

(*iv*) For circular section :

$$h^2 = \frac{d^2}{16} + \frac{1}{128}\frac{d^4}{R^2} + ...$$

d = Diameter of circular section,
R = Radius of curvature of curved bar.

4. Resultant stress in a curved bar subject to direct stress and bending stress is given by

$$\sigma_r = \sigma_0 + \sigma_b$$

where $\sigma_b = \frac{M}{A \times R}\left[1 + \frac{R^2}{h^2}\left(\frac{y}{R + y}\right)\right]$.

y is positive and maximum at top layer and is –ve and maximum at bottom layer. Also
y = distance of any layer from centroidal axis.

5. Resultant stress in a circular ring :

$$\sigma_r = \sigma_0 + \sigma_b \quad \text{where} \quad \sigma_0 = \frac{P}{2A} \times \sin\theta$$

and $$\sigma_b = \frac{M}{A \times R}\left[1 + \frac{R^2}{h^2}\left(\frac{y}{R + y}\right)\right]$$

where M = bending moment

$$= P \times R\left[\frac{1}{\pi} - \frac{1}{2}\sin\theta\right].$$

6. Resultant stress in a chain link

$$\sigma_r = \sigma_0 + \sigma_b$$

where $\sigma_0 = \frac{P}{2A} \times \sin\theta$ ... for circular portion

$= \frac{P}{2A}$ ... for straight portion

σ_b = bending stress ...for circular portion

$$= \frac{M}{A \times R}\left[1 + \frac{R^2}{h^2} \times \left(\frac{y}{R + y}\right)\right].$$

and M = bending moment

$$= \frac{P \times R}{2}\left[\frac{2R + L}{\pi R + L} - \sin\theta\right] \quad \text{... For circular portion}$$

σ_b = for straight portion

$$= \frac{M}{I} \times y$$

where $M = \frac{P \times R}{2}\left[\frac{2R + L}{\pi R + L} - 1\right]$ and I = M.O. Inertia.

EXERCISE 10

(A) Theoretical Questions

1. (*a*) What are the assumptions made in the derivation of stresses in a curved bar which is subjected to bending moments ?
   (*b*) Find an expression for bending stress produced in a curved bar which is subjected to bending moment.
2. Write-down the expression for Winkler-Bach formula.
3. Prove that the position of neutral axis from centroid axis is given by

$$y_0 = -\frac{R \times h^2}{(R^2 + h^2)}$$

where R = radius of curvature of curved bar or distance or axis of curvature from centroidal axis
h^2 = a constant which depends upon the shape of cross-section.
4. Find an expression for h^2 for the following cross-sections
   (*i*) rectangular section (*ii*) circular section and (*iii*) trapezoidal section.
5. Write an expression for resultant stresses in a curved member subjected to direct stress and bending stress.
6. Find an expression for the bending moment in a circular ring which is subjected to a tensile load P along the diameter.
7. Prove that bending moment (M) is a simple chain link for curved portion is given by,

$$M = \frac{P \times R}{2}\left(\frac{2R + L}{\pi R + L} - \sin\theta\right)$$

8. Explain in details the procedure of finding resultant stress in case of :
   (*a*) circular ring and
   (*b*) chain link
   When they are subjected to a tensile load along the longitudinal axis.

(B) Numerical Problems

1. Determine : (*i*) location of neutral axis, (*ii*) maximum and minimum stresses and (*iii*) ratio of maximum and minimum stress,
   When a curved beam of rectangular cross-section of width 10 mm and depth 20 mm is subjected to pure bending of moment + 50 Nm. The beam is curved in a plane parallel to depth. The mean radius of curvature is 25 mm. Also plot the variation of stresses across the section.
   [**Ans.** (*i*) y_0 = – 1.39 mm, (*ii*) 1.76, (*iii*) 103 N/mm^2 (comp.) and 58.45 N/mm^2 (tensile)]
2. Determine : (*i*) location of neutral axis and (*ii*) maximum and minimum stresses, when a curved beam of trapezoidal section of bottom width 45 mm, top width 30 mm and height 60 mm is subjected to pure bending moment of + 1350 Nm. The bottom width is towards the centre of curvature. The radius of curvature is 75 mm and beam is curved in a plane parallel to depth. Also plot the variation of stresses across the section.
   [**Ans.** (*i*) y_0 = – 3.98 mm, (*ii*) 76.98 N/mm^2 (comp.), 50.65 N/mm^2 (tensile)]
3. Solve the problem 2, assuming the section to be circular of diameter 60 mm. The mean radius of curvature of the beam is 75 mm.
   [**Ans.** (*i*) y_0 = – 3.1 mm, (*ii*) 91.95 N/mm^2 (comp.) and 48.64 N/mm^2 (tensile)
4. A central horizontal section of a hook is symmetrical trapezium 90 mm deep. The inner width being 90 mm and outer being 45 mm. The hook carries a load of 67.5 kN, the load line passes at a distance of 40 mm from the inside edge of the section. The centre of curvature is in the load line. Calculate the extreme intensities of stress. Also plot the stress distribution across the section.
   [**Ans.** $(\sigma_r)_0$ = 53.15 N/mm^2 (comp) and $(\sigma_r)_i$ = 106.5 N/mm^2 (tensile)]

5. A central horizontal section of a hook is a symmetrical trapezium of inner width = 67.5 mm and of outer width = 22.5 mm. The depth of the section is 90 mm. The hook carries a load of 37.5 kN. The load line passes through the centre of curvature. The radius of the hook is 52.5 mm. Determine the maximum compressive and tensile stresses in the section of the hook.

[**Ans.** (*i*) 83.70 MN/m^2 (tensile) and (*ii*) 43.2 MN/m^2 (comp)]

6. A central horizontal section of a hook is a trapezium with inner width = 80 mm, outer width = 50 mm and depth 150 mm. The centre of curvature of the section is at a distance of 120 mm from the inner fibre and the load line is 100 mm from the inner fibre. What maximum load, the hook will carry if maximum stress is not to exceed 120 MN/m^2 [**Ans.** 122.5 kN]

7. Determine the maximum tensile and compressive stresses produced in a closed ring when the ring is subjected to a pull of 11.25 kN, the line of action of which passes through the centre of the ring. The mean radius of curvature of the closed ring is 90 mm. The ring is circular in cross-section with diameter equal to 30 mm. [**Ans.** 20.35 N/mm^2 (comp.) and 13.6 N/mm^2 (tensile)]

8. A pull of 75 kN is applied at the ends of a simple link which consists of semi-circular ends of mean radius of curvature 60 mm and middle straight portion of length 60 mm. The link is circular in cross-section with radius = 30 mm. Determine the stresses in the link.

[**Ans.** (*i*) At $\theta = 0$, 93.65 N/mm^2 (comp) and 40.75 N/mm^2 (tensile)

(*ii*) At $\theta = 90°$, 45.56 N/mm^2 (tensile) and 5.6 N/mm^2 (compressive)

(*iii*) Straight portion, 31.95 N/mm^2 (tensile) and 12 N/mm^2 (compressive)]

9. Find : (*i*) the position of neutral axis and (*ii*) greatest bending stresses, when a curved bar of rectangular section 30 mm wide by 37.5 mm deep initially unstressed is subjected to bending moment of 281.25 Nm which tends to straighten the bar. The mean radius of curvature is 75 mm.

[**Hint.** $b = 30$ mm ; $d = 37.5$ mm ; $M = -281.25$ Nm $= -281.25 \times 10^3$ Nmm (the bending moment M is –ve as it decreases the curvature of the bar or tends to straighten the bar) ; $R = 75$ mm

$$A = b \times d = 30 \times 37.5 = 1125 \text{ mm}^2$$

$$h^2 = \frac{R^3}{d}\left[\log_e\left(\frac{2R+d}{2R-d}\right)\right] - R^2$$

$$= \frac{75^3}{37.5}\left[\log_e\left(\frac{2\times 75+37.5}{2\times 75-37.5}\right)\right] - 75^2$$

$$= 11250[0.5108] - 5625$$

$$= 5746.78 - 5625 = 121.8 \text{ mm}^2$$

$$(i)\ y_0 = \frac{-Rh^2}{(R^2+h^2)} = \frac{-75\times 121.8}{(75^2+121.8)} = \frac{-75\times 121.8}{5625+121.8} = -1.59 \text{ mm}$$

$$(ii)\ \sigma_b = \frac{M}{A\times R}\left[1+\frac{R^2}{h^2}\left(\frac{y}{R+y}\right)\right]$$

$$= \frac{-281.25\times 10^3}{1125\times 75}\left[1+\frac{75^2}{121.8}\left(\frac{y}{75+y}\right)\right]$$

$$= -3.33\left[1+46.182\left(\frac{y}{75+y}\right)\right]$$

At $y = \frac{d}{2} = \frac{37.5}{2} = 18.75$ mm, we get

$$(\sigma_b)_0 = -3.33\left[1 + 46.182 \times \frac{18.75}{(75 + 18.75)}\right] = -34.03 \text{ N/mm}^2 \text{ [compressive]}$$

At $y = -\frac{d}{2} = -18.75$ mm,

$$(\sigma_b)_i = -3.33\left[1 + 46.182 \times \frac{(-18.75)}{75 - 18.75}\right] = +47.93 \text{ N/mm}^2 \text{ (tensile)]}$$

11

Unsymmetrical Bending

11.1. INTRODUCTION

In the chapter of bending stresses, it is assumed that neutral axis of the cross-section of the beam is perpendicular to the plane of loading. This means that the plane of loading is parallel to the plane containing the principal centroidal axis of the inertia of cross-section of the beam. This type of bending is known symmetrical bending.

If the plane of loading or the plane of bending does not lie in (or parallel to) a plane that contains the principal centroidal axis of the cross-section, that bending is known as unsymmetrical bending. In case of unsymmetrical bending, the neutral axis is not perpendicular to the plane of bending.

The unsymmetrical bending will be when

(*i*) section is symmetrical (such as rectangular, circular, I-sections) but load line is inclined to both the principal axes.

(*ii*) the section is unsymmetrical (such as angle section or channel section etc.) and load line is along any centroidal axis.

11.2. PROPERTIES OF BEAM CROSS-SECTION

The integral $\int xy\, dA$ is known as product of inertia and pair of axes, for which it is zero, are known as principal axes of the cross-section. The moments of inertia of an area about its principal axes are known as principal moments of inertia. The bending moment about any other axis is known as unsymmetrical bending.

11.2.1. Principal Moments of Inertia. The principal axes of any area are those axes about which the product of inertia (I_{xy}) is zero. Axes of symmetry through centroid are automatically principal axes, as the product moments for opposite quadrant concelling each other out.

If *OX* and *OY* are two perpendicular axes through centriod and *OU* and *OV* are principal axes at an angle θ with axes *OX* and *OY* in anticlockwise direction as shown in Fig. 11.1. then :

(*i*) Condition for principal axes is

$$\tan 2\theta = \frac{2I_{xy}}{I_{yy} - I_{xx}} \qquad ...(11.1)$$

(*ii*) Principal moments of inertia about axes OU and OV are :

$$I_{UU} = \frac{1}{2}(I_{xx} + I_{yy}) + \frac{1}{2}(I_{xx} - I_{yy}) \cos 2\theta - I_{xy} \sin 2\theta \text{ also} \quad ...(11.1A)$$

$$I_{UU} = \frac{1}{2}(I_{xx} + I_{yy}) + \frac{1}{2}(I_{xx} - I_{yy}) \sec 2\theta \quad ...(11.1B)$$

$$I_{VV} = \frac{1}{2}(I_{xx} + I_{yy}) - \frac{1}{2}(I_{xx} - I_{yy}) \sec 2\theta \quad ...(11.2)$$

Also $\quad I_{UU} + I_{VV} = I_{xx} + I_{yy}$ ...(11.2A)

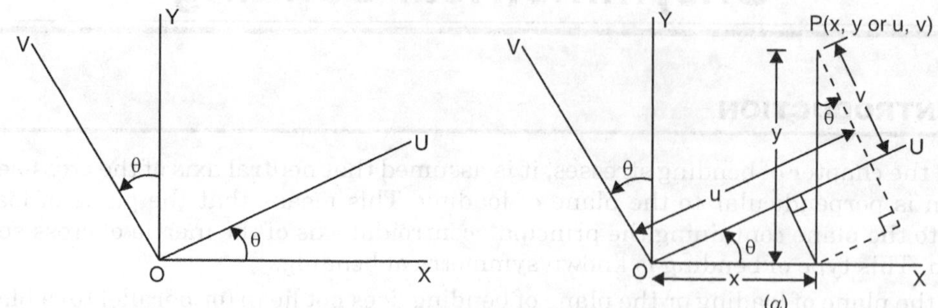

Fig. 11.1

(*iii*) The co-ordinates u, v relative to OU, OV and x, y relative to OX, OY of any point P are given by [Refer to Fig. 11.1(*a*)]

$$\left.\begin{array}{l} u = x \cos\theta + y \sin\theta \text{ and} \\ v = y \cos\theta - x \sin\theta \end{array}\right] \quad ...(11.2B)$$

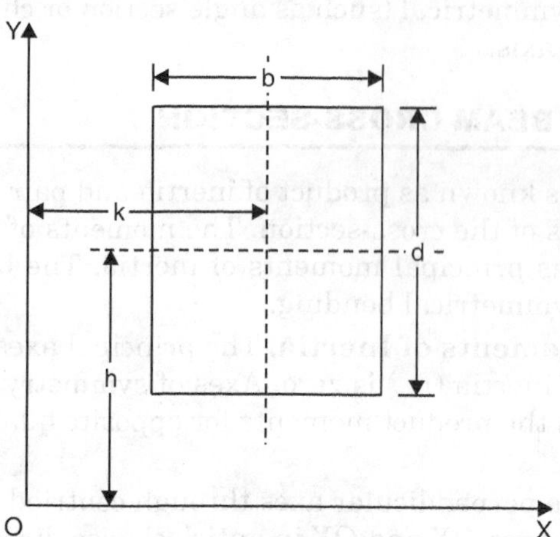

Fig. 11.2

(*iv*) If I_{xx} and I_{yy} are equal, then 2θ will be 90° and $\cos 2\theta = 0$. And $\sec 2\theta = \frac{1}{\cos 2\theta} = \frac{1}{0} = \infty$. Hence equation (11.1*B*) will not give correct result for I_{UU}. Hence to find I_{UU} in such cases, the equation (11.1*A*) should be used.

(v) A rectangle of width b and depth d is shown in Fig. 11.2. The sides of the rectangle are parallel to the principal axes. The product of inertia I_{xy} will be

$$I_{xy} = \int xy\, dA = \iint xy\, dy\, dx = \left[\frac{x^2}{2}\right]_{k-\frac{b}{2}}^{k+\frac{b}{2}} \times \left[\frac{y^2}{2}\right]_{h-\frac{d}{2}}^{h+\frac{d}{2}}$$

$$= \left[\frac{\left(k+\frac{b}{2}\right)^2}{2} - \frac{\left(k-\frac{b}{2}\right)^2}{2}\right] \times \left[\frac{\left(h+\frac{d}{2}\right)^2}{2} - \frac{\left(h-\frac{d}{2}\right)^2}{2}\right]$$

$$= kb \times hd = bd \times hk = Ahk \qquad (\because \; b \times d = A) \quad ...(11.2C)$$

Hence the product of inertia of a rectangle whose sides are parallel to the axes is equal to area of rectangle × distance of its C.G. from one axis × distance of its C.G. from other axis.

11.3. STRESS IN UNSYMMETRICAL BENDING

The cross-section of a beam subjected to a bending moment M in the plane of Y-Y is shown in Fig. 11.3. The co-ordinate axes XX and YY pass through the centroid G of the section.

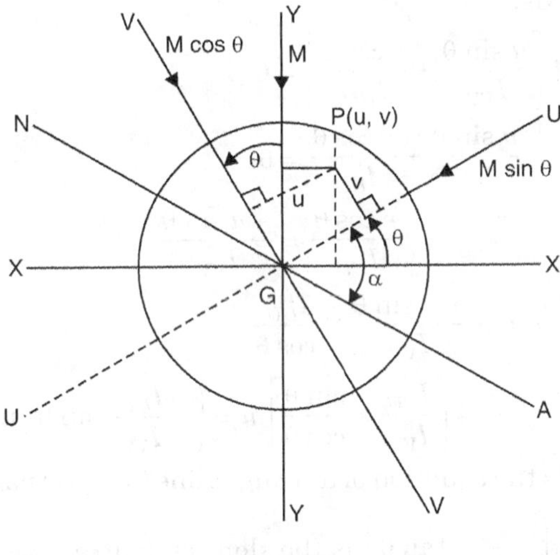

Fig. 11.3

Let UU, VV = principal axes passing through G and inclined at an angle θ to XX and YY axes respectively.

It is required to find the resultant stress at any point P having co-ordinates (x, y) with respect to axes XX, YY and (u, v) with respect to axes UU, VV.

To find the stress distribution over the section the moment M in plane YY is resolved into components in the plane UU and VV. (*i.e.*, in the plane of principal axes UU and VV)

The moment in plane $UU = M \sin\theta$

Moment in plane $VV = M \cos\theta$

The moment in plane UU, will bend the beam about axis VV. The bending stress (σ_b) due to this moment will be equal to $\frac{(M \sin\theta)}{I_{VV}} \times u$ $\qquad \left(\frac{M}{I} = \frac{\sigma}{y} \quad \text{or} \quad \sigma = \frac{M}{I} \times y\right).$

Similarly the moment in plane *VV*, will bend the beam about axis *UU*. The bending stress due to this moment will be equal to $\frac{(M \cos \theta)}{I_{UU}} \times v$.

Hence the resultant bending stress at any point $P(u, v)$ will be given by,

$$\sigma_b = \frac{(M \sin \theta) \times u}{I_{VV}} + \frac{(M \cos \theta) \times v}{I_{UU}}$$

$$= M\left[\frac{u \sin \theta}{I_{VV}} + \frac{v \cos \theta}{I_{UU}}\right] \quad \text{...(11.3)}$$

In the above equation, the signs of u and v will determine the nature of bending stress. If the co-ordinates of a point with respect of *XX*, *YY* axes are known then the co-ordinates of the same point with respect of *UU*, *VV* axes will be given by equation (11.2*B*) as

$$u = x \cos \theta + y \sin \theta$$

and

$$v = y \cos \theta - x \sin \theta$$

where θ = Inclination of principal axes *UU*, *VV* with axes *XX*, *YY*.

11.3.1. Neutral Axis (N.A.). At the neutral axis, the resultant bending stress is zero. Hence the equation of neutral axis is obtained by substituting the value of σ_b equal to zero in equation (11.3).

Hence for neutral axis,

$$M\left[\frac{u \sin \theta}{I_{VV}} + \frac{v \cos \theta}{I_{UU}}\right] = 0$$

or

$$\frac{u \sin \theta}{I_{VV}} + \frac{v \cos \theta}{I_{UU}} = 0$$

or

$$\frac{v \cos \theta}{I_{UU}} = \frac{-u \sin \theta}{I_{VV}}$$

or

$$v = \frac{-u \sin \theta}{I_{VV}} \times \frac{I_{UU}}{\cos \theta}$$

$$= -\left[\frac{I_{UU}}{I_{VV}} \times \frac{\sin \theta}{\cos \theta}\right] u = \left(-\frac{I_{UU}}{I_{VV}} \tan \theta\right) u \quad \text{...(11.4)}$$

The equation (11.4) is the equation of a straight line ($y = mx$) passing through the centroid *G* of the section. Here $m = \left(-\frac{I_{UU}}{I_{VV}} \tan \theta\right)$ is the slope of neutral axis.

11.3.2. Slope of Neutral Axis. Let α = Angle made by neutral axis with axis *UU*.

Then $\tan \alpha$ = slope of the neutral axis.

From equation (11.4), the slope of neutral axis is given as

$$\tan \alpha = m = \left(-\frac{I_{UU}}{I_{VV}} \tan \theta\right) \quad \text{...(11.4A)}$$

$$\therefore \quad \alpha = \tan^{-1}\left(-\frac{I_{UU}}{I_{VV}} \tan \theta\right) \quad \text{...(11.4B)}$$

Note. (*i*) The nature of stress on one side of N.A. will be same whereas on the other side of N.A., the stress will be of opposite nature.

(*ii*) The stress will be maximum at a point which is having maximum distance from N.A.

11.4. DEFLECTION OF BEAMS IN UNSYMMETRICAL BENDING

A transverse section of a beam with centroid G is shown in Fig. 11.4 along with rectangular co-ordinate axes XX' and YY'. The principal axes UU' and VV' inclined at an angle θ to the XY set of co-ordinates axes are also shown. W is the load acting along line YY'. This load can be resolved into two components, $W \sin \theta$ along UG and $W \cos \theta$ along VG.

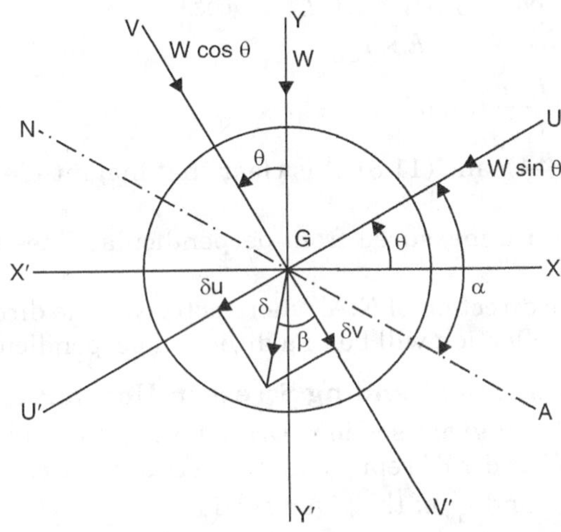

Fig. 11.4

The component $W \sin \theta$ will bend the beam about VV' axis whereas $W \cos \theta$ will bend the beam about UU' axis.

Let δu = deflection due to load $W \sin \theta$ along line UU' and

δv = deflection due to load $W \cos \theta$ along line VV'

The values of δu and δv are given as

$$\delta u = \frac{K(W \sin \theta)\, L^3}{E \times I_{VV}} \text{ and} \qquad \text{...(11.5)}$$

$$\delta v = \frac{K(W \cos \theta)\, L^3}{E \times I_{UU}} \qquad \text{...(11.6)}$$

where K = A constant depending upon the end conditions of the beam and position of load along the beam

L = Length of beam

The resultant deflection 'δ' is obtained as

$$\delta = \sqrt{\delta u^2 + \delta v^2} \qquad \text{...(11.6A)}$$

$$= \sqrt{\left[\frac{K(W \sin \theta)\, L^3}{E \times I_{VV}}\right]^2 + \left[\frac{K\,(W \cos \theta) L^3}{E \times I_{UU}}\right]^2}$$

$$= \frac{KWL^3}{E}\sqrt{\frac{\sin^2 \theta}{I_{VV}^2} + \frac{\cos^2 \theta}{I_{UU}^2}} \qquad \text{...(11.7)}$$

$K = \frac{1}{48}$ for a simply supported beam carrying a point load at the centre.

The angle β made by resultant deflection δ with the line GV' is given by,

$$\tan \beta = \frac{\delta u}{\delta v} = \frac{\left(\frac{K\,(W \sin \theta) L^3}{E \times I_{VV}}\right)}{\frac{K\,(W \cos \theta) L^3}{E \times I_{UU}}} = \frac{\sin \theta}{\cos \theta}\,\frac{I_{UU}}{I_{VV}}$$

$$= \frac{I_{UU}}{I_{VV}} \tan \theta \qquad \text{...(11.8)}$$

From equations (11.4*A*) and (11.8), it is clear that magnitude of angles α and β are the same $\left(= \frac{I_{UU}}{I_{VV}} \tan \theta\right)$. They are measured from perpendicular lines *GU* and *GV* in the same direction. Angle α gives the direction of N.A. whereas β gives the direction of resultant deflection. Hence the resultant deflection will be in a direction perpendicular to neutral axis.

11.4.1. Method for finding Bending Stress in Unsymmetrical Bending

(*i*) Find the C.G. of the given section. Draw horizontal and vertical lines *XGX'* and *YGY'* through *G*. Then *XX'* and *YY'* represents the rectangular co-ordinate axes.

(*ii*) Determine I_{xx}, I_{yy} and I_{xy} of the given section.

(*iii*) Calculate the value of θ from $\tan 2\theta = \frac{2I_{xy}}{I_{yy} - I_{xx}}$. If the value of θ is +ve, then the principal axes *UU'* will be in the counter-clockwise direction with *X*-axis. Now find the location of *VV'* axis which is at right angle to *UU'* axis.

(*iv*) Find the values of I_{UU} and I_{VV}. If $I_{xx} = I_{yy}$, then value of I_{UU} will be obtained from equation (11.1*A*).

(*v*) Find *M* and its components along principal axes *GU* and *GV*.

(*vi*) Find the resultant bending stress using

$$\sigma_b = \frac{(M \sin \theta) \times u}{I_{VV}} + \frac{(M \cos \theta) \times v}{I_{UU}}$$

$$= M\left[\frac{u \sin \theta}{I_{VV}} + \frac{v \cos \theta}{I_{UU}}\right]$$

Problem 11.1. *Fig. 11.5 shows an unequal angle of dimensions 100 mm by 60 mm and 10 mm thick. Determine :*

(*i*) *position of the principal axes and*

(*ii*) *magnitude of the principal moments of inertia for the given angle.*

Sol. Let us first find the location of centroid *G* of the given angle.

Divide the angle into two rectangles 1 and 2. Then

$$A_1 = 60 \times 10 = 600 \text{ mm}^2$$

$$A_2 = (100 - 10) \times 10 = 90 \times 10 = 900 \text{ mm}^2$$

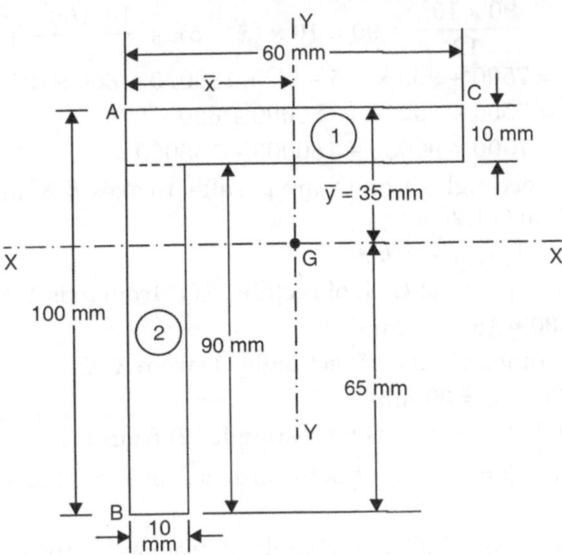

Fig. 11.5

Take horizontal distances from line AB and vertical distances from line AC vertically downwards. Let (x_1, y_1) and (x_2, y_2) are the co-ordinates of the C.G. of rectangles (1) and (2).

Then $$x_1 = \frac{60}{2} = 30 \text{ mm}, \quad y_1 = \frac{10}{2} = 5 \text{ mm}$$

$$x_2 = \frac{10}{2} = 5 \text{ mm}, \quad y_2 = 10 + \frac{90}{2} = 55 \text{ mm}$$

Now the centroid G, is obtained as

$$\bar{x} = \frac{A_1 x_1 + A_2 x_2}{A_1 + A_2} = \frac{600 \times 3 + 900 \times 5}{600 + 900} = \frac{18000 + 4500}{1500}$$
$$= 12 + 3 = 15 \text{ mm and}$$

$$\bar{y} = \frac{A_1 y_1 + A_2 y_2}{A_1 + A_2} = \frac{600 \times 5 + 900 \times 55}{600 + 900} = \frac{3000 + 49500}{1500}$$
$$= 2 + 33 = 35 \text{ mm}$$

Hence the centroid G is below line AC at a distance of $\bar{y} = 35$ mm and towards right of line AB at a distance of $\bar{x} = 15$ mm.

Draw a horizontal line X-X and vertical line Y-Y through G. Then X-X and Y-Y are the axes through centroid.

Now the moments of inertia about X-axis, about Y-axis and product of Inertia of the given angle is obtained as

$$I_{XX} = \frac{60 \times 10^3}{12} + (60 \times 10) \times \left(\bar{y} - \frac{10}{2}\right)^2 + \frac{10 \times (90^3)}{12} + 10 \times 90 \times (65 - 45)^2$$
$$= 5000 + 600 \times (35 - 5)^2 + 607500 + 900 \times 20^2$$
$$= 5000 + 540000 + 607500 + 360000$$
$$= 1512500 \text{ mm}^4$$

$$I_{YY} = \frac{90 \times 10^3}{12} + 90 \times 10 \times (\bar{x} - 5)^2 + \frac{10 \times 60^3}{12} + 10 \times 60 \times (30 - \bar{x})^2$$

$$= 7500 + 900 \times (15 - 5)^2 + 180000 + 600 \times (30 - 15)^2$$

$$= 7500 + 90000 + 180000 + 600 \times 225$$

$$= 7500 + 90000 + 180000 + 135000 = 412500 \text{ mm}^4 \text{ and}$$

As the sides of the rectangles 1 and 2 are parallel to axes *X-X* and *Y-Y*, hence product of inertia is given by equation (11.2*C*)as

$$I_{xy} = A_1 h_1 k_1 + A_2 h_2 k_2$$

where k_1 = Horizontal distance of C.G. of rectangle (1) from axis *Y-Y*

$= (30 - \bar{x}) = 30 - 15 = 15$ mm

h_1 = Vertical distance of C.G. of rectangle (1) from *X-X*

$= (\bar{y} - 5) = (35 - 5) = 30$ mm

k_2 = Horizontal distance of C.G. of rectangle (2) from *Y-Y*

$= (\bar{x} - 5) = 15 - 5 = 10$ mm. It is towards left of *Y-Y*. Hence –ve.

$= - 10$ mm

h_2 = Vertical distance of C.G. of rectangle (2) from *X-X* = (65 – 45) = 20 mm. It is below *X-X* and hence is –ve.

$= - 20$ mm

$\therefore \quad I_{xy} = 600 \times 15 \times 30 + 900 \times (-10)(-20) \qquad (\because A_1 = 600 \text{ and } A_2 = 900)$

$= 270000 + 180000 = 450000 \text{ mm}^4$

(*i*) Position of Principal Axes

For principal axes,

$$\tan 2\theta = \frac{2I_{xy}}{I_{yy} - I_{xx}}$$

$$= \frac{2 \times 450000}{412500 - 1512500} = -0.818$$

tan 2θ is –ve in 2nd quadrant

$$\therefore \quad 2\theta = \tan^{-1}(-0.818) = -39.28° = 180 - 39.28 = 140.72°$$

$$\therefore \quad \theta = \frac{140.72}{2} = \mathbf{70.36°.} \quad \textbf{Ans.}$$

The axis *U-U* will be obtained by drawing *U-U* at an angle of 70.36° with axis *X-X* through *G* in anticlockwise as shown in Fig. 11.5(*a*). The axis *V-V* is at right angle to *U-U* through *G*. The axes *U-U* and *V-V* are principal axes. **Ans.**

(*ii*) Value of Principal Moment of Inertia

Moment of inertia about principal axes *U-U* is given by,

$$I_{UU} = \frac{I_{xx} + I_{yy}}{2} + \frac{(I_{xx} - I_{yy})}{2} \sec 2\theta$$

$$= \frac{1512500 + 412500}{2} + \frac{1512500 - 412500}{2} \sec 140.72°$$

$$= 962500 + 550000 \times \left(-\frac{1}{0.774}\right) = 962500 - 710594$$

$$= \mathbf{251905.6 \text{ mm}^4.} \quad \textbf{Ans.}$$

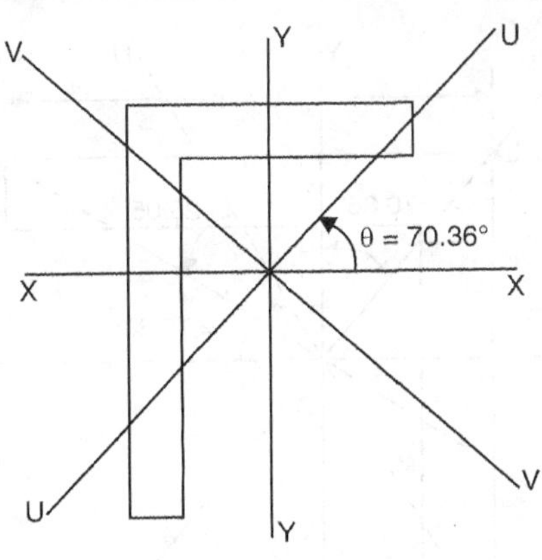

Fig. 11.5 (*a*)

and $$I_{VV} = I_{XX} + I_{YY} - I_{UU}$$
$$= 1512500 + 412500 - 251906 = \mathbf{1673094\ mm^4}.\quad \textbf{Ans.}$$

Problem 11.2. *A cantilever of length 1 m carries a point load of 2000 N at the free-end. The cross-section of the cantilever is an unequal angle of dimensions 100 mm by 60 mm and 10 mm thick. The small leg of the angle (i.e., 60 mm) is horizontal. The load passes through the centroid of the cross-section. Determine:*

(*i*) *position of neutral axis and*

(*ii*) *the magnitude of maximum stress set up, at the fixed section of the cantilever.*

Sol. The position of centroid G has been obtained in Problem 11.1. The inclination of the principal axes UU' and VV' is also calculated.

For a cantilever, the maximum bending moment is at the fixed end. Hence maximum B.M. is 2000 × 1000 = 2,000,000 Nmm. This bending moment is in the plane of YY'. As the given bending moment is not in the plane of principal axes, hence it is a case of unsymmetrical bending.

Resolving this bending moment in the planes of principal axes VV' and UU' respectively, we get

B.M. in the plane of UU' = 2,000,000 sin 70.36° = 1880320 Nmm

and in the plane of VV' = 2,000,000 cos 70.36° = 643320 Nmm.

The moment in plane UU' will bend the beam about axis VV'. Hence bending moment about axis $VV' = M_v$ = 2,000,000 × sin 70.36° = 1880320 Nmm

Similarly bending moment about axis $UU' = M_u$ = 2,000,000 × cos 70.36° = 643320 Nmm.

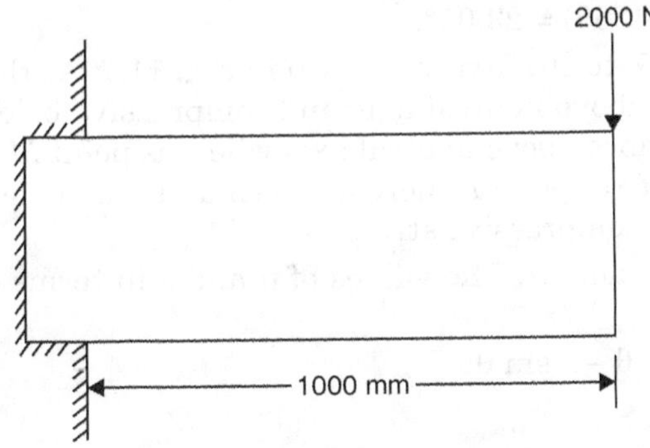

Fig. 11.6 (*a*)

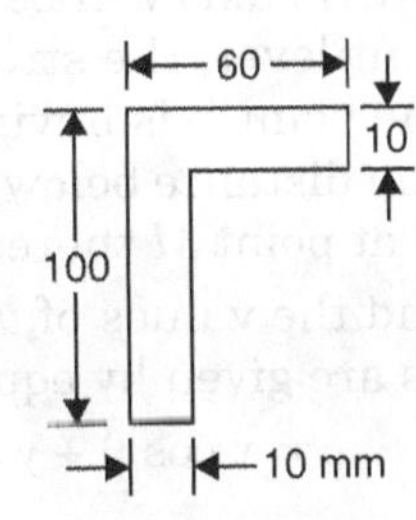

Fig. 11.6 (*b*) Cross-section of cantilever

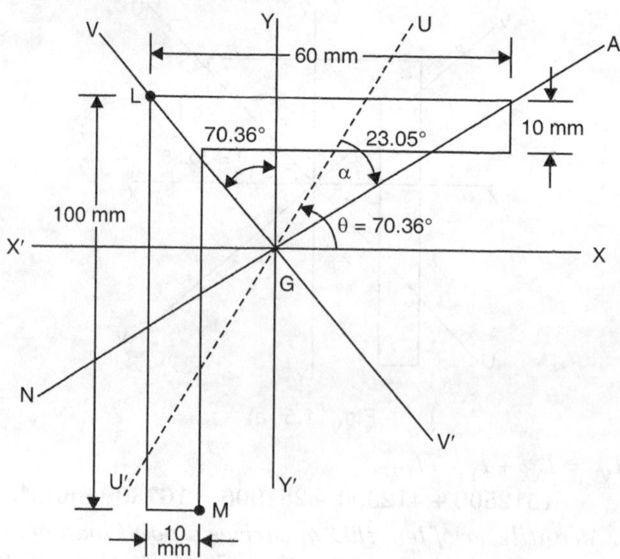

Fig. 11.6 (c) Enlarged view of cross-section

The combined bending stress at any point (u, v) is given by

$$\sigma = \frac{M_v \times u}{I_{VV}} + \frac{M_u \times v}{I_{UU}} \qquad \text{[Here } u \text{ and } v \text{ are positive in quadrant UGV]}$$

$$= \frac{1880320 \times u}{1673094} + \frac{643320 \times v}{251905.6} \qquad [\because \text{ From Problem 11.1, } I_{VV} = 1673094 \text{ mm and } I_{UU} = 251905.6 \text{ mm}^4]$$

$$= 1.128u + 2.65v \qquad \ldots(i)$$

At the neutral axis, $\sigma = 0$. Hence equation of neutral axis is obtained by putting $\sigma = 0$ in equation (*i*).

$\therefore$ Equation of neutral axis is

$$1.128u + 2.65v = 0$$

$$\therefore \qquad v = \frac{-1.128}{2.65} u = -0.4256u$$

This is equation of a straight line passing through G (*i.e.*, $y = mx$) with $m = -0.425$

or $\qquad \tan \alpha = -0.4256$ or $\alpha = \tan^{-1}(-0.4256) = 23.05°$.

Hence neutral axis will be inclined to $-23.05°$ to UU' axis as shown in Fig. 11.7. As this is the case of cantilever, the stress will be tensile above neutral axis and compressive below neutral axis. The point L is having maximum distance above neutral axis whereas point M is having maximum distance below neutral axis. Hence at point L there will be maximum tensile stress, whereas at point M there will be maximum compressive stress.

Let us find the values of u and v at points L and M. The values of u and v in terms of x, y co-ordinates are given by equation (11.2*B*) as

$$u = x \cos \theta + y \sin \theta \text{ and } v = y \cos \theta - x \sin \theta$$

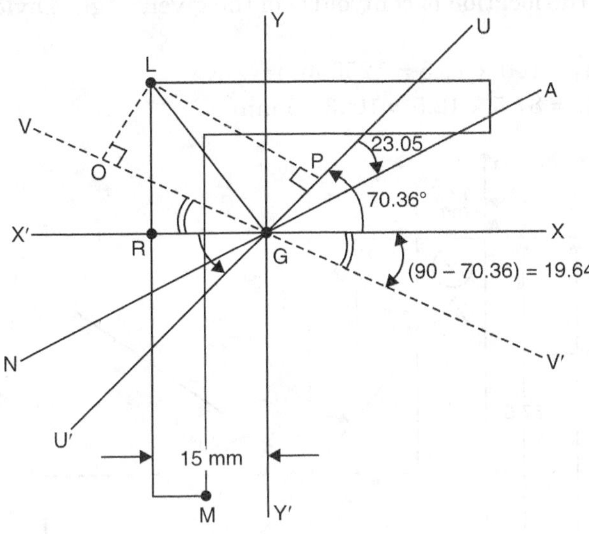

Fig. 11.7

For point *L*

$$x = -\bar{x} = -15 \text{ mm}$$

$$y = \bar{y} = 35 \text{ mm}$$

$$\therefore \quad u = -15 \cos 70.36° + 35 \sin 70.36° = -5.04 + 32.97 = 27.93$$

$$v = 35 \cos 70.36° - (-15) \sin 70.36° = 11.76 + 14.13 = 25.89$$

∴ Resultant bending stress at point L is

$$\sigma = 1.128\, u + 2.65\, v$$

$$= 1.128 \times 27.93 + 2.65 \times 25.89$$

$$= 31.504 + 68.608 = \mathbf{100.112 \text{ N/mm}^2 \text{ (Tensile)}}. \quad \textbf{Ans.}$$

For point *M*

$$x = -(\bar{x} - 10) = -(15 - 10) = -5 \text{ mm}$$

$$y = -(100 - \bar{y}) = -(100 - 35) = -65 \text{ mm}$$

$$\therefore \quad u = x \cos \theta + y \sin \theta = -5 \cos 70.36 + (-65) \sin 70.36$$

$$= -5 \times 0.336 - 65 \times 0.942 = -1.68 - 61.23 = -62.91 \text{ mm}$$

$$v = y \cos \theta - x \sin \theta = -65 \cos 70.36° - (-5) \sin 70.36°$$

$$= -65 \times 0.336 + 5 \times 0.942 = -21.84 + 4.71 = -17.13$$

and bending stress becomes as

$$\sigma = 1.128u + 2.65v$$

$$= 1.128 \times (-62.91) + 2.65 \times (-17.13) = -70.96 - 45.39$$

$$= \mathbf{-116.35 \text{ N/mm}^2 \text{ (Compressive)}}. \quad \textbf{Ans.}$$

Problem 11.3. *A simply supported beam of span 3 m carries a load of 500 N at its centre. The section of the beam is an equal angle of size 100 mm by 100 mm and 12.5 mm thick as shown in Fig. 11.8. The load line passes through centroid G of the section and is along line YG. Determine: (i) Stresses at points A, B and C of mid-section of the beam, (ii) Deflection of beam at the mid-section and (iii) Position of neutral axis. Take E = 200000 N/mm²*.

Sol. Let us find the location of centroid G of the given angle. Divide the given angle into two rectangles 1 and 2.

$$A_1 = 100 \times 12.5 = 1250 \text{ mm}^2$$
$$A_2 = 87.5 \times 12.5 = 1093.75 \text{ mm}^2$$

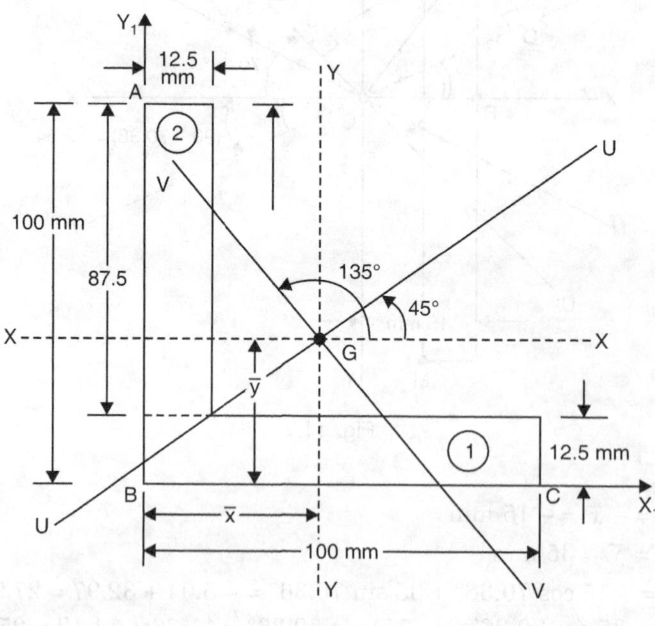

Fig. 11.8

Let $(\bar{x}, \bar{y})$ are the co-ordinates of the centroid G with respect to rectangular axes BX_1 and BY_1.

Then x_1 = Distance of C.G. of rectangle (1) from $BY_1 = \dfrac{100}{2} = 50$ mm

y_1 = Distance of C.G. of rectangle (1) from $BX_1 = \dfrac{12.5}{2} = 6.25$ mm

x_2 = Distance of C.G. of rectangle (2) from $BY_1 = \dfrac{12.5}{2} = 6.25$ mm

y_2 = Distance of C.G. of rectangle (2) from $BX_1 = 12.5 + \dfrac{87.5}{2} = 56.25$ mm

Now $\bar{x} = \dfrac{A_1x_1 + A_2x_2}{A_1 + A_2} = \dfrac{1250 \times 50 + 1093.75 \times 6.25}{1250 + 1093.75} = \dfrac{62500 + 6836}{2343.75}$

$= \dfrac{69336}{2343.75} = 29.58$ mm

and $\bar{y} = \dfrac{A_1y_1 + A_2y_2}{A_1 + A_2} = \dfrac{1250 \times 6.25 + 1093.75 \times 56.25}{1250 + 1093.75} = \dfrac{7812.5 + 61523}{2343.75}$

$= 29.58$ mm.

Now draw X-X axis and Y-Y axis through G. Then the moment of inertia about X-axis, about Y-axis and product of inertia of the given angle is obtained as

$$I_{XX} = \left[\frac{100 \times 12.5^3}{12} + 100 \times 12.5 \times \left(\bar{y} - \frac{12.5}{2}\right)^2\right]$$

$$+ \left[\frac{12.5 \times 87.5^3}{12} + 12.5 \times 87.5 \times \left(\frac{87.5}{2} - \bar{y}\right)^2\right]$$

$= [16274 + 1250 \times (29.58 - 6.25)^2] + [697835 + 1093.75\ (56.25 - 29.58)^2]$

$= [16274 + 680361] + [697835 + 777972]$

$\left[\because \text{ Distance between C.G. of rectangle (2) and C.G. of the angle} = \left(12.5 + \frac{87.5}{2}\right) - \bar{y} = 56.25 - \bar{y}\right]$

$= 2172442 \text{ mm}^4$

$I_{yy} = I_{xx}$. As the given section has equal dimensions.

To calculate the product of inertia (I_{xy}), we find first the co-ordinates of the C.G. of rectangle (1) and rectangle (2) with respect to axes X-X and Y-Y.

As the sides of the rectangle 1 and 2 are parallel to axes X-X and Y-Y, hence product of inertia is given by equation (11.2C) as

$$I_{xy} = A_1 h_1 k_1 + A_2 h_2 k_2$$

where k_1 = Horizontal distance of C.G. of rectangle (1), from Y-Y = $\left(\frac{100}{2} - \bar{x}\right)$ = 50 – 29.58 = 20.32 mm. It is towards right of Y-Y and hence +ve = 20.32.

h_1 = Vertical distance of C.G. of rectangle (1) from X-X axis = $\left(\bar{y} - \frac{12.5}{2}\right)$ = 29.58 – 6.25 = 23.33. It is below X-X axis and hence –ve = – 23.33.

k_2 = Horizontal distance of C.G. of rectangle (2) from Y-Y = $\bar{x} - \frac{12.5}{2}$ = 29.58 – 6.25 = 23.33. It is towards left of Y-Y and hence it is –ve = – 23.33.

h_2 = Vertical distance of C.G. of rectangle (2) from X-X axis = $\left(12.5 + \frac{87.5}{2}\right) - \bar{y}$ = (12.5 + 43.75) – 29.58 = 26.67. It is above X-X axis and hence +ve = 26.67 mm.

$\therefore \quad I_{xy} = 1250 \times 20.32 \times (-23.33) + 1093.75 \times (-23.33) \times 26.67$

$= -592582 - 680543 = -1273125 \text{ mm}^4$

Principal Axes

Let us now find the **principal axes.** For principal axes, we have

$$\tan 2\theta = \frac{2I_{xy}}{I_{yy} - I_{xx}} = \infty \quad (\text{as } I_{xx} = I_{yy})$$

$$= \tan 90°$$

$\therefore\ \theta_1 = 45°$ and $\theta_2 = 90 + 45 = 135°$ gives the position of principal axes UU' and VV' through G respectively as shown in Fig. 11.8.

Principal Moment of Inertia

In this case $2\theta = 90°$, hence principal moment of inertia about principal axis *U-U* will be obtained from equation (11.1*A*),

$$I_{UU} = \frac{1}{2}(I_{xx} + I_{yy}) + \frac{1}{2}(I_{xx} - I_{yy}) \cos 2\theta - I_{xy} \sin 2\theta$$

$$\left[\text{Here donot use } I_{UU} = \frac{1}{2}(I_{xx} + I_{yy}) + \frac{1}{2}(I_{xx} - I_{yy}) \sec 2\theta\right]$$

$$= \frac{1}{2}[2172442 + 2172442] + 0 - (-1273125) \sin 90°$$

$$= 2172442 + 1273125 = 3445567 \text{ mm}^4$$

and
$$I_{VV} = I_{XX} + I_{YY} - I_{UU}$$
$$= 2172442 + 2172442 - 3445567$$
$$= 899317 \text{ mm}^4.$$

(*i*) **Stresses at the points A, B, C.** For a simply supported beam of span *L* carrying a point load *W* at the mid-span, the B.M. at the mid-section is given by,

$$M = \frac{W \times L}{4} = \frac{500 \times 3}{4} \text{ Nm} = \frac{500 \times 3}{4} \times 1000 = 375000 \text{ Nmm}$$

This B.M. is acting in the plane *Y-Y* and will bend the beam in *X-X* plane. Let us resolve the bending moment *M* into its components in the planes of principal axes *UU′*, and *VV′*.

(*i*) Component of moment (*M*) in the plane $UU' = M \sin \theta$. This component will bend the given angle in plane *VV′*. Hence bending stress due to this component at any point *P* having co-ordinates (*u*, *v*) will be $\dfrac{(M \sin \theta) \times u}{I_{VV}}$.

(*ii*) Component of moment (*M*) in the plane $VV' = M \cos \theta$. This component will bend the given angle in plane *UU′*. Hence bending stress due to this component at point *P*(*u*, *v*) will be $\dfrac{(M \cos \theta) \times v}{I_{UU}}$.

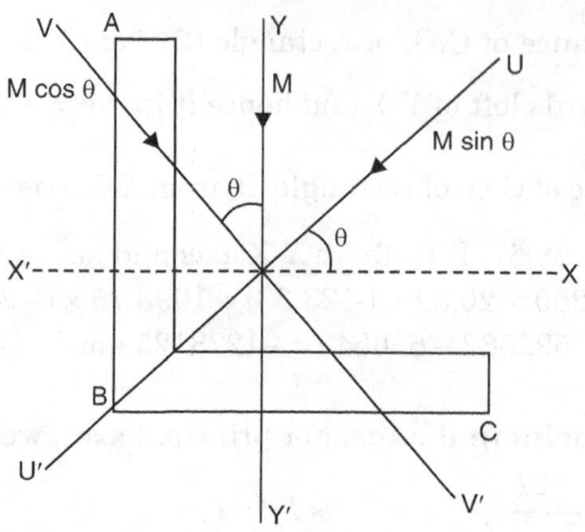

Fig. 11.8 (*a*)

∴ Resultant bending stress at any point $P(u, v)$ is given by,

$$\sigma_b = \frac{(M \sin \theta) \times u}{I_{VV}} + \frac{(M \cos \theta) \times v}{I_{UU}}$$

$$= \frac{375000 \times \sin 45° \times u}{I_{VV}} + \frac{375000 \times \cos 45° \times v}{I_{UU}}$$

$$= 375000 \times 0.707 \left(\frac{u}{I_{VV}} + \frac{v}{I_{UU}} \right) \qquad [\because \sin 45° = \cos 45° = 0.707]$$

$$= 375000 \times 0.707 \left(\frac{u}{899317} + \frac{v}{3445567} \right) \qquad \text{...(A)}$$

Let us find the values of u and v at points A, B and C. The values of u and v in terms of x, y co-ordinates are given by equation (11.2B) as

$$u = x \cos \theta + y \sin \theta \text{ and}$$
$$v = y \cos \theta - x \sin \theta$$

For Point A

$$x = -\bar{x} = -29.58 \text{ mm}$$
$$y = 100 - \bar{y} = 100 - 29.58 = 70.42 \text{ mm}$$

∴ $$u = -29.58 \times \cos 45° + 70.42 \sin 45° = 28.87 \text{ mm and}$$
$$v = 70.42 \cos 45° - (-29.58) \sin 45° = 70.7 \text{ mm}$$

Substituting these values in equation (A), we get

$$\sigma_b \text{ at point } A = 375000 \times 0.707 \left(\frac{28.87}{899317} + \frac{70.7}{3445567} \right)$$
$$= 8.51 + 5.44 = \mathbf{13.95 \text{ N/mm}^2}. \quad \textbf{Ans.}$$

For Point B

$$x = -\bar{x} = -29.58 \text{ mm and } y = -\bar{y} = -29.58 \text{ mm}$$

∴ $$u = x \cos \theta + y \sin \theta = -29.58 \cos 45° + (-29.58) \sin 45° = -41.83 \text{ mm}$$
$$v = y \cos \theta - x \sin \theta = -29.58 \cos 45° - (-29.58) \sin 45° = 0$$

Substituting these values in equation (A), we get

$$\sigma_b \text{ at point } B = 375000 \times 0.707 \left(-\frac{41.83}{899317} + 0 \right) = \mathbf{-12.33 \text{ N/mm}^2}. \quad \textbf{Ans.}$$

For Point C

$$x = 100 - \bar{x} = 100 - 29.58 = 70.42 \text{ and } y = -\bar{y} = -29.58 \text{ mm}$$

∴ $$u = x \cos \theta + y \sin \theta = 70.42 \times \cos 45° + (-29.58) \sin 45° = 28.87 \text{ mm}$$
$$v = y \cos \theta - x \sin \theta = -29.58 \cos 45° - 70.42 \sin 45° = -70.7 \text{ mm}$$

Substituting these values in equation (A), we get

$$\sigma_b \text{ at point } C = 375000 \times 0.707 \left(\frac{28.87}{899317} + \frac{-70.7}{3445567} \right)$$
$$= 8.51 - 5.44 = \mathbf{3.07 \text{ N/mm}^2}. \quad \textbf{Ans.}$$

(*ii*) **Deflection of the beam at mid-section (δ).** The deflection is given by equation (11.6A) as

$$\delta = \sqrt{\delta u^2 + \delta v^2} \qquad ...(B)$$

where δ_u = deflection along UU' due to load component $W \sin\theta$ (which is along UU' axis) for the bending of beam about VV' axis

$$= \frac{K\,(W \sin\theta)\,L^3}{E \times I_{VV}}$$

δ_V = deflection along VV' due to load component $W \cos\theta$ (which is along VV' axis) for the bending of beam about UU' axis $= \dfrac{K\,(W\cos\theta)\,L^3}{E \times I_{UU}}$

where K = a constant $= \dfrac{1}{48}$ for a simply supported beam carrying a point load W at the centre

L = length of beam

Now the equation (*B*) becomes as

$$\delta = \sqrt{\left[\frac{K\,(W\sin\theta)\,L^3}{E \times I_{VV}}\right]^2 + \left[\frac{K\,(W\cos\theta)\,L^3}{E \times I_{UU}}\right]^2}$$

$$= \frac{KWL^3}{E}\sqrt{\frac{\sin^2\theta}{(I_{VV})^2} + \frac{\cos^2\theta}{(I_{UU})^2}}$$

Here $W = 500$ N, $L = 3$ m $= 3000$ mm, $K = \dfrac{1}{48}$ and $E = 200{,}000$ N/mm^2

$$\therefore \quad \delta = \frac{1}{48} \times \frac{500 \times (3000)^3}{200{,}000}\sqrt{\frac{\sin^2 45°}{(899317)^2} + \frac{\cos^2 45°}{(3445567)^2}}$$

$$= \frac{5 \times 27}{96} \times 10^6\,\sqrt{6.182 \times 10^{-13} + 4.211 \times 10^{-14}}$$

$$= \frac{5 \times 27}{96} \times 10^6\,\sqrt{61.82 \times 10^{-14} + 4.211 \times 10^{-14}}$$

$$= \frac{5 \times 27}{96} \times 10^6 \times 8.125 \times 10^{-7} = \mathbf{1.1425\ mm.}\quad \textbf{Ans.}$$

The angle β made by the deflection with axis $V'V$ clockwise is given by

$$\tan\beta = \frac{I_{UU}}{I_{VV}}\tan\theta$$

$$= \frac{3445567}{899317}\tan 45° = 3.831$$

$$\therefore \quad \beta = \tan^{-1} 3.831 = \mathbf{75.37°.}\quad \textbf{Ans.}$$

(*iii*) **Position of the neutral axis.** For neutral axis, resultant stress should be zero *i.e.*, σ_b in equation (A) should be zero

$$\therefore \quad 0 = 375000 \times 0.707\left(\frac{u}{899317} + \frac{v}{3445567}\right) = 0$$

or $$\frac{u}{899317} + \frac{v}{3445567} = 0$$

or $$v = -\frac{u}{899317} \times 3445567 = -3.831u$$

This is a equation of straight line, passing through G. The slope of this line is

$$m = \tan \alpha = -3.831$$

$$\therefore \quad \alpha = \tan^{-1} - 3.831 = \mathbf{-75.37.} \quad \textbf{Ans.}$$

Problem 11.4. *Calculate the maximum tensile stress in the I-section of a simply supported beam of span 3 m which carries a load of 7.5 kN at the centre of the beam. The load-line in inclined at an angle of 30° with the vertical as shown in Fig. 11.9 and passes through the centroid of the section. The dimensions of the I-section are shown in Fig. 11.9.*

Sol. The given section is symmetrical about XX' and YY' axes. The product of inertia about these axes will be zero. Hence these axes (*i.e.*, X-X and Y-Y) will also be principal axes (*i.e.*, U-U and V-V) as shown in Fig. 11.9. The centroid G of the section will be as shown in Fig. 11.9 *i.e.*, at the centre of the web. The principal axes will be XX' and YY'. (*i.e.*, $UU = XX'$ and $VV = YY'$)

$$W = 7.5 \text{ kN} = 7500 \text{ N}$$

$$L = 3 \text{ m} = 3000 \text{ mm.}$$

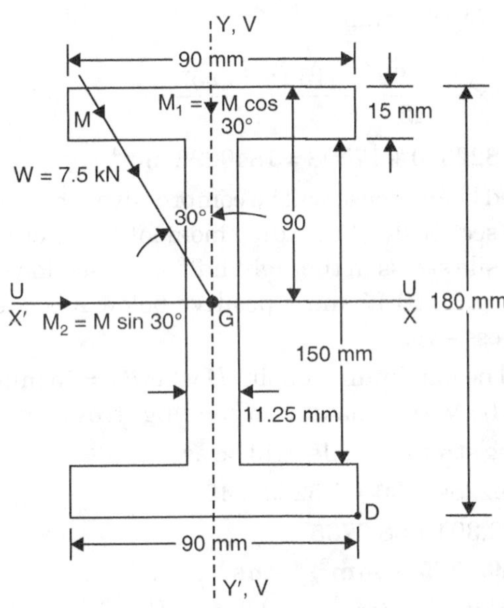

Fig. 11.9

The load is acting at the centre of the simply supported beam. The bending moment at the centre is maximum and is equal to $\frac{W \times L}{4} = \frac{7500 \times 3000}{4} = \frac{22500000}{4} = 5625000$ Nmm.

$$\therefore \quad M = 5625000 \text{ Nmm.}$$

This bending moment is along a line which is inclined at an angle of 30° with vertical line YG as shown in Fig. 11.9. This bending moment is resolved into components in the plane of principal axes V-V and U-U *i.e.*, one along YG and other along $X'G$. They are $M \cos 30°$ and $M \sin 30°$.

The component along YG

$$M_1 = M \cos 30° = 5625000 \times 0.866 = 4871250 \text{ Nmm}$$

The component along $X'G$

$$M_2 = M \sin 30° = 5625000 \times 0.5 = 2812500 \text{ Nmm}$$

The component along YG will bend the beam about XX' and component along $X'G$ will bend the beam about YY'. Hence the resultant bending stress at any point (x, y) will be given as

$$\sigma = \frac{M_1}{I_{xx}} \times y + \frac{M_2}{I_{yy}} \times x$$

$$= \frac{4871250}{21591561} y + \frac{2812500}{1840298} x$$

$$= 0.2256\, y + 1.5283\, x \qquad ...(i)$$

where $I_{xx} = 2\left[\frac{90 \times 15^3}{12} + 90 \times 15 \times (90 - 7.5)^2\right]$ for the flanges $+ \frac{11.25 \times 150^3}{12}$ for the web

$$= 2\,[25312.5 + 9188437] + 3164062$$

$$= 21591561 \text{ mm}^4$$

and $$I_{yy} = 2 \times \frac{15 \times 90^3}{12} + \frac{150 \times 11.25^3}{12}$$

$$= 1822500 + 17798 = 1840298 \text{ mm}^4$$

For a simply supported beam, these will be compressive stress in the upper half and tensile stress in the lower half of the section due to bending moment M_1 whereas there will be compressive stress in the left half and tensile stress in the right half of the section due to bending moment M_2. Taking x positive towards right of YY' and y positive below XX', we will get tensile stress as positive and compressive stress – ve.

The tensile stress will be maximum at point D where $x = 45$ mm and $y = 90$ mm. Substituting these values in equation (i), we get maximum bending stress.

∴ Maximum bending stress (tensile) will be as

$$\sigma = 0.2256 \times 90 + 1.5283 \times 45$$

$$= 20.304 + 68.7735$$

$$= \mathbf{89.0775 \text{ N/mm}^2}. \quad \textbf{Ans.}$$

Problem 11.5. *Calculate the stresses at the corners of the rectangular section of a simply supported beam of span 5 m which carries a load of 4 kN at the centre of the span. The load line is inclined at an angle of 30° to the vertical longitudinal plane as shown in Fig. 11.10 and passes through the centroid of the section. The dimensions of the section are shown in Fig. 11.10.*

Sol. Given: $W = 4$ kN, $L = 5$ m. The given section is symmetrical about XX' and YY' axes. The product of inertia about these axes will be zero. Hence XX' and YY' axes will be principal axes (*i.e.*, UU and VV axes) also. The centroid G of the section is at the centre of the rectangular section. The principal axes are XX' and YY'. Let us calculate moment of inertia I_{XX} and I_{YY}.

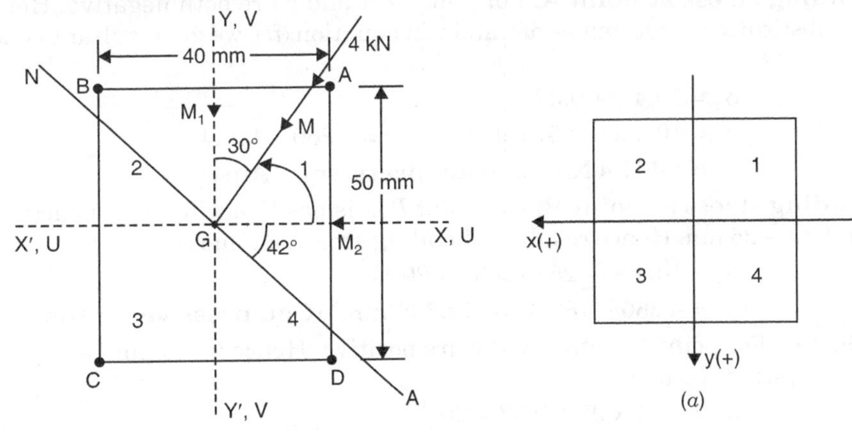

Fig. 11.10

$$I_{XX} = \frac{40 \times 50^3}{12} = \frac{1250000}{3} = 4.167 \times 10^5 \text{ mm}^4$$

$$I_{YY} = \frac{50 \times 40^3}{12} = 2.667 \times 10^5 \text{ mm}^4$$

For a simply supported beam, carrying a point load at the centre, the B.M. is maximum at the centre. It is given by,

$$M = \frac{W \times L}{4} = \frac{4 \times 5}{4} = 5 \text{ kNm} = 5000 \text{ Nm} = 5 \times 10^6 \text{ Nmm.}$$

This bending moment may be resolved in two components, one along YG and other along XG.

Component along YG, $M_1 = M \times \cos 30°$

$= (5 \times 10^6) \times 0.866 = 4.33 \times 10^6$ Nmm

(This component will bend the beam about axis XX'. For a simply supported beam, there will be compressive stress in the upper half and tensile stress in the lower half of the section due to this component).

Component along XG, $M_2 = M \times \sin 30° = (5 \times 10^6) \times 0.5 = 2.5 \times 10^6$ Nmm.

(This component will bend the beam about axis YY'. There will be compressive stress in the right half of the section and tensile stress in the left half due to this component.)

The resultant bending stress one due to bending about axis XX' and other due to bending about axis YY' at any point (x, y) will be given as

$$\sigma = \frac{M_1}{I_{xx}} \times y + \frac{M_2}{I_{yy}} \times x \qquad ...(i)$$

$$\sigma = \frac{4.33 \times 10^6}{4.167 \times 10^5} y + \frac{2.5 \times 10^6}{2.667 \times 10^5} x$$

$$= 10.4y + 9.37x \qquad ...(ii)$$

There will be tensile stress due to bending about axis XX' and due to bending about axis YY' in third quadrant of the Fig. 11.10. Hence taking x positive towards left of YY' and y positive below XX', we will get tensile stress as positive in third quadrant.

(*i*) **Bending stress at point A.** For point A, x and y are both negative. Hence $x = -20$ and $y = -25$. Substituting these values of x and y in equation (*ii*), we get resultant bending stress at A, as

$$\sigma_A = 10.4\,y + 9.37x$$
$$= 10.4 \times (-25) + 9.37\,(-20) = -260 - 187.4$$
$$= \mathbf{-447.4\ N/mm^2\ (Compressive).}\quad \textbf{Ans.}$$

(*ii*) **Bending stress at point B.** For point B, x is positive whereas y is negative. Hence $x = 20$ mm and $y = -25$ mn. Hence resultant bending stress becomes as

$$\sigma_B = 10.4 \times (-25) + 9.37 \times 20$$
$$= -260 + 187.4 = \mathbf{-72.6\ N/mm^2\ (Compressive).}\quad \textbf{Ans.}$$

(*iii*) **Point C.** For point C, both x and y are positive. Hence $x = 20$ mm and $y = 25$ m. The resultant bending stress becomes as

$$\sigma_C = 10.4 \times 25 + 9.37 \times 20$$
$$= 260 + 187.4 = \mathbf{447.4\ N/mm^2\ (Tensile).}\quad \textbf{Ans.}$$

(*iv*) **Point D.** For point D, x is negative and y is positive. Hence $x = -20$ mm and $y =$ 25 mm. The resultant bending stress becomes as

$$\sigma_D = 10.4 \times (+25) + 9.37\,(-20)$$
$$= 260 - 187.4 = \mathbf{72.6\ N/mm^2\ (Tensile).}\quad \textbf{Ans.}$$

Equation of N.A.

At the neutral axis (N.A.) the bending stress is zero. Hence $\sigma = 0$ in equation (*ii*).

or $$0 = 10.4y + 9.37x$$

or $$\tan\alpha = \frac{y}{x} = -\frac{9.37}{10.4} = -0.9$$

$$\therefore \quad \alpha = \tan^{-1} - 0.9 = -42°$$

As angle α is negative, the N.A. should be drawn in such a way that it passes through the quadrant in which either x or y is negative. This is only possible in 2nd and 4th quadrant. Hence a line making an angle of 42° with XX' axis in clockwise direction and passing through G, gives the position of N.A. This is shown in Fig. 11.10. Above the N.A., there will be compressive stress and below it tensile stress.

Problem 11.6. *Calculate the maximum load which a simply supported beam of T-section can carry at the centre. The span of the beam is 6 m and dimensions of T-section are shown in Fig. 11.11. The load is inclined to 30° to the Y-axis and passes through the centroid of T-section. The maximum tensile and compressive stresses are not allowed to exceed 30 N/mm² and 60 N/mm² respectively.*

Sol. The given section is symmetrical about axis Y-Y. The centroid will lie on axis Y-Y. The product of inertia will be zero about this axis. Hence it will be the principal axis. The other principal axis will be X-X passing through centroid G. The principal axes will be XX and YY axes. Let us now calculate the position of centroid G.

Let $\bar{y}$ = Distance of centroid G of the section from axis $X'X'$ (*i.e.*, from horizontal line BA.)

Then $$\bar{y} = \frac{A_1 y_1 + A_2 y_2}{(A_1 + A_2)}$$

where A_1 = Area of flange = $60 \times 10 = 600\ \text{mm}^2$

A_2 = Area of web = $80 \times 10 = 800\ \text{mm}^2$

$$y_1 = \text{Distance of C.G. of flange from } X'X' = \frac{10}{2} = 5 \text{ mm}$$

$$y_2 = \text{Distance of C.G. of web from } X'X' = 10 + \frac{80}{2} = 50 \text{ mm}$$

$$\therefore \quad \bar{y} = \frac{600 \times 5 + 800 \times 50}{600 + 800} = \frac{3000 + 40000}{1400} = \frac{43000}{1400}$$

$$= 30.714 \text{ mm} \simeq 30.7 \text{ mm}$$

Now let us find I_{xx} and I_{yy}.

$$I_{xx} = \left[\frac{60 \times 10^3}{12} + 60 \times 10 \times (\bar{y} - 5)^2 \text{ for flange}\right]$$

$$+ \left[\frac{10 \times 80^3}{12} + 10 \times 80 \times (y_2 - \bar{y})^2 \text{ for web}\right]$$

$$= [5000 + 600 \times (30.7 - 5)^2] + [426667 + 800\,(50 - 30.7)^2]$$

$$= 5000 + 396294 + 426666.7 + 297992 = 1125953 \text{ mm}^4$$

and

$$I_{yy} = \frac{10 \times 60^3}{12} + \frac{80 \times 10^3}{12}$$

$$= 180000 + 6666.7 = 186666.7 \text{ mm}^4 \simeq 186667 \text{ mm}^4$$

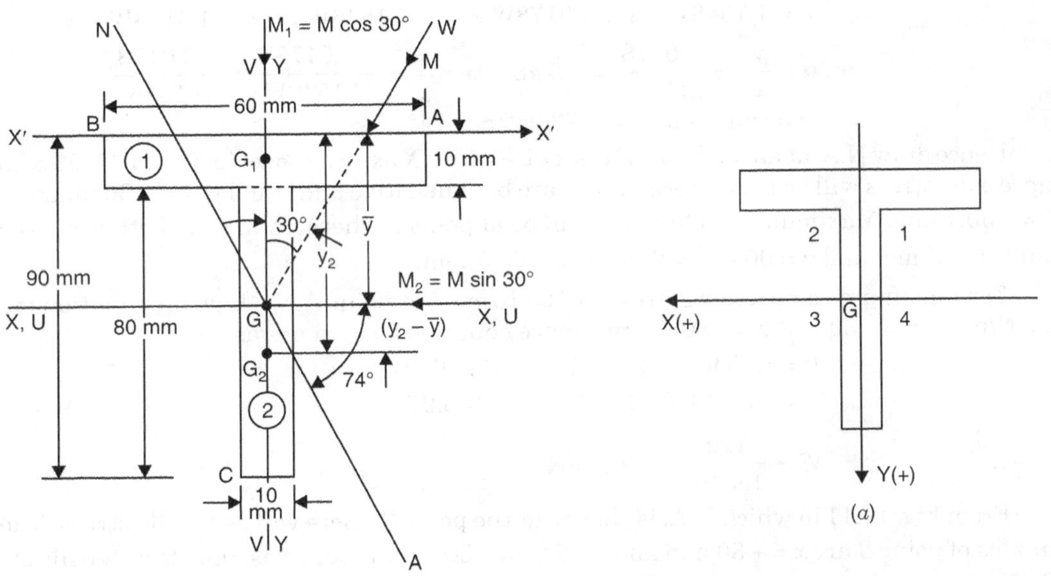

Fig. 11.11

For a simply supported beam, carrying a point load W at the centre, the maximum bending moment at the centre is given by,

$$M = \frac{W \times L}{4} = \frac{W \times 6}{4} = 1.5W \text{ where } W = \text{load at centre in kN}$$

Resolving the above moment in two components.

The moment in the plane of YG,

$$M_1 = M \times \cos 30°$$
$$= 1.5W \times 0.866 = 1.3W \text{ kNm}$$
$$= 1.3W \times 1000 \times 1000 \text{ Nmm} = 1.3 \times 10^6 W \text{ Nmm}$$

(This component will bend the beam about axis X-X)

For a simply supported beam, there will be compressive stress in the upper half and tensile stress in the lower half due to moment M_1.

The moment in the plane of XG, $M_2 = M \times \sin 30°$

or $$M_2 = 1.5\ W \times 0.5 = 0.75W \text{ kNm} = 0.75 \times 10^6 W \text{ Nmm}$$

(This component will bend the beam about axis Y-Y.)

Hence there will be compressive stress in the right half and tensile stress in the left half due to moment M_2. The resulting bending stress one due to bending about axis XX and other due to bending about axis Y-Y at any point (x, y) will be given as

$$\sigma = \frac{M_1}{I_{xx}} \times y + \frac{M_2}{I_{yy}} \times x$$
$$= \frac{1.3 \times 10^6 W}{1125953} y + \frac{0.75 \times 10^6 W}{186667} \times x$$

or $$\sigma = 1.1546W \times y + 4.0178\ W \times x \quad ...(i)$$

For neutral axis, $\sigma = 0$. Hence equation of N.A. is

$$0 = 1.1546W \times y + 4.0178W \times x \text{ or } 1.1546W \times y = -4.0178W \times x$$

or $$\tan \alpha = \frac{y}{x} = \frac{-4.0178}{1.1546} = -3.48 \quad \text{or} \quad \frac{y}{x} = \frac{-4.0178W}{1.1546W} = \frac{-4.0178}{1.1546}$$

$\therefore$ $$\alpha = \tan^{-1} - 3.48 = -73.96° \simeq -74°$$

Hence draw N.A. at an angle of 74° clockwise with XX axis as shown in Fig. 11.11. Maximum compressive stress will be at A where x and y are both negative and equal to $x = -30$ mm and $y = -\bar{y} = -30.7$ mm. Maximum tensile stress will be at point C where x and y are both positive and equal to $x = 5$ mm and $y = 90 - \bar{y} = 90 - 30.7 = 59.3$ mm.

For maximum compressive stress of 120 N/mm^2 at point A, we have in equation (i) $\sigma = -120$ N/mm^2, $x = -30$ and $y = -30.7$ mm. Hence equation (i) becomes as

$$-120 = 1.1546W \times (-30.7) + 4.0178W \times (-30)$$
$$= -35.44W - 120.53W = -155.97W$$

$\therefore$ $$W = \frac{120}{155.97} = 0.769 \text{ kN}$$

From Fig. 11.11 in which N.A. is shown, at the point B, there will be tensile stress. The co-ordinates of point B are $x = +30$ mm and $y = -\bar{y} = -30.7$ mm. Let us assume that tensile stress 60 N/mm^2 is maximum at point B. Then load W will be as given by equation (i) in which $\sigma = 60$ N/mm^2, $x = 30$ mm and $y = -30.7$ mm.

$\therefore$ Hence equation (i) becomes as

$$60 = 1.1546W \times (-30.7) + 4.0178W \times 30$$
$$= -35.44W + 120.53W = 85.09W$$

$\therefore$ $$W = \frac{60}{85.09} = 0.705 \text{ kN}$$

From maximum tensile stress of 60 N/mm^2 at point C, we have in equation (i) $\sigma = 60$ N/mm^2, $x = 5$ mm and $y = 59.3$ mm. Hence equation (i) becomes as

$$60 = 1.1546W \times (59.3) + 4.0178W \times (5)$$
$$= 68.47W + 20.09W = 88.56W$$

$$\therefore \quad W = \frac{60}{88.56} = 0.677 \text{ kN}$$

Comparing all these cases, we find that maximum load carried by the beam will be 0.677 kN so that the max. compressive and max. tensile stress are within limit.

11.5 SHEAR CENTRE

Shear centre is a point (in or outside a section) through which the applied shear force produces no torsion or twist of the member. If the load is not applied through the shear centre, there will be twisting of the beam due to unbalanced moment caused by the shear force acting on the section.

- The shear centre lies on the axis of symmetry, if the section is symmetrical about one axis. The shear centre does not coincide with the centroid in this case.
- For sections having two axes of symmetry, the shear centre lies on the intersection of these axes and thus coincides with the centroid.
- Shear centre is also known as centre of twist.

11.6. DETERMINATION OF SHEAR CENTRE FOR CHANNEL SECTION

Fig. 11.12 shows a channel section. The given section is symmetrical about the axis *X-X*. Hence shear centre will lie on this axis.

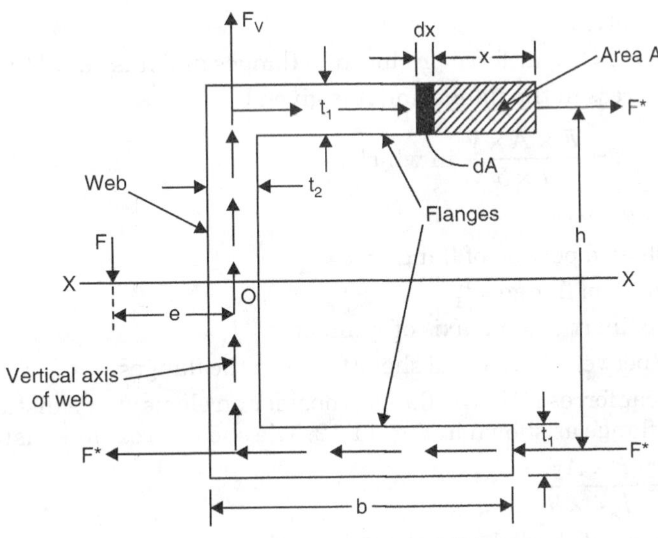

Fig. 11.12

Let F = Applied shear force on the section

t_1 = Thickness of flanges.

t_2 = Thickness of web.

F^* = Shear force produced in flanges.

F_V = Vertical shear force produced in the web.

b = Length of flange.

h = Distance between lines of action of F^*.

We know that shear stress follows the direction of boundary and hence shear stress distribution is horizontal in flanges and vertical in web. The shear forces due to these shear stresses will be horizontal in flanges and vertical in the web. The total shear force in the web must be equal to the applied vertical shear force. Hence the vertical force is balanced. But the shear forces in the flanges are unbalanced. They are equal and opposite and hence produces a clockwise couple of magnitude $F^* \times h$.

If the applied force F acts through the vertical axis of the web and passes through O (*i.e.*, the centre of the web) then there will be no moments due to vertical forces (*i.e.*, due to applied force F and due to shear forces produced in the web). But there is a clockwise moment $F^* \times h$ which is unbalanced and can twist the section of the channel.

Now if the line of application of the applied vertical force F is displaced to the left by a distance of e from the vertical axis of the web as shown in Fig. 11.12, then the unbalanced clockwise moments $F^* \times h$ can be balanced by the counter-clockwise moments due to applied force F and vertical force produced in the web.

Taking moments of all forces about point O *i.e.*, centre of the web

$$F \times e = F^* \times h$$ [The moment of force F_V produced in the web is zero as it passes through O]

$$\therefore \quad e = \frac{F^* \times h}{F}$$

The above equation gives the location of shear centre.

In the above equation,

F^* = Shear force produced in flanges and it is equal to $\int \tau \times dA$

where τ is the shear stress in the flange and it is given by,

$$\tau = \frac{F \times A \times \bar{y}}{I \times b} \text{ in which}$$

F = Applied shear force

A = Area of shaded portion of flange = $x \times t_1$

b = Actual width of flange = t_1

I = Moment of inertia about axis of symmetry = I_{XX}

h = Distance between horizontal shear forces in the flanges.

For finding shear forces F^* in the flange, consider an element at a distance x from the right hand edge of the top flange as shown in Fig. 11.12. The shear stress at a distance x is given by,

$$\tau = \frac{F \times A\bar{y}}{I_{XX} \times b}$$

where $A\bar{y}$ = Moment of shaded area about X-X axis

$$= (x \times t_1) \times \frac{h}{2} \quad \text{and} \quad b = t_1 \qquad \left[\because A = x \times t_1 \text{ and } \bar{y} = \frac{h}{2}\right]$$

$$\therefore \quad \tau = \frac{F \times (x \times t_1) \times \frac{h}{2}}{I_{XX} \times t_1} = \frac{F \times x \times h}{2I_{XX}}$$

The shear force in the elementary area dA, is given by

$$= \tau \times dA = \tau \times dx \times t_1 \qquad [\because \; dA = dx \times t_1]$$

The total shear force in the top flange will be

$$F^* = \int_0^b \tau \times dA = \int_0^b \tau \times dx \times t_1 \qquad [\because \; dA = dx \,.\, t_1]$$

The limits are from O to total length of flange *i.e.*, b

$$\therefore \qquad F^* = \int_0^b \frac{F \,.\, x \,.\, h}{2I_{XX}} \,.\, t_1 \,.\, dx = \frac{F \,.\, h \,.\, t_1}{2I_{XX}} \int_0^b x\, dx$$

$$= \frac{F \,.\, h \,.\, t_1}{2I_{XX}} \left[\frac{x^2}{2}\right]_0^b = \frac{F \,.\, h \,.\, t_1 \,.\, b^2}{2I_{XX} \times 2} = \frac{F \,.\, h \,.\, t_1 \,.\, b^2}{4I_{XX}}$$

The shear force in the bottom flange will be also equal to F^*, but it will be in opposite direction as shown in Fig. 11.12.

Let us now find the location of shear centre.

Let e = distance of shear centre along the axis of symmetry *i.e.*, axis X-X from the centre of the web *i.e.*, from point O.

Taking moments of all forces about the centre O of the web, we get

$$F \times e = F^* \times h$$

[The moment of force F_V about point O is zero as this force passes through O]

$$\therefore \qquad e = \frac{F^* \times h}{F}$$

$$= \frac{F \,.\, h \,.\, t_1 \,.\, b^2}{4I_{XX}} \times \frac{h}{F} \qquad \left[\because \; F^* = \frac{F \,.\, h \,.\, t_1 \,.\, b^2}{4I_{XX}}\right]$$

$$= \frac{h^2 \,.\, t_1 \,.\, b^2}{4I_{XX}} \qquad \text{...(11.9)}$$

where $I_{XX} = 2\left[\frac{b \times t_1^{\,3}}{12} + bt_1 \,.\, \left(\frac{h}{2}\right)^2 \text{ for two flanges}\right] + \frac{t_2 \times h^3}{12}$

$$\text{for web} = \frac{b \times t_1^{\,3}}{6} + \frac{bt_1 h^2}{2} + \frac{t_2 h^3}{12}$$

For a section shown in Fig. 11.13, the shear centre is at O, the common point of intersection of applied vertical force F and shear forces F^* produced in the legs of the channel due to shear stress distribution in the channel section. If applied force F is acting at O, the resultant moment on the section is zero as all forces pass through O. Hence there will be no twisting of the section.

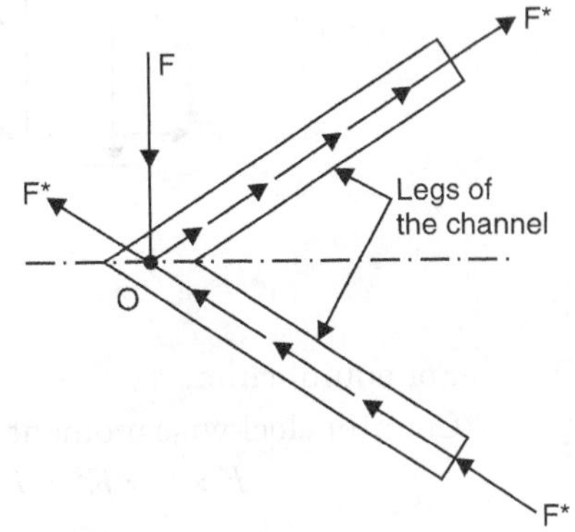

Fig. 11.13

Problem 11.7. *Determine the position of the shear centre for a channel section of 120 mm by 120 mm outside and 10 mm thick.*

Sol. Given :

Outside dimensions = 120 mm × 120 mm

Thickness of flange and web = 10 mm

The given section is symmetrical about the axis X-X. Hence shear centre lies on this axis.

Let F = applied force on the section vertically downwards.

The shear stress produced by the applied force in the channel section, has to follow the direction of the boundary, the shear stress distribution will be horizontal in the flanges and vertical in the web. The vertical upward shear force in the web will be equal to vertical downward applied shear force. Hence the vertical force is balanced. The horizontal force in the flanges will be equal and opposite. This pair of forces in the flanges will produce a twist in the cross-section. To balance this twist, the applied force should act at such a point so that the moment of the applied force and the couple due to horizontal forces in the flanges are balanced. The point at which the applied force acts is known as *shear centre.*

If applied force F acts through the vertical axis of the web, there will be no moment from the vertical forces. But the horizontal forces in the flanges will give the clockwise couple, which can twist the cross-section of the channel. If the line of action of the vertical downward applied force F is displaced to the left at a distance 'e' from the vertical axis of the web, then there will be a counter-clockwise couple which will balance the clockwise couple and there will be no twist in the cross-section.

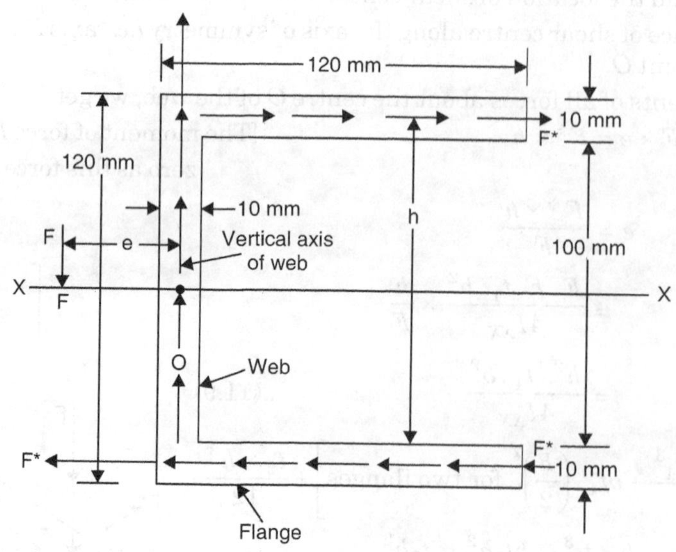

Fig. 11.14

For equilibrium,

Counter clockwise moment about O = clockwise moment (or couple)

$$F \times e = F^* \times h$$ [The vertical force in the web passes through O and hence moment of this force is zero]

$\therefore$ $$e = \frac{F^* \times h}{F},$$ where F^* = shear force due to shear stress is flanges.

To find the value of F*

Shear force $F^* = \tau \times$ Area. Let us find the shear stress in flange.

Shear stress in flanges are equal. It is given by

$$\tau = \frac{F \times A\bar{y}}{I \times b} \qquad ...(i)$$

where F = applied force on the section.

Consider a length x and thickness dx from the tip of the flange as shown by section lines in the Fig. 11.14 (a).

For Flange

Area of flange upto length x,

$$A = x \times 10 = 10x$$

$\bar{y}$ = Distance of C.G. of area A from N.A.

$$= \frac{120}{2} - \frac{10}{2} = 60 - 5 = 55$$

I = M.O.I. of the complete section about X-X

$$= 2\left[\frac{120 \times 10^3}{12} + 120 \times 10 \times 55^2\right] + \frac{10 \times 100^3}{12}$$

(For flanges) (For web)

$$= 2\,[10000 + 3630000] + 833333 = 8113333 \text{ mm}^4$$

b = Actual width at the section considered = 10 mm

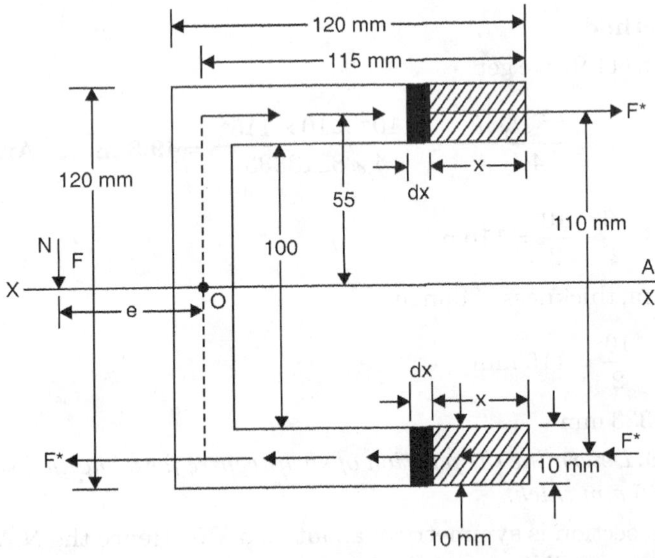

Fig. 11.14 (a)

Substituting these values in (*i*)

$$\therefore \qquad \tau = \frac{F\,(10x)\,55}{8113333 \times 10} = 6.778 \times 10^{-6}\, F \times x$$

Shear force on area dA, $dF^* = \tau \times dA = \tau \times b \times dx$

Total shear force in each flange becomes as

$$F^* = \int \tau \times dA = \int_0^{115} \tau \times b \times dx \qquad [\because \; dA = b \times dx]$$

$$= \int_0^{115} (6.778 \times 10^{-6}\, F \times x)\, 10 dx$$

$$= 6.778 \times 10^{-5}\, F \int_0^{115} x dx = 6.778 \times 10^{-5}\, F \left[\frac{x^2}{2}\right]_0^{115}$$

$$= 6.778 \times 10^{-5}\, F \times \frac{115^2}{2} = 0.448F$$

Let the applied force F acts at a distance 'e' from O as shown in Fig. 11.14(a). For no twist in the section, we must have

$$F \times e = F^* \times 110$$

$$\therefore \qquad e = \frac{F^* \times 110}{F} = \frac{0.448\, F \times 110}{F} = \mathbf{49.28\ mm.} \quad \textbf{Ans.}$$

Alternate Method

Using equation (11.9), we get

$$e = \frac{h^2 \,.\, t_1 \,.\, b^2}{4\, I_{XX}} = \frac{110^2 \times 10 \times 115^2}{4 \times 8113333} = \mathbf{49.3\ mm.} \quad \textbf{Ans.}$$

where $\qquad h = 120 - \dfrac{10}{2} - \dfrac{10}{2} = 110$ mm

$t_1 = 10$ mm, thickness of flange

$b = 120 - \dfrac{10}{2} = 115$ mm

$I_{XX} = 8113333\ \text{mm}^4$.

Problem 11.8. *Determine the position of shear centre for a channel section of 400 mm by 200 mm outside and 5 mm thick.*

Sol. The given section is symmetrical about axis X-X. Hence the N.A. will coincide with X-X axis and shear centre will lie on axis X-X.

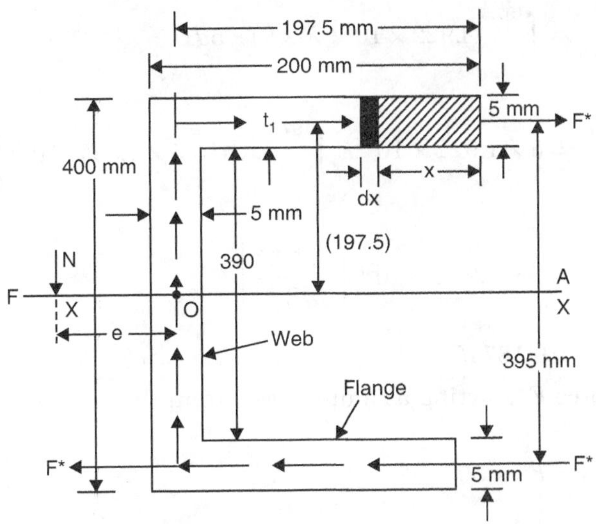

Fig. 11.15

Let F = applied external force on the section.

Let us find I_{XX}.

$$I_{XX} = 2\left[\frac{200 \times 5^3}{12} + 200 \times 5 \times 197.5^2\right] \text{ for flange} + \frac{5 \times 390^3}{12} \text{ for web}$$

$$= 2[2083 + 39006250] + 24716250 = 102732916 \text{ mm}^4$$

Let us now find the shear stress in the flanges.

Consider a length x and thickness dx from one tip of the flange as shown in Fig. 11.15.

The shear stress (τ) at a distance x is given by,

$$\tau = \frac{FA\overline{y}}{I_{XX} \times b} = \frac{F \times (5x) \times 197.5}{102732916 \times 5} = 1.922 \times 10^{-6}\, F \times x$$

where A = shaded area = $x \times 5 = 5x$ mm^2

$\overline{y}$ = Distance of C.G. of the shaded area from N.A. $= 200 - \dfrac{5}{2} = 197.5$ mm

b = Width of section considered = 5 mm

Shear force in the elementary area dA

$$= \tau \times dA = \tau \times b \times dx \qquad [\because\ dA = bdx]$$

$$= \tau \times 5 \times dx = 5\tau \times dx$$

Total shear force F^* in each flange,

$$F^* = \int \tau \times dA = \int_0^{197.5} \tau \times 5dx$$

$$= \int_0^{197.5} (1.922 \times 10^{-6}\ F \times x) \times 5dx$$

$$= 5 \times 1.922 \times 10^{-6} F \int_0^{197.5} x dx = 5 \times 1.847 \times 10^{-3} \left[\frac{x^2}{2}\right]_0^{197.5}$$

$$= 5 \times 1.922 \times 10^{-6} \left[\frac{197.5^2}{2}\right] F$$

$$= 0.1874F$$

If the applied force F is acting at a distance e from the centre line of the web, then for equilibrium, we have

$$F \times e = F^* \times 395$$

$$\therefore \qquad e = \frac{F^* \times 395}{F} = \frac{0.1874F \times 395}{F} = 74.02 \simeq \mathbf{74\ mm} \quad \textbf{Ans.}$$

Alternate Method

Using equation (11.9), we get

$$e = \frac{h^2 \times t_1 \times b^2}{4I_{XX}} = \frac{395^2 \times 5 \times 197.5^2}{4 \times 102732916} = \mathbf{74.05\ mm} \quad \textbf{Ans.}$$

where $h = 400 - \frac{5}{2} - \frac{5}{2} = 395$ mm

t_1 = Thickness of flange = 5 mm

b = Length of flange upto centre line of web = $200 - \frac{5}{2} = 197.5$ mm

$I_{XX} = 102732916\ \text{mm}^4$ (calculated)

11.7. DETERMINATION OF SHEAR CENTRE FOR I-SECTION

In case of a section having two axes of symmetry, the shear centre coincides with the centroid. But in case of a section having one axis of symmetry, the shear centre does not coincide with the centroid, but lies on the axis of symmetry. The equal I-section is symmetrical about X-X axis and Y-Y axis. Hence shear centre coincides with centroid.

But in case of unequal I-section, the section is symmetrical about X-X axis hence shear centre does not coincide with the centroid but lies on the X-X axis.

We know that shear stress follows the direction of boundary and hence shear stress distribution is horizontal in flanges and vertical in web. The corresponding shear forces due to these stresses will be horizontal in flanges and vertical in web. The total shear force in the web will be equal to the applied vertical shear force. Hence the vertical force is balanced. But the shear forces in the flanges are unbalanced. They are equal and opposite and hence produces couples as shown in Fig. 11.16.

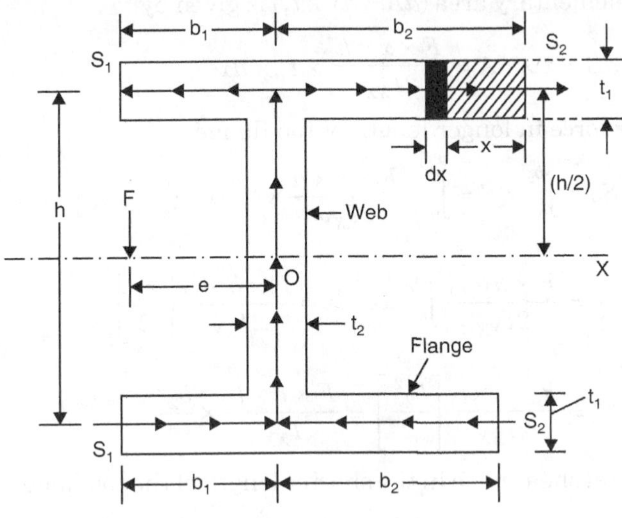

Fig. 11.16

The unequal I-section is shown in Fig. 11.16. It is symmetrical about *X-X* axis, hence shear centre will lie on this axis.

Let F = Vertical downward applied shear force

$b_1 + b_2$ = Total length of flange

t_1 = Thickness of flange

t_2 = Thickness of web

S_1 = Shear force in shorter length of top flange

S_2 = Shear force in longer length of top the flange.

There will be equal and opposite shear forces in the bottom flange also as shown in Fig. 11.16.

The vertical upward shear force in the web will be balanced by vertical downward applied shear force.

To find S_1 and S_2 consider an element at a distance x from the right hand edge of the top flange. The shear stress at a distance x is given by

$$\tau = \frac{FA\bar{y}}{I \times b}$$

$$= \frac{F \times (x \times t_1) \times \frac{h}{2}}{I_{XX} \times t_1}$$

$$= \frac{F \times x \times h}{2I_{XX}}$$

where F = Applied shear force,

$A\bar{y}$ = Moment of shaded area about axis of symmtery *i.e.*, about axis X-$X = (x \times t_1) \times \frac{h}{2}$

I = Moment of inertia about axis *X-X*

b = Width of the shaded section = t_1

Shear force in elementary area ($dA = dx \times t_1$) is given by

$$dS = \tau \times dA = \frac{F \times x \times h}{2I_{XX}} \times t_1 \times dx$$

$\therefore$ Total shear force in longer length of top flange

$$S_2 = \int_0^{b_2} dS = \int_0^{b_2} \frac{F \times x \times h}{2I_{XX}} \times t_1\, dx$$

$$= \frac{F \times h \times t_1}{2I_{XX}} \int_0^{b_2} x \times dx = \frac{F \times h \times t_1}{2I_{XX}} \left[\frac{x^2}{2}\right]_0^{b_2}$$

$$= \frac{F \times h \times t_1}{2I_{XX}} \left[\frac{{b_2}^2}{2}\right] = \frac{F \times h \times t_1}{I_{XX}} \times \frac{{b_2}^2}{4}$$

Similarly the total shear force in the shorter length of the top flange will be

$$S_1 = \int_0^{b_1} dS = \int_0^{b_1} \frac{F \times x \times h}{2I_{XX}} \times t_1\, dx = \frac{F \times h \times t_1}{2I_{XX}} \left[\frac{x^2}{2}\right]_0^{b_1}$$

$$= \frac{F \times h \times t_1}{I_{XX}} \times \frac{{b_1}^2}{4}.$$

The total shear force in the bottom flange for shorter and longer length will S_1 and S_2 respectively as shown in Fig. 11.16.

The couple due to S_2 will be clockwise whereas the couple due to S_1 will be anticlockwise. Also couple due to S_2 is having larger valve and hence net moment due to these couples will be clockwise. To balance this net moment the applied downward shear force F should act to the left of point O on the axis X-X to give anticlockwise moment.

Let e = distance of point of application of F from point O. Taking moments of all forces about point O, we get

$S_2 \times h - S_1 \times h = F \times e$ [Moment of shear force in the web is zero as the shear force passes through O]

$$\therefore \quad e = \frac{(S_2 - S_1) \times h}{F} = \left(\frac{F \times h \times t_1}{I_{XX}} \times \frac{{b_2}^2}{4} - \frac{F \times h \times t_1}{I_{XX}} \times \frac{{b_1}^2}{4}\right) \times \frac{h}{F}$$

or

$$e = \frac{h^2 \times t_1}{4I_{XX}} ({b_2}^2 - {b_1}^2) \qquad \text{...(11.10)}$$

Problem 11.9. *Determine the position of shear centre for unequal I-section shown in Fig. 11.17.*

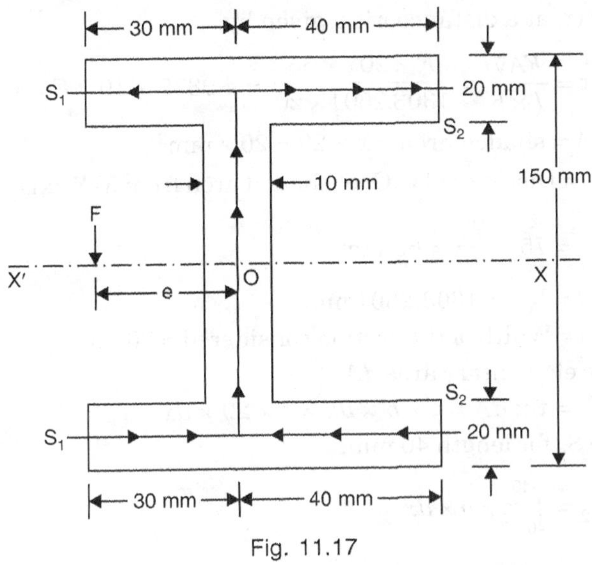

Fig. 11.17

Sol. The unequal *I*-section shown in Fig. 11.17 is symmetrical about *XX*-axis. Let us find the value of I_{XX} (moment of inertia of *I*-section about axis *XX*).

$$I_{XX} = 2\left[\frac{70 \times 20^3}{12} + (70 \times 20)\,(65)^2\right] + \frac{10 \times 110^3}{12}$$

$$= 2[46667 + 5915000] + 1109167 = 1303.2501 \text{ mm}^4$$

Fig. 11.17 (*a*)

Let F = Shear force applied on the section.

S_1 = Shear force produced by external applied shear force F on one part of the flange.

S_2 = Shear force produced on the other part of flange.

To find S_2

Consider a length x and thickness dx from one tip of the flange as shown in Fig. 11.17 (a). The shear stress (τ) at a distance x is a given by,

$$\tau = \frac{FA\bar{y}}{I \times b} = \frac{F \times 20x \times 65}{1303.2501 \times 20} = 4.9875 \times 10^{-6} F \times x$$

Here A = shaded area = $x \times 20 = 20\,x$ mm^2

$\bar{y}$ = Distance of C.G. of shaded area from X-X axis

$$= 75 - \frac{20}{2} = 65 \text{ mm}$$

$I = I_{XX} = 1303.2501$ mm^4

b = Width of the section considered = 20 mm

Shear force in the elementary area dA

$$= \tau \times dA = \tau \times b \times dx = \tau \times 20 \times dx \qquad [\because \; b = 20 \text{ mm}]$$

Total shear force S_2 for length 40 mm,

$$S_2 = \int_0^{40} \tau \times b \times dx$$

$$= \int_0^{40} (4.9875 \times 10^{-6} \times F \times x) \times 20dx \qquad [\because \; b = 20 \text{ mm}]$$

$$= (4.9875 \times 20 \times 10^{-6} \times F) \left[\frac{x^2}{2}\right]_0^{40}$$

$$= 9.975 \times 10^{-5} F \times \left[\frac{1600}{2}\right] = 7.98 \times 10^{-2} F.$$

To find S_1

The total shear force S_1 for the flange length of 30 mm will be

$$S_1 = \int_0^{30} \tau \times b \times dx = \int_0^{30} (4.9875 \times 10^{-6} \times F \times x) \times 20 \times dx$$

$$= (4.9875 \times 20 \times 10^{-6} \times F) \left[\frac{x^2}{2}\right]_0^{30}$$

$$= 4.9875 \times 20 \times 10^{-6} F \times \left(\frac{900}{2}\right)$$

$$= 4.488 \times 10^{-2} F$$

Let e = distance of shear centre from the centre O of the web. Hence it is point where external shear force F is acting.

Taking moments of shear forces S_1, S_2 and F about the centre O of the web,

$$S_2 \times 130 - S_1 \times 130 = F \times e$$

or $$(S_2 - S_1) \times 130 = F \times e$$

$$\therefore \qquad e = \frac{(S_2 - S_1) \times 130}{F} = \frac{(7.98 \times 10^{-2}\ F - 4.488 \times 10^{-2} F) \times 130}{F}$$

$$= (7.98 - 4.488) \times 10^{-2} \times 130 = \mathbf{4.5396\ mm} \simeq \mathbf{4.54\ mm} \quad \textbf{Ans.}$$

Alternative Method

Using equation (11.10), we get

$$e = \frac{h^2 \times t_1 \times (b_2{}^2 - b_1{}^2)}{4 I_{XX}}$$

where h = Height of I-section $- \frac{t_1}{2} - \frac{t_1}{2} = 150 - \frac{20}{2} - \frac{20}{2} = 130$ mm

t_1 = Thickness of flange = 20 mm

b_2 = Longer length of flange upto centre line of web = 40 mm

b_1 = Shorter length of flange upto centre line of web = 30 mm

I_{XX} = 13032501 mm^4 (calculated)

$$= \frac{130^2 \times 20 \times [40^2 - 30^2]}{4 \times 13032501} = \frac{130^2 \times 20 \times 700}{4 \times 13032501} = \mathbf{4.538\ mm.} \quad \textbf{Ans.}$$

Problem 11.10. *Determine the position of shear centre for the C-section shown in Fig. 11.18.*

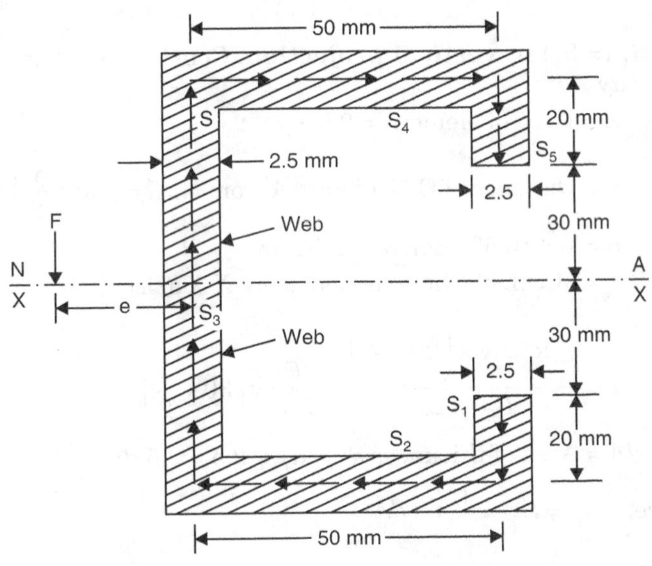

Fig. 11.18

Sol. The given section is symmetrical about axis X-X. Hence N.A. coincides with X-X axis.

Let $\quad F$ = applied shear force vertically downwards.

S_3 = Shear force in the web in the upward direction.

S_1, S_2, S_4, S_5 = Shear forces in different portions.

Also $S_1 = S_5$ and $S_2 = S_4$ (due to symmetry)
Shear stress in any layer,

$$\tau = \frac{FA\bar{y}}{I_{XX} \times b}$$

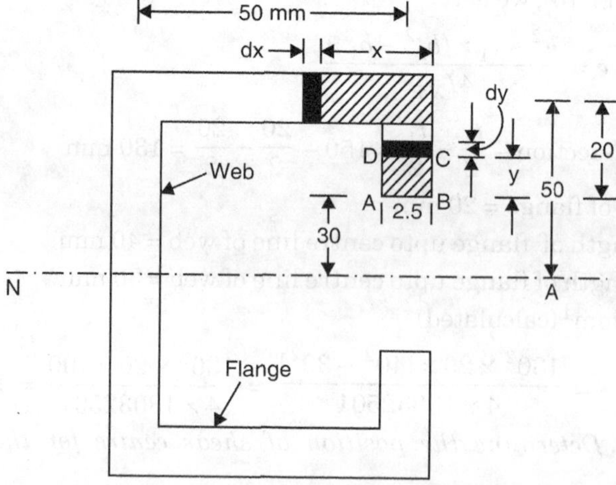

Fig. 11.19

Shear force S_1 (= S_5) in vertical projection. Consider an element of length y width 2.5 mm and thickness dy.

A = Area of element = $2.5 \times y = 2.5y$

$\bar{y}$ = Distance of C.G. of area A from N.A. = $\left(30 + \frac{y}{2}\right)$

b = Width of element = 2.5 mm

I_{XX} = M.O.I. of whole section about X-X axis

$$\therefore \quad \tau = \frac{F \times 2.5y \times \left(30 + \frac{y}{2}\right)}{I_{XX} \times 2.5} = \frac{F}{I_{XX}} y \left(30 + \frac{y}{2}\right)$$

dA = Area of thickness $dy = dy \times 2.5 = 2.5\, dy$

$\therefore$ Shear force, $S_1 (= S_5) = \int \tau \times dA$

$$= \int_0^{20} \frac{F}{I_{XX}} y \left(30 + \frac{y}{2}\right) \times 2.5\, dy = \frac{2.5\, F}{I_{XX}} \int_0^{20} \left(30y + \frac{y^2}{2}\right) dy$$

$$= \frac{2.5F}{I_{XX}} \left[\frac{30y^2}{2} + \frac{y^3}{6}\right]_0^{20} = \frac{2.5F}{I_{XX}} \left[15 \times 20^2 + \frac{20^3}{6}\right]$$

$$= \frac{2.5F}{I_{XX}} [6000 + 1333.33] = \frac{2.5 \times 7333.33\, F}{I_{XX}} = 18333.3 \frac{F}{I_{XX}}$$

Shear force S_2 (= S_4) in top and bottom flanges

Consider an element of length x width 2.5 mm and thickness dx as shown in Fig. 11.19 (a).

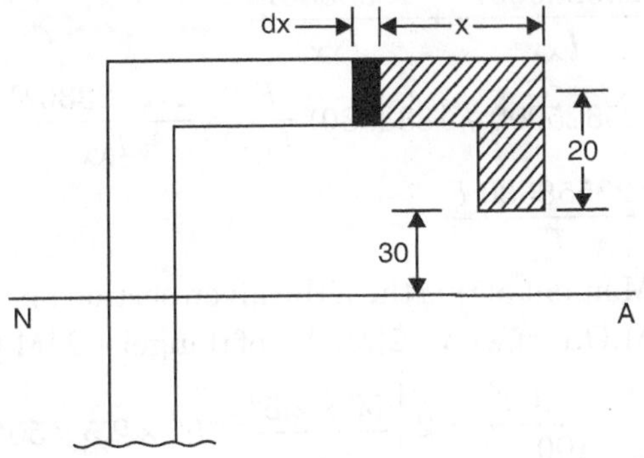

Fig. 11.19 (a)

The shear stress in the flange at a distance x is given by

$$\tau = \frac{FA\bar{y}}{I_{xx} \times b}$$

Here $A\bar{y}$ is made up of two parts *i.e.*, moment of flange area upto length x about N.A. plus moment of shaded area of vertical projection about N.A.

$\therefore$ $A\bar{y}$ = Moment of flange area upto length x about N.A. + Moment of shaded area of vertical projection about N.A.

$$= (2.5 \times x) \times (30 + 20) + (2.5 \times 20) \times \left(30 + \frac{20}{2}\right)$$

$$= 125x + 2000$$

and $b = 2.5$

$$\therefore \qquad \tau = \frac{F \times (125x + 2000)}{I_{XX} + 2.5}$$

Area of thickness dx, $dA = 2.5 \times dx$

$$\therefore \quad \text{Shear force } S_4 \, (= S_2) = \int \tau \times dA$$

$$= \int_0^{50} \frac{F(125x + 2000)}{2.5\, I_{XX}} \times 2.5 \times dx = \int_0^{50} \frac{F(125x + 2000)}{I_{XX}} dx$$

$$= \frac{F}{I_{XX}} \left[\frac{125\, x^2}{2} + 2000x\right]_0^{50}$$

$$= \frac{F}{I_{XX}} \left[\frac{125}{2} \times 50^2 + 2000 \times 50\right] = \frac{F}{I_{XX}} [156250 + 100000]$$

$$= \frac{F \times 256250}{I_{XX}}$$

Taking the moments of all forces about the centre of web, we get

$$F \times e = S_4 \times 100 + (S_5 + S_1) \times 50$$

$$= \frac{F \times 256250}{I_{XX}} \times 100 + \frac{2 \times 18333.3F}{I_{XX}} \times 50$$

$$= \frac{25625000F}{I_{XX}} + \frac{1833330F}{I_{XX}}$$

$$= (25625000 + 1833330)\frac{F}{I_{XX}} = \frac{27558330F}{I_{XX}}$$

or $$e = \frac{27558330\,F}{I_{XX}}$$

Now I_{XX} = Moment of inertia of the given section about axis X-X

= M.O.I. of web + 2[M.O.I. of flange] + 2 [M.O.I. of vertical projection]

$$= \frac{2.5 \times 100^3}{100} + 2\left[\frac{50 \times 2.5^3}{12} + 50 \times 2.5 \times 50^2\right] + 2\left[\frac{2.5 \times 20^3}{12} + 2.5 \times 20 \times \left(50 - \frac{20}{2}\right)^2\right]$$

$$= 208333 + 2[65.1 + 312500] + 2[1666 + 80000]$$

$$= 208333 + 625730 + 163332 = 997395$$

$\therefore$ $e = \frac{27558330}{997395}$ = **27.63 mm. Ans.**

HIGHLIGHTS

1. The unsymmetrical bending will be when :
   (*i*) section is symmetrical but load line is inclined to both the principal axes,
   (*ii*) section is unsymmetrical (such as angle section or channel section) and load line is long any centroidal axis.
2. Principal axes are the axes about which product of inertia is zero. The moment of inertia about principal axes is known as principal moment of inertia.
3. The condition for principal axes is

$$\tan 2\theta = \frac{2I_{xy}}{I_{yy} - I_{XX}}$$

where θ = angle in anti-clockwise direction of principal axes [*UU* and *VV*] with axes *XX* and *YY*.

4. The principal moments of inertia about principal axes (*UU* and *VV*) will be

$$I_{UU} = \cos^2\theta \times I_{XX} + \sin^2\theta\, I_{yy} - \sin 2\theta \times I_{xy}$$

$$= \frac{1}{2}[I_{XX} + I_{yy}] + \frac{1}{2}[I_{xx} - I_{yy}] \sec 2\theta$$

and $$I_{VV} = \frac{1}{2}[I_{XX} + I_{yy}] - \frac{1}{2}[I_{xx} - I_{yy}] \sec 2\theta$$

5. The position of Neutral Axis (N.A.) is given by

$$\tan\alpha = \left(-\frac{I_{UU}}{I_{VV}}\tan\theta\right)$$

where α = angle made by neutral axis with principal axis UU in clockwise direction

θ = Angle made by principal axis UU with XX axis in anti-clockwise direction.

I_{UU} and I_{VV} = Principal moment of inertia about principal axes UU and VV respectively.

6. Shear centre is the point (in or outside a section) through which the applied shear force produces no torsion or twist of the member.

7. The location of shear centre is given for different sections as :

(*a*) **For channel section**

$$e = \frac{h^2 \times t_1 \times b^2}{4I_{XX}}$$

where h = Distance between the centre lines of the flanges

t_1 = Thickness of flange

b = Length of flange upto the centre line of the web

I_{XX} = Moment of inertia of the channel about axis of symmetry (*X-X* axis)

e = Horizontal distance of applied vertical downward shear force from the centre of the web.

(*b*) **For unequal I-section**

$$e = \frac{h^2 \times t_1(b_2^2 - b_1^2)}{4I_{XX}}$$

where t_1 = Thickness of the flanges of *I*-section

b_2 = Longer length of *I*-section from vertical centre line of web

b_1 = Shorter length of *I*-section from vertical centre line of web

I_{XX} = Moment of inertia of *I*-section about axis of symmetry (*X-X* axis)

8. For sections having symmetry about the two-axes, the shear centre coincides with the centroid, whereas for sections having symmetry about one axis, the shear centre lies on the axis of symmetry but does not coincides with centroid.

EXERCISE 11

(A) Theoretical Questions

1. Define and explain the terms: unsymmetrical bending and shear centre.
2. State the condition for the two axes to be principal axes for the given two perpendicular axes.
3. How will you find the resultant stress in unsymmetrical bending ?
4. Write a short note of deflection of beams in unsymmetrical bending.
5. Define shear centre. What is the importance of shear centre ?
6. How will you determine the shear centre for channel section and *I*-section ?

(B) Numerical Problems

1. Determine the maximum stress setup and position of the neutral axis for an angle of dimensions 50 mm by 30 mm by 5 mm. The given angle is used as a cantilever of length 500 mm with 30 mm leg horizontal. A load of 1 kN is applied at the free end.

   [**Ans.** (*i*) 203 N/mm² (tensile) and 233 N/cm² (Comp.), (*ii*) $\alpha = -23.08°$]

2. Find the position of principal axes and the values of the principal moments of inertia for an unequal angle 75 mm by 45 mm by 7.5 mm.

   [**Ans.** (*i*) $\theta = 70.36°$, (*ii*) $I_{UU} = 8.05 \times 10^8$ mm⁴, $I_{VV} = 52.8 \times 10^8$ mm⁴]

3. A simply supported beam of span 3.6 m carries a load of 600 N at its centre. The section of the beam is an equal angle of size 120 mm by 120 mm by 15 mm as shown in Fig. 11.20. The load line passes through the centroid (G) of the section and is along line YG. Determine :

(*i*) stresses at points *A*, *B* and *C* of mid-section of the beam
(*ii*) deflection of beam at mid-section and
(*iii*) position of neutral axis. Take $E = 2 \times 10^5$ N/mm².

[Ans. (*i*) $\sigma_A = 11.65$ N/mm², $\sigma_B = -10.3$ N/mm², $\sigma_C = 2.586$ N/mm²
(*ii*) $\delta = 1.143$ mm (*iii*) $\alpha = -75.37°$]

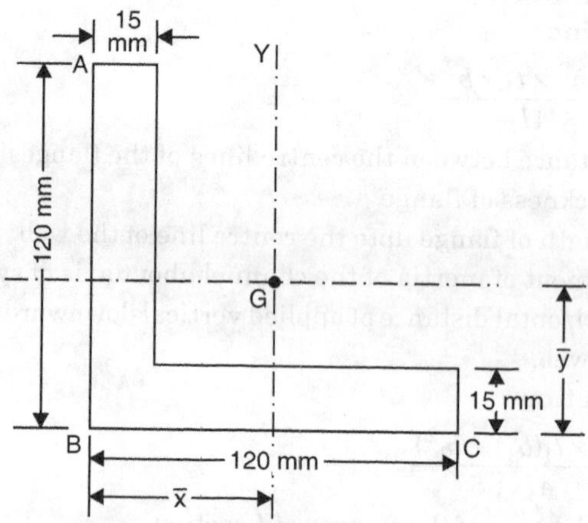

Fig. 11.20

4. **Determine the position of shear centre of a channel having dimensions : flanges 120 mm × 20 mm and web 160 mm × 10 mm.** **[Ans.** $e = 50.84$ mm**]**

5. **Determine the position of shear centre for the unequal *I*-section shown in Fig. 11.21.** **[Ans.** $e = 9.08$ mm**]**

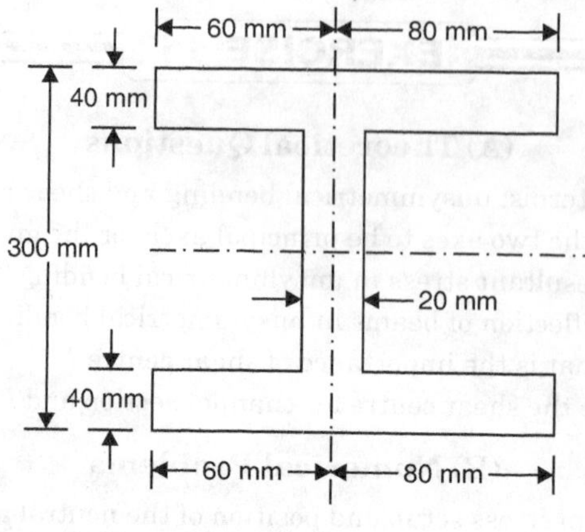

Fig. 11.21

EXAMINATION PAPER AND OBJECTIVE TYPE QUESTIONS WITH SOLUTIONS

EXAMINATION PAPER

(EME-302)

B. TECH

(SEM. III) ODD SEMESTER THEORY EXAMINATION 2009–2010

STRENGTH OF MATERIALS

Time : 3 Hours *Total Marks : 100*

Note : (1) *Attempt* ***all*** *questions:*

(2) *Assume any* ***data*** *required, but not given.*

1. Attempt any **two** parts of the following : **10 × 2 = 20**

(*a*) At a point in a strained material, there are normal stresses of 30 N/mm^2, tension and 20 N/mm^2, compression on two planes at right angles to one another, together with shearing stresses of 15 N/mm^2 on the same planes. If the loading on the material is increased so that the stresses reach values of K times those given, find the maximum permissible value of K if the maximum direct stress in the material is not to exceed 80 N/mm^2, and maximum shear stress is not to exceed 50 N/mm^2.

(*b*) A shaft of 15 cm diameter is subjected to a maximum torque of 20 kNm and a maximum bending moment of 18 kNm. Find the factor of safety (*i*) according to the maximum shear stress theory; (*ii*) according to the maximum strain energy theory if the elastic limit in simple tension is 240 MN/m^2. Take $\mu = 0.3$.

(*c*) Write short notes on any two of the following:

(*i*) Castigliano's theorem

(*ii*) Compatibility equations

(*iii*) Three-dimensional stresses.

2. Attempt any **two** parts of the following : **10 × 2 = 20**

(*a*) A timber joist of 6 metre span has to carry a load of 15 kN/metre. Find the dimensions of the joist if the maximum permissible stress is limited to 8 N/mm^2. The depth of the joist has to be twice the width.

(*b*) A beam, simply supported at ends A and B is loaded with two point loads of 60 kN and 50 kN at distance 1 metre and 3 metre respectively from end A. Determine the position and magnitude of maximum deflection. Take $E = 2 \times 10^5$ N/mm^2 and $I = 8500$ cm^4.

(*c*) Find the internal and external diameters required for a hollow shaft, which is to transmit 40 kW of power at 240 rev/minute. The shear stress is to be limited to 100 MN/m^2. Take outside diameter to be twice the inside diameter.

3. Attempt any **two** parts of the following : **10 × 2 = 20**

(*a*) A leaf spring has 12 plates each 50 mm wide and 5 mm thick, the longest plate being 600 mm long. The greatest bending stress is not to exceed 180 N/mm^2 and the central deflection is 15 mm. Estimate the magnitude of the greatest central load that can be applied to the spring. $E = 0.206 \times 10^6$ N/mm^2.

(*b*) Determine the section of a cast iron hollow cylindrical column 5 metre long with ends firmly built-in if it carries an axial load of 300 kN. The ratio of internal to external diameter is 3/4. Use factor of safety of 8. Take $\sigma_c = 567$ N/mm^2 and Rankine's constant $a = 1/1600$.

(*c*) From the first principles drive the expression for the critical buckling for a column having one end fixed and one end hinged.

4. Attempt any **two** parts of the following: **10 × 2 = 20**

(*a*) A cylindrical vessel 1.5 metre in diameter, 2 metre long and 1.5 cm thick is closed at both the ends by rigid plates and this cylinder is filled with water at atmospheric pressure. Find how much additional amount of water is required to be pumped so as to make the final pressure in the cylinder as 70 bar. Take E = 210 GN/m^2 and $\mu = 0.3$ for the material of the cylinder. Bulk modulus of the water is 2.4 GN/m^2.

(*b*) Calculate the thickness of metal necessary for a cylindrical shell of internal diameter of 80 mm to withstand an internal pressure of 25 N/mm^2, if the maximum permissible tensile stress is 125 N/mm^2.

(*c*) Write short notes on any **two** of the following:

(*i*) Lame's theory of thick cylinders

(*ii*) Compound cylinders

(*iii*) Radial, axial and circumferential stresses in thick cylinders.

5. Attempt any **two** parts of the following: **10 × 2 = 20**

(*a*) A curved bar of square section, 3 cm sides and mean radius of curvature 4.5 cm is initially unstressed. If a bending moment of 300 Nm is applied to the bar tending to straighten it, find the stresses at the inner and outer faces.

(*b*) A 60 mm × 40 mm × 6 mm unequal angle is placed with the longer leg vertical, and is used as a beam. It is subjected to a bending moment of 12 kN-cm acting in the vertical plane through the centroid of the section. Determine the maximum bending stress induced in the section.

(*c*) Write short notes on any **two** of the following:

(*i*) Centroidal Principal Axes

(*ii*) Assumptions for the theory of curved beams

(*iii*) Application of curved beams with large initial curvature.

SOLUTION OF EXAMINATION PAPER

1. (*a*) $\sigma_x = 30$ N/mm^2, $\sigma_y = -20$ N/mm^2

$\tau = 15$ N/mm^2

Principal stresses are given by,

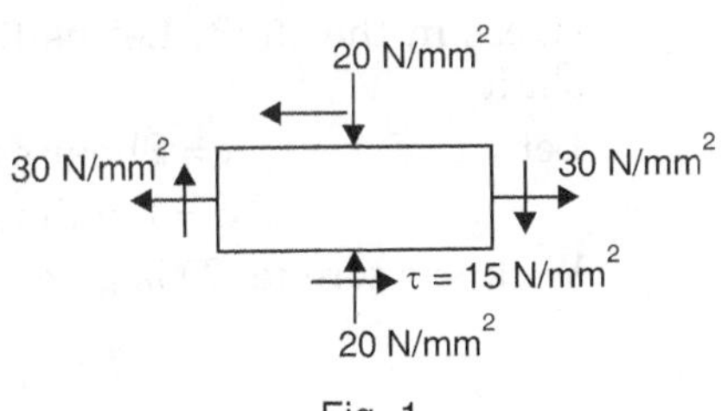

Fig. 1

$$\text{Principal stresses } = \frac{\sigma_x + \sigma_y}{2} \pm \sqrt{\left(\frac{\sigma_x - \sigma_y}{2}\right)^2 + \tau^2}$$

and max. shear stress is given by,

$$\tau_{max} = \sqrt{\left(\frac{\sigma_x - \sigma_y}{2}\right)^2 + \tau^2}$$

Let the load is increased on the material such that stresses become K times the given stresses.

Then, $\sigma_x = 30K$ N/mm^2, $\sigma_y = -20K$ N/mm^2 and $\tau = 15K$ N/mm^2.

(*i*) *The value of K for maximum direct stress of 80 N/mm²*.

Max. direct (or principal) stress

$$= \frac{\sigma_x + \sigma_y}{2} + \sqrt{\left(\frac{\sigma_x - \sigma_y}{2}\right)^2 + \tau^2}$$

$$= \frac{30K + (-20K)}{2} + \sqrt{\left(\frac{30K - (-20K)}{2}\right)^2 + (15K)^2}$$

$$= 5K + \sqrt{(25K)^2 + (15K)^2} = 5K + \sqrt{625K^2 + 225K^2}$$

$$= 5K + \sqrt{850K^2} = 5K + 29.15K = 34.15K$$

But max. direct stress given = 80 N/mm^2

$\therefore$ $80 = 34.15K$

or $K = \dfrac{80}{34.15} = \mathbf{2.34}$ ...(*i*)

(*ii*) *Value of K for maximum shear stress of 50 N/mm²*

$$\tau_{max} = \sqrt{\left(\frac{\sigma_x - \sigma_y}{2}\right)^2 + \tau^2} = \sqrt{\left(\frac{30K - (-20K)}{2}\right)^2 + (15K)^2}$$

$$= \sqrt{(25K)^2 + (15K)^2} = \sqrt{850K^2} = 29.15K$$

But τ_{max} (given) = 50 N/mm^2

$\therefore$ $50 = 29.15K$

or $K = \dfrac{50}{29.15} = \mathbf{1.71}$ ...(*ii*)

To satisfy both the condition of maximum direct stress of 80 N/mm^2 and maximum shear stress of 50 N/mm^2, the maximum permissible value of K will be minimum of the two values of K *i.e.*, **1.71. Ans.**

1. (*b*) $d = 15$ cm $= 0.15$ m ; Torque, $T = 20$ kNm $= 20{,}000$ Nm ; Bending moment, $M = 18$ kNm $= 18000$ Nm. In simple tension, maximum stress $= 240$ MN/m^2. Due to torque, there will be shear stress whereas due to bending moment, there will be bending stress in the shaft. Let us first find the bending stress and shear stress set up in the shaft.

Let τ = Shear stress

σ = Bending stress

We know torque (T) is given by,

$$T = \frac{\pi}{16} \times \tau \times d^3$$

$$\therefore \quad \tau = \frac{16T}{\pi d^3} = \frac{16 \times 20000}{\pi \times (0.15)^3} = \frac{16 \times 20000}{0.0106}$$

$$= 30188679 \text{ N/m}^2 = 30.188 \text{ MN/m}^2 \simeq 30.19 \text{ MN/m}^2$$

The bending stress is obtained from bending equation as given by,

$$\frac{M}{I} = \frac{\sigma}{y}$$

or

$$\sigma = \frac{M}{I} \times y, \quad \text{where} \quad I = \frac{\pi}{64} d^4 \text{ and } y = \frac{d}{2}$$

$$= \frac{M}{\frac{\pi}{64} d^4} \times \frac{d}{2} = \frac{32M}{\pi d^3} = \frac{32 \times 18000}{\pi \times (0.15)^3}$$

$$= 54.339 \times 10^6 \text{ N/m}^2 = 54.34 \text{ MN/m}^2$$

The bending stress is tensile (or compressive) in nature.

At any point on the shaft, there will be shear stress and tensile (or compressive) stress.

Let us find now principal stresses (σ_1 and σ_2) due to bending stress and shear stress.

The principal stresses are given by

$$(\sigma_1 \text{ and } \sigma_2) = \frac{\sigma_x + \sigma_y}{2} \pm \sqrt{\left(\frac{\sigma_x - \sigma_y}{2}\right)^2 + \tau^2}$$

$$= \frac{\sigma + 0}{2} \pm \sqrt{\left(\frac{\sigma - 0}{2}\right)^2 + \tau^2} \qquad (\because \ \sigma_x = \sigma, \sigma_y = 0)$$

$$= \frac{54.34}{2} \pm \sqrt{\left(\frac{54.34}{2}\right)^2 + 30.19^2}$$

$$= 27.17 \pm \sqrt{27.17^2 + 30.19^2} = 27.17 \pm \sqrt{738.2 + 911.4}$$

$$= 27.17 \pm 40.61$$

$$= 27.17 + 40.61 \text{ and } 27.17 - 40.61$$

$$= 67.78 \text{ and } -13.44 \text{ MN/m}^2$$

(*i*) *Factor of safety according to maximum shear stress theory.*

According maximum shear stress theory,

Maximum shear stress = Half of difference between principal stresses

$$= \frac{1}{2}(\sigma_1 - \sigma_2) = \frac{1}{2}[67.78 - (-13.44)] = \frac{1}{2}[67.78 + 13.44]$$

$$= 40.61 \text{ MN/m}^2$$

In simple tension, the principal stresses are 240, 0, 0 and maximum shear stress

$$= \frac{1}{2}[240 - 0] = 120 \text{ MN/m}^2$$

$$\therefore \quad \text{Factor of safety} = \frac{\text{Maximum shear stress in simple tension}}{\text{Maximum shear stress due to } \sigma_1 \text{ and } \sigma_2}$$

$$= \frac{120}{40.61} = 2.95 \simeq \mathbf{3.} \quad \mathbf{Ans.}$$

(*ii*) *Factor of safety according to maximum strain energy theory.*

According to maximum strain energy theory,

$$\text{Maximum strain energy} = \frac{1}{2E}[\sigma_1^2 + \sigma_2^2 + \sigma_3^2 - 2\mu(\sigma_1\sigma_2 + \sigma_2\sigma_3 + \sigma_3\sigma_1)]$$

$$= \frac{1}{2E}[67.78^2 + (-13.44)^2 + 0^2 - 2 \times 0.3 \times \{67.78 \times (-13.44) + 0 + 0\}] \quad (\because \sigma_3 = 0)$$

$$= \frac{1}{2E}[67.78^2 + 13.44^2 + 2 \times 0.3 \times 67.78 \times 13.44]$$

$$= \frac{1}{2E}[4594 + 180.6 + 546.6] = \frac{5321.2}{2E}$$

In simple tension, maximum strain energy

$$= \frac{1}{2E}[240^2] \quad [\text{Here } \sigma_2 = 0, \sigma_3 = 0]$$

$$= \frac{57600}{2E}$$

$$\therefore \quad \text{Factor of safety} = \frac{\text{Maximum strain energy in simple tension}}{\text{Maximum strain energy developed in shaft}}$$

$$= \frac{\left(\frac{57600}{2E}\right)}{\left(\frac{5321.2}{2E}\right)} = \frac{57600}{5321.2} = 10.82 \simeq \mathbf{11.} \quad \mathbf{Ans.}$$

1. (*c*) (*i*) Refer to page 69 Article 2.5.

(*ii*) **Compatibility equations** are the equations which are used for converting a statically indeterminate problem to a statically determinate problem. For compatibility equations the geometric considerations, among the deformed parts of a body and at the boundaries, must be satisfied.

Let us apply the compatibility equations to a statically indeterminant problem.

An elastic stepped bar is loaded as shown in Fig. 1(*A*). Find the reactions R_1 and R_2 caused by the application of the force P_1 at the junction *B*. The area of cross-section of upper part is A_1 whereas of lower part is A_2.

From statics, we have

$$R_1 + R_2 = P_1 \qquad ...(1)$$

In the above equation, P_1 is given. R_1 and R_2 are unknown. Hence there is only one equation, but unknown are two. Hence this system is statically indeterminate. Now let us see how this system can be converted into statically determinate system.

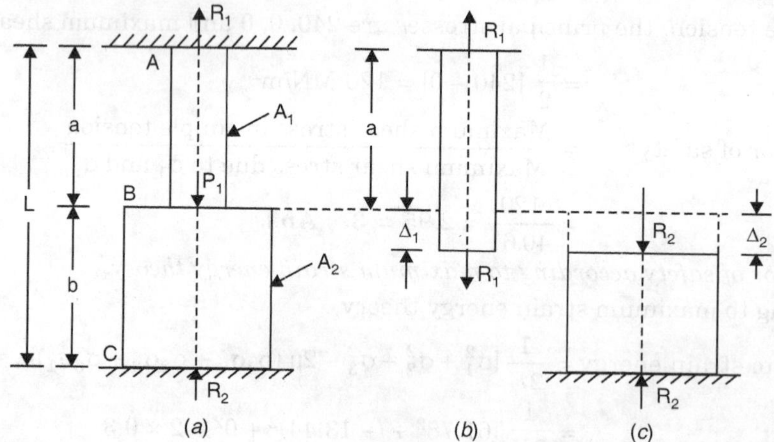

Fig. 1(A)

Consider the bar to be divided in two parts AB and BC. The upper part is subjected to a tensile force R_1. Hence there is elongation Δ_1 in this part due to tensile force R_1. The lower part is subjected to a compressive force R_2. Hence there is contraction Δ_2 in this part due to compressive force R_2. Due to geometric consideration these deflections must be equal, or Δ_1 should be equal to Δ_2.

Hence compatibility equation becomes as

$$\Delta_1 = \Delta_2$$

But $\quad \Delta_1 = \dfrac{R_1 \times a}{A_1 E} \quad \left(\because E = \dfrac{\text{Stress}}{\text{Strain}} = \dfrac{\left(\dfrac{P}{A}\right)}{\left(\dfrac{\Delta L}{L}\right)} = \dfrac{P \times L}{A \times \Delta L} \right.$

or $\Delta L = \dfrac{P \times L}{AE}$. Hence for AB, $P = R_1$, $L = a$,

$A = A_1$ and $\Delta L = \Delta_1$

$$\left. \therefore \quad \Delta_1 = \frac{R_1 \times a}{A_1 E} \right)$$

Similarly $\quad \Delta_2 = \dfrac{R_2 \times b}{A_2 \times E}$

But $\quad \Delta_1 = \Delta_2$

$$\therefore \quad \frac{R_1 \times a}{A_1 E} = \frac{R_2 \times b}{A_2 \times E} \qquad ...(2)$$

Now there are two equations (1) and (2) and there are two unknowns R_1 and R_2. They can be solved. Hence the problem has been converted into statically determinate problem

From equation (2), $\quad R_1 = \dfrac{R_2 \times b}{A_2 \times E} \times \dfrac{A_1 \times E}{a} = \dfrac{R_2 \times A_1 \times b}{A_2 \times a} \qquad ...(3)$

Substituting the value of R_1 in equation (1), we get

$$\frac{R_2 \times A_1 \times b}{A_2 \times a} + R_2 = P_1$$

or $$R_2 \left[\frac{A_1 \times b}{A_2 \times a} + 1\right] = P_1$$

$\therefore$ $$R_2 = \frac{P_1}{\left(\frac{A_1 \times b}{A_2 \times a}\right) + 1} = \frac{P_1}{\left(1 + \frac{A_1 \times b}{A_2 \times a}\right)}.$$ **Ans.**

Substituting the value of R_2 in equation (3), we get

$$R_1 = \left[\frac{P_1}{\left(\frac{A_1 \times b}{A_2 \times a}\right) + 1}\right] \times \frac{A_1 \times b}{A_2 \times a} = \frac{P_1 \times A_1 \times b}{(A_1 \times b) + (A_2 \times a)}$$

$$= \frac{P_1}{1 + (A_2 \times a)/(A_1 \times b)}.$$ **Ans.**

(*iii*) Refer to page 55 Article 2.1.

2. (*a*) Given :

$L = 6$ m, load/m, $w = 15$ kN $= 15000$ N

Max. stress, $\sigma = 8$ N/mm^2 $= 8 \times 10^6$ N/m^2

Depth $= 2 \times$ Width or $d = 2b$

Max. bending moment will be at the centre and it is equal to $\frac{W \times L}{8}$

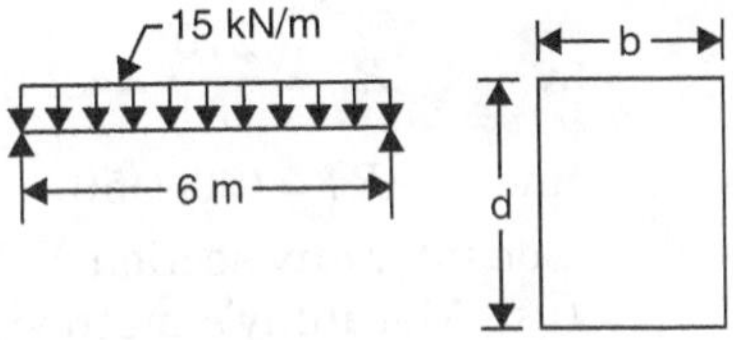

where $W = w \times L = 15000 \times 6 = 90{,}000$ N

Fig. 2

$\therefore$ Max. bending moment,

$$M = \frac{W \times L}{8} = \frac{90000 \times 6}{8} \text{ Nm}$$

Using equation, $\frac{M}{I} = \frac{\sigma}{y}$, we get

$$\sigma = \frac{M}{I} \times y \qquad ...(i)$$

where $$I = \frac{bd^3}{12} = \frac{b \times (2b)^3}{12} \qquad (\because \ d = 2b)$$

$$= \frac{b \times 8 \times b^3}{12} = \frac{2}{3} b^4$$

$$y = \frac{d}{2} = \frac{2b}{2} = b \qquad (\because \ d = 2b)$$

$$M = \frac{90000 \times 6}{8} \text{ Nm}$$

$\sigma = 8 \times 10^6$ N/m^2

Substituting these values in equation (*i*), we get

$$8 \times 10^6 = \frac{\left(\frac{90000 \times 6}{8}\right)}{\left(\frac{2}{3} b^4\right)} \times b = \frac{90000 \times 6 \times 3}{8 \times 2} \times \frac{1}{b^3}$$

or

$$b^3 = \frac{90000 \times 6 \times 3}{16} \times \frac{1}{8 \times 10^6}$$

$$= \frac{90000 \times 9}{64 \times 10^6} = \frac{90 \times 9 \times 10^3}{64 \times 10^6} = \frac{810}{64 \times 10^3}$$

$$\therefore \quad b = \left(\frac{810}{64 \times 10^3}\right)^{1/3} = \frac{(810)^{1/3}}{4 \times 10} = \frac{9.32}{40} = 0.233 \text{ m} = \mathbf{233 \text{ mm.}} \quad \textbf{Ans.}$$

and $d = 2b = 2 \times 233 = \mathbf{466 \text{ mm.}}$ **Ans.**

2. (*b*) In this question, span of beam is not given. Assume it to be 5 m.

$\therefore \quad L = 5$ m.

$E = 2 \times 10^5 \text{ N/mm}^2, I = 8500 \text{ cm}^4$

or $\quad I = 8500 \times 10^4 \text{ mm}^4$.

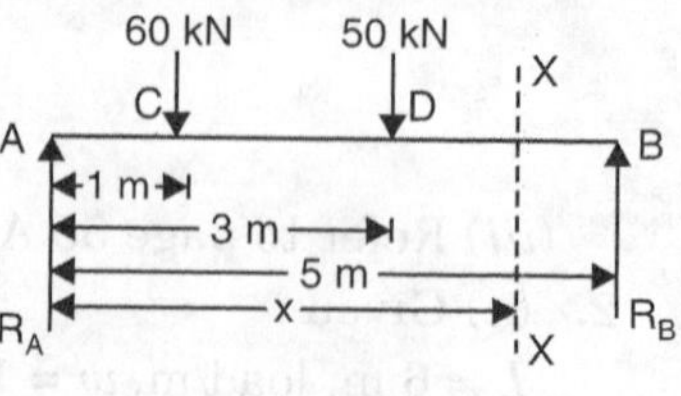

Fig. 3

Find : (*i*) Max. deflection.

(*ii*) Position of max. deflection.

First, find the reactions R_A and R_B

$$\Sigma M_A = 0, 5 \times R_B = 60 \times 1 + 50 \times 3 = 60 + 150 = 210$$

$$\therefore \quad R_B = \frac{210}{5} = 42 \text{ kN}$$

and $\quad R_A = (60 + 50) - R_B = 110 - 42 = 68$ kN

Consider any section *X-X* at a distance *x* from *A*, the bending moment at *X-X* is given by (use Macaulay's method)

$$M_x = EI \frac{d^2y}{dx^2} = R_A \times x \;\vdots\; - 60\,(x - 1) \;\vdots\; - 50\,(x - 3)$$

Integrating, we get

$$EI \frac{dy}{dx} = \frac{68 \times x^2}{2} + C_1 \;\vdots\; \frac{-60(x-1)^2}{2} \;\vdots\; - \frac{50(x-3)^2}{2} \qquad (\because \; R_A = 68)$$

$$= 34x^2 + C_1 \;\vdots\; - 30\,(x-1)^2 \;\vdots\; - 25\,(x-3)^2 \qquad ...(i)$$

Integrating again, we get

$$EIy = \frac{34x^3}{3} + C_1 x + C_2 \;\vdots\; \frac{-30(x-1)^3}{3} \;\vdots\; - \frac{25(x-3)^3}{3}$$

$$= 11.33x^3 + C_1 x + C_2 \mid - 10(x-1)^3 \mid - 8.33(x-3)^3 \qquad ...(ii)$$

To find C_1 and C_2, use boundary conditions.

When $x = 0, y = 0$ $\quad \therefore \quad 0 = 0 + 0 + C_2 \quad \therefore \quad C_2 = 0$

When $x = 5$ m, $y = 0$ $\quad \therefore \quad 0 = 11.33 \times 5^3 + C_1 \times 5 + 0 - 10 \times (5-1)^3 - 8.33 \times (5-3)^3$

$$= 1416.25 + 5C_1 - 10 \times 4^3 - 8.33 \times 2^3$$

$$= 1416.25 + 5C_1 - 640 - 66.64$$

$\therefore \quad 5C_1 = -1416.25 + 640 + 66.64 = -709.61$

or $\quad C_1 = \frac{-709.61}{5} = -141.9$

Substituting the values of C_1 and C_2 in equation (*ii*), we get deflection at any section as

$$EIy = 11.33x^3 + (-141.9)x + 0 \;\vdots\; -10(x-1)^3 \;\vdots\; -8.33(x-3)^3 \quad ...(iii)$$

Position of maximum deflection

The maximum deflection will be at a section between C and D. For maximum deflection, $\frac{dy}{dx}$ should be zero. Hence from equation (*i*), we have

$$EI\frac{dy}{dx} = 34x^2 + C_1 \;\vdots\; -30(x-1)^2 \text{ ... upto point } D$$

For maximum deflection, $\frac{dy}{dx} = 0.$

Hence substituting $\frac{dy}{dx} = 0$, in the above equation, we get

$$0 = 34x^2 + C_1 - 30(x-1)^2$$

$$= 34x^2 - 141.9 - 30(x-1)^2 \qquad (\because \; C_1 = -141.9)$$

$$= 34x^2 - 141.9 - 30(x^2 + 1 - 2x) = 34x^2 - 141.9 - 30x^2 - 30 + 60x$$

$$= (34-30)x^2 + 60x - 141.9 - 30 = 4x^2 + 60x - 171.9$$

or $\quad 0 = x^2 + 15x - 42.975$

or $\quad x = \frac{-15 \pm \sqrt{15^2 + 4 \times 42.975}}{2}$

$$= \frac{-15 + \sqrt{225 + 171.9}}{2} \qquad \text{(–ve value is not possible)}$$

$$= \frac{-15 + 19.92}{2} = \mathbf{2.46\ m.} \quad \textbf{Ans.}$$

The deflection will be maximum at a distance of 2.46 m from end A.

Max. Deflection

Substituting $x = 2.46$ in equation (*iii*), we get maximum deflection.

$$\therefore \quad EIy_{max} = 11.33(2.46)^2 - 141.9 \times 2.46 - 10(2.46-1)^3$$

[Do not consider last term]

$$= 68.56 - 349 - 31.12 = -311.56$$

$$\therefore \quad y_{max} = \frac{-311.56}{EI} \qquad (\because \; EI = 2 \times 10^5 \text{ N/mm}^2 \times 8500 \times 10^4 \text{ mm}^4$$

$$= 2 \times 8500 \times 10^9 \text{ Nmm}^2$$

$$= 2 \times 8500 \times 10^9 \times \frac{1}{1000}(\text{kN}) \times \frac{1}{10^6}\text{m}^2$$

$$= 2 \times 8500 \text{ kN m}^2)$$

$$= \frac{-311.56}{2 \times 8500}\text{m} = \frac{-311.56 \times 1000}{2 \times 8500}\text{mm}$$

$$= -18.327 \simeq \mathbf{-18.33\ mm.} \quad \textbf{Ans.}$$

(–ve sign means the deflection is downwards)

2. (*c*) Power, $P = 40$ kW $= 40 \times 10^3$ W; N = 240 r.p.m. ; $\tau = 100$ MN/m^2 $= 100 \times 10^6$ N/m^2 ; $D_0 = 2D_i$

Using, $$P = \frac{2\pi NT}{60}$$

$\therefore$ $$40 \times 10^3 = \frac{2\pi \times 240 \times T}{60}$$

$\therefore$ $$T = \frac{40 \times 10^3 \times 60}{2\pi \times 240} \text{ Nm} = 1591.55 \text{ Nm}$$

The torque in case of hollow shaft is given by

$$T = \frac{\pi}{16D_0}(D_0^4 - D_i^4) \times \tau, \text{ where } D_i = \frac{D_0}{2}$$

$\therefore$ $$1591.55 = \frac{\pi}{16D_0}\left[D_0^4 - \left(\frac{D_0}{2}\right)^4\right] \times 100 \times 10^6$$

$$= \frac{\pi}{16D_0}\left[D_0^4 - \frac{D_0^4}{16}\right] \times 100 \times 10^6 = \frac{\pi}{16D_0} \times \frac{15D_0^4}{16} \times 100 \times 10^6$$

$$= \frac{15\pi \times D_0^3}{256} \times 100 \times 10^6$$

or $$D_0^3 = \frac{256 \times 1591.55}{15\pi \times 100 \times 10^6} \text{ m}^3 = \frac{256 \times 1591.55}{15\pi \times 100 \times 10^6} \times 10^9 \text{ mm}^3$$

$$= \frac{256 \times 1591.55}{1.5 \times \pi \times 10^9} \times 10^9 = \frac{256 \times 1591.55}{1.5 \times \pi} = 86460.77$$

$\therefore$ $D_0 = (86460.77)^{1/3} =$ **44.22 mm. Ans.**

and $D_i = \frac{D_0}{2} = \frac{44.22}{2} =$ **22.11 mm. Ans.**

3. (*a*) No. of plates, $n = 12$; Width, $b = 50$ mm ; thickness, $t = 5$ mm, $L = 600$ mm
$\sigma_b = 180$ N/mm^2 ; $\delta = 15$ mm and $E = 0.206 \times 10^6$ N/mm^2
Find the load at the centre.
Deflection (δ) is given by,

$$\delta = \frac{3WL^3}{8n \times b \times t^3 \times E}$$

or $$15 = \frac{3 \times W \times 600^3}{8 \times 12 \times 50 \times 5^3 \times 0.206 \times 10^6}$$

(Each side of this equation is in mm)

or $$W = \frac{15 \times 8 \times 12 \times 50 \times 5^3 \times 0.206 \times 10^6}{3 \times 600^3}$$

$$= \frac{120 \times 600 \times 125 \times 0.206 \times 10^6}{3 \times 6^3 \times 10^6}$$

$$= \frac{40 \times 100}{36} \times 125 \times 0.206 = \mathbf{2861.11\ N} \quad \text{...}(i)$$

The bending stress is given by,

$$\sigma_b = \frac{3W \times L}{2nbt^2}$$

or $$180 = \frac{3W \times 600}{2 \times 12 \times 50 \times 5^2} \quad \text{(Each side of this equation is in N/mm}^2\text{)}$$

$\therefore$ $$W = \frac{180 \times 12 \times 12 \times 50 \times 5^2}{3 \times 600} = 3000 \text{ N} \quad ...(ii)$$

(*i*) If W = 3000 N, then bending stress will be 180 N/mm^2 but deflection will be more than 15 mm as for 15 mm deflection, load is 2861.11 N.

(*ii*) If W = 2861.11 N, then bending stress will be less than 180 N/mm^2 and deflection is 15 mm.

Hence the maximum central load will be **2861.11 N. Ans.**

3. (*b*) L = 5 m, W = 300 kN = 300 × 10^3 N

Both ends are fixed, and $\frac{D_i}{D_0} = \frac{3}{4} \quad \therefore \quad D_i = \frac{3}{4} D_0$

Factor of safety S.F. = 8, σ_c = 567 N/mm^2

Rankine constant, $a = \frac{1}{1600}$

The safe load (P) is given by Rankine's formula as

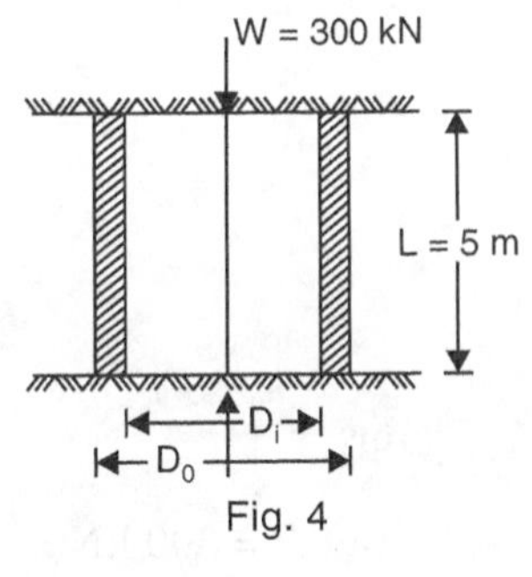

Fig. 4

$$P = \frac{1}{\text{S.F.}} \frac{\sigma_c \times A}{\left[1 + a\left(\frac{l}{k}\right)^2\right]} = \frac{1}{8} \frac{567 \times A}{\left[1 + a\left(\frac{l}{k}\right)^2\right]}$$

(Each side of the equation is in N $\therefore$ P is in N and A is in mm^2)

where A = Area of cross-section

$$= \frac{\pi}{4}\left[D_0^2 - D_i^2\right]$$

a = Rankine's constant = $\frac{1}{1600}$

l = Actual length of column = 5 m

k = Least radius of gyration

$$= \sqrt{\frac{I}{A}} = \sqrt{\frac{\frac{\pi}{64}[D_0^4 - D_i^4]}{\frac{\pi}{4}[D_0^2 - D_i^2]}} = \sqrt{\frac{1}{16}(D_0^2 + D_i^2)}$$

$\therefore$ $$P = \frac{70.875 \times \frac{\pi}{4}\left[D_0^2 - D_i^2\right]}{1 + \frac{1}{1600}\left(\frac{l^2}{k^2}\right)}$$

$$= \frac{70.875 \times \frac{\pi}{4}\left[D_0^2 - D_i^2\right]}{1 + \frac{1}{1600} \times \frac{5 \times 5}{\frac{1}{16}(D_0^2 + D_i^2)}}$$

$$= \frac{70.875 \times \frac{\pi}{4}\left[D_0^2 - \left(\frac{3}{4}D_0\right)^2\right]}{1 + \frac{1}{100} \times \frac{25}{\left[D_0^2 + \left(\frac{3}{4}D_0\right)^2\right]}} \qquad \left[\because \; D_i = \frac{3}{4}D_0\right]$$

$$= \frac{\frac{70.875 \times \pi}{4}\left(D_0^2 - \frac{9}{16}D_0^2\right)}{1 + \frac{1}{4} \times \frac{1}{\left(D_0^2 + \frac{9}{16}D_0^2\right)}} = \frac{\frac{70.875 \times \pi}{4} \times \frac{7D_0^2}{16}}{1 + \frac{1}{4} \times \frac{1}{\frac{25}{16}D_0^2}}$$

$$= \frac{24.35\,D_0^2}{1 + \frac{0.16}{D_0^2}} = \frac{24.35\,D_0^2}{\left(\frac{D_0^2 + 0.16}{D_0^2}\right)}$$

or $$P = \frac{24.35\,D_0^4}{D_0^2 + 0.16}$$

But P = 300 kN = 300 × 1000 N

$$\therefore \qquad 300 \times 1000 = \frac{24.35 D_0^4}{D_0^2 + 0.16}$$

or $300000\,(D_0^{\,2} + 0.16) = 24.35 D_0^{\,4}$

or $300000\,D_0^{\,2} + 48000 = 24.35 D_0^{\,4}$

or $24.35\,D_0^{\,4} - 300000\,D_0^{\,2} - 48000 = 0$

This is a quadratic equation in $D_0^{\,2}$.

$$\therefore \qquad D_0^{\,2} = \frac{300000 \pm \sqrt{(300000)^2 + 4 \times 24.35 \times 48000}}{2 \times 24.35}$$

$$= \frac{300000 \pm \sqrt{9 \times 10^{10} + 4675200}}{48.70} = \frac{300000 \pm 300007.79}{48.7}$$

$$= 12320 \text{ mm}^2$$

$\therefore \qquad D_0 = \sqrt{12320}$ = 110.99 mm = **111 mm. Ans.**

and $\qquad D_i = \frac{3}{4}D_0$ = 0.75 × 111 = **83.25 mm. Ans.**

3. (*c*) Refer to page 387 Article 7.12.

4. (*a*) d = 1.5 m = 1500 mm ; L = 2 m = 2000 mm ; t = 1.5 cm = 15 mm ;

p = 70 bar = 70 × 10^5 N/m^2 ; 7 × 10^6 N/m^2 = $\frac{7 \times 10^6 \text{ N}}{10^6 \text{ mm}^2}$ = 7 N/mm^2;

$E = 210$ GN/m^2 = 210×10^9 N/m^2 = $210 \times 10^3 \times 10^6$ N/m^2 = 210×10^3 N/mm^2

$\mu = 0.3$ and bulk modulus $K = 2.4$ GN/m^2 = 2.4×10^9 N/m^2 = 2.4×10^3 N/mm^2

Let us first calculate the stresses set up in the material of cylinder due to internal pressure p

Hoop stress, $$\sigma_1 = \frac{p \times d}{2t} = \frac{7 \times 1500}{2 \times 15} = 350 \text{ N/mm}^2$$

Longitudinal stress, $$\sigma_2 = \frac{p \times d}{4t} = 175 \text{ N/mm}^2$$

Now calculate the strains in the material of the cylinder.

Hoop strain, $$e_1 = \frac{\sigma_1}{E} - \frac{\mu\sigma_2}{E} = \frac{1}{E}[\sigma_1 - \mu\sigma_2]$$

$$= \frac{1}{210000}[350 - 0.3 \times 175] \qquad [\because \ E = 210 \times 10^3 \text{ N/mm}^2]$$

$$= \frac{1}{210000} \times 175\,[2 - 0.3] = \frac{175 \times 1.7}{210{,}000}$$

Longitudinal strain, $$e_2 = \frac{\sigma_2}{E} - \frac{\mu\sigma_1}{E} = \frac{1}{E}(\sigma_2 - \mu\sigma_1)$$

$$= \frac{1}{210000}[175 - 0.3 \times 350]$$

$$= \frac{1}{210000} \times 175\,[1 - 0.6] = \frac{175 \times 0.4}{210000}$$

Increase in capacity of the cylinder due to strains

$$= (2e_1 + e_2) \times \text{Volume of cylinder}$$

$$= \left[\frac{2 \times 175 \times 1.7}{210000} + \frac{175 \times 0.4}{210000}\right] \times \frac{\pi}{4} d^2 \times L$$

$$= \left[\frac{175 \times 3.4}{210000} + \frac{175 \times 0.4}{210000}\right] \times \frac{\pi}{4} \times 1500^2 \times 2000 \text{ mm}^3$$

$$= \frac{175 \times 3.8}{210000} \times \frac{\pi}{4} \times 1500^2 \times 2000$$

$$= 11191924 \text{ mm}^3 = 11.19 \times 10^6 \text{ mm}^3 \qquad \ldots(i)$$

Due to bulk modulus, there will be decrease in volume of water which is equal to

$$= \frac{p}{K} \times \text{Volume} \qquad \left[\because \ K = \frac{p}{-\left(\frac{dV}{V}\right)} = \frac{-p \times V}{dV}. \quad \therefore \ dV = \frac{-p \times V}{K}\right]$$

$$= \frac{7\,(\text{N/mm}^2)}{2400\,(\text{N/mm}^2)} \times \frac{\pi}{4} \times 1500^2 \times 2000$$

$$= 10308350 = 10.3 \times 10^6 \text{ mm}^3 \qquad \ldots(ii)$$

$\therefore$ Additional amount of water required to be pumped

$= (i) + (ii) = 11.19 \times 10^6 + 10.3 \times 10^6 =$ **21.49×10^6 mm^3. Ans.**

4. (*b*) Internal dia., $d_1 = 80$ mm, $\therefore$ Internal radius, $r_1 = 40$ mm

Internal fluid pressure, $p = 25$ N/mm^2

Max. permissible tensile stress, $\sigma_1 = 125$ N/mm^2

Calculate thickness of metal required.

The maximum permissible tensile stress is the hoop stress (σ_1) which is at the inner radius of cylindrical shell.

Let r_2 = External radius

r_1 = Internal radius

The radial pressure (p) and hoop stress (σ_1) at any radius (r) are given by **Lame's Equation** as

$$p = \frac{b}{r^2} - a \qquad ...(i)$$

and
$$\sigma_1 = \frac{b}{r^2} + a \qquad ...(ii)$$

where a and b are constants. Their values are obtained from boundary conditions. Boundary conditions are :

At $r = 40$ mm, $p = 25$ N/mm^2 and $\sigma_1 = 125$ N/mm^2. Then equations (i) and (ii) becomes as

$$25 = \frac{b}{40^2} - a \qquad (\because \; p = 25 \text{ N/mm}^2 \text{ and } r = 40 \text{ mm})$$

$$125 = \frac{b}{40^2} + a \qquad (\because \; \sigma_1 = 125 \text{ and } r = 40 \text{ mm})$$

Solve the above equations for a and b. Adding these equations, we get

$$25 + 125 = \frac{b}{40^2} + \frac{b}{40^2} = \frac{2b}{40^2}$$

or
$$150 = \frac{2b}{1600}$$

$\therefore$
$$b = \frac{150 \times 1600}{2} = 120000$$

and
$$a = \frac{b}{40^2} - 25 = \frac{120000}{1600} - 25 \qquad (\because \; b = 120000)$$

$$= 75 - 25 = 50$$

Substituting values $a = 50$ and $b = 120000$ in equation (i), we get

$$p = \frac{120000}{r^2} - 50$$

At $r = r_2$, pressure is zero (*i.e.*, atmospheric pressure) at outer surface.

$\therefore$ $p = 0$ at $r = r_2$. The above equation becomes as

$$0 = \frac{120000}{r_2^2} - 50$$

Find r_2 from this equation

$\therefore$
$$50 = \frac{120000}{r_2^2}$$

or
$$r_2^2 = \frac{120000}{50} = 2400$$

$\therefore$
$$r_2 = \sqrt{2400} = 48.989 \simeq 49 \text{ mm}$$

$\therefore$ Thickness of metal required $= r_2 - r_1 = 49 - 40 =$ **9 mm. Ans.**

4. (*c*) (*i*) Refer to page 460 Article 9.1.

(*ii*) Refer to page 468 Article 9.4.

(*iii*) Refer to page 460 Article 9.2.

5. (*a*) Square section of 3 cm (= 30 mm) sides. ∴ Area, $A = 30 \times 30 = 900$ mm^2.

Mean readius of curvature, $R = 4.5$ cm = 45 mm

Bending moment, $M = 300$ Nm = 300 × 1000 Nmm

The bending moment is –ve as it tends to decrease the curvature when applied

∴ $M = -30000$ Nmm.

To find stresses to inner and outer faces of the section.

The stress at any layer in the cross-section at a distance '*y*' from centroidal axis is given by

$$\sigma = \frac{M}{R \times A}\left\{1 + \frac{R^2}{h^2}\left(\frac{y}{R+y}\right)\right\} \qquad ...(i)$$

The value of 'h^2' for a rectangular section is given by

$$h^2 = \frac{R^3}{d}\left[\log_e\left(\frac{2R+d}{2R-d}\right)\right] - R^2$$

where $b = d = 30$ mm for square section

$$\therefore \quad h^2 = \frac{45^3}{30}\left[\log_e\left(\frac{2\times45+30}{2\times45-30}\right)\right] - 45^2 \qquad [\because \; R = 45 \text{ mm}]$$

$$= 3037.5\left[\log_e\left(\frac{120}{60}\right)\right] - 2025$$

$$= 3037.5\,[\log_e 2.0] - 2025 = 3037.5 \times 0.693 - 2025$$

$$= 80.43$$

Hence equation (*i*) becomes as

$$\therefore \quad \sigma = \frac{M}{R\times A}\left[1 + \frac{45^2}{80.43}\left(\frac{y}{45+y}\right)\right]$$

$$= \frac{-300000}{45\times900}\left[1 + 25.177\left(\frac{y}{45+y}\right)\right] \qquad ...(ii)$$

(*i*) *Stress at outer face of the section, where* $y = \frac{d}{2} = \frac{30}{2} = 15$ *mm*

$$\sigma = \frac{-300000}{45\times900}\left[1 + 25.177\left(\frac{15}{45+15}\right)\right]$$

$$= -7.4\left[1 + \frac{25.177}{4}\right] = \mathbf{-53.97\ N/mm^2\ (compressive)} \quad \textbf{Ans.}$$

(*ii*) *Stress at the inner face, where* $y = \frac{-d}{2} = \frac{-30}{2} = -15$ *mm.*

From equation (*ii*), we have

$$\sigma = \frac{-300000}{45\times900}\left[1 + 25.177\left(\frac{-15}{45-15}\right)\right]$$

$$= -7.4\left[1 - \frac{25.177 \times 15}{30}\right] = -7.4\,[1 - 12.58]$$

$$= -7.4 \times (-11.58) = \mathbf{85.69\ N/mm^2\ (tensile).}\quad \textbf{Ans.}$$

5. (*b*) Dimensions of angle 60 mm × 40 mm × 6 mm

Binding moment, $M = 12$ kN cm = 12000 N cm

= 120 Nm

Calculate G (centroid)

$$\bar{x} = \frac{A_1x_1 + A_2x_2}{A_1 + A_2}$$

where $A_1 = 54 \times 6 = 324$

$A_2 = 40 \times 6 = 240$

$x_1 = 3,\ x_2 = 20$

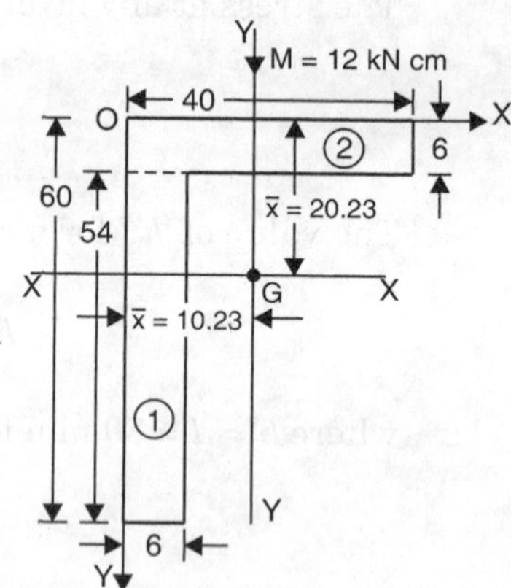

Fig. 5

$$= \frac{324 \times 3 + 240 \times 20}{324 + 240}$$

$$= \frac{972 + 4800}{564} = \frac{5772}{564} = 10.23 \text{ mm}$$

$$\bar{y} = \frac{A_1y_1 + A_2y_2}{A_1 + A_2}$$

where $y_1 = 6 + \frac{54}{2} = 33,\quad y_2 = 3$

$$= \frac{324 \times 33 + 240 + 3}{564}$$

$$= \frac{10692 + 720}{564} = 20.23 \text{ mm.}$$

Now, let us find I_{XX}, I_{YY} and I_{XY}

$$I_{XX} = \frac{40 \times 6^3}{12} + (40 \times 6) \times (20.23 - 3)^2 + \frac{6 \times 54^3}{12} + (6 \times 54) \times (33 - 20.23)^2$$

$$= 720 + 71249 + 78732 + 52835 = 203536 \text{ mm}^4$$

$$I_{YY} = \frac{6 \times 40^3}{12} + (6 \times 40) \times (20 - 10.23)^2 + \frac{54 \times 6^3}{12} + 54 \times 6 \times (10.23 - 3)^2$$

$$= 32000 + 22908 + 972 + 16936 = 72186 \text{ mm}^4$$

$$I_{XY} = A_1\,h_1\,k_1 + A_2\,h_2\,k_2$$

where $h_1 = 33 - 20.23 = 12.77,\ k_1 = 10.23 - 3 = 7.23$

$h_2 = 20.23 - 3 = 17.23,\qquad k_2 = 20 - 10.23 = 9.77$

$\therefore\quad I_{XY} = 324 \times 12.77 \times 7.23 + 240 \times 17.23 \times 9.77 = 29914 + 40401$

$= 70315 \text{ mm}^4$

For principal axes,

$$\tan 2\theta = \frac{2I_{XY}}{I_{YY} - I_{XX}} = \frac{2 \times 70315}{72186 - 203536} = -1.07065$$

$\therefore \quad 2\theta = \tan^{-1}(-1.07065) \simeq -47°$

$= 180 - 47° = 133°$

$\therefore \quad \theta = 66.5°$

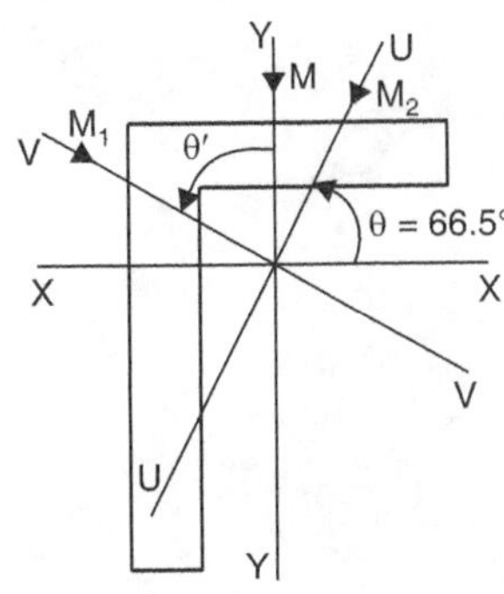

Fig. 5 (*a*)

$$I_{UU} = \frac{1}{2}[I_{XX} + I_{YY}] + \frac{1}{2}[I_{XX} - I_{YY}] \sec 2\theta$$

$$= \frac{1}{2}[203536 + 72186] + \frac{1}{2}[203536 - 72186] \sec(133°)$$

$$= 137861 + 65675 \times (-1.466) = 41581 \text{ mm}^4$$

Now, $I_{VV} = I_{XX} + I_{YY} - I_{UU} = 203536 + 72186 - 41581$
$= 234141 \text{ mm}^4$

Bending moment in the vertical plane along YY, $M = 12$ kN cm

$M = 12$ kN cm $= 12 \times 1000 \times 10$ N mm $= 120000$ N mm

Resolving this moment along VV and UU, we get

Moment along VV, $M_1 = M \cos\theta = 120000 \times \cos 66.5° = 47849.8 \simeq 47850$ Nm.

Moment along UU, $M_2 = M \sin\theta = 120000 \times \sin 66.5° = 110047$ Nmm.

The moment M_1 will bend the angle about UU whereas the moment M_2 will bend about VV. The combined bending stress at any point will be given by

$$\sigma = \frac{M_1 \times v}{I_{UU}} + \frac{M_2 \times u}{I_{VV}}$$

$$= \frac{47850}{41581} \times v + \frac{110047}{234141} \times u$$

or $\quad \sigma = 1.15\,v + 0.47u \quad$ ...(A)

The equation of neutral axis (N.A.) is $\sigma = 0$.

For neutral axis

$\therefore \quad 0 = 1.15\,v + 0.47u$

or $\quad 1.15v = -0.47u$ or $v = \dfrac{-0.47}{1.15}u = -0.408u$

$\therefore$ Slope of neutral axis,

$\tan\alpha = -0.408$

or $\quad \alpha = \tan^{-1}(-0.408) = -22.195 \simeq -22.2°$.

This slope is clockwise with UU. Hence N.A. is inclined at an angle of $-22.2°$ to UU as shown in Fig. 5 (*b*).

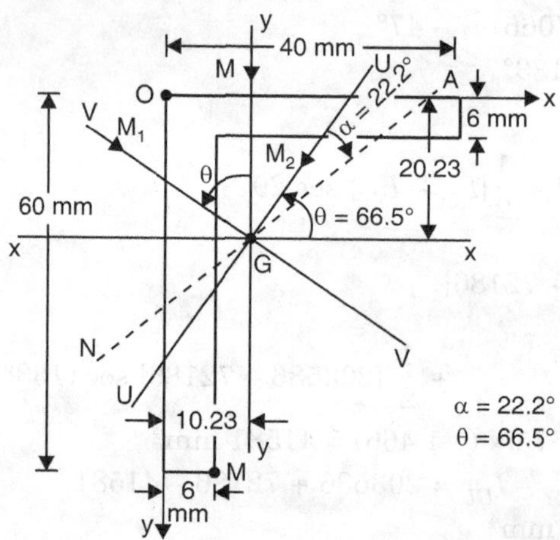

Fig. 5 (*b*)

For a beam, the stress will be compressive above the N.A. and tensile below N.A. The maximum compressive stress will be at *O* and maximum tensile stress will be at point *M*.

The values of *u* and *v* at point *O* and *M* are obtained from equations

$$u = x \cos \theta + y \sin \theta$$

and $$v = y \cos \theta - x \sin \theta, \text{ where } \theta = 66.5°$$

For point O, $x = -\bar{x} = -10.23,\ y = +\bar{y} = 20.23$

$\therefore$ $u = -10.23 \times \cos 66.5° + 20.23 \sin 66.5°$

$= -4.08 + 18.55 = 14.47$

and $v = y \cos \theta - x \sin \theta = 20.23 \times \cos 66.5° + 10.23 \sin 66.5$

$= 8.066 + 9.38 = 17.44$

$\therefore$ Stress at point *O* from equation (*A*) becomes

$$\sigma_O = 1.15v + 0.47u = 1.15 \times 17.44 + 0.47 \times 14.47$$

$= 20.05 + 6.8 =$ **26.85 N/mm² (Compressive). Ans.**

For point M,

$$x = (\bar{x} - 6) = -(10.23 - 6) = -4.23$$

and $$y = -(60 - \bar{y}) = -(60 - 20.23) = -39.77$$

Now, $u = -4.23 \times \cos 66.5° + (-39.77) \sin 66.5°$

$= -1.686 - 36.47 = -38.16$

and $v = -39.77 \cos 66.5° - (-4.23) \sin 66.5° = -15.86 + 3.88 = -11.98$

$\therefore$ Stress at point *O* from equation (*A*) becomes as

$$\sigma_M = 1.15(-11.98) + 0.47(-38.16)$$

$= -13.77 - 17.93 =$ **– 31.7 N/mm² (Tensile). Ans.**

5. (*c*) (*i*) Refer to page 567 Article 11.2.

(*ii*) Refer to page 521 Article 10.2.

(*iii*) Refer to page 521 Article 10.1.

OBJECTIVE TYPE QUESTIONS FROM COMPETITIVE EXAMINATIONS

Tick mark the most appropriate answer:

1. For the state of plane stress shown in Fig. 6, the maximum and minimum principal stresses are:

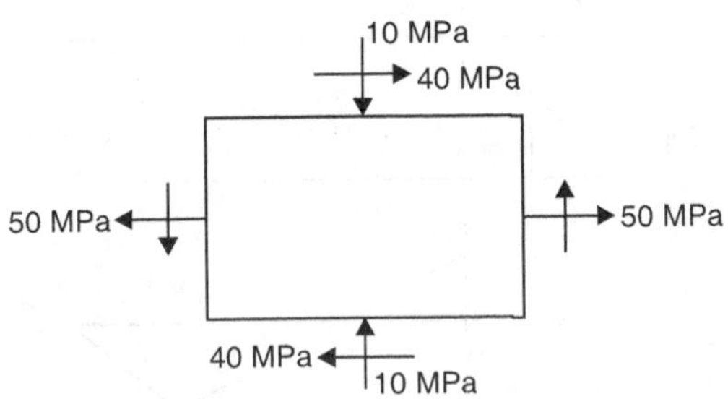

Fig. 6

(*a*) 60 MPa and 30 MPa ☐ (*b*) 50 MPa and 10 MPa ☐

(*c*) 40 MPa and 20 MPa ☐ (*d*) 70 MPa and 30 MPa. ☐

2. Which of the following is true? (μ = Poisson's ratio)

(*a*) $0 < \mu < -\frac{1}{2}$ ☐ (*b*) $1 < \mu < 0$ ☐

(*c*) $1 < \mu < -1$ ☐ (*d*) $\infty < \mu < -\infty$ ☐

3. For most brittle materials, the ultimate strength in compression is much large than the ultimate strength in tension. This is mainly due to

(*a*) presence of flaws and microscopic cracks or cavities ☐

(*b*) necking in tension ☐

(*c*) severity of tensile stress as compared to compressive stress ☐

(*d*) non-linearity of stress-strain diagram. ☐

4. Bending moment *M* and torque *T* is applied on a solid circular shaft. If the maximum bending stress equals to maximum shear stress developed, then *M* is equal to

(*a*) $\frac{T}{2}$ ☐ (*b*) T ☐

(*c*) $2T$ ☐ (*d*) $4T$. ☐

5. When bending moment *M* and torque *T* is applied on a shaft then equivalent torque is

(*a*) $M + T$ ☐ (*b*) $\sqrt{M^2 + T^2}$ ☐

(*c*) $\frac{1}{2}\sqrt{M^2 + T^2}$ ☐ (*d*) $\frac{1}{2}(M + \sqrt{M^2 + T^2})$. ☐

6. The bending moment diagram for the case shown in Fig. 7 below.

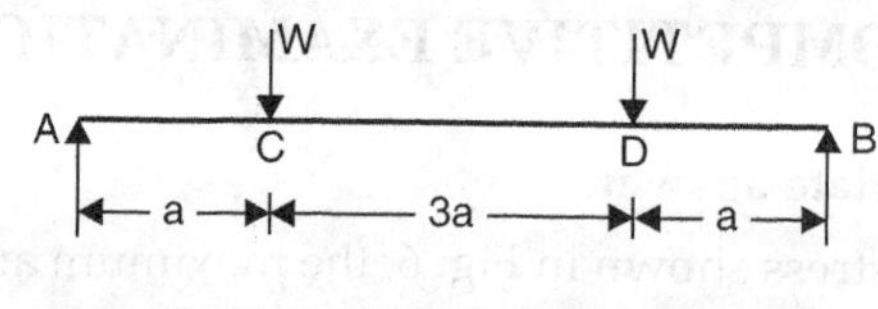

(a)

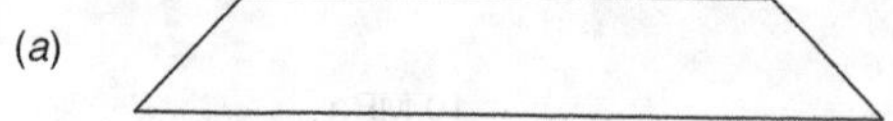

(b)

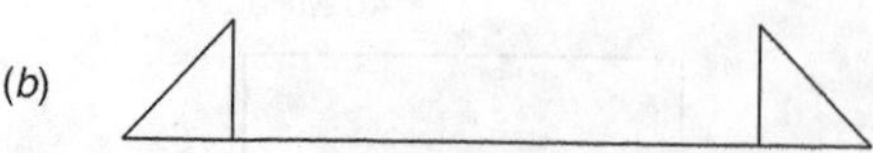

(c)

(d)

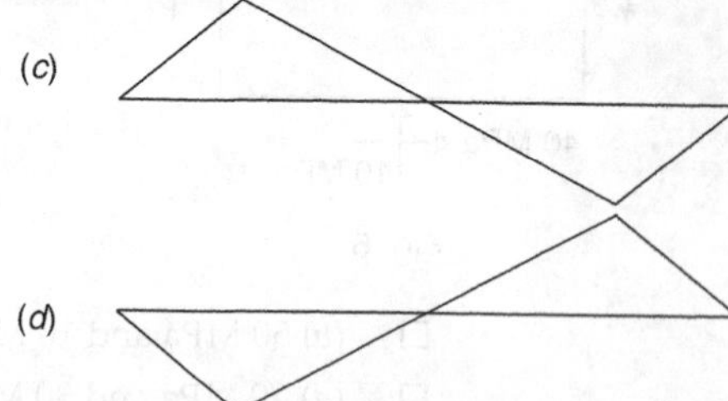

Fig. 7

7. If a prismatic bar be subjected to an axial tensile stress σ, the shear stress induced at a plane inclined at θ with the axis will be

(a) $\frac{\sigma}{2}\sin 2\theta$ ☐ (b) $\frac{\sigma}{2}\cos 2\theta$ ☐

(c) $\frac{\sigma}{2}\cos^2\theta$ ☐ (d) $\frac{\sigma}{2}\sin^2\theta$. ☐

8. A vertical hanging bar of length L and weighing w N/unit length carries a load W at the bottom as shown in Fig. 8. The tensile force in the bar at a distance y from the support will be given by

(a) $W + w \times L$ ☐

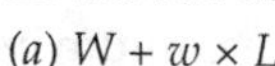

(b) $W + w(L - y)$ ☐

(c) $(W + w) \times \frac{y}{L}$ ☐

(d) $W + \frac{W}{w}(L - y)$. ☐

Fig. 8

9. In case of bi-axial state of normal stresses, the normal stress on 45° plane is equal to

(a) sum of normal stresses ☐ (b) difference of normal stresses ☐

(c) half the sum of normal stresses ☐ (d) half the difference of normal stresses. ☐

10. The temperature stress is a function of

1. Co-efficient of linear expansion

2. Temperature rise

3. Modulus of elasticity. The correct answer is

(*a*) 1 and 2 only ☐ (*b*) 1 and 3 only ☐

(*c*) 2 and 3 only ☐ (*d*) 1, 2 and 3 only. ☐

11. When two springs of equal lengths are arranged to form a cluster spring then which of the following statements are true :

1. Angle of twist in both the springs will be equal
2. Deflection of both the springs will be equal
3. Load taken by each spring will be half the total load
4. Shear stress in each spring will be equal

(*a*) 1 and 2 only ☐ (*b*) 2 and 3 only ☐

(*c*) 3 and 4 only ☐ (*d*) 1, 2 and 4 only. ☐

12. When σ and Young's modulus of elasticity *E* remains constant, the energy absorbing capacity of part subjected to dynamic forces, is a function of its

(*a*) length ☐ (*b*) cross-section ☐

(*c*) volume ☐ (*d*) none of the above. ☐

13. If the principal stresses corresponding to a two-dimensional state of stress are σ_1 and σ_2 when σ_1 is greater than σ_2 and both are tensile, then which one of the following would be correct criterion for failure by yielding, according to maximum shear stress theory ?

(*a*) $\frac{\sigma_1 - \sigma_2}{2} = \pm \frac{\sigma_{yp}}{2}$ ☐ (*b*) $\frac{\sigma_1}{2} = \pm \frac{\sigma_{yp}}{2}$ ☐

(*c*) $\frac{\sigma_2}{2} = \pm \frac{\sigma_{yp}}{2}$ ☐ (*d*) $\sigma_1 = \pm 2\sigma_{yp}$. ☐

14. The buckling load will be maximum for a column, if

(*a*) one end of the column is clamped and other end is free ☐

(*b*) both ends of the column are clamped ☐

(*c*) both ends of the column are hinged ☐

(*d*) one end of the column is hinged and other end is free. ☐

15. A length of 10 mm diameter steel wire is coiled to a close-coiled helical spring having 8 coils of 75 mm mean diameter, and the spring has a stiffness *k*. If the same length of the wire is coiled to 10 coils of 60 mm mean diameter, then the spring stiffness will be

(*a*) *k* ☐ (*b*) 1.25 *k* ☐

(*c*) 1.56 *k* ☐ (*d*) 1.95 *k*. ☐

16. A metal pipe of 1 m diameter contains a fluid having a pressure of 100 N/cm². If the permissible tensile stress in the metal is 2 kN/cm², then the thickness of the metal required for making the pipe would be

(*a*) 5 mm ☐ (*b*) 10 mm ☐

(*c*) 20 mm ☐ (*d*) 25 mm. ☐

17. Circumferential and longitudinal strains in cylindrical boiler under internal steam pressure, are e_1 and e_2 respectively. Change in volume of the boiler cylinder per unit volume will be

(a) $e_1 + 2e_2$ ☐ (b) $e_1e_2^2$ ☐

(c) $2e_1 + e_2$ ☐ (d) $e_1^2 e_2$. ☐

18. In the assembly of pulley, key and shaft

(a) pulley is made the weakest ☐

(b) key is made the weakest ☐

(c) key is made the strongest ☐

(d) all the three are designed for equal strength. ☐

19. Shear stress distribution diagram of a beam of rectangular cross-section, subject to transverse loading will be

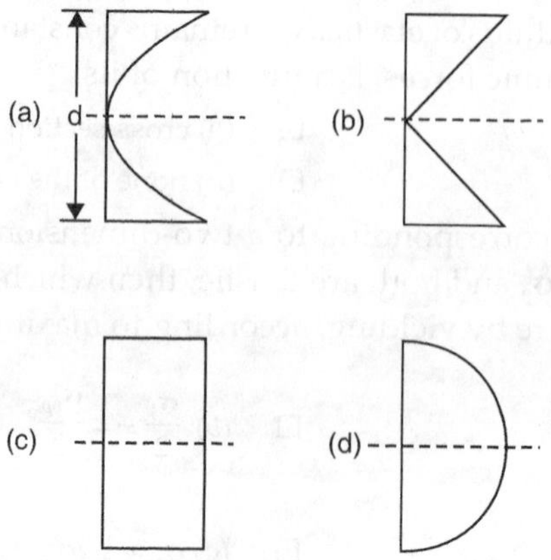

(a) d (b)

(c) (d)

Fig. 9

20. A horizontal beam with square cross-section is simply supported with sides of the square horizontal and vertical and carries a distributed loading that produces maximum bending stress σ in the beam. When the beam is placed with one of the diagonals horizontal, the maximum bending stress will be

(a) $\frac{\sigma}{\sqrt{2}}$ ☐ (b) σ ☐

(c) $\sqrt{2} \times \sigma$ ☐ (d) 2σ. ☐

21. A shaft was initially subjected to bending moment and then was subjected to torsion. If the magnitude of the bending moment is found to be the same as that of the torque, then the ratio of maximum bending stress to shear stress would be

(a) 0.25 ☐ (b) 0.50 ☐

(c) 2.0 ☐ (d) 4.0. ☐

22. A simply supported beam of rectangular section 4 cm by 6 cm carries a mid-span concentrated load such that 6 cm side lies parallel to the line of action of loading; deflection under

the load is δ. If the beam is now supported with the 4 cm side parallel to the line of action of loading, the deflection under the load will be

(*a*) 0.44 × δ ☐ (*b*) 0.67 × δ ☐
(*c*) 1.50 × δ ☐ (*d*) 2.25 × δ. ☐

23. A beam *AB* is hinge-supported at its ends and is loaded by a couple *P* × *C* as shown in Fig. 10. The magnitude of shearing force at a section *x* of the beam is

(*a*) 0 ☐
(*b*) *P* ☐
(*c*) $\frac{P}{2L}$ ☐
(*d*) $\frac{P \times C}{2L}$ ☐

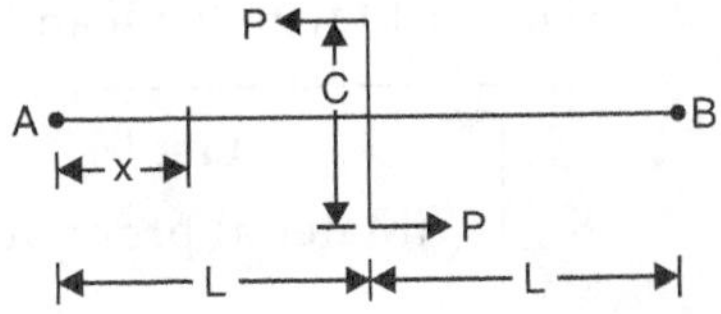

Fig. 10

24. The frictional torque for square thread at mean radius (*R*) while raising load (*W*) is given by

(*a*) *WR* × tan (ϕ – α) ☐ (*b*) *WR* × tan (ϕ + α) ☐
(*c*) *WR* × tan α ☐ (*d*) *WR* × tan ϕ ☐

where ϕ = Angle of friction and α = Helix angle.

25. Design of shafts made of brittle materials is based on

(*a*) Guest's theory ☐ (*b*) Rankine's theory ☐
(*c*) St. Venant's theory ☐ (*d*) Von Mises theory. ☐

26. Principal stresses at a point in plane stressed element are $\sigma_x = \sigma_y$ = 5000 N/cm². Normal stress on the plane inclined at 45° to the *x*-axis will be

(*a*) 0 ☐ (*b*) 5000 N/cm² ☐
(*c*) 7070 N/cm² ☐ (*d*) 10000 N/cm². ☐

27. State of stress in plane element is shown in Fig. 11. Which one of the following figures shown in Fig. 12 is the correct sketch of Mohr's circle of the state of stress ?

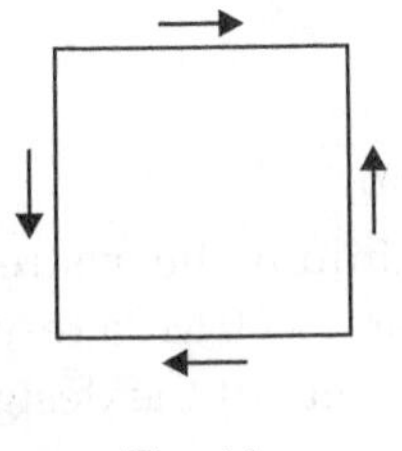

(a) (b) (c) (d)

Fig. 11 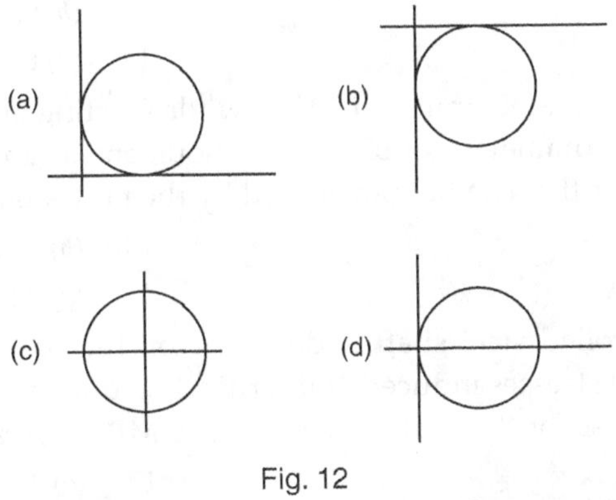 Fig. 12

28. A steel rod of 1 sq. cm cross-sectional area is 100 cm long and has a Young's modulus of elasticity 20 × 10⁶ N/cm². It is subjected to an axial pull of 20 kN. The elongation of the rod will be

(*a*) 0.05 cm ☐ (*b*) 0.1 cm ☐
(*c*) 0.15 cm ☐ (*d*) 0.20 cm. ☐

29. If the area of cross-section of a wire is circular and if the radius of this wire decreases to half its original value due to stretch to the wire by a load, then modulus of elasticity of the wire be

(*a*) one-fourth of its original value ☐ (*b*) halved ☐
(*c*) doubled ☐ (*d*) unaffected ☐

E depends upon the material. It is independent of area, load etc.

30. Match list I with list II and select the correct answer using codes given below the lists:

List I (**Material properties**)	**List II** (**Test of determine material properties**)
A. Ductility	1. Impact test
B. Toughness	2. Fatigue test
C. Endurance limit	3. Tension test
D. Resistance to penetration	4. Hardness test

Codes:

	A	*B*	*C*	*D*	
(*a*)	3	2	1	4	☐
(*b*)	4	2	1	3	☐
(*c*)	3	1	2	4	☐
(*d*)	4	1	2	3	☐

31. If a material had a modulus of elasticity of 21×10^6 N/cm² and a modulus of rigidity of 8×10^6 N/cm², then approximate value of the Poisson's ratio of the material would be

(*a*) 0.26 ☐ (*b*) 0.31 ☐
(*c*) 0.47 ☐ (*d*) 0.5. ☐

32. A shaft can safely transmit 90 kW while rotating at a given speed. If this shaft is replaced by a shaft of diameter double the previous one and rotated at half the speed of the previous, the power that can be transmitted by the new shaft is

(*a*) 90 kW ☐ (*b*) 180 kW ☐
(*c*) 360 kW ☐ (*d*) 720 kW. ☐

33. A cold rolled steel shaft is designed on the basis of maximum shear stress theory. The principal stresses induced at its critical section are 60 MPa and – 60 MPa respectively. If the yield stress for the shaft material is 360 MPa, the factor of safety of the design is

(*a*) 2 ☐ (*b*) 3 ☐
(*c*) 4 ☐ (*d*) 5. ☐

34. An eccentrically loaded riveted joint is shown in Fig. 13 with 4 rivets at *P*, *Q*, *R* and *S*. Which of the rivets are the most loaded?

(*a*) *P* and *Q* ☐ (*b*) *Q* and *R* ☐
(*c*) *R* and *S* ☐ (*d*) *S* and *P*. ☐

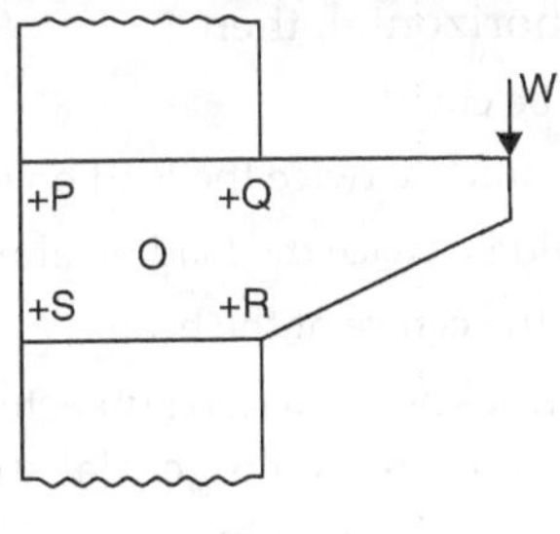

Fig. 13

35. When a helical compression spring is cut into two equal halves, the stiffness of each of the resulting springs will be

(*a*) unaltered ☐ (*b*) double ☐

(*c*) one-half ☐ (*d*) one-fourth. ☐

36. While calculating the stress induced in a close-coiled helical spring, Wahl's factor must be considered to account for

(*a*) the curvature and stress concentration effect ☐

(*b*) shock loading ☐

(*c*) poor service conditions ☐

(*d*) fatigue loading. ☐

37. A straight bar is fixed at the edges *A* and *B* as shown in Fig. 14. Its elastic modulus is *E* and cross-section is *A*. There is a load *P* = 120 N acting at *C*. Determine the reactions at the ends

(*a*) 60 N at *A*, 60 N at *B* ☐

(*b*) 30 N at *A*, 90 N at *B* ☐

(*c*) 40 N at *A*, 80 N at *B* ☐

(*d*) 80 N at *A*, 40 N at *B*. ☐

A, C, P = 120 N, B, L, 2L

Fig. 14

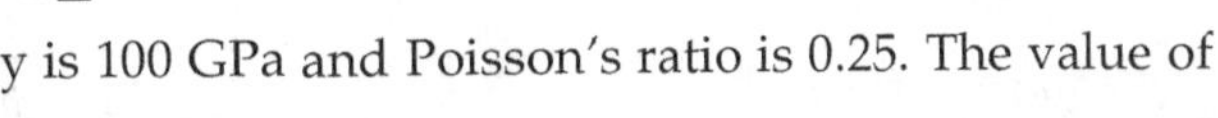

38. For a material, the modulus of rigidity is 100 GPa and Poisson's ratio is 0.25. The value of modulus of elasticity in GPa is

(*a*) 125 ☐ (*b*) 150 ☐

(*c*) 200 ☐ (*d*) 250. ☐

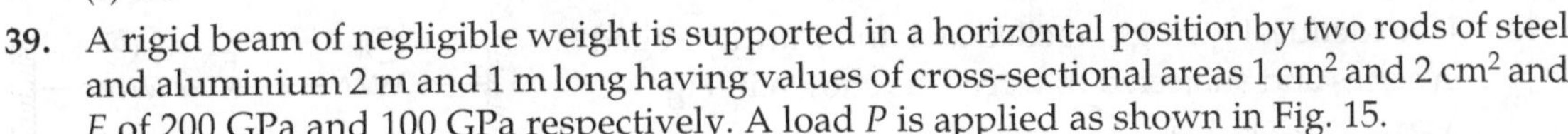

39. A rigid beam of negligible weight is supported in a horizontal position by two rods of steel and aluminium 2 m and 1 m long having values of cross-sectional areas 1 cm^2 and 2 cm^2 and *E* of 200 GPa and 100 GPa respectively. A load *P* is applied as shown in Fig. 15.

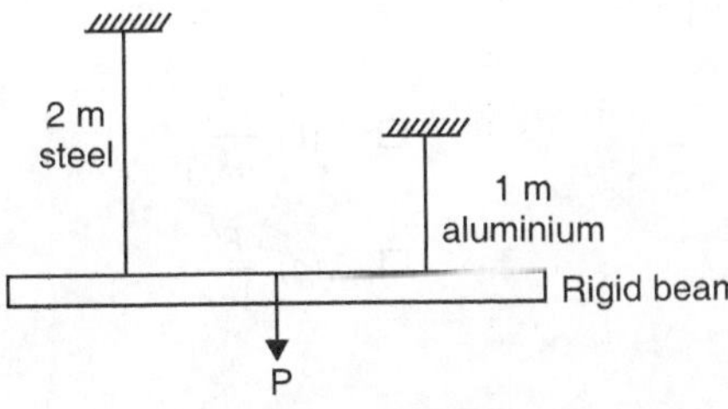

Fig. 15

If the rigid beam is to remain horizontal, then

(*a*) the load on both rods should be equal ☐

(*b*) the load on aluminium rod should be twice the load on steel ☐

(*c*) the load on the steel rod should be twice the load on aluminium ☐

(*d*) the load P must be applied at the centre of the beam. ☐

40. A thin cylinder of radius r and thickness t when subjected to an internal hydrostatic pressure p causes a radial displacement u, then the tangential strain caused is

(*a*) $\frac{du}{dr}$ ☐ (*b*) $\frac{1}{r}\cdot\frac{du}{dr}$ ☐

(*c*) $\frac{u}{r}$ ☐ (*d*) $\frac{2u}{r}$. ☐

41. Determine the stiffness of the beam shown in Fig. 16 given below :

When $I = 375 \times 10^{-4}$ m^4

$L = 0.5$ m

$E = 200$ GPa

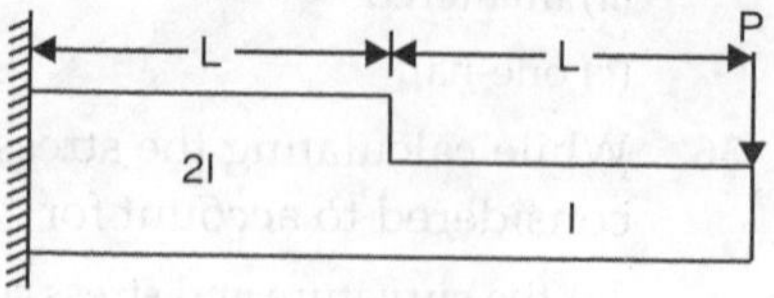

Fig. 16

The stiffness is given by

(*a*) 12×10^{10} N/m ☐ (*b*) 10×10^{10} N/m ☐

(*c*) 4×10^{10} N/m ☐ (*d*) 8×10^{10} N/m. ☐

42. The strain energy stored in a body of volume V subjected to uniform stress σ is

(*a*) $\frac{\sigma \times E}{V}$ ☐ (*b*) $\frac{\sigma E^2}{V}$ ☐

(*c*) $\frac{\sigma \times V^2}{E}$ ☐ (*d*) $\frac{\sigma^2}{2E} \times V$. ☐

43. For the same internal diameter, wall thickness, material and internal pressure, the ratio of maximum stress, induced in a thin cylindrical and in a thin spherical vessel will be

(*a*) 2 ☐ (*b*) $\frac{1}{2}$ ☐

(*c*) 4 ☐ (*d*) $\frac{1}{4}$. ☐

44. Two identical springs labelled as 1 and 2 are arranged in series and subjected to force F as shown in Fig. 17. Assume that each spring constant is k. The strain energy stored in spring 1 is

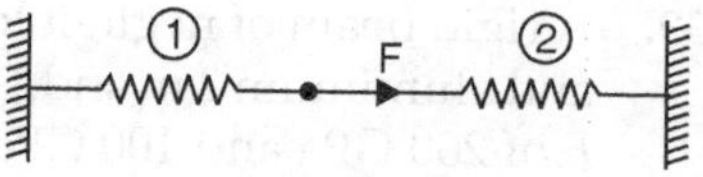

Fig. 17

(*a*) $\frac{F^2}{2k}$ ☐ (*b*) $\frac{F^2}{4k}$ ☐

(*c*) $\frac{F^2}{8k}$ ☐ (*d*) $\frac{F^2}{16k}$. ☐

45. A rod having cross-sectional area 100×10^{-6} m^2 is subjected to a tensile load. Based on the Tresca failure criterion, if the uniaxial yield stress of the material is 200 MPa, the failure load is

(*a*) 10 kN ☐ (*b*) 20 kN ☐
(*c*) 100 kN ☐ (*d*) 200 kN. ☐

46. Wire diameter, mean coil diameter and number of turns of a closely-coiled steel spring are *d*, *D* and *N* respectively and stiffness of the spring is *k*. A second spring is made of the same steel but with wire diameter, mean coil diameter and number of turns as 2*d*, 2*D* and 2*N* respectively. The stiffness of the new spring is

(*a*) *k* ☐ (*b*) 2*k* ☐
(*c*) 4*k* ☐ (*d*) 8*k*. ☐

47. If the diameter of a long column is reduced by 20%, the percentage of reduction in Euler's buckling load is

(*a*) 4 ☐ (*b*) 36 ☐
(*c*) 49 ☐ (*d*) 59. ☐

48. With one fixed end and other free end, a column of length *L* buckles at load P_1. Another column of same length and same cross-section fixed at both ends buckles at load P_2. Then P_2/P_1 is

(*a*) 1 ☐ (*b*) 2 ☐
(*c*) 4 ☐ (*d*) 16. ☐

49. The principal stresses σ_1, σ_2 and σ_3 at a point respectively are 80 MPa, 30 MPa and – 40 MPa. The maximum shear stress is

(*a*) 25 MPa ☐ (*b*) 35 MPa ☐
(*c*) 55 MPa ☐ (*d*) 60 MPa. ☐

50. The Poisson's ratio of a material which has Young's modulus of 120 GPa and shear modulus of 50 GPa, is

(*a*) 0.1 ☐ (*b*) 0.2 ☐
(*c*) 0.3 ☐ (*d*) 0.4. ☐

ANSWERS

Answers to Objective Type Questions from Competitive Examinations with Explanations

1. (*d*)	**2.** (*a*)	**3.** (*a*)	**4.** (*a*)	**5.** (*b*)	**6.** (*a*)
7. (*a*)	**8.** (*b*)	**9.** (*c*)	**10.** (*d*)	**11.** (*c*)	**12.** (*c*)
13. (*b*)	**14.** (*b*)	**15.** (*c*)	**16.** (*d*)	**17.** (*c*)	**18.** (*b*)
19. (*d*)	**20.** (*c*)	**21.** (*c*)	**22.** (*d*)	**23.** (*d*)	**24.** (*b*)
25. (*b*)	**26.** (*b*)	**27.** (*c*)	**28.** (*b*)	**29.** (*d*)	**30.** (*c*)
31. (*b*)	**32.** (*c*)	**33.** (*b*)	**34.** (*d*)	**35.** (*b*)	**36.** (*a*)
37. (*d*)	**38.** (*d*)	**39.** (*b*)	**40.** (*c*)	**41.** (*c*)	**42.** (*d*)

43. (*a*) **44.** (*c*) **45.** (*b*) **46.** (*a*) **47.** (*d*) **48.** (*d*)
49. (*d*) **50.** (*b*).

EXPLANATIONS

1. Here

$$\sigma_x = 50 \text{ MPa}$$
$$\sigma_y = -10 \text{ MPa}$$
$$\tau_{xy} = 40 \text{ MPa}$$

The principal stresses σ_1 and σ_2 are given as

$$\sigma_1, \sigma_2 = \frac{\sigma_x + \sigma_y}{2} \pm \sqrt{\left(\frac{\sigma_x - \sigma_y}{2}\right)^2 + \tau_{xy}{}^2}$$

$$= \frac{50 + (-10)}{2} \pm \sqrt{\left[\frac{50 - (-10)}{2}\right]^2 + 40^2}$$

$$= 20 \pm \sqrt{30^2 + 40^2}$$

$$= 20 \pm 50,$$

$\therefore$ $\sigma_1 = 20 + 50 = 70$ MPa

and $\sigma_2 = 20 - 50 = -30$ MPa = **30 MPa. (Compressive). Ans.**

2. μ = Poisson's ratio = $\dfrac{\text{Lateral strain}}{\text{Longitudinal strain}}$

The lateral strain is opposite to longitudinal strain. This means if longitudinal strain is tensile, then lateral strain is compressive. Hence μ is negative. For most of the material μ lies between -0.25 to -0.40.

4. Bending stress, $\sigma_b = \dfrac{M}{I} \times y$

$\therefore$
$$(\sigma_b)_{max} = \frac{M}{I} \times \frac{d}{2} \quad \text{where,} \quad I = \frac{\pi}{64} d^4$$

$$= \frac{M}{\frac{\pi}{64} d^4} \times \frac{d}{2} = \frac{32\,M}{\pi d^3}$$

$$T = \frac{\pi}{16} \times d^3 \times \tau_{max}$$

$\therefore$
$$\tau_{max} = \frac{16\,T}{\pi d^3}$$

$\therefore$ If
$$\sigma_{max} = \tau_{max}$$

Then
$$\frac{32\,M}{\pi d^3} = \frac{16\,T}{\pi d^3} \quad \text{or} \quad M = \frac{T}{2}. \quad \textbf{Ans.}$$

5. Due to bending moment (*M*), the bending stress will be produced in the shaft. This bending stress (σ_b) is given by

$$\sigma_b = \frac{32\,M}{\pi d^3}$$

Due to torque (T), the shear stress will be produced. This shear stress (τ) is given by

$$\tau = \frac{16\,T}{\pi d^3}$$

$\therefore$ The principal stresses due to bending and shear stresses are

$$\sigma_1 \text{ and } \sigma_2 = \frac{\sigma_b}{2} \pm \sqrt{\left(\frac{\sigma_b}{2}\right)^2 + \tau^2}$$

$$= \frac{1}{2} \times \frac{32\,M}{\pi d^3} \pm \sqrt{\left(\frac{16M}{\pi d^3}\right)^2 + \left(\frac{16\,T}{\pi d^3}\right)^2}$$

For finding equivalent torque (T_e) when the shaft is subjected to bending moment and torque, we should determine the maximum shear produced by the principal stresses.

$\therefore$ Max. shear stress due to principal stresses is given by

$$\tau_{max} = \frac{\sigma_1 - \sigma_2}{2}, \qquad \text{where } \sigma_1 = \frac{16M}{\pi d^3} + \sqrt{\left(\frac{16M}{\pi d^3}\right)^2 + \left(\frac{16\,T}{\pi d^3}\right)^2}$$

$$= \sqrt{\left(\frac{16M}{\pi d^3}\right)^2 + \left(\frac{16\,T}{\pi d^3}\right)^2} \qquad \text{and } \sigma_2 = \frac{16M}{\pi d^3} - \sqrt{\left(\frac{16M}{\pi d^3}\right)^2 + \left(\frac{16\,T}{\pi d^3}\right)^2}$$

$$= \frac{16}{\pi d^3}\sqrt{M^2 + T^2}$$

Equivalent torque is

$$T_e = \frac{\pi}{16} \times d^3 \times \tau_{max.}$$

$$= \frac{\pi}{16} \times d^3 \times \left[\frac{16}{\pi d^3} \times \sqrt{M^2 + T^2}\right] \qquad \left(\because \tau_{max} = \frac{16}{\pi d^3} \times \sqrt{M^2 + T^2}\right)$$

$$= \sqrt{\mathbf{M^2 + T^2}}\ \textbf{. Ans.}$$

7.

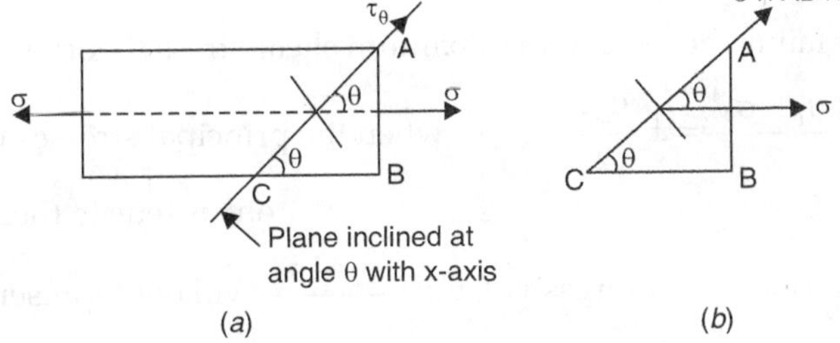

Fig. 18

Force on $AB = \sigma \times$ Area

$= \sigma \times AB \times 1$ (Thickness = unity)

Force on $AC = (\sigma \times AB) \times \cos \theta$

$\therefore$ τ_θ = Shear stress on the plane AC

$$= \frac{\text{Shear force}}{\text{Area}} = \frac{\sigma \times AB \times \cos \theta}{AC \times 1} = \sigma \times \sin \theta \times \cos \theta \left(\because \frac{AB}{AC} = \sin \theta\right)$$

$$= \frac{\sigma}{2} \times 2 \sin \theta \cos \theta = \frac{\sigma}{2} \times \sin 2\theta. \quad \textbf{Ans.}$$

8. The tensile force in the bar at a distance y from the support

= Weight at the lower end + Weight of bar for a length $(L - y)$

= **W + w (L – y). Ans.**

9. The normal stress on the inclined plane in case of biaxial stress system is given by

$$\sigma_\theta = \frac{1}{2}(\sigma_x + \sigma_y) + \frac{1}{2}(\sigma_x - \sigma_y) \cos 2\theta$$

$$\therefore \quad \sigma_{45°} = \frac{1}{2}(\sigma_x + \sigma_y) + \frac{1}{2}(\sigma_x - \sigma_y) \cos 90°$$

$$= \frac{1}{2}(\sigma_x + \sigma_y). \quad \textbf{Ans.}$$

10. Temperature stress = $\alpha \times (\Delta t) + E$

where α = Co-efficient of linear expansion,

Δt = Temperature rise, and

E = Modulus of elasticity.

Hence temperature stress depends upon all the three and (*d*) is the answer.

12. The energy absorbed by a part subject to dynamic force is given by

$$U = \frac{\sigma^2}{2E} \times \text{volume}$$

when σ and E are constant then

$$U \propto \text{volume}$$

Hence (*c*) is the answer.

13. The criterion of failure according to maximum shear stress theory is

$$\frac{\sigma_1 - \sigma_2}{2} = \pm \frac{\sigma_{yp}}{2}$$

when the principal stresses σ_1 and σ_2 are opposite *i.e.,* one is tensile then other is compressive

But if both are tensile (or compressive), then $\frac{\sigma_1 - \sigma_2}{2}$ will not represent the maximum shear stress. It will represent the stress less than maximum shear stress. But $\frac{\sigma_1}{2}$ will represent the maximum shear stress. Hence criterion of failure is

$$\frac{\sigma_1}{2} = \pm \frac{\sigma_{yp}}{2}. \quad \textbf{Ans.}$$

14. Buckling load for column with different end condition are:

$$P_E = \frac{\pi^2 EI}{L^2} \quad \text{...both ends hinged}$$

$$= \frac{\pi^2 EI}{4L^2} \quad \text{...one end is fixed (or clamped) and other is hinged}$$

$$= \frac{2\pi^2 EI}{L^2} \quad \text{...one end is fixed (clamped) and other is hinged}$$

$$= \frac{4\pi^2 EI}{L^2} \quad \text{...both ends fixed (or clamped)}$$

$\therefore$ When both ends are clamped, the buckling load is maximum.

15. The spring stiffness (k) for a close-coiled helical spring in terms of dia. of wire (d), mean radius of coil (R), no. of turns (n) and modulus of rigidity (C) is given by

$$k = \frac{W}{\delta} = \frac{Cd^4}{64R^3 \times n}$$

When the dia. of wire (d) and material of coil is same, then

$$k \propto \frac{1}{R^3 \times n} \quad [\because \text{ For the same material, } C \text{ is constant}]$$

$\therefore$ $k \times R^3 \times n = \text{constant}$

or $k_1 \times R_1^3 \times n_1 = k_2 \times R_2^3 \times n_2$

or $$k_2 = k_1 \times \left(\frac{R_1}{R_2}\right)^3 \times \left(\frac{n_1}{n_2}\right) = k_1 \times \left(\frac{D_1}{D_2}\right)^3 \times \left(\frac{n_1}{n_2}\right)$$

$$= k_1 \times \left(\frac{75}{60}\right)^3 \times \left(\frac{8}{10}\right) \quad \left[\because \begin{array}{l} D_1 = 75 \text{ mm}, n_1 = 8 \\ D_2 = 60 \text{ mm}, n_2 = 10 \end{array}\right]$$

$= \mathbf{1.56k_1}$. **Ans.**

$\therefore$ Answer is (c).

16. Here $d = 1 \text{ m} = 100 \text{ cm}$

$p = 100 \text{ N/cm}^2$

Max. permissible tensile stress = $2 \text{ kN/cm}^2 = 2000 \text{ N/cm}^2$

For thin cylinder, the maximum stress is circumferential (or hoop) stress.

Hence here $\sigma_c = 2000 \text{ N/cm}^2$

Let t = thickness.

Then $$\sigma_c = \frac{p \times d}{2t}$$ [Here p and σ_c should have same unit. Then d and t will have the same unit]

or $$t = \frac{p \times d}{2 \times \sigma_c} = \frac{100 \times 100}{2 \times 2000} = 2.5 \text{ cm} = \mathbf{25 \text{ mm}}.$$ **Ans.**

17. Original volume, $V = \frac{\pi}{4} D^2 \times L$

Change in volume, dV will be obtained by taking differential as

$$dV = \frac{\pi}{4} D^2 \times dL + \frac{\pi}{4} Ld(2D^2)$$

$$= \frac{\pi}{4} D^2 \times dL + \frac{\pi}{4} L \times 2D\, d(D)$$

$\therefore$ Volumetric strain, $$\frac{dV}{V} = \frac{\frac{\pi}{4} D^2 \times dL + \frac{\pi}{4} L \times 2D \times d(D)}{\frac{\pi}{4} D^2 \times L} = \frac{dL}{L} + \frac{2d(D)}{D}$$

$$\frac{dL}{L} = \text{Longitudinal strain} = e_2$$

$$\frac{d(D)}{D} = \text{Circumferential strain} = e_1 \qquad \left[\frac{\pi(D + dD) - \pi D}{\pi D} = \frac{dD}{D}\right]$$

$$\therefore \quad \frac{dV}{V} = e_2 + 2e_1 = 2e_1 + e_2 \qquad \left[\frac{dV}{V} = \text{Change in volume per unit volume}\right]$$

18. The part which is cheapest in overall cost and can be easily replaced when there is some damage, is made the weakest part. The key in comparison to pulley and shaft can be replaced easily.

19. The shear stress distribution in a beam of rectangular cross-section is parabolic and having maximum value at the neutral axis. Hence the answer is (*d*).

Shear stress is given by

$$\tau = \frac{FA \times \overline{y}}{I \times b}, \qquad \text{where } A \times \overline{y} \text{ for rectangular section} = \frac{b}{2}\left(\frac{d^2}{4} - y^2\right)$$

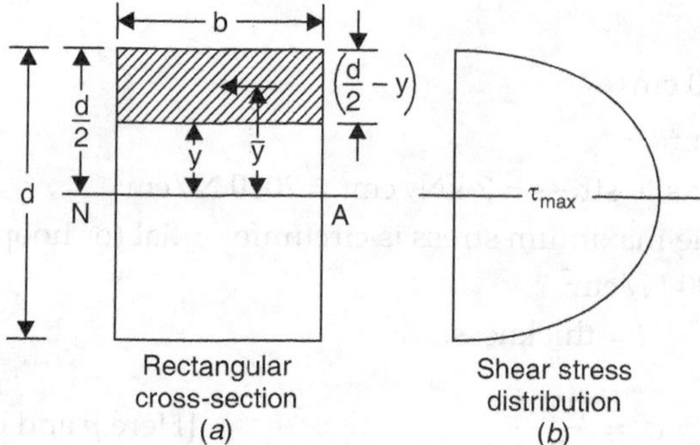

Fig. 19

where A = Shaded area = $b \times \left(\frac{d}{2} - y\right)$

y = Section where shear stress is τ

$\bar{y}$ = Distance of C.G. of shaded area from N.A. = $y + \frac{1}{2}\left(\frac{d}{2} - y\right)$

I = M.O. Inertia

b = Width at section y which is b here.

$$\therefore \quad \tau = \frac{F \times \frac{b}{2}\left(\frac{d^2}{4} - y^2\right)}{I \times b} = \frac{F}{2I}\left(\frac{d^2}{4} - y^2\right)$$

which is the equation of a parabola.

20. When beam is placed as shown in Fig. 20(a) 1st position, then bending stress is given by

$$\sigma_b = \frac{M}{I} \times y$$

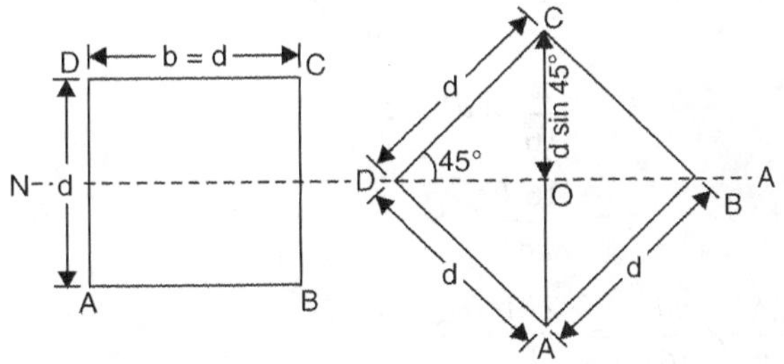

(a) 1st position (b) 2nd position

Fig. 20

and
$$(\sigma_b)_{max} = \frac{M}{I} \times \frac{d}{2} = \frac{M}{\frac{bd^3}{12}} \times \frac{d}{2} \qquad \left[\because I = \frac{bd^3}{12} = \frac{d \times d^3}{12} = \frac{d^4}{12}\right]$$

$$= \frac{6M}{bd^2} = \frac{6M}{d \times d^2} = \frac{6M}{d^3} \qquad (\because b = d \text{ being square}) \quad ...(i)$$

The bending stress when beam is placed as shown in Fig. 20(b) 2nd position is given by

$$(\sigma_b)^*_{max} = \frac{M}{I^*} \times y^*$$

where y^* = Distance of top layer from N.A. *i.e.*, distance OC

= Half of diagonal of the square = OC

= $d \times \sin 45° = \frac{d}{\sqrt{2}}$

I^* = M.O.I, of 2nd position about N.A.

$$= \frac{M}{\left(\frac{d^4}{12}\right)} \times \frac{d}{\sqrt{2}}$$

$$= \frac{12M}{d^3 \times \sqrt{2}}$$

$$= 2 \times \frac{bh^3}{12} \quad \text{where } b = BD = 2 \times \frac{d}{\sqrt{2}}$$

$$h = \frac{d}{\sqrt{2}} = \sqrt{2} \times d$$

$$\therefore \quad \frac{(\sigma_b)^*_{max}}{(\sigma_b)_{max}} = \frac{\left(\dfrac{12\,M}{d^3 \times \sqrt{2}}\right)}{\left(\dfrac{6M}{d^3}\right)}$$

$$= \frac{2 \times \sqrt{2}d \times \left(\dfrac{d}{\sqrt{2}}\right)^3}{12} = \frac{1}{6} \times \frac{d^4}{2} = \frac{d^4}{12}$$

$$= \frac{2}{\sqrt{2}} = \sqrt{2}$$

$$\therefore \quad (\sigma_b)^*_{max} = \sqrt{2} \times (\sigma_b)_{max}$$

Hence the answer is (*c*).

21. $$\sigma_b = \frac{M}{I} \times y = \frac{M}{\frac{\pi}{64} \times d^4} \times \frac{d}{2} = \frac{32\,M}{\pi d^3}$$

$$\therefore \quad T = \frac{\pi}{16} \times d^3 \times \tau$$

$$\therefore \quad \tau = \frac{16\,T}{\pi d^3}$$

$$\therefore \quad \frac{\sigma_b}{\tau} = \frac{32M}{\pi d^3} \times \frac{\pi d^3}{16T} = \frac{2M}{T} = 2. \qquad (\because \; M = T)$$

22. For a simply supported beam carrying a point load at the centre, the deflection (δ) is given by

$$\delta = \frac{PL^3}{48\,EI}$$

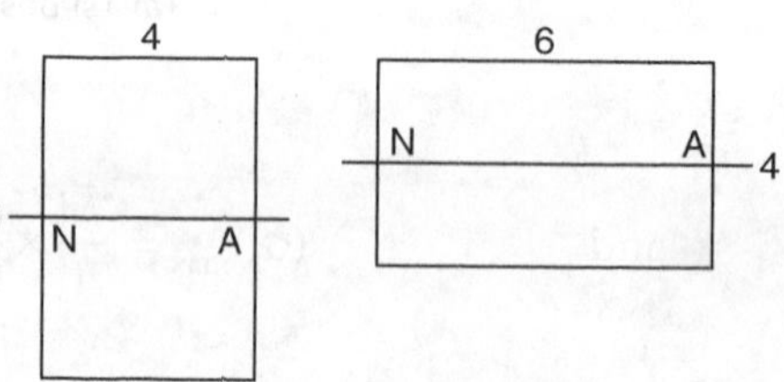

(*a*) 1st position (*b*) 2nd position

Fig. 21

Here load *P*, span of beam *L* and *E* is same for both positions,

$$\therefore \quad \delta \propto \frac{1}{I}$$

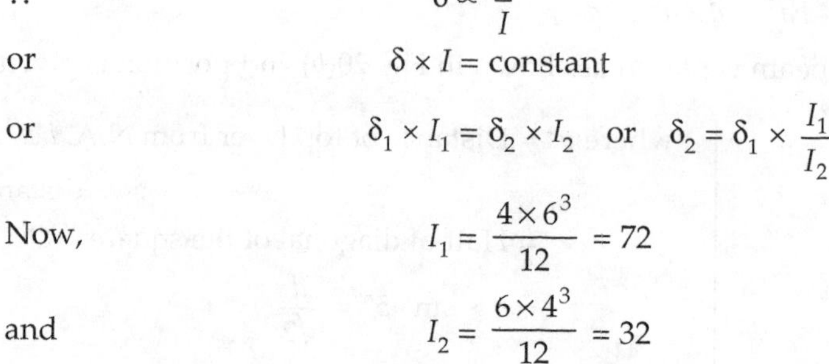

or $$\delta \times I = \text{constant}$$

or $$\delta_1 \times I_1 = \delta_2 \times I_2 \quad \text{or} \quad \delta_2 = \delta_1 \times \frac{I_1}{I_2}$$

Now, $$I_1 = \frac{4 \times 6^3}{12} = 72$$

and $$I_2 = \frac{6 \times 4^3}{12} = 32$$

$$\therefore \quad \delta_2 = \delta_1 \times \frac{72}{32} = \mathbf{2.25\,\delta_1.} \quad \textbf{Ans.}$$

Hence answer is (*d*).

23. Couple acting on beam = $P \times C$ anti-clockwise moment at any point should be zero

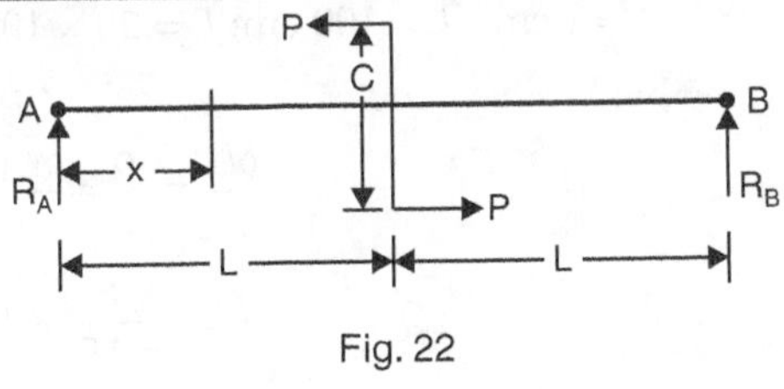

Fig. 22

$$\therefore \quad M_A = 0$$

or $$R_B \times 2L + P \times C = 0$$

or $$R_B = \frac{-P \times C}{2L}$$

(–ve sign shows that reaction R_B is acting downwards)

There is no load on the beam

$$\therefore \quad R_A + R_B = 0 \quad \text{or} \quad R_A = -R_B = \frac{P \times C}{2L}$$

The reaction R_A is acting upwards.

Shear force at x, $\quad F_x = R_A = \frac{\mathbf{P \times C}}{\mathbf{2L}}$. **Ans.**

24. Resolving forces along the inclined plane and normal to the plane, we get

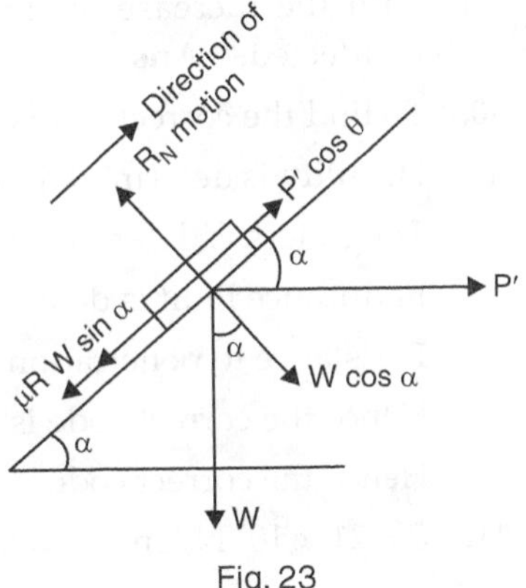

Fig. 23

$$P' \cos \alpha = \mu R_N + W \sin \alpha \quad ...(1)$$

$$R_N = W \cos \alpha + P' \sin \alpha$$

Substituting the value of R_N in (1), we get

$$P' \cos \alpha = \mu \, [W \cos \alpha + P' \sin \alpha] + W \sin \alpha$$

Taking $$\mu = \tan \phi = \frac{\sin \phi}{\cos \phi}$$

and substituting in the above equation and simplifying, we get

$$P' = W \tan (\alpha + \phi)$$

$\therefore$ Frictional torque at mean radius (R) is = $R \times P' = R \times W \tan (\alpha + \phi)$.

26.
$$\sigma_\theta = \frac{\sigma_x + \sigma_y}{2} + \frac{\sigma_x - \sigma_y}{2} \times \cos 2\theta$$

$$\sigma_{45^\circ} = \frac{\sigma_x + \sigma_y}{2} + \frac{\sigma_x - \sigma_y}{2} \times \cos 90^\circ$$

$$= \frac{\sigma_x + \sigma_y}{2}$$

$$= \frac{5000 + 5000}{2} = \mathbf{5000\ N/cm^2}. \quad \textbf{Ans.}$$

27. There is no normal stress either in x-direction or in y-direction.

$\therefore$ $\sigma_x = 0$ and $\sigma_y = 0$. While drawing Mohr's circle of stresses, σ_x and σ_y are taken along x-axis from origin and shear stress (τ) along y-axis. Hence shear stress (τ) will be taken on the origin of axis along y-direction upward and downwards. Hence answer is (c).

28. $A = 1\ \text{cm}^2; L = 100\ \text{cm}; E = 20 \times 10^6\ \text{N/cm}^2, P = 20\ \text{kN} = 20 \times 1000\ \text{N}$

$$\frac{\delta L}{L} = \frac{\sigma}{E} = \frac{\left(\frac{P}{A}\right)}{E}$$

$$\therefore \quad \delta L = \frac{P}{AE} \times L = \frac{20,000 \times 100}{1 \times 20 \times 10^6} = \mathbf{0.1\ cm.} \quad \textbf{Ans}$$

29. $$A_1 = \pi r_1^2,\ A_2 = \pi\left(\frac{r_1}{2}\right)^2 = \frac{\pi r_1^2}{4}$$

The modulus of elasticity for a material is independent of area and load applied etc. Hence with the increase or decrease of area (*i.e.*, radius), the modulus of elasticity will be unaffected. **Ans.**

30. To find the correct answer proceed as given below

Ductility is determined by tension test (3)

Toughness is determined by Impact test (1)

Endurance limit is determined by Fatigue Test (2)

Resistance to penetration is determined by Hardness test (4)

Hence the correct code is which contains 3, 1, 2, 4.

Hence the correct code is (*c*).

31. $E = 21 \times 10^6\ \text{N/cm}^2, C = 8 \times 10^6\ \text{N/cm}^2$

$$E = 2C(1 + \mu)$$

$$\therefore \quad \mu = \frac{E}{2C} - 1$$

$$= \frac{21 \times 10^6}{2 \times 8 \times 10^6} - 1 = \frac{21}{16} - 1 = 1.3125 - 1 = \mathbf{0.3125.} \quad \textbf{Ans.}$$

32. $P = 2\pi NT$, where

$P_1 = 2\pi N_1 T_1$

$P_2 = 2\pi N_1 T_1$

$= 2\pi \times \left(\frac{N_1}{2}\right) \times T_1$

$= 2\pi N_1 \times T_1 \times 4$

$= P_1 \times 4 = 90 \times 4 = \mathbf{360\ kW.}$ **Ans.**

$$T = \frac{\pi}{16} \times d^3 \times \tau \quad \therefore \quad T_1 = \frac{\pi}{16} \times d_1^3 \times \tau$$

$$N_2 = \frac{N_1}{2} \qquad T_2 = \frac{\pi}{6} \times (2d_1)^3 \times \tau$$

$$= \frac{\pi}{16} \times 8 \times d_1^3 \times \tau$$

$$= 8 \times T_1$$

33. According to maximum shear stress theory for design purpose, we have the equation $(\sigma_1 - \sigma_2) = \sigma_t$ where σ_t = Permissible stress in simple tension or $60 - (-60) = \sigma_t$ or $60 + 60 = \sigma_t$

or $\sigma_t = 120$ and safety factor $= \frac{\sigma_t^*}{\sigma_t} = \frac{360}{120} = \mathbf{3}$. **Ans.**

34. Here each rivet is subjected to direct stress due to load W and bending stress due to bending moment. Bending moment is equal to $W \times e$, where e = eccentricity.

The value of e is maximum for rivets P and S. The direct stress is same for all the rivets. Bending stress is maximum when e is maximum. Eccentricity is maximum for rivets P and S. Hence rivets P and S are having maximum bending stress. Hence rivets P and S are most loaded. **Ans.**

35. The stiffness of a helical compression spring is given by,

$$k = \frac{Cd^4}{64\,R^3 \times n}$$

where C = Modulus of rigidity

d = Dia. of wire

n = No. of turns

R = Mean radius of coil.

When spring is cut into two equal halves, only no. of turns will be effected it will become half. Other value such as C, d and R will be same.

$$\therefore \qquad k_2 = \frac{Cd^4}{64 \times R^3 \times \left(\frac{n}{2}\right)} = \frac{2Cd^4}{64R^3 \times n} = \mathbf{2 \times k}. \quad \textbf{Ans.}$$

36. For close-coiled helical spring,

$$\tau = \frac{16\,WR}{\pi d^3}$$

While deriving this equation, the effect of curvature of spring and stress concentration effect are neglected. Hence the correct expression for shear stress will be

$$\tau = \frac{16\,WR}{\pi d^3} \times K$$

where K = Wahl's correction factor

$$= \frac{4S - 1}{4S - 4} + \frac{0.615}{S}$$

where $S = \text{Spring index} = \frac{D}{d} = \frac{\text{Mean dia. of coil}}{\text{Dia. of wire}}$.

37. Free body diagram through C

$$R_1 + R_2 = 120 \text{ N} \qquad \text{...}(i)$$

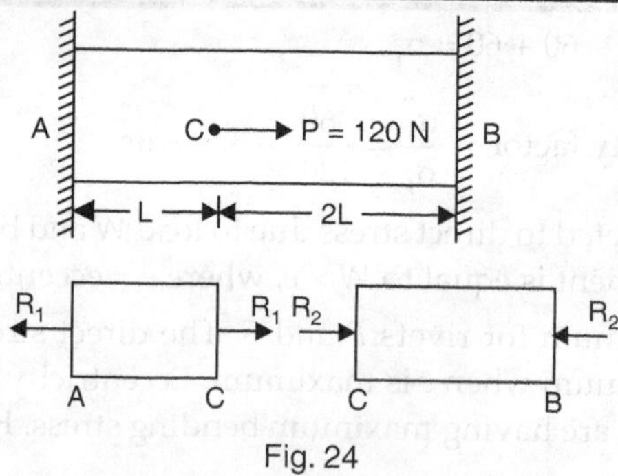

Fig. 24

Extension in AC = Compression in CB

For $\quad AC: \dfrac{\delta L_1}{L_1} = \dfrac{\sigma_1}{E} \quad$ or $\quad \delta L_1 = \dfrac{\sigma_1}{E} \times L_1 = \dfrac{R_1}{AE} \times L_1 \qquad \left(\because \quad \sigma_1 = \dfrac{R_1}{A}\right)$

For $\quad CB: \dfrac{\delta L_2}{L_2} = \dfrac{\sigma_2}{E} \quad$ or $\quad \delta L_2 = \dfrac{\sigma_2}{E} \times L_2 = \dfrac{R_2}{AE} \times L_2 \qquad \left(\because \quad \sigma_2 = \dfrac{R_2}{A}\right)$

As $\quad \delta L_1 = \delta L_2$

$\therefore \quad \dfrac{R_1 \times L_1}{AE} = \dfrac{R_2 \times L_2}{AE}$

or $\quad R_1 \times L_1 = R_2 \times L_2 \quad$ But $\quad L_1 = L \quad$ and $\quad L_2 = 2L$

$\therefore \quad R_1 \times L = R_2 \times 2L$

or $\quad R_1 = 2R_2 \qquad ...(ii)$

From (*i*) and (*ii*), $\quad 3R_2 = 120$ N $\quad \therefore \quad$ **$R_2 = 40$ N. Ans.**

and $\quad R_1 = 120 - 40 =$ **80 N. Ans.**

38. $C = 100$ G Pa $= 100 \times 10^9$ N/m^2

$\mu = 0.25$

$E = 2C(1 + \mu)$

$= 2 \times 100 \times 10^9 (1 + 0.25) = 200 \times 1.25 \times 10^9$

$= 250 \times 10^9$ N/m^2 = **250 GPa. Ans.**

39. Steel rod $\qquad$ Aluminium rod

$L_1 = 2$ m $\qquad L_2 = 1$ m

$A_1 = 1$ cm^2 $\qquad A_2 = 2$ cm^2

$E_1 = 200$ GPa $\qquad E_1 = 100$ GPa

The rigid beam will be horizontal if:

Extension of steel rod = Extension of aluminium rod

$$\delta L_1 = \delta L_2$$

$$\frac{P_1 \times L_1}{A_1 \times E_1} = \frac{P_2 \times L_2}{A_2 \times E_2}$$

or $$P_1 = P_2 \times \frac{A_1}{A_2} \times \frac{E_1}{E_2} \times \frac{L_2}{L_1} = P_2 \times \frac{1}{2} \times \frac{200}{100} \times \frac{1}{2} = \frac{P_2}{2}$$

or $$2P_1 = 2P_2 \quad \text{or} \quad P_2 = \mathbf{2P_1}. \quad \textbf{Ans.}$$

or load on aluminium rod = 2 times the load on steel rod.

40. Radial displacement $= u$

Initial radius $= r$

Final radius $= r + u$

Tangential strain = Circumferential strain = $\dfrac{\text{Final circumference} - \text{Initial}}{\text{Initial circumference}}$

$$= \frac{2\pi(r+u) - 2\pi r}{2\pi r} = \frac{2\pi u}{2\pi r} = \frac{\mathbf{u}}{\mathbf{r}}. \quad \textbf{Ans.}$$

41. Stiffness = $\dfrac{\text{Load}}{\text{Deflection}} = \dfrac{P}{\text{Deflection under } P}$

Let us find deflection under load P. This can be done by conjugate Beam Method. In this method, the beam carries the $\dfrac{M}{EI}$ load corresponding to actual load. The deflection at any section will be equal to B. M. at that section due the load carried by conjugate beam. Refer to Fig. 25 (*b*).

Load for conjugate beam is $\dfrac{M}{EI}$.

B.M. at $A = 0$, hence value of $\dfrac{M}{EI}$ at $A = 0$

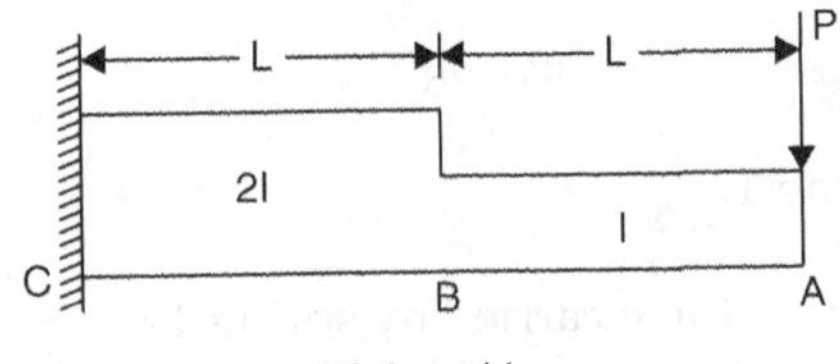

(*a*) Actual beam

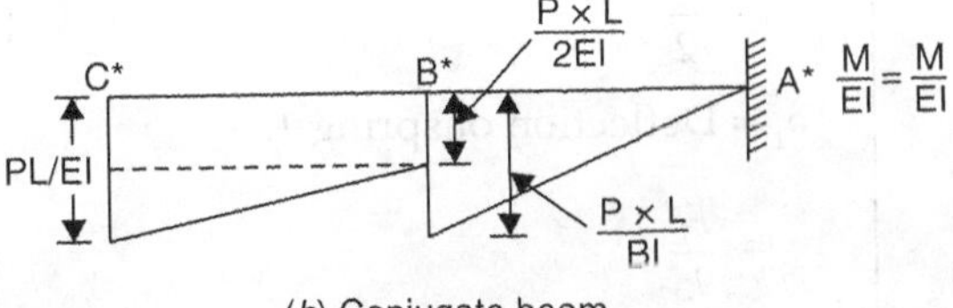

(*b*) Conjugate beam

Fig. 25

B.M. at $C = P \times 2L$, hence value of $\frac{M}{EI}$ at $C = \frac{P \times 2L}{E \times (2I)} = \frac{PL}{EI}$

B.M. at $B = P \times L$, hence value of $\frac{M}{EI}$ at B for $AB = \frac{P \times L}{EI}$, for $BC = \frac{P \times L}{EI} = \frac{PL}{2EI}$

Deflection (δ) at A = B.M. at A^* due to load carried by conjugate beam

$$= \left(\frac{1}{2} \times \frac{PL}{EI} \times L\right) \times \frac{2L}{3} + \left(\frac{PL}{2EI} \times L\right) \times 1.5L + \left(\frac{1}{2} \times \frac{PL}{2EI} \times L\right) \times \frac{5L}{3}$$

$$= \frac{PL^3}{EI}\left(\frac{1}{3} + 0.75 + \frac{5}{12}\right) = \frac{PL^3}{EI} \times \frac{18}{12} = \frac{1.5 \times PL^3}{EI}$$

$$\therefore \quad \text{Stiffness} = \frac{P}{\delta} = \frac{P}{\left(\frac{1.5 \times PL^3}{EI}\right)} = \frac{EI}{1.5 \times L^3} = \frac{(200 \times 10^9) \times (375 \times 10^{-6})}{1.5 \times (0.5)^3}$$

$$= \frac{200 \times 10^9 \times 375 \times 10^{-4} \times 10^3}{1.5 \times 125} = \mathbf{4 \times 10^{10} \text{ N/m.}} \quad \textbf{Ans.}$$

42. The maximum stress induced in a thin cylinder is hoop stress (σ_c). It is given by

$$\sigma_c = \frac{P \times d}{2t}$$

The maximum hoop stress produced in spherical vessel $= \frac{P \times d}{4t}$

$$\therefore \quad \frac{\text{Max. stress in cylindrical vessel}}{\text{Max. stress in spherical vessel}} = \frac{\frac{pd}{2t}}{\frac{p \times d}{4t}} = 2.0.$$

44.

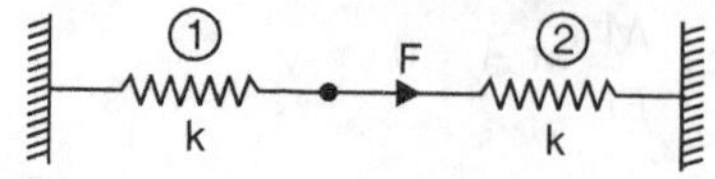

Fig. 26

Strain energy stored in spring 1,

$U_1 = \frac{1}{2} \times F_1 \times \delta_1$ where F_1 = force carried by spring 1

$= \frac{1}{2} \times \frac{F}{2} \times \frac{F}{2k}$ | $= \frac{F}{2}$ ($\because$ both springs are identical)

δ_1 = Deflection of spring 1.

$= \frac{\mathbf{F^2}}{\mathbf{8k}}$. **Ans.** | $= \frac{F_1}{k}$

$= \frac{F_1}{2k}$ $\left(\because F_1 = \frac{F}{2}\right)$

45. $A = 100 \times 10^{-6}$ m^2

Let P = tensile load at failure

$\therefore$ For one-dimensional stress system,

we have stresses as $\left(\frac{P}{A}, 0, 0\right)$

Tensile stress due to load P, $\sigma_1 = \frac{P}{A}$

This stress is in one-direction only *i.e.*, $(\sigma, 0, 0)$

Max. shear stress due to stress system $(\sigma_1, 0, 0) = \frac{1}{2}(\sigma_1 - 0) = \frac{\sigma_1}{2}$

Uniaxial yield stress, $\sigma_t^* = 200$ MPa $= 200 \times 10^6$ N/m^2.

$\therefore$ For uniaxial yield stress, we have stress system as $(\sigma_t^*, 0, 0)$

Max. shear stress due to uniaxial yield stress $= \frac{1}{2}(\sigma_t^* - 0) = \frac{\sigma_t^*}{2}$

According to Tresca failure criterion,

Max. shear stress developed = Max. shear due to yield stress

i.e., $\frac{\sigma_1}{2} = \frac{\sigma_t^*}{2}$ or $\sigma_1 = \sigma_t^*$

or $\frac{P}{A} = 200 \times 10^6$ N/m^2 $\left(\because \sigma_t^* = 200 \times 10^6 \text{ N/m}^2 \text{ and } \sigma_1 = \frac{P}{A}\right)$

or $P = 200 \times 10^6 \times A$

$= (200 \times 10^6) \times (100 \times 10^{-6}) = 20000 =$ **20 kN. Ans.**

46. We know that stiffness of a close-coiled helical spring is given by

$$k = \frac{Cd^4}{64\, R^3 \times n}$$

1st case, $k_1 = \frac{Cd^4}{64\left(\frac{D}{2}\right)^3 \times N}$ $\left(\text{here } R = \frac{D}{2} \text{ and } n = N\right)$

2nd case, $k_2 = \frac{C(2d)^4}{64\left(\frac{2D}{2}\right)^3 \times 2N}$

(here dia. of wire = $2d$; Mean coil dia. = $2D$ and number of turns = $2N$)

$$= \frac{Cd^4 \times 16}{64 \times \left(\frac{D}{2}\right)^3 \times 8 \times 2N} = \frac{Cd^4}{64\left(\frac{D}{2}\right) \times N} = \mathbf{k_1} \quad \left(\because k_1 - \frac{Cd^4}{64\left(\frac{D}{2}\right)^3 \times N}\right)$$ **Ans.**

47. Euler's buckling load,

$$P = \frac{\pi^2 EI}{L^2}, \quad \text{where} \quad I = \frac{\pi}{64} d^4$$

For circular column when diameter is reduced by 20%, then dia. of new column = 0.8 d

$\therefore$ New moment of inertia, $I^* = \frac{\pi}{4} \times (0.8\, d)^4 = \frac{\pi}{4} \times d^4 \times (0.8)^4$

Initial buckling load, $P = \frac{\pi^2 E}{L^2} \times \frac{\pi}{64} d^4$

New buckling load, $P^* = \frac{\pi^2 E}{L^2} \times \frac{\pi}{64} \times d^4 \times 0.8^4$

$\therefore$ % reduction in load $= \frac{P - P^*}{P} \times 100$

$$= \frac{\left(\frac{\pi^2 E}{L^2} \times \frac{\pi}{64} d^4 - \frac{\pi^2 E}{L^2} \times \frac{\pi}{64} d^4 \times 0.8^4\right)}{\frac{\pi^2 E}{L^2} \times \frac{\pi}{4} d^4} \times 100$$

$$= \frac{1 - 0.8^4}{1} = (1 - 0.4096) \times 100 = 59\%. \quad \textbf{Ans.}$$

48. For a column of one end fixed and other free, the bucking load is

$$P_1 = \frac{\pi^2 EI}{4L^2}$$

For a second column of same length and same cross-sectional area when both ends are fixed, the buckling load is

$$P_2 = \frac{4\pi^2 EI}{L^2}$$

$$\therefore \quad \frac{P_2}{P_1} = \frac{\left(\frac{4\pi^2 EI}{L^2}\right)}{\left(\frac{\pi^2 EI}{4L^2}\right)} = 4 \times 4 = 16. \quad \textbf{Ans.}$$

49. $\sigma_{max} = \frac{1}{2} [\sigma_1 - \sigma_3] = \frac{1}{2} [80 - (-40)] = $ **60 MPa. Ans.**

50. $E = 2C(1 + \mu)$

$\therefore \mu = \frac{E}{2C} - 1 = \frac{120}{2 \times 50} - 1 = 1.2 - 1.0 = $ **0.2. Ans.**